Accessing the E-book edition

Using the VitalSource® ebook

Access to the VitalBook™ ebook accompanying this book is via VitalSource® Bookshelf – an ebook reader which allows you to make and share notes and highlights on your ebooks and search across all of the ebooks that you hold on your VitalSource Bookshelf. You can access the ebook online or offline on your smartphone, tablet or PC/Mac and your notes and highlights will automatically stay in sync no matter where you make them.

1. **Create a VitalSource Bookshelf account at** ***https://online.vitalsource.com/user/new*** or log into your existing account if you already have one.

2. **Redeem the code provided in the panel below to get online access to the ebook.** Log in to Bookshelf and select **Redeem** at the top right of the screen. Enter the redemption code shown on the scratch-off panel below in the **Redeem Code** pop-up and press **Redeem**. Once the code has been redeemed your ebook will download and appear in your library.

No returns if this code has been revealed.

DOWNLOAD AND READ OFFLINE

To use your ebook offline, download BookShelf to your PC, Mac, iOS device, Android device or Kindle Fire, and log in to your Bookshelf account to access your ebook:

On your PC/Mac

Go to ***https://support.vitalsource.com/hc/en-us*** and follow the instructions to download the free **VitalSource Bookshelf** app to your PC or Mac and log into your Bookshelf account.

On your iPhone/iPod Touch/iPad

Download the free **VitalSource Bookshelf** App available via the iTunes App Store and log into your Bookshelf account. You can find more information at ***https://support.vitalsource.com/hc/en-us/categories/200134217-Bookshelf-for-iOS***

On your Android™ smartphone or tablet

Download the free **VitalSource Bookshelf** App available via Google Play and log into your Bookshelf account. You can find more information at ***https://support.vitalsource.com/hc/en-us/categories/200139976-Bookshelf-for-Android-and-Kindle-Fire***

On your Kindle Fire

Download the free **VitalSource Bookshelf** App available from Amazon and log into your Bookshelf account. You can find more information at ***https://support.vitalsource.com/hc/en-us/categories/200139976-Bookshelf-for-Android-and-Kindle-Fire***

N.B. The code in the scratch-off panel can only be used once. When you have created a Bookshelf account and redeemed the code you will be able to access the ebook online or offline on your smartphone, tablet or PC/Mac.

SUPPORT

If you have any questions about downloading Bookshelf, creating your account, or accessing and using your ebook edition, please visit ***http://support.vitalsource.com/***

ACTUARIAL MODELS

The Mathematics of Insurance

Second Edition

ACTUARIAL MODELS

The Mathematics of Insurance

Second Edition

VLADIMIR I. ROTAR

CRC Press
Taylor & Francis Group
Boca Raton London New York

CRC Press is an imprint of the
Taylor & Francis Group, an **informa** business
A CHAPMAN & HALL BOOK

CRC Press
Taylor & Francis Group
6000 Broken Sound Parkway NW, Suite 300
Boca Raton, FL 33487-2742

Printed in the UK by Severn, Gloucester on responsibly sourced paper

MIX
Paper | Supporting responsible forestry
FSC® C022174

Version Date: 20140609

International Standard Book Number-13: 978-1-4822-2706-2 (Pack - Book and Ebook)

Visit the Taylor & Francis Web site at
http://www.taylorandfrancis.com

and the CRC Press Web site at
http://www.crcpress.com

Preface

As mentioned in the first edition, this book is not a monograph, it is a textbook. Its goal is to give a comprehensive exposition of the basic models of insurance processes. A supporting goal is to present some mathematical frameworks and methods used in Actuarial Modeling.

The format of the book. The material is presented in the form of *three nested "routes."* Route 1 contains the basic material designed for a one-semester course. This material is *self-contained* and has a moderate level of difficulty.

Route 2 contains all of Route 1, offers a more complete exposition, and is suited for a two-semester course or self-study. It is *slightly* more challenging but should be approachable for any reader familiar with primary concepts of calculus and linear algebra.

Route 3 (more precisely, the part that is not included in Routes 1-2) is designed primarily for graduate study.

The routes are explicitly designated. To assist in navigating the text, we use *markers similar to road signs.* We also provide the Table of Contents with similar signs, so the reader has a general outline of the book's structure.

Potential audience. The book is intended for a large audience: students, actuaries, mathematicians, and anyone conducting research in areas related to the subject of insurance processes. Some parts of the book may be useful for studying economic and social models of a more general nature. The author believes that the *main audience* is students who are majoring in mathematics, statistics, economics or finance and who are taking courses in Actuarial Modeling.

For the most part, this text is based on the course "Actuarial Mathematics" which the author has taught many times over the last fifteen years in the Department of Mathematics at the University of California at San Diego (UCSD) and in the Department of Mathematics and Statistics at San Diego State University (SDSU).

Prerequisites are two semesters of calculus and one upper division course in probability for undergraduate students. There is no need for any specialized mathematical background that differs from the standard introductory topics. The only possible exception is the big O and little o notation which is not always introduced in standard Calculus courses. This simple and convenient notation is defined and illustrated by examples in two pages in the Appendix.

To facilitate reading, the main text is preceded by *an introductory Chapter 0* containing a digest of basic facts from Probability Theory and the Theory of Interest. Ideally, the reader will not have to refer to outside sources for background material; everything is under one cover and is presented in a uniform notation and style.

More on the contents. Our aim is to give an explicit exposition of main ideas and basic mathematical models. Therefore, we sometimes leave details important merely from an economic point of view to courses in the economics of insurance.

This also applies to counting and computation issues. The book contains many computational examples, but these serve as illustrations of results rather than instruction on computational aspects of the theory.

More specifically, the book includes examples and exercises on *numerical calculations with use of Microsoft Excel.* The goal here is not to teach numerical methods in insurance—this requires software more powerful than Excel—but to assist the reader in developing an appreciation of particular formulas and to demonstrate practical possibilities and restrictions of different approaches under consideration.

Mainly, this textbook contains the standard material taught in courses in Actuarial Modeling. Nevertheless, it also contains several topics which the author feels would make the material more modern or/and somewhat deeper. Cases in point are the modern *theory of risk evaluation*; a *generalization of Arrow's theorem*; the classification of distributions with regard to *light- and heavy-tails*; *the accuracy of normal and Poisson approximations*, and consequently, a more accurate estimation of some characteristics of insurance processes; a sufficiently detailed description of *cash flows in the Markov environment*; a model with *payments of dividends*; a systematic presentation of *reinsurance models*; and the application of the *martingale technique* to Ruin Theory.

Regarding the last topic, to make the use of this book possible for instructors and readers who do not wish to discuss martingales, the exposition of Ruin Theory is organized in a way that the knowledge of martingales is not required. This part of the exposition is self contained. However in an additional section, the reader who is familiar with martingales, will be able to enjoy short proofs and a unified representation for the discrete and continuous time cases.

On exams of actuarial societies. The book can be used as a source for preparing for exams on actuarial models. The author purposefully reviewed the programs and problems of the corresponding exams of the the Society of Actuaries (SOA) and the Casualty Actuarial Society (CAS). As a result, nearly all topics from these exams on actuarial models are included in Routes 1 or 2 of the book.

To the same end, with the kind permission of the CAS, a significant number of problems given in the previous sessions of the CAS Exams were included in the book as examples. In the first edition, these problems concerned the 2003-2005 examinations. These problems are still relevant and by no means obsolete, so it made sense to include many of them again. However, new problems from the the 2011-2013 examinations have been also included.

Certainly, anyone interested in taking actuarial exams should study the syllabus and problems given in previous examinations. However, the author believes that upon the completion of the text and its exercises, the preparation for these exams will be relatively effortless.

It is also worthwhile to mention two topics from the actuarial exam syllabi, which are not included in this book. First, we do not discuss data analysis. This topic should be considered separately in a course in actuarial statistics. Secondly, although we touch on simulation questions, we do not discuss them in detail. In particular, some technical simulation methods were not included in the book.

What is new in the second edition. During the years since the first edition was published, the book has been adopted as a text at a number of universities, and the author himself has taught courses in Actuarial Mathematics at both universities mentioned above using this book. The result has been great feedback from students, teaching assistants, colleagues in lecturing, and from individual readers who reached out with questions. This feedback allowed to eliminate many typos and a few more substantial (though still minor) flaws. It also has provided better understanding of which material went smoothly for students and which material required a more detailed explanation or—in rare occasions—even had to be eliminated or presented in another way.

All of this has led to *significant editing and revamping* of all chapters. This especially concerns Chapter 6 (in the first edition, Chapter 7), "Global Characteristics of the Surplus Process". Chapter 3 on conditional expectation from the first edition has been shortened and the material has been moved to Chapter 0, "On Preliminary Facts from Probability and Interest.

Another goal was to take into account some new results and current trends in teaching actuarial modeling (which may be seen in some new textbooks that were published during these years). In particular, pension fund modeling is becoming more topical, and thus a *new chapter* on pensions models has been added to the text.

Overall, the author believes that these modifications have led to significant improvement, and the second edition provides a more robust and polished exposition of the material.

Textbook web site. Possible *additional remarks*, more detailed *answers to the exercises*, and *errata* will be posted at *http://actuarialtextrotar.sdsu.edu*.

Acknowledgments

My sincere and deep thanks go to my colleagues for useful discussions of various scientific and pedagogical questions relevant to this book. This concerns

Caroline Bennet, San Diego State University, USA
Eric Bieri, US Insurance Pricing, John Hancock Insurance Company, USA
Paul Brock, San Diego State University, USA
John Elwin, San Diego State University, USA
Patrick Fitzsimmons, University of California at San Diego, USA
Victor Korolev, Moscow State University, Russia
Luis Latorre, an actuary, Madrid, Spain
Jeffrey Liese, California Polytechnic State University, San Luis Obispo, USA
R. Duncan Luce, University of California at Irvine, USA
Donald Lutz, San Diego State University, USA
Mukul Majumdar, Cornel University, USA
Michael O'Sullivan, San Diego State University, USA
Yosef Rinott, Hebrew University, Jerusalem, Israel
Alexey Sholomitskii, The High Economic School, Russia
Sergey Shorgin, The Institute for Problems of Informatics, RAS, Russia
Alexander Slastnikov, The Central Economics and Mathematics Institute, RAS, Russia
Lee Van de Wetering, San Diego State University, USA
Hans Zwiesler, Ulm University, Germany

I am thankful to the referees of the first and second editions as well for their useful remarks.

My special thanks go to *Mark Dunster, Robert Grone, Collen Kelly, David Lesley, David Macky, Helen Noble, Eugene Pampiga, Steven Pierce, Peter Salamon*, and *Arthur Springer* for help in English editing.

I am also thankful to my former students *Xin-Wei Du* and *Esteban Mansilla* who, when taking courses on Actuarial Mathematics, read a draft of the corresponding chapters and also provided useful feedback.

My thanks go also to *Sarah Borg* and *Sara Zarei* for their permission to use material from their master theses (references can be found in the corresponding sections).

Sunil Nair is the editor of already the third book of mine, and it is a pleasure to acknowledge that I always felt comfortable with all matters concerning the preparation of the book.

Additional thanks go to the editorial staff at Taylor & Francis and in particular to *Shashi Kumar* for help in the preparation of the final file.

Last but not least, I am grateful to the Casualty Actuarial Society, and especially to *Ms. Elizabeth Smith*, for the kind permission to reprint some problems given in previous sessions of the CAS exams on actuarial models.

V. R.

Contents

Flag-signs in the margins designate a route: either Route 1 (which is entirely contained in Route 2), or Route 2 (which is entirely contained in Route 3), or Route 3; see the Preface and Introduction for a more detailed description of these routes. A new flag-sign is posted only at the moment of a route change.

Introduction

1,2,3

We begin with what may be viewed as a small miracle. Two investors, Ann and David, expect random incomes amounting to random variables (r.v.) X_1 and X_2, respectively. We do not exclude the case where the X's may take on negative values, which corresponds to losses. For simplicity, suppose X_1 and X_2 are independent with the same probability distribution. Then X_1 and X_2 have the same expected value $m = E\{X_i\}$ and variance $\sigma^2 = Var\{X_i\}$.

Assume that Ann and David evaluate the riskiness of their investments by the variance of income, and being risk averse, they want to reduce the riskiness of their future incomes. To this end, Ann and David decide to divide the total income into equal shares, so each will have the random income

$$Y = \frac{1}{2}(X_1 + X_2).$$

Then for both, Ann and David, the expected value of the new income will be

$$E\{Y\} = \frac{1}{2}(m+m) = m,$$

that is, the same as before sharing the risk. On the other hand (see Chapter 0 for more details), the variance

$$Var\{Y\} = \frac{1}{4}(\sigma^2 + \sigma^2) = \frac{\sigma^2}{2},$$

is half as large.

Although this result is easy to prove, this is a key fundamental fact. And it is indeed quite astonishing. The riskiness of the system as a whole did not change, the r.v.'s X_1 and X_2 remained as they were, but the level of risk faced by *each* participant has decreased.

Now, consider n participants of a mutual risk exchange, and denote their random incomes by $X_1, ..., X_n$. Assume again that the X's are independent and identically distributed, and set $m = E\{X_i\}$ and $\sigma^2 = Var\{X_i\}$. If the participants divide their total income into equal shares, then the income for each is

$$Y = \frac{X_1 + ... + X_n}{n}.$$

In this case (see again details in Chapter 0),

$$E\{Y\} = m, \quad \text{while} \quad Var\{Y\} = \frac{\sigma^2}{n},$$

and for large n, the variance is close to zero. Thus, for a large number of participants, the risk of each separate participant may be reduced nearly to zero.

The phenomenon we observed in its simplest form is called *redistribution of risk*. It is at the heart of most stabilization financial mechanisms, certainly including insurance.

People use insurance because they can redistribute the risk, making it small for each if the number of participants is large. Insurance companies play the role of organizers of such a redistribution. Of course, they do it for profit, although there are non-profit organizations of mutual insurance.

With some exaggeration, one may say that the theory we study in this book deals with various generalizations of the scheme above. To have a general picture, let us consider a brief outline of the book.

Chapter 0 contains basic facts from Probability Theory and the Theory of Interest we use in the book.

In *Chapter 1*, we will see that variance is not the only possible characteristic of riskiness, and as a matter of fact, is far from being the best. There are more sophisticated and flexible risk measures, and in Chapter 1 we study several of the most important ones.

In *Chapter 2* we built the first relatively complete model of insurance. First, we consider just *one client or insured.* The object of study here is the random future payment X of the company to the client.

Once we know how the random variable (r.v.) X appears, we consider a *group of n clients*, or a *risk portfolio*. In this case, we study the r.v. of the total payment

$$S_n = X_1 + \ldots + X_n, \tag{1}$$

where X_i is the payment to the ith client. For small n, we can compute the distribution of S_n directly by certain methods studied in Chapter 2. For large n, we apply approximation methods.

To cover clients' risks, the company collects premiums. Denote by π_n the total premium corresponding to the risk portfolio above. It is natural to expect—and we prove it—that for the company to function with financial stability, the premium π_n needs to be larger than the mean total payment $E\{S_n\}$ which is usually called a *net premium*. The difference $\Delta_n = \pi_n - E\{S_n\}$, called a *security loading*, is the additional payment for the risk the insurer incurs.

The determination of a value of Δ_n acceptable for the company is one of the main, if not the main, tasks of Actuarial Modeling, and in the course of the book, we consider several basic results on this point.

In *Chapter 3*, we generalize the previous model and instead of (1), consider the sum

$$S_N = X_1 + \ldots + X_N, \tag{2}$$

where not only the X's but the number of terms in the sum, N, is also random.

The most important for insurance interpretation of this scheme concerns the situation where we deal with a portfolio as a whole, and we are interested in the total claim the company will have to pay out. Here, N is the number of future *claims* to be received by the company during a certain period, and the X's are the sizes of payments corresponding to these claims.

In Chapter 3, we explore possible probability distribution of the r.v. N and the aggregate payment S_N itself.

Both models above are static. *Chapters 4-6* concern dynamic models. The main object of study here is a *surplus process* R_t, where t stands for time and the process itself may be defined, for example, as follows.

Let N_t be the number of claims received by time t. We consider N_t at all moments t from some time interval, and hence we view N_t as a *random process*.

By analogy with (2), the total aggregate claim paid by the company by time t is the r.v.

$$S_{N_t} = X_1 + ... + X_{N_t},$$

where X_i is the size of the ith claim. The process S_{N_t} is called a *claim process.* In *Chapters 4-5*, we study various types of the processes N_t and S_{N_t}.

Along with the flow of claims, we consider the cash flow of premiums the company is receiving. Let c_t be the total premium collected by time t. Suppose also that at time $t = 0$, the company has an initial surplus u. Then the surplus of the company at time t is the r.v.

$$R_t = u + c_t - S_{N_t}. \tag{3}$$

In *Chapter 6*, we consider global characteristics of the process R_t. In one way or another, they reflect the "quality" of the process R_t or, in other words, the extent the future surplus process will meet the goals of the company. These characteristics are relevant to either the profitability of insurance operations or to their viability, i.e., the degree of protection against adversity.

Chapters 7-10 address life insurance and annuities. (The last term means a series of regular payments; for example, pension payments.) Risk redistribution continues to play an important role in this case, but this type of insurance product has an additional special feature: an essential time lag between the moment when the client pays a premium and the time when the company pays the corresponding benefit.

The company somehow invests the premiums paid, and during the lifetime period mentioned the amount invested is growing at some rate. Therefore, the total amount of the premium sufficient for the company to fulfill its obligation may be (and should be) less than the size of the benefit. To determine how much less is the main task of the actuary in this case. The models are rather sophisticated and cover various types of insurance and annuity contracts.

In *Chapter 11*, we explore particular but important models of pension plans.

In *Chapter 12*, we return to what we started with, that is, to risk redistribution. However, we consider it at another level, namely, at the level of reinsurance when companies redistribute the risk incurred between themselves. Such a risk redistribution may be even more flexible than that at the first level—the companies may share individual risks or redistribute the total accumulated risk in different ways.

* * *

As mentioned in the preface, the material is presented in the form of three nested "routes". Route 1 consists of the material designed for a one semester course. Route 2 is intended for a broader and deeper study, perhaps, for a two-semester course. Route 3 (more precisely, the part that is not included in Routes 1-2) is designed primarily for graduate study. Route 1 is completely contained in Route 2, and both of them are included in Route 3.

All routes are self-contained.

The special "road signs" will help the reader to continue in the chosen route. For example, the sign

Route 1 $\Rightarrow$ *page 111*

2,3 indicates that the readers who chose Route 1 should advance to p.111 to continue the route. Below this sign, a small "flag" in the margin designates which route runs now (as in the margin here). In other words, if the reader does not switch to the page mentioned (as p.111 above), she/he will have entered Route 2 (and hence, Route 3 too).

1,2,3 If the reader goes to the page mentioned (as p.111 above), then in the margin of this page, she/he will see a sign confirming that this is the right place to move to, and showing a particular location on the page where Route 1 picks up (as in the margin here).

To see a general picture of the book's structure, the corresponding flag-signs are also placed in the Table of Contents. (The rare cases where we switch routes even inside a subsubsection are not reflected in Contents.)

Exercise sections are excluded from this routing system. If the main text of a route continues in another chapter, the road sign directs the reader to this chapter though the exercise section of the current chapter may contain exercises from the current route. We hope that if the reader wants to do exercises, she/he will visit the exercise section anyway.

In the exercises, the problems belonging to Route 2 are marked by an asterisk *, problems from Route 3 by **. However, if a whole section belongs to Route 2 or 3, in the exercises, we mark by * or ** only the title of this section. In the exercises for chapters belonging entirely to Route 2 or 3, we naturally do not mark anything.

Occasionally, purely technical proofs or additional remarks are enclosed by the signs ► ◄. This material may be omitted in the first reading.

If we have not used a definition or fact recently or are using it for the first time, then the corresponding references are given. This is being done *just in case*. If the reader is already familiar with a referred item, it makes sense to ignore the reference and move ahead.

Certainly, when moving along Route 1, the reader is welcome to look around or venture into areas that are not included in the first route. However, the reader should not be discouraged if something seems difficult. Route 2 is indeed slightly more involved than Route 1, but only slightly, and requires just a more in-depth reading.

Another matter is that if you are taking a one-semester course, then it may be reasonable to postpone the material of Route 2—at least, most of this material—for a while, and return to it when you have more time and experience. Enticing topics you skipped on the way will await you.

More technical remarks. The symbol ■ indicates the end of a proof, while the symbol □ marks the end of an example or a series of examples.

The numbering of sections and formulas in each chapter is self-contained. The adopted system of references is clear from the following examples.

Section 2.3 is the third subsection of the second section of the *current* chapter.

The formula (2.3.4) is the fourth formula of the third subsection of the second section of the *current* chapter.

Example 2.3-4 is the fourth example from Section 2.3 of the *current* chapter.

In each chapter, theorems, propositions, and corollaries are being enumerated in a linear fashion through the whole chapter: the theorem that appears after Proposition 2 is Theorem 3, and the corollary following Theorem 3 is Corollary 4.

If we refer to a formula, section, example, etc., from another chapter, we write the number of the chapter to which we are referring in bold font. For instance, Section **1**.2.3 is the third subsection of the second section of the *first* chapter. Formula (**1**.2.3.4) is the formula (2.3.4) of the *first* chapter. Theorem **1**.2 is Theorem 2 of Chapter 1, etc.

The following *abbreviations* are used throughout the entire book.

APV—actuarial present value;
CLT—central limit theorem;
c.d.f.—cumulative distribution function;
c.v.—coefficient of variation;
d.f.—distribution function (we omit here the adjective "cumulative");
EU—expected utility;
EUM—expected utility maximization;
FSD—first stochastic dominance
iff—if and only if;
i.i.d.—independent and identically distributed;
LLN—law of large numbers;
l.e.r.—loss elimination ratio;
l.-h.s.—left-hand side;
m.g.f.—moment generating function;
p.d.f.—probability density function;
RDEU—rank dependent expected utility;
r.v.—random variable;
r.vec.—random vector;
r.-h.s.—right-hand side.
SSD—second stochastic dominance

Chapter 0

Preliminary Facts from Probability and Interest

1,2,3

This chapter's primary purpose is for further reference. Nevertheless, it is recommended that the reader at least skims through the chapter before starting to read the main text.

We deal here mainly with definitions and basic notions to which we will repeatedly refer throughout the book. Most facts are given without proof, but we discuss their significance and plausibility. We touch briefly on simple or standard notions, and pay more attention either to notions that are less traditional but necessary or that are more difficult; for example, moment generating functions, and conditional expectations. For the last topic, we even give exercises at the end of the chapter.

Sections 1-7 concern Probability Theory. In Section 8, we consider elements of the theory of interest.

1 PROBABILITY AND RANDOM VARIABLES

1.1 Sample space, events, probability measure

When building a model of any experiment, we first specify the space of all possible outcomes which may occur as a result of the experiment. We call such a space a *sample space*, or a *space of elementary outcomes*. A traditional notation for a sample space is Ω, and for its elements, the individual outcomes, is ω. So, $\Omega = \{\omega\}$.

We denote the standard set operations with sets $A, B, ...$ of elements from Ω: complement A^c, union $A \cup B$, and intersection $A \cap B$ of sets A, B. For $A \cap B$ we use a shorter notation AB, and call it the *product* of A and B. The reader is invited to verify on her/his own the set identities

$$(A \cap B)^c = A^c \cup B^c, \ (A \cup B)^c = A^c \cap B^c. \tag{1.1.1}$$

(A proof may be found in practically any probability textbook, e.g., [102], [116], [122].)

Next, we specify a collection, or a class, $\mathcal{A}$ of sets A which we will consider. Sets from $\mathcal{A}$ are called *events*.

For the theory to be complete and non-contradictory, the class $\mathcal{A}$ should be sufficiently rich; more precisely, we assume the following properties to be true:

(a) if $A \in \mathcal{A}$, then the complement A^c also belongs to $\mathcal{A}$;
(b) if events $A_1, A_2, ...$ are from $\mathcal{A}$, then their union $\cup_i A_i$ also belongs to $\mathcal{A}$. (1.1.2)

From (1.1.2) it follows that

(c) an empty set $\emptyset$ and the whole space Ω belong to $\mathcal{A}$;
(d) for any events $A_1, A_2, \ldots$ from $\mathcal{A}$, their intersection $\cap_i A_i \in \mathcal{A}$.

(The last property follows from properties (a), (b) and (1.1.1). To prove (c), we take any set $A \in \mathcal{A}$, and write $\emptyset = A \cap A^c \in \mathcal{A}$ by property (d), and $\Omega = A \cup A^c \in \mathcal{A}$ by property (b).)

We call a space Ω *discrete*, if it consists of a finite or countable number of points: $\Omega = \{\omega_1, \omega_2, \ldots\}$. Otherwise, we call the space *non-discrete* or *uncountable.*
If a space Ω is discrete, we can consider as $\mathcal{A}$ the class of *all* subsets of Ω.

► If Ω is uncountable, for example, if it may be identified with the real line, we cannot consider all subsets of Ω. The reason is that in general it is impossible to define probabilities simultaneously for all sets. So, some sets should be excluded from consideration. Fortunately, it suffices to exclude only some very exotic sets, which by no means will prevent us from building models of real phenomena. (For detail, see any advanced textbook on Measure Theory or Probability, e.g., [27], [70], [120], [129].) ◄

Throughout the book we assume that the class $\mathcal{A}$ of events under consideration is such that we are able to define probabilities of all events as we do in the following definition.
We call a *probability distribution*, or a *probability measure,* a function $P(A)$ of sets A from $\mathcal{A}$ such that

$$\begin{aligned} &(i) \quad 0 \leq P(A) \leq 1 \text{ for all } A \in \mathcal{A}; \\ &(ii) \quad P(\Omega) = 1; \\ &(iii) \quad P(\cup_i A_i) = \textstyle\sum_i P(A_i) \quad \text{for any } \textit{disjoint} \text{ events } A_1, A_2, \ldots \in \mathcal{A}. \end{aligned} \tag{1.1.3}$$

The value of $P(A)$ is called the probability of event A.
In particular, from definition (1.1.3) it follows that $P(\emptyset) = 0$. Indeed, $1 = P(\Omega) = P(\emptyset \cup \Omega) = P(\emptyset) + P(\Omega) = P(\emptyset) + 1$. So, $1 = P(\emptyset) + 1$, which implies that $P(\emptyset) = 0$.
Note also that in Property (i) we might require just $P(A) \geq 0$. Indeed, by Properties (iii) and (ii), $P(A) \leq P(A) + P(A^c) = P(\Omega) = 1$. We presented Property (i) in the above form for the completeness of the picture.

Next, we state the following two elementary properties of $P(A)$.

- For any events $A_1, A_2, \ldots \in \mathcal{A}$, *not necessarily disjoint,*

$$P(\cup_i A_i) \leq \sum_i P(A_i). \tag{1.1.4}$$

- For any $A, B \in \mathcal{A}$

$$P(A \cup B) = P(A) + P(B) - P(AB).$$

1.2 Independence and conditional probabilities

Events A_1 and A_2 are said to be *independent* if

$$P(A_1 A_2) = P(A_1) P(A_2). \tag{1.2.1}$$

We say that events $A_1, A_2, ..., A_n$ are *mutually independent* if for any sample of integers $(i_1, i_2, , ..., i_k)$ from $(1, ..., n)$

$$P(A_{i_1} \cdots A_{i_k}) = P(A_{i_1}) \cdots P(A_{i_k}).$$

For example, events A_1, A_2, A_3 are mutually independent if they are pairwise independent, that is,

$$P(A_1A_2) = P(A_1)P(A_2),\; P(A_1A_3) = P(A_1)P(A_3),\; P(A_2A_3) = P(A_2)P(A_3),$$

and

$$P(A_1A_2A_3) = P(A_1)P(A_2)P(A_3).$$

The conditional probability of event A given event B is

$$P(A \,|\, B) = \frac{P(AB)}{P(B)},$$

provided $P(B) \neq 0$. From (1.2.1) it follows that

> If A and B are independent, then $P(A \,|\, B) = P(A)$.

The following formula, in spite of its simplicity, proves to be a very useful tool in solving a great many probability problems.

The law of total probability, or **the formula for total probability.** Consider a collection of *disjoint* events H_i such that $\cup_i H_i = \Omega$. Such a collection is called a *partition.* Assume that $P(H_i) \neq 0$ for all i's. Then for any event A,

$$P(A) = \sum_i P(A \,|\, H_i)P(H_i). \tag{1.2.2}$$

In the next formula, we interpret H_i 's defined above as the events corresponding to different *hypotheses* about the nature of an experiment, or the different possible causes of an observable event A. The event A itself is viewed as a particular result of the experiment. In this case, $P(H_i)$ is called the *prior probability* that the ith hypothesis is true (that is, before the experiment is carried out). The probability $P(H_i \,|\, A)$ is called *posterior.* It is the probability that the ith hypothesis is true given that A occurred as a result of the experiment.

The Bayes formula (or rule). For any event A such that $P(A) \neq 0$, and events H_i defined above, for each i,

$$P(H_i \,|\, A) = \frac{P(A \,|\, H_i)}{P(A)} = \frac{P(A \,|\, H_i)}{\sum_k P(A \,|\, H_k)}. \tag{1.2.3}$$

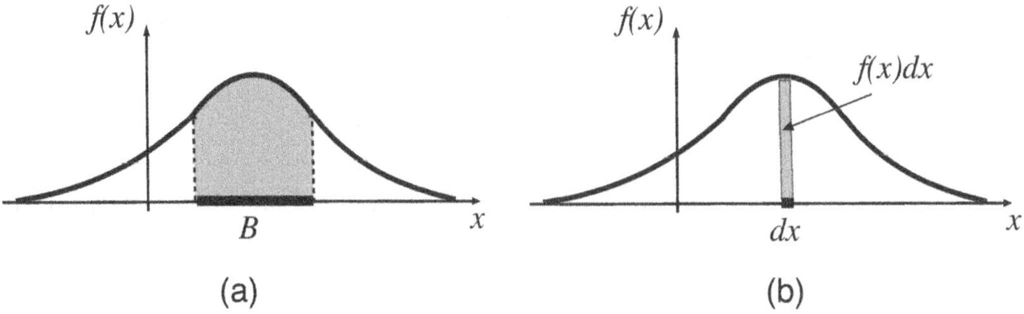

FIGURE 1.

1.3 Random variables, random vectors, and their distributions

1.3.1 Random variables

Below, we will use the symbol B for subsets of *the real line*.

A *random variable* (r.v.) $X = X(\omega)$ is a function defined on the space Ω. The function of sets $F_X(B) = P(X \in B)$ is said to be the *probability distribution* (briefly, distribution) of X. More precisely, $F_X(B)$ is the probability of the set $\{\omega : X(\omega) \in B\}$.

In other words, $F_X(B)$ is the probability that the value of X will fall in the set B. If it cannot cause misunderstanding, we will omit the index X in $F_X(B)$.

► Speaking more rigorously, for the theory to be non-contradictory, we should consider only sets B for which we can define the notion of length. Such sets are called Borel sets. Non-Borel sets exist but are very exotic, so we can exclude them from consideration. (A rigorous definition and/or examples may be found in advanced textbooks on Measure Theory or Probability; see, e.g., [27], [70], [120], [124], [129].)

Similar remarks may be made regarding the r.v. X itself. For the probability $P(X \in B)$ to be well defined, the set $\{\omega : X(\omega) \in B\}$ should belong to the class $\mathcal{A}$ of events A for which the probability $P(A)$ is defined. So formally, we define a r.v. as a function $X(\omega)$ for which the set $\{\omega : X(\omega) \in B\} \in \mathcal{A}$ for any (Borel) set B. (For more detail see, e.g., [27], [59], [70], [76], [120], [129].) In this book, we do not consider this issue.◄

A r.v. X and its distribution $F(B)$ are called *discrete* if X assumes a finite or countably infinite number of values $x_1, x_2, \ldots$. We say that the distribution $F(B)$ is concentrated at points $x_1, x_2, \ldots$.

A r.v. X and its distribution $F(B)$ are called *absolutely continuous,* if there exists a non-negative function $f(x)$ such that for any B

$$F(B) = \int_B f(x)dx, \tag{1.3.1}$$

that is, $P(X \in B)$ is the area above the set B and under $f(x)$; see Fig.1a. The function $f(x)$ is called the *probability density function* of the r.v. X. Briefly, we call $f(x)$ the *density* of X.

Setting $B = (-\infty, \infty)$, we have

$$\int_{-\infty}^{\infty} f(x)dx = 1. \tag{1.3.2}$$

Since a point is an interval of zero length, from (1.3.1) it follows that

For any absolutely continuous r.v. X, and any number a,
$$P(X = a) = 0. \tag{1.3.3}$$

For an infinitesimally small interval $[x, x+dx]$, the probability $P(x \le X \le x+dx)$ may be represented as $f(x)dx$; see Fig.1b.

The two types of distributions described above do not exhaust all possible distributions. For example, we can consider a mixture of discrete and continuous distributions; see for more detail Section 1.3.5. In this book, we omit the adjective 'absolute' and refer to absolutely continuous distributions as continuous.

► The point is that there exist distributions for which (1.3.3) is true (and hence these distributions are not discrete), but which cannot be represented in the form of (1.3.1), that is, they do not have densities. Such distributions are called continuous but non-absolutely continuous; see, e.g., [27], [120], [129]. These distributions are rather exotic, and we do not consider them here. ◄

1.3.2 Random vectors

Let $\mathbf{X} = (X_1, ..., X_k)$, where X_i's are r.v.'s. We call $\mathbf{X}$ a k-dimensional *random vector* (r.vec.). The function of sets $F(B) = F_{\mathbf{X}}(B) = P(\mathbf{X} \in B)$, where now B is a subset of the k-dimensional space $\mathbb{R}^k$, is the distribution of $\mathbf{X}$.

A r.vec. $\mathbf{X} = (X_1, ..., X_k)$ and its distribution $F(B)$ are said to be discrete if all coordinates X_i are discrete. A r.vec. $\mathbf{X}$ and its distribution $F(B)$ are said to be continuous if there exists a non-negative function $f(\boldsymbol{x})$, where $\boldsymbol{x} = (x_1, ..., x_k)$, such that for any B from $\mathbb{R}^k$,

$$F(B) = \int \cdots \int_B f(\boldsymbol{x}) d\boldsymbol{x}. \tag{1.3.4}$$

The integral in (1.3.4) is a k-dimensional integral, and the differential $d\boldsymbol{x} = dx_1 \cdots dx_k$. Setting $B = \mathbb{R}^k$ we have

$$\int \cdots \int_{\mathbb{R}^k} f(\boldsymbol{x}) d\boldsymbol{x} = 1. \tag{1.3.5}$$

R.v.'s $X_1, ..., X_k$ are said to be *mutually independent* if for any sets $B_1, ..., B_k$ from the real line, the events $\{X_1 \in B_1\}, ..., \{X_k \in B_k\}$ are mutually independent.

Consider now the case $k = 2$, and $\mathbf{X} = (X_1, X_2)$. Let r.v.'s X_i, take on values $x_{i1}, x_{i2}, ...$ (where $i = 1, 2$), and $f_{ij} = P(X_1 = x_{1i}, X_2 = x_{2j})$. We say that the probabilities f_{ij} specify the *joint distribution* of (X_1, X_2), and call f_{ij} *joint probabilities*.

The probabilities $f_i^{(1)} = P(X_1 = x_i)$ and $f_j^{(2)} = P(X_2 = x_j)$ are called *marginal*; they characterize the distributions of the coordinates separately. The collections $f^{(1)} = \left(f_1^{(1)}, f_2^{(1)}, ...\right)$ and $f^{(2)} = \left(f_1^{(2)}, f_2^{(2)}, ...\right)$ are called *marginal distributions*. The joint distribution completely determines the marginal distributions:

$$f_i^{(1)} = \sum_j f_{ij}, \; f_j^{(2)} = \sum_i f_{ij}.$$

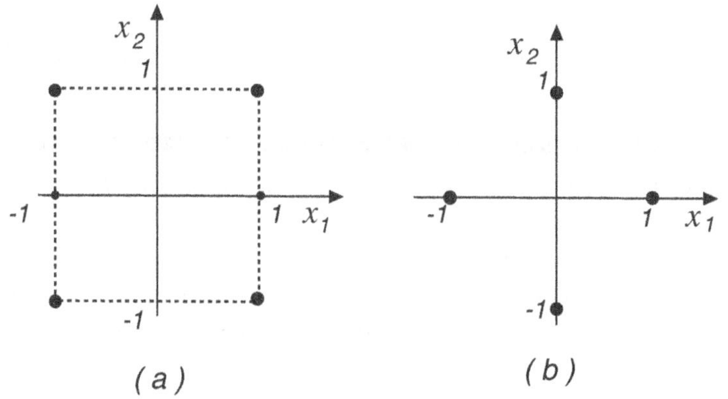

FIGURE 2.

The converse assertion is not true in general. To specify the joint distribution, we should know the marginal distributions and the structure of dependency between X_1 and X_2. In the case of independency the situation is simpler.

Proposition 1 *Discrete r.v.'s X_1, X_2 are independent if and only if*

$$f_{ij} = f_i^{(1)} f_j^{(2)} \text{ for all } i, j.$$

EXAMPLE 1. Let a r.vec. $\mathbf{X} = (X_1, X_2)$ take on four vector-values corresponding to four points in Fig.2a with equal probabilities $1/4$. In this case, it is convenient to set $f_{ij} = P(X_1 = i, X_2 = j)$, where $i = \pm 1, j = \pm 1$. One may guess that in this case, X_1, X_2 are independent. Indeed, $f_1^{(1)} = P(X_1 = 1) = \frac{1}{4} + \frac{1}{4} = \frac{1}{2}$, and similarly, all other marginal probabilities $f_{-1}^{(1)} = f_1^{(2)} = f_{-1}^{(2)} = \frac{1}{2}$. Hence, $f_{ij} = \frac{1}{4} = \frac{1}{2} \cdot \frac{1}{2} = f_i^{(1)} f_j^{(2)}$ for all $i = \pm 1, j = \pm 1$.

Now let $\mathbf{X}$ take on values corresponding to the four points in Fig.2b with equal probabilities. In this case, X_1 and X_2 take on values $0, +1, -1$ with probabilities $\frac{1}{2}, \frac{1}{4}, \frac{1}{4}$, respectively, and are certainly dependent. If, for instance, X_1 takes on the value 1, then X_2 may be only zero, while if $X_1 = 0$, then the r.v. X_2 may be either 1 or -1.

Certainly, the fact that X_1, X_2 are dependent follows from Proposition 1 also. For example, $f_{11} = P(X_1 = 1, X_2 = 1) = 0$, while $f_1^{(1)} f_1^{(2)} = \frac{1}{4} \cdot \frac{1}{4} \neq 0$. □

In the case where $k > 2$, the results are similar, but the notation is a bit cumbersome. We define the joint probabilities $f_{i_1 i_2 \dots i_k} = P(X_1 = x_{1i_1}, \dots, X_k = x_{i_k})$, and the marginal probabilities $f_j^{(m)} = P(X_m = x_{mj})$. Then

$$f_{i_m}^{(m)} = \sum\nolimits^{(m)} f_{i_1 \dots i_k},$$

where in $\sum^{(m)}$ summation is over all $i_1, \dots, i_{m-1}, i_{m+1}, \dots, i_k$, that is, i_m is fixed.

The independence case is described by

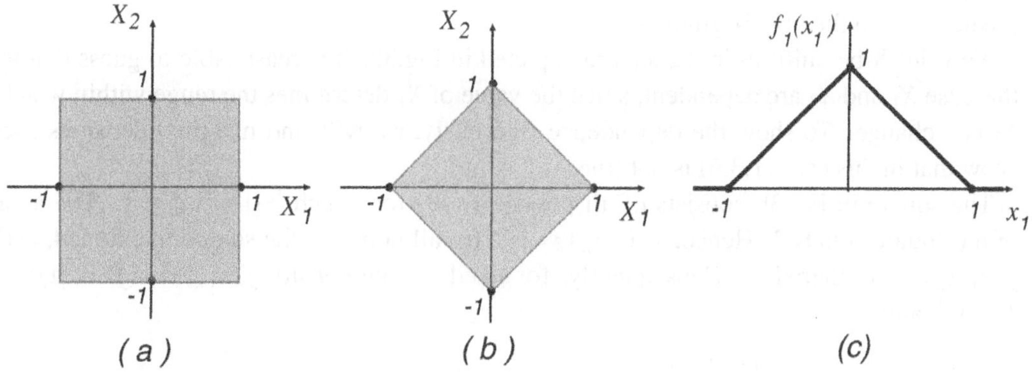

FIGURE 3.

Proposition 2 *Discrete r.v.'s $X_1,...,X_k$ are mutually independent if and only if*

$$f_{i_1 i_2 ... i_k} = f^{(1)}_{i_1} f^{(2)}_{i_2} \cdot ... \cdot f^{(k)}_{i_k} \text{ for all } i_1,...,i_k.$$

Now let a r.vec. $\mathbf{X} = (X_1, X_2)$ be continuous, $f(\boldsymbol{x}) = f(x_1,x_2)$ be its density, and let $f_1(x_1)$ and $f_2(x_2)$ be the separate marginal densities of the r.v.'s X_1 and X_2, respectively. Then

$$f_1(x_1) = \int_{-\infty}^{\infty} f(x_1,x_2)dx_2, \; f_2(x_2) = \int_{-\infty}^{\infty} f(x_1,x_2)dx_1.$$

If $k > 2$ and the joint density is a function $f(\boldsymbol{x}) = f(x_1,...,x_k)$, then the marginal density of X_m is given by

$$f_m(x_m) = \int f(x_1,...,x_{m-1},x_m,x_{m+1},...,x_k)dx_1...dx_{m-1}dx_{m+1}...dx_k,$$

where the integral is a $(k-1)$-dimensional integral with respect to $x_1,...,x_{m-1},x_{m+1},...x_k$.

Proposition 3 *Continuous r.v.'s $X_1,...,X_k$ are mutually independent if and only if*

$$f(x_1,...,x_k) = f_1(x_1) \cdot ... \cdot f_k(x_k) \quad \text{for any } x_1,...,x_k. \tag{1.3.6}$$

EXAMPLE 2. We use the same idea as in Example 1 for the case of continuous distribution. Let a r.vec. $\mathbf{X} = (X_1, X_2)$ take on values from the square in Fig.3a, and the joint density $f(x_1,x_2) = 1/4$ for all points $\mathbf{x} = (x_1,x_2)$ from this square, and $f(x_1,x_2) = 0$ otherwise.

(The total integral of f over the square should be one; see (1.3.5). Hence, if we set f equal to a constant, this constant should be equal to one divided by the area of the square.)

This type of distribution is called *uniform*, and we say that all points from the square are equiprobable.

For $|x_1| \leq 1$ and $|x_2| \leq 1$,

$$f_1(x_1) = \int_{-1}^{1} f(x_1,x_2)dx_2 = \int_{-1}^{1} \frac{1}{4}dx_2 = \frac{1}{2}, \; f_2(x_2) = \int_{-1}^{1} f(x_1,x_2)dx_1 = \frac{1}{2},$$

and $f(x_1,x_2) = \frac{1}{4} = \frac{1}{2}\cdot\frac{1}{2} = f_1(x_1)f_2(x_2)$. Hence, X_1, X_2 are independent, which could be predicted from the very beginning.

Now, let $\mathbf{X}$ be uniform in the square depicted in Fig.3b. It is reasonable to guess that in this case X_1 and X_2 are dependent, since the value of X_1 determines the range within which X_2 can change. To show the dependence rigorously, we will find marginal densities and show that in this case (1.3.6) is not true.

The square in Fig.3b consists of all points (x_1,x_2) for which $|x_1|+|x_2| \leq 1$. The area of the square equals 2. Hence, $f(x_1,x_2) = 1/2$ for all points in the square mentioned, and $f(x_1,x_2) = 0$ otherwise. Consequently, for fixed x_1, the density $f(x_1,x_2) = 0$ if $|x_2| > 1-|x_1|$, and

$$f_1(x_1) = \int_{-1+|x_1|}^{1-|x_1|} \frac{1}{2}dx_2 = 1-|x_1|, \text{ if } |x_1| \leq 1, \text{ and } = 0 \text{ otherwise.}$$

Similarly, $f_2(x_2) = 1-|x_2|$, if $|x_2| \leq 1$, and $= 0$ otherwise. The graph of $f_1(x_1)$ is given in Fig.3c. The distribution with this density is called *triangular*; it will appear in this book repeatedly.

Certainly, $f_1(x_1)f_2(x_2) = (1-|x_1|)(1-|x_2|) \neq \frac{1}{2} = f(x_1,x_2)$, and by Proposition 3, X_1 and X_2 are dependent. □

1.3.3 Cumulative distribution functions

The *cumulative distribution function,* or simply the *distribution function (d.f.)* of a r.v. X, is the function $F_X(x) = P(X \leq x)$. As usual, when it does not lead to confusion, we will omit the index X.

Note that if $F(B) = P(X \in B)$, the distribution of X, then the d.f. $F(x) = F((-\infty, x])$. Thus, if we know the distribution $F(B)$, we know the distribution function $F(x)$. We will see later that the converse assertion is also true: the distribution function completely determines $F(B)$ for any set B.

Any distribution function $F(x)$ is non-decreasing, $F(-\infty) = 0$, and $F(\infty) = 1$.

By definition, for any interval $(a,b]$ and r.v. X with a d.f. $F(x)$,

$$P(a < X \leq b) = P(X \leq b) - P(X \leq a) = F(b) - F(a).$$

From this it follows, in particular, that

If $F(x)$ is constant on an interval $(a,b]$, then $P(X \in (a,b]) = 0$.

(1.3.7)

See also Fig.4.

Let us consider the limit of $F(x) = P(X \leq x)$ as x converges to a number c from the left, that is, if $x < c$. Since for each $x < c$, the event $\{X \leq x\}$ does not include the point c,

$$\lim_{x\to c, x<c} P(X \leq x) = P(X < c).$$

Clearly, $P(X < c)$ is not equal to $P(X \leq c)$ if X assumes the value c with a positive probability. Set

$$F(c-0) = P(X < c)$$

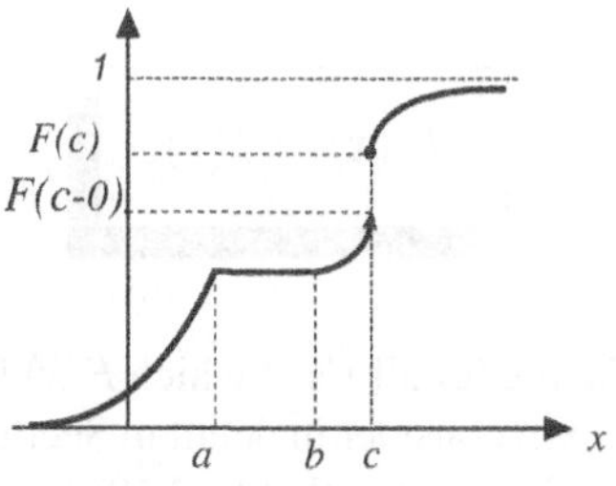

FIGURE 4.

(see Fig.4). Since $P(X = c) = P(X \leq c) - P(X < c)$, we have

$$P(X = c) = F(c) - F(c-0),$$

or, in other words,

For any c, the probability $P(X = c)$ equals the jump of $F(x)$ at the point c. (1.3.8)

See again Fig.4.

From (1.3.7)-(1.3.8) it follows that if a discrete r.v. X assumes values $x_1, x_2, \ldots$ with probabilities $f_1, f_2, \ldots$, respectively, then its d.f. $F(x)$ is constant in all intervals (x_i, x_{i+1}), and makes a jump of f_i at the point x_i, $i = 1, 2, \ldots$; see Fig.5.

Now consider the continuous case. Let $f(x)$ be the density of a r.v. X. Then by the definitions of density and d.f.,

$$F(x) = P(X \leq x) = \int_{-\infty}^{x} f(u)du. \tag{1.3.9}$$

(Certainly, once we wrote x as a limit of integration, we should use another letter inside the integral.)

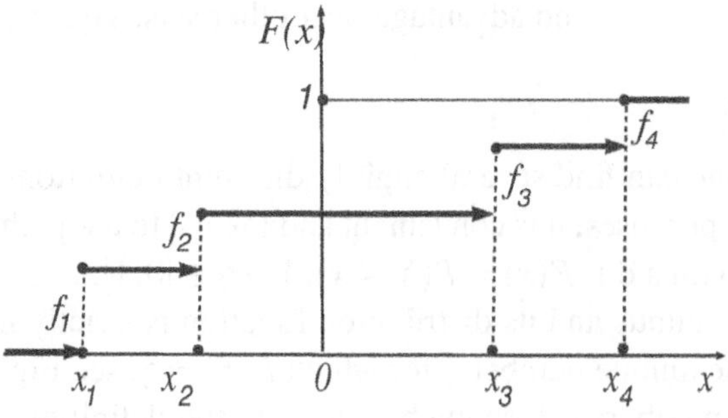

FIGURE 5. The distribution function of a discrete random variable.

From (1.3.9) it immediately follows that

$$\boxed{F'(x) = f(x).} \tag{1.3.10}$$

Rigorously speaking, (1.3.10) is true for all x's at which $F(x)$ is differentiable. A good example is the uniform distribution which we will consider in detail in Section 3.2.1. The d.f. and density in this case are graphed in Fig.10 on page 31. We see that (1.3.10) is true for all x's except 0 and 1, where $F'(x)$ does not exist. As a matter of fact, the function $F(x)$ is not differentiable at points where $f(x)$ is not continuous. The reader may look up this fact in practically any textbook on Calculus, e.g., [136]. We skip these formalities since in all schemes we consider in this book, we can exclude from consideration x's for which (1.3.10) does not hold.

It is worth realizing also that the relations (1.3.9)-(1.3.10) are those between a function and its antiderivative, and are strongly related to the second fundamental theorem of Calculus.

In conclusion, we briefly touch on the multidimensional case. Let $\mathbf{X} = (X_1,...,X_k)$, where X_i's are r.v.'s., and $\boldsymbol{x}$ stands for a non-random vector $(x_1,...,x_k)$. The d.f. of $\mathbf{X}$ is the function

$$F(\boldsymbol{x}) = F(x_1,...,x_k) = P(X_1 \le x_1,...,X_k \le x_k).$$

This is the probability of a "corner" (see Fig.6).

Suppose that $\mathbf{X}$ has a density $f(\boldsymbol{x})$. By virtue of (1.3.4), the multidimensional counterpart of (1.3.9) in this case is the relation

$$F(x_1,...,x_k) = \int_{-\infty}^{x_1} ... \int_{-\infty}^{x_k} f(u_1,...,u_k)du_1...du_k. \tag{1.3.11}$$

Differentiating (1.3.11), we get the counterpart of (1.3.10):

$$f(x_1,...,x_k) = \frac{\partial^k}{\partial x_1 ... \partial x_k} F(x_1,...,x_k) \tag{1.3.12}$$

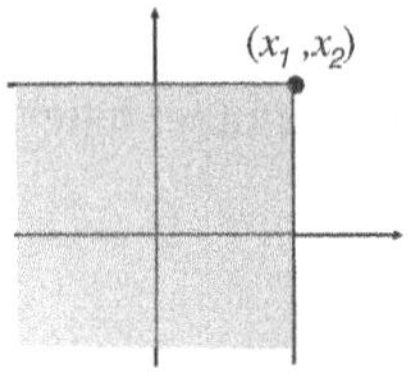

FIGURE 6.

The last formula turns out to be useful in many problems. However, in general, d.f.'s in the multidimensional case do not play as important a role as they do in the one-dimensional case. They represent the probabilities of "corners", and these sets in the multidimensional framework have no advantage over other sets, say, circles.

1.3.4 Quantiles

In the literature, one can find several slightly different definitions of a quantile (see below). For our further purposes, it is convenient and logical to adopt the following definition.

Consider a r.v. X with a d.f. $F(x) = P(X \le x)$. Let $\gamma \in [0,1]$.

If the r.v. X is continuous and its distribution function is strictly increasing, then q_γ, the γ-*quantile* of X, is the unique number q for which $F(q) = \gamma$; see Fig.7a.

If there are many numbers q for which $F(q) = \gamma$, the definition we adopt, chooses the right end point of the interval where $F(x) = q$; see Figures 7b,c. In the literature, one can

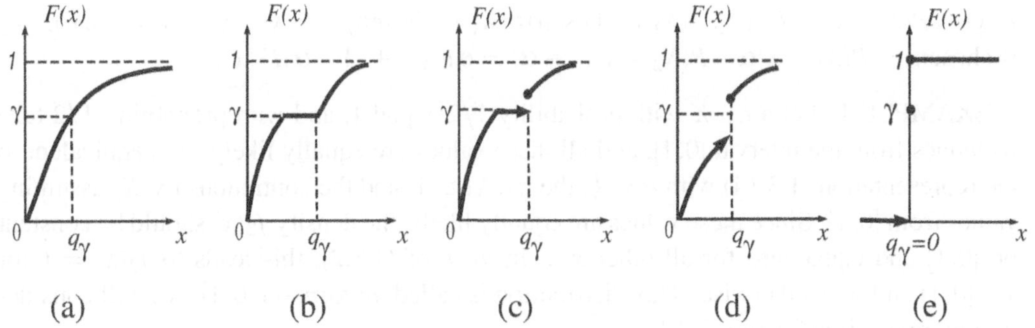

FIGURE 7. Quantiles.

find definitions where the γ-quantile is the left end point, or the middle, or even any point from this interval. The difference is not essential.

If the r.v. takes on some values with positive probabilities (and hence the d.f. has "jumps"), it may happen that there is no number q such that $F(q) = \gamma$; see Fig.7d. Then we choose the point at which $F(x)$ "jumps" over the level γ.

In particular, if $X = 0$ with probability one, the point 0 is the γ-quantile for all $\gamma \in [0, 1)$ (see Fig.7e).

The 0.5-quantile is called a *median*. If the distribution is continuous, then $P(X \le q_{0.5}) = P(X \ge q_{0.5}) = 0.5$. (The r.v. is as likely to be larger than the median as it is to be smaller.) In general, since it may happen that $P(X = q_{0.5}) > 0$, we have $P(X < q_{0.5}) \le 0.5$ and $P(X > q_{0.5}) \le 0.5$.

Another term in use for a γ-quantile is a 100γth *percentile*.

Formally, the above definitions may be unified as follows. The γ-quantile q_γ is the number such that $F(q_\gamma - \varepsilon) \le \gamma$ and $F(q_\gamma + \varepsilon) > \gamma$ for any arbitrarily small $\varepsilon > 0$.

The reader familiar with the notion of supremum may realize that the γ-quantile above may be also defined as $q_\gamma = \sup\{x : F(x) \le \gamma\}$.

1.3.5 Mixtures of distributions

Let $F_1(B)$ and $F_2(B)$ be two distributions, and let $\alpha \in [0, 1]$. Consider the distribution

$$F^{(\alpha)}(B) = \alpha F_1(B) + (1 - \alpha)F_2(B). \tag{1.3.13}$$

We call the distribution $F^{(\alpha)}(B)$ a *mixture of distributions* F_1, F_2. In particular, the d.f.

$$F^{(\alpha)}(x) = \alpha F_1(x) + (1 - \alpha)F_2(x).$$

Such a definition admits an explicit interpretation. Let F_1 and F_2 be the distributions of r.v.'s X_1 and X_2, respectively, and a r.v.

$$X = \begin{cases} X_1 & \text{with probability } \alpha, \\ X_2 & \text{with probability } 1 - \alpha. \end{cases} \tag{1.3.14}$$

In other words, we choose X_1 with probability α, and X_2 with probability $1 - \alpha$. (We skip formalities of defining the r.v.'s X, X_1, X_2 on the same sample space.) Then the distribution of X is the linear combination (1.3.13) of F_1 and F_2, which we call here a mixture. Indeed, $F(x) = P(X \le x) = P(X \le x \,|\, X_1 \text{ is chosen})P(X_1 \text{ is chosen}) + P(X \le x \,|\, X_2 \text{ is chosen})P(X_2$ is chosen$) = P(X_1 \le x)\alpha + P(X_2 \le x)(1-\alpha) = \alpha F_1(x) + (1-\alpha)F_2(x)$.

EXAMPLE 1. Let a r.v. X with probability $1/2$ equal 1, and with probability $1/2$ take on values from the interval $[0,1]$, and all these values are equally likely. It is equivalent to the representation (1.3.14) with $\alpha = \frac{1}{2}$, the r.v. $X_1 \equiv 1$, and the continuous r.v. X_2 assuming values from $[0,1]$. Since these values are equally likely, the density $f_2(x)$ should be constant on $[0,1]$ and equal zero for all other x's. In view of (1.3.2), this leads to $f_2(x) = 1$ for $x \in [0,1]$ and $= 0$, otherwise. This distribution is called *uniform* on $[0,1]$; we will consider it in more detail in Section 3.2.1.

By (1.3.9), $F_2(x) = x$ for $x \in [0,1]$, $F_2(x) = 0$ for $x < 0$, and $F_2(x) = 1$ for $x > 1$. For more detail on the d.f. of a uniform distribution see Section 3.2.1.

The r.v. X_1 takes on the value 1 with probability one. So, $F_1(x) = 0$ for $x < 1$, $F_1(x) = 1$ for $x \ge 1$ (graph $F_1(x)$ by analogy with Fig.7d). Then

$$F(x) = \frac{1}{2}F_1(x) + \frac{1}{2}F_2(x), \tag{1.3.15}$$

which amounts to

$$F(x) = \begin{cases} 0 & \text{if } x < 0, \\ \frac{1}{2}x & \text{if } 0 \le x < 1, \\ 1 & \text{if } x \ge 1; \end{cases} \tag{1.3.16}$$

the graph is given in Fig.8. Note that the last distribution is neither continuous nor discrete. It is a mixture of continuous and discrete distributions. □

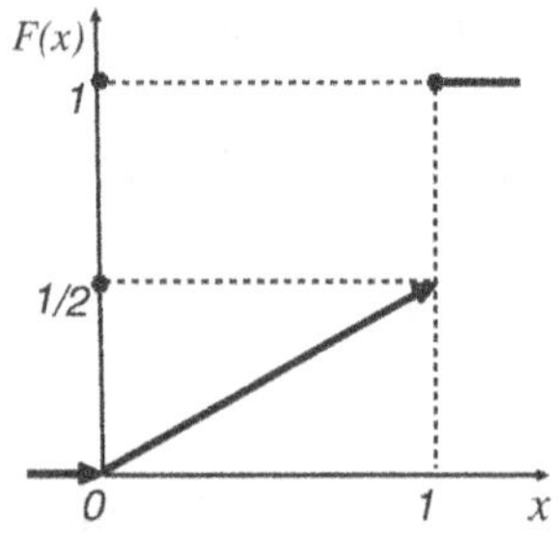

FIGURE 8.

2 EXPECTATION

2.1 Definitions

Below, unless stated otherwise, we assume that all series or integrals under consideration exist and are finite.

The expected value of a r.v. is often defined in textbooks for the discrete and continuous cases separately. If a r.v. X is discrete and takes on values $x_1, x_2, \ldots$ with probabilities $f_1, f_2, \ldots$, respectively, the *expected value*, or the *mean value*, by definition, is

$$E\{X\} = \sum_i x_i f_i \tag{2.1.1}$$

(where the summation is taken over all possible values i), provided that the above series converges.

If $f(x)$ is the density of a r.v. X, its expected value

$$E\{X\} = \int_{-\infty}^{\infty} xf(x)dx, \tag{2.1.2}$$

provided that the above integral exists and is finite.

It may be proved that, similar to (2.1.1), for a function $u(x)$,

$$E\{u(X)\} = \sum_i u(x_i)f_i, \tag{2.1.3}$$

where the summation is over all possible values of i.

If $f(x)$ is the density of a r.v. X, the expected value

$$E\{u(X)\} = \int_{-\infty}^{\infty} u(x)f(x)dx. \tag{2.1.4}$$

Clearly, (2.1.1) and (2.1.2) follow from (2.1.3) and (2.1.4) respectively, if we set $u(x) = x$.

We can unify and generalize the definitions above by writing

$$E\{u(X)\} = \int_{-\infty}^{\infty} u(x)dF(x), \tag{2.1.5}$$

where $F(x)$ is the d.f. of X. The last integral is called a *Riemann-Stieltjes integral* and is defined as follows:

- If X has a density $f(x)$, then $dF(x) = f(x)dx$, and the integral above coincides with (2.1.4).
- If $F(x)$ is constant on an interval $(a,b]$, that is, $P(X \in (a,b]) = 0$, we again define dF on this interval as a usual differential and have $dF(x) = 0$. (In this case the interval $(a,b]$ is automatically excluded from integration in (2.1.5).)
- If at a point c, the d.f. $F(x)$ has a jump of $\delta = F(c) - F(c-0)$, that is, $P(X = c) = \delta$, then we define $dF(c)$ as the jump at c, that is, $dF(c) = \delta$. This is a definition, but it is quite natural for the following reason. The differential $dF(x)$ means the change of $F(x)$ in the infinitesimally small interval $[x, x+dx]$, but if $F(x)$ jumps at c, the change is not small and equals $F(c) - F(c-0)$.

  So, the part of the integral (2.1.5) at the point c above is $u(c)\delta = u(c)[F(c) - F(c-0)]$.
- In particular, if X is discrete, and assumes values $x_1, x_2, \ldots$ with probabilities $f_1, f_2, \ldots$, respectively, we set $dF(x_j) = f_j$ for all j, and $dF(x) = 0$ in all intervals between x_j's. So defined, the integral $\int_{-\infty}^{\infty} u(x)dF(x) = \sum_j u(x_j)f_j$; that is, we have come to the definition (2.1.3).

In any case,

> We view $dF(x)$ as $P(x \le X \le x+dx)$, the probability that X will assume a value from an infinitesimally small interval $[x, x+dx]$.

In particular, definition (2.1.5) allows us to consider mixtures of continuous and discrete distributions.

EXAMPLE 1. Let us return to Example 1.3.5-1. Let $u(x) = x^2$. By (1.3.16), we can write

$$E\{u(X)\} = \int_{-\infty}^{\infty} u(x)dF(x) = \int_0^1 u(x)\cdot\frac{1}{2}dx + u(1)[F(1)-F(1-0)]$$
$$= \int_0^1 x^2\frac{1}{2}dx + 1^2\cdot\frac{1}{2} = \frac{2}{3}.$$

Another way to solve the same problem is to use (1.3.15) by writing $dF(x) = \frac{1}{2}dF_1(x) + \frac{1}{2}dF_2(x)$ and substituting it into (2.1.5). We have

$$E\{u(X)\} = \int_{-\infty}^{\infty} u(x)dF(x) = \frac{1}{2}\int_{-\infty}^{\infty} u(x)dF_1(x) + \frac{1}{2}\int_{-\infty}^{\infty} u(x)dF_2(x)$$
$$= \frac{1}{2}u(1) + \frac{1}{2}\int_0^1 u(x)dx = \frac{1}{2}1^2 + \frac{1}{2}\int_0^1 x^2dx = \frac{2}{3}. \ \square$$

Next, we consider a set B in the real line, and the function

$$I_B(x) = \begin{cases} 1 \text{ if } x \in B, \\ 0 \text{ if } x \notin B. \end{cases}$$

The function $I_B(x)$ is called the *indicator* of B.

Let $u(x)$ in (2.1.5) be equal to $I_B(x)$. Then the r.v. $u(X) = I_B(X) = 1$ if $X \in B$, and 0 otherwise. Hence, $E\{u(X)\} = 1\cdot P(X \in B) + 0\cdot P(X \notin B) = P(X \in B) = F_X(B)$. On the other hand, in view of (2.1.5),

$$E\{u(X)\} = \int_{-\infty}^{\infty} I_B(x)dF(x) = \int_B dF(x)$$

(since $I_B(x) = 0$ for $x \notin B$). Thus,

$$F_X(B) = \int_B dF(x), \tag{2.1.6}$$

and we see that, indeed, the d.f. $F(x)$ determines $F_X(B)$ for all B.

In conclusion, we state without proofs two elementary properties of expectation.

(i) For any numbers c_1, c_2 and any r.v.'s X_1, X_2,

$$E\{c_1X_1 + c_2X_2\} = c_1E\{X_1\} + c_2E\{X_2\}. \tag{2.1.7}$$

(ii) For any independent r.v.'s X_1, X_2,

$$E\{X_1X_2\} = E\{X_1\}E\{X_2\}, \tag{2.1.8}$$

provided that the expectations above exist.

2.2 Integration by parts and a formula for expectation

An essential advantage of the representation (2.1.5) is that it points out the possibility of integration by parts. Consider, for simplicity, a positive r.v. X, so its d.f. $F(x) = 0$ for $x < 0$. Let $u(x)$ be a differentiable function such that the integral $\int_0^\infty u(x)dF(x)$ exists and is finite. Set, also for simplicity, $u(0) = 0$. Then

$$\begin{aligned} E\{u(X)\} &= \int_0^\infty u(x)dF(x) = -\int_0^\infty u(x)d(1-F(x)) \\ &= -u(\infty)[1-F(\infty)] + u(0)[1-F(0)] + \int_0^\infty (1-F(x))du(x). \end{aligned}$$

Since $u(0) = 0$, we have $u(0)[1-F(0)] = 0$. Because $F(\infty) = 1$, if $u(\infty) = \infty$, then in the expression $u(\infty)[1-F(\infty)]$, we have the indeterminate form $\infty \cdot 0$. As a matter of fact, we can set $u(\infty)[1-F(\infty)] = 0$. To prove this, one should show that $|u(x)|[1-F(x)] \to 0$ as $x \to \infty$. We will prove it in Section 2.5. So, eventually

$$E\{u(X)\} = \int_0^\infty (1-F(x))du(x). \tag{2.2.1}$$

Setting $u(x) = x$, we obtain from (2.2.1) a useful formula for the expected value of a positive r.v.:

$$E\{X\} = \int_0^\infty (1-F(x))dx. \tag{2.2.2}$$

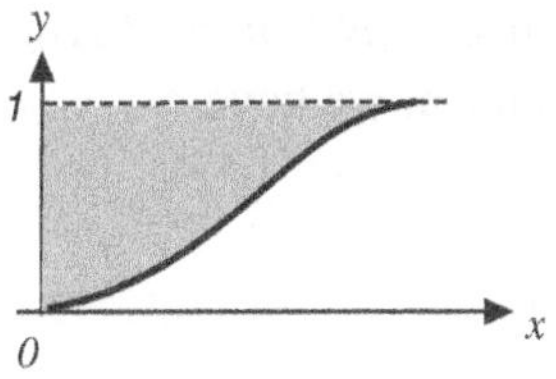

FIGURE 9.

Some examples are considered in Section 3.

In particular, from (2.2.2) it follows that the expected value equals the area between the graph of $F(x)$ and the line $y = 1$ (see Fig.9).

Consider now an integer valued r.v. X assuming only values $0, 1, 2, \ldots$. In this case, the d.f. $F(x)$ is constant on intervals $[n, n+1)$ for $n = 0, 1, \ldots$ (see also Fig.5). Then, for any $x \in [n, n+1)$, we have $1 - F(x) = 1 - F(n) = P(X > n)$, and from (2.2.2) it follows that

$$E\{X\} = \sum_{n=0}^{\infty} \int_n^{n+1} (1-F(x))dx = \sum_{n=0}^{\infty} P(X > n) \int_n^{n+1} dx = \sum_{n=0}^{\infty} P(X > n) \cdot 1.$$

Thus,

$$E\{X\} = \sum_{n=0}^{\infty} P(X > n). \tag{2.2.3}$$

We will use formulas (2.2.2) and (2.2.3) repeatedly in this book.

2.3 Can we encounter an infinite expected value in models of real phenomena?

Certainly, we can construct a r.v. X for which $E\{X\} = \infty$. Let, say, X take on values 2, 4, 8, ..., 2^k, ..., and so on, with probabilities $\frac{1}{2}$, $\frac{1}{4}$, ..., 2^{-k}, ..., respectively. Then $E\{X\} =$

$2 \cdot \frac{1}{2} + 2 \cdot \frac{1}{2} + ... + 2^k \cdot 2^{-k} + ... = 1 + 1 + ... + 1 + ... = \infty$. We consider this r.v. in Section 1.2.2 in connection with the so called St. (Saint) Petersburg's paradox.

This example, however, is somewhat artificial, and the question is whether we can encounter infinite expectations in models of real phenomena. In this book, we will see several such examples. The first is considered below.

EXAMPLE 1. (*Record values.*) Let $X_1, X_2, ...$ be a sequence of independent identically distributed (i.i.d.) r.v.'s, and N be the number of the first X_i which is larger than X_1. Formally, $N = \min\{n : X_n > X_1\}$. Let, for instance, X_1 be the first payment made by an insurance company in the new year, $X_2, X_3, ...$ be next consecutive payments, and N be the number of the first payment among $X_2, X_3, ...$ which was larger than X_1. We show that $E\{N\} = \infty$.

Let the event $A_{in} = \{X_i \geq X_1, ..., X_i \geq X_n\}$, $i = 1, ..., n$. So, if A_{in} occurs, then among the first n r.v.'s, the ith r.v. assumes a *record* value in the sense that it is not less than the values of the other $n-1$ r.v.'s. Then the event $\{N > n\} = A_{1n}$. In view of symmetry, $P(A_{in})$ is the same for all i. Indeed, because the X's are i.i.d., among the r.v.'s $X_1, ..., X_n$ each has "an equal chance to take on a value not smaller than the values of the others." On the other hand, at least one A_{in} should occur, which means that $\cup_{i=1}^n A_{in} = \Omega$. Then, by (1.1.4),

$$1 = P(\Omega) = P(\cup_{i=1}^n A_{in}) \leq \sum_{i=1}^n P(A_{in}) = nP(A_{1n}).$$

In the last equality, we used that $P(A_{in}) = P(A_{1n})$ for all i. Hence, $P(N > n) = P(A_{1n}) \geq 1/n$ for all $n = 1, 2, ...$ Using (2.2.3) and neglecting in it the first term, we have

$$E\{N\} = \sum_{n=0}^{\infty} P(N > n) \geq \sum_{n=1}^{\infty} \frac{1}{n}.$$

As we remember from Calculus, the last series (called harmonic) diverges, that is, equals infinity. □

To make the definition of an infinite expected value more precise, we say that for a non-negative r.v. X with a d.f. F, the expected value $E\{X\} = \infty$ if the integral $\int_0^\infty x dF(x) = \infty$. If X is non-positive, we say that $E\{X\} = -\infty$ if $\int_{-\infty}^0 x dF(x) = -\infty$.

When dealing with r.v.'s taking positive and negative values, we should be more cautious. Any r.v. X may be represented as $X = X^+ - X^-$, where X^+ and X^- are positive and negative parts of X, respectively. More precisely, $X^+ = \max\{X, 0\}$, and $X^- = \max\{-X, 0\}$. Note that both quantities, X^+, X^-, are positive. Let us write $E\{X\} = E\{X^+\} - E\{X^-\}$.

- If $E\{X^+\} < \infty$, and $E\{X^-\} < \infty$, we define $E\{X\}$ in the usual way, and we say that $E\{X\}$ is finite. Certainly, in this case, $E\{X\} = E\{X^+\} - E\{X^-\}$.
- If $E\{X^+\} = \infty$, and $E\{X^-\} < \infty$, we say that $E\{X\} = \infty$.
- If $E\{X^+\} < \infty$, and $E\{X^-\} = \infty$, we say that $E\{X\} = -\infty$.
- If $E\{X^+\} = \infty$, and $E\{X^-\} = \infty$, we say that $E\{X\}$ does not exist.

2.4 Moments of r.v.'s. Correlation

2.4.1 Variance and other moments

Assuming that $E\{X^2\} < \infty$, we define the *variance* of X as the quantity

$$Var\{X\} = E\{(X - E\{X\})^2\}.$$

Skipping proofs, we list some properties of variance. First, it is easy to verify that

$$Var\{X\} = E\{X^2\} - (E\{X\})^2. \tag{2.4.1}$$

Secondly, for any constant c,

$$Var\{X + c\} = Var\{X\}, \;\; Var\{cX\} = c^2 Var\{X\}. \tag{2.4.2}$$

Furthermore, for any *independent* r.v.'s X and Y,

$$Var\{X + Y\} = Var\{X\} + Var\{Y\}. \tag{2.4.3}$$

Note that $Var\{X - Y\}$ is equal to $Var\{X\} + Var\{Y\}$ rather than to $Var\{X\} - Var\{Y\}$ (!). Indeed, by (2.4.2),

$$\begin{aligned} Var\{X - Y\} &= Var\{X + (-Y)\} = Var\{X\} + Var\{-Y\} = Var\{X\} + (-1)^2 Var\{Y\} \\ &= Var\{X\} + Var\{Y\}. \end{aligned}$$

The quantity $\sigma_X = \sqrt{Var\{X\}}$ is called the *standard deviation* of X.

For a r.v. X, the quantities $m_k = E\{X^k\}$ for a natural k and $\overline{m}_k = E\{|X|^k\}$ for any $k \geq 0$ are called the kth *moment* and the kth *absolute moment*, respectively. So, the expectation $E\{X\}$ is the first moment.

Set $m = m_1$. The quantities $\mu_k = E\{(X - m)^k\}$ and $\bar{\mu}_k = E\{|X - m|^k\}$ are called the kth *central moment* and the kth *absolute central moment*, respectively. So, the variance is the second central moment.

Before considering the notions of covariance and correlation, we present a fundamental inequality.

2.4.2 The Cauchy-Schwarz inequality

For any r.v.'s ξ and η with finite second moments,

$$(E\{\xi\eta\})^2 \leq E\{\xi^2\}E\{\eta^2\}. \tag{2.4.4}$$

Proof. Let t be a real number. Then

$$0 \leq E\{(\xi - t\eta)^2\} = E\{\xi^2\} - 2tE\{\xi\eta\} + t^2 E\{\eta^2\}. \tag{2.4.5}$$

Denote by $Q(t)$ the r.-h.s. of (2.4.5). As a function of t, this is a quadratic function. Since $Q(t) \geq 0$ for all t, the discriminant of $Q(t)$ should be non-positive. The discriminant is equal to $4(E\{\xi\eta\})^2 - 4E\{\xi^2\}E\{\eta^2\} = 4[(E\{\xi\eta\})^2 - E\{\xi^2\}E\{\eta^2\}]$. We see that it is non-positive if and only if (2.4.4) is true. ∎

It is straightforward to verify that, if $\xi = t_0\eta$ for some number t_0, then (2.4.4) becomes a strict *equality*.

Conversely, let $(E\{\xi\eta\})^2 = E\{\xi^2\}E\{\eta^2\}$. Then the discriminant above equals zero. Consequently, there exists only one root—say, t_0, of the equation $Q(t) = 0$. In this case, $E\{(\xi - t_0\eta)^2\} = 0$. The r.v. $(\xi - t_0\eta)^2$ is non-negative. Its expectation may be equal to zero only if $\xi - t_0\eta = 0$ with probability one. Thus,

$$(E\{\xi\eta\})^2 = E\{\xi^2\}E\{\eta^2\} \text{ iff } \xi = t_0\eta \text{ for some } t_0 \text{ with probability one.} \tag{2.4.6}$$

2.4.3 Covariance and correlation

Covariance is a measure of dependency between two r.v.'s. Let X_1, X_2 be r.v.'s with finite second moments, $m_i = E\{X_i\}$, $\sigma_i^2 = Var\{X_i\} > 0$. (Unlike above, here the symbol m_i denotes the mean value of X_i.) We call the covariance between X_1, X_2 the quantity

$$Cov\{X_1, X_2\} = E\{(X_1 - m_1)(X_2 - m_2)\}.$$

Multiplying the variables in the parentheses above, one may prove that

$$Cov\{X_1, X_2\} = E\{X_1X_2\} - m_1m_2 \tag{2.4.7}$$

(compare with (2.4.1)).

Note also that $Cov\{X_1, X_1\} = E\{(X_1 - m_1)^2\} = Var\{X_1\}$.

If X_1, X_2 are independent, $Cov\{X_1, X_2\} = 0$. Indeed, in this case, by virtue of (2.1.8), $Cov\{X_1, X_2\} = E\{X_1 - m_1\}E\{X_2 - m_2\} = 0$, since $E\{X_1 - m_1\} = E\{X_1\} - m_1 = m_1 - m_1 = 0$. (Certainly, $E\{X_2 - m_2\}$ also equals zero, but it suffices to consider $E\{X_1 - m_1\}$.)

Thus,

> If $Cov\{X_1, X_2\} \neq 0$, then the r.v.'s X_1, X_2 are dependent.

The converse to the above assertion is not true.

EXAMPLE 1. Consider a r.vec. $\mathbf{X} = (X_1, X_2)$ whose distribution is presented in Fig.2b in Example 1.3.2-1. As was noted there, the r.v.'s X_1, X_2 in this case are dependent. On the other hand, as is easy to compute, $E\{X_1\} = E\{X_2\} = 0$, and in view of (2.4.7), $Cov\{X_1, X_2\} = E\{X_1X_2\}$. In our example, for any outcome, one of variables, X_1 or X_2, equals zero, so $X_1X_2 = 0$. Hence, $Cov\{X_1, X_2\} = 0$. □

Thus, independence implies zero covariance, but not vice versa.

It is straightforward to verify that for any r.v.'s X_1, X_2,

$$Var\{X_1 + X_2\} = Var\{X_1\} + Var\{X_2\} + 2Cov\{X_1, X_2\}.$$

Thus, for (2.4.3) to be true, we only need $Cov\{X_1, X_2\}$ to be zero, which, as we saw, is a weaker property than independence.

The *correlation coefficient* of r.v. X_1, X_2, briefly the *correlation*, is the quantity

$$Corr\{X_1, X_2\} = \frac{Cov\{X_1, X_2\}}{\sigma_1\sigma_2}. \tag{2.4.8}$$

(We have assumed in the beginning of this section that $\sigma_i > 0$.) Note that correlation is a dimensionless characteristic. Say, if we measure X_1, X_2 in dollars, the dimension of $Cov\{X_1, X_2\}$ and $Var\{X_i\}$ is dollar^2, while $Corr\{X_1, X_2\}$, as follows from (2.4.8), does not have a dimension. For this reason, correlation, which may be viewed as the normalized covariance, is a more adequate measure of dependence.

The following properties are true.

1. $-1 \leq Corr\{X_1, X_2\} \leq 1$ for all X_1, X_2.

2. If $Corr\{X_1, X_2\} = 1$, then $X_2 = a + bX_1$ for some a and $b > 0$.

3. If $Corr\{X_1, X_2\} = -1$, then $X_2 = a + bX_1$ for some a and $b < 0$.

R.v.'s X_1, X_2 for which $Corr\{X_1, X_2\} = 0$ are called *non-correlated* or *uncorrelated.* R.v.'s for which $Corr\{X_1, X_2\} > 0$, are called *positively correlated.* If $Corr\{X_1, X_2\} < 0$, the r.v.'s X_1, X_2 are *negatively correlated.* R.v.'s for which $Corr\{X_1, X_2\} = \pm 1$, are called *perfectly correlated.*

A good exercise is to derive Property 1 from (2.4.4), and Properties 2-3 from (2.4.6). (Set $\xi = X_1 - m_1$, $\eta = X_2 - m_2$.)

2.5 Inequalities for deviations

For any r.v. X, the probability $P(X > x) \to 0$ as $x \to \infty$. Our next goal is to estimate how fast it is vanishing. For large x's, the probability $P(X > x)$ is often called a *tail* of the distribution of X, or the probability of *large deviation.* Similarly, one can consider $P(X < x)$ as $x \to -\infty$. The latter is referred as a left tail, and the former—in this context—as a right tail.

Proposition 4 *(An inequality for deviations). Let a function $u(x)$ be non-negative and non-decreasing. Then for any x and any r.v. X,*

$$P(X \geq x) \leq \frac{E\{u(X)\}}{u(x)}, \tag{2.5.1}$$

provided that the r.-h.s. of (2.5.1) is finite.

Replacing X in (2.5.1) by $|X|$, and setting $u(x) = x^k$ for $x > 0$ and $k > 0$, we come to

Corollary 5 *(Markov's inequality). For any $x > 0$,*

$$P(|X| \geq x) \leq \frac{E\{|X|^k\}}{x^k} = \frac{\overline{m}_k}{x^k}. \tag{2.5.2}$$

Let $m = E\{X\}$. Setting $k = 2$ in (2.5.2) and replacing X by $X - m$, we come to

Corollary 6 *(Chebyshev's inequality). For any* $x > 0$,

$$P(|X-m| \geq x) \leq \frac{E\{(X-m)^2)\}}{x^2} = \frac{Var\{X\}}{x^2}. \tag{2.5.3}$$

Proof of Proposition 4. Let $F(x)$ be the d.f. of X. Using consecutively in the inequalities below the facts that $u(x)$ is non-negative and non-decreasing, we have

$$E\{u(X)\} = \int_{-\infty}^{\infty} u(z)dF(z) \geq \int_{x}^{\infty} u(z)dF(z) \geq \int_{x}^{\infty} u(x)dF(z) \geq$$
$$= u(x)\int_{x}^{\infty} dF(z) = u(x)P(X \geq x). \blacksquare \tag{2.5.4}$$

► Let X be non-negative, and $x > 0$. In (2.5.4), we have shown that

$$\int_{x}^{\infty} u(z)dF(z) \geq u(x)P(X \geq x). \tag{2.5.5}$$

If the integral $\int_0^\infty u(x)dF(x)$ is finite, the l.-h.s. of (2.5.5) converges to zero as $x \to \infty$. Then so does the r.-h.s., which we promised to prove in Section 2.2. ◄

2.6 Linear transformations of r.v.'s. Normalization

Let X be a r.v. with a d.f. $F(x)$. Denote by $f(x)$, m, and σ^2 the density, the expected value, and the variance of X, respectively. Let a r.v. $Y = a + bX$, where a and b are numbers, $b > 0$. Then the d.f. of Y is the function

$$F_Y(x) = P(Y \leq x) = P(a+bX \leq x) = P(X \leq (x-a)/b) = F((x-a)/b).$$

If the density $f(x)$ exists, we can use (1.3.10), which implies that the density of Y is $f_Y(x) = \frac{d}{dx}F_Y(x) = \frac{1}{b}f((x-a)/b)$. Using rules (2.1.7) and (2.4.2) for means and variances, we summarize all of this in the following table.

	the d.f.	the density	the mean	the variance
X	$F(x)$	$f(x)$	m	σ^2
$a+bX,\ b>0$	$F((x-a)/b)$	$\frac{1}{b}f((x-a)/b)$	$a+bm$	$b^2\sigma^2$

(2.6.1)

Let $X' = X - m$. Then

$$E\{X'\} = E\{X\} - m = m - m = 0.$$

The r.v. X' is called *centered*, and the operation itself—*centering*.
Next, assuming $\sigma > 0$, consider the r.v.

$$X^* = \frac{X-m}{\sigma}. \tag{2.6.2}$$

The reader can verify directly or by making use of rule (2.6.1) that

$$E\{X^*\} = 0, \; Var\{X^*\} = 1. \tag{2.6.3}$$

We call such a r.v. *normalized*, and the operation (2.6.2)—standard normalization, or simply *normalization*. We will use this operation repeatedly in this book. It is worth emphasizing that X^* is the same r.v. X, but considered, so to speak, in a standard scale.

EXAMPLE 1. Let $S_n = X_1 + ... + X_n$, where X_i's are i.i.d. r.v.'s with the same distribution as X above. Sometimes such r.v.'s are called independent replicas of X. By virtue of rules (2.1.7) and (2.4.2),

$$E\{S_n\} = mn, \; Var\{S_n\} = \sigma^2 n,$$

and for increasing n, the last characteristics are increasing. We can rescale S_n by defining

$$S_n^* = \frac{S_n - mn}{\sigma\sqrt{n}}.$$

For S_n^*, we have $E\{S_n^*\} = 0$ and $Var\{S_n^*\} = 1$. □

---

3 SOME BASIC DISTRIBUTIONS

In this section, we list some distributions playing an important role in theory and applications. Table 1 in Appendix, Section 1 presents a summary of the distributions we discuss below. In next chapters, we will extend this list. Proofs of facts given below may be found in practically any textbook on Probability; see, e.g., [102], [116], [120], [122].

3.1 Discrete distributions

3.1.1 The binomial distribution

Let $n \geq 1$ be an integer and $p \in [0,1]$. The *binomial distribution* with parameters n and p is the distribution of a r.v. X taking values $0, 1, ..., n$, and such that

$$f_k = P(X = k) = \binom{n}{k} p^k q^{n-k}, \tag{3.1.1}$$

where $q = 1 - p$, and

$$\binom{n}{k} = \frac{n!}{k!(n-k)!} = \frac{n(n-1)\cdots(n-k+1)}{k!}, \tag{3.1.2}$$

the number of ways to choose k objects from n distinct objects.

For future use, let us observe that the quantity $\binom{r}{k}$ may be defined for any real number r and integer $k \geq 0$, if we adopt as a definition the last expression in (3.1.2), setting

$$\binom{r}{k} = \frac{r(r-1)\cdots(r-k+1)}{k!}. \tag{3.1.3}$$

The binomial distribution with parameters n and p above usually appears in applications as the distribution of the number of successes in the sequence of n independent trials if the probability of success in a separate trial is equal to p. Adopting this interpretation, we can represent X as follows.

Consider the r.v.'s $X_1, ..., X_n$ such that $X_i = 1$ if the ith trial is successful, and $X_i = 0$ otherwise. Then the total number of successes is given by

$$X = X_1 + ... + X_n. \tag{3.1.4}$$

The r.v. X_i's defined above are called *Bernoulli's variables.* We have assumed that $P(X_i = 1) = p$, and since the trials are independent, the r.v.'s X_i are independent. Because $E\{X_i\} = p$ and $Var\{X_i\} = pq$ (the reader is encouraged to show this), from (3.1.4) it follows that

$$E\{X\} = np \text{ and } Var\{X\} = npq.$$

The next distribution we consider is multivariate, and represents a natural generalization of the binomial distribution.

3.1.2 The multinomial distribution

Assume that each of n independent trials may result in any of l possible outcomes with respective probabilities $p_1, ..., p_l$ such that $\sum_{i=1}^{l} p_i = 1$. Denote by K_i the number of trials with outcome i. Consider the *joint distribution* of the r.v.'s $K_1, ..., K_l$. Let $m_1, ..., m_l$ be non-negative integers such that $m_1 + ... + m_l = n$. Then for any such integers,

$$P(K_1 = m_1, ..., K_l = m_l) = \frac{n!}{m_1! \cdot ... \cdot m_l!} p_1^{m_1} \cdot ... \cdot p_l^{m_l} \tag{3.1.5}$$

(see, e.g., [102], [116], [122].) It is worthwhile to emphasize that the marginal distributions of K_i's are binomial; for example,

$$P(K_1 = k) = \binom{n}{k} p_1^k (1 - p_1)^{n-k}. \tag{3.1.6}$$

Formally, relation (3.1.6) follows from (3.1.5) if we set $m_1 = k$ and add up all probabilities (3.1.5) over all possible values of $m_2, ..., m_l$. However, we can justify (3.1.6) without calculations if we consider n independent trials above, and call a trial successful if it results in the first outcome. Then K_1 is the number of successful trials.

3.1.3 The geometric distribution

Two closely related distributions are called *geometric* in the literature. First, this is the distribution of a r.v. N taking values $1, 2, ...$ and such that

$$f_k = P(N = k) = pq^{k-1}, \tag{3.1.7}$$

where the parameter $p \in [0, 1]$, and $q = 1 - p$.

A classical example is the distribution of the number N of the first successful trial in a sequence of independent trials with the same probability of success p. In this case, the event $\{N > k\}$ occurs if the first k trials are not successful, which implies that

$$P(N > k) = q^k. \tag{3.1.8}$$

In some applications it is more convenient to use the term 'geometric distribution' not for the distribution above but for the distribution of the r.v. $K = N - 1$ which, naturally, assumes values $0, 1, 2, \ldots$. Then

$$P(K = k) = pq^k,\ k = 0, 1, 2, \ldots. \tag{3.1.9}$$

It follows from (3.1.8) that

$$P(K > k) = q^{k+1}. \tag{3.1.10}$$

The first version of the geometric distribution has the following property: for any integers m and k

$$P(N > m + k \,|\, N > k) = P(N > m). \tag{3.1.11}$$

We may clarify this in the following way. Assume that there was no success during the first k trials. The property (3.1.11) means that this fact has no effect on how long we will wait for a success after k trials: the probability that it will happen after an additional m trials does not depend on k.

Such a property is called the *memoryless* or the *lack of memory* property. In comparison with the exponential distribution that we consider later, it is worth emphasizing that (3.1.11) is true only for integers m and k. As a good exercise, the reader can check that, for instance, for $m = k = 2.5$ the relation (3.1.11) is not true.

It is easy to double check that for the r.v. K, (3.1.11) should be slightly changed:

$$P(K > m + k \,|\, K > k) = P(K > m - 1).$$

It may be computed that

$$E\{N\} = \frac{1}{p},\ E\{K\} = \frac{q}{p},\ \text{and}\ Var\{N\} = Var\{K\} = \frac{q}{p^2}. \tag{3.1.12}$$

(See practically any textbook on Probability, e.g., [102], [116], [122, p.67], and also Section 4.4 where we compute moments of the negative binomial distribution.)

In this book, we will primarily consider the geometric distribution in the sense of K.

3.1.4 The negative binomial distribution

Consider again a sequence of independent trials each having probability p of being a success. Suppose that the trials are performed until a total of ν successes is accumulated, and let N_ν be the number of the trials required. In other words, N_ν is the number of the trial at which the νth success occurs. Then N_ν takes on values $\nu, \nu+1, \ldots$. The νth success may occur at the mth trial only if the mth trial is successful, and among the previous $m-1$ trials there are exactly $\nu - 1$ successes. Then, as it follows from (3.1.1),

$$P(N_\nu = m) = p\binom{m-1}{\nu-1}p^{\nu-1}q^{m-\nu} = \binom{m-1}{\nu-1}p^\nu q^{m-\nu},\ m = \nu, \nu+1, \ldots. \tag{3.1.13}$$

(see also, e.g., [102], [116], [122]).

Denote by T_i the number of trials after the $(i-1)$th success until the ith success occurs. For $i = 1$, we set $T_1 = N_1$. Clearly, $N_\nu = T_1 + T_2 + \ldots + T_\nu$.

Since the trials are independent, the r.v.'s T_i are independent. For the same reason, to wait for the ith success after the $(i-1)$th success has occurred, is the same as to wait for the first success. So, all T_i's have the geometric distribution. Then, in view of (3.1.12),

$$E\{N_\nu\} = \frac{\nu}{p}, \; Var\{N_\nu\} = \frac{\nu q}{p^2}. \tag{3.1.14}$$

As in the case of the geometric distribution, we consider an alternative definition of the negative binomial distribution. This is the distribution of the r.v. $K_\nu = N_\nu - \nu$, which takes on values $0, 1, 2, ...$ Since $P(K_\nu = m) = P(N_\nu = m + \nu)$, it follows from (3.1.13) that

$$P(K_\nu = m) = \binom{\nu + m - 1}{\nu - 1} p^\nu q^m, \; m = 0, 1, 2, ...,$$

or, due to the formula $\binom{n}{k} = \binom{n}{n-k}$,

$$P(K_\nu = m) = \binom{\nu + m - 1}{m} p^\nu q^m, \; m = 0, 1, 2, ... \; . \tag{3.1.15}$$

From (3.1.14), we get that

$$E\{K_\nu\} = \frac{\nu}{p} - \nu = \frac{\nu q}{p}, \; Var\{K_\nu\} = \frac{\nu q}{p^2}. \tag{3.1.16}$$

Distribution (3.1.15) appears in many applications including those which are not relevant to a sequence of trials and numbers of successes. Moreover, in some applications, the parameter ν in (3.1.15) is positive but not necessarily an integer. The last instance requires clarification.

Consider a r.v. K_ν, not connected with a sequence of trials, whose distribution is formally defined by (3.1.15). The parameter ν in (3.1.15) is positive, and the coefficients $\binom{\nu+m-1}{m}$ are defined in accordance with the formula (3.1.3). Since $\nu > 0$, all these coefficients are positive, and hence all probabilities in (3.1.15) are also positive. Then, to show that the distribution is well defined, it remains to prove that the sum of these probabilities equals one. To this end, we use the Taylor expansion (4.2.9) from Appendix for the function $(1-x)^{-\alpha}$, which gives

$$\sum_{m=0}^{\infty} P(K_\nu = m) = p^\nu \sum_{m=0}^{\infty} \binom{\nu + m - 1}{m} q^m = p^\nu (1-q)^{-\nu} = p^\nu p^{-\nu} = 1.$$

The distribution (3.1.15) is called negative binomial with parameters p and ν. If ν is an integer, the distribution of the r.v. $K_\nu + \nu$ coincides with the distribution of the νth success. If $\nu = 1$, this is the geometric distribution with parameter p.

Formulas (3.1.16) remain true in the general case of an arbitrary positive ν. An easy way to show this, is to use moment generation functions, which we will do in Section 4.4.

3.1.5 The Poisson distribution

For reasons discussed later in Sections **3**.2.1 and **5**.2, this distribution plays a very important role in theory and applications. Formally, this is the distribution of a r.v. N taking values $0, 1, 2, ...$ and such that

$$f_k = P(N = k) = e^{-\lambda} \frac{\lambda^k}{k!}, \; k = 0, 1, 2, ... \; .$$

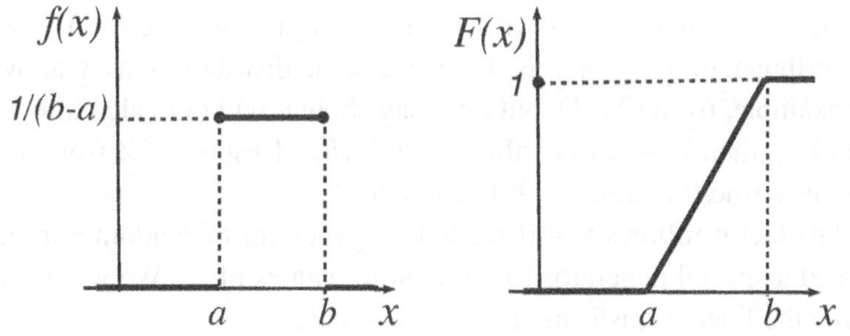

FIGURE 10. The density and the distribution function of a random variable uniform on $[a, b]$.

The positive parameter λ has a simple sense:

$$E\{N\} = \lambda. \tag{3.1.17}$$

The variance coincides with the mean value:

$$Var\{N\} = \lambda. \tag{3.1.18}$$

This is not an accident and may be explained by making use of properties of the Poisson process we consider in Section **4**.2 devoted to random processes. Proofs of (3.1.17)-(3.1.18) use expansion (4.2.5) in Appendix, and may be found in practically any textbook on Probability.

In this book, we will repeatedly return to properties of the Poisson distribution. In particular, we will prove in Section **2**.2.1.2 that, if N_1 and N_2 are independent Poisson r.v.'s with parameters λ_1 and λ_2, respectively, then the sum $N_1 + N_2$ has the Poisson distribution with parameter $\lambda_1 + \lambda_2$.

3.2 Continuous distributions

3.2.1 The uniform distribution and simulation of r.v.'s

We already considered this distribution. Now, we explore it in more detail. Let a continuous r.v. X assume all values from an interval $[a, b]$ and only from $[a, b]$. The last fact implies that $f_X(x)$, the density of X, should be equal to zero for $x \notin [a, b]$. If we assume that all values from $[a, b]$ are equally likely, we must set $f(x)$ equal to a constant c for $x \in [a, b]$, since otherwise these values will not be equally likely. On the other hand, we should have $\int_a^b f_X(x) = 1$, which implies that c should be $1/(b - a)$. Thus, we have arrived at

$$f_X(x) = \begin{cases} \frac{1}{b-a} & \text{if } x \in [a, b], \\ 0 & \text{if } x \notin [a, b]; \end{cases} \tag{3.2.1}$$

see Fig.10. Substituting (3.2.1) into (1.3.9), it is not difficult to derive that the distribution function of X is

$$F_X(x) = \begin{cases} 0 & \text{if } x < a, \\ \frac{x-a}{b-a} & \text{if } a \leq x \leq b, \\ 1 & \text{if } x > b; \end{cases} \tag{3.2.2}$$

see again Fig.10.

Note that, if a r.v. Z is uniformly distributed on $[0,1]$, then the r.v. $Y = a + (b-a)Z$ is uniformly distributed on $[a,b]$, that is, Y has the same distribution as X above. Formally, it follows, for example, from (2.6.1), but one may show it without calculations. If Z assumes values from $[0,1]$, then Y assumes values from $[a,b]$. If values of Z from $[0,1]$ are equally likely, the same is true for values of Y from $[a,b]$.

From the last fact it follows that if we have a generator of random numbers from $[0,1]$, we do not need a special generator for simulating values of X. We would simulate values of Z and apply the linear transformation $a + (b-a)Z$.

Making use of (2.1.5) and (2.4.1), we have

$$E\{Z\} = \int_0^1 x dx = \frac{1}{2},\ E\{Z^2\} = \int_0^1 x^2 dx = \frac{1}{3},\ Var\{Z\} = \frac{1}{12}.$$

Hence,

$$E\{X\} = E\{Y\} = a + (b-a)\frac{1}{2} = \frac{a+b}{2},$$

which is to be expected. Indeed, since all values are equally likely, the mean should be in the middle of $[a,b]$. By (2.6.1),

$$Var\{X\} = Var\{Y\} = (b-a)^2 Var\{Z\} = \frac{(b-a)^2}{12}.$$

It is worthwhile to warn the reader against the following widespread but incorrect reasoning.

Assume that we know that a r.v. ξ takes on values from, say, $[0,1]$, but we have no additional information about the distribution of ξ. Since we equally do not know anything about chances of values from $[0,1]$, we may consider these values equally likely. Hence, we can assume that ξ is uniform. Then, by (3.2.2), the d.f. of ξ is $F_\xi(x) = x$ for $x \in [0,1]$.

But the r.v. ξ^2 also takes on values from $[0,1]$, and we do not have any additional information about its distribution either. Then, on the same grounds, we can set $F_{\xi^2}(x) = x$, and write $x = F_\xi(x) = P(\xi \leq x) = P(\xi^2 \leq x^2) = F_{\xi^2}(x^2) = x^2$. So, we have arrived at a false assertion that $x = x^2$.

The above reasoning was faulty because the absence of information does not allow us to jump to any particular conclusion about the distribution of ξ. Knowing nothing means that we cannot say anything about the distribution except that it is concentrated on $[0,1]$. Whereas the assertion on uniformity should be based on the rather concrete information that all values from $[0,1]$ are *equally* likely.

Simulation. The inverse distribution function method. The following property of the uniform distribution is very important for the simulation of r.v.'s.

Let $F(x)$ be a d.f., and let $F^{-1}(y)$ be its inverse. If for some y, there are many x's for which $F(x) = y$ or there is no such an x, then we define $F^{-1}(y)$ similar to what we did in Section 1.3.4. Namely, $F^{-1}(y)$ is a number x such that $F(x-\varepsilon) \leq y$ and $F(x+\varepsilon) > y$ for

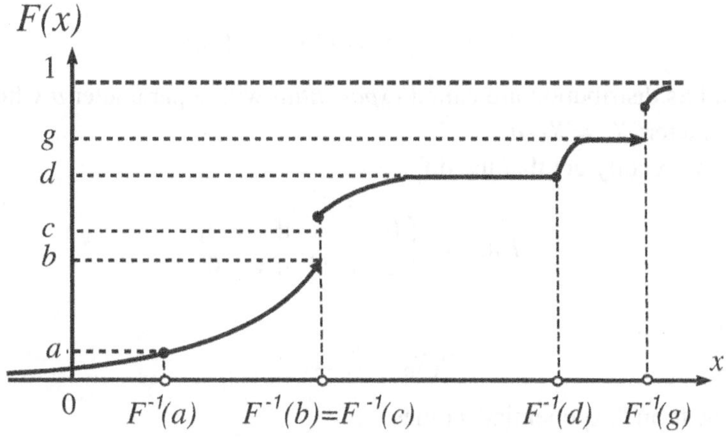

FIGURE 11. The inverse of a distribution function. (The letter *g* is chosen since *e* stands for the natural e, and *f*—for the density.)

any $\varepsilon > 0$. The definition is illustrated in Fig.11.

The reader familiar with the notion of supremum may realize that the above inverse $F^{-1}(y) = \sup\{x : F(x) \leq y\}$.

Proposition 7 *Let Z be uniformly distributed on* $[0,1]$. *Then the r.v.* $X = F^{-1}(Z)$ *has the distribution function* $F(x)$.

Proof. As we saw above, $P(Z \leq z) = z$ for $z \in [0,1]$. On the other hand, for any x we have $P(X \leq x) = P(F^{-1}(Z) \leq x) = P(Z) \leq F(x)) = F(x)$, since $0 \leq F(x) \leq 1$. ■

EXAMPLE 1. Let X take on values from $[0,2]$ and have the d.f. $F(x) = x^3/8$ for $x \in [0,2]$. We want to simulate values of X, for instance, using Excel. Since $F^{-1}(y) = 2y^{1/3}$, we may represent X as $X = F^{-1}(Z) = 2Z^{1/3}$, where Z is uniform on $[0,1]$. In the Excel worksheet in Fig.12, five values of Z are simulated in Column A with use of the Excel random number generator. The corresponding values of X are in Column B. For example, B1=2*A1^(1/3). □

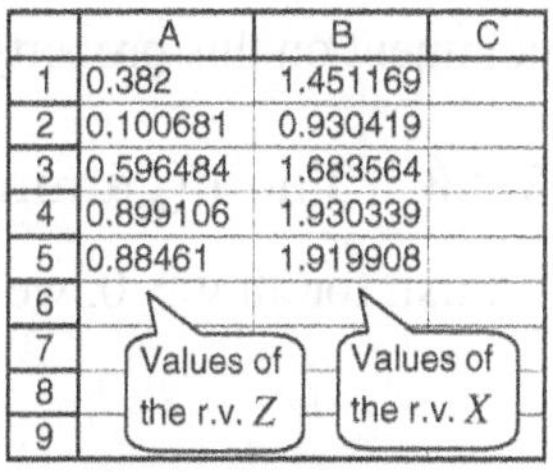

	A	B	C
1	0.382	1.451169	
2	0.100681	0.930419	
3	0.596484	1.683564	
4	0.899106	1.930339	
5	0.88461	1.919908	
6			
7			
8			
9			

FIGURE 12.

3.2.2 The exponential distribution

We call a continuous *positive* r.v. X_1 and its distribution *standard exponential* if the corresponding density $f_1(x) = e^{-x}$ for $x \geq 0$. It is straightforward to compute that $E\{X_1\} = \int_0^\infty xe^{-x}dx = 1$, $E\{X_1^2\} = \int_0^\infty x^2 e^{-x}dx = 2$, and hence, $Var\{X_1\} = 1$.

Consider now the r.v. $X = X_a = X_1/a$ for a positive a. In accordance with (2.6.1), the density of X_a is

$$f_a(x) = \begin{cases} 0 & \text{if } x < 0, \\ ae^{-ax} & \text{if } x \geq 0, \end{cases} \tag{3.2.3}$$

and

$$E\{X_a\} = 1/a, \; Var\{X_a\} = 1/a^2. \tag{3.2.4}$$

Such a r.v. and its distribution are called *exponential* with a parameter a which, as we see, is a scale parameter: $X_a = X_1/a$.

By (1.3.9), we readily get that the d.f.

$$F_a(x) = \begin{cases} 0 & \text{if } x < 0, \\ 1 - e^{-ax} & \text{if } x \geq 0. \end{cases} \tag{3.2.5}$$

Consequently,

$$P(X_a > x) = e^{-ax}, \tag{3.2.6}$$

from which the term "exponential" comes.

The exponential distribution has the unique

Lack-of-Memory (or Memoryless) Property: for any $x, y \geq 0$,

$$P(X > x + y \,|\, X > x) = P(X > y). \tag{3.2.7}$$

Indeed,

$$P(X > x + y \,|\, X > x) = \frac{P(X > x + y, X > x)}{P(X > x)} = \frac{P(X > x + y)}{P(X > x)}. \tag{3.2.8}$$

Then, in view of (3.2.6),

$$P(X > x + y \,|\, X > x) = \frac{e^{-a(x+y)}}{e^{-ax}} = e^{-ay} = P(X > y).$$

Thus, if X has exceeded a certain level x, the overshoot (over x) has the same distribution as the r.v. X itself, and does not depend on the particular level x that was exceeded. We comment on this property in more detail in Section **2**.1.1.

3.2.3 The Γ(gamma)-distribution

First, for all $\nu > 0$, we define the *Γ(gamma)-function*

$$\Gamma(\nu) = \int_0^\infty x^{\nu-1} e^{-x} dx.$$

It is easy to verify that $\Gamma(1) = \int_0^\infty e^{-x} dx = 1$ and, using integration by parts, that

$$\Gamma(\nu + 1) = \nu\Gamma(\nu). \tag{3.2.9}$$

Then for an integer $k \geq 1$, we have $\Gamma(k+1) = k\Gamma(k) = k(k-1)\Gamma(k-1) = ... = k(k-1) \cdot ... \cdot 1 \cdot \Gamma(1) = k!$. Thus,

$$\Gamma(k+1) = k!$$

for $k = 0, 1, ...$ So, the Γ-function may be viewed as a generalization of the notion of factorial.

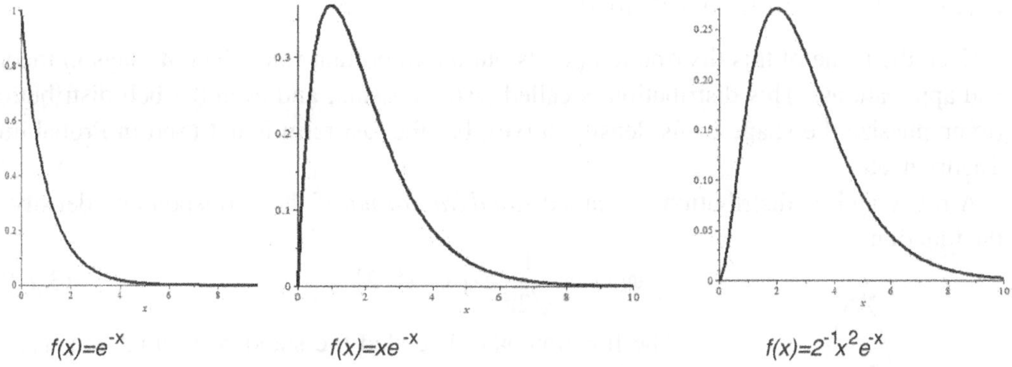

FIGURE 13. The Γ-densities for $\nu = 1, 2, 3$.

Consider, first, a continuous r.v. $X_{1\nu}$ whose density is the function

$$f_{1\nu}(x) = \frac{1}{\Gamma(\nu)} x^{\nu-1} e^{-x} \text{ for } x \geq 0, \text{ and } = 0 \text{ otherwise.}$$

Due to $\Gamma(\nu)$ in the denominator, $\int_0^\infty f_{1\nu}(x)dx$ is indeed equal to one. For $\nu = 1$, this is the standard exponential density. The parameter ν characterizes the type of the distribution. In Fig.13, we demonstrate how this type depends on ν by sketching the graphs of $f_{1\nu}(x)$ for $\nu = 1, 2, 3$.

The kth moment

$$\begin{aligned} E\{X_{1\nu}^k\} &= \frac{1}{\Gamma(\nu)} \int_0^\infty x^k x^{\nu-1} e^{-x} dx = \frac{1}{\Gamma(\nu)} \int_0^\infty x^{k+\nu-1} e^{-x} dx \\ &= \frac{\Gamma(\nu+k)}{\Gamma(\nu)} = \frac{(\nu+k-1)\Gamma(\nu+k-1)}{\Gamma(\nu)} = \frac{(\nu+k-1)(\nu+k-2)\Gamma(\nu+k-2)}{\Gamma(\nu)} = \dots \\ &= \frac{(\nu+k-1)\cdot \ldots \cdot \nu\Gamma(\nu)}{\Gamma(\nu)} = (\nu+k-1)\cdot \ldots \cdot \nu, \end{aligned}$$

by virtue of (3.2.9). In particular,

$$E\{X_{1\nu}\} = \nu, \; E\{X_{1\nu}^2\} = (\nu+1)\nu, \text{ and hence } Var\{X_{1\nu}\} = \nu. \tag{3.2.10}$$

Now, let $a > 0$ and the r.v. $X_{a\nu} = X_{1\nu}/a$. By (2.6.1), the density of $X_{a\nu}$ is

$$f_{a\nu}(x) = \frac{a^\nu}{\Gamma(\nu)} x^{\nu-1} e^{-ax} \text{ for } x \geq 0, \text{ and } = 0 \text{ otherwise.} \tag{3.2.11}$$

The distribution defined and its density $f_{a\nu}(x)$ are called the Γ(Gamma)-distribution and Γ-density, respectively, with parameters a and ν. As we saw, a is just a scale parameter, while parameter ν may be called essential since it specifies the type of the distribution.

From (3.2.10) it follows that

$$E\{X_{a\nu}\} = \frac{\nu}{a}, \; E\{X_{a\nu}^2\} = \frac{(\nu+1)\nu}{a^2}, \text{ and } Var\{X_{a\nu}\} = \frac{\nu}{a^2}. \tag{3.2.12}$$

3.2.4 The normal distribution

Even the name of this distribution points out the important role which it plays in theory and applications. This distribution is called also Gaussian, and even the bell distribution (to emphasize the shape of its density curve), but the last term is not used in Probability Theory itself.

A r.v. X and its distribution are called *standard normal* if the corresponding density is the function

$$\varphi(x) = \frac{1}{\sqrt{2\pi}} \exp\{-x^2/2\}. \tag{3.2.13}$$

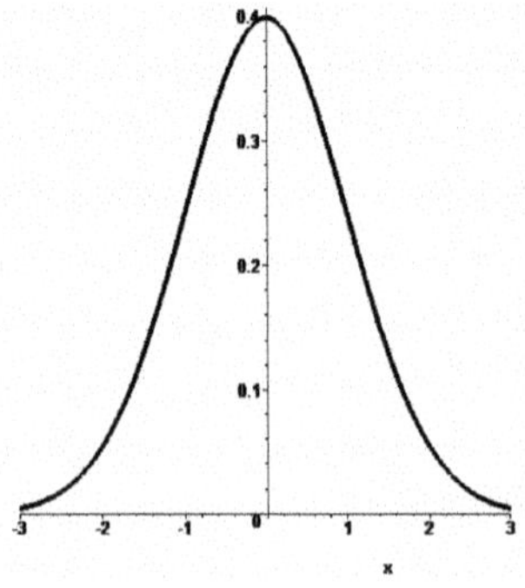

FIGURE 14.

The function $\varphi(x)$ is called the standard normal density; its graph for $-3 \le x \le 3$ is presented in Fig.14.

Since density (3.2.13) is an even function, the integrand in the integral $\int_{-\infty}^{\infty} x\varphi(x)dx$ is an odd function, and hence this integral equals zero. Integrating by parts, one can compute that $\int_{-\infty}^{\infty} x^2\varphi(x)dx = 1$. Thus,

$$E\{X\} = 0, \; Var\{X\} = 1, \tag{3.2.14}$$

from which the term "standard" comes.

Let m be an arbitrary number, and let $\sigma > 0$. The r.v. $Y = m + \sigma X$ and its distribution are called (m, σ^2)*-normal.* In view of (3.2.14) and (2.6.1),

$$E\{Y\} = m, \; Var\{Y\} = \sigma^2, \tag{3.2.15}$$

which justifies the choice of the notation m and σ^2. By the same rule (2.6.1), the density of Y is the function

$$\varphi_{m\sigma}(x) = \frac{1}{\sqrt{2\pi}\sigma} \exp\left\{-\frac{(x-m)^2}{2\sigma^2}\right\}. \tag{3.2.16}$$

In accordance with (1.3.9), the d.f. of X is equal to

$$\Phi(x) = \frac{1}{\sqrt{2\pi}} \int_{-\infty}^{x} e^{-u^2/2} du.$$

The function $\Phi(x)$ is called the *standard normal d.f.* The integral above cannot be computed analytically, but any particular value of $\Phi(x)$ may be computed to a high degree of accuracy using the numerical integration technique. (See also Table 1 in Section 2 in Appendix.)

Using again (2.6.1), we get that the d.f. of Y is the function

$$\Phi_{m\sigma}(x) = \Phi\left(\frac{x-m}{\sigma}\right). \tag{3.2.17}$$

In Section **2**.2.1.2, we show that the sum of independent (m_1, σ_1^2)- and (m_2, σ_2^2)-normal r.v.'s is $(m_1+m_2, \, \sigma_1^2+\sigma_2^2)$-normal.

The importance of the normal distribution is explained, first of all, by the central limit theorem that we consider in Section 6.2.

We present also the following well known and useful estimate for $\Phi(x)$ (see, e.g., [38], [102],[116], [122]): for any $x > 0$

$$x^{-1}(1 - x^{-2})\varphi(x) \le 1 - \Phi(x) = \Phi(-x) \le x^{-1}\varphi(x). \tag{3.2.18}$$

4 MOMENT GENERATING FUNCTIONS

4.1 Laplace transform

The *Laplace transform* of a r.v. X and its distribution F is the function

$$M_X(z) = E\{e^{zX}\} = \int_{-\infty}^{\infty} e^{zx} dF(x), \tag{4.1.1}$$

defined for all z for which the above expectation exists. In general, the argument z is a complex number. When it cannot cause confusion, we will omit the index X in $M_X(z)$.

If z is a real number, the Laplace transform is also called a *moment generating function* (m.g.f.). We will use the latter term since it is customary in actuarial modeling. The letter M comes from the word "moment".

The terminology chosen is related to the following fact. As above, denote by m_k the kth moment $E\{X^k\}$. Making use of the Taylor expansion of the exponential function (see (4.2.5) in Appendix), we can write

$$M_X(z) = E\{e^{zX}\} = E\left\{\sum_{k=0}^{\infty} \frac{(zX)^k}{k!}\right\} = \sum_{k=0}^{\infty} \frac{z^k E\{X^k\}}{k!} = \sum_{k=0}^{\infty} \frac{m_k}{k!} z^k. \tag{4.1.2}$$

(We omit the formal justification of passing the expectation operation inside the sum.)

Thus, the m.g.f. $M_X(z)$ may be expanded into the series in powers of z, where the kth coefficient equals the kth moment divided by $k!$. Thus, $M^{(k)}(0) = m_k$. We consider this issue in more detail in Sections 4.4-4.5.

It is worth emphasizing that for real z's, the integral in (4.1.1) may not exist. Therefore, m.g.f.'s exist not for all r.v.'s and/or not for all values of z; see examples in Section 4.3 below. Certainly for $z = 0$, we can always write that

$$M_X(0) = E\{e^{0 \cdot X}\} = E\{1\} = 1. \tag{4.1.3}$$

If $z = it$, where the imaginary $i = \sqrt{-1}$, the Laplace transform is equal to

$$K_X(t) = E\{e^{itX}\} = \int_{-\infty}^{\infty} e^{itx} dF(x),$$

and is called the *characteristic function* of a r.v. X. The same function is called also the characteristic function of F, or the *Fourier transform* of F. Since $|e^{itx}| = 1$, the characteristic function always exists. Indeed, for the last integral

$$\left|\int_{-\infty}^{\infty} e^{itx} dF(x)\right| \le \int_{-\infty}^{\infty} |e^{itx}| \, dF(x) = \int_{-\infty}^{\infty} dF(x) = 1,$$

that is, the integral is finite for any z.

If X assumes values $0, 1, \dots$, and $p_k = P(X = k)$, it is convenient to consider real $z \le 0$, and set $s = e^z$ which is not larger than one. Then $E\{e^{zX}\} = E\{s^X\}$, where $0 < s \le 1$. Denoting the last function by $G_X(s)$, we can write

$$G_X(s) = E\{s^X\} = s^0 p_0 + s^1 p_1 + s^2 p_2 + \dots = \sum_{k=0}^{\infty} p_k s^k.$$

The function $G_X(s)$ is called a *probability generating function* (or simply a generating function). The coefficients in its expansion in powers of s are just the corresponding probabilities p_k.

The Laplace transform in its different versions has proven to be a powerful weapon for solving a wide variety of problems. There are several reasons for this, but possibly the primary reason is connected with the following property:

For any independent r.v.'s X_1 and X_2,
$$M_{X_1+X_2}(z) = M_{X_1}(z)M_{X_2}(z) \tag{4.1.4}$$
for all z for which the Laplace transforms above are well defined.

Indeed, by property (2.1.8),

$$M_{X_1+X_2}(z) = E\{e^{z(X_1+X_2)}\} = E\{e^{zX_1}e^{zX_2}\} = E\{e^{zX_1}\}E\{e^{zX_2}\} = M_{X_1}(z)M_{X_2}(z).$$

Before considering examples, we establish two elementary and one non-elementary property.

A. *The Laplace transform of a linear transformation.*

$$\text{For } Y = a + bX, \quad \text{the Laplace transform } M_Y(z) = e^{az}M_X(bz). \tag{4.1.5}$$

Indeed, $M_Y(z) = E\{e^{z(a+b)X}\} = E\{e^{az}e^{bzX}\} = e^{az}E\{e^{bzX}\} = e^{az}M_X(bz)$.

B. *The Laplace transform of a mixture (or linear combination) of distributions is equal to the mixture (or the linear combination) of the Laplace transforms.* More precisely, let F_1, F_2 be two distributions, and $F = \alpha F_1 + (1-\alpha)F_2$, where $0 \le \alpha \le 1$ [see Section 1.3.5]. Let $M_1(z), M_2(z), M(z)$ be the Laplace transforms of the distributions F_1, F_2, F, respectively. Then

$$M(z) = \alpha M_1(z) + (1-\alpha)M_2(z).$$

This immediately follows from (4.1.1):

$$M(z) = \int_{-\infty}^{\infty} e^{zx}dF(x) = \int_{-\infty}^{\infty} e^{zx}d(\alpha F_1(x) + (1-\alpha)F_2(x))$$
$$= \alpha\int_{-\infty}^{\infty} e^{zx}dF_1(x) + (1-\alpha)\int_{-\infty}^{\infty} e^{zx}dF_2(x).$$

The non-elementary property mentioned is the uniqueness property: r.v.'s with different distributions have different Laplace transforms. The situation is similar when we consider only real z's, that is, moment generating functions. However, in this case, we will always assume that the m.g.f.'s under consideration exist at least for all z such that $|z| < c$, where c is a positive constant.

Theorem 8 *R.v.'s with distinct distributions have distinct m.g.f.'s.*

In other words, if for two r.v.'s, X and Y, with the d.f.'s $F_X(x)$ and $F_Y(x)$, respectively, $M_X(z) = M_Y(z)$ for all z in a neighborhood of zero, then $F_X(x) = F_Y(x)$ for all x.

4.2 An example when a m.g.f. does not exist

Let a r.v. X have the density $f(x) = 1/(2x^2)$ for $|x| \geq 1$, and $= 0$ otherwise. The reader may verify that indeed $\int_{-\infty}^{\infty} f(x)dx = 1$. If the m.g.f. had existed, it would have been equal to

$$\int_{-\infty}^{\infty} e^{zx} f(x)dx = \frac{1}{2}\int_{-\infty}^{-1} e^{zx}\frac{1}{x^2}dx + \frac{1}{2}\int_{1}^{\infty} e^{zx}\frac{1}{x^2}dx.$$

If $z > 0$, the first integral converges, because $zx < 0$, and hence $e^{zx} \leq 1$. However, the second integral diverges, since for large x the function $e^{zx}\frac{1}{x^2}$ grows faster than, say, $e^{zx/2}$. The proof may be found practically in any Calculus textbook; see, e.g., [136]. We omit the details. If $z < 0$, the second integral is finite, while the first diverges. So, the m.g.f. exists only for $z = 0$.

4.3 The m.g.f.'s of basic distributions

Next, we consider the m.g.f.'s of the basic distribution from Section 3. Table 2 in Appendix, Section 1 presents a summary of the results below.

In all formulas below, z is real.

4.3.1 The binomial distribution

The easiest way to find the m.g.f. of a binomial r.v. X is to use the representation

$$X = X_1 + ... + X_n,$$

where the r.v.'s X_i are independent and equal 1 or 0 with probabilities p and q, respectively; see Section 3.1.1. The m.g.f. of each X_i is

$$M_{X_i}(z) = e^{z\cdot 1}p + e^{z\cdot 0}q = e^z p + q = 1 + p(e^z - 1),$$

since $q = 1 - p$. By property (4.1.4), the m.g.f. of X is the product of the m.g.f.'s of X_i's, so

$$M_X(z) = (e^z p + q)^n = (1 + p(e^z - 1))^n.$$

For the corresponding generating function, setting $s = e^z$, we have

$$G_X(s) = (sp + q)^n.$$

4.3.2 The geometric and negative binomial distributions

For the geometric distribution in the form (3.1.9),

$$M_K(z) = \sum_{k=0}^{\infty} e^{kz} pq^k = p\sum_{k=0}^{\infty} (e^z q)^k = \frac{p}{1 - qe^z}. \tag{4.3.1}$$

(In the last step, we used formula (4.2.10) for a geometric series.)

To get the m.g.f. for the r.v. $N = K + 1$ (see Section 3.1.3), we use (4.1.5), which leads to $M_N(z) = e^z M_K(z)$; see also Table 2 in Section 1 in Appendix.

In general, for the negative binomial distribution (3.1.15), using expansion (4.2.9) from Appendix, we have

$$M_{K_\nu}(z) = \sum_{m=0}^{\infty} e^{mz}\binom{\nu+m-1}{m}p^\nu q^m = p^\nu \sum_{m=0}^{\infty}\binom{\nu+m-1}{m}(e^z q)^m$$
$$= p^\nu(1-qe^z)^{-\nu} = \left(\frac{p}{1-qe^z}\right)^\nu. \tag{4.3.2}$$

The generating function

$$G_{K_\nu}(s) = \left(\frac{p}{1-qs}\right)^\nu.$$

To get the result for $N_\nu = K_\nu+\nu$, we again use (4.1.5), which leads to $M_{N_\nu}(z) = e^{z\nu}M_{K_\nu}(z)$.

4.3.3 The Poisson distribution

For the Poisson distribution with parameter λ, its m.g.f.

$$M(z) = \sum_{k=0}^{\infty} e^{zk}e^{-\lambda}\frac{\lambda^k}{k!} = e^{-\lambda}\sum_{k=0}^{\infty}\frac{(e^z\lambda)^k}{k!} = e^{-\lambda}e^{\lambda e^z} = \exp\{\lambda(e^z-1)\}, \tag{4.3.3}$$

by virtue of the Taylor expansion (4.2.5) from Appendix.

Next we consider continuous distributions.

4.3.4 The uniform distribution

For the distribution with the density (3.2.1), the m.g.f.

$$M(z) = \int_a^b e^{zx}\frac{1}{b-a}dx = \frac{e^{bz}-e^{az}}{z(b-a)} \tag{4.3.4}$$

if $z \neq 0$. For $z = 0$ the last expression is not defined, but we can set $M(0) = 1$ in view of (4.1.3). Note that the limit of the r.-h.s. of (4.3.4) as $z \to 0$ equals one (one can use, for example, L'Hôpital's rule).

4.3.5 The exponential and gamma distributions

For a r.v. X, let the density $f(x) = ae^{-ax}$ for $x \geq 0$. Then the m.g.f.

$$M(z) = \int_0^\infty e^{zx}ae^{-ax}dx = \frac{a}{a-z} = \frac{1}{1-z/a} \tag{4.3.5}$$

for $z < a$. It is important to emphasize that for $z \geq a$ the m.g.f. does not exist.

In general, for the density (3.2.11) and $z < a$, making the change of variable $y = (a-z)x$, we have

$$M(z) = \int_0^\infty e^{zx}f_{a\nu}(x)dx = \frac{a^\nu}{\Gamma(\nu)}\int_0^\infty e^{zx}x^{\nu-1}e^{-ax}dx = \frac{a^\nu}{\Gamma(\nu)}\int_0^\infty x^{\nu-1}e^{-(a-z)x}dx$$
$$= \frac{a^\nu}{\Gamma(\nu)}\frac{1}{(a-z)^\nu}\int_0^\infty y^{\nu-1}e^{-y}dy = \frac{a^\nu}{\Gamma(\nu)}\frac{1}{(a-z)^\nu}\Gamma(\nu) = \left(\frac{a}{a-z}\right)^\nu = \frac{1}{(1-z/a)^\nu},$$

again provided that $z < a$.

4.3.6 The normal distribution

For a standard normal r.v. X,

$$M_X(z) = \int_{-\infty}^{\infty} e^{zx}\frac{1}{\sqrt{2\pi}}e^{-x^2/2}dx = \frac{1}{\sqrt{2\pi}}\int_{-\infty}^{\infty} \exp\{zx - x^2/2\}dx$$
$$= (2\pi)^{-1/2}\int_{-\infty}^{\infty} \exp\left\{\frac{1}{2}\left(z^2-(x-z)^2\right)\right\}dx = (2\pi)^{-1/2}e^{z^2/2}\int_{-\infty}^{\infty} \exp\left\{-\frac{1}{2}(x-z)^2\right\}dx$$

(in the third step we completed the square). With the change of variable $y = x - z$, we have

$$M_X(z) = (2\pi)^{-1/2}e^{z^2/2}\int_{-\infty}^{\infty} \exp\left\{-y^2/2\right\}dy = e^{z^2/2}\int_{-\infty}^{\infty} (2\pi)^{-1/2}\exp\left\{-y^2/2\right\}dy.$$

The integrand in the last integral is the standard normal density. Hence, the integral itself equals one, and

$$M_X(z) = e^{z^2/2}.$$

For the (m, σ^2)-normal r.v. $Y = m + \sigma X$, by virtue of (4.1.5),

$$M_Y(z) = e^{mz}M_X(\sigma z) = \exp\{mz + \sigma^2 z^2/2\}. \tag{4.3.6}$$

4.4 The moment generating function and moments

Consider a r.v. X with a d.f. $F(x)$, and assume that for some $c_0 > 0$,

$$E\{e^{|zX|}\} < \infty \quad \text{for } |z| < c_0. \tag{4.4.1}$$

Since $E\{e^{zX}\} \leq E\{e^{|zX|}\}$, condition (4.4.1) ensures that the m.g.f.

$$M(z) = E\{e^{zX}\} = \int_{-\infty}^{\infty} e^{zx}dF(x)$$

is well defined for all z from the interval $(-c_0, c_0)$.

Clearly, $M(0) = 1$.

It may be proved that under condition (4.4.1), we can differentiate $M(z)$ an arbitrary number of times, and we can do that by passing the operation of differentiation through the integral.

In particular, differentiating $M(z)$ once, we get

$$M'(z) = E\{Xe^{zX}\} = \int_{-\infty}^{\infty} xe^{zx}dF(x). \tag{4.4.2}$$

Hence,

$$M'(0) = E\{X\}.$$

Differentiating (4.4.2), we have

$$M''(z) = E\{X^2e^{zX}\} = \int_{-\infty}^{\infty} x^2e^{zx}dF(x), \tag{4.4.3}$$

and

$$M''(0) = E\{X^2\}.$$

Continuing in the same fashion, we get that the kth derivative

$$M^{(k)}(z) = E\{X^k e^{zX}\} = \int_{-\infty}^{\infty} x^k e^{zx} dF(x), \tag{4.4.4}$$

and for all k

$$M^{(k)}(0) = m_k = E\{X^k\}, \tag{4.4.5}$$

the kth moment of X.

From (4.4.3) it follows, in particular, that $M''(z) = E\{X^2 e^{zX}\} \geq 0$ for all z, and hence

$$\boxed{M(z) \text{ is always convex.}} \tag{4.4.6}$$

► As a matter of fact, to show the convexity of $M(z)$, we do not have to consider derivatives. Since e^{zX}, as a function of z, is convex, we can use the counterpart (for convexity) of Definition 1 from Section 4.3 in Appendix, and write that

$$\begin{aligned} M(z_1) + M(z_2) &= E\{e^{z_1 X} + e^{z_2 X}\} \geq E\left\{2\exp\left\{\left(\frac{z_1 X + z_2 X}{2}\right)\right\}\right\} \\ &= 2E\left\{\exp\left\{\left(\frac{z_1 + z_2}{2}\right)X\right\}\right\} = 2M\left(\frac{z_1 + z_2}{2}\right), \end{aligned}$$

which implies the convexity of $M(z)$. ◄

EXAMPLE 1. As was promised in Section 3.1.4, we compute the mean and the variance of the negative binomial r.v. K_ν. By (4.3.2), the corresponding m.g.f. $M(z) = [p/(1-qe^z)]^\nu$. Then

$$M'(z) = p^\nu \nu q e^z/(1-qe^z)^{\nu+1},$$

and

$$M''(z) = p^\nu \nu q \left[\frac{e^z}{(1-qe^z)^{\nu+1}} + \frac{e^{2z} q(\nu+1)}{(1-qe^z)^{\nu+2}}\right].$$

Since $q = 1-p$, we have $E\{K_\nu\} = M'(0) = p^\nu \nu q/(1-q)^{\nu+1} = \nu q/p$, $E\{K_\nu^2\} = M''(0) = p^\nu \nu q[p^{-\nu-1} + q(\nu+1)p^{-\nu-2}] = \nu q p^{-2}[p + q(\nu+1)]$. This readily implies that $Var\{K_\nu\} = E\{K_\nu^2\} - (E\{K_\nu\})^2 = \nu q p^{-2}$, so we have arrived at (3.1.16). □

4.5 Expansions for m.g.f.'s

4.5.1 Taylor's expansions for m.g.f.'s

The reader may look over general facts on Taylor's expansions in Section 4.2 in Appendix.

We also use below, for the first time in this book, the common Calculus notation $o(x)$ ("little o" notation). Since it is not always used in introductory Calculus courses, we explain its significance in two pages of a special Section 4.1 in Appendix. If the reader is not familiar with the $o(\cdot)$ notation, it is worth the time required to understand it anyhow. It is convenient, explicit, and saves time in a great many calculations.

In short, $o(x)$ is a function such that $[o(x)/x] \to 0$ as $x \to 0$. We view $o(x)$ as a term negligible in comparison with x for small x's.

First, we state that under condition (4.4.1), the expansion

$$M(z) = \sum_{k=0}^{\infty} \frac{m_k}{k!} z^k \tag{4.5.1}$$

is true, provided $|z| < c_0$. We skip the proof of this fact and the other formal operations below.

Secondly, making use of the general Taylor formula (see (4.2.3) in Appendix) and (4.4.5), we have

$$M(z) = 1 + m_1 z + \frac{m_2 z^2}{2!} + \ldots + \frac{m_k z^n}{n!} + o(z^n). \tag{4.5.2}$$

In particular,

$$M(z) = 1 + m_1 z + \frac{m_2 z^2}{2} + o(z^2). \tag{4.5.3}$$

4.5.2 Cumulants

Set $K(z) = \ln(M(z))$. If $M(z)$ admits the Taylor expansion, so does $K(z)$; we skip here formalities. Since $K(0) = \ln(M(0)) = 0$, the Taylor expansion for $K(z)$ will look as follows:

$$K(z) = \ln(M(z)) = \sum_{i=1}^{\infty} \frac{\varkappa_i z^i}{i!}, \tag{4.5.4}$$

where $\varkappa$'s are the corresponding coefficients, more precisely, the corresponding derivatives of $K(z)$ at zero. The quantity $\varkappa_i$ is called the ith *cumulant* of the r.v. X (and its distribution). The significance of these characteristics and their usefulness is connected with the following instance.

Let $M_1(z)$ and $M_2(z)$ be the m.g.f.'s of independent r.v.'s X_1 and X_2, and $M(z)$ be the m.g.f. of $X_1 + X_2$. Denote by $K_1(z)$, $K_2(z)$, and $K(z)$ the logarithms of the corresponding m.g.f.'s. Then taking the logarithms in (4.1.4), we get

$$K(z) = K_1(z) + K_2(z). \tag{4.5.5}$$

The last relation turns out to be convenient in many problems. Combining (4.5.4) and (4.5.5), we see that

$$\varkappa_i = \varkappa_{i1} + \varkappa_{i2}, \tag{4.5.6}$$

where $\varkappa_{i1}, \varkappa_{i2}$, and $\varkappa_i$ are the ith cumulants of X_1, X_2, and $X_1 + X_2$, respectively.

Let us now return to (4.5.4). As a simple example, consider the case $n = 2$. It will also be a good exercise on the use of the notation $o(\cdot)$. By virtue of (4.2.8) in Appendix and

(4.5.3), for a r.v. X and its m.g.f. $M(z)$,

$$K(z) = \ln[M(z)] = \ln[1+(M(z)-1)] = (M(z)-1) - \frac{1}{2}(M(z)-1)^2 + o\left((M(z)-1)^2\right)$$
$$= m_1z + \frac{m_2z^2}{2} + o(z^2) - \frac{1}{2}\left(m_1z + \frac{m_2z^2}{2} + o(z^2)\right)^2 + o\left(\left(m_1z + \frac{m_2z^2}{2} + o(z^2)\right)^2\right)$$
$$= m_1z + \frac{m_2z^2}{2} + o(z^2) - \frac{1}{2}(m_1z)^2 + o(z^2) + o(z^2) = m_1z + \frac{m_2 - m_1^2}{2}z^2 + o(z^2).$$

Since $m_2 - m_1^2 = \sigma^2$, the variance of X, we have

$$K(z) = mz + \frac{\sigma^2 z^2}{2} + o(z^2), \tag{4.5.7}$$

where $m = m_1 = E\{X\}$.

Comparing it with (4.5.4), for the first two cumulants we have

$$\varkappa_1 = m, \ \varkappa_2 = \sigma^2.$$

In this case, rule (4.5.6) coincides with the corresponding rules for the mean and the variance of sums of r.v.'s.

A good exercise is, using the same little-o-technique, to show that the third cumulant

$$\varkappa_3 = \mu_3 = E\{(X-m)^3\}.$$

---

5 CONVERGENCE OF RANDOM VARIABLES AND DISTRIBUTIONS

Consider a sample space $\Omega = \{\omega\}$, a class of events $\mathcal{A}$, a probability measure $P(A)$ defined on events A from $\mathcal{A}$, and r.v.'s X which were defined as functions $X(\omega)$ on Ω.

We say that a sequence of r.v.'s $X_n = X_n(\omega)$ converges to a r.v. $X = X(\omega)$ *almost surely*, or *with probability one*, if

$$P(X_n \to X) = 1.$$

More precisely, this means that the set of all ω's for which $X_n(\omega) \to X(\omega)$ has probability one. In this case, we write $X_n \xrightarrow{\text{a.s.}} X$.

In many models, such convergence either does not take place, or if it does, it is difficult to prove. On the other hand, in many applications, it is sufficient to consider a weaker type of convergence when we just require $X_n - X$ to be small for large n with a probability close to one. We will now translate this heuristic definition into mathematical terms.

Saying that $X_n - X$ is small, we mean that $|X_n - X|$ is less than a sufficiently small positive number ε. So, we want $|X_n - X|$ to be less than ε for large n with a large probability.

Saying that a probability is large, we mean that it is close to one, and saying "for large n" we mean that $n \to \infty$. We are ready to give a formal definition.

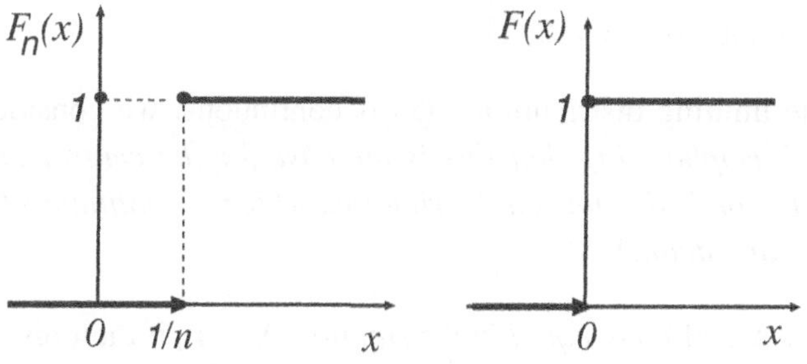

FIGURE 15.

A sequence of r.v.'s X_n converges to a r.v. X *in probability* if for any arbitrary small $\varepsilon > 0$

$$P(|X_n - X| < \varepsilon) \to 1 \text{ as } n \to \infty. \tag{5.1}$$

In this case, we write $X_n \xrightarrow{P} X$. Note that (5.1) is equivalent to the relation

$$P(|X_n - X| \geq \varepsilon) \to 0 \text{ as } n \to \infty. \tag{5.2}$$

It may be proved that convergence almost surely implies convergence in probability; see, e.g., [27], [129], [122, pp 95-100].

Later, we will also need the definition of convergence to infinity. Following the same logic, we say that $X_n \xrightarrow{P} \infty$, *converges to infinity in probability*, if for any arbitrary large number $k > 0$,

$$P(X_n > k) \to 1 \text{ as } n \to \infty. \tag{5.3}$$

We say that $X_n \xrightarrow{P} -\infty$, *converges to negative infinity in probability*, if for any arbitrary large $k > 0$,

$$P(X_n < -k) \to 1 \text{ as } n \to \infty. \tag{5.4}$$

The next type of convergence is weaker and is connected with the proximity of distributions.

We say that a sequence of distributions F_n *weakly converges* to a distribution F, writing it as $F_n \xrightarrow{w} F$, or simply $F_n \to F$, if the d.f.'s

$$F_n(x) \to F(x) \text{ for all } x \text{ at which the d.f. } F(x) \text{ is continuous.}$$

We consider only points of continuity of $F(x)$ because otherwise the above definition would not cover very natural types of convergence.

EXAMPLE 1. Let F_n be the distribution of a r.v. $X_n(\omega) \equiv \frac{1}{n}$, and let F be the distribution of a r.v. $X(\omega) \equiv 0$ (so each r.v. assumes only one value for all ω's). The graphs of the d.f.'s $F_n(x)$ and $F(x)$ are given in Fig.15. Clearly, $X_n(\omega) \to X(\omega)$ for all ω's. The reader can readily verify also that $F_n(x) \to F(x)$ for any $x \neq 0$. However, $F_n(0) = 0$ while $F(0) = 1$, so $F_n(0) \not\to (0)$. Since it would be very unnatural to claim in this case that the distribution

F_n does not converge to F, we should just exclude the point zero from consideration. Note that $F(x)$ is not continuous at $x = 0$. □

Certainly, if the limiting distribution $F(x)$ is continuous, we consider convergence for all x's. *It is worth emphasizing that this is the case for the central limit theorem where the limiting distribution is the normal distribution, which is continuous (see Section 6.2 for precise statements and detail).*

We say that r.v.'s X and Y are *equal in distribution*, $X \stackrel{d}{=} Y$, if their distributions are equal.

We say that r.v.'s X_n *converge to* X *in distribution*, writing $X_n \xrightarrow{d} X$, if the distributions $F_{X_n} \to F_X$.

It may be proved that if X assumes only one value, the convergencies $X_n \xrightarrow{d} X$ and $X_n \xrightarrow{P} X$ are equivalent, but in general this is not true. (See details, e.g., in [122, Sec.7-3.1].)

To understand the above statement, observe the simple fact that the equality of the distributions of r.v.'s does not imply the equality of the r.v.'s themselves.

EXAMPLE 2. You play a game with a friend in which you pay \$1 to your friend with probability $1/2$, and with the same probability your friend pays \$1 to you (say, you toss a coin). Clearly, your gain is the r.v. $X = 1$ or -1 with equal probabilities $1/2$. The gain of your friend is the r.v. $Y = -1$ or 1 with the same probabilities. The probability distributions of the r.v.'s X and Y are, clearly, the same: $X \stackrel{d}{=} Y$, but the r.v.'s themselves are not equal to each other since $Y = -X$.

If we set now $X_n = Y$, we will have $F_{X_n} = F_X$ (and hence $F_{X_n} \to F_X$) but the r.v. X_n themselves does not converge to X. □

▶ Weak convergence of distributions does not exhaust all interesting types of convergence. Consider distributions $F_n(B) = P(X_n \in B)$ and $F(B) = P(X \in B)$, where sets B are subsets of the real line. (See also Section 1.3.1 for definitions and a remark concerning Borel sets.) When considering the convergence of d.f.'s, we deal with sets $B = (\infty, x]$.

Note that the convergence of distribution functions may not imply the convergence for all sets.

EXAMPLE 3. We divide the interval $[0,1]$ into n equal parts, and set $x_{kn} = k/n$, $k = 0, 1, ..., n-1$. Let X_n be a r.v. taking values x_{kn} with equal probabilities $1/n$, and X be a r.v. uniformly distributed on $[0,1]$. For any $x \in [0,1]$, the d.f. $F_n(x) = P(X_n \le x) = \frac{1}{n} \times$ [the number of points $x_{kn} \le x$].

A good and not very difficult exercise is to prove that $F_n(x) \to x$, that is, $F_n(x)$ converges to the length of the interval $[0,x]$, which, in turn, is the d.f. $F(x) = P(X \le x)$. Thus, $F_n \xrightarrow{w} F$.

Let now B be the set of all possible points x_{kn}, where $n = 1, 2, ...,$ and $k = 0, 1, ..., n-1$. Certainly, $F_n(B) = 1$, since all possible values of X_n are in B. On the other hand, $F(B) = 0$. Indeed, the uniform distribution is continuous, so for each point x_{kn}, the probability $P(X = x_{kn}) = 0$. Since the number of all such points is countable, the probability $P(X \in B)$ is the sum of all these zero probabilities, that is, zero. □

So, if we want to have the *convergence of distributions over all sets*, we should require more from the model. Assume, for example, that r.v.'s X_n and X have densities $f_n(x)$ and

$f(x)$, respectively. Then for any set B,

$$|F_n(B) - F(B)| = \left| \int_B f_n(x)dx - \int_B f(x)dx \right| = \left| \int_B [f_n(x) - f(x)]dx \right|$$
$$\leq \int_B |f_n(x)dx - f(x)|\, dx \leq \int_{-\infty}^{\infty} |f_n(x)dx - f(x)|\, dx. \tag{5.5}$$

Thus, in the case of continuous r.v.'s, the convergence over all sets is connected with the convergence of densities. In particular, for $F_n(B) \to F(B)$ for all B, it is sufficient that the last integral in (5.5) converges to zero.

We restrict ourselves to a simple example.

EXAMPLE 4. Let X_n be uniformly distributed on $[\frac{1}{n}, 1 + \frac{1}{n}]$, and let X be uniform on $[0, 1]$. We prove that in this case we have convergence over all sets. Indeed, $f_n(x) = 1$ if $x \in [\frac{1}{n}, 1 + \frac{1}{n}]$, and $= 0$ otherwise, while $f(x) = 1$ if $x \in [0, 1]$, and $= 0$ for other x's. Then

$$\int_{-\infty}^{\infty} |f_n(x)dx - f(x)|\, dx = \int_0^{1/n} + \int_{1/n}^{1} + \int_1^{1+1/n} = \int_0^{1/n} |0 - 1|\, dx + \int_{1/n}^{1} |1 - 1|\, dx$$
$$+ \int_1^{1+1/n} |1 - 0|\, dx = \frac{2}{n} \to 0 \text{ as } n \to \infty.$$

From this and (5.5), it follows that $F_n(B) \to F(B)$ for all B. □ ◀

6 LIMIT THEOREMS

6.1 The Law of Large Numbers (LLN)

Let $X_1, X_2, ...$ be a sequence of independent identically distributed (i.i.d.) r.v.'s. Let $S_n = X_1 + ... + X_n$, and $\overline{X}_n = S_n/n$. Set $m = E\{X_i\}$, provided that it exists. It does not depend on i since X's are identically distributed. The LLN says that though $\overline{X}_n$ is random for each particular n, this randomness vanishes as n gets larger, and $\overline{X}_n$ approaches m. The point here, however, is that since $\overline{X}_n$ is not a sequence of numbers but of random variables, the very notion of convergence should be defined properly. We use the notions of convergence almost surely (with probability one) and in probability discussed in Section 5.

Theorem 9 *(The strong LLN)*
(a) Suppose that $E\{|X_i|\}$ is finite. Then $\overline{X}_n \xrightarrow{a.s.} m$ (that is, $\overline{X}_n$ converges to m almost surely). More specifically,

$$P\left(\overline{X}_n \to m\right) = 1. \tag{6.1}$$

(b) If for some c

$$P\left(\overline{X}_n \to c\right) = 1, \tag{6.2}$$

then $E\{|X_i|\}$ is finite, and $c = m$.

As was told in Section 5, almost surely convergence implies convergence in probability, so we state

Corollary 10 *(The weak LLN) Suppose that $E\{|X_i|\}$ is finite. Then $\overline{X}_n \xrightarrow{P} m$ (that is, $\overline{X}_n$ converges to m in probability). More specifically, for any $\varepsilon > 0$,*

$$P\left(|\overline{X}_n - m| \geq \varepsilon\right) \to 0 \quad as \quad n \to \infty. \tag{6.3}$$

The LLN is a mathematical theorem but *it may be viewed as a fundamental law of nature.* First of all, due to this law, the random behavior of a great many of real processes in the long run exhibits a sort of stability. Consider, for instance, the consecutive values of daily income of a company in the long run. These values—denote them by $X_1, X_2, \ldots$.—may be essentially random, uncertain. However, if the X's are i.i.d., the average income per day for n days, that is, $\overline{X}_n = \frac{1}{n}(X_1 + \ldots + X_n)$, for large n, in the long run, is practically certain, being close to the non-random value equal to the expected value of the X's.

Not of less importance is that the LLN allows us to estimate the mean values of r.v.'s. Suppose, for example, that we want to estimate the mean of the highest August temperature in a particular area. The only thing we can do is to review such a temperature, say, in the last 100 years and compute the average. Our intuition tells us that the average will be close to the mean value (though, perhaps, not exactly equal to), and the LLN confirms it and explains in which sense it is true.

Proofs of the LLN and detailed discussions may be found in many textbooks on Probability; see, e.g., [27], [38], [116], [120], [122], [129].

6.2 The Central Limit Theorem (CLT)

Let $E\{X_i^2\}$ be finite, and $\sigma^2 = Var\{X_i\}$. Since the X's are i.i.d., we have $E\{S_n\} = mn$ and $Var\{S_n\} = \sigma^2 n$. Consider the normalized sum

$$S_n^* = \frac{S_n - E\{S_n\}}{\sqrt{Var\{S_n\}}} = \frac{S_n - mn}{\sigma\sqrt{n}}.$$

It is worth emphasizing that the normalized r.v. S_n^* is just the same sum S_n considered in an appropriate scale: after normalization, $E\{S_n^*\} = 0$, and $Var\{S_n^*\} = 1$ (see Example 2.6-1).

Theorem 11 *(The CLT) For any x,*

$$P(S_n^* \leq x) \to \Phi(x) \quad as \quad n \to \infty,$$

where $\Phi(x) = \frac{1}{\sqrt{2\pi}} \int_{-\infty}^{x} e^{-u^2/2} du$, the standard normal distribution function.

Corollary 12 *For any a and b,*

$$P(a \leq S_n^* \leq b) \to \frac{1}{\sqrt{2\pi}} \int_a^b e^{-x^2/2} dx \quad as \quad n \to \infty.$$

In spite of its simple statement, the CLT deals with a deep and important fact. The theorem says that as the number of the terms in the sum S_n is getting larger, the influence of separate terms is diminishing and the distribution of S_n is getting close to a standard distribution (namely, normal) *regardless of which distribution the separate terms have.* It may be continuous or discrete, or neither of these, uniform, exponential, binomial, or anything else—provided that the variance of the terms is finite, the distribution of the sum S_n for large n may be well approximated by a normal distribution.

The theorem has an enormous number of applications because it allows one to estimate probabilities concerning sums of r.v.'s in situations where the distribution of the terms X's is not known; more precisely, when one knows or has estimated only the most "rough" characteristics: the mean and standard deviation.

The same theorem may be useful when formally the distribution of the separate terms is known, but it is difficult (if possible) to present a tractable representation for the distribution of the sum.

Proofs, with use of different methods, and generalizations of this theorem may be found in many textbooks. Often, to make proofs more transparent, in textbooks some additional unnecessary conditions are imposed. Complete proofs with conditions close to necessary may be found, for example, in [27], [38], [120], [122], [129].

---

7 CONDITIONAL EXPECTATIONS. CONDITIONING

Usually, standard courses in Probability do not pay much attention to conditional expectations. For us this topic is quite important; so we consider it in detail and even give exercises on conditioning in Section 9.

7.1 Conditional expectation given a r.v.

Let X and Y be r.v.'s. Our immediate goal is to define the quantity which we will denote by $E\{Y\,|\,X=x\}$, and which we will understand exactly as it sounds: the mean value of Y given that X took on a particular value x.

7.1.1 The discrete case

Let X be a discrete r.v. or a r.vec. —we will see that it does not matter in our case—taking values $x_1, x_2, \ldots$. Below we will often omit the index i of x_i, writing just x but keeping in mind that we consider only x's which coincide with one of the values of X. In particular, this means that for such x's, $P(X=x)\neq 0$.

We define the conditional cumulative distribution function (or simply conditional d.f.) of Y given $X=x$ as the function

$$F_Y(y\,|\,X=x) = P(Y\le y\,|\,X=x) = \frac{P(Y\le y,\, X=x)}{P(X=x)}. \tag{7.1.1}$$

By analogy with the standard representation (2.1.5), let us set

$$E\{Y\,|\,X=x\} = \int_{-\infty}^{\infty} y\,dF_Y(y\,|\,X=x). \tag{7.1.2}$$

For $E\{Y\,|\,X=x\}$ so defined, we will also use the notation $m_{Y|X}(x)$. The function $m_{Y|X}(x)$ is often called a *regression function* of Y on X. When it does not cause misunderstanding, we omit the index $Y|X$ in $m_{Y|X}(x)$.

If Y is also discrete and takes on values $y_1, y_2, \ldots$, then the definition (7.1.2) may be written as

$$E\{Y\,|\,X=x\} = \sum_j y_j P(Y=y_j\,|\,X=x). \tag{7.1.3}$$

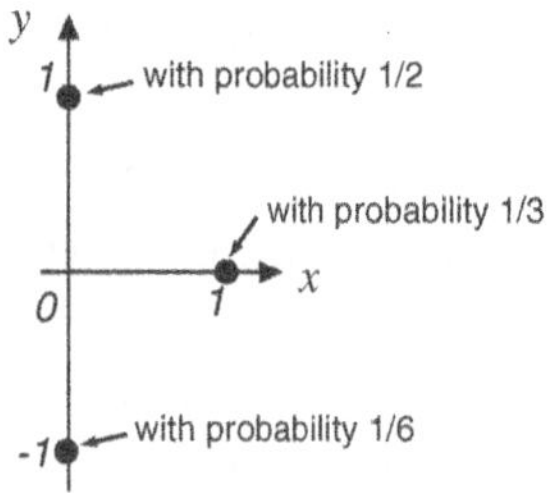

FIGURE 16.

EXAMPLE 1. Let a r.vec. (X,Y) take on vector-values $(0,1)$, $(1,0)$, $(0,-1)$ with probabilities $\frac{1}{2}$, $\frac{1}{3}$, $\frac{1}{6}$, respectively; see also Fig.16. If $X=1$, then Y takes on only one value 0, so $P(Y=0\,|\,X=1)=1$, and

$$m(1) = E\{Y\,|\,X=1\} = 0.$$

In accordance with (7.1.3),

$$m(0) = E\{Y\,|\,X=0\} = 1\cdot P(Y=1\,|\,X=0) + 0\cdot P(Y=0\,|\,X=0) + (-1)\cdot P(Y=-1\,|\,X=0)$$
$$= 1\cdot\frac{1/2}{(1/2)+(1/6)} + 0\cdot 0 + (-1)\cdot\frac{1/6}{(1/2)+(1/6)} = \frac{1}{2}.$$

EXAMPLE 2 is classical. Let N_1 and N_2 be independent Poisson r.v.'s with parameters λ_1 and λ_2, respectively. Find $E\{N_1\,|\,N_1+N_2=n\}$.

Thus, N_1 plays the role of Y, and N_1+N_2 plays the role of X; we also replaced x by n. For example, there are two groups of clients of an insurance company, and N_i is the number of claims coming from the ith group. We are interested in the mean number of claims coming from the first group *given* the total number of claims.

It is known that the conditional distribution of N_1 given N is binomial. More precisely,

$$P(N_1=k\,|\,N_1+N_2=n) = \binom{n}{k} p^k (1-p)^{n-k},$$

where

$$p = \frac{\lambda_1}{\lambda_1+\lambda_2}.$$

We give a formal proof of this fact in Section **3**.3.1.2, where we consider this phenomenon in detail. Now, to give an example on conditional expectation, it is enough to take this fact for granted.

Since the mean value of the binomial distribution is np,

$$E\{N_1\,|\,N_1+N_2=n\} = \frac{n\lambda_1}{\lambda_1+\lambda_2}. \ \square \tag{7.1.4}$$

Let us return to the regression function $m(x) = m_{Y|X}(x)$. This is the mean value of Y given $X=x$. Since X is a random variable, its values x may be different, random. To reflect this circumstance, let us replace in $m(x)$ the argument x by the r.v. X itself, that is, consider the r.v. $m(X)$. This is a function of X, and its significance is the same as above: it is the conditional mean value of Y given a value of X. However, since X is random, the conditional mean value of Y given X is also random, and the value of $m(X)$ depends on which value X will assume.

The r.v. $m(X)$ has a special notation: $E\{Y\,|\,X\}$, and is called the *conditional expectation* of Y given X. It is important to keep in mind that this is a r.v., and since $E\{Y\,|\,X\}$ is a function of X, its value is completely determined by the value of X.

EXAMPLE 3. In the situation of Example 1, X takes on two values: 0 and 1 with probabilities $\frac{2}{3}$ and $\frac{1}{3}$, respectively; see also Fig.16. Hence, $m(X)$ takes on the values $m(0) = \frac{1}{2}$ and $m(1) = 0$ with the above probabilities. Consequently, $E\{Y\,|\,X\}$ is a r.v. taking values $\frac{1}{2}$ and 0 with respective probabilities $\frac{2}{3}$ and $\frac{1}{3}$.

EXAMPLE 4. Let us return to Example 2 and set $N = N_1 + N_2$, $\lambda = \lambda_1 + \lambda_2$. By virtue of (7.1.4), $m_{N_1\,|\,N}(n) = \dfrac{\lambda_1}{\lambda}n$, and hence

$$E\{N_1\,|\,N\} = \frac{\lambda_1}{\lambda}N. \ \square$$

7.1.2 The case of continuous distributions

If the denominator in (7.1.1) equals zero, we cannot write this representation as is. For example, this is the case for all x's in the continuous case. So, we should define $F_Y(y\,|\,X = x)$ and $E\{Y\,|\,X = x\}$ in another way. In this section, we assume that the vector (X,Y) has a joint density $f(x,y)$. Denote by $f_X(x)$ the marginal density of X and consider the function

$$f_{Y|X}(y\,|\,x) = \frac{f(x,y)}{f_X(x)}, \tag{7.1.5}$$

provided $f_X(x) \neq 0$. If $f_X(x) = 0$, we set $f_{Y|X}(y\,|\,x) = 0$ by definition.

We call $f_{Y|X}(y\,|\,x)$ *the conditional density of* Y given $X = x$. When it cannot cause confusion, we omit the index $Y\,|\,X$ in $f_{Y|X}(y\,|\,x)$.

To clarify the significance of definition (7.1.5), let us consider infinitesimally small intervals $[x, x+dx]$, $[y, y+dy]$ and assume $f_X(x) \neq 0$. Reasoning somewhat heuristically, we represent the probability $P(x \leq X \leq x+dx)$ as $f_X(x)dx$; see also Section 1.3.1. Similarly, we write $P(y \leq Y \leq y+dy,\, x \leq X \leq x+dx) = f(x,y)dxdy$. Then

$$P(y \leq Y \leq y+dy\,|\,x \leq X \leq x+dx) = \frac{P(y \leq Y \leq y+dy,\, x \leq X \leq x+dx)}{P(x \leq X \leq x+dx)}$$

$$= \frac{f(x,y)dxdy}{f_X(x)dx} = f(y\,|\,x)dy. \tag{7.1.6}$$

Since the interval $[x, x+dx]$ is infinitesimally small, we can view $P(y \leq Y \leq y+dy\,|\,x \leq X \leq x+dx)$ as $P(y \leq Y \leq y+dy\,|\,X = x)$, which leads to

$$P(y \leq Y \leq y+dy\,|\,X = x) = f(y\,|\,x)dy.$$

Thus, $f(y\,|\,x)$ indeed plays the role of the density of Y when X assumes a value x, which justifies the formal definition (7.1.5). Note that the reasoning above is just a clarification. Formally, we adopt (7.1.5) as a definition.

Now, again as a definition, we set the conditional d.f.

$$F_Y(y\,|\,X=x) = \int_{-\infty}^{y} f(t\,|\,x)dt, \tag{7.1.7}$$

and, by analogy with (7.1.2),

$$m(x) = m_{Y\,|\,X}(x) = E\{Y\,|\,X=x\} = \int_{-\infty}^{\infty} yf(y\,|\,x)dy. \tag{7.1.8}$$

Note also that by virtue of (7.1.7), $dF_Y(y\,|\,X=x) = f(y\,|\,x)dy$ (the conditional density is the derivative of the conditional d.f.). Hence, we can write (7.1.8) as

$$m(x) = m_{Y\,|\,X}(x) = E\{Y\,|\,X=x\} = \int_{-\infty}^{\infty} ydF(y\,|\,x)dy. \tag{7.1.9}$$

Comparing it with (7.1.2), we see that the latter representation is true for the discrete and continuous cases as well.

Now, as above, we define the expected value of Y given X as

$$E\{Y\,|\,X\} = m(X).$$

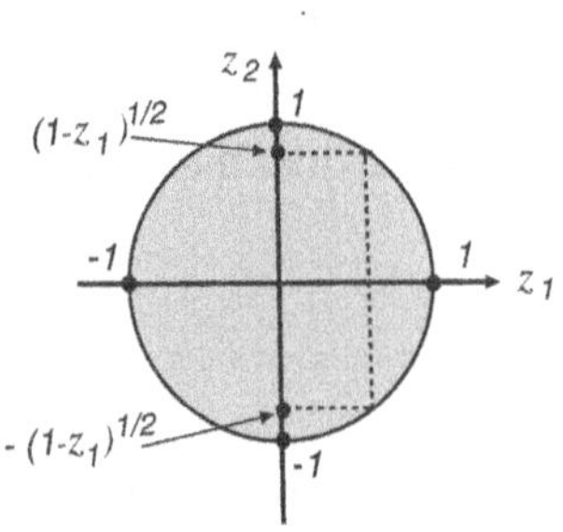

FIGURE 17.

EXAMPLE 1. Let a vector $\mathbf{Z} = (Z_1, Z_2)$ be uniformly distributed on the unit disk $O_1 = \{z_1^2 + z_2^2 \leq 1\}$. The joint density of $\mathbf{Z}$ is $f(z_1, z_2) = 1/\pi$ if $(z_1, z_2) \in O_1$, and $= 0$ otherwise; see also Fig.17.

First, we find $f_{Z_2\,|\,Z_1}(z_2\,|\,z_1)$, so the role of X above is played by Z_1, and the role of Y by Z_2.

If $Z_1 = z_1$, then values of Z_2 lie on the interval $\left[-\sqrt{1-z_1^2}, \sqrt{1-z_1^2}\right]$, see again Fig.17. Since the joint distribution is uniform, we can guess that the same is true for the conditional distribution of Z_2 given Z_1. Let us show it rigorously.

For the marginal density of Z_1, we have

$$f_1(z_1) = \int_{-\infty}^{\infty} f(z_1, z_2)dz_2 = \int_{-\sqrt{1-z_1^2}}^{\sqrt{1-z_1^2}} \frac{1}{\pi}dz_2 = \frac{2\sqrt{1-z_1^2}}{\pi}. \tag{7.1.10}$$

Then, by definition,

$$f_{Z_2\,|\,Z_1}(z_2\,|\,z_1) = \frac{f(z_1, z_2)}{f_1(z_1)} = \frac{1}{2\sqrt{1-z_1^2}} \text{ if } z_2 \in \left[-\sqrt{1-z_1^2}, \sqrt{1-z_1^2}\right], \text{ and } = 0 \text{ otherwise.}$$

For a fixed z_1, as a function of z_2, the last density is constant. Hence, the conditional distribution under consideration is indeed uniform on $\left[-\sqrt{1-z_1^2}, \sqrt{1-z_1^2}\right]$.

Let us consider conditional expectations, say, $E\{Z_2 \,|\, Z_1\}$ and $E\{Z_2^2 \,|\, Z_1\}$. We can use the general formula (7.1.8), but it is not necessary.

Indeed, for a r.v. ξ uniformly distributed on $[-a,a]$ for some a, we have $E\{\xi\} = 0$ and $E\{\xi^2\} = a^2/3$; see Section 3.2.1. As we now know, given Z_1 the r.v. Z_2 is uniformly distributed on $[-(1-Z_1^2)^{1/2}, (1-Z_1^2)^{1/2}]$. Hence, $E\{Z_2 \,|\, Z_1\} = 0$ (which could be predicted by using the symmetry argument), and

$$E\{Z_2^2 \,|\, Z_1\} = (1-Z_1^2)/3.$$

EXAMPLE 2. Let X_1, X_2 be independent r.v.'s with densities $f_1(x)$, $f_2(x)$, respectively, and let $S = X_1 + X_2$. In many particular problems, it is important to realize that

$$f_{X_1 \,|\, S}(x \,|\, s) = \frac{f_1(x) f_2(s-x)}{f_S(s)}. \tag{7.1.11}$$

Since the denominator is exactly what the definition (7.1.5) requires, to prove (7.1.11) it suffices to show that the joint density of the vector (X_1, S) is

$$f(x,s) = f_1(x) f_2(s-x). \tag{7.1.12}$$

Heuristically, it is almost obvious. For any r.v. X, its density $f(x)$ is connected with the probability that X will take on a value close to x. In our case, if $X_1 \approx x$ and $S \approx s$, then $X_2 \approx s - x$. Since X_1, X_2 are independent, $P(X_1 \approx x, S \approx s) = P(X_1 \approx x, X_2 \approx s - x) = P(X_1 \approx x) P(X_2 \approx s - x)$. This is reflected in (7.1.12). A rigorous proof may be found, e.g., in [122, p.197].

EXAMPLE 3. We apply the result of the previous example to the particular case when X_1 and X_2 are exponential with the same parameter a. It suffices to consider $f_{X_1 \,|\, S}(x \,|\, s)$ for $x \in [0,s]$. Indeed, X_1 and X_2 are positive, and given their sum equals s, both terms are not greater than s.

Later, in Proposition 4 from Section **2**.2.1.2, we will prove that $f_S(s) = f_{a,2}(s)$, where $f_{a,2}$ is the Γ-density with parameters $(a,2)$. So, $f_S(s) = a^2 s e^{-as}$, and by (7.1.11), for $x \le s$

$$f_{X_1 \,|\, S}(x \,|\, s) = \frac{ae^{-ax} ae^{-a(s-x)}}{a^2 s e^{-as}} = \frac{1}{s}.$$

Thus, the conditional distribution is uniform on $[0,s]$. The result is nice: though the values of the exponential r.v. X_1 are not equally likely, given the information that the sum $X_1 + X_2$ is s, the r.v. X_1 may take on any value from $[0,s]$ with equal likelihood. This is strongly connected with the fact that X's are exponential; see also Exercise 4.

Once we know the conditional distribution, we can compute various expectations. In particular, $E\{X_1 \,|\, S\} = S/2$ (as the mean of the distribution uniform on $[0,S]$). If we define the conditional variance as the variance of the corresponding conditional distribution, we can write that $Var\{X_1 \,|\, S\} = S^2/12$. $\square$

7.2 Properties of conditional expectations

1. Let us recall that the conditional expectation $E\{Y\,|\,X\}$ is a r.v. The main and extremely important property of conditional expectation is that for any X and for any Y with a finite $E\{Y\}$,

$$\boxed{E\{E\{Y\,|\,X\}\} = E\{Y\}.} \tag{7.2.1}$$

Thus,

> If we "condition Y on X" and then compute the expected value of the conditional expectation, we come back to the original unconditional expectation $E\{Y\}$.

We call (7.2.1) the *formula for total expectation* or the *law of total expectation*. In the next section, we consider some examples of applications of this formula, and later on we will apply it repeatedly.

The validity of (7.2.1) is based on the fact that the part in the left member of (7.2.1) that concerns X cancels. To demonstrate it, consider the continuous case. Since $E\{Y\,|\,X\} = m(X)$, by virtue of (7.1.5) and (7.1.8), we have

$$\begin{aligned} E\{E\{Y\,|\,X\}\} &= E\{m(X)\} = \int_{-\infty}^{\infty} m(x) f_X(x)dx = \int_{-\infty}^{\infty}\left(\int_{-\infty}^{\infty} yf(y\,|\,x)dy\right) f_X(x)dx \\ &= \int_{-\infty}^{\infty}\int_{-\infty}^{\infty} y\left(\frac{f(x,y)}{f_X(x)}\right) f_X(x)dydx = \int_{-\infty}^{\infty}\int_{-\infty}^{\infty} yf(x,y)dydx \\ &= \int_{-\infty}^{\infty} y\left(\int_{-\infty}^{\infty} f(x,y)dx\right)dy = \int_{-\infty}^{\infty} yf_Y(y)dy = E\{Y\}, \end{aligned}$$

where $f_Y(y)$ is the marginal density of Y.

The proof in the discrete case is similar; we should just replace integrals by the corresponding sums.

The next properties are straightforward and quite plausible from a heuristic point of view. We omit proofs.

2. For any number c and r.v.'s X, Y, Y_1, and Y_2,

$$E\{cY\,|\,X\} = cE\{Y\,|\,X\} \quad \text{and} \quad E\{Y_1 + Y_2\,|\,X\} = E\{Y_1|X\} + E\{Y_2\,|\,X\}. \tag{7.2.2}$$

3. If r.v.'s X and Y are independent, then $E\{Y\,|\,X\}$ is not random and equals $E\{Y\}$.

4. Consider $Y = g(X)Z$, where Z is a r.v. and $g(x)$ is a function. Then

$$E\{g(X)Z\,|\,X\} = g(X)E\{Z\,|\,X\}. \tag{7.2.3}$$

In particular,

$$E\{g(X)\,|\,X\} = g(X), \tag{7.2.4}$$

and $E\{X\,|\,X\} = X$. Intuitively, it is quite understandable. When conditioning on X, we view X as a constant, and hence $g(X)$ may be brought outside of the conditional expectation.

EXAMPLE 1. (*The "beta" of a security*). In Finance, the "beta" (β) of a security (say, a stock or a portfolio of stocks) is a characteristic describing the relation of the security's return (that is, the income per \$1 investment) with that of the financial market as a whole. In other words, β shows how the return of a security depends on the situation in the market on the average.

Let Y be the (future) return of a security, and let X be the future value of a market index, i.e., a global characteristic monitoring either the value of the market as a whole, or an essential part of it. (Typical examples are Dow Jones or S&P indices.)

Let us adopt the following simple model that sometimes works well: $Y = \xi X + \varepsilon$, where coefficients ξ and ε are *random* but do not depend on X.

Loosely put, ε characterizes the "random factors" that are relevant to the security and are not associated with the market. The random coefficient ξ reflects the impact of the global market situation on the value of the security.

In view of Properties 2-4, $E\{Y\,|\,X\} = E\{\xi X + \varepsilon\,|\,X\} = E\{\xi X\,|\,X\} + E\{\varepsilon\,|\,X\} = X\,E\{\xi|X\} + E\{\varepsilon\,|\,X\} = X\,E\{\xi\} + E\{\varepsilon\}$.

The mean value $E\{\xi\}$ is denoted by β and called "beta." We set $a = E\{\varepsilon\}$, and eventually write $E\{Y\,|\,X\} = \beta X + a$.

Note that β may be negative. For example, as a rule, the price of gold is growing when the market is dropping.

By definition, the market itself has a beta of 1. Indeed, in this case $Y = X$, and hence $\xi = 1$, $\varepsilon = 0$.

If the variation of the stock return is larger on the average than that of the market, the absolute value of beta is greater than 1, whereas for a stock whose return varies to a less extent than the market's return, $|\beta| < 1$. The reader may find particular values of β for many stocks in Web sites concerning the stock market.

EXAMPLE 2. Let X_1, X_2 be i.i.d. r.v.'s. Find $E\{X_1\,|\,X_1 + X_2\}$. (What is the mean value of one term (addend) given the value of the sum?) By the symmetry argument, we can guess that the answer is simple:

$$E\{X_1\,|\,X_1 + X_2\} = \frac{1}{2}(X_1 + X_2).$$

This is, indeed, true, and may be shown as follows. Since X_1, X_2 are independent and have the same distribution, $E\{X_1\,|\,X_1 + X_2\} = E\{X_2\,|\,X_1 + X_2\}$ by symmetry. Then, using the properties above, we have $2E\{X_1\,|\,X_1 + X_2\} = E\{X_1\,|\,X_1 + X_2\} + E\{X_2\,|\,X_1 + X_2\} = E\{X_1 + X_2\,|\,X_1 + X_2\} = X_1 + X_2$. $\square$

7.3 Conditioning and some useful formulas

The formula for total expectation

$$E\{Y\} = E\{E\{Y\,|\,X\}\} \tag{7.3.1}$$

proves to be a very useful tool in many problems including those where the original setup does not involve any conditioning.

Formulas below present some modifications or particular versions of (7.3.1), which we will repeatedly use throughout the whole book.

7.3.1 A formula for variance

Let us consider a counterpart of (7.3.1) for variances. We show that

$$Var\{Y\} = E\{Var\{Y\,|\,X\}\} + Var\{E\{Y\,|\,X\}\}, \tag{7.3.2}$$

where $Var\{Y\,|\,X\}$ is the variance of Y with respect to the conditional distribution of Y given X. In particular, we can write that $Var\{Y\,|\,X\} = E\{Y^2\,|\,X\} - (E\{Y\,|\,X\})^2$.

To memorize (7.3.2), notice that in this formula the order of the operations $E\{\cdot\}$ and $Var\{\cdot\}$ alternates. To prove (7.3.2), we make use of (7.3.1) in the following way:

$$\begin{aligned} Var\{Y\} &= E\{Y^2\} - (E\{Y\})^2 = E\{E\{Y^2\,|\,X\}\} - (E\{E\{Y\,|\,X\}\})^2 \\ &= E\left\{E\{Y^2\,|\,X\} - (E\{Y\,|\,X\})^2\right\} + E\left\{(E\{Y\,|\,X\})^2\right\} - (E\{E\{Y\,|\,X\}\})^2 \\ &= E\left\{Var\{Y\,|\,X\}\right\} + Var\left\{E\{Y\,|\,X\}\right\}. \end{aligned}$$

We skip here particular examples; in the following chapters, we use (7.3.2) repeatedly.

7.3.2 More detailed representations of the formula for total expectation

We use representation (7.1.9) which is true for both continuous and discrete cases. Since $E\{Y\,|\,X\} = m(X)$, from (7.3.1) and (7.1.9) it follows that

$$E\{Y\} = E\{m(X)\} = \int_{-\infty}^{\infty} m(x)dF_X(x) = \int_{-\infty}^{\infty} \left(\int_{-\infty}^{\infty} y dF_Y(y\,|\,X=x)\right) dF_X(x). \tag{7.3.3}$$

In the case when all densities exist, we can rewrite it as

$$E\{Y\} = \int_{-\infty}^{\infty} m(x)f_X(x)dx = \int_{-\infty}^{\infty} \left(\int_{-\infty}^{\infty} yf(y\,|\,x)dy\right) f_X(x)dx. \tag{7.3.4}$$

Let us now recall that all r.v.'s we consider are defined on some sample space $\Omega = \{\omega\}$, i.e., each r.v. X is a function $X(\omega)$. Consider an event $A \subseteq \Omega$, and the *indicator* of A, that is, the r.v.

$$\mathbf{1}_A = \mathbf{1}_A(\omega) = \begin{cases} 1 \text{ if } \omega \in A, \\ 0 \text{ if } \omega \notin A. \end{cases} \tag{7.3.5}$$

Thus, $\mathbf{1}_A$ assumes the value 1 if A occurs, and the value 0 otherwise. Clearly, $E\{\mathbf{1}_A\} = 1 \cdot P(A) + 0 \cdot P(A)$, so

$$E\{\mathbf{1}_A\} = P(A). \tag{7.3.6}$$

By analogy with (7.3.6), we define the quantity $P(A\,|X) = E\{\mathbf{1}_A\,|X\}$. The conditional probability $P(A\,|X)$ should be understood as it is written: this is the probability of A given the r.v. X. Note that $P(A\,|X)$ is a function of X, and it is a random variable.

Setting $Y = \mathbf{1}_A$ in (7.3.1), we have $P(A) = E\{\mathbf{1}_A\} = E\{E\{\mathbf{1}_A\,|X\}\} = E\{P(A\,|X)\}$. Thus,

$$P(A) = E\{P(A\,|X)\}. \tag{7.3.7}$$

In this particular case, $m(x) = P(A\,|X = x)$, and by virtue of (7.3.3), we have

$$P(A) = \int_{-\infty}^{\infty} P(A\,|X = x)dF_X(x). \tag{7.3.8}$$

We can view it as a version of the formula for total probability.

If X has a density $f_X(x)$, the last relation may be rewritten as

$$P(A) = \int_{-\infty}^{\infty} P(A\,|X = x)f_X(x)dx. \tag{7.3.9}$$

EXAMPLE 1 is classical. Let ξ_1, ξ_2 be independent exponential r.v.'s with respective parameters a_1, a_2. Find $P(\xi_2 > \xi_1)$. A particular illustration is considered in Exercise 9. By (7.3.9) and by virtue of the independence condition,

$$P(\xi_2 > \xi_1) = \int_{-\infty}^{\infty} P(\xi_2 > \xi_1\,|\,\xi_1 = x)f_{\xi_1}(x)dx = \int_0^{\infty} P(\xi_2 > x\,|\,\xi_1 = x)a_1e^{-a_1x}dx$$

$$= \int_0^{\infty} P(\xi_2 > x)a_1e^{-a_1x}dx = \int_0^{\infty} e^{-a_2x}a_1e^{-a_1x}dx = a_1\int_0^{\infty} e^{-(a_1+a_2)x}dx = \frac{a_1}{a_1+a_2}. \;\square$$

Let us return to (7.3.8) and set $A = \{Z \le z\}$, where Z is a r.v. and z is a number. After such a substitution, we get that the d.f.

$$F_Z(z) = P(Z \le z) = \int_{-\infty}^{\infty} P(Z \le z\,|X = x)dF_X(x) = \int_{-\infty}^{\infty} F_Z(z\,|X = x)dF_X(x), \tag{7.3.10}$$

where $F_Z(z\,|X = x)$ is the conditional d.f. of Z given $X = x$. If all densities exist, differentiating (7.3.10) in z, we obtain that the density of Z is

$$f_Z(z) = \int_{-\infty}^{\infty} f_{Z|X}(z\,|x)f_X(x)dx. \tag{7.3.11}$$

See also Exercise 8.

EXAMPLE 2. Find the distribution of the ratio $Z = \xi_1/\xi_2$ for independent standard normal r.v.'s ξ_1 and ξ_2. Certainly, Z is well defined only if $\xi_2 \ne 0$, but since $P(\xi_2 = 0) = 0$, we can eliminate this case from consideration.

Consider the conditional density $f_{Z|\xi_2}(z\,|x)$. We will omit the index $Z\,|\,\xi_2$, writing just $f(z\,|x)$. Because ξ_1 and ξ_2 are independent, the density of Z given $\xi_2 = x \ne 0$ is the density of the r.v. ξ_1/x. If $x > 0$, this is a normal r.v. with zero mean and variance $1/x^2$. Furthermore, a normal r.v. with zero mean is symmetric and when multiplying it by -1, we do not change its distribution. Hence, for $x < 0$, we again have a normal r.v. with zero mean and variance $1/x^2$. Thus, in both cases, $f(z\,|\,x) = (2\pi)^{-1/2}|x|\exp\{-z^2x^2/2\}$.

Now, by (7.3.11),

$$f_Z(z) = \int_{-\infty}^{\infty} f_{Z|\xi_2}(z\,|\,x) f_{\xi_2}(x)dx = \int_{-\infty}^{\infty} (2\pi)^{-1/2}|x|\exp\{-z^2x^2/2\}(2\pi)^{-1/2}\exp\{-x^2/2\}dx$$
$$= (2\pi)^{-1}\int_{-\infty}^{\infty} |x|\exp\{-x^2(1+z^2)/2\}dx.$$

The integrand is an even function, so $\int_0^\infty = \int_{-\infty}^0$. By the variable change $y = x\sqrt{1+z^2}$, we have

$$f_Z(z) = \frac{1}{2\pi}\cdot\frac{1}{1+z^2}\cdot 2\int_0^\infty y\exp\{-y^2/2\}dy = \frac{1}{\pi(1+z^2)}\int_0^\infty d\left(-\exp\{-y^2/2\}\right)$$
$$= \frac{1}{\pi(1+z^2)}.$$

The distribution we have arrived at is called the *Cauchy distribution*. □

7.4 Conditional expectation given a random vector

We will see that in the multidimensional case, we can proceed practically in the same fashion.

7.4.1 General definitions

Let $\mathbf{X} = (X_1, ..., X_k)$ be a r.vec., and Y be a r.v. Set $\boldsymbol{x} = (x_1, ..., x_k)$. If the distribution of $\mathbf{X}$ is discrete and $\boldsymbol{x}$ is a possible value of $\mathbf{X}$, then $P(\mathbf{X} = \boldsymbol{x}) \neq 0$, and we can again consider the conditional d.f.

$$F_Y(y\,|\,\mathbf{X} = \boldsymbol{x}) = P(Y \leq y\,|\,\mathbf{X} = \boldsymbol{x}) = \frac{P(Y \leq y, \mathbf{X} = \boldsymbol{x})}{P(\mathbf{X} = \boldsymbol{x})}.$$

Then we can again set

$$E\{Y\,|\,\mathbf{X} = \boldsymbol{x}\} = \int_{-\infty}^{\infty} y dF_Y(y\,|\,\mathbf{X} = \boldsymbol{x})$$

and proceed exactly as we did in Section 7.1.1.

Let the r.vec. $(\mathbf{X}, Y) = (X_1, ..., X_n, Y)$ be continuous, let $f(\boldsymbol{x}, y)$ be its joint density, and $f_{\mathbf{X}}(\boldsymbol{x})$ be the joint density of the r.vec. $\mathbf{X}$. With respect to $(\mathbf{X}, Y)$, the density $f_{\mathbf{X}}(\boldsymbol{x})$ is marginal. By analogy with what we did in Section 7.1.2, we define the conditional density as

$$f(y\,|\,\boldsymbol{x}) = f_{Y|\mathbf{X}}(y\,|\,\boldsymbol{x}) = \frac{f(\boldsymbol{x}, y)}{f_{\mathbf{X}}(\boldsymbol{x})},$$

provided $f_{\mathbf{X}}(\boldsymbol{x}) \neq 0$. Then, exactly as in (7.1.8), we define

$$m(\boldsymbol{x}) = \int_{-\infty}^{\infty} y f(y\,|\,\boldsymbol{x})dy,$$

and set $E\{Y\,|\,\mathbf{X}\} = m(\mathbf{X})$.

EXAMPLE 1 is similar to Example 7.1.2-1. Let a vector $\mathbf{Z} = (Z_1, Z_2, Z_3)$ be uniformly distributed in the unit ball $O_1 = \{z_1^2 + z_2^2 + z_3^2 \leq 1\}$. We will condition Z_3 on (Z_1, Z_2).

Set $\boldsymbol{z} = (z_1, z_2, z_3)$. The joint density of $\mathbf{Z}$ is $f(\boldsymbol{z}) = \frac{3}{4\pi}$ for $\boldsymbol{z} \in O_1$, and $= 0$ otherwise. (For the total integral of $f(\boldsymbol{z})$ to be one, the density in this case should be equal to one divided by the volume of the ball.) If the r.vec. (Z_1, Z_2) assumes a value (z_1, z_2), then values of Z_3 lie in the interval $\left[-\sqrt{1-z_1^2-z_2^2}, \sqrt{1-z_1^2-z_2^2}\right]$. The density of (Z_1, Z_2) is marginal with respect to the total density $f(\boldsymbol{z})$. So,

$$f_{(Z_1,Z_2)}(z_1, z_2) = \int_{-\infty}^{\infty} f(z_1, z_2, z_3) dz_3 = \int_{-\sqrt{1-z_1^2-z_2^2}}^{\sqrt{1-z_1^2-z_2^2}} \frac{3}{4\pi} dz_3 = \frac{3\sqrt{1-z_1^2-z_2^2}}{2\pi}. \tag{7.4.1}$$

Then

$$f_{Z_3\,|\,(Z_1,Z_2)}(z_3\,|\,z_1, z_2) = \frac{f(z_1, z_2, z_3)}{f_{(Z_1,Z_2)}(z_1, z_2)} = \frac{1}{2\sqrt{1-z_1^2-z_2^2}}$$

if $z_3 \in \left[-\sqrt{1-z_1^2-z_2^2}, \sqrt{1-z_1^2-z_2^2}\right]$, and $= 0$ otherwise.

As a function of z_3, the last density is constant, so the conditional distribution under consideration is uniform on $\left[-\sqrt{1-z_1^2-z_2^2}, \sqrt{1-z_1^2-z_2^2}\right]$.

Now, similar to what we did in Example 7.1.2-1, we can get that $E\{Z_3\,|\,Z_1, Z_2\} = 0$, and $E\{Z_3^2\,|\,Z_1, Z_2\} = (1 - Z_1^2 - Z_2^2)/3$. □

Proposition 13 *The conditional expectation $E\{Y\,|\,\mathbf{X}\}$ defined above satisfies Properties 1-4 from Section 7.2.*

7.4.2 On conditioning in the multi-dimensional case

In this case, the formulas from Section 7.3.2 continue to be true with the natural replacement X by $\mathbf{X}$, and x by $\boldsymbol{x}$. In particular, we can write the following counterpart of (7.3.11):

$$f_Z(z) = \int_{-\infty}^{\infty} \cdots \int_{-\infty}^{\infty} f_{Z\,|\,\mathbf{X}}(z\,|\,\boldsymbol{x}) f_{\mathbf{X}}(\boldsymbol{x}) d\boldsymbol{x}, \tag{7.4.2}$$

where integration in the multidimensional integral above is carried out over all values of the vector $\boldsymbol{x}$.

EXAMPLE 1. This nice example is a generalization of Example 7.3.2-2. Consider the system of linear equations

$$\mathbf{AZ} = \mathbf{b}, \tag{7.4.3}$$

where $\mathbf{A} = \{a_{ij}\}$ is a $n \times n$ matrix, $\mathbf{b} = (b_1, ..., b_n)$ is a n-dimensional vector, and the n-dimensional vector $\mathbf{Z} = (Z_1, ..., Z_n)$ is the vector of unknowns, that is, $\mathbf{Z}$ is a solution to (7.4.3).

Assume that all a_{ij} and b_i are independent standard normal r.v.'s. We prove that in this case, as in Example 7.3.2-2, each Z_i has the Cauchy distribution.

In view of symmetry, it suffices to consider Z_1. Let A_j be the cofactor of the element a_{1j}. By the well known formula for solutions of the system of linear equations,

$$Z_1 = \frac{b_1A_1 + b_2A_2 + ... + b_nA_n}{a_{11}A_1 + a_{22}A_2 + ... + a_{1n}A_n}.$$

Since all r.v.'s a_{1j} and $(A_1, ..., A_n)$ have continuous distributions, the probability that the denominator will be equal to zero is zero, so we may exclude this case from consideration. For the same reason, we can divide the numerator and the denominator by $\sqrt{A_1^2 + ... + A_n^2}$ and write

$$Z_1 = \frac{(b_1A_1 + b_2A_2 + ... + b_nA_n)/\sqrt{A_1^2 + ... + A_n^2}}{(a_{11}A_1 + a_{22}A_2 + ... + a_{1n}A_n)/\sqrt{A_1^2 + ... + A_n^2}}. \tag{7.4.4}$$

Consider the conditional distribution of Z_1 given the vector $\overline{A} = (A_1, ..., A_n)$. Note that the b's and a_{1j}'s are independent of each other and do not depend on $\overline{A}$. Then, once $(A_1, ..., A_n)$ is given, the numerator and the denominator in (7.4.4) are independent.

Furthermore, given $\overline{A}$, the r.v. $b_1A_1 + b_2A_2 + ... + b_nA_n$ is the sum of independent normal r.v.'s. Hence, this sum is normal. Its mean is zero (since each a_{1j} has zero mean), and given $(A_1, ..., A_n)$, the conditional variance of this sum is $A_1^2 + ... + A_n^2$. Consequently, dividing by $\sqrt{A_1^2 + ... + A_n^2}$ we normalize the sum (see also Section 2.6), making its variance equal to one. Thus, given $\overline{A} = (A_1, ..., A_n)$, the numerator in (7.4.4) is a standard normal r.v.

By the same argument, the denominator in (7.4.4) is also standard normal. Thus, given $\overline{A}$, the conditional distribution of Z_1 is the distribution of the ratio of two independent standard normal r.v.'s. By the result of Example 7.3.2-2, this is the Cauchy distribution.

So, we have arrived at a remarkable fact. The conditional distribution of Z_1 given $\overline{A}$ does not depend on $\overline{A}$ at all and is equal to the Cauchy distribution. Then the unconditional distribution of Z_1 must also be equal to the same Cauchy distribution. Formally, we can prove it as follows. By virtue of (7.4.2),

$$f_{Z_1}(x) = \int_{\boldsymbol{u}} f_{Z_1|\overline{A}}(x\,|\,\boldsymbol{u}) f_{\overline{A}}(\boldsymbol{u}) d\boldsymbol{u},$$

where $\boldsymbol{u} = (u_1, ..., u_n)$, and integration is over all possible values of $\boldsymbol{u}$. As was proved, $f_{Z_1|\overline{A}}(x\,|\,\boldsymbol{u}) = 1/[\pi(1+x^2)]$. Consequently,

$$f_{Z_1}(x) = \int_{\boldsymbol{u}} \frac{1}{\pi(1+x^2)} \cdot f_{\overline{A}}(\boldsymbol{u}) d\boldsymbol{u} = \frac{1}{\pi(1+x^2)} \int_{\mathbf{u}} f_{\overline{A}}(\boldsymbol{u}) d\boldsymbol{u} = \frac{1}{\pi(1+x^2)}$$

because the integral of any density equals one. □

7.4.3 On the infinite-dimensional case

The situation becomes more complicated when $\mathbf{X}$ is an infinite-dimensional vector. For example, in Chapters 4 and 5, we study random processes X_t where time t is continuous. In many problems, we need to consider the conditional expectations of a r.v. Y given a realization of a process X_t until time T, which may be written as $E\{Y\,|\,X_t,\ 0 \leq t \leq T\}$. Note that the "conditioning part" involves *all* X_t for $t \in [0, T]$.

The significance of such a conditional expectation is absolutely the same as above, and it may be treated as above, including Properties 1-4 from Section 7.2. However, in a mathematically rigorous exposition, we cannot proceed in the same way as before. One reason is that the notion of a density $f(\boldsymbol{x})$ presupposes that $\int_{\boldsymbol{x}} f(\boldsymbol{x}) = 1$, where integration is over all $\boldsymbol{x}$'s. Thus, we must define what integration in an infinite-dimensional space means. This requires additional constructions.

In this case, practically the only way out is to appeal to a general definition of conditional expectation based on the measure theory. However, it is important to realize that for applications, we need only two things:

— the very fact that the conditional expectation exists, and

— that it satisfies the main properties 1-4 from Section 7.2.

If we take these facts for granted, then there will be no problem in using the notion of conditional expectation in applied models.

---

8 ELEMENTS OF THE THEORY OF INTEREST 1,2,3

In all examples below we choose, for certainty, a year as the fundamental unit of time.

8.1 Compound interest

We begin with the notion of rate. Let us view a variable C_t as the capital of an investor at time t. For an interval $[t, t+h]$, set $\Delta C_t = C_{t+h} - C_t$, the absolute change of C_t over this interval. If the *relative* change

$$\frac{\Delta C_t}{C_t} = kh, \tag{8.1.1}$$

we call k the *rate* of change over this interval; more precisely, the average rate.

Now, consider the case when C_t changes only m times a year at regular moments in time at an average *constant* rate δ. Then, we may divide each year into intervals of length $\frac{1}{m}$, and in accordance with (8.1.1), for each such interval we can write

$$\frac{\Delta C_t}{C_t} = \delta \frac{1}{m}. \tag{8.1.2}$$

It is noteworthy that, while we consider intervals whose lengths are less than one year, the coefficient δ in (8.1.2) continues to be a rate per unit of time; that is, an annual rate.

Assume, for example, that we deposit an amount C_0 into a bank account, and the bank credits (or compounds) interest monthly proceeding from an annual rate δ equal, say, to 5%. The bank will carry out crediting or *compounding* in the following way.

The bank will divide the rate δ by 12, and at the end of the first month the interest $C_0 \frac{\delta}{12}$ will be credited. This corresponds to the general rule (8.1.2) except for the fact that months have slightly different durations.

So, the initial capital C_0 increases to $C_0(1+\delta/12)$. Next month, this amount will again be multiplied by $(1+\delta/12)$, the total amount will equal $C_0(1+\delta/12)^2$, and so on, up to the end of the year when the final amount will be equal to $C_0(1+\delta/12)^{12} = C_0(1+0.05/12)^{12} \approx C_0 \cdot 1.0512$. We see that the real annual profit per one unit of money will be equal to $i = (1+\delta/12)^{12} - 1 \approx 0.0512 = 5.12\%$, which is a bit larger than the rate 5%.

If, *keeping the same annual rate* δ, the bank compounds interest not twelve but n times a year, it will lead to the amount $C_0(1+\delta/n)^n$.

The quantity $i = (1+\delta/n)^n - 1$ is called an (annual) *effective interest rate* and shows the real growth over a year. Another term for the same quantity is *yield*. The reader can find both words, rate and yield, in her/his bank statement and compare the numbers there.

It is noteworthy that when changing the number of the moments of compounding (or the number of *conversion periods*), we maintain the same rate δ. In Section 8.2, we will come back to this issue and consider the case when i is fixed while the rate is changing.

As is well known from Calculus, the function $(1+\delta/n)^n$ is increasing in n. So, the yield is larger than the interest rate, and the more often the interest is compounded, the better for the investor.

Furthermore, $(1+\delta/n)^n \to e^{\delta}$ as $n \to \infty$. (The reader remembers that e is defined as $\lim_{n\to\infty}(1+1/n)^n$.) If $n = \infty$, we say that the interest is *compounded continuously*. The annual growth factor in this case is e^{δ}, the effective interest (or yield) is $i = e^{\delta} - 1$, and at the end of the year, the capital invested becomes equal to $C_1 = C_0 e^{\delta}$.

In the case of continuous compounding, we call the characteristic δ also the *force of interest*. The interest rate and the force of interest are synonyms in this case.

At the end of t years, the capital will become equal to $C_t = C_0 e^{\delta t}$. One may show this reasoning as above, but a better and more general way is to proceed as follows.

Let us come back to the relation (8.1.1) and consider an infinitesimally small interval $[t, t+dt]$. Then (8.1.1) may be written as

$$\frac{dC_t}{C_t} = \delta dt. \tag{8.1.3}$$

This may be viewed as a differential equation for C_t. The reader remembers (or can readily verify) that a solution is

$$C_t = C_0 e^{\delta t}, \tag{8.1.4}$$

where C_0 is the value of the function C_t at $t = 0$.

From (8.1.4) it again follows that the annual interest rate $\frac{C_1 - C_0}{C_0}$ is equal to

$$i = e^{\delta} - 1. \tag{8.1.5}$$

Making use of the differential equation (8.1.3) allows to consider the case when the rate $\delta = \delta(t)$ is changing in time, which is closer to reality. Replacing in (8.1.3) the letter t by s, we can write it as

$$\frac{dC_s}{C_s} = \delta(s)ds. \tag{8.1.6}$$

Since $\dfrac{dC_s}{C_s} = d(\ln C_s)$, integrating both sides of (8.1.6) from 0 to t, we come to the relation

$$\ln C_t - \ln C_0 = \int_0^t \delta(s)ds.$$

From this it follows that

$$C_t = C_0 \exp\left\{\int_0^t \delta(s)ds\right\}. \tag{8.1.7}$$

If $\delta(s)$ equals a constant δ, we come back to (8.1.4).

EXAMPLE 1. Let the rate $\delta(s)$ increase linearly during a year from 4% to 6%. Find the annual effective interest (or yield). In this case, $\delta(s) = 0.04 + 0.02s$ for $0 \le s \le 1$. Then $\int_0^1 \delta(s)ds = \int_0^1 (0.04 + 0.02s)ds = 0.05$. So, the yield equals

$$\exp\left\{\int_0^1 \delta(s)ds\right\} - 1 = e^{0.05} - 1 \approx 0.0512. \tag{8.1.8}$$

Assume now that the interest is growing within the *same limits*, but the "speed" of this growth is not a constant, and is slower in the beginning. For example, $\delta(s) = 0.04 + 0.02s^2$. Then, as is easy to compute, $\int_0^1 \delta(s)ds = 0.0466...$, and the yield is equal to $\exp\{0.0466...\} - 1 \approx 0.0477$ [(compare with (8.1.8)]. □

In general, the varying rate $\delta(s)$ depends on the market and, consequently, is random. In this case, $\delta(s)$ is a random process, and C_t is a random process too.

An interesting question is whether we underestimate or overestimate the expected income $E\{C_t\}$ if we replace the random rate δ by its mean value $E\{\delta\}$.

The answer follows from Jensen's inequality which we will prove and discuss in detail in Section **1**.3.4.2. In particular, from this inequality it follows that if $u(x)$ is a convex function, than for any r.v. X

$$E\{u(X)\} \ge u(E\{X\})$$

(provided that all quantities above are finite). Since e^x is a convex function,

$$E\{C_t\} = C_0 E\left\{\exp\left\{\int_0^t \delta(s)ds\right\}\right\} \ge C_0 \exp\left\{E\left\{\int_0^t \delta(s)ds\right\}\right\} = C_0 \exp\left\{\int_0^t E\{\delta(s)\}ds\right\}.$$

Thus, we underestimate the mean value of the yield replacing δ by $E\{\delta\}$. More precisely, since the annual yield is $(C_1/C_0) - 1$, the expected annual yield is

$$E\left\{\frac{C_1}{C_0} - 1\right\} \ge \exp\left\{\int_0^1 E\{\delta(s)\}ds\right\} - 1.$$

EXAMPLE 2. Assume that for a "good" year, the annual rate is 8%, while for a "bad" year, the rate is 2%. Suppose that both scenarios are equally likely. In this case, the r.v. $\delta(s)$ does not depend on s, and if $C_0 = 1$, then

$$E\{C_1\} = \frac{1}{2}e^{0.02} + \frac{1}{2}e^{0.08} \approx 1.0517, \text{ while } e^{E\{\delta\}} = e^{0.05} = 1.0513.$$

The difference is not large but may be significant for large investments. □

8.2 Nominal rate

Next, we look at the same phenomenon from another point of view. Assume that the effective annual interest rate i is fixed, and the bank compounds interest m times a year. With what annual rate should the bank compound interest in order to maintain the *given* efficient interest rate (yield) i ?

Such an annual rate is usually denoted by $i^{(m)}$ and is called a *nominal annual interest rate* (more precisely, an annual rate of interest payable (or convertible) m-thly). As follows from (8.1.2) and the discussion at the beginning of Section 8.1, the bank will divide the rate $i^{(m)}$ by m, and at the end of each period, the current capital will be multiplied by $1+i^{(m)}/m$. In order for such a compounding to lead to the annual interest i, we should have

$$\left(1+\frac{i^{(m)}}{m}\right)^m = 1+i.$$

This implies that

$$i^{(m)} = m[(1+i)^{1/m} - 1]. \tag{8.2.1}$$

Continuous compounding corresponds again to the case $m \to \infty$. The limit

$$\lim_{m\to\infty} i^{(m)} = \lim_{m\to\infty} (m[(1+i)^{1/m} - 1]) = \ln(1+i). \tag{8.2.2}$$

(For example, one can set $x = 1/m$ and apply the L'Hôpital rule.)

So, continuous compounding corresponds to the annual rate $\delta = \ln(1+i)$, which is consistent with (8.1.5).

Note also that $i^{(m)}$ is decreasing from $i^{(1)} = i$ to $i^{(\infty)} = \delta = \ln(1+i)$ as m is growing from 1 to infinity. (To prove that $i^{(m)}$ is decreasing, it suffices to compute the derivative of $i^{(m)}$ in m.)

8.3 Discount and annuities

We define a *discount factor* or simply *discount* v_t as the value of one unit of money to be paid at time t if evaluation is carried out from the standpoint of the present time $t = 0$. In another terminology, v_t is the *present value* of one unit of money to be paid at time t.

The question of how to calculate a discount is rather complicated, if it is even solvable, since it involves many issues such as inflation, different choices of investment, randomness of investment results, etc.

Roughly, v_t may be defined as the amount of money which one should invest into a risk-free security at time $t = 0$, in order to have a unit of money at time t.

Consider a unit time interval (say, a year), and assume that the risk-free efficient interest rate for this period is well defined and equals i. Then, investing $\frac{1}{1+i}$ units of money, the investor will obtain $\frac{1}{1+i} \cdot (1+i) = 1$ at the end of the period. Thus, to get one unit at the end, one should invest $\frac{1}{1+i}$ at the beginning. Hence,

$$v_1 = v = \frac{1}{1+i}.$$

Let time be discrete, that is, $t = 0, 1, \ldots$, and let the risk-free interest in each period $[t-1, t]$ be the same and equal i. Then the present value of one unit to be paid at the end of the period $[t-1, t]$, that is, at time t, is

$$v_t = \left(\frac{1}{1+i}\right)^t = v^t, \tag{8.3.1}$$

where t is an integer.

Assume that an investor expects a future cash flow during n periods of time with payments at the beginning of each period. Such a sequence of payments is called an *annuity-due*. Denote by c_t the cash at time t; that is, at the beginning of the period $[t, t+1)$. The present value of this payment is $v^t c_t$. The total cash flow is the sequence $c_0, c_1, c_2, \ldots, c_{n-1}$, and the present value of this flow is the number

$$C_n = c_0 + vc_1 + v^2 c_2 + \ldots + v^{n-1} c_{n-1} = \sum_{t=0}^{n-1} v^t c_t.$$

In the theory of interest, for the particular case of $c_t = 1$, the quantity C_n is denoted by $\ddot{a}_{\overline{n}|}$, and we adopt this notation in this book. (The dots in $\ddot{a}$ indicate that payments are provided in the beginning of each period. See also Sections **9**.1, **9**.2.1 where we discuss annuities in much more detail.) Thus,

$$\ddot{a}_{\overline{n}|} = 1 + v + \ldots + v^{n-1} = \frac{1-v^n}{1-v}, \tag{8.3.2}$$

provided $v \neq 1$. This quantity is called the present value of an *annuity-due with unit rate,* or simply an annuity-due. (More precisely, a *certain annuity-due*, because for now we deal with a non-random cash flow.)

For $v < 1$ and an infinite horizon $n = \infty$, the limit in (8.3.2) is equal to

$$\ddot{a}_{\overline{\infty}|} = \frac{1}{1-v}. \tag{8.3.3}$$

This quantity is also called the present values of a *perpetuity-due* or, making it shorter, a perpetuity-due.

If payments are made at moments $t = 1, \ldots, n$, that is, at the ends of the periods, the present value of the cash flow equals

$$v + v^2 + \ldots + v^n = v\frac{1-v^n}{1-v}.$$

This quantity is denoted by $a_{\overline{n}|}$. It is called the present value of an *annuity-immediate* (or *payable in arrears*), or simply an annuity-immediate. The *perpetuity-immediate* is given by

$$a_{\overline{\infty}|} = \frac{v}{1-v},$$

provided $v < 1$.

Now, let us consider the continuous time case. Assume that an investment grows in accordance with (8.1.7). Then, to find v_t, we should set in (8.1.7) $A_t = 1$ and $A_0 = v_t$, which implies that

$$v_t = \exp\left\{-\int_0^t \delta(s)ds\right\}. \tag{8.3.4}$$

If the rate $\delta(s)$ equals a constant δ, then (8.3.4) implies that

$$v_t = e^{-\delta t},$$

and we come to

$$v_t = v^t, \tag{8.3.5}$$

where $v = e^{-\delta}$.

The relation (8.3.5) is adopted in many models not necessarily based on continuous compounding of interest—that is, v is not necessarily presented as $e^{-\delta}$. The term *discount* is usually applied to the characteristic v.

8.4 Accumulated value

Consider a series of payments of a unit of money at discrete time moments $t = 0, 1, ...,$ $n-1$. The unit paid at moment t, "starts to grow" and at time n becomes equal to $(1+i)^{n-t}$, where i is the (annual) interest, or more precisely, the effective interest rate. Hence, the total *accumulated value* at time n is equal to the quantity

$$\ddot{s}_{\overline{n}|} = \sum_{t=0}^{n-1}(1+i)^{n-t} = (1+i)\cdot\frac{(1+i)^n-1}{i}. \tag{8.4.1}$$

Since $v = \frac{1}{1+i}$, we can write

$$\ddot{s}_{\overline{n}|} = \frac{(1/v)^n-1}{1-v} = v^{-n}\cdot\frac{1-v^n}{1-v}. \tag{8.4.2}$$

Note that we could come to the same formula just dividing the annuity-due $\ddot{a}_{\overline{n}|}$ from (8.3.2) by v^n. Indeed, $\ddot{a}_{\overline{n}|}$ is the value of the cash flow from the standpoint of the time $t = 0$, while $\ddot{s}_{\overline{n}|}$ is the value of the *same* cash flow from the standpoint of the time $t = n$. The former value is equal to the latter value times the discount factor v^n.

8.5 Effective and nominal discount rates

Let i be an effective annual interest rate. Then the discount factor $v = \frac{1}{1+i}$ [see (8.3.1)]. A person investing a unit of money at the beginning of a year will be credited i units at the end of the year. The present value of the interest payment i from the standpoint of the initial time is $vi = \frac{1}{1+i}i$. The quantity

$$d = \frac{i}{1+i} \tag{8.5.1}$$

is called an *effective rate of discount* or a *rate of interest-in-advance*. The latter term is connected to the following interpretation. Assume that the interest is paid in advance at the beginning of the year. For this to be equivalent to the payment of i at the end of the year, the payment in advance should be equal to the present value of the amount i paid at the end of the year. The present value mentioned is equal to $\frac{1}{1+i}\cdot i = \frac{i}{1+i}$.

It is easy to see also that

$$d = 1 - v.$$

The last relation may be interpreted as follows. The quantity v is what you should invest at the beginning of the year to get one unit at the end of the year. So, d is an investment profit, more precisely, a profit rate since we are talking about a unit of money.

Let $d^{(m)}$ be the equivalent *annual nominal rate of interest-in-advance* credited m times in a year. Another term is a *nominal rate of discount* convertible $m$thly. In other words, $d^{(m)}$ is an *annual* discount rate which leads to the effective annual interest rate i, and hence to the effective annual rate of discount d, if interest is compounded m times a year.

Since $d^{(m)}$ is an annual *rate*, in accordance with rule (8.1.2) which we apply now to payments in advance, the rate for each period of length $\frac{1}{m}$ should be $\tilde{d}^{(m)} = d^{(m)}/m$. As we know (see Section 8.2), in order to get the interest i at the end of the year, the effective interest rate in *each such period* should be equal to $\tilde{\imath}^{(m)} = i^{(m)}/m$, where $i^{(m)}$ is the nominal *annual* interest rate. Then, in accordance with (8.5.1),

$$\tilde{d}^{(m)} = \frac{\tilde{\imath}^{(m)}}{1+\tilde{\imath}^{(m)}}, \text{ and } d^{(m)} = m\tilde{d}^{(m)} = \frac{i^{(m)}}{1+i^{(m)}/m}.$$

Substituting $i^{(m)} = m[(1+i)^{1/m} - 1]$ from (8.2.1), after simple algebra we get that

$$d^{(m)} = m(1-(1+i)^{-1/m}) = m(1-v^{1/m}). \tag{8.5.2}$$

Furthermore,

$$d^{(m)} \to \delta \text{ as } m \to \infty, \tag{8.5.3}$$

where $\delta = \ln(1+i)$. This is certainly not surprising since the case $m = \infty$ corresponds to continuous compounding with rate δ. (To prove (8.5.3) one can set $x = 1/m$, $v = e^{-\delta}$ and apply L'Hôpital's rule.)

---

9 EXERCISES

Exercises in this chapter concern only Section 7.

1. We say that Y does not depend on an event A if Y and $\mathbf{1}_A$ are independent. Show that in this case $E\{Y \mid A\} = E\{Y\}$. (*Advice:* Start with $E\{Y \mid A\} = \frac{1}{P(A)}E\{Y; A\} = \frac{1}{P(A)}E\{Y\mathbf{1}_A\}$.)

2. The simple assertion of Exercise 1 may be generalized in the following way. For two events A_1 and A_2, if the vector $(Y, \mathbf{1}_{A_1})$ does not depended on $\mathbf{1}_{A_2}$, then $E\{Y \mid A_1A_2\} = E\{Y \mid A_1\}$. Show this. (*Advice:* Start with $E\{Y \mid A_1A_2\} = \frac{1}{P(A_1A_2)}E\{Y\mathbf{1}_{A_1}\mathbf{1}_{A_2}\}$.)

3. Graph the marginal density (7.1.10) from Example 7.1.2-1. Do you see without calculations that the integral of the function in (7.1.10) is indeed equal to one? Find $E\{Z_2^4 \mid Z_1\}$. Write $E\{Z_2^3 \mid Z_1\}$.

4. In the situation of Example 7.1.2-2, find the conditional density $f_{X_1 \mid S}(x \mid s)$ when

   (a) X_1 and X_2 are standard normal;

(b) X_1 and X_2 have the Γ-distributions with parameters (a, ν_1) and (a, ν_2), respectively. In this case, the distribution we will get is called the B(beta)-distribution. Write a particular formula for the conditional density for $\nu_1 = \nu, \nu_2 = 1$, and $E\{X_1 \mid S\}$ in this case. Does it grow when ν is increasing? To what does it converge as $\nu \to \infty$? Interpret this with regard to the form of the Γ-density $f_{a\nu}(x)$.

5. Graph the marginal density (7.4.1) from Example 7.4-1. Do you see without calculations that the total integral of this function (with respect to both arguments) is indeed equal to one? Find $E\{Z_3^4 \mid Z_1, Z_2\}$. Write $E\{Z_3^3 \mid Z_1, Z_2\}$.

6. Let the joint density of (X,Y) be $f(x,y) = x+y$ for $0 < x \leq 1, 0 \leq y \leq 1$, and $= 0$ otherwise. Find the conditional density $f(y \mid x)$, and $E\{Y \mid X\}$. At which x does the function $m(x)$ attain its maximum? Find $E\{X \mid Y\}$.

7. Let the joint density of (X,Y) be $f(x,y) = 2ye^{-x}/x^2$ for $0 < y \leq x$, and $= 0$ otherwise. Find the conditional density $f(y \mid x)$ and $E\{Y \mid X\}$.

8. Derive the formula (7.3.11) directly from the definition of $f(z \mid x)$. (*Advice*: Rewrite the definition of $f(z \mid x)$ as $f(x,z) = f(z \mid x) f_X(x)$ and integrate in x.)

9. A clerk in the claim department of an insurance company is waiting for the next customer call. It is known that the waiting time for the next man is exponentially distributed with mean m_1, while for a woman this r.v. is exponential with mean m_2. Men and women call independently. Find the probability that the first call will be from a man.

10. Let X_1 and X_2 be i.i.d. r.v.'s. Assume that we know values of $X_{\min} = \min\{X_1, X_2\}$, and $X_{\max} = \max\{X_1, X_2\}$. What can we say about the expected value of, for instance, X_1 given this information? (*Advice*: Note that by symmetry $E\{X_1 \mid X_{\min}, X_{\max}\} = E\{X_2 \mid X_{\min}, X_{\max}\}$.)

11. Let r.v.'s X, Y, and Z be mutually independent. Suppose X and Y are standard normal, while the distribution of Z is arbitrary. Find the distribution of the r.v. $\dfrac{X+YZ}{\sqrt{1+Z^2}}$.

12. Suppose that the random number generator in your computer is perfect.

(a) You simulate one value of a r.v. X uniformly distributed on $[0,1]$. After that you simulate n independent trials with the probability of success equal to the value of X you got. Denote by Y the number of successes. Write $E\{Y\}$ and $Var\{Y\}$. Find the distribution of Y. (*Hint*: While the first two questions are not very complicated, the third requires some calculations. In particular, we should know that $\int_0^1 x^k (1-x)^m dx = k!m!/(k+m+1)!$.)

(b) You simulate one value of a r.v. N having the Poisson distribution with parameter λ. After that, you simulate independent trials with a fixed probability p of success. The number of trials equals the value of N you got. Let Y be the number of successes. Find $E\{Y\}$, $Var\{Y\}$ and the distribution of Y. In this book, we will consider this classical problem several times using different methods. Now our goal is to practice in conditioning.

Chapter 1

Comparison of Random Variables. Preferences of Individuals

1,2,3

This chapter concerns various rules of comparison and subsequent selection among risky alternatives.

1 A GENERAL FRAMEWORK AND FIRST CRITERIA

1.1 Preference order

What do we usually do when we choose an investment strategy in the presence of uncertainty? Consciously or not, we compare random variables (r.v.'s) of the future income, corresponding to different possible strategies, and we try to figure out which of these r.v.'s is the "best".

Suppose you are one of 2 million of people who buy a lottery ticket to win a *single* one million dollar prize. Your gain is a random variable (r.v.)

$$\xi = \begin{cases} 1{,}000{,}000 & \text{with probability } 1/2{,}000{,}000, \\ 0 & \text{with probability } 1 - 1/2{,}000{,}000. \end{cases}$$

If the ticket's price is \$1, then your random profit is $\xi - 1$. If you have decided to buy the ticket, it means that, when comparing the r.v.'s $X = \xi - 1$ and $Y = 0$ (the profit if you do not buy a ticket), you have decided, perhaps at an intuitive level, that X is better for you than Y.

The fact that the mean value $E\{X\} = E\{\xi\} - 1 = \frac{1}{2} - 1 = -\frac{1}{2}$ is negative does not say that the decision is unreasonable. You pay for hope or for fun.

Suppose you buy auto insurance against a possible future loss ξ. Assume that with probability 0.9 the r.v. $\xi = 0$ (nothing happened), and with probability 0.1, the loss ξ takes on values between \$0 and \$2000, and all these values are equally likely. In this case, $E\{\xi\} = 0.1 \cdot 1000 = 100$. If the premium c you pay is equal, say, to \$110, it means that the loss of ξ is worse for you than the loss of the certain amount $c = 110$. The fact that you pay \$10 more than the mean loss, again, does not necessarily mean that you made a mistake. The additional \$10 may be viewed as a payment for stability.

For the insurance company the decision in this case is, in a sense, the opposite. The company gets your premium c, and it will pay you a random payment ξ. The company compares the r.v. $\widetilde{X} = c - \xi$ with the r.v. $Y = 0$, and if the company signs the insurance contract, it means that it has decided that the random income $\widetilde{X}$ is better than zero income.

In the reasoning above, we assumed that decision did not depend on the total wealth of the client or the company but just on the r.v.'s under comparison. Suppose that in the case of insurance you also take into account your total wealth or a part of it, which we denote by w. Then the r.v.'s under comparison are $w - \xi$ (your wealth if you do not insure the loss) and $w - c$ (your wealth if you do insure the loss for the premium c).

In the examples we considered, one of the variables under comparison was non-random. Certainly, this is not always the case. For example, if you decide to insure only half of the future loss ξ for a lower premium c', then the r.v.'s we should consider are $X = w - \xi$ (you do not buy an insurance) and $Y = w - \frac{\xi}{2} - c'$ (you insure half of the loss for c').

This chapter addresses various criteria for the comparison of risky alternatives. As a rule, we will talk about possible values of future income. In this case, while the criteria may vary, they usually have one feature in common. When choosing a possible investment strategy, we have competing interests: we want the income to be large, but we also want the risk to be low. As a rule, we can reach a certain level of stability only by sacrificing a part of the income: we should pay for stability. So, our decision becomes a trade-off between the possible growth and stability.

Let us consider the general framework where we deal with a fixed class $\mathcal{X} = \{X\}$ of r.v.'s X. We assume that r.v.'s from $\mathcal{X}$ are all defined on some sample space $\Omega = \{\omega\}$ (see Section **0**.1.3.1). That is, $X = X(\omega)$.

Defining a rule of comparison on the class $\mathcal{X}$ means that for pairs (X,Y) of r.v.'s from $\mathcal{X}$, we should determine whether X is better than Y, or X is worse than Y, or these two random variables are equivalent for us.

Formally, this means that among possible pairs (X,Y) (the order of the r.v.'s in the pair (X,Y) is essential), we specify a collection of those pairs (X,Y) for which X is preferable or equivalent to Y. In other words, "X is not worse than Y", and as a rule we will use the latter terminology. We will write it as $X \succsim Y$.

If (X,Y) does not belong to the collection mentioned, we say "X is worse than Y" or "Y is better than X", writing $X \prec Y$ or $Y \succ X$, respectively. If simultaneously $X \succsim Y$ and $Y \succsim X$, we say that "X is equivalent to Y", writing $X \simeq Y$.

Not stating it each time explicitly, we will always assume that the relation $\succsim$ satisfies the following two properties.

(i) *Completeness*: For any X and Y from $\mathcal{X}$, either $X \succsim Y$, or $Y \succsim X$. (As was mentioned, these relations may hold simultaneously);

(ii) *Transitivity*: For any X, Y, and Z from $\mathcal{X}$, if $X \succsim Y$ and $Y \succsim Z$, then $X \succsim Z$.

The rule of comparison so defined is called a *preference order on the class* $\mathcal{X}$.

Before discussing examples, we state one general requirement on preference orders. This requirement is quite natural when we view X's as the r.v.'s of future income.

The monotonicity property:

$$\text{If } X,Y \in \mathcal{X} \quad \text{and} \quad P(X \geq Y) = 1, \text{ then } X \succsim Y. \tag{1.1.1}$$

This requirement reflects the rule "the larger, the better". If a random income $X = X(\omega)$ is greater than or equal to a random income $Y = Y(\omega)$ for *all* ω's or, at least, with probability one, then for us X is not worse than Y.

It makes sense to emphasize that in (1.1.1) we consider not all r.v.'s but only those from the class $\mathcal{X}$ under consideration. We will see later that this is an important circumstance.

It is natural to consider also

The strict monotonicity property:

$$\text{If } X, Y \in \mathcal{X},\ P(X \geq Y) = 1, \text{ and } P(X > Y) > 0, \text{ then } X \succ Y. \tag{1.1.2}$$

The significance of this requirement is also clear. If a random income $X = X(\omega)$ is not smaller than $Y = Y(\omega)$ with probability one, and with a positive probability $X(\omega)$ is larger than $Y(\omega)$, then we prefer X to Y.

In this book, we will accept only preference orders which are monotone in the class of the r.v's under consideration. However, we will not always require strict monotonicity; see for example, the VaR criterion in Section 1.2.2. Nevertheless, if a rule of comparison is not strictly monotone, this says about some non-flexibility of this rule, and it makes sense, at least, to recheck to what extent it meets our goals.

EXAMPLE 1. Let two r.v.'s, $X = X(\omega)$ and $Y = Y(\omega)$, be defined on a sample space Ω consisting of only two outcomes: ω_1 and ω_2. The probabilities of the outcomes are equal to $1/2$. We may view X, Y as the random income corresponding to two investment strategies, and ω_1 , ω_2 as two states of the future market. Let

$$X(\omega_1) = 1,\ X(\omega_2) = 3,$$
$$Y(\omega_1) = 1,\ Y(\omega_2) = 2.$$

Clearly, $X(\omega) \geq Y(\omega)$ for both ω's and for any monotone order $\succeq$, we will have $X \succeq Y$, i.e., X is not worse than Y, which is natural. Suppose, however, that for an individual, her/his preference order $\succeq$ is monotone but is not strictly monotone, and though $P(X > Y) = \frac{1}{2} > 0$, the r.v. X is equivalent to Y. This means that the individual is indifferent whether to choose X or Y. The only way to interpret it, is to say that the individual needs at most two units of money, and does not need more. If, as a matter of fact, it is not true, the individual's preferences should be described in a more flexible way. □

Let $V(X)$ be a function taking on *numerical* values. We say that an order $\succsim$ is preserved, or completely characterized, by $V(X)$ if for any $X, Y \in \mathcal{X}$,

$$X \succsim Y \Leftrightarrow V(X) \geq V(Y), \tag{1.1.3}$$

where the symbol $\Leftrightarrow$ means if and only if, also abbreviated iff.

The function $V(X)$ may be viewed as a measure of the "quality" of X: the larger $V(X)$, the better X, and X is not worse than Y iff $V(X) \geq V(Y)$.

Below, we consider various examples; but first it is worthwhile to note that in the case of (1.1.3), the *monotonicity property* may be restated as

$$\text{If } X,Y \in \mathcal{X} \text{ and } P(X \geq Y) = 1, \text{ then } V(X) \geq V(Y). \tag{1.1.4}$$

The *strict monotonicity* is equivalent in this case to the property

$$\text{If } X,Y \in \mathcal{X},\ P(X \geq Y) = 1, \text{ and } P(X > Y) > 0, \text{ then } V(X) > V(Y). \tag{1.1.5}$$

Let us turn to examples.

1.2 Several simple criteria

We will talk about preferences of economic agents—separate individuals, companies, etc.—using also, for brevity, the term "investor".

1.2.1 The mean-value criterion

The investor cares only about the mean values of r.v.'s, that is,

$$X \succsim Y \Leftrightarrow E\{X\} \geq E\{Y\}.$$

In this case, the collection of all pairs (X,Y) mentioned above is just the collection of all pairs (X,Y) for which $E\{X\} \geq E\{Y\}$, and in (1.1.3), $V(X) = E\{X\}$.

Clearly, this criterion is strictly monotone; the reader is invited to show it on her/his own. Note, however, that from the mean-value criterion's point of view, for example, r.v.'s

$$X = \begin{cases} 100 & \text{with probability } 1/2 \\ 0 & \text{with probability } 1/2 \end{cases}, \quad \text{and} \quad Y = 50 \tag{1.2.1}$$

are equivalent. This might not reflect people's real preferences; so, the criterion may occur to be too simple, non-flexible. However, as we will see, in some situations quite reasonable comparison rules may turn out to be close to the mean-value criterion.

1.2.2 Value-at-Risk (VaR)

Another term in use is the *capital-at-risk criterion.* For a r.v. X, denote by $q_\gamma = q_\gamma(X)$ its γ-quantile or the 100γ-th percentile. The reader is recommended to look up the rigorous definition in Section **0**.1.3.4, and especially Fig.**0**-7 there. Loosely speaking, it is the largest number x for which $P(X \leq x) \leq \gamma$. If the r.v. X is continuous and its distribution function (d.f.) $F(x)$ is increasing, q_γ is just the number for which $F(q_\gamma) = \gamma$. See also Fig.**0**.7a. The discrete distribution case is illustrated in Fig.**0**.7de.

Let γ be a fixed level of probability, viewed as sufficiently small. Assume that an investor does not take into consideration events whose probabilities are less than γ. Then, for such an investor the worst, smallest conceivable level of the income is q_γ.

Let, for instance, $\gamma = 0.05$. Then $q_{0.05}$ is the smallest value of the income among all values which may occur with 95% probability. One may say that $q_{0.05}$ is the value at 5% risk. Note that q_γ may be negative, which corresponds to losses.

The VaR criterion is defined as

$$X \succsim Y \Leftrightarrow q_\gamma(X) \geq q_\gamma(Y),$$

i.e., we set $V(X) = q_\gamma(X)$ in (1.1.3).

In applications of VaR, for the γ-quantile of X, the notation $VaR_\gamma(X)$ is frequently used; we will keep the notation $q_\gamma(X)$.

The particular choice of $\gamma = 0.05$ is very common, but it has rather a psychological explanation: 0.01 is "too small", while 0.1 is "too large". As a matter of fact, whether a particular value of γ should be viewed as small or not depends on the situation. We can view a probability of 0.05 as small if it is the probability that there will be a rain tomorrow. However, the same number should be considered very large if it is the probability of being involved in a traffic accident: it would mean that on the average you are likely to be involved in an accident one out of twenty times you are in traffic.

EXAMPLE 1. Let a r.v. X (say, a random income) take on values $0, 10$ with probabilities 0.1 and 0.9, respectively, and let a r.v. Y take on the same values with respective probabilities 0.07 and 0.93. The reader is suggested to check that for $\gamma = 0.05$, we have $q_\gamma(X) = q_\gamma(Y) = 0$ (look, for example at Fig.7d in Section **0**.1.3.4). So, X and Y are equivalent under the VaR criterion. However, for $\gamma = 0.08$ we have $q_\gamma(X) = 0$ while $q_\gamma(Y) = 10$, that is, X is worse than Y. So, the result of comparison depends on γ. □

The VaR criterion is monotone. Indeed, if $X \geq Y$ with probability one, than $P(X \leq x) \leq P(Y \leq x)$, so in this case $q_\gamma(X) \geq q_\gamma(Y)$. However, *VaR is not strictly monotone.*

EXAMPLE 2. Let Y be uniform on $[0,2]$, and

$$X = \begin{cases} Y & \text{if} Y \leq 1, \\ 2 & \text{if } Y > 1. \end{cases}$$

We see that $P(X > Y) = P(1<Y<2) = \frac{1}{2} > 0$. However, if $x \leq 1$, then $P(X \leq x) = P(Y \leq x)$, and hence $q_\gamma(X) = q_\gamma(Y)$ if $\gamma \leq \frac{1}{2}$. □

The fact that the VaR criterion is not strictly monotone does not provide sufficient grounds to reject the VaR in any case; we discuss it in more detail in Section 1.2.3. However, we should be aware that this is not a flexible criterion since it does not take into account all values of r.v.'s, as we saw in the example above.

EXAMPLE 3. Let X be normal with mean m and variance σ^2. Since the d.f. of X is $\Phi\left(\frac{x-m}{\sigma}\right)$ (see, e.g., Section **0**.3.2.4), the γ-quantile of X is a solution to the equation $\Phi\left(\frac{q-m}{\sigma}\right) = \gamma$. Denote by $q_{\gamma s}$ the γ-quantile of the *standard* normal distribution, i.e., $\Phi(q_{\gamma s}) = \gamma$. Then we can rewrite the equation mentioned as $\frac{q-m}{\sigma} = q_{\gamma s}$, and

$$q_\gamma(X) = m + q_{\gamma s}\sigma.$$

The coefficient $q_{\gamma s}$ depends only on γ. Usually people choose $\gamma < 0.5$, and in this case $q_{\gamma s} < 0$. For example, if $\gamma = 0.05$, then $q_{\gamma s} \approx -1.64$ (see Table 2 in Appendix, Section 2), and the VaR criterion is preserved by the function $q_\gamma(X) \approx m - 1.64\sigma$. Criteria of the type

$$V(X) = m - k\sigma, \qquad (1.2.2)$$

where k is a positive number, are frequently used in practice, and not only for normal r.v.'s; as we will see in Section 1.2.5, maybe too frequently. The expression in (1.2.2) can be interpreted as follows. If we view X as a future income and variance or standard deviation as a measure of riskiness, then we want the mean m to be as large as possible and σ as small as possible. This is reflected by the minus sign in (1.2.2). The number k may be viewed as a weight we assign to variance.

EXAMPLE 4. There are $n = 10$ assets with random *returns* $X_1, ..., X_n$. The term "return" means the income per \$1 investment. For example, if the today price of a stock is \$11, while the yesterday price was \$10, the return for this one-day period is $\frac{11}{10} = 1.1$. Note that a return X may be less than one, and in this case we face a loss.

Assume $X_1, ..., X_n$ to be independent and their distributions to be closely approximated by the normal distribution with mean m and variance σ^2.

Let us compare two strategies of investing n million dollars: either investing the whole sum in one asset, for example, in the first, or distributing the investment sum equally between n assets. We proceed from the VaR criterion with $\gamma = 0.05$.

For the first strategy, the income will be the r.v. $Y_1 = nX_1 = 10X_1$. The mean $E\{Y_1\} = nm$, and $Var\{Y_1\} = n^2\sigma^2$, so to compute $q_\gamma(Y_1)$ we should replace in (1.2.2) m by nm, and σ by $n\sigma$. Replacing $q_{\gamma s}$ by its approximate value -1.64, we have

$$q_\gamma(Y_1) = mn - 1.64n\sigma = 10m - 16.4\sigma.$$

For the second strategy, the income is the r.v. $Y_2 = X_1 + ... + X_n$. Hence, $E\{Y_2\} = nm$, $Var\{Y_2\} = n\sigma^2$, and

$$q_\gamma(Y_2) = mn - 1.64\sqrt{n}\sigma \approx 10m - 5.2\sigma.$$

Thus, the second strategy is preferable, which might be expected from the very beginning. Nevertheless, in the next example we will see that if the X_i's have a distribution different from normal, we may jump to a different conclusion.

EXAMPLE 5[1]. There are ten *independent* assets such that investment into each with 99% probability gives 4% profit, and with 1% probability the investor loses the whole investment. Assume that we invest \$10 million and compare the same two strategies as in Example 4. Let us again apply the VaR criterion with $\gamma = 0.05$.

If we invest all \$10 million into the first asset, we will get \$10.4 million with probability 0.99, and in the notation of the previous example $q_\gamma(10X_1) = 10.4$.

For the second strategy, the number of successful investments has the binomial distribution with parameters $p = 0.99$, $n = 10$. If the number of successes is k, the income is

[1]This example is very close to an example from [1] presented also in [8, p.14] with the corresponding reference.

$k \times 1\text{million} \times 1.04$. The d.f. $F(x)$ of the income is given in the table below. The values of $F(x)$ are the values of the binomial d.f. with the parameters mentioned.

k	≤ 6	7	8	9	10
The income x	≤ 6.24	7.28	8.32	9.36	10.4
The d.f. $F(x)$	$\leq 2.002 \cdot 10^{-6}$	0.000114	0.004266	0.095618	1

The 0.05-quantile of this distribution is $9.36 < 10.4$. Therefore, following VaR, we should choose the first investment strategy.

Note, however, that if we choose as γ a number slightly smaller than 0.01, for example $\gamma = 0.0095$, then the result will be different. In this case, $q_\gamma(10X_1) = 0$, while the 0.0095-quantile of the distribution presented in the table, is again 9.36.

Certainly, the results of the comparison above should not be considered real recommendations. On the contrary, the last example indicates a limitation of the application of VaR, and shows that this criterion is quite sensitive for the choice of γ. □

The reader can find more about the VaR criterion, for example, in [60], [66], [64]. Some references may be found also in *http://www.riskmetrics.com* and *http://www.gloriamundi.org*.

1.2.3 An important remark: risk measures rather than criteria

This simple but important remark concerns the two criteria above and practically all other criteria we will consider in this chapter. The point is that we do not have to limit ourselves to using only one criteria each time. On the contrary, we can combine them.

For example, when considering a random income X, we may compute its expectation $E\{X\}$ and its quantile $q_\gamma(X)$. In this case, we will know what we can expect on the average, and what is the worst conceivable (or likely) outcome. When comparing two r.v.'s, we certainly may take into account both characteristics. How we will do this depends on our preferences. The simplest way is to consider the linear combination $\alpha E\{X\} + \beta q_\gamma(X)$, where α and β play the role of weights we assign to the mean and to the quantile. The larger β, the more cautious we are.

Under such an approach to risk assessment, various functions $V(X)$ present rather possible characteristics of the random income X than criteria. In this case, we call $V(X)$ a *risk measure*.

Route 1 ⇒ page 79

1.2.4 Tail-Value-at-Risk (TailVaR)

2,3

Another term for this criterion is Tail conditional expectation (TCE). This is a modification of VaR. The motivation is illustrated by the following example.

EXAMPLE 1. Consider two r.v.'s X and Y of the future income such that

X takes on

values	-2	-1	10	20
with probabilities	0.01	0.02	0.47	0.5

Y takes on

values	$-2\cdot 10^6$	-1	10	20
with probabilities	0.01	0.01	0.48	0.5

The probabilities that the income will be negative in both cases are small: 3% and 2%. For $\gamma = 0.025$, we would have $q_\gamma(X) = -1$ and $q_\gamma(Y) = 10$. So, under the VaR criterion, Y is preferable, which does not look natural. While we may neglect negative values of the income in the first case, this may be unreasonable in the second: a loss of 2 million can be too serious to ignore, even if such an event occurs with a small probability of 1%. □

In situations as above, we speak about the possibility of *large deviations*, or a *heavy tail* of the distribution (for the term "tail", see also Section **0**.2.5). We compare the "tails" of different distributions in more detail in Section **2**.1.1. For now, we introduce a criterion which involves the mean values of large deviations.

First, consider the function

$$V(X;\,t) = E\{X\,|\,X \le t\}, \tag{1.2.3}$$

the mean value of X *given* that the income X did not exceed a level t.

(Formally, the right member of (1.2.3) is defined by

$$E\{X\,|\,X \le t\} = \frac{1}{P(X \le t)}\int_{-\infty}^{t} x\,dF(x) = \frac{1}{F(t)}\int_{-\infty}^{t} x\,dF(x), \tag{1.2.4}$$

where $F(x)$ is the d.f. of X. The formula covers the cases of discrete and continuous r.v.'s simultaneously if we understand the integral above as in (**0**.2.1.5) from Section **0**.2.1. We consider conditional expectations in detail in Section **0**.7; though now, it is sufficient for us just to use definition (1.2.4).)

If we are interested only in losses, then it suffices to consider $t \le 0$. Then $V(X;\,t)$ is negative.

Note that in such situations, people often consider not the income but the losses directly, that is, instead of the r.v. X, the r.v. $\widetilde{X} = -X$. Negative values of X correspond to positive values of $\widetilde{X}$ and vice versa. In this case, $E\{X\,|\,X \le t\} = -E\{\widetilde{X}\,|\,\widetilde{X} \ge |t|\}$ if $t \le 0$. The risk measure $E\{\widetilde{X}\,|\,\widetilde{X} \ge s\}$ is the expected value of the loss given that it has exceeded a level s. In insurance, it is called an *expected policyholder deficit*. See also Exercise 9.

Let us come back to $V(X;t)$ and take, as the level t, the γ-quantile $q_\gamma(X)$. Accordingly, we set

$$V_{\text{tail}}(X) = E\{X\,|\,X \le q_\gamma(X)\},$$

and define the rule of comparison of r.v.'s by the relation

$$X \succsim Y \Leftrightarrow V_{\text{tail}}(X) \ge V_{\text{tail}}(Y).$$

EXAMPLE 2. Consider the r.v.'s from Example 1 for $\gamma = 0.025$. As we already saw, in this case, $q_\gamma(X) = -1$ and $q_\gamma(Y) = 10$. If the r.v. X took a value less or equal to -1, then it can take on only values -1 and -2. Hence,

$$V_{\text{tail}}(X) = E\{X \mid X \le -1\} = (-2)P(X = -2 \mid X \le -1) + (-1)P(X = -1 \mid X \le -1)$$
$$= (-2)\cdot\frac{0.01}{0.03} + (-1)\cdot\frac{0.02}{0.03} = -\frac{4}{3}.$$

For Y, computing in the same manner, we have

$$V_{\text{tail}}(Y) = (-2\cdot 10^6)\cdot\frac{0.01}{0.5} + (-1)\cdot\frac{0.01}{0.5} + 10\cdot\frac{0.48}{0.5} = -39990.42,$$

which is much less than $-4/3$. So, $X \succsim Y$. □

Route 2 ⇒ page 78

3

Now, assume, for simplicity, that the income consists of a fixed non-random positive part and a random loss ξ. Since the positive part is certain, we can exclude it from consideration, setting the income $X = -\xi$. Let us denote by $G(x)$ the d.f. of ξ, and set $\overline{G}(x) = P(\xi > x) = 1 - G(x)$, the tail of the distribution of ξ. Suppose $G(x)$ is continuous. For $t \le 0$, we have

$$P(X \le t) = P(\xi \ge -t) = P(\xi \ge |t|) = \overline{G}(|t|), \tag{1.2.5}$$

and

$$E\{X \mid X \le t\} = E\{-\xi \mid -\xi \le t\} = -E\{\xi \mid \xi \ge |t|\} = \frac{-1}{P(\xi \ge |t|)}\int_{|t|}^{\infty} x dG(x)$$
$$= \frac{1}{\overline{G}(|t|)}\int_{|t|}^{\infty} x d(1 - G(x)) = \frac{1}{\overline{G}(|t|)}\int_{|t|}^{\infty} x d\overline{G}(x).$$

Integration by parts implies that

$$E\{X \mid X \le t\} = \frac{1}{\overline{G}(|t|)}\left(-|t|\overline{G}(|t|) - \int_{|t|}^{\infty}\overline{G}(x)dx\right) = -\left(|t| + \frac{1}{\overline{G}(|t|)}\int_{|t|}^{\infty}\overline{G}(x)dx\right). \tag{1.2.6}$$

EXAMPLE 3. (a) Let $\overline{G}(x) = e^{-x}$, that is, ξ is a standard exponential r.v. We may avoid long calculations if we recall that ξ has the memoryless property (see, e.g., Section **0**.3.2.2). By this property, if ξ has exceeded a level x, the overshoot (over x) has the same distribution as the r.v. ξ itself. Since $E\{\xi\} = 1$, we can write

$$E\{\xi \mid \xi \ge t\} = t + 1.$$

Hence,

$$|E\{X \mid X \le q\}| = E\{\xi \mid \xi \ge |q|\} = |q| + 1,$$

and

$$\gamma = P(X \le q) = P(\xi > |q|) = \overline{G}(|q|) = e^{-|q|}.$$

Hence, and $|q| = \ln \frac{1}{\gamma}$, provided $\gamma > 0$. Thus, for $\gamma > 0$,

$$|E\{X\,|\,X \le q_\gamma\}| = \ln\frac{1}{\gamma} + 1, \text{ and } V_{\text{tail}}(X) = E\{X\,|\,X \le q_\gamma\} = -\ln\frac{1}{\gamma} - 1.$$

(b) Now, let $\overline{G}(x) = 1/(1+x)^2$ for $x \ge 0$. This is a particular case of the Pareto distribution that we consider in more detail in Section **2**.1.1. The tail of this distribution is viewed as "heavy".

To find $q = q_\gamma$, we make use of (1.2.5) and write

$$\gamma = P(X \le q) = \overline{G}(|q|) = \frac{1}{(1+|q|)^2}, \tag{1.2.7}$$

which implies

$$|q_\gamma| = \frac{1}{\sqrt{\gamma}} - 1, \tag{1.2.8}$$

again provided $\gamma > 0$. From (1.2.6) it follows that

$$\begin{aligned} |E\{X\,|\,X \le q)| &= |q| + \frac{1}{\overline{G}(|q|)}\int_{|q|}^{\infty} \overline{G}(x)dx = |q| + (1+|q|)^2 \int_{|q|}^{\infty} \frac{1}{(1+x)^2}dx \\ &= |q| + (1+|q|)^2 \frac{1}{(1+|q|)} = |q| + (1+|q|) = 2|q| + 1. \end{aligned}$$

Substituting (1.2.8), we have

$$|E\{X\,|\,X \le q_\gamma\}| = \frac{2}{\sqrt{\gamma}} - 1.$$

The absolute value above corresponds to the losses. For the (negative) income X

$$V_{\text{tail}}(X) = E\{X\,|\,X \le q_\gamma\} = 1 - \frac{2}{\sqrt{\gamma}}.$$

(c) Let us compare the two cases above. In the first, $P(\xi > x) = e^{-x}$; in the second, $P(\xi > x) = 1/(1+x)^2$. The latter function converges to zero much more slowly than the former. One may say that the tail in the latter case is much "heavier". It means that the probability to have essential losses is larger in the latter case, and we should expect that under the TailVaR criterion the exponential distribution is "better".

This is indeed the case for all γ's. To show it, we should prove that the difference

$$\left(\frac{2}{\sqrt{\gamma}} - 1\right) - (\ln\frac{1}{\gamma} + 1) = \frac{2}{\sqrt{\gamma}} - \ln\frac{1}{\gamma} - 2$$

is positive for *all* positive γ's. Denote this difference by $C(\gamma)$. Note that $C(1) = 0$, and $C'(\gamma) = -\frac{1}{\gamma^{3/2}} + \frac{1}{\gamma} = \frac{1}{\gamma^{3/2}}(\sqrt{\gamma} - 1) < 0$ for all $\gamma \in [0,1)$. Hence, $C(\gamma) > 0$ for $\gamma < 1$. □

2,3 Next, we test the TailVaR on monotonicity. It turns out that ingeneral, *the TailVaR criterion is not monotone.*

EXAMPLE 4. Though we are interested in losses, to make the example illustrative, we will consider non-negative r.v.'s $X = X(\omega)$ and $Y = Y(\omega)$. (Subtracting from both r.v.'s a large number c, we may come to r.v.'s with negative values, but the result of comparison of $X - c$ and $Y - c$ will be the same as for X and Y.)

Let the space of elementary outcomes $\Omega = \{\omega_1, \omega_2, \omega_3\}$, and the probabilities and values of X and Y be as follows:

	ω_1	ω_2	ω_3
$P(\omega) =$	0.1	0.4	0.5
$X(\omega) =$	0	10	20
$Y(\omega) =$	0	10	10

Clearly, $P(X \geq Y) = 1$ and $P(X > Y) = 0.5 > 0$, so it is quite reasonable to prefer X to Y.

Set, however, $\gamma = 0.2$. Then, as we can see from the table (or by graphing the d.f.'s of X and Y), the quantiles $q_\gamma(X) = q_\gamma(Y) = 10$. Now, $V_{\text{tail}}(X) = E\{X \,|\, X \leq 10\} = 0 \cdot \frac{1}{5} + 10 \cdot \frac{4}{5} = 8$, while $V_{\text{tail}}(Y) = E\{Y \,|\, Y \leq 10\} = E\{Y\} = 0 \cdot 0.1 + 10 \cdot 0.9 = 9$. Thus, with respect to the TailVaR, Y is better than X, which contradicts common sense. □

Nevertheless, the TailVaR criterion arose as a result of reasonable argumentation. Therefore, it makes sense not to reject it but realize each time in what situations the monotonicity property is fulfilled. Note also the following.

First of all, the TailVaR is monotone in the class of continuous r.v.'s. An advice on how to show this is given in Exercise 7.

It may be also shown that if Ω is finite and all ω's are equiprobable, then under some mild conditions on r.v.'s or for a slightly modified criterion, monotonicity does take place. We consider it in more detail in Section 1.3. The discussion there points out how to redefine the TailVar to make it monotone.

1.2.5 The mean-variance criterion 1,2,3

This criterion is, in a sense, the same as (1.2.2), but the motivation and derivation are different. Consider an investor expecting a random income X. Set $m_X = E\{X\}, \sigma_X^2 = Var\{X\}$. Suppose the investor measures the riskiness of X by its variance, and wishes the mean income m_X to be as large as possible and the variance σ_X^2—as small as possible. The quality of the r.v. X for such an investor is determined by a function of m_X and σ_X^2. In a simplest case, it is a linear function of m_X and σ_X, and we can write it as

$$V(X) = \tau m_X - \sigma_X, \tag{1.2.9}$$

where the minus reflects the fact that the quality decreases as the variance increases.

The positive parameter τ plays the role of a weight assigned to m_X: a larger τ indicates that the investor values mean more highly. This parameter is usually called a *tolerance to risk*.

We assigned—unlike in (1.2.2)—a weight to the mean rather than to the standard deviation, merely following a tradition in Finance; it does not matter which parameter is endowed by a coefficient. For example, we can write $V(X) = \tau(m_X - \frac{1}{\tau}\sigma_X)$. Then a weight is assigned to the standard deviation, while the factor τ in the very front does not change the comparison rule.

Note also that often, instead of (1.2.9), people consider $V(X) = \tau m_X - \sigma_X^2$, but the difference is also non-essential: both criteria proceed from the same measure of riskiness. The choice of one of them is rather the matter of convenience.

The function $V(X)$ in (1.2.9) preserves the corresponding preference order $\succsim$ among r.v.'s:

$$X \succsim Y \Leftrightarrow \tau m_X - \sigma_X \geq \tau m_Y - \sigma_Y.$$

It is noteworthy that when presenting (1.2.9), we did not assume r.v.'s under consideration to be normal, which we did when deriving criterion (1.2.2). We will see that this may cause problems. The mean-variance criteria (in slightly different forms) are very popular, especially in Finance, and at first glance look quite natural. However, there are situations where the choice of such criteria may contradict common sense.

EXAMPLE 1. Let $X = 0$, a number $a \geq 1$, and

$$Y = \begin{cases} a \text{ with probability } \frac{1}{a}, \\ 0 \text{ with probability } 1 - \frac{1}{a}. \end{cases}$$

Clearly, $E\{Y\} = 1$ and $Var\{Y\} = E\{Y^2\} - (E\{Y\})^2 = a - 1$. Then,

$$V(Y) = \tau \cdot 1 - \sqrt{a-1} = \tau - \sqrt{a-1}, \quad \text{while } V(X) = \tau \cdot 0 - 0 = 0. \tag{1.2.10}$$

So, whatever τ is, we can choose a sufficiently large a for which $V(Y) < 0$. On the other hand, $V(X) = 0$, and under the mean-variance criterion, Y is worse than X, whereas $P(X \leq Y) = 1$. Clearly, if we replace in (1.2.10) the standard deviation $\sqrt{a-1}$ by the variance $a - 1$, then the difference will be even more dramatic. □

Note also that it would be a mistake to think that the example above is contrived, and in practice problems, we do not watch such cases. To show this, consider

EXAMPLE 2. Let X take on values from $[1, \infty)$, and $P(X > x) = 1/x^\alpha$ for all $x \geq 1$ and some $\alpha > 2$. This is a version of the Pareto distribution we discuss in Section **2**.1.1. The Pareto distribution, in different versions, is used in many applications including actuarial modeling. It is not difficult to compute that

$$m_X = \frac{\alpha}{\alpha - 1}, \text{ and } \sigma_X^2 = \frac{\alpha}{(\alpha-2)(\alpha-1)^2}.$$

The reader can check it right away or wait until Section **2**.1.1.

Let Y be uniformly distributed on $[0, 1]$. Obviously, $X \geq Y$ with probability one.

In accordance with (1.2.9),

$$V(X) = \tau \cdot \frac{\alpha}{\alpha - 1} - \frac{\sqrt{\alpha}}{\sqrt{\alpha - 2}\,(\alpha - 1)}, \text{ and } V(Y) = \tau \cdot \frac{1}{2} - \frac{1}{\sqrt{12}} \tag{1.2.11}$$

(for the standard deviation of the uniform distribution, see Section **0**.3.2.1).

We see from (1.2.11) that, whatever τ is, if α approaches 2, the function $V(X)$ converges to $-\infty$. Consequently, for *any* τ, we can choose α (and hence a r.v. X) such that $V(X) < V(Y)$.

Thus, under the mean-variance criterion, X is worse than Y, and consequently the criterion (1.2.9) is not monotone. □

It is also worth noting that the linearity of the function in the r.-h.s. of (1.2.9) is not an essential circumstance, neither is the choice of particular r.v.'s X and Y.

> One may observe the same phenomenon for $V(X)$ equal to almost *any* function $g(m_X, \sigma_X)$ of the mean and standard deviation. Moreover, for *any* r.v. X, we may point out a r.v. Y such that $P(Y \geq X) = 1$ whereas $V(Y) < V(X)$.

More precisely, it may look as follows. Let $V(X) = g(m_X, \sigma_X)$. To avoid cumbersome formulations, assume that $g(x,y)$ is smooth. Since we want the mean to be large and the variance to be small, it is natural to assume that the partial derivatives $g_1(x,y) = \frac{\partial}{\partial x} g(x,y) > 0$, $g_2(x,y) = \frac{\partial}{\partial y} g(x,y) < 0$.

Proposition 1 *Assume, in addition, that the partial derivatives $g_1(x,y)$ and $g_2(x,y)$ are continuous functions. Then for any r.v. X with a finite variance, there exists a r.v. Y such that $P(Y \geq X) = 1$, while*

$$g(m_X, \sigma_X) > g(m_Y, \sigma_Y).$$

We will prove it in the end of this section.

Proposition 1 is a strong argument against using variance as a measure of risk. However, *if we restrict ourselves to a sufficiently narrow class of r.v.'s, the monotonicity property may hold.*

In particular, this is true if we consider only normal r.v.'s because there are no two normal r.v.'s, X and Y with different variances and such that $X \leq Y$ with probability one.

To show it rigorously, assume that the normal r.v.'s X and Y mentioned exist. We have $P(X \leq x) = \Phi((x - m_X)/\sigma_X)$ and $P(Y \leq x) = \Phi((x - m_Y)/\sigma_Y)$. Since $P(Y \geq X) = 1$, it is true that $P(Y \leq x) \leq P(X \leq x)$, and hence $\Phi((x - m_Y)/\sigma_Y) \leq \Phi((x - m_X)/\sigma_X)$ for any x.

The function $\Phi(x)$ is strictly increasing. Therefore, from the last inequality, it follows that $\frac{x - m_Y}{\sigma_Y} \leq \frac{x - m_X}{\sigma_X}$ for *all* x. Certainly, this cannot be true if $\sigma_Y \neq \sigma_X$ because two lines with different slopes intersect and at only one point.

On the other hand, if $\sigma_Y = \sigma_X$, the comparison is trivial: $Y \succsim X$ if $m_Y \geq m_X$.

The case of normal r.v.'s is simple because the normal distribution is characterized only by two parameters: mean m and standard deviation σ. Each normal distribution may be identified with a point (m, σ) in a plane, and the rule of comparison will be equivalent to a rule of comparison of points in this plane.

If we consider a family of distributions with three or more parameters but still compare these distributions proceeding from their means and variances, we may come to paradoxes similar to what we saw above.

We touch on one more example of the violation of monotonicity. Consider the family of r.v.'s $c + X_{a\nu}$, where c is a parameter and $X_{a\nu}$ has the Γ-distribution with parameters a, ν (see Section **0**.3.2.3 and Fig.**0**.13 there). The distribution of $c + X_{a\nu}$ is called a translated Γ-distribution or a Γ-distribution with a shift; it is widely used in many areas including insurance as we will see in this book repeatedly. The distribution is asymmetric, and it may be only very roughly characterized by its mean and variance. So, in this case. it is possible to build an example of violation of the monotonicity property. We skip details; the first such example was suggested by K. Borch [15].

In conclusion, it is worth again emphasizing that the reasoning above *does not mean that we should not use mean-variance criteria, but it does mean that we should be cautious.*

► ***Proof of Proposition 1*** uses the Taylor expansion for functions of two variables. Let $E\{X\} = m$, $Var\{X\} = \sigma^2$, and a number $\varepsilon \in (0,1)$. Set $Y = X + \xi_\varepsilon$, where the r.v. ξ_ε is independent of X, and

$$\xi_\varepsilon = \begin{cases} \varepsilon^{-1} & \text{with probability } \varepsilon^3, \\ 0 & \text{with probability } 1 - \varepsilon^3. \end{cases}$$

Obviously, $P(Y \geq X) = 1$. Furthermore, $E\{\xi_\varepsilon\} = \varepsilon^2$, $Var\{\xi_\varepsilon\} = \varepsilon - \varepsilon^4$, and hence, $E\{Y\} = m + \varepsilon^2$, and $Var\{Y\} = \sigma^2 + \varepsilon - \varepsilon^4$.

Then $\sigma_Y = \sqrt{\sigma^2 + \varepsilon - \varepsilon^4}$. Applying Taylor's expansion for this function of ε (see the Appendix, (4.2.3)), and assuming $\sigma \neq 0$, we get that $\sigma_Y = \sigma + \frac{1}{2\sigma}\varepsilon + o(\varepsilon)$, where here and below $o(\varepsilon)$ stands for a remainder negligible as $\varepsilon \to 0$. (See also the Appendix, Section 4.1.)

By the Taylor expansion for $g(x,y)$, we have $g(m_Y, \sigma_Y) = g\left(m + \varepsilon^2, \sigma + \frac{1}{2\sigma}\varepsilon + o(\varepsilon)\right) =$

$$g(m,\sigma) + g_1(m,\sigma)\varepsilon^2 + g_2(m,\sigma)\left(\frac{1}{2\sigma}\varepsilon + o(\varepsilon)\right) + o(\varepsilon) = g(m,\sigma) + \frac{1}{2\sigma}g_2(m,\sigma)\,\varepsilon + o(\varepsilon).$$

Because $g_2(m,\sigma^2) < 0$ and the remainder $o(\varepsilon)$ is negligible for small ε, there exists $\varepsilon > 0$ such that $\frac{1}{2\sigma}g_2(m,\sigma^2)\varepsilon + o(\varepsilon) < 0$.

For such an ε, we have $g(m_Y, \sigma_Y^2) < g(m, \sigma^2) = g(m_X, \sigma_X^2)$.

In the case $\sigma = 0$, we have $\sigma_Y = \sqrt{\varepsilon - \varepsilon^4} = \sqrt{\varepsilon} + o(\varepsilon)$, and the proof is similar. ■ ◄

Routes 1 and 2 ⇒ page 86

3

1.3 On coherent measures of risk

In this section, we discuss some desirable properties of risk measures. It is important to emphasize, however, that we should not expect these properties to hold in all situations, especially simultaneously. The properties themselves have long been known, but they attracted a great deal of attention due to the paper [8] which had given deeper insight into the nature of some useful criteria. See also a further discussion in [9], the monograph [33], and "an exposition for the lay actuary" with some examples in [91].

We describe properties below in terms of $V(X)$ preserving $\succsim$.

I. *Subadditivity.* For all $X, Y \in \mathcal{X}$,

$$V(X+Y) \geq V(X) + V(Y). \tag{1.3.1}$$

This requirement concerns the diversification of portfolios. Let us view X and Y as the random results of the investments into two assets, and $V(X)$ and $V(Y)$ as the values of the corresponding investments. Then the left member of (1.3.1) is the value of the portfolio consisting of the two investments mentioned, while the right member is the sum of the values of X and Y, considered separately.

Note also that if (1.3.1) is true for two r.v.'s, it is true for any number of r.v.'s.

Thus, under a preference with this property, it is reasonable to have many risks in one portfolio (when risks may, in a sense, compensate each other) rather than to deal with these risks separately.

II. *Positive Homogeneity.* For any $\lambda \geq 0$ and $X \in \mathcal{X}$,

$$V(\lambda X) = \lambda V(X).$$

III. *Translation Invariance.* For any number c and $X \in \mathcal{X}$,

$$V(X+c) = V(X) + c.$$

Properties II-III establish invariance with respect to the change of scale. For example, if we decide to measure income not in dollars but in cents, under the requirement II, the value of investment (if this value is measured in money units) should be multiplied by 100. If we add to a random income a certain amount c, in accordance with III, the value of the income should increase by c.

Note at once that the value of investment may be measured not only in money units. This is the case, for example, when we apply the utility theory which we discuss in detail in Section 3. Properties II-III are not so innocent as they might seem, and many criteria we consider later, do not satisfy them. However, if (II-III) hold, it certainly "makes life better".

Since in this setup, in general, $V(X)$ is not connected with some particular probability measure, we call $V(X)$ monotone if $V(X) \geq V(Y)$ when $X(\omega) \geq Y(\omega)$ for all ω.

Criteria satisfying I-III together with the monotonicity property are called *coherent*.

Because the mean-variance criterion is monotone only in special situations, consider, as examples, the first three criteria from Section 1.2.

The mean-value function $V(X) = E\{X\}$ satisfies all three criteria above, as is easy to see. (For example, $E\{X+Y\} = E\{X\} + E\{Y\}$, and similarly one can check the other properties.) This is true not because this criterion is very good, but because it is very simple.

The VaR and TailVaR satisfy II-III. Assume, for simplicity, that X is a continuous r.v. Then $P(X \leq q_\gamma(X)) = \gamma$. To show that $q_\gamma(\lambda X) = \lambda q_\gamma(X)$, we should prove that $P(\lambda X \leq \lambda q_\gamma(X)) = \gamma$. But it is obvious since λ cancels out.

To prove that, for example, $V_{\text{tail}}(\lambda X) = \lambda V_{\text{tail}}(X)$, it suffices to write $V_{\text{tail}}(\lambda X) = E\{\lambda X \mid \lambda X \leq q_\gamma(\lambda X)\} = \lambda E\{X \mid \lambda X \leq \lambda q_\gamma(X)\} = \lambda E\{X \mid X \leq q_\gamma(X)\} = \lambda V_{\text{tail}}(\lambda X)$.

Property III is considered similarly.

It remains to check the main (and most sophisticated) property I. The VaR does not satisfy this property in general as is shown in

EXAMPLE 1. Let us revisit Example 1.2.2-5. Note that if Property I holds for two r.v.'s, then it holds for any number of r.v.'s. In Example 1.2.2-5, we computed that $q_\gamma(X_1 + ... + X_{10}) = 9.36$ if $\gamma = 0.05$. Since $P(X_1 = 0) = 0.01$, for the same γ, the quantile $q_\gamma(X_1) = 1.04$. Then $q_\gamma(X_1) + ... + q_\gamma(X_n) = 10 \cdot 1.04 = 10.4 > 9.36$, and hence Property I does not hold. □

In general, the TailVaR criterion does not satisfy Property I either, and, as we know, it is not even monotone. Nevertheless, in Example 2c below, we consider some conditions under which both properties hold. As was mentioned in Section 1.2.4, in particular, it concerns the case where the space Ω is finite and all ω's are equally likely.

Since in the scheme of equiprobable ω's, the r.v.'s themselves may assume various values, the requirement that all ω's are equally likely is not very strong, and the TailVaR criterion may prove to be efficient in many situations. Nevertheless, it is worthwhile to make the following two remarks.

The goal of the TailVaR criterion is to exclude, as far as it is possible, strategies which could lead to large losses, but it does not take into account possibilities of other values of income, large or moderate. One may say it is a pessimistic criterion.

Secondly, the TailVaR criterion and other criteria we considered are normative, that is, invented by people. These criteria are applied consciously by companies for explicitly stated goals and in explicitly designated situations. When we deal with separate people, the picture may be different. Real individuals are not always pessimistic, often make decisions at an intuitive level, and sometimes are quite sophisticated. To describe their behavior, we should proceed from qualitatively different principles. An introduction to the corresponding theory is given in Sections 3-4.

Next, we give an implicit representation of the whole class of criteria satisfying Properties I-III together with monotonicity. Consider r.v.'s $X = X(\omega)$ defined on a sample space Ω. Denote by $E_P\{X\}$ the expected value of X with respect to a probability measure P defined on sets from Ω.

It was shown in [8] that functions $V(X)$ satisfying all properties mentioned are functions which may be represented as

$$V(X) = \min_{P \in \mathcal{P}} E_P\{X\}, \tag{1.3.2}$$

where $\mathcal{P} = \{P\}$ is a family of probability measures P on Ω. In other words, each function $V(\cdot)$ corresponds to a family $\mathcal{P}$, and vice versa.

For the reader familiar with the notion of infimum, note that in general the minimum above may be not attainable, and more rigorously, a necessary and sufficient condition is the existence of a family $\mathcal{P}$ such that $V(X) = \inf_{P \in \mathcal{P}} E_P\{X\}$.

EXAMPLE 2. (a) Let P_0 be the probability measure representing the "actual" probabilities of the occurrence of events ω, and let $\mathcal{P}$ consist of *only one* measure P_0. Then (1.3.2) implies that $V(X) = E_{P_0}\{X\}$, and we deal with the mean-value criterion.

(b) Let $\Omega = \{\omega_1, ..., \omega_n\}$ be finite, and let $\mathcal{P}$ consist of *all* probability measures on events from Ω. Denote by $P(\omega)$ the probability of ω corresponding to measure P. The expected value with respect to P is

$$E_P\{X\} = \sum_{i=1}^{n} X(\omega_i)P(\omega_i).$$

To minimize the last expression, we should choose P which assigns the probability one to the minimum value of $X(\omega)$. For such a measure, $E_P\{X\} = \min_\omega X(\omega)$. So,

$$V(X) = \min_\omega X(\omega).$$

(c) Let again $\Omega = \{\omega_1, ..., \omega_n\}$. Assume that the "actual" probability measure P_0 assigns the equal probabilities $\frac{1}{n}$ to each ω. We show that the TailVaR criterion admits the representation (1.3.2).

To make our reasoning simpler, consider only r.v.'s $X(\omega)$ taking different values for different ω's.

We fix a γ and denote by $k = k(\gamma)$ the integer such that $\frac{k-1}{n} \leq \gamma < \frac{k}{n}$; that is, $k = [n\gamma] + 1$, where $[a]$ denotes the integer part of a. Consider all sets A from Ω containing exactly k points. Let $P_A(\omega)$ be the measure assigning the probability $\frac{1}{k}$ to each point from A, and zero probability to all other $n-k$ points. Let $\mathcal{P}$ consist of all such measures P_A.

Consider now a r.v. $X(\omega)$ and set $x_i = X(\omega_i)$. We have assumed x_i's to be different. Without loss of generality, we can suppose that $x_1 < x_2 < ... < x_n$, since otherwise we can renumerate the ω's. The reader is invited to verify that with respect to the original measure P_0, first, $q_\gamma(X) = x_k$, where $k = k(\gamma)$ chosen above, and second, that

$$TailVar(X) = E\{X \mid X \leq q_\gamma\} = \frac{1}{k}(x_1 + ... + x_k). \tag{1.3.3}$$

On the other hand, for any A consisting of k points, say, points $\omega_{i_1}, ..., \omega_{i_k}$,

$$E_{P_A}\{X\} = X(\omega_{i_1})\frac{1}{k} + ... + X(\omega_{i_k})\frac{1}{k} = \frac{1}{k}(x_{i_1} + ... + x_{i_k}) \geq \frac{1}{k}(x_1 + ... + x_k) = E\{X \mid X \leq q_\gamma\},$$

because $x_1, ..., x_k$ are the k least values of X. Thus, the minimum in (1.3.2) is attained at P_{A_0}, where $A_0 = \{\omega_1, ..., \omega_k\}$, and this minimum is equal to $TailVar(X)$.

Note that if x_i's are not different, formally we cannot reason as above since in this case (1.3.3) may be not true, and we may construct an example close to Example 1.2.4-4. However, in this case, we may modify the TailVar criterion itself defining it as in (1.3.3). In the case of different x_i's, it will coincide with the "usual" TailVar. □

In conclusion, note that we should not, certainly, restrict ourselves only to coherent measures. Often, it is reasonable to sacrifice some properties mentioned above in order to deal with more flexible characteristics of distributions. It concerns, in particular, criteria we consider in following sections of this chapter.

2 COMPARISON OF R.V.'S AND LIMIT THEOREMS

In this section, we return to the mean-value and VaR criteria and look at them from another point of view. We saw that they were not very sophisticated—especially the former, and do not satisfy all desirable properties—at least, the latter. Nevertheless, reasonable decision making rules may occur to be close to the criteria mentioned when the corresponding decision acts are made repeatedly. To understand this, we will proceed from the limit theorems of Probability Theory presented in Section **0**.6.

2.1 A simple model of insurance with many clients

Consider an insurance company dealing with n clients. Let X_i, $i = 1, ..., n$, be the random value of the payment to the ith client. We assume the X's to be independent and identically distributed (i.i.d.), which may be interpreted as if the clients come from a *homogeneous group*. We keep all notations from Section **0**.6.

Let $m = E\{X_i\}$, and $c = m + \varepsilon$, where $\varepsilon > 0$, be the premium for each client. Thus, we assume that the premium is, at least a bit, larger than m. The total profit of the company equals $nc - S_n$, where $S_n = X_1 + ... + X_n$. Set $\overline{X}_n = S_n/n$.

The probability that the company will not suffer a loss is equal to

$$P(nc - S_n \geq 0) = P(S_n - mn \leq n\varepsilon) = P(\overline{X}_n - m \leq \varepsilon) \geq P(|\overline{X}_n - m| \leq \varepsilon)$$
$$= 1 - P(|\overline{X}_n - m| > \varepsilon) \to 1 \quad \text{as} \quad n \to \infty,$$

for *any* arbitrarily small $\varepsilon > 0$, by Corollary **0**.10 to the LLN in Section **0**.6.

Thus, if n is "large", for the company not to suffer a loss, the premium c should be "just a little bit" larger than m. In this case, the company would prefer, with regard to each client, the profit $c - X_i$ rather than zero.

We see that for large n the necessary premium c may be close to the expected value m, and accordingly the criterion of the choice of a premium is close to the mean-value criterion.

The "little bit" mentioned, that is, the value of ε, is one of the main objects of study in Actuarial Modeling, and we will return to it repeatedly. Here we will make just preliminary observations.

First, note that though ε can be small, it cannot be zero.

Indeed, let $\sigma^2 = Var\{X_i\}$. Assume $\sigma > 0$; otherwise X's take on just one value and the situation is trivial. As in Section **0**.6, set $S_n^* = \dfrac{S_n - mn}{\sigma\sqrt{n}}$. Then, if $\varepsilon = 0$,

$$P(nc - S_n \geq 0) = P(mn - S_n \geq 0) = P(S_n - mn \leq 0) = P\left(\frac{S_n - mn}{\sigma\sqrt{n}} \leq 0\right)$$
$$= P(S_n^* \leq 0) \to \Phi(0) = \frac{1}{2} \quad \text{as} \quad n \to \infty,$$

by the Central Limit Theorem (CLT) **0**.11 in Section **0**.6. So, in this case, the probability that the company will not suffer a loss is close only to $1/2$.

Note also that since the limiting normal distribution is continuous, the probability that S_n equals exactly some value is close to zero. Therefore, the probability of making a profit and the probability of not suffering a loss asymptotically, for large n, are the same.

Now, let $\varepsilon > 0$. The same CLT provides the first heuristic approximation for a reasonable value of ε. Assume that the company specifies the lowest acceptable level β for the probability of not suffering a loss. For instance, the company wishes the mentioned probability to be not less than $\beta = 0.95$, in the worst case—to equal 0.95.

Set $\varepsilon = a\sigma/\sqrt{n}$, where the number a is what we want to estimate. Let c be an acceptable premium for the company. Then

$$\beta \le P(nc - S_n \ge 0) = P(S_n - mn \le n\varepsilon) = P\left(\frac{S_n - mn}{\sigma\sqrt{n}} \le \frac{\varepsilon\sqrt{n}}{\sigma}\right)$$
$$= P\left(\frac{S_n - mn}{\sigma\sqrt{n}} \le a\right) = P(S_n^* \le a).$$

By the CLT, $P(S_n^* < a) \approx \Phi(a)$ for large n. Thus, up to normal approximation, $\beta \le \Phi(a)$, and hence for the premium to be acceptable, a should be not less than $q_{\beta s}$, the β-quantile of the standard normal distribution. In the boundary case, the least acceptable premium

$$c \approx m + \frac{q_{\beta s}\sigma}{\sqrt{n}}. \tag{2.1.1}$$

The sign $\approx$ indicates that the answer is true within the accuracy of normal approximation.

For $\beta = 0.95$, we have $q_{\beta s} = 1.64...$, and $c \approx m + 1.64\sigma/\sqrt{n}$.

EXAMPLE 1. A special insurance pays $b = \$150$ to passengers of an airline in the case of a serious flight delay. Assume that for each of 10,000 clients who bought such an insurance, the probability of a delay is $p = 0.1$. In this case,

$$X_i = \begin{cases} b & \text{with probability } p, \\ 0 & \text{with probability } 1-p, \end{cases}$$

$m = bp = 15$, $\sigma = b\sqrt{p(1-p)} = 45$ (recall the formulas for the mean and the variance of a binomial r.v.). Then for $\beta = 0.95$, by (2.1.1), $c \approx 15 + \frac{1.64 \cdot 45}{100} \approx 15.74$. So, a premium of \$16 would be enough for the company. □

Note that the choice of c in (2.1.1) is closely related to the VaR criterion. For each premium c, the company compares its random profit $nc - S_n$ with the r.v. $Y \equiv 0$, the profit in the case when the company does not sell the insurance product. For the c chosen, up to normal approximation, $\beta = P(nc - S_n > 0)$ and hence $P(nc - S_n \le 0) = 1 - \beta$. Thus, zero is the $(1-\beta)$-quantile for the r.v. $nc - S_n$. On the other hand, Y takes on only one value—zero, and this singular value is the γ-quantile for any γ, including $\gamma = 1 - \beta$. (See again the definition of quantile in Section **0**.1.3 and Fig.7e there.)

Thus, for the least acceptable c in (2.1.1), $(1-\beta)$-quantiles of the r.v.'s $nc - S_n$ and Y coincide, that is, $nc - S_n$ is equivalent to Y in the sense of the VaR criterion. For c larger than the value in (2.1.1), $nc - S_n$ will be better than $Y = 0$.

The approach based on limit theorems is, however, far from being universal. First of all, the acts of making decisions are not always repeated a large number of times. We can say so about an insurance company when it deals with a large number of clients, but a separate client may make decisions rarely enough, and the law of large numbers (LLN) in this case may not work well.

Second—and this is also important—even when limit theorems formally might work, real people in real situations may proceed from preferences not connected with means or variances.

For example, when comparing r.v.'s as in (1.2.1) even repeatedly, people rarely consider such r.v.'s equivalent. Usually, the less risky alternative (as Y in (1.2.1)) is preferred to the more risky (as X in (1.2.1)); see Section 3.4 for more detail. So, the LLN argument does not work here.

The same concerns the CLT. Assume, for instance, that an individual proceeds—perhaps unconsciously—from the same argument based on the CLT, as we used above. Then adding a small amount ε of money to X from (1.2.1) would have made a difference: the r.v. $X+\varepsilon$ would have been better than $Y=50$. However, usually people—not companies but separate individuals—do not exhibit such behavior.

We consider now an old celebrated example when the application of the LLN leads to a conclusion that is inconsistent with usual human behavior.

2.2 St. Petersburg's paradox

The problem below was first investigated by Daniel Bernoulli in his paper [13] published in 1738 when D. Bernoulli worked in Saint Petersburg. Consider a game of chance consisting of tossing a regular coin until a head appears. Suppose that if the first head appears right away at the first toss, the payment equals 2, say, dollars if we update the problem to the present day. If the first head appears at the second toss, the payment equals 4, and so on; namely, if a head appears at the first time at the kth toss, the payment equals 2^k. It is easy to see that the payment in this case is a r.v. X taking values $2,4,8,...,2^k,...$ with probabilities $\frac{1}{2},\frac{1}{4},\frac{1}{8},...,\frac{1}{2^k},...$, respectively, and $E\{X\}=2\cdot\frac{1}{2}+4\cdot\frac{1}{4}+8\cdot\frac{1}{8}+...=1+1+1+...=\infty$. By the LLN, this means that if the game is played repeatedly, and X_j is the payment in the jth game, then with probability one

$$\frac{X_1+..+X_n}{n}\to\infty \quad \text{as} \quad n\to\infty.$$

Thus, in the long run, the average payment will be greater than *an arbitrary large* number.

Then if a player had proceeded from the LLN, she/he would have agreed to pay *any, arbitrary large, entry price* for participating in each play. Certainly, it does not reflect preferences of real people: most would not agree to pay each time, for example, \$100 if even they are guaranteed to participate in a large number of plays. (Would the reader agree to pay \$100 each time?) □

There exists a purely mathematical solution to this paradox based on the fact that in this particular case, $\frac{X_1+...+X_n}{n\log_2 n}\to 1$ with probability one. A not very short proof may be found, e.g., in [38], [120, p.57]. Thus, if the entry price for each play depends on the number of plays n and equals $c = \log_2 n$, then for large n, the total payment for participating in n plays will be close to the total gain, and the price c would be "fair".

This solution is strongly connected with the particular problem under consideration. Fortunately, D. Bernoulli did not know the fact mentioned and suggested a general solution that had proved to be very fruitful and, in the twentieth century, had led to a developed theory we consider in the next section.

---

3 EXPECTED UTILITY

3.1 Expected utility maximization (EUM)

3.1.1 Utility function

D. Bernoulli proceeded from the simple observation that the "degree of satisfaction" of having capital, or in other words, the "utility of capital", depends on the particular amount of capital in a nonlinear way. For example, if we give \$1000 to a person with a wealth of \$1,000,000, and the same \$1000 to a person with zero capital, the former will feel much less satisfied than the latter.

To model this phenomenon, D. Bernoulli assumed that the satisfaction of possessing a capital x, or the "utility" of x, may be measured by a function $u(x)$ that, as a rule, is not linear. Such a function is called a *utility function*, or a utility of money function. The word "satisfaction" would possibly reflect the significance of the definition better, but the term "utility" has been already adopted.

The utility function, if it exists, can be viewed as a characteristic of the individual, as if the individual is endowed by this function; so to speak, it is "built into the mind". To some extent, we can talk about the utility function of a company too. In this case, it reflects the preferences of the company.

D. Bernoulli himself suggested as a good candidate for the "natural" utility function $u(x) = \ln x$, assuming that the increment of the utility is proportional not to the absolute but to the relative growth of the capital. More specifically, if capital x is increased by a small dx, then the increment of the utility, $du(x)$, is proportional to dx/x, that is,

$$du = k\frac{dx}{x} \tag{3.1.1}$$

for a constant k. The solution to this equation is $u(x) = k\ln x + C$, where C is another constant. We will see soon that the values of k and C depend just on the choice of units in measuring utility, and hence do not matter.

Consider now a random income X. In this case, the utility of the income is the r.v. $u(X)$. Bernoulli's suggestion was to proceed from the expected utility $E\{u(X)\}$.

EXAMPLE 1. Assume that the utility function of the player in St. Petersburg's paradox is $u(x) = \ln x$. Then the expected utility

$$E\{u(X)\} = \sum_{k=1}^{\infty} u(2^k)2^{-k} = \sum_{k=1}^{\infty} \ln(2^k)2^{-k} = (\ln 2)\sum_{k=1}^{\infty} k2^{-k} = 2\ln 2,$$

and, unlike $E\{X\}$, the expected utility is finite. (To realize that $\sum_{k=1}^{\infty} k2^{-k} = 2$, one may compute it directly, or observe that this is the expected value of the geometric r.v. with the parameter $p = 1/2$. See Section **0**.3.1.3.) □

Next, we consider the general case. Clearly, we can restrict ourselves to non-decreasing utility functions, which reflects the rule "the larger, the better or at least not worse".

3.1.2 Expected utility maximization criterion

By definition, this criterion corresponds to the preference order $\succsim$ for which

$$X \succsim Y \Leftrightarrow E\{u(X)\} \geq E\{u(Y)\} \tag{3.1.2}$$

for a utility function u. Not stating it each time explicitly, we will always assume that $u(x)$ is defined on an interval (which may be the whole real line).

The relation (3.1.2) means that among two r.v.'s, we prefer the r.v. with the larger expected utility. In particular,

$$\text{if } E\{u(X)\} = E\{u(Y)\}, \text{ we say that } X \simeq Y,$$

X is *equivalent* to Y.

If $u(x)$ is non-decreasing (as we agreed), the rule (3.1.2) is monotone. (If $X \geq Y$ with probability one, then $u(X) \geq u(Y)$ with probability one too, which immediately implies that $E\{u(X)\} \geq E\{u(Y)\}$.) In Exercise 11, we discuss strict monotonicity.

The investor who follows (3.1.2) is called an *expected utility maximizer* (EU maximizer; we will use also the same abbreviation EUM when it does not cause misunderstanding).

It is worth emphasizing that when we are talking about an EU maximizer, we mean that the person's preferences *may be described by* (3.1.2), or in other words that the person behaves as if she/he were an EU maximizer. However, this does not imply in any way that calculations in (3.1.2) are really running in the mind. A good image illustrating this was suggested in [82]. A thrown ball exhibits a trajectory described as the solution to a certain equation, but no one thinks that the ball "has and itself solves" this equation. People do not get confused about the ball but they sure do about models of other people.

The first property of EUM criterion. *The preference order (3.1.2) does not change if $u(x)$ is replaced by any function $u^*(x) = bu(x) + a$, where b is a positive and a is an arbitrary number.*

Indeed, if we replace in (3.1.2) u by u^*, then b and a will cancel.

Thus, u may be defined up to a linear transformation, and the scale in which we measure utility may be chosen at our convenience. In particular, there is nothing wrong or strange if u assumes negative values.

EXAMPLE 1. Consider (3.1.1) with $u(x) = k \ln x + C$ as above. We see now that constants k and C indeed do not matter, and we can restrict ourselves to $u(x) = \ln x$.

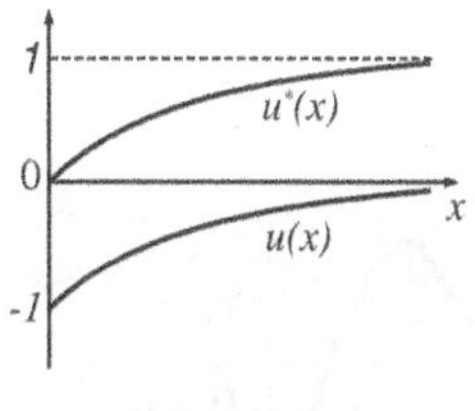

FIGURE 1.

EXAMPLE 2. Let $u(x) = -\frac{1}{1+x}$, $x \geq 0$; see Fig.1. Should the fact that $u(x)$ is negative for all x's make us uncomfortable? Not at all. Consider $u^*(x) = u(x) + 1 = \frac{x}{1+x}$. The new function is positive but reflects the same preference order. The sign of $u(x)$ does not matter; what matters when we compare X and Y is whether $E\{u(X)\}$ is larger than $E\{u(Y)\}$ or not. □

Consider an example of the comparison of r.v.'s.

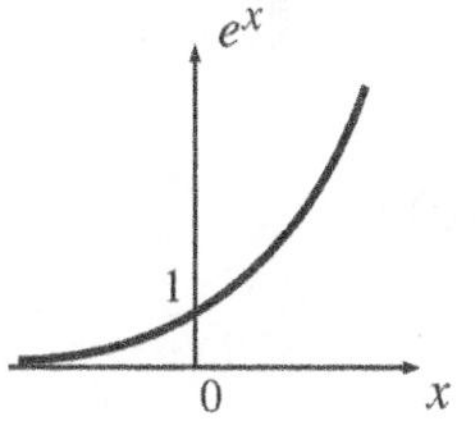

FIGURE 2.

EXAMPLE 3. (a) *A reckless gambler.* Let a gambler's utility function $u(x) = e^x$. Negative x's correspond to losses, and positive—to gains. The values of $u(x)$ for large negative x's practically do not differ, while in the region of positive x's the function $u(x)$ grows very fast; see Fig.2. We may interpret it as if the gambler is not concerned about possible losses and is highly enthusiastic about large gains. Note that the function e^x is convex, and as we will see later, in Section 3.4, the convexity of the utility function corresponds to the inclination to risk.

Consider a game in which the gambler wins a dollars with a probability of p, and loses the same amount a with the probability $q = 1 - p$. So, we deal with $X = \pm a$ with the mentioned probabilities. Assume $p < q$.

In our case, $E\{u(X)\} = e^a p + e^{-a} q$. The gambler will participate in such a game if X is better than a r.v. $Y \equiv 0$, which amounts to $E\{u(X)\} > u(0) = 1$. This is equivalent to $e^a p + e^{-a} q > 1$.

If we set $e^a = y$, the last inequality may be reduced to the quadratic inequality $py^2 - y + q > 0$. One root of the corresponding quadratic equality is one, the other is q/p. Since $y \geq 1$, the solution is $y > q/p$, and consequently, $a > \ln(q/p)$. Thus, the gambler is inclined to bet large stakes, and will participate in the game only if $a > \ln(q/p)$. For instance, if $p = \frac{1}{4}$, the lowest acceptable stake for the gambler is $\ln 3 \approx 1.1$.

(b) *A cautious gambler.* Consider now a gambler who views the loss of a unit of money as a disaster. What this unit of money is equal to, \$1,000,000 or just \$100, depends on the gambler. On the other hand, the gambler does not mind taking some risk and participating in a game with a moderate stake. The utility function of such a gambler may look as in Fig.3a: $u(x) \to -\infty$ as $x \to -1$, and $u(x)$ is growing as a convex function for positive x's. For instance, the function

$$u(x) = \begin{cases} kx^2 & \text{for } x \geq 0, \\ \ln(1-x^2) & \text{for } -1 < x < 0 \end{cases}$$

has a similar graph. We wrote x^2 in $\ln(1-x^2)$ to make the function smooth at zero. The parameter k indicates the gambler's inclination to risk. The larger k, the steeper $u(x)$ for positive x's.

Consider the same r.v. X as in the example above. Then $E\{u(X)\} = p \cdot ka^2 + q \cdot \ln(1-a^2)$. Denote the r.-h.s. by $g(a)$. The gambler will participate in the game if $E\{u(X)\} > u(0) = 0$, which amounts to $g(a) > 0$. The reader can readily verify that if $k \leq q/p$, the graph of $g(a)$

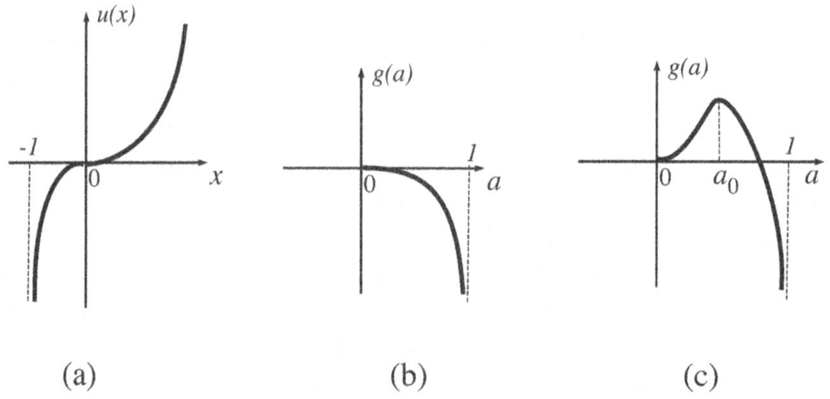

FIGURE 3.

looks as in Fig.3b, and $g(a)$ does not assume positive values. In this case, the person we consider will refuse to play.

The graph for $k > q/p$ is sketched in Fig.3c. In this case, $g(a)$ is positive in a neighborhood of zero. The maximum of $g(a)$ is attained at a_0 such that $a_0^2 = 1 - \frac{q}{kp}$. For example, for $p = \frac{1}{4}$, and $k = 4$, we get $a_0 = \frac{1}{2}$, so the gambler's optimal behavior is to bet half of the unit of money. □

The next notion we introduce is a *certainty equivalent*. First, note that any number c may be viewed as a r.v. taking only one value c. Consider a preference order $\succsim$ not necessarily connected with expected utility maximization. Assume that for a r.v. X, we can find a number $c = c(X)$ such that $c \simeq X$ with respect to the order $\succsim$. That is, c is equivalent to X, or the decision maker is indifferent whether to choose c or X. It may be said that the decision maker considers c an "adequate price of X".

The number $c(X)$ so defined is called a *certainty equivalent* of X.

Now let us consider an EU maximizer with a utility function u. For such a person, in accordance with (3.1.2), the relation $c \simeq X$ is equivalent to $E\{u(X)\} = E\{u(c)\}$, and since c is not random, $E\{u(X)\} = u(c)$. If u is a one-to-one function, there exists the inverse function $u^{-1}(y)$, and

$$c(X) = u^{-1}(E\{u(X)\}).$$

EXAMPLE 4. (a) In the situation of Example 3a, $u^{-1}(x) = \ln x$. So, the certainty equivalent $c(X) = \ln(e^a p + e^{-a} q)$. For example, for $p = \frac{1}{4}$ and $a = 10$, we would have $c(X) = \ln(\frac{1}{4}e^{10} + \frac{3}{4}e^{-10}) \approx 8.614$, which is close to 10. It is not surprising: the gambler does not care much about losses.

(b) Consider Example 3b for $p = \frac{1}{4}$, $k = 4$. Here the situation is quite different. The gambler bets $a = a_0 = \frac{1}{2}$. In this case, $E\{u(X)\} = g\left(\frac{1}{2}\right) = \frac{1}{4} \cdot 4 \cdot \frac{1}{4} + \frac{3}{4} \cdot \ln(\frac{3}{4}) \approx 0.034$. On the other hand, $u^{-1}(y) = \sqrt{y/k}$ for positive y's. So, in our case the certainty equivalent $c(X) \approx \sqrt{0.034/4} \approx 0.0922$. □

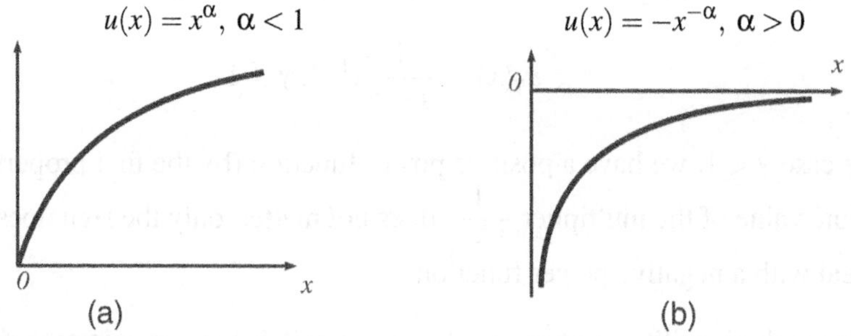

FIGURE 4. Positive- and negative-power utility functions.

Note that the certainty equivalent of a certain number a is, of course, this number: $c(a) = u^{-1}(E\{u(a)\}) = u^{-1}(u(a)) = a$.

3.1.3 Some "classical" examples of utility functions

1. *Positive-power functions.* Let $u(x) = x^\alpha$ for all $x \geq 0$ and some $\alpha > 0$; see Fig.4a. The expected utility in this case is considered only for positive r.v.'s, and $E\{u(X)\} = E\{X^\alpha\}$, the moment of X of the order α. If $\alpha = 1$, then $E\{u(X)\} = E\{X\}$, and the EUM criterion coincides with the mean-value criterion. For $\alpha < 1$ the function $u(x)$ is concave (downward), for $\alpha > 1$ - convex (concave upward). We will see soon that this is strongly connected with the attitude of the investor to risk. For $u(x)$ we are considering, the certainty equivalent of a r.v. X is $c(X) = (E\{X^\alpha\})^{1/\alpha}$. In the simplest case $\alpha = 1$, the certainty equivalent $c(X) = E\{X\}$.

   EXAMPLE 1. Let $X = b > 0$ or 0 with equal probabilities. Then $c(X) = \left(\frac{1}{2}b^\alpha\right)^{1/\alpha} = 2^{-1/\alpha}b$. The smaller α is, the smaller the certainty equivalent. We will interpret this fact later when we consider the notion of risk aversion.

   EXAMPLE 2. Let X be uniform on $[0, b]$. Then $c(X) = \left(\int_0^b x^\alpha \frac{1}{b} dx\right)^{1/\alpha} = \left(\frac{1}{1+\alpha} b^\alpha\right)^{1/\alpha}$ $= \left(\frac{1}{1+\alpha}\right)^{1/\alpha} b$. Because $(1+\alpha)^{1/\alpha}$ is decreasing in α, again the smaller α, the smaller the certainty equivalent. □

2. *Negative-power functions.* Next, consider $u(x) = -1/x^\alpha$ for all $x > 0$ and some $\alpha > 0$; see Fig.4b. We again deal only with positive r.v.'s, and $E\{u(X)\} = -E\{X^{-\alpha}\}$. The fact that $u(x)$ is negative does not matter, but the fact that $u(x) \to -\infty$, as $x \to 0$, is meaningful: now the investor is much more "afraid" of being ruined than in the previous case when $u(x) \to 0$ as $x \to 0$. We see also that $u(x) \to 0$ as $x \to +\infty$, which may be interpreted as the saturation effect: the investor does not distinguish much large values of the capital. Compare it with the previous case of the positive power where $u(x) \to +\infty$ as $x \to +\infty$.

Both cases above may be described by the unified formula

$$u_\gamma(x) = \frac{1}{1-\gamma}x^{1-\gamma},\ \gamma \neq 1. \tag{3.1.3}$$

In the case $\gamma < 1$, we have a positive power function (by the first property above, the absolute value of the multiplier $\frac{1}{1-\gamma}$ does not matter, only the sign does). For $\gamma > 1$, we deal with a negative power function.

3. *The logarithmic utility function,* $u(x) = \ln x$, $x > 0$, is in a sense intermediate between the two cases above and has been already discussed.

4. *Quadratic utility functions.* Consider $u(x) = 2ax - x^2$, where parameter $a > 0$; the multiplier 2 is written for convenience. Certainly, such a utility function is meaningful only for $x \le a$ when the function is increasing. Hence, in this case, we consider only r.v.'s X such that $P(X \le a) = 1$. Negative values of X are interpreted as the case when the investor loses or owes money. We have $E\{u(X)\} = 2aE\{X\} - E\{X^2\} = 2aE\{X\} + (E\{X\})^2 - Var\{X\}$. Thus, the expected utility is a quadratic function of the mean and the variance.

5. *Exponential utility functions.* Let $u(x) = -e^{-\beta x}$, where parameter $\beta > 0$, and the function is considered for all x's. The graph is depicted in Fig.5. Since $u(x) \to 0$, as $x \to \infty$, faster than any power function, the saturation effect in this case is stronger than in Case 2. The expected utility $E\{u(X)\} = -E\{e^{-\beta X}\} = -M(-\beta)$, where $M(z) = E\{e^{zX}\}$.

   The function $M(z)$, we also use the notation $M_X(z)$ to emphasize that it depends on the choice of the r.v. X, is the *moment generating function* of X. (See a definition and examples in Section **0**.4.)

   In Exercise 15, we show that the certainty equivalent

$$c(X) = -\frac{1}{\beta}\ln(M_X(-\beta)).$$

   Consider a negative β, setting $\beta = -a$ for some $a > 0$. Then

$$c(X) = \frac{1}{a}\ln(M_X(a)) = \frac{1}{a}\ln(E\{e^{aX}\}). \tag{3.1.4}$$

   This is the *Masset criterion* popular in Economics. When X is a loss, the same expression appears as the premium for the coverage of X in accordance with the so called *exponential principle*. We consider it later in Section **2**.4 and in Exercise **2**.65. In particular, we compare there the cases $\beta > 0$ and $\beta < 0$. In Exercise 14 we consider two important properties of the Masset criterion.

EXAMPLE 3. Let X be distributed exponentially with parameter a. Then $M_X(z) = a/(a-z)$. (See Section **0**.4.) Now calculations lead to $c(X) = \frac{1}{\beta}[\ln(a+\beta) - \ln a)]$. □

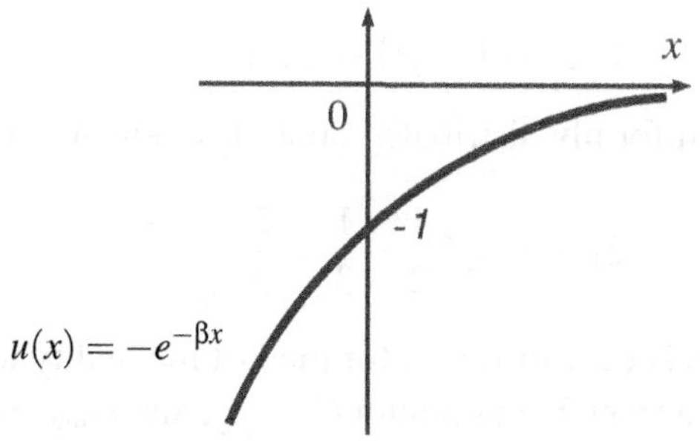

FIGURE 5. The exponential utility function.

In the case of exponential utility, EU maximization has an important property stated in

Proposition 2 *Let $u(x) = -e^{-\beta x}$ and, under the EUM criterion with this utility function, $X \succsim Y$. Then $w+X \succsim w+Y$ for any number w.*

The number w above may be interpreted as the initial wealth, and X and Y as random incomes corresponding to two investment strategies. Proposition 2 claims that in the exponential utility case, the preference relation between X and Y does not depend on the initial wealth.

Proof is straightforward. By definition, $w+X \succsim w+Y$ iff $E\{u(w+X)\} \geq E\{u(w+Y)\}$. For the particular u above, $E\{u(w+X)\} = -E\{e^{-\beta(w+X)}\} = -e^{-\beta w}E\{e^{-\beta X}\}$, and the same is true for Y. So, in the last inequality, the common multiplier $-e^{-\beta w}$ cancels out, and the validity of the relation $E\{u(w+X)\} \geq E\{u(w+Y)\}$ does not depend on w. Hence, if this relation is true for $w = 0$, it is true for all w. ■

3.2 Utility and insurance

Consider an individual with a wealth of w, facing a possible random loss ξ. Assume that the individual is an EU maximizer with a utility function $u(x)$. What premium G would the individual be willing to pay to insure the risk?

The individual's wealth after paying the premium will become $X = w - G$, while if she/he does not buy the insurance, the wealth will equal the r.v. $Y = w - \xi$.

Then in accordance with the principle (3.1.2), a premium G will be acceptable for the person under consideration only if

$$u(w-G) \geq E\{u(w-\xi)\}. \tag{3.2.1}$$

For the maximal accepted premium $G_{\max}$,

$$u(w-G_{\max}) = E\{u(w-\xi)\}. \tag{3.2.2}$$

($G_{\max}$ exists if, say, u is continuous and increasing; we skip formalities.)

EXAMPLE 1. Let $u(x) = 2x - x^2$, $w = 1$, and let ξ be uniformly distributed on $[0,1]$. Because $w - \xi = 1 - \xi \leq 1$, we deal only with x's for which $u(x)$ increases. Let $y = w - G$.

By (3.2.1),

$$2y - y^2 \geq 2E\{(1-\xi)\} - E\{(1-\xi)^2\}.$$

Observing that $1-\xi$ is also uniformly distributed on $[0,1]$ (show it!), we have

$$2y - y^2 \geq 2\frac{1}{2} - \frac{1}{3} = \frac{2}{3}.$$

We are interested in $y \leq 1$. As is easy to verify, for the last inequality to be true, we should have $y \geq 1 - \frac{1}{\sqrt{3}}$. Hence, any acceptable premium $G \leq \frac{1}{\sqrt{3}}$, and $G_{\max} = \frac{1}{\sqrt{3}} \approx 0.57$.

In the example we consider, the loss is positive with probability one. In Exercise 17, we provide similar calculations for the case when $\xi = 0$ with probability 0.9, and ξ is uniformly distributed on $[0,1]$ with probability 0.1. This corresponds to the typical situations when the loss equals zero with a large probability. Nevertheless, it is worth noting that situations when the loss is a positive (or practically positive) r.v. are not rare, especially when we deal with an aggregate loss concerning a large group of clients. For example, if a university provides medical insurance for its employees as one insurance contract, the total loss may be considered a positive r.v. The same remark concerns other examples in this section. □

Next, we consider not an insured but an insurer. The latter offers the complete coverage of a loss ξ for a premium H which, in general, may be different from G above. Assume that the insurer is an EU maximizer with a utility function $u_1(x)$ and a wealth of w_1. (Actually, it is more natural to interpret w_1 as an additional reserve kept by the insurer to fulfill its obligations.) Following a similar logic, we obtain that an acceptable premium H for the insurer must satisfy the inequality

$$u_1(w_1) \leq E\{u_1(w_1 + H - \xi)\}, \tag{3.2.3}$$

and hence for the minimal accepted premium $H_{\min}$

$$u_1(w_1) = E\{u_1(w_1 + H_{\min} - \xi)\}. \tag{3.2.4}$$

EXAMPLE 2. Let $u_1(x) = x^\alpha$, $w_1 = 1$, and ξ be the same as in Example 1. Taking again into account that $\eta = 1 - \xi$ is uniformly distributed on $[0,1]$, we derive from (3.2.3) that

$$1 \leq E\{(H+1-\xi)^\alpha\} = E\{(H+\eta)^\alpha\} = \int_0^1 (H+x)^\alpha dx = \frac{1}{\alpha+1}[(H+1)^{\alpha+1} - H^{\alpha+1}].$$

Hence, $H_{\min}$ is a solution to the equation

$$(H+1)^{\alpha+1} - H^{\alpha+1} = \alpha + 1.$$

For example, when $\alpha = 1/2$, it is easy to calculate—using even a simple calculator— that $H_{\min} \approx 0.52$. □

Clearly, for a premium P to be acceptable for both sides, the insurer and the insured, we should have

$$H_{\min} \leq P \leq G_{\max}.$$

Hence, if $H_{\min} > G_{\max}$, insurance is impossible. If $H_{\min} \le G_{\max}$, the premium will be chosen from the interval $[H_{\min}, G_{\max}]$. For instance, in the situation of Examples 1-2, we have $0.52 \le P \le 0.57$.

If for example, the insurer has a sort of monopoly in the market, the premium will be close to $G_{\max}$. In the case of competition or if a law imposes restrictions on the size of premiums, we can expect the premium to be closer to $H_{\min}$.

It is worth emphasizing that in general and in the examples above, premiums depend on the initial wealth w. Now we consider the special case of exponential utility, when premiums do not depend on wealth.

Let $u(x) = -e^{-\beta x}$. Due to Proposition 2, we can set $w = 0$ in (3.2.1)-(3.2.4). Hence, (3.2.2) is equivalent to $-e^{\beta G_{\max}} = -E\{e^{\beta\xi}\}$, and $G_{\max} = \frac{1}{\beta}\ln(E\{e^{\beta\xi}\})$. We see in the r.-h.s. the moment generating function (m.g.f.) $M_\xi(z) = E\{e^{z\xi}\}$. Thus,

$$G_{\max} = \frac{1}{\beta}\ln M_\xi(\beta). \tag{3.2.5}$$

The same formula is true for the insurer: in a similar way, we derive from (3.2.4) that in the case $u_1(x) = -e^{-\beta_1 x}$,

$$H_{\min} = \frac{1}{\beta_1}\ln M_\xi(\beta_1). \tag{3.2.6}$$

It may be proved (see Exercise 33 and an advice there for detail) that the r.-h.s. of (3.2.5) is non-decreasing in β. Consequently, for $H_{\min} \le G_{\max}$, we should require $\beta_1 \le \beta$. This fact will be interpreted when we consider the notion of risk aversion.

EXAMPLE 3. Assume that the random loss ξ may be well approximated by a normal r.v. with mean m and variance σ^2. In this case, the m.g.f. $M(z) = \exp\{mz + \sigma^2 z^2/2\}$; see Section **0**.4.3. The reader is invited to provide simple calculations leading in this case to

$$G_{\max} = m + \beta\frac{\sigma^2}{2}, \quad H_{\min} = m + \beta_1\frac{\sigma^2}{2}. \tag{3.2.7}$$

The answer looks nice and natural: the larger β and/or the variance, the more the premium exceeds the expected value of the loss. For $H_{\min} \le G_{\max}$, we indeed should have $\beta_1 \le \beta$. □

The same logic can be applied to more complicated forms of insurance. Let, for example, a client be willing to insure only half of a possible loss ξ. Then the corresponding equation for the maximal premium will be

$$E\{u(w - \frac{1}{2}\xi - G_{\max})\} = E\{u(w - \xi)\}. \tag{3.2.8}$$

In conclusion, it is worth noting that the expected utility analysis can work well when we deal rather with the preferences of individual clients. This does not mean that we cannot apply the EUM criterion to the description of the behavior of companies, but one should do it with caution. As we will see in later chapters, the behavior of companies may be determined by principles qualitatively different from those based on expected utility.

3.3 How to determine the utility function in particular cases

In Section 3.5.5, we will see that in the EUM case, when one considers r.v.'s taking only n fixed values, to completely determine the preference order, it suffices to specify $n-1$ equivalent distributions. At least theoretically it may be done by questioning the individual.

Another way is to determine certainty equivalents, which may be illustrated by the following

EXAMPLE 1. We believe that Chris is an EU maximizer, and we try to determine his utility function $u(x)$. In view of the first property from Section 3.1.2, the scale in measuring utility does not matter, so we can set, say, $u(0) = 0$ and $u(100) = 1$, where money is measured in convenient units (for example, not in \$1 but in \$100).

You invite Chris to compare a game with prizes $X = 100$ or 0, each with probability $1/2$, with a payment of 50 for sure. Chris finds X worse than a certain payment of 50. Then you reduce 50 to $49, 48, ...,$ and so on, up to the moment when Chris starts to hesitate. Assume that it happens at $c = 40$. Then we can view c as the certainty equivalent of X. This means that $u(c) = E\{u(X)\} = \frac{1}{2}u(100) + \frac{1}{2}u(0) = \frac{1}{2}$. Hence $u(40) = 0.5$, and we know the value of $u(x)$ at one more point. You can continue such a process, for example, figuring out how much Chris values a r.v. $X_1 = 100$ and 40 with equal probabilities. Assume that Chris's answer is 60. Then $u(60) = \frac{1}{2}u(100) + \frac{1}{2}u(40) = \frac{3}{4}$, etc.

Similar questioning may involve insurance premiums. Suppose, for example, that Chris's initial wealth is 100 units of money. (To make an example meaningful we should certainly assume that the units are substantially larger than \$1.) Assume that, when facing a possible loss of the whole wealth with a probability of 0.1, Chris is willing to pay a premium of at most 25 to insure the loss. In view of (3.2.2), it means that $u(75) = \dfrac{1}{10}u(0) + \dfrac{9}{10}u(100) = 9/10$. □

Unfortunately, in real life it works not so well as in nice theoretical examples. The problem is not in mathematical modeling but in making results of such an inquiry reliable, reflecting the real preferences of the individual. This is a psychological rather than mathematical question. The difficulty is that answers depend on the situation, on the form in which the questions are asked, whether the questioning involves real money or the experiment is virtual, and on many other psychological and social issues. These problems are beyond the scope of this book on mathematical modeling. For a corresponding discussion see, e.g., [56], [72], [85], [144], [147], and references therein.

3.4 Risk aversion

3.4.1 A definition

Below, by the symbol Z_ε we will denote a r.v.

$$Z_\varepsilon = \begin{cases} \varepsilon & \text{with probability } 1/2, \\ -\varepsilon & \text{with probability } 1/2, \end{cases}$$

where $\varepsilon > 0$. We will talk about the *risk aversion* of an individual with a preference order $\succsim$ if the following condition holds.

Condition Z: For any r.v. X, any $\varepsilon > 0$, and any r.v. Z_ε independent of X, it is true that $X \succsim X + Z_\varepsilon$.

Condition Z reflects the rule "the less stable, the worse". An investor with preferences satisfying this property would not accept an offer resulting in either an additional income with probability $1/2$ or a loss of the same amount and with the same probability.

It is important to emphasize that *Condition Z concerns an arbitrary preference order*, not only the EUM criterion.

An individual whose preference order satisfies Condition Z is called a *risk averter*. If $X \precsim X + Z_\varepsilon$ for any X, any $\varepsilon > 0$, and any Z_ε independent of X, then we call such an individual a *risk lover* or *risk taker*.

The fact that we consider in Condition Z a non-strict relation $\succsim$ is not essential. We do it to avoid below some superfluous constructions. Formally, the above definition does not exclude the case when an individual is simultaneously a risk averter and a risk lover, that is, $X \simeq X + Z_\varepsilon$ for all X and ε. In this case, we say that the individual is *risk neutral*.

Certainly, a person may be neither a risk averter nor a risk lover. For example, it may happen that for some particular X and ε, it is true that $X \succsim X + Z_\varepsilon$, and for another r.v., say, X^*, it may turn out that $X^* \precsim X^* + Z_\varepsilon$.

Next, we consider the EUM criterion and figure out when *this particular criterion* satisfies Condition Z.

Proposition 3 *Let $\succsim$ be a EUM order defined in (3.1.2). Then Condition Z holds iff $u(x)$ is concave.*

We will prove this proposition at the end of this section; now we turn to examples and comments.

Usually we deal with smooth utility functions, so to check whether an EU maximizer with a utility function u is a risk averter, it suffices to check the second derivative u''.

For example, for $u = x^\alpha$, we have $u''(x) = \alpha(\alpha - 1)x^{\alpha-2}$. Thus, $u''(x) < 0$ for $\alpha < 1$, which corresponds to the risk aversion case, while for $\alpha > 1$ we deal with a risk lover. The case $\alpha = 1$ when $E\{u(X)\} = E\{X\}$ may be assigned to both types: the person is *risk neutral*. Other utility functions are considered in Exercise 26.

Whether a person is a risk averter or a risk lover (or neither) depends, of course, not only on her/his personality but on the particular situation. You may be a risk averter in routine life but if you have decided to spend some time in a casino, you are definitely a risk lover.

There is also strong evidence based on experiments that many people incline to behave as risk averters when concerned with future gains (positive values of X), and as risk lovers when facing losses.

For example, a person may choose \$500 for sure rather than \$1,000 with probability $1/2$. However, the same person may prefer to take a risk of losing \$1,000 with probability $1/2$ rather than to lose (only) \$500 for sure. A utility function in this case may look as in Fig.6.

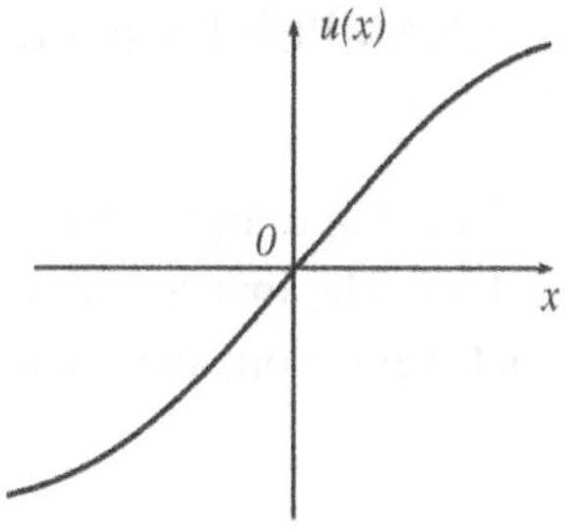

FIGURE 6.

Certainly, the utility function may be more complicated or—better to say—more sophisticated. *For example*, in the region of moderate x's the function may be concave and in the region of large income values—convex.

The following inequality clarifies why the concavity of utility functions is relevant to risk aversion.

3.4.2 Jensen's inequality

We assume all expectations below to be finite.

Proposition 4 *Let a r.v. X take on values from a finite or infinite interval I, and let a function $u(x)$ be concave on I. Then*

$$E\{u(X)\} \le u(E\{X\}). \tag{3.4.1}$$

If u is convex (concave upward), then

$$E\{u(X)\} \ge u(E\{X\}). \tag{3.4.2}$$

The proof is relegated to Section 3.4.4.

Being purely mathematical assertions, inequalities (3.4.1)-(3.4.2) are relevant to the basic question of insurance: why is it possible?

Assume that a client of an insurance organization is a EU maximizer and consider relation (3.2.2). If the client is a risk averter (which is natural to assume since the client is willing to pay to insure a risk), then $u(x)$ is concave, and by Jensen's inequality,

$$u(w - G_{\max}) = E\{u(w-\xi)\} \le u(E\{w-\xi\}) = u(w - E\{\xi\}).$$

Since u is non-decreasing, it implies that $w - G_{\max} \le w - E\{\xi\}$, or

$$G_{\max} \ge E\{\xi\}.$$

Thus, the maximum premium the client agrees to pay is larger than (or, for the boundary case, equals) the average coverage of the risk, $E\{\xi\}$.

So, *the company will get on the average more than it will pay,* which means that the company can function.

To the contrary, if the client had been a risk lover, from Jensen's inequality it would have followed that $G_{\max} \le E\{\xi\}$, and insurance would have been impossible.

EXAMPLE 1. Consider Example 3.2-1. We computed $G_{\max} = \frac{1}{\sqrt{3}} \approx 0.57$, while $E\{\xi\} = 0.5$. □

The same argument may be applied to the certainty equivalent of a r.v. X (see Section 3.1.1). The inverse of an increasing function is increasing. From this and (3.4.1) it follows that if u is concave, then $c(X) = u^{-1}(E\{u(X)\}) \le u^{-1}(u(E\{X\})) = E\{X\}$. Thus,

> In the case of risk aversion, $c(X) \le E\{X\}$.

For the risk lover a similar argument leads to $c(X) \geq E\{X\}$.

EXAMPLE 2. Let X be exponential with parameter a, and $u(x) = -e^{-\beta x}$, $\beta > 0$. The function is concave, and the person with such a utility function is a risk averter. Continuing the calculations from Example 3.1.3-3, we have

$$c(X) = \frac{1}{\beta}[\ln(a+\beta) - \ln a)] = -\frac{1}{\beta}(\ln a)\left[1 - \frac{\ln(a+\beta)}{\ln a}\right] = \frac{1}{\beta}\left(\ln\frac{1}{a}\right)\left[1 - \frac{\ln(a+\beta)}{\ln a}\right]. \tag{3.4.3}$$

Let X be "large"; formally let $a \to 0$. Then $E\{X\} = \frac{1}{a} \to \infty$. Since the third factor in (3.4.3) converges to one, $c(X) \sim \frac{1}{\beta}\ln\left(\frac{1}{a}\right)$. (The relation $u \sim v$ means $\frac{u}{v} \to 1$.) Thus, in our case

$$c(X) \sim \frac{1}{\beta}\ln(E\{X\}).$$

Since $\ln x$ is much smaller than x for large x's, the certainty equivalent is much smaller than the mean value $E\{X\}$. We interpret this as saying that the individual is a strong risk averter.
□

Route 1 ⇒ page 129

2,3

3.4.3 How to measure risk aversion in the EUM case

We use below the Calculus notation $o(\varepsilon)$ for a function $o(\varepsilon)$ such that $\frac{o(\varepsilon)}{\varepsilon} \to 0$ as $\varepsilon \to 0$. The reader who is not familiar with this very convenient and simple notation is recommended to look at the detailed explanation in Appendix, Section 4.1.

Let $x > 0$ be a fixed capital. Its certainty equivalent is the same number x. Let us consider the r.v. $x + Z_\varepsilon$ and compute its certainty equivalent for small ε.

Lemma 5 *Suppose the second derivative u'' exists and is continuous, and $u'(x) > 0$ for x chosen above. Then the certainty equivalent*

$$c(x + Z_\varepsilon) = x - \frac{1}{2}R(x)\varepsilon^2 + o(\varepsilon^2), \tag{3.4.4}$$

where

$$R(x) = -\frac{u''(x)}{u'(x)}. \tag{3.4.5}$$

The proof is given in Section 3.4.4; now we will discuss the significance of (3.4.4).

By definition of $o(\varepsilon^2)$, we view the third term in (3.4.4) as negligible with respect to the second.

The function $R(x)$ may be considered a characteristic of the concavity of u. In particular, if u is concave (risk aversion!), then $R(x) \geq 0$.

If u is concave, by Proposition 4, the r.v.

$$x + Z_\varepsilon \precsim x,$$

and for the corresponding certainty equivalents we have

$$c(x+Z_\varepsilon) \le x.$$

The difference $x - c(x+Z_\varepsilon)$ may be viewed as a "price for risk", a "measure of riskiness". By Lemma 5, this measure is proportional to $R(x)$ up to the negligible remainder $o(\varepsilon^2)$. The characteristic $R(x)$ is called an *absolute risk aversion function,* or the *Arrow-Pratt index of risk aversion.*

For a r.v. X, the expectation $E\{R(X)\}$ may be called an *expected absolute risk aversion.*

If x is measured in dollars, the dimension of $R(x)$ is dollar^{-1}. We define the *relative risk aversion function* as $R_r(x) = |x|R(x)$. This function does not have dimension. We call $E\{R_r(X)\}$ an *expected relative risk aversion.*

EXAMPLE 1. Let $u(x) = -e^{-\beta x}$. Then, as is easy to calculate by substituting into (3.4.5), the absolute risk aversion function $R(x) = \beta$ and does not depend on x. In light of this, formulas (3.2.7) from Example 3.2-3 look very nice and understandable: the larger the risk aversion characteristics β and β_1, the more the premiums $G_{\max}$ and $H_{\min}$ exceed the expected value m. The differences $G_{\max} - m$ and $H_{\min} - m$ are proportional to β and β_1, respectively. The insurance is possible if the insurer is not more risk averse than the insured ($\beta_1 \le \beta$). This reflects reality. The insurance company can afford to be less risk averse. It deals with many clients, the number of those is usually essentially larger than the number of claims, and payments are compensated by premiums. The corresponding rigorous model will be considered in Section **2**.3.

EXAMPLE 2. Let $u(x) = x^\alpha$, $x \ge 0$, $\alpha > 0$. Then $R(x) = (1-\alpha)/x$ (compute on your own), and the relative aversion $R_r(x) = 1-\alpha$. It is non-negative iff u is concave ($\alpha < 1$).

For $u(x) = -x^{-\alpha}$, $x > 0$, $\alpha > 0$, we have $R(x) = (1+\alpha)/x$, and the relative aversion $R_r(x) = 1+\alpha$ and is positive for all $\alpha > 0$ (which is consistent with the fact that u is concave for all α).

In Examples 3.1.3-1 and 2, we considered two types of a particular r.v. X. In the first example, X took on two values, $b{>}0$ and 0, with equal probabilities; in the second, X was uniform on $[0,b]$. The certainty equivalents proved to be $c(X) = 2^{-1/\alpha}b$ and $c(X) = (1+\alpha)^{-1/\alpha}b$, respectively. In both cases, the less the risk aversion α, the less the certainty equivalent.

EXAMPLE 3. For $u(x) = \ln(x)$, $x > 0$, we have $R(x) = 1/x$, and $R_r(x) = 1$. □

In Exercise 37, the reader is invited to prove that the cases considered above exhaust all cases with constant risk aversion functions.

EXAMPLE 4. Let $u(x) = x/(1+|x|)$. The graph is given in Fig.7a. We see that the person with such a utility function is a risk averter in the region of gains and a risk lover in the case of losses (negative x's). We observe also the saturation effect in both sides for $x \to \pm\infty$. So, we expect that at $\pm\infty$ the absolute aversion should vanish. It is true since, as is easy to compute, $R(x) = 2/(1+|x|)$ for $x > 0$, and $R(x) = -2/(1+|x|)$ for $x < 0$. See Fig.7b.

Since $u''(0)$ does not exist, $R(0)$ is not defined. However, we can write that $R_r(x) = |x|R(x) = 2x/(1+|x|)$ for $x \ne 0$. Then, by continuity, we may set $R_r(0) = 0$. □

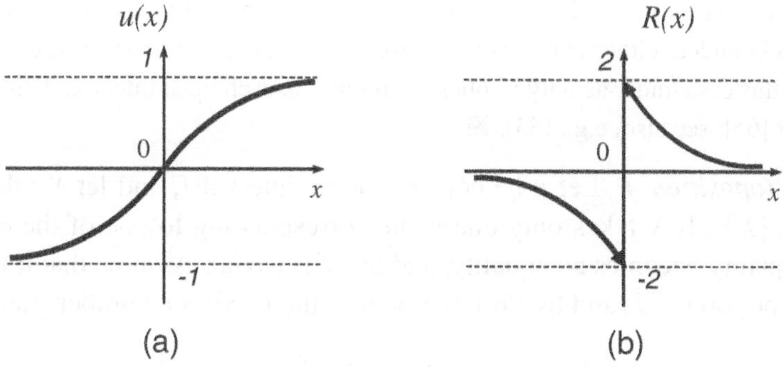

FIGURE 7.

3.4.4 Proofs

To make constructions below simpler, we assume additionally that functions $u(x)$ are continuous. Since we deal with concave functions which are continuous on any open interval, it is a very minor assumption.

Proof of Proposition 3. Sufficiency. Note that if $u(x)$ is concave, then setting $\lambda = \frac{1}{2}$ in the definition (**0**.4.3.1), we get that for any x_1, x_2 from the domain of $u(x)$,

$$\frac{u(x_1)+u(x_2)}{2} \leq u\left(\frac{x_1+x_2}{2}\right). \tag{3.4.6}$$

Now, let $F_X(x)$ and $F_{Z_\varepsilon}(x)$ be the d.f. of X and Z_ε, respectively. Since X and Z_ε are independent,

$$E\{u(X+Z_\varepsilon)\} = \int_{-\infty}^{\infty}\int_{-\infty}^{\infty} u(x+z)dF_X(x)dF_{Z_\varepsilon}(z) = \int_{-\infty}^{\infty}\left(\int_{-\infty}^{\infty} u(x+z)dF_{Z_\varepsilon}(z)\right)dF_X(x)$$
$$= \int_{-\infty}^{\infty}\left(u(x+\varepsilon)\frac{1}{2}+u(x-\varepsilon)\frac{1}{2}\right)dF_X(x) = E\left\{\frac{1}{2}u(X+\varepsilon)+\frac{1}{2}u(X-\varepsilon)\right\}.$$

By virtue of (3.4.6), the r.v. in $\{.\}$ above is not greater than $u(X)$. Hence,

$$E\{u(X+Z_\varepsilon)\} \leq E\{u(X)\},$$

and, consequently, $X+Z_\varepsilon \preceq X$.

(In the proof above we avoided conditional expectations (Section **0**.7). Otherwise, we could just condition on X, writing $E\{u(X+Z_\varepsilon)\} = E\{E\{u(X+Z_\varepsilon)\,|\,X\}\} = E\{\frac{1}{2}u(X+\varepsilon) + \frac{1}{2}u(X-\varepsilon)\}$.)

Necessity. Let x_1, x_2 be two numbers such that $x_1 < x_2$. Set $x_0 = \dfrac{x_1+x_2}{2}$, $\varepsilon = \dfrac{x_2-x_1}{2}$, and $X \equiv x_0$. By definition of risk aversion, $E\{u(X+Z_\varepsilon)\} \leq E\{u(X)\}$. For the particular X above, it means that $\frac{1}{2}u(x_0+\varepsilon)+\frac{1}{2}u(x_0-\varepsilon) \leq u(x_0)$. By the choice of x_0 and ε, this implies (3.4.6). The last property is called midconcavity, and formally it does not imply concavity, that is, (**0**.4.3.1) for all $\lambda \in [0,1]$. However, it does if $u(x)$ is continuous (see, e.g., [55]), which we have assumed. This proves necessity.

As a matter of fact we do not have to assume continuity. The utility function u is non-decreasing and hence it is bounded, either from above or from below, on any interval where it is defined. It is known that in this case midconcavity implies continuity on each open interval. This fact is traced to Jensen's paper [65]; see also, e.g., [55]. ■

Proof of Proposition 4. Let u be concave on an interval I, and let X take values from I. Set $m = E\{X\}$. If X takes only one value corresponding to one of the endpoints of I, Jensen's inequality becomes an equality, and proof is trivial. Assume that it is not so. Then m is an interior point of I, and by Proposition **0**-1, there exists a number c such that

$$u(x) - u(m) \leq c(x-m) \quad \text{for any } x \in I. \tag{3.4.7}$$

(We can include the endpoints since u is assumed to be continuous. Note that as a matter of fact this assumption is not necessary, and we made it for simplicity.)

Setting $x = X$, we write $u(X) - u(m) \leq c(X-m)$. Computing the expectations of both sides, we have $E\{u(X)\} - u(m) \leq c(m-m) = 0$, which amounts to (3.4.1).

To prove (3.4.2), it suffices to consider the function $-u(x)$ which is concave if $u(x)$ is convex. ■

Proof of Lemma 5. By Taylor's expansion, $u(x+Z_\varepsilon) = u(x) + u'(x)Z_\varepsilon + \frac{1}{2}u''(x)Z_\varepsilon^2 + o(Z_\varepsilon^2)$. Note also that $E\{Z_\varepsilon\} = 0$, $E\{Z_\varepsilon^2\} = \varepsilon^2$. Hence, $E\{u(x+Z_\varepsilon)\} = u(x) + \frac{1}{2}u''(x)\varepsilon^2 + o(\varepsilon^2)$.

The function u' is continuous, and $u'(x) > 0$. Hence, $u'(y) > 0$ for y's from a neighborhood of x. Then we can consider $u^{-1}(y)$ and apply the Taylor expansion. Since, $(u^{-1}(y))' = 1/u'(u^{-1}(y))$, we have $u^{-1}(y+s) = u^{-1}(y) + (u^{-1}(y))'s + o(s) = u^{-1}(y) + \dfrac{1}{u'(u^{-1}(y))}s + o(s)$ for small s. Eventually, for small ε, the certainty equivalent

$$\begin{aligned} c(x+Z_\varepsilon) &= u^{-1}(E\{u(x+Z_\varepsilon)\}) = u^{-1}\left(u(x) + \frac{1}{2}u''(x)\varepsilon^2 + o(\varepsilon^2)\right) \\ &= u^{-1}(u(x)) + \frac{1}{u'(u^{-1}(u(x)))}\left(\frac{1}{2}u''(x)\varepsilon^2 + o(\varepsilon^2)\right) + o\left(\frac{1}{2}u''(x)\varepsilon^2 + o(\varepsilon^2)\right) \\ &= x + \frac{1}{2}\frac{u''(x)}{u'(x))}\varepsilon^2 + \frac{1}{u'(x)}o(\varepsilon^2) + o\left(\frac{1}{2}u''(x)\varepsilon^2 + o(\varepsilon^2)\right). \end{aligned}$$

The last two terms are $o(\varepsilon^2)$. ■

3.5 A new perspective: EUM as a linear criterion

3.5.1 Preferences on distributions

Let us recall the original Probability Theory framework with a sample space $\Omega = \{\omega\}$ and a probability measure $P(A)$ on it. Random variables $X = X(\omega)$ are functions on Ω.

Let $F_X(B) = P(X \in B)$, where B is a set from the real line. As in Section **0**.1.3.1, we call the function of sets $F_X(B)$ the distribution of X. In particular, the d.f. $F_X(x) = F((-\infty, x])$. Certainly, if we know $F_X(B)$ for all B, we know $F_X(x)$ for all x. The converse assertion is also true: knowing the d.f. $F(x)$, we can determine $F_X(B)$ for any set B. (See (**0**.2.1.6) and explanations on this point in Sections **0**.1.3.1, **0**.1.3.3.)

We fix a class $\mathcal{X} = \{X\}$ of r.v.'s X and consider a preference order $\succsim$ on $\mathcal{X}$.

Suppose that, when comparing r.v.'s, we take into account not the structure of r.v.'s (that is, how they depend on ω) but only the information about the possible values of r.v.'s and the corresponding probabilities. In other words, we proceed only from distributions F_X and compare not r.v.'s themselves but rather their distributions. For example, this is the case in the EUM framework. Indeed, $E\{u(X)\}$ is completely determined by the distribution F_X, and therefore, when comparing r.v.'s, as a matter of fact, we compare the corresponding distributions.

In the situation we have described, instead of the preference order on the class $\mathcal{X} = \{X\}$, it suffices to consider a preference order $\succsim$ on the set $\mathcal{F} = \{F_X\}$ of the distributions of the r.v.'s X from $\mathcal{X}$. This order is determined by the rule

$$F_X \succsim F_Y \iff X \succsim Y. \tag{3.5.1}$$

Because any distribution and its d.f. completely determine each other, it does not matter where we define the preference rule $\succsim$: among distributions or distribution functions. Below, when it does not cause misunderstanding, we use the symbol F for both. The reader can even understand the word "distribution" as "distribution function".

Conversely, if we agreed to compare only distributions, and if we defined somehow a preference order on a class $\mathcal{F} = \{F\}$ of distributions F, then we have defined, by the same relation (3.5.1), the preference order among all r.v.'s having the distributions from $\mathcal{F}$.

3.5.2 The first stochastic dominance

Next, we specify in terms of distributions the rule "the larger, the better". In Section 1.1, we introduced the natural monotonicity property: if $P(X \geq Y) = 1$, then $X \succsim Y$. However, if we proceed only from distributions, such a rule does not cover all situations when X is "obviously better" than Y.

EXAMPLE 1. Let Ω consist of two outcomes, ω_1 and ω_2, and $P(\omega_1)=1/3$, $P(\omega_2)=2/3$. Let

$$X(\omega_1) = 10, \quad X(\omega_2) = 20,$$
$$Y(\omega_1) = 20, \quad Y(\omega_2) = 10.$$

For example, we may view X and Y as the prices for two stocks and ω_1, ω_2 as two possible states of the future financial market. With a positive probability of $1/3$ the value of X will be less than the value of Y. Nevertheless, because $P(\omega_1) < P(\omega_2)$, if for us it is not important which ω will occur, but merely what income we will have, we will certainly prefer X to Y. □

The purpose of the next definition is to take such cases into account.

We say that the distribution F dominates the distribution G in the sense of the *first stochastic dominance* (FSD) if

$$F(x) \leq G(x) \quad \text{for } any \ x. \tag{3.5.2}$$

(That is, whatever x is, for the distribution F, the probability of having an income *not larger* than x, is *not larger* than that for the distribution G.)

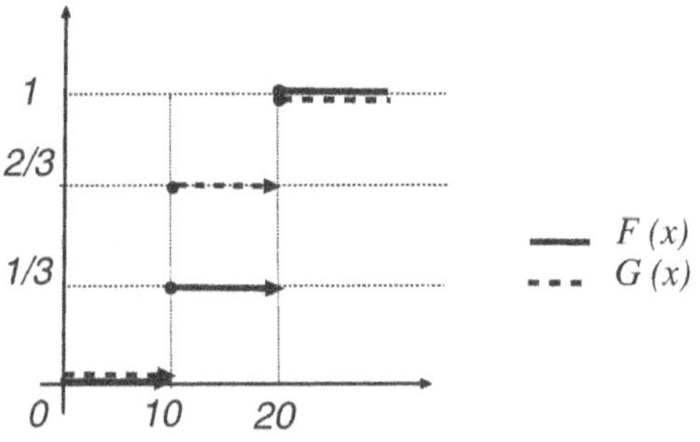

FIGURE 8.

Certainly, if $P(X \geq Y) = 1$, then $F_X(x) \leq F_Y(x)$ for *any* x. Indeed, $F_X(x) = P(X \leq x) \leq P(Y \leq x) = F_Y(x)$. The converse assertion is not true.

EXAMPLE 2. Let us revisit Example 1. Let $F(x)$ and $G(x)$ be the d.f.'s of X and Y, respectively. Their graphs are given in Fig.8. We see that (3.5.2) is true, though $P(X < Y) = \frac{1}{3} > 0$. $\square$

In the case of the comparison of distributions, the rule "the larger, the better" is reflected in the following definition.

> A preference order $\succsim$ on a set $\mathcal{F} = \{F\}$ is said to be *monotone with respect to the first stochastic dominance* (FSD), if $F \succsim G$ for any pair of distributions $F, G \in \mathcal{F}$ with property (3.5.2).

For brevity, in this case we will also say that $\succsim$ *satisfies the FSD rule*.

A preference order $\succsim$ is said to be *strictly monotone* with respect to the FSD, if $F \succ G$ (i.e., F is better than G), once (3.5.2) is true and at least for one x the inequality in (3.5.2) is strict.

EXAMPLE 3. Consider the situation of Example 2. If a preference order $\succsim$ satisfies the FSD rule, then for the distributions F_X and F_Y from this example, we have $F_X \succsim G_Y$. Moreover, if $\succsim$ is strictly monotone, then $F \succ G$.

EXAMPLE 4. Ann's future income X has the exponential distribution with a mean of $m > 0$, and Paul's income Y is uniformly distributed on $[0,1]$. For which m is Ann's position better in the sense of the FSD rule?

Let $F(x)$ and $G(x)$ be the respective d.f.'s. We should figure out when (3.5.2) is true. We have $F(x) = 1 - e^{-x/m}$ for $x \geq 0$, and $G(x) = x$ for $x \in [0,1]$. Furthermore, $F(0) = G(0) = 0$, $F(1) < 1$, and $G(1) = 1$; see Fig.9.

Since $F(x)$ is concave, it may coincide with $G(x)$ at no more than one point besides the origin. The derivative $F'(x) = e^{-x/m}/m$, and $G'(x) = 1$.

Hence, if $m \geq 1$, then $F'(0) \leq G'(0)$, and $F'(x) < F'(0) \leq 1$. From this it follows that $F(x) < G(x)$ for all $x > 0$; see Fig.9ab.

If $m < 1$, then $F'(0) > 1$, and $F(x) > G(x)$ for x's from some interval; see Fig.9c.

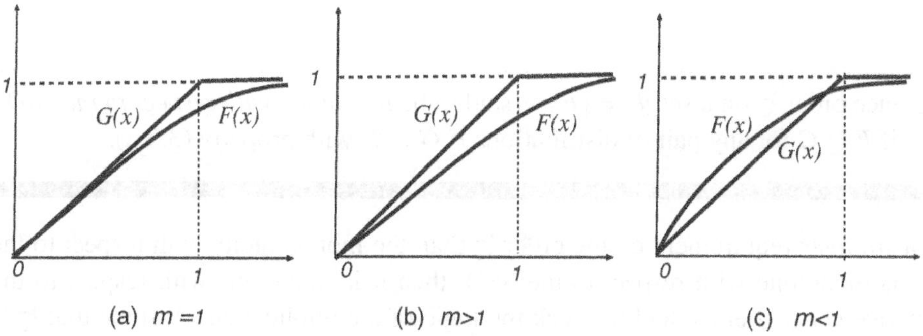

FIGURE 9.

Thus, if we follow the natural FSD rule, for $m \geq 1$ we prefer X to Y. For $m < 1$, the comparison is not so obvious, and to make a decision, we should determine the preference order in more detail. □

3.5.3 The second stochastic dominance

This section concerns a property reflecting the rule "the riskier, the worse". We say that F dominates G in the sense of the *second stochastic dominance* (SSD) if

$$\int_{-\infty}^{t} (F(x) - G(x))dx \leq 0 \text{ for any } t, \tag{3.5.3}$$

provided that the integral above is finite for any t. Clearly, if F dominates G in the sense of the FSD, that is, (3.5.2) is true, then (3.5.3) is also true. (Since in this case, the integrand in (3.5.3) is non-positive.)

EXAMPLE 1. To clarify (3.5.3), let us recall the definition of risk aversion from Section 3.4.1. Let a r.v. X have a distribution F, and let G_ε be the distribution of the r.v. $X_\varepsilon = X + Z_\varepsilon$, as it was defined in Section 3.4.1. Then

$$\begin{aligned} G_\varepsilon(x) &= P(X_\varepsilon \leq x) = P(X_\varepsilon \leq x \,|\, Z_\varepsilon = \varepsilon)\frac{1}{2} + P(X_\varepsilon \leq x \,|\, Z_\varepsilon = -\varepsilon)\frac{1}{2} \\ &= \frac{1}{2}(P(X + \varepsilon \leq x) + P(X - \varepsilon \leq x)) = \frac{1}{2}(F(x-\varepsilon) + F(x+\varepsilon)). \end{aligned} \tag{3.5.4}$$

The distribution F dominates G_ε in the sense of the SSD. Indeed, set $Q(t) = \int_{-\infty}^{t} F(x)dx$. Note that $\int_{-\infty}^{t} F(x+\varepsilon)dx = \int_{-\infty}^{t+\varepsilon} F(x)dx = Q(t+\varepsilon)$, and similarly $\int_{-\infty}^{t} F(x-\varepsilon)dx = Q(t-\varepsilon)$. Then, in view of (3.5.4),

$$\int_{-\infty}^{t} (F(x) - G_\varepsilon(x))dx = Q(t) - \frac{1}{2}(Q(t-\varepsilon) + Q(t+\varepsilon)).$$

To show that the last expression is non-positive for any t, it remains to prove that $Q(t)$ is a convex function; see Appendix, Section 4.3. Assume, for simplicity, that $F(x)$ has the density $f(x) = F'(x)$. Then $Q'(t) = F(t)$, $Q''(t) = f(t) \geq 0$, since f is a density. As a matter of fact, the smoothness of $f(x)$ is not necessary: it suffices to take into account that $F(x)$ is not decreasing. □

A preference order $\succsim$ on a set $\mathcal{F} = \{F\}$ is said to be *monotone with respect to the SSD*, if $F \succsim G$ for any pair of distributions $F, G \in \mathcal{F}$ with property (3.5.3).

This is a *stronger* requirement on the order $\succsim$ than the monotonicity with respect to the FSD: if $\succsim$ is monotone with respect to the SSD, then it is monotone with respect to the FSD. (Not vice versa! Let us double check the logic of the implication. Assume that $\succsim$ is monotone with respect to the SSD. Let F dominate G in the sense of the FSD. Then (3.5.3) is true. Then $F \succsim G$. Hence, $\succsim$ is monotone with respect to the FSD also.)

We say that an individual is a risk averter in the sense of the SSD if her/his preference order $\succsim$ is monotone with respect to the SSD.

From Example 1 it follows that if somebody is a risk averter in the sense of the SSD, then she/he is a risk averter in the sense of Section 3.4.1.

3.5.4 The EUM criterion

We return to the EUM criterion. For a r.v. X, the expected value $E\{u(X)\}$ may be written as

$$E\{u(X)\} = \int_{-\infty}^{\infty} u(x)dF(x), \tag{3.5.5}$$

where $F(x)$ is the *distribution function* of X.

In more detail, this formula is discussed in Section **0**.1.3. In particular, if X has a probability density function $f(x)$, then $dF(x) = f(x)dx$, and the integral above may be understood in the "usual" way as $\int_{-\infty}^{\infty} u(x)f(x)dx$.

If X is discrete and assumes values $x_1, x_2, \ldots$ with probabilities $f_1, f_2, \ldots$, respectively, then we set $dF(x_j) = f_j$ for all j, and $dF(x) = 0$ for all other x's. This will lead to $\int_{-\infty}^{\infty} u(x)dF(x) = \sum_j u(x_j)f_j$, that is, to the definition of expected value in the discrete case.

As was already mentioned, since $E\{u(X)\}$ is completely determined by the d.f. F of X, we may restrict ourselves to the corresponding order $\succsim$ on a set $\mathcal{F} = \{F\}$ of distributions.

Let us *fix* a utility function u and set

$$U(F) = \int_{-\infty}^{\infty} u(x)dF(x), \tag{3.5.6}$$

assuming that the last integral is finite for all $F \in \mathcal{F}$.

We will call $U(F)$ a *utility functional*. The word *functional* is used in Mathematics when the argument in a function (as in $U(F)$) is not a number but an object of a more general nature (as a distribution F) but the values of the function are real numbers. It is convenient to use this common mathematical term here in order to distinguish the utility *functional* $U(F)$ from the utility *function* $u(x)$.

The EUM preference order $\succsim$ in $\mathcal{F}$ may be defined as

$$F \succsim G \Leftrightarrow U(F) \geq U(G),$$

that is, $\succsim$ is *preserved* by U.

Certainly, the criterion we have defined is the same criterion as above, just presented in terms of distributions.

Consider the difference $U(F)-U(G)$. Integrating by parts, we have

$$U(F)-U(G)=\int_{-\infty}^{\infty}u(x)d(F(x)-G(x))=\int_{-\infty}^{\infty}(G(x)-F(x))du(x). \tag{3.5.7}$$

(The differences $F(\infty)-G(\infty)=1-1=0$, $F(-\infty)-G(-\infty)=0-0=0$; the limits $u(x)(F(x)-G(x))$ at $\pm\infty$ equal zero; see, e.g., (**0**.2.5.5) and the argumentation there.)

Let F dominate G in the sense of the FSD, and the utility function $u(x)$ be non-decreasing. Then $G(x)-F(x)\geq 0$, $du(x)\geq 0$, and (3.5.7) implies that $U(F)\geq U(G)$. Thus, we can state the following:

> If $u(x)$ is non-decreasing, then $U(F)$ is monotone with respect to the FSD.

The above condition is also necessary. Indeed, let $x_1 > x_2$, a r.v. X_1 assume just one value x_1, and a r.v. X_2 assume only one value x_2. Let F and G be the d.f.'s of X_1 and X_2, respectively. Then F dominates G in the sense of the FSD (show why). If $U(F)$ is monotone with respect to the FSD, than $U(F)\geq U(G)$. On the other hand, in the EUM case, $U(F)=E\{u(X_1)\}=u(x_1)$, and $U(G)=u(x_2)$. So, $u(x_1)\geq u(x_2)$.

► Consider now risk aversion. Suppose that F dominates G in the sense of the SSD and the utility function $u(x)$ is concave. For simplicity, assume that u is sufficiently smooth. Integrating (3.5.7) by parts one time more, we have

$$\begin{aligned}U(F)-U(G)&=\int_{-\infty}^{\infty}(G(x)-F(x))u'(x)dx=\int_{-\infty}^{\infty}u'(x)d\left(\int_{-\infty}^{x}(G(s)-F(s))ds\right)\\&=\lim_{A\to\infty}u'(A)\left(\int_{-\infty}^{A}(G(s)-F(s))ds\right)+\int_{-\infty}^{\infty}\left(\int_{-\infty}^{x}(F(s)-G(s))\right)u''(x)dx\\&\geq\int_{-\infty}^{\infty}\left(\int_{-\infty}^{x}(F(s)-G(s))ds\right)u''(x)dx.\end{aligned} \tag{3.5.8}$$

(We took into account that $u'(A)\geq 0$ and $\int_{\infty}^{A}$ above are non-negative.) Furthermore, because $\int_{-\infty}^{x}(F(s)-G(s))ds\leq 0$ and $u''(x)\leq 0$, we have $U(F)\geq U(G)$. Thus,

> If $u(x)$ is concave and non-decreasing, then $U(F)$ is monotone with respect to the SSD.

It may be proved that in the case of EUM, the definitions of risk aversion in the sense of Condition Z from Section 3.4.1 and in the sense of the SSD are equivalent. In view of Proposition 3, from this it follows, in particular, that the above concavity condition is also necessary. ◄

3.5.5 Linearity of the utility functional

Let F_1 and F_2 be two distributions, and let a number $\alpha \in [0,1]$. We call the distribution $F^{(\alpha)}$ a *mixture of distributions* F_1, F_2 if

$$F^{(\alpha)} = \alpha F_1 + (1-\alpha)F_2, \tag{3.5.9}$$

that is, the d.f. $F^{(\alpha)}(x) = \alpha F_1(x) + (1-\alpha)F_2(x)$, and in general, $F^{(\alpha)}(B) = \alpha F_1(B) + (1-\alpha)F_2(B)$.

The reader is recommended to look up this notion and comments on it in Section **0**.1.3.5. In particular, we mentioned there that the r.v. X having the distribution $F^{(\alpha)}$ may be represented as

$$X = \begin{cases} X_1 & \text{with probability } \alpha, \\ X_2 & \text{with probability } 1-\alpha, \end{cases}$$

where X_1, X_2 have the distributions F_1, F_2, respectively. For a particular example, see also Exercise 41.

Now, let us consider the concept of mixture from a geometric point of view.

EXAMPLE 1. Consider r.v.'s taking only three fixed values, x_1,x_2, and x_3, such that $x_1 < x_2 < x_3$. Any distribution F of such a r.v. may be identified with the probability vector (p_1, p_2, p_3), where p_i is the probability of the value x_i. Since $p_1 + p_2 + p_3 = 1$, to specify the distribution, it is enough to know two probabilities from these three. For us, it will be convenient to choose (p_1, p_3). All points (p_1, p_3) lie in the triangular region $\Delta = \{(p_1, p_3) : p_1, p_3 \geq 0,\, p_1 + p_3 \leq 1\}$ depicted in Fig.10a.

Any distribution F (from those we are considering) may be identified with a point in the triangle Δ. The origin corresponds to the probability vector $(0,1,0)$, the "North-East" border to vectors $(p_1, 0, p_3)$, etc. If F is the distribution of income, the "best" point is certainly the point $(0,1)$.

Consider two points in Δ, say, $\mathbf{p}^1$ and $\mathbf{p}^2$, corresponding to some distributions F_1 and F_2. Then the mixture $\alpha F_1 + (1-\alpha)F_2$ will correspond to the point $\mathbf{p}^{(\alpha)} = \alpha\mathbf{p}^1 + (1-\alpha)\mathbf{p}^2$ lying in the segment connecting $\mathbf{p}^1$ and $\mathbf{p}^2$, and such that the distances from this point to points $\mathbf{p}^1$ and $\mathbf{p}^2$ are in the proportion $\alpha/(1-\alpha)$. In particular, $\mathbf{p}^{(1/2)}$ lies in the middle of this segment. See also Fig.10a.

Certainly, this scheme may be generalized to the case of n values $x_1, ..., x_n$ such that $x_1 < ... < x_n$, whose probabilities are $p_1, ..., p_n$, respectively. Since $p_1 + ... + p_n = 1$, it suffices to consider $n-1$ probabilities, say, $p_2, ..., p_n$, and the counterpart of Δ is the set $\Delta^{(n)} = \{(p_2, ..., p_n) : p_i \geq 0,\, p_2 + ... + p_n \leq 1\}$. For $n = 4$, it is a three-dimensional tetrahedral; see Fig.10b. For $n > 4$ it is a multidimensional prism. Mixtures of two distributions lie in a line in $\Delta^{(n)}$. □

For brevity, we call this illustrative scheme the Δ-*scheme*; we will repeatedly return to it.

In the general framework, a mixture admits the same interpretation. If we view distributions F as points in a space $\mathcal{F}$, then for fixed points F_1 and F_2, the collection of points $F^{(\alpha)} = \alpha F_1 + (1-\alpha)F_2$ for $\alpha \in [0,1]$ may be viewed as the segment connecting F_1 and F_2. When α runs from 0 to 1, the distribution F_α varies from F_2 to F_1.

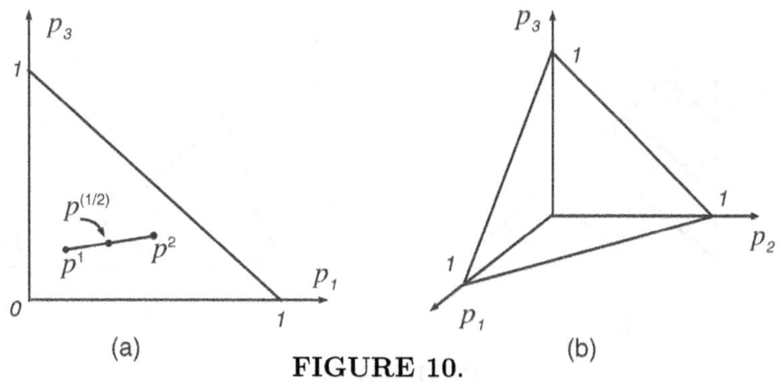

FIGURE 10.

The main property of the EUM criterion:

$$U(\alpha F_1 + (1-\alpha)F_2) = \alpha U(F_1) + (1-\alpha)U(F_2). \tag{3.5.10}$$

That is, *the utility of a mixture is equal to the mixture of the utilities,* or, in other terms, U is a *linear functional.* (More precisely, functionals for which (3.5.10) holds only for $\alpha \in [0,1]$ are called affine; we keep the more explicit term 'linear'.)

To prove (3.5.10), it suffices to replace F in (3.5.6) by the mixture (3.5.9) and to write

$$U\left(F^{(\alpha)}\right) = \int_{-\infty}^{\infty} u(x)dF^{(\alpha)}(x) = \int_{-\infty}^{\infty} u(x)d(\alpha F_1(x) + (1-\alpha)F_2(x))$$
$$= \alpha \int_{-\infty}^{\infty} u(x)dF_1(x) + (1-\alpha)\int_{-\infty}^{\infty} u(x)dF_2(x) = \alpha U(F_1) + (1-\alpha)U(F_2).$$

EXAMPLE 2. Consider again the Δ-scheme for $n=3$. Any rule of comparison of distributions in this particular case amounts to a rule of comparison of points in Δ. Identifying distributions F with vectors (p_1, p_2, p_3), we write the expectation $U(F)$ as

$$U(F) = u(x_1)p_1 + u(x_2)p_2 + u(x_3)p_3 = u(x_1)p_1 + u(x_2)(1-p_1-p_3) + u(x_3)p_3$$
$$= [u(x_1) - u(x_2)]p_1 + [u(x_3) - u(x_2)]p_3 + u(x_2) = ap_1 + bp_3 + h,$$

where $a = u(x_1) - u(x_2)$, $b = u(x_3) - u(x_2)$, $h = u(x_2)$.
Thus, $U(F)$ is a linear function of p_1 and p_3. □

Let us fix an order $\succsim$ on $\mathcal{F}$. We call a set $\mathbf{F} \subseteq \mathcal{F}$ of distributions an *equivalence set,* or *equivalence class,* if $F \simeq G$ for any $F, G \in \mathbf{F}$. In other words, any equivalence set contains only distributions equivalent to each other.

We see from (3.5.10) that under an EUM criterion, such a set is linear in the sense that if $F, G \in \mathbf{F}$, then any mixture $\alpha F + (1-\alpha)G$ also belongs to $\mathbf{F}$. Indeed, let $U(F) = U(G)$. Then $U(\alpha F + (1-\alpha)G) = \alpha U(F) + (1-\alpha)U(G) = \alpha U(F) + (1-\alpha)U(F) = U(F)$.

EXAMPLE 3. In the Δ-scheme, $n=3$, for points (p_1, p_3) from Δ to be equivalent, the corresponding values of $U(F) = ap_1 + bp_3 + h$ should be equal to the same constant. In

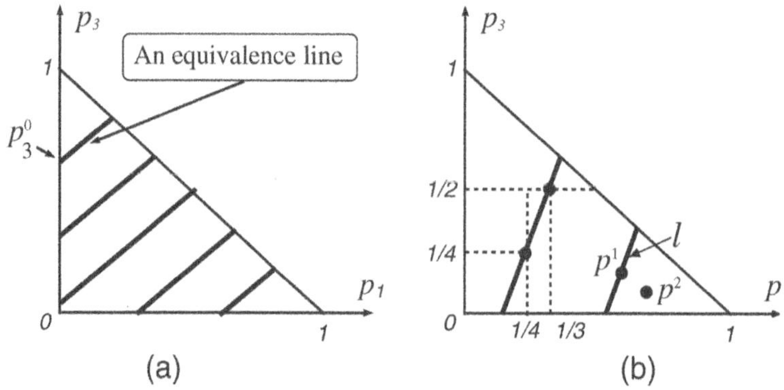

FIGURE 11.

other words, an equivalence set is a set of points

$$ap_1 + bp_3 + h = d, \tag{3.5.11}$$

where d is a constant. This is a line, or more precisely, the part of the line (3.5.11) lying in Δ. This part is, certainly, a segment; see Fig.11a. All equivalence lines are parallel with the slope

$$k = -\frac{a}{b} = \frac{u(x_2) - u(x_1)}{u(x_3) - u(x_2)}. \tag{3.5.12}$$

EXAMPLE 4. Let $u(x)$ be strictly increasing. Consider in Δ two parallel lines with the slope k from (3.5.12). Points in each line are equivalent, points lying in different lines are not. Which line is better?

Answer: The higher. Intuitively it is clear that the closer to the best point $(0,1)$, the better. For a rigorous proof, consider a point $\mathbf{p}^0 = (0, p_3^0)$ in the vertical axis and denote by F^0 the corresponding distribution (see Fig.11a). Clearly, $U(F^0) = bp_3^0 + h = [u(x_3) - u(x_2)]p_3^0 + u(x_2)$. The function u is increasing, and hence $u(x_3) > u(x_2)$ because $x_3 > x_1$. Then $U(F^0)$ is increasing in p_3^0, and the higher the point $\mathbf{p}^0$ is, the better it is. On the other hand, the higher $\mathbf{p}^0$, the higher the line starting from this point with the slope k.

EXAMPLE 5. Ann is an EU maximizer. You ask her to compare two random variables both taking the same three values x_1, x_2, x_3; the first r.v.–with probabilities $(\frac{1}{4}, \frac{1}{2}, \frac{1}{4})$, respectively, the second—with probabilities $(\frac{1}{3}, \frac{1}{6}, \frac{1}{2})$. It has turned out that Ann found these distributions equivalent for her. Is this information enough to completely determine Ann's preferences among ALL random variables taking the same values?

Answer: Yes. Since the points mentioned are equivalent, the line going through these points and having the slope $(\frac{1}{2} - \frac{1}{4})/(\frac{1}{3} - \frac{1}{4}) = 3$ is an equivalence line. Hence, $k = 3$. Consider two other points, $\mathbf{p}^1$ and $\mathbf{p}^2$, in Δ, and draw the line l with the same slope k going through $\mathbf{p}^1$; see Fig.11b. If the point $\mathbf{p}^2$ turns out to be below l, then $\mathbf{p}^2$ is worse than $\mathbf{p}^1$; if $\mathbf{p}^2$ is above l, then $\mathbf{p}^2$ is better. □

Certainly, the particular probabilities in Example 5 do not matter: *in the Δ-scheme for $n = 3$, in order to completely determine the preference order, it suffices to find two equivalent distributions.*

EXAMPLE 6. Assume that in the previous example, the values x_1, x_2, x_3 are equally spaced: $x_2 - x_1 = x_3 - x_2$. For instance, $x_1 = 0, x_2 = 20, x_3 = 40$. Then, for a concave $u(x)$, we have $k \geq 1$. Indeed, in our case, $x_2 = (x_1 + x_3)/2$, and by property (3.4.6), $u(x_2) \geq [u(x_1) + u(x_3)]/2$. This implies that $u(x_2) - u(x_1) \geq u(x_3) - u(x_2)$. Hence, in the risk aversion case, the slope of equivalence lines corresponds to an angle at least 45°. □

The reader certainly sees that the above reasoning may be easily generalized to the Δ-scheme for $n > 3$. In this case, equivalence sets are planes for $n = 4$, and hyper-planes in R^{n-1} for $n > 4$. We come to a general conclusion:

> To completely determine the preference order among all distributions concentrated at n fixed points, it is enough to know $n-1$ equivalent distributions.

In Exercise 44, we justify it rigorously. Risk aversion is again connected with the position of equivalence hyper-planes; we skip details.

3.5.6 An axiomatic approach

D. Bernoulli did not derive the EUM criterion from some original assumptions: he simply suggested it and gave an argument for why this criterion seems natural. The modern approach to such problems is more sophisticated. We do not point out a good solution from the very beginning, but first establish desirable properties of the solution, called axioms. After that, we try to figure out which solutions satisfy the properties established.

In Utility Theory such an approach was first applied in 1944, more than 200 years after D. Bernoulli's paper [13], by J. von Neumann and O. Morgenstern in [96].

The basic axiom. Consider a preference order $\succsim$ on a set $\mathcal{F} = \{F\}$ of distributions F.

Axiom 6 *(the Independence Axiom). Let F, G, and H belong to $\mathcal{F}$, and $F \succsim G$. Then for any $\alpha \in [0,1]$,*

$$\alpha F + (1-\alpha)H \succsim \alpha G + (1-\alpha)H.$$

The axiom sounds quite plausible: if you mix F and G with the *same* distribution H, then the relation between the mixtures will be the same as for the original distributions F and G. (We will see, however, in Section 4 that this is far from being always true.)

A geometric illustration is given in Fig.12. Let a "point" F be "better" than G. Consider the "segments" connecting F and H, and G and H. Let us choose two points, one in each segment, in a way that their positions between F and H, and G and H, respectively, would be in the same "proportion". Then these points are in the same relation as the original points F and G.

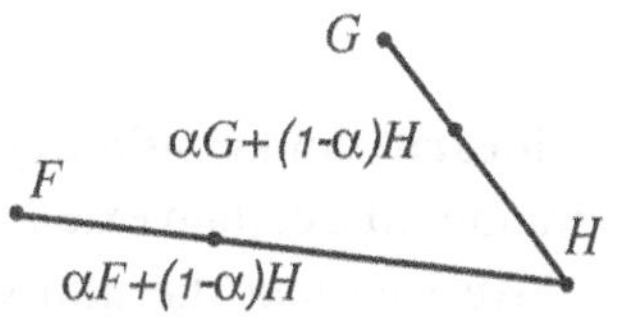

FIGURE 12.

EXAMPLE 1. John, when comparing two random variables,

$$X_1 = \begin{cases} \$100 & \text{with probability } 0.2, \\ \$0 & \text{with probability } 0.8, \end{cases} \quad \text{and } X_2 = \begin{cases} \$50 & \text{with probability } 0.4, \\ \$0 & \text{with probability } 0.6, \end{cases}$$

has decided that for him $X_2 \succ X_1$. (John seems to be a strong risk averter.) After that John is offered to play one of two games. In both games, a coin will be tossed, and in the case of a head, John will get \$100. In the case of a tail, in the first game, John will get a random prize amounting to the r.v. X_1, while in the second game—to the r.v. X_2. Thus, eventually, the prizes for the games are

$$Y_1 = \begin{cases} \$100 & \text{with probability } 0.5 + 0.5 \cdot 0.2 = 0.6, \\ \$0 & \text{with probability } 0.4, \end{cases}$$

$$Y_2 = \begin{cases} \$100 & \text{with probability } 0.5, \\ \$50 & \text{with probability } 0.5 \cdot 0.4 = 0.2, \\ \$0 & \text{with probability } 0.3, \end{cases}$$

respectively. If John, consciously or not, follows the Independence Axiom, he would prefer Y_2 to Y_1. □

Proposition 7 *The EUM criterion satisfies the Independence Axiom.*

Proof is practically immediate. In the EUM case, if $F \succsim G$, then $U(F) \geq U(G)$. Hence, by (3.5.10), $U(\alpha F + (1-\alpha)H) = \alpha U(F) + (1-\alpha)U(H) \geq \alpha U(G) + (1-\alpha)U(H) = U(\alpha G + (1-\alpha)H)$. Consequently, $\alpha F + (1-\alpha)H \succsim \alpha G + (1-\alpha)H$. □

The main result of classical utility theory is that under some additional, more technical, conditions, the converse claim is also true:

The Independence Axiom implies the EUM principle.

We skip here details and a rigorous formulation; for a detailed exposition see, e.g., [82].

Route 2 ⇒ page 115

3 **Linearity and continuity lead to EUM.** Consider a somewhat weaker proposition illustrating, to a certain extent, the situation.

We saw that the utility functional (3.5.6) was linear, that is, $U(\alpha F_1 + (1-\alpha)F_2) = \alpha U(F_1) + (1-\alpha)U(F_2)$. The question we discuss here is whether any linear utility functional admits the representation (3.5.6). (That is, instead of the Independence Axiom, we consider the linearity property itself.) The answer is "yes, if $U(F)$ is in a certain sense continuous".

To specify what this means, first note that convergence of distributions may be defined in different ways. We consider two.

1. **Weak convergence.** We say that a sequence F_n converges to a distribution F weakly, and write it as $F_n \xrightarrow{w} F$, if $F_n(x) \to F(x)$ for any x at which $F(x)$ is continuous.

2. **Convergence for all sets.** We say that a sequence F_n converges to a distribution F for all sets, and write it as $F_n \overset{c}{\to} F$, if $F_n(B) \to F(B)$ for any B.

For more detail on convergence of r.v.'s and distributions, see Section **0**.5.

Clearly, convergence in the later case implies convergence in the former.

Accordingly, we consider two definitions of continuity of a functional $U(F)$.

Condition C1. (*Weak continuity*): $U(F_n) \to U(F)$ provided $F_n \overset{w}{\to} F$.

Condition C2. (*Continuity with respect to convergence for all sets*): $U(F_n) \to U(F)$ provided $F_n \overset{c}{\to} F$.

Condition C1 is stronger than Condition C2 since convergence $U(F_n) \to U(F)$ in the former case takes place under a weaker (!) requirement on the convergence of F_n.

Theorem 8 *Suppose $U(F)$ is defined on the set of all distributions and is linear, i.e., (3.5.10) is true. Then, if Condition C2 holds, there exists a bounded function $u(x)$ such that*

$$U(F) = \int_{-\infty}^{\infty} u(x)dF(x). \tag{3.5.13}$$

If Condition C1 holds, the function $u(x)$ in (3.5.13) is continuous.

(A proof may be found in [40], [120].)

---

4 NON-LINEAR CRITERIA

2,3

The EUM approach may be considered as a first approximation to the description of people's preferences. Over the years, there has been a great deal of discussion about the adequacy of this approach. Many experiments have been provided and a number of examples have been constructed, showing that the EUM approach is far from being efficient in all situations. The existence of such examples is not surprising; on the contrary, it would have been surprising if the behavior of such sophisticated (and sometimes strange) creatures as human beings had been always well described by simple *linear* functions. This section concerns some elements of modern utility theory.

4.1 Allais' paradox

The following example considered by M. Allais [2] is probably the most famous. Though being contrived, it is very illustrative. Consider distributions F_1, F_2, F_3, F_4 of a random income with values \$0, \$10 millions, or \$30 millions. The corresponding probabilities are given in the following table.

	\$0	\$10 million	\$30 million
F_1	0	1	0
F_2	0.01	0.89	0.1
F_3	0.9	0	0.1
F_4	0.89	0.11	0

Apparently, a majority of people would prefer F_1 to F_2, reasoning as follows. Ten million dollars is a lot of money, for ordinary people as inconceivable as thirty million. So, it is better to get ten for sure than to go in pursuit of thirty million at the risk of receiving nothing (even if the probability of this is very small). Thus, $F_1 \succ F_2$.

The situation with F_3 and F_4 is different. Now the probabilities of receiving nothing are large—and hence one should be ready to lose, and these probabilities are practically the same. Then it is reasonable to choose the variant with the larger prize. So, $F_3 \succ F_4$.

Let us consider the mixtures $\frac{1}{2}F_1 + \frac{1}{2}F_3$ and $\frac{1}{2}F_2 + \frac{1}{2}F_4$. If the preference $\succ$ had been preserved by a utility functional (3.5.5), in the light of the linearity property (3.5.10), we would have had $\dfrac{1}{2}F_1 + \dfrac{1}{2}F_3 \succ \dfrac{1}{2}F_2 + \dfrac{1}{2}F_4$.

However, as a matter of fact, as is easy to calculate, $\dfrac{1}{2}F_1 + \dfrac{1}{2}F_3 = \dfrac{1}{2}F_2 + \dfrac{1}{2}F_4$.

Next, we address several directions in which the EUM criterion can be generalized. (In particular, the situation in Allais's paradox may be described by using the schemes below; we will skip concrete calculations.)

Note also that examples in this section do not aim to justify criteria we are introducing; justification comes from empirical evidence and qualitative reasoning based on axioms we will discuss. The goal of these examples is more modest—to demonstrate how criteria introduced will work *if we accept them*, and what new features they can raise in comparison with the classical EUM criterion.

4.2 Weighted utility

The criterion below is based on two functions: $u(x)$ which we view as a utility function, and a non-negative function $w(x)$ which we call a weighting function. Consider the functional

$$W(F) = \frac{\int_{-\infty}^{\infty} u(x)w(x)dF(x)}{\int_{-\infty}^{\infty} w(x)dF(x)}, \tag{4.2.1}$$

assuming that the denominator in (4.2.1) is not zero.

Note that if F is the distribution of a r.v. X, then (4.2.1) may be rewritten as

$$W(F) = \frac{E\{u(X)w(X)\}}{E\{w(X)\}}.$$

The difference between the classical expected utility scheme and the last case is that now we assign to different values x weights $w(x)$. If all weights $w(x) \equiv 1$, the denominator in (4.2.1) equals one, and we deal with the EUM case.

Since we compare here rather distributions than r.v.'s themselves, it is convenient to define a preference order $\succsim$ on the set $\mathcal{F} = \{F\}$ of distributions F for which $W(F)$ is well defined. Let an order $\succsim$ be *preserved* by the functional W:

$$F \succsim G \Longleftrightarrow W(F) \geq W(G).$$

When comparing r.v.'s, we will say that $X \succsim Y$ if $F_X \succsim G_X$, where F_X and G_X are the distributions of X and Y, respectively.

Following tradition, we denote by δ_c the distribution of a non-random $X \equiv c$. Then the certainty equivalent $c(F)$ is defined as a number c such that $\delta_c \simeq F$.

In the particular case of this section, $W(\delta_c) = u(c)w(c)/w(c) = u(c)$. Thus, for the certainty equivalent $c = c(F)$ of a r.v. X with a distribution F, we have $u(c) = W(F)$, and

$$c(F) = u^{-1}(W(F)). \tag{4.2.2}$$

EXAMPLE 1. Let $u(x) = x^{\alpha}$, and $w(x) = x^{\beta}$. We assume $\alpha > 0$. As to the parameter β, depending on the situation, it may be either positive (the larger a value, the larger its weight), or negative (the larger a value, the less its weight), or zero. In the last case, $\beta = 0$, we deal with the EUM criterion.

For a positive r.v. X with a distribution F, by definition (4.2.1),

$$W(F) = \frac{E\{X^{\alpha+\beta}\}}{E\{X^{\beta}\}}. \tag{4.2.3}$$

In particular, for $\alpha = \beta = 1$, we have

$$W(F) = \frac{E\{X^2\}}{E\{X\}} = \frac{E\{X^2\}}{(E\{X\})^2}E\{X\}.$$

It is well known that $(E\{X\})^2 \leq E\{X^2\}$ (see, for example, Exercise 33b, or recall that $Var\{X\} = E\{X^2\} - (E\{X\})^2 \geq 0$). Hence, $W(F) \geq E\{X\}$. Because in the last case $u(x) = x$ and, consequently, $u^{-1}(x) = x$, the certainty equivalent $c(F) = W(F) \geq E\{X\}$, so we deal with a risky person. It is not surprising since large values have large weights.

EXAMPLE 2. Let X in the previous example be uniformly distributed on $[0,d]$. The reader is invited to verify that in this case, for $\beta > -1$,

$$W(F) = \frac{1+\beta}{1+\alpha+\beta}d^{\alpha}.$$

Following (4.2.2), for the certainty equivalent we have

$$c(F) = (W(F))^{1/\alpha} = \left(\frac{1+\beta}{1+\alpha+\beta}\right)^{1/\alpha} d. \tag{4.2.4}$$

If $\beta = 0$, we come to the result of Example 3.1.3-2, which is not surprising since in this case $w(x) \equiv 1$, and we deal with expected utility. The larger β, the greater $c(F)$, and this is also understandable: the larger β, the larger weights for large x's, so the certainty equivalent must grow as β increases. □

Next, we generalize the model from Section 3.2. Consider the maximal premium a client with a wealth of w is willing to pay to insure a risk ξ. As in Section 3.2, we write that

$$w - G_{\max} \simeq w - \xi. \tag{4.2.5}$$

Since the r.-h.s. of (4.2.5) is certain, (4.2.5) is equivalent to $w - G_{\max} = c(F_{w-\xi})$, where as usual, the symbol F_X stands for the distribution of a r.v. X. So,

$$G_{\max} = w - c(w-\xi) = w - u^{-1}(W(F_{w-\xi})).$$

EXAMPLE 3. As in Examples 1-2, let $u(x) = x^{\alpha}$ and $w(x) = x^{\beta}$. Let ξ be uniformly distributed on $[0,d]$, and $w = d$. Then the r.v. $w - \xi$ is also uniformly distributed on $[0,d]$, and it follows from (4.2.4) that

$$G_{\max} = w - c(w-\xi) = d\left(1 - \left(\frac{1+\beta}{1+\alpha+\beta}\right)^{1/\alpha}\right).$$

The classical EUM case corresponds to $\beta = 0$. We see that for $\beta > 0$, the maximum premium becomes smaller. □

Next, we discuss the linearity issue (see Section 3.5.5). If we replace F in (4.2.1) by a mixture $F^{(\alpha)} = \alpha F_1 + (1-\alpha)F$ (compare with Section 3.5.5), we get

$$W(F^{(\alpha)}) = \frac{\alpha\int_{-\infty}^{\infty} u(x)w(x)dF_1(x) + (1-\alpha)\int_{-\infty}^{\infty} u(x)w(x)dF_2(x)}{\alpha\int_{-\infty}^{\infty} w(x)dF_1(x) + (1-\alpha)\int_{-\infty}^{\infty} w(x)dF_2(x)}. \tag{4.2.6}$$

In general, this quantity is certainly not equal to $\alpha W(F_1) + (1-\alpha)W(F_2)$ (see also Exercise 57). Hence, in this case, the linearity property and the Independence Axiom do not hold.

EXAMPLE 4. Consider the Δ-scheme described in Section 3.5.5 with r.v.'s taking only three fixed values x_1, x_2, x_3. Let $x_1 < x_2 < x_3$. Any distribution F of such a r.v. is identified with the probability vector (p_1, p_2, p_3), where p_i is the probability of the value x_i. Consider points in $\Delta = \{(p_1, p_3) : p_1, p_3 \geq 0,\ p_1 + p_3 \leq 1\}$. Setting $p_2 = 1 - p_1 - p_3$ we have

$$\begin{aligned} W(F) &= \frac{u(x_1)w(x_1)p_1 + u(x_2)w(x_2)p_2 + u(x_3)w(x_3)p_3}{w(x_1)p_1 + w(x_2)p_2 + w(x_3)p_3} \\ &= \frac{ap_1 + bp_3 + h}{\widetilde{a}p_1 + \widetilde{b}p_3 + \widetilde{h}}, \end{aligned} \tag{4.2.7}$$

where

$$\begin{aligned} &a = u(x_1)w(x_1) - u(x_2)w(x_2),\ b = u(x_3)w(x_3) - u(x_2)w(x_2),\ h = u(x_2)w(x_2), \\ &\widetilde{a} = w(x_1) - w(x_2),\ \widetilde{b} = w(x_3) - w(x_2),\ \widetilde{h} = w(x_2). \end{aligned}$$

Since we consider only distributions for which $\int_{-\infty}^{\infty} w(x)dF(x) > 0$, we consider only those points in Δ for which the denominator in (4.2.7) is positive. All points for which $W(F)$ equals a constant d are points for which $ap_1 + bp_3 + h = d(\widetilde{a}p_1 + \widetilde{b}p_3 + \widetilde{h})$, or

$$(a - d\widetilde{a})p_1 + (b - d\widetilde{b})p_3 + h - d\widetilde{h} = 0. \tag{4.2.8}$$

This is a line, or more precisely the part of the line (4.2.8) lying in Δ. However, the slope of this line depends on d, so lines corresponding to different d's are not parallel (!).

Thus, although the Independence Axiom and the corresponding linearity property are not true in this case, we have a sort of the linearity property when we consider equivalent distributions.

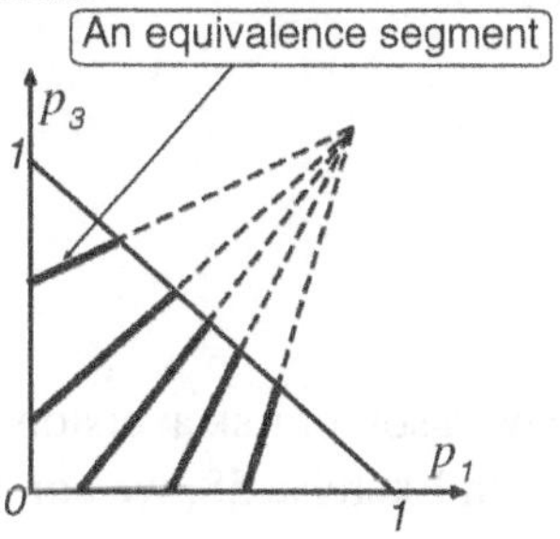

FIGURE 13.

It is interesting that all lines (4.2.8) intersect at the same point. This point is the intersection of two lines: l_0 defined by the equation $ap_1 + bp_3 + h = 0$ (points where $W(F) = 0$), and l_∞ defined by $\widetilde{a}p_1 + \widetilde{b}p_3 + \widetilde{h} = 0$ (points where the denominator in (4.2.7) is zero, and hence $W(F)$ is not defined). Advice on how to show it rigorously is given in Exercise 56, a typical picture is depicted in Fig.13.

EXAMPLE 5. In the Δ-scheme, let $x_1 = 0, x_2 = 1, x_3 = 2$; $u(x) = \sqrt{x}$, and $w(x) = 1/(1+x)$. Then, as is easy to verify, $a = -\frac{1}{2}, b = \frac{\sqrt{2}}{3} - \frac{1}{2} \approx -0.029, h = \frac{1}{2}, \widetilde{a} = \frac{1}{2}, \widetilde{b} = -\frac{1}{6}$ and $\widetilde{h} = \frac{1}{2}$. Thus, up to the third digit, the point of the intersection of all lines (4.2.8) is the intersection of the lines $-0.5p_1 - 0.029p_3 + 0.5 = 0$ and $0.5p_1 - 0.166p_3 + 0.5 = 0$. The approximate solution is the point (0.704, 5.111). □

The next scheme generalizes the approach of this section.

4.3 Implicit or comparative utility

4.3.1 Definitions and examples

In this section, we consider not a utility function $u(x)$ but a function $v(x,y)$ which we will call an *implicit utility* or *comparative utility* function, and interpret it as a function indicating to what extent income x is preferable to income y. One may say that $v(x,y)$ is the comparative utility of x with respect to y. In light of this, we assume $v(x,x) = 0$, $v(x,y) \geq 0$ if $x \geq y$, and $v(x,y) \leq 0$ if $x \leq y$. Sometimes one can assume $v(x,y) = -v(y,x)$ but in general it may be false: x may be "better" than y to a smaller extent than y is "worse" than x; see also Example 3 below.

It is natural to assume that $v(x,y)$ is non-decreasing in x and is non-increasing in y, which again reflects the property "the larger, the better".

EXAMPLE 1. Let

$$v(x,y) = \frac{x-y}{1+|x|+|y|}. \tag{4.3.1}$$

In this case, for small x and y the comparative utility almost equals the difference $x-y$, while for large x's and y's the measure $v(x,y)$ may be viewed as a relative difference: $x-y$ is divided by $1+|x|+|y|$. □

We define the certainty equivalent of a r.v. X as a solution to the equation

$$E\{v(X,c)\} = 0, \tag{4.3.2}$$

provided that this solution exists and is unique. The interpretation is clear: $c(X)$ is the certain amount whose comparative utility with respect to X equals zero on the average.

EXAMPLE 2. Let $v(x,y)$ be given by (4.3.1), and let $X = d > 0$ or 0 with equal probabilities. Then (4.3.2) is equivalent to $\frac{1}{2}v(d,c) + \frac{1}{2}v(0,c) = 0$. Obviously, c should be between d and 0. So, $c \geq 0$ and the equation may be written as

$$\frac{1}{2} \cdot \frac{d-c}{1+d+c} + \frac{1}{2} \cdot \frac{0-c}{1+0+c} = 0.$$

This is a quadratic equation. Its positive solution is

$$c = \frac{\sqrt{1+2d}-1}{2}.$$

As is easy to verify, c above is less than $E\{X\} = d/2$, so we have a sort of risk aversion. For large d, we have $c \sim \sqrt{d/2}$, which is much smaller than $d/2$. In Exercise 58, we compute the maximal accepted premium. □

Once we have defined what is certainty equivalent in this case, we can define the corresponding preference order by the rule

$$X \succsim Y \iff c(X) \geq c(Y).$$

First of all, note that this scheme includes the classical EU maximization as a particular case. Indeed, let $v(x,y) = u(x) - u(y)$, where u is a utility function. Assume that u is increasing, so its inverse u^{-1} exists. In this case, (4.3.2) implies $E\{u(X)\} - E\{u(c)\} = 0$, and since c is certain, we have $E\{u(X)\} = u(c)$. Hence, $c(X) = u^{-1}(E\{u(X)\})$, as in the classical case. Because $u^{-1}(x)$ is increasing, the relation $c(X) \geq c(Y)$ is equivalent to the relation $E\{u(X)\} \geq E\{u(Y)\}$.

Furthermore, the weighted utility scheme is also a particular case of the comparative utility. To show it, set

$$v(x,y) = w(x)[u(x) - u(y)].$$

In this case, for the certainty equivalent c of a r.v. X with a distribution F we write

$$\begin{aligned} 0 = E\{v(X,c)\} &= \int_{-\infty}^{\infty} v(x,c)dF(x) = \int_{-\infty}^{\infty} w(x)[u(x) - u(c)]dF(x) \\ &= \int_{-\infty}^{\infty} u(x)w(x)dF(x) - u(c)\int_{-\infty}^{\infty} w(x)dF(x). \end{aligned}$$

From this it follows that $u(c) = W(F)$, where $W(F)$ is the same as in (4.2.1). If $u(x)$ is strictly increasing, then $c = u^{-1}(W(F))$. Since u^{-1} is also strictly increasing, $c(F) \geq c(G)$ iff $W(F) \geq W(G)$.

Let now $v(x,y)$ be concave with respect to x. Consider a r.v. X and set $m = E\{X\}$ and $c = c(X)$. By definition, $v(m,m) = 0$. Then, by Jensen's inequality,

$$v(m,m) = 0 = E\{v(X,c)\} \leq v(E\{X\},c) = v(m,c).$$

Thus, $v(m,m) \leq v(m,c)$. Since $v(x,y)$ is non-increasing in y, this implies that

$$c(X) \leq E\{X\}.$$

A good example for such a function v is

$$v(x,y) = g(x-y),$$

where $g(s)$ is a concave increasing function such that $g(0) = 0$. Note that in this case we should not expect $v(x,y) = -v(x,y)$.

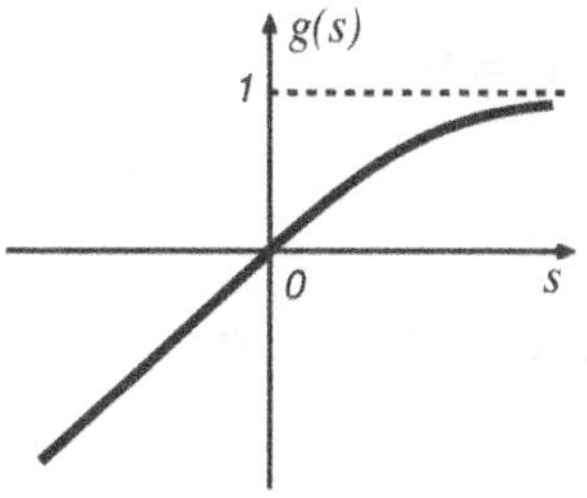

FIGURE 14.

EXAMPLE 3 (*a jealous person*). Let Mr. J.'s implicit utility function $v(x,y) = g(x-y)$, where

$$g(s) = \begin{cases} \dfrac{s}{1+s} & \text{if } s \geq 0, \\ s & \text{if } s < 0. \end{cases}$$

The function $g(s)$ is concave, its graph is given in Fig.14.

Mr. J. may be characterized as pretty jealous. Assume that x is Mr. J.'s wealth, and he compares it with a Mr. A.'s wealth y. If x is much larger than y, the comparative utility $v(x,y)$ is, nevertheless, not large, and $v(x,y) \to 1$ as $x - y \to \infty$. (Mr. J. does not think that his wealth is much more valuable than that of Mr. A.)

On the other hand, if x is much smaller than y, the comparative utility is negative with a large absolute value, and $v(x,y) \to -\infty$, as $x - y \to -\infty$. (Now Mr. J. considers himself much less happy than Mr. A.)

Let a r.v. $X = d > 0$ or 0 with equal probabilities. In this case, equation (4.3.2) is reduced to $\frac{1}{2}g(d-c) + \frac{1}{2}g(-c) = 0$, which leads to

$$\frac{d-c}{1+d-c} = c.$$

This is a quadratic equation. The solution that lies between d and 0 is

$$c = c(X) = \frac{d}{1 + (d/2) + \sqrt{1 + (d^2/4)}}.$$

The denominator is greater than two. So, $c < (d/2) = E\{X\}$. □

4.3.2 In what sense the implicit utility criterion is linear

As we saw, the equation (4.3.2) may be written as

$$\int_{-\infty}^{\infty} v(x,c)dF(x) = 0, \tag{4.3.3}$$

where F is the d.f. of X. Consequently, the solution to this equation depends not on the r.v. X itself but on its d.f. F. That is, $c(X)$ is a function (or functional) of F, and it should not cause confusion if we use also the notation $c(F)$ defining it as a solution to (4.3.3).

Consider a set $\mathcal{F} = \{F\}$ of distributions. Assume that the function $v(x,y)$ is such that $c(F)$ exists and is unique for each $F \in \mathcal{F}$. Let us define a preference order $\succsim$ in $\mathcal{F}$ by

$$F \succsim G \Longleftrightarrow c(F) \geq c(G).$$

Proposition 9 *Let $F \simeq G$ (F is equivalent to G.) Then for any $\alpha \in [0,1]$ the mixture $F^{(\alpha)} = \alpha F + (1-\alpha)G \simeq F$.*

Proof is short. If $F \simeq G$, then F and G have the same certainty equivalent. Denote it by c. By definition (4.3.3),

$$\int_{-\infty}^{\infty} v(x,c)dF(x) = 0, \quad \text{and} \quad \int_{-\infty}^{\infty} v(x,c)dG(x) = 0.$$

Then

$$\int_{-\infty}^{\infty} v(x,c)dF^{(\alpha)}(x) = \alpha \int_{-\infty}^{\infty} v(x,c)dF(x) + (1-\alpha)\int_{-\infty}^{\infty} v(x,c)dG(x) = 0.$$

Hence, c is the certainty equivalent for $F^{(\alpha)}$ too. ■

Geometrically, it means that equivalent points still lie in the same line, but equivalency lines may not be parallel.

EXAMPLE 1. Consider again the Δ-scheme for $n = 3$. In this case, equation (4.3.3) may be written as

$$v(x_1,c)p_1 + v(x_2,c)p_2 + v(x_3,c)p_3 = 0.$$

Since $p_2 = 1 - p_1 - p_3$, we can rewrite it as

$$a(c)p_1 + b(c)p_3 + h(c) = 0, \tag{4.3.4}$$

where $a(c) = v(x_1,c) - v(x_2,c)$, $b(c) = v(x_3,c) - v(x_2,c)$, $h(c) = v(x_2,c)$.

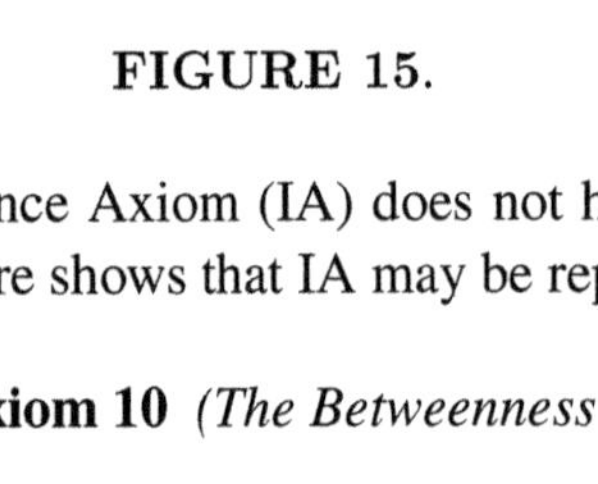

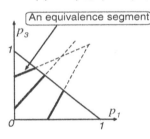

FIGURE 15.

Let us compare this with what we did in Section 3.5.5, where a, b, and h did not depend on c. For a fixed c, all points that satisfy (4.3.4) lie in a line, and the slope of this line depends on c. Unlike the case of weighted utility, the dependence of the slope on c is rather arbitrary, so we should not expect that all these lines intersect at the same point. A typical picture is given in Fig.15. □

As we saw, and as we see now again, the Independence Axiom (IA) does not hold in this case. A general theory which we do not present here shows that IA may be replaced by the following weaker

Axiom 10 *(The Betweenness Axiom.) Let $F, G \in \mathcal{F}$, and $F \simeq G$. Then for any $\alpha \in [0,1]$,*

$$\alpha F + (1-\alpha)G \simeq G.$$

We see that, unlike the IA, the Betweenness Axiom (BA) deals not with the case when $F \succsim G$ but only with equivalent F and G. The fact that BA is weaker than IA is presented in

Proposition 11 *If IA holds, then BA holds too.*

Proof. Set $H = G$ in the formulation of the Independence Axiom 6. Let the IA hold, and let $F \succsim G$. Then $\alpha F + (1-\alpha)G \succsim \alpha G + (1-\alpha)G = G$, that is, $\alpha F + (1-\alpha)G \succsim G$. Now let $F \simeq G$. Then $F \succsim G$, and $G \succsim F$ simultaneously. Consequently, $\alpha F + (1-\alpha)G \succsim G$ and $\alpha F + (1-\alpha)G \precsim G$, simultaneously. That is, $\alpha F + (1-\alpha)G \simeq G$. ■

It proves that together with some more technical assumptions,

The Betweenness Axiom implies the implicit (or comparative) utility principle.

4.4 Rank Dependent Expected Utility

4.4.1 Definitions and examples

The next approach essentially differs from those of previous sections. For simplicity, we restrict ourselves to non-negative r.v.'s.

Consider probability distributions F on $[0,\infty)$ and two functions: $u(x)$ viewed again as a *utility* function, and a function $\Psi(p)$ defined on $[0,1]$, which we call a *transformation function*. We assume $\Psi(p)$ to be non-decreasing, $\Psi(0) = 0$, $\Psi(1) = 1$. We consider an individual (or investor) whose preferences are preserved by the function (or functional)

$$R(F) = \int_0^\infty u(x)d\Psi(F(x)). \tag{4.4.1}$$

The *transformation* (or *weighting*) function Ψ reflects the attitude of the individual to different probabilities. The individual, when perceiving information about the distribution F, "transforms" the actual distribution function $F(x)$ into another one, $F_\Psi(x) = \Psi(F(x))$, underestimating or overestimating real probabilities.

First, let us show that $F_\Psi(x)$ is indeed a distribution function.[2] First, since we consider non-negative r.v.'s, we have $F(x) = 0$ for $x < 0$. Second, due to properties of Ψ, the function $F_\Psi(x) = \Psi(F(x))$ is non-decreasing. Moreover, for $x < 0$ we have $F_\Psi(x)=\Psi(F(x))=\Psi(0)=0$, and $F_\Psi(\infty)=\Psi(F(\infty))=\Psi(1) = 1$. Hence, $F_\Psi(x)$ is a d.f.

Note also that in (4.4.1), we transform a distribution function, not a density. The latter transformation would lead to non-desired consequences. For example, a transformation $\Psi(f(x))$ of a density f may be not a density, since $\int_0^\infty \Psi(f(x))dx$ might not equal one; so we would no longer deal with a probability distribution.

The quantity (4.4.1) is referred to as a *Rank Dependent Expected Utility (RDEU)*. Another term in use is *distorted expectation*; see, e.g., [33].

The corresponding preference order $\succsim$ is preserved by the function $R(F)$, that is,

$$F \succsim G \Longleftrightarrow R(F) \geq R(G).$$

A simple example is $\Psi(p) = p^\beta$. If $\beta = 1$, the subject perceives F as is and deals with the "usual" expected utility. If $\beta < 1$, then $p^\beta \geq p$, and the investor overestimates the

[2] We skip the question of continuity from the right. Since we required $\Psi(0) = 0$, we cannot require $\Psi(p)$ to be right continuous at $p = 0$, and hence $\Psi(F(x))$ may be not right continuous everywhere. However, it does not influence the integral in (4.4.1), so it is not an essential circumstance.

probability for the income to be less that a fixed value: the investor is "security-minded". In the case $\beta > 1$, the investor underestimates the probability mentioned, being "potential-minded" (or "opportunity-minded") .

EXAMPLE 1. Let X be uniformly distributed on $[0,1]$. Its distribution function

$$F(x) = \begin{cases} 0 & \text{if } x < 0, \\ x & \text{if } x \in [0,1], \\ 1 & \text{if } x > 1, \end{cases} \quad \text{and} \quad F_\Psi(x) = \begin{cases} 0 & \text{if } x < 0, \\ x^\beta & \text{if } x \in [0,1], \\ 1 & \text{if } x > 1. \end{cases}$$

Then the density of the transformed distribution,

$$f_\Psi(x) = F'_\Psi(x) = \begin{cases} 0 & \text{if } x < 0, \\ \beta x^{\beta-1} & \text{if } x \in [0,1], \\ 0 & \text{if } x > 1. \end{cases}$$

For example, if $\beta > 1$, the density $f_\Psi(x)$ is increasing, and while for the original distribution all values of X are equally likely, in the "investor's mind" it is not so: smaller values are less likely.

To the contrary, if $\beta < 1$, the density $f_\Psi(x) \to \infty$ (!) as $x \to 0$, that is, the investor strongly overestimates the probability to get nothing.

The case $\beta = 0$ corresponds to an "absolutely pessimistic" investor: $F_\Psi(x) = 0$ for $x < 0$, and $= 1$ for $x > 0$, that is, F_Ψ is the distribution of a r.v. $X \equiv 0$. In this case, the investor expects that she/he will get nothing for sure.

EXAMPLE 2. Assume that an investor does not distinguish small values of the income. For instance, hoping for an income equal to \$100,000 on the average, the investor considers income values of \$1 or even \$100 as too small and, consciously or not, identifies them with zero income.

Denote by F the distribution of the investor's income and assume that the investor identifies with zero all values which are less then the γ-quantile $q_\gamma(F)$ for some small fixed γ.

Suppose the same is true for "inconceivable" large values. For instance, the same investor may consider \$1million or \$10 million an improbable luck, and (consciously or not) identify these numbers. More precisely, choosing for simplicity the same level γ, assume that the investor identifies with $q_{1-\gamma}(F)$ all values which are larger than $q_{1-\gamma}(F)$.

In both cases, we may talk about the existence of a *perception threshold*. Such a situation may be described by the *truncation* transforming function Ψ such that

$$\Psi(0) = 0, \quad \text{and} \quad \Psi(p) = \begin{cases} \gamma \text{ if } 0 < p < \gamma, \\ p \text{ if } \gamma \le p < 1-\gamma, \\ 1 \text{ if } p \ge 1-\gamma. \end{cases}$$

In this case,

$$F_\Psi(x) = \begin{cases} \gamma & \text{if } 0 \le x < q_\gamma(F), \\ F(x) & \text{if } q_\gamma(F) \le x < q_{1-\gamma}(F), \\ 1 & \text{if } x \ge q_{1-\gamma}(F). \end{cases}$$

Hence,

$$R(F) = u(0) \cdot \gamma + \int_{q_\gamma(F)}^{q_{1-\gamma}(F)} u(x) dF(x) + u(q_{1-\gamma}(F)) \cdot \gamma. \tag{4.4.2}$$

The functional (4.4.2) is not linear and should be distinguished from the naive criterion when truncation is carried out at a fixed, perhaps large, value not depending on F.

EXAMPLE 3. Let F be the distribution of a r.v. taking only two values, say, a and $b > a$ with respective probabilities p and $1-p$. Then $F(x) = 0$ if $x \in [0,a)$, $F(x) = p$ if $x \in [a,b)$, and $F(x) = 1$ if $x \in [b,\infty)$. Consequently, $\Psi(F(x)) = 0$ if $x \in [0,a)$, $\Psi(F(x)) = \Psi(p)$ if $x \in [a,b)$, and $\Psi(F(x)) = 1$ if $x \in [b,\infty)$. Then, by (4.4.1),

$$R(F) = u(a)\Psi(p) + u(b)[1-\Psi(p)]. \tag{4.4.3}$$

In this case, $\Psi(\cdot)$ "transforms" just one probability p.

Note also that if a r.v. $X \equiv c$, then its d.f. $\delta_c(x) = 1$ or 0 for $x \geq c$, and $x < c$, respectively. Hence, $\Psi(\delta_c(x))$ also equals 1 or 0 for $x \geq c$, and $x < c$, respectively, and by (4.4.1) or (4.4.3),

$$R(\delta_c) = u(c). \tag{4.4.4}$$

EXAMPLE 4. Let a person having, for instance, the utility function $u(x) = \sqrt{x}$, choose between one of two retirement plans: either the annual pension is equal to $X = \$100,000$, or it is equal to r.v. $Y = \$50,000$ or $\$200,000$ with probabilities $1/2$.

For the numbers above, the expected utility criterion leads to a slight preference for the latter plan ($E\{u(X)\} \approx 316$ and $E\{u(Y)\} \approx 335$), which does not looks very realistic. One would expect most people to choose the plan X.

On the other hand, by (4.4.4), we have $R(F_X) = u(10^5)$ and, by (4.4.3), $R(F_Y) = u(5 \cdot 10^4)\Psi(1/2) + u(2 \cdot 10^5)[1-\Psi(1/2)]$.

It is easy to calculate that $R(F_X) > R(F_Y)$, that is, the individual would prefer X, if $\Psi(1/2) > 0.59$. This means that such a person slightly overestimates the probability of the unlucky event to get $\$50,000$ (since this probability equals $1/2$). So, one can expect $\Psi(p)$ to be concave for large p's. Certainly, this naive example is given merely for illustration. □

For the certainty equivalent $c = c(F)$ of a distribution F, we have $u(c) = R(F)$, and

$$c(F) = u^{-1}(R(F)). \tag{4.4.5}$$

EXAMPLE 5. Let F be the uniform distribution on $[0,b]$, $u(x) = x^{\alpha}$, and $\Psi(p) = p^{\beta}$. Then, by (4.4.5),

$$c(F) = \left(\int_0^b x^{\alpha} d(x/b)^{\beta}\right)^{1/\alpha} = \left(\beta \frac{b^{\alpha}}{\beta+\alpha}\right)^{1/\alpha} = \left(\frac{\beta}{\beta+\alpha}\right)^{1/\alpha} b.$$

For $\beta = 1$, we have $c(F) = [1/(1+\alpha)]^{1/\alpha} b$, which corresponds to the EUM case considered in Example 3.1.3-2. For $\beta > 1$, the certainty equivalent gets larger; for $\beta < 1$ – smaller. It makes sense: for $\beta > 1$, the individual underestimates the probabilities of "bad events", so the certainty equivalent is larger in comparison with the case when the probabilities mentioned are perceived correctly. The case $\beta < 1$ is the opposite. □

4.4.2 Application to insurance

Consider the insurance model from Section 3.2, keeping the same notation for the wealth w, the random loss ξ, and the premium G. Following the same logic, we see that for G to be acceptable, the certain quantity $w-G$ should be preferable to the r.v. $w-\xi$.

For the maximal accepted premium $G_{\max}$, the r.v. $w-G_{\max}$ should be equivalent to $w-\xi$. In the RDEU framework, this means that

$$u(w-G_{\max}) = R(F_{w-\xi}), \tag{4.4.6}$$

where, as usual, F_X denotes the distribution of a r.v. X.

EXAMPLE 1. As in Example 3.2-1, let $u(x) = 2x - x^2$, $w = 1$, and the r.v. ξ be uniformly distributed on $[0,1]$. Let $\Psi(p) = p^{\beta}$. Since $1-\xi$ is also uniformly distributed on $[0,1]$,

$$R(F_{1-\xi}) = \int_0^1 (2x - x^2)dx^{\beta} = C_{\beta},$$

where $C_{\beta} = \dfrac{\beta(3+\beta)}{(1+\beta)(2+\beta)}$. Let $y = w - G_{\max}$. Then $2y - y^2 = C_{\beta}$. As in Example 3.2-1, we obtain from this that

$$G_{\max} = \sqrt{1 - C_{\beta}} = \sqrt{\frac{2}{(1+\beta)(2+\beta)}}.$$

If $\beta = 0$ (the absolutely pessimistic investor from Example 4.4.1-1), $G_{\max}$ is equal to one, that is, to the maximal possible loss. (The investor feels that the maximal loss will happen.) The larger β, the less $G_{\max}$, which is natural. For $\beta = 1$, we have $G_{\max} = 1/\sqrt{3}$ as in the expected utility case in Example 3.2-1. □

4.4.3 Further discussion and the main axiom

Next, we consider some possible forms of the transformation function $\Psi(p)$. To clarify the classification below, note that when saying that a subject underestimates the probability of an event, we mean that the subject perceives the likelihood of this event to be less than it really is. In the extreme case, the subject neglects the possibility of such an event. The four cases we discuss below are illustrated in Fig.16.

- $\Psi(p) \geq p$ and is concave. For any certain level of income, the subject overestimates the probabilities that the income will not reach this level. The subject is "security minded".

- $\Psi(p) \leq p$ and is convex: the opposite case. The subject is "potential-minded".

- $\Psi(p)$ is S-shaped. The subject underestimates the probabilities of very large and very small values and, consequently, proceeds from moderate values of income.

- $\Psi(p)$ is inverse-S-shaped: "cautiously hopeful". Roughly speaking, the subject overestimates the probabilities of "very large" and "very small" values.

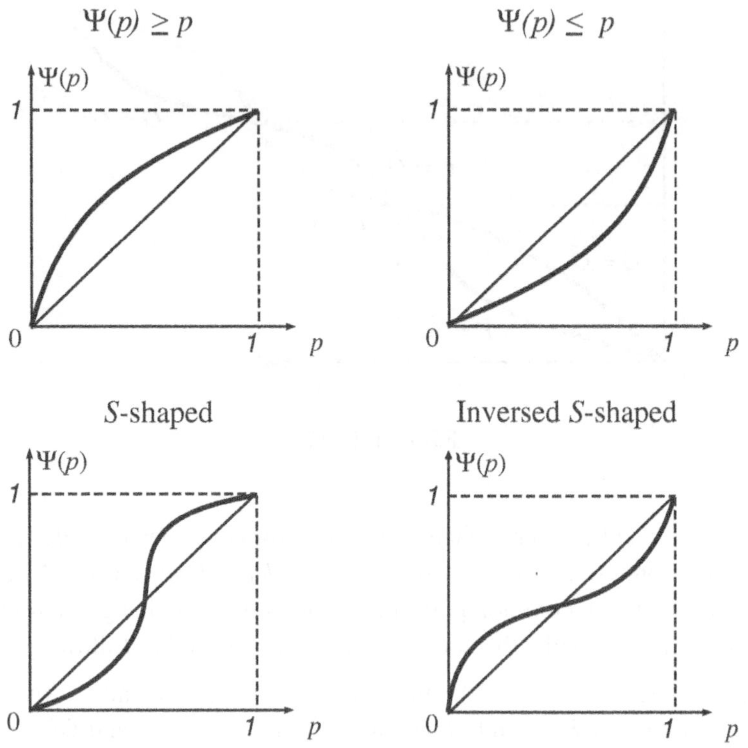

FIGURE 16. Transformation functions.

Many experiments presented in the literature testify to the inverse S-shaped pattern (see, e.g., [83]). However, one should realize that these experiments, usually dealing with students, concern one-time gains or investments, and the amounts of money involved are not large. In such situations, it is not surprising that people count to some extent on large values of the income, overestimating real probabilities of their occurrence.

In long run investment, when dealing with significant amounts of money and in situations when these amounts really matter for the investor (say, in the case of a retirement plan), the investor may exhibit a different behavior, proceeding from moderate values of the income rather than from the possibilities of large deviations. In such situations, an S-shaped transformation may be more adequate.

An interesting theory on possible forms of the transformation function Ψ may be found, e.g., in [83] and [103].

In conclusion, we discuss the main axiom connected with RDEU.

Consider an investor with a preference order $\succsim$ and two d.f.'s, $F(x)$ and $G(x)$. Assume that $F(x) = G(x)$ for all x's from a set A, and suppose that for the investor, $F \succsim G$.

Now, assume that we change $F(x)$ and $G(x)$ in a way such that

- all changes concern only values of $F(x)$ and $G(x)$ at x's from A,
- when changing $F(x)$ and $G(x)$, we keep them equal to each other for $x \in A$.

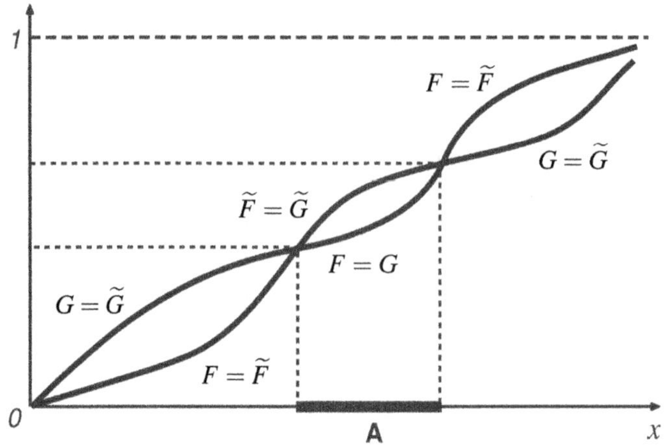

FIGURE 17.

If for any F and G, after such a change, the investor continues to prefer F to G, we say that the investor's preferences satisfy the *ordinal independence axiom*. The idea of such axioms (if you change the common part, the relation does not change) is referred to as the *sure-thing-principle*. Formally, the above axiom may be stated as follows.

For any two d.f.'s, $F(x)$ and $G(x)$, such that $F(x) = G(x)$ for all x's from some set A, consider two other d.f.'s, $\widetilde{F}(x)$ and $\widetilde{G}(x)$, such that $F(x) = \widetilde{F}(x)$ and $G(x) = \widetilde{G}(x)$ for all $x \notin A$, and $\widetilde{F}(x) = \widetilde{G}(x)$ for $x \in A$; see also Fig.17.

Consider the set $\mathcal{F} = \{F\}$ of all distributions F and a preference order $\succsim$ on $\mathcal{F}$.

Axiom 12 *(The ordinal independence). For any pair F, G from $\mathcal{F}$, if $F \succsim G$, then for any distributions $\widetilde{F}$ and $\widetilde{G}$ with the mentioned above properties, $\widetilde{F} \succsim \widetilde{G}$.*

One can prove that along with some more technical assumptions,

> *The ordinal independence axiom implies the RDEU principle.*

To conclude the whole Section 4, it is worth pointing again to one special feature of all models we considered: we identified random variables with their distributions. In other words, two r.v.'s with the same distributions are viewed in these models as equivalent. Much evidence has been accumulated showing that this is not always the case. Sometimes people distinguish outcomes with the same probabilities if the ways leading to these outcomes are different. The question touched on is connected with the so called coalescing property in the modern theory of gambles; see, for example, [83]. This interesting issue, however, is beyond the scope of this book.

1,2,3

5 OPTIMAL PAYMENT FROM THE STANDPOINT OF AN INSURED

5.1 Arrow's theorem

We consider here the following problem.

An individual with a wealth of w is facing a random loss X with a mean $m > 0$. To protect her/himself against at least a part of the risk, the individual appeals to an insurer.

The insurer, having many clients, when specifying the corresponding premium g, proceeds merely from the mean value of the future payment.

As a *particular example*, we may consider the situation when if the mean payment is λ, the insurer agrees to sell the coverage for the premium $g = (1+\theta)\lambda$ for a fixed $\theta > 0$. The coefficient θ is called a *relative security loading coefficient.* For instance, if $\theta = 0.1$, the insurer adds 10% to the mean payment. In following chapters, we will consider the characteristic θ repeatedly and in detail.

However, the case of such a determination of the premium is merely an example. Here, for us it is only important that there is a strict correspondence between g and λ, and once λ is given, the premium g is fixed.

If the coverage is complete, the mean payment is equal to the mean loss, that is, $\lambda = m$. (In the particular case of security loading, the premium would be $g_{\text{complete}} = (1+\theta)m$).

However, the premium in the case of complete coverage may be too large for the individual (or she/he may be just not willing to pay it). In this case, the individual buys a non-complete coverage with a mean payment $\lambda < m$. In this case, the policy is specified by a *payment function* $r(x)$, the amount that the insurer will pay if the loss X assumes a value x. Since the coverage is not complete, $0 \le r(x) \le x$. Note right away, that this implies that $0 \le r(0) \le 0$; that is $r(0) = 0$.

As we have assumed, the insurer requires only one condition on $r(x)$ to hold:

$$E\{r(X)\} = \lambda. \tag{5.1.1}$$

In this case, the premium g is completely specified by λ, and the individual can choose any $r(x)$ provided that (5.1.1) is true. The question is which $r(x)$ is the best.

Before considering examples of possible payment functions, note that (5.1.1) may be rewritten as follows.

Assume that $r(x)$ is non-decreasing, which is natural. Let $F_0(x)$ be the d.f. of X. Then, by virtue of (**0**.2.2.1),

$$E\{r(X)\} = \int_0^\infty (1 - F_0(x))dr(x).$$

Thus, condition (5.1.1) may be rewritten as

$$\int_0^\infty (1 - F_0(x))dr(x) = \lambda. \tag{5.1.2}$$

Consider particular examples of the payment function $r(x)$.

EXAMPLE 1 (*Proportional insurance* or *quota share insurance*). In this case, : $r(x) = kx$, $k \le 1$. Then, $E\{r(X)\} = E\{kX\} = km$, and (5.1.1) implies that $k = \lambda/m$.

EXAMPLE 2 (*Excess-of-loss or stop-loss insurance*). We will call it also *insurance with a deductible*. In this case,

$$r(x) = r_d(x) = \begin{cases} 0 & \text{if } x \leq d, \\ x-d & \text{if } x > d, \end{cases} \tag{5.1.3}$$

where the number d is called a deductible. In this case, payment is carried out only if the loss exceeds the level d, and if it happens, the insurer pays the overshoot. The term "excess-of-loss" is used when such a rule concerns each contract separately, "stop-loss" —when it concerns a whole risk portfolio.

Inserting (5.1.3) into (5.1.2), we have

$$\int_d^\infty (1 - F_0(x))dx = \lambda. \tag{5.1.4}$$

The last relation is an equation for d given λ. Simple particular examples are relegated to Exercise 61.

EXAMPLE 3 (*Insurance with a limit coverage*). In this case,

$$r(x) = \begin{cases} x & \text{if } x \leq s, \\ s & \text{if } x > s, \end{cases}$$

where s is the maximum the insurer will pay. Again using (5.1.2), we see that restriction (5.1.1) may be written as

$$\int_0^s (1 - F_0(x))dx = \lambda,$$

which is an equation for s. □

We return to the optimization problem. Assume that the preferences of the individual are preserved by a function $U(F)$. For the reader who skipped Section 4 on non-linear criteria, we start with the EUM case where

$$U(F) = \int_0^\infty u(x)dF(x), \tag{5.1.5}$$

and u is a non-decreasing utility function. In the end of the section, we consider the non-linear case.

Denote by $F_{(r)}(x)$ the distribution function of the r.v.

$$X_{(r)} = w - g - X + r(X),$$

the wealth of the individual under the choice of a payment function $r(x)$. Our goal is to find a function r for which $F_{(r)}$ is the best. More rigorously, we a looking for a function r^* which maximizes the function $Q(r) = U(F_{(r)})$.

It is remarkable that under certain conditions, the optimal payment function does not depend on the particular form of the utility function $u(x)$ and on the premium g. More precisely, r^* has the type (5.1.3) with the deductible d specified in (5.1.4).

Let us state it rigorously. For a fixed $\lambda \in (0, m]$, consider the set of all function $r(x)$ satisfying (5.1.2), that is, the set

$$\mathcal{R}_\lambda = \{r(x) : E\{r(X)\} = \lambda\}.$$

The theorem below belongs to K.Arrow; see, e.g., [5], [7].

Theorem 13 *Let $u(x)$ in (5.1.5) be concave, and $r^*(x) = r_d(x)$, where d satisfies (5.1.4). Then for any function $r(x)$ from $\mathcal{R}_\lambda$,*

$$Q(r) \le Q(r^*). \tag{5.1.6}$$

So,

> The optimal payment is the *same* for *any* concave utility function.

EXAMPLE 4. Two people facing the same loss X but having different utility functions, say $\sqrt{x}$ and $\ln x$, will prefer the *same* deductible policy and with the same deductible, provided that they choose the same mean payment λ. □

Proof of Theorem 13. As we know from Calculus, the l.-h.s. of (5.1.4), as a function of d, is continuous. By the general formula (**0**.2.2.2), this function is equal to m at $d = 0$, and it converges to zero as $d \to \infty$. Hence, for any $\lambda \in (0, m]$, there exists a number d for which the l.-h.s. of (5.1.4), that is, $E\{r_d(X)\}$, is equal to λ.

In particular, this means that $r_d \in \mathcal{R}_\lambda$.

(As a matter of fact, since $m > 0$, the number d for which (5.1.4) is true, is unique. Indeed, let x_0 be the smallest point at which $F_0(x) = 1$. If $F_0(x) < 1$ for all x's, we set $x_0 = \infty$. Because $m > 0$, the point $x_0 > 0$. Then $F_0(x) < 1$ for all $x < x_0$, and $1 - F_0(x) > 0$ on $[0, x_0)$. Consequently, the l.-h.s. of (5.1.4) is strictly decreasing for $d \in [0, x_0)$.)

Now, let us fix $r \in \mathcal{R}_\lambda$ and set $F(x) = F_{(r)}(x)$. Denote by $F^*(x)$ the d.f. of the r.v. $X^* = X_{(r^*)}$, i.e., the final wealth in the case (5.1.3) with d satisfying (5.1.4). Set $a = w - g - d$, and $b = w - g$.

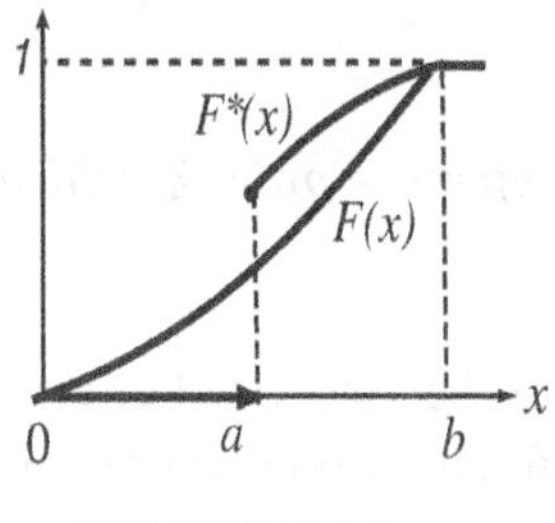

FIGURE 18.

Note that for both d.f.'s, $F(b) = F^*(b) = 1$. Since $P(X^* < a) = 0$, and $P(X^* = a) = P(X \ge d)$, the d.f. $F^*(x) = 0$ for $x < a$, and has a jump of $P(X \ge d)$ at the point a; see Fig.18. For $x > a$, we have $F^*(x) = P(w - g - X \le x)$ because $r(X) = 0$ for $X \le d$.

On the other hand, $F(x) = P(w - g - X + r(X) \le x)$, and since $r(X) \ge 0$, we have

$$F(x) \le F^*(x) \text{ for } x \ge a. \tag{5.1.7}$$

A typical picture is given in Fig.18.

We need also the following relations. Because $E\{r(X)\} = E\{r^*(X)\}$, we have $E\{X_{(r)}\} = E\{X^*\}$. Therefore, integrating by parts, we get that

$$\int_{-\infty}^{b} [F^*(z) - F(z)]\,dz = \int_{-\infty}^{b} z\,d\,[F(z) - F^*(z)] = E\{r(X)\} - E\{r^*(X)\} = 0. \tag{5.1.8}$$

(How to prove that $\lim_{z \to -\infty} z\,[F(z) - F^*(z)] = 0$ is shown in Section **0**.2.5.) From (5.1.8) it follows that $\int_{-\infty}^{x} [F^*(z) - F(z)]\,dz = -\int_{x}^{b} [F^*(z) - F(z)]\,dz$ for $x \le b$. Then, in view of

(5.1.7), for $a \le x \le b$,

$$\int_{-\infty}^{x} [F^*(z) - F(z)]\,dz \le 0. \tag{5.1.9}$$

On the other hand, since $F^*(x) = 0$ for $x < a$, inequality (5.1.9) is true for $x < a$ also, and hence it is true for all $x \le b$. Note also that, since $F^*(z) - F(z) = 1 - 1 = 0$ for $x > b$, eventually (5.1.9) is true for all x's.

Let us proceed to a direct proof. Assuming for simplicity that u is sufficiently smooth, integrating by parts, and taking into account that $F^*(b) = F(b)$, we have

$$\begin{aligned} U(F^*) - U(F) &= \int_{-\infty}^{b} u(x)d[F^*(x) - F(x)] = \int_{-\infty}^{b} [F(x) - F^*(x)]u'(x)dx \\ &= \int_{-\infty}^{b} u'(x)d\left(\int_{-\infty}^{x} [F(z) - F^*(z)]dz\right). \end{aligned} \tag{5.1.10}$$

Making use of (5.1.8), we integrate by parts in (5.1.10) one more time, which leads to

$$U(F^*) - U(F) = \int_{-\infty}^{b} \left(\int_{-\infty}^{x} [F^*(z) - F(z)]dz\right) u''(x)dx. \tag{5.1.11}$$

Since $u''(x) \le 0$, (5.1.11) and (5.1.9) implies that $U(F^*) \ge U(F)$. ■

Route 1 ⇒ page 141

2,3

5.2 A generalization

Now, let us realize that (5.1.9) means that F^* dominates F in the sense of the second stochastic dominance (SSD); see Section 3.5.3. Consequently, we have proved above the following much more general theorem.

Theorem 14 *The payment function $r^*(x)$ is optimal for any preference order $\succsim$ which is monotone with respect to the SSD.*

EXAMPLE 1. Consider the case of implicit utility described in Section 4.3. Let $v(x,y)$ be a given function as it is defined in this section, and let the preference order be preserved by the certainty equivalent $c(F)$ defined in (4.3.3). Assume that $v(x,y)$ is concave with respect to x and non-increasing in y. We show that the corresponding preference order is monotone with respect to the SSD.

To simplify calculations, assume that $v(x,y)$ is sufficiently smooth. For two distributions, F and G, similarly to (3.5.8), we have

$$\int_{-\infty}^{\infty} v(x,c)d[F(x) - G(x)] \ge \int_{-\infty}^{\infty} \left(\int_{-\infty}^{x} [F(z) - G(z)]dz\right) v''_{xx}(x,c)dx,$$

where

$$v''_{xx}(x,c) = \frac{\partial^2}{\partial x^2} v(x,c).$$

Consequently, if $v''_{xx} \le 0$ and F dominates G in the sense of the SSD, then

$$\int_{-\infty}^{\infty} v(x,c)d[F(x) - G(x)] \ge 0 \tag{5.2.1}$$

for all c. Let $c = c(F)$. Then $\int_{-\infty}^{\infty} v(x,c(F))dF(x) = 0$, and hence $\int_{-\infty}^{\infty} v(x,c(F))dG(x) \le 0$. On the other hand, by definition, $\int_{-\infty}^{\infty} v(x,c(G))dG(x) = 0$. Since $v(x,y)$ is non-increasing in y, from this it easily follows that $c(G) \le c(F)$, and hence $F \succsim G$. □

5.3 Historical remarks regarding the whole chapter

The remarks below are far from being comprehensive. The modern theory of comparing risky alternatives was established by J. von Neumann and O. Morgenstern in [96]. A rather complete exposition of this theory and generalizations may be found in [82] by R.D. Luce and H. Raifa.

There is a very rich literature on further developments of this theory. Many of them are reflected in the monographs by M. Denuit, J. Dhaene, M. Goovaerts and P. Kaas [33], P. Fishburn [39], R.D. Luce [83], J. Quiggin [106], P. Wakker [140]. The reader can find there also historical notes and a rich bibliography. A bibliography with comments may be found also in P.Wakker's web-site http://people.few.eur.nl/wakker/refs/webrfrncs.doc

To the author's knowledge, the weighted utility concept was first considered by S.H. Chew and K.R. MacCrimmon (1979, [25], [26]) and H. Bühlmann (1980, [21]). The theory with corresponding axioms was later developed by S.H. Chew (1989, [24]). Other axioms leading to close criteria were considered by P. Fishburn (1988, [39]).

The implicit or comparative utility and the betweenness axiom or axioms quite similar to it were suggested and explored independently by A.Ya. Kiruta (1980, [73]) , S.A. Smolyak (1983, [133]; see also [134]) and E. Dekel (1986, [32]). Various results on criteria close to comparative utility were considered in the mentioned monograph by P. Fishburn [39], and on the betweenness axiom—in the mentioned S.H. Chew's paper [24].

The full RDEU model, including a set of axioms, was first suggested by J. Quiggin [107], though some earlier works of J. Quiggin had already contained some relevant ideas; see references in J. Quiggin [106]. Some models including weighting functions, say, in the case of binary gambles were considered earlier in the prospect theory of D. Kahneman and A. Tversky [68]. A special case of RDEU was independently considered in the "dual model" of M. Yaari (1987, [146]) and developed further by A. Roell (1987, [114]). To the author's knowledge, an axiomatic system for the most general case including continuous distributions, was considered by P. Wakker (1994, [140]).

The idea of the above proof of Theorem 13 and Theorem 14 belongs to C. Gollier and H. Schlesinger; see [48] and also [125]. Other generalizations of Arrow's theorem (mostly using optimization technique) may be found, e.g., in papers by E. Karni [71], A. Raviv [109], and I. Zilcha and S.H. Chew [149]. See also references therein.

6 EXERCISES

Sections 1 and 2

1. Make sure that you indeed understand why if $E\{X\} = m$ and $Var\{X\} = \sigma^2 \neq 0$, then for the normalized r.v. $X^* = (X - m)/\sigma$, we have $E\{X^*\} = 0$, $Var\{X^*\} = 1$.

2. Find the 0.2-quantile of a r.v. X taking values $0, 3, 7$ with probabilities $0.1, 0.3$, 0.6, respectively.

3. This exercise concerns the VaR criterion with a parameter γ. R.v.'s X with or without indices correspond to an income.

   (a) Let X_1 take on values $0, 1, 2, 3$ with probabilities 0.1, 0.3, 0.4, 0.2, respectively, and X_2 take on the same values with probabilities 0.1, 0.4, 0.2, 0.3. Find all γ's for which $X_1 \succsim X_2$.

   (b) Let r.v. X_1 be uniform on $[0, 1]$, and X_2 be exponential with $E\{\xi_2\} = m$. When is the relation $X_2 \succsim X_1$ true for all γ's? Let $m = 1/2$. Find all γ's for which $X_1 \succsim X_2$.

   (c) Let r.v.'s $X_1 = 1 - \xi_1$, $X_2 = 1 - \xi_2$, where the loss ξ_1 is uniform on $[0, 1]$, and ξ_2 is exponential with $E\{\xi_2\} = m$. When is the relation $X_1 \succsim X_2$ true for all γ ? Let $m = 1/2$. Find all γ's for which $X_1 \succsim X_2$. (*Advice*: Compare just ξ's and observe that $q_\gamma(X) = 1 - q_{1-\gamma}(\xi)$.)

   (d) Let X_1 be uniform on $[0, 3]$, and let X_2 be uniform on $[1, 2]$. Find all γ's for which $X_1 \succsim X_2$.

4.* Solve Exercises 3a,d for the case of the TailVaR criterion.

5. It is known that if $X_1, ..., X_n$ are independent exponential r.v.'s with unit means, then $S_n = X_1 + ... + X_n$ has the Γ-distribution with parameters $(1, n)$. We will prove it in Section **2**.2.2.

   Let $n = 10$, and let r.v.'s $X_1, ..., X_{10}$ be defined as above. Suppose that $Y_i = 1.1 \cdot X_i$ for $i = 1, ..., 10$, represent the returns for the investments in 10 independent assets. (For the notion of "return" see Example 1.2.2-4.) Thus, since $E\{X_i\} = 1$, an investment into each asset gives 10% profit on the average. Proceeding from the VaR criterion, figure out what is more profitable: to invest \$10 into one asset, or split it between 10 assets. You are recommended to use Excel (or another software); the corresponding command in Excel for quantiles is =GAMMAINV$(p, \nu, 1/a)$ where a is the scale parameter.

6. Consider two assets. The investment of a unit of money into the ith asset leads to an income of $1 + \xi_i$ units; $i = 1, 2$. Assume that ξ's have a joint normal distribution, $E\{\xi_i\} = 0$, $Corr\{\xi_1, \xi_2\} = \rho$. Suppose that you invest some money into each asset, K_i is the *profit* of the investment into the ith asset, and K is the total *profit*. Prove that

$$q_\gamma(K) = \sqrt{q_\gamma^2(K_1) + q_\gamma^2(K_2) + 2\rho q_\gamma(K_1) q_\gamma(K_2)}.$$

   The last formula is relevant to the JP Morgan RiskMetics^{TM} methodology; see, e.g. [64]. Some references may be found also in *http://www.riskmetrics.com* and *http://www.gloriamundi.org*.

7.* (a) Proceeding from (1.2.4) and using integration by parts, prove that $E\{X \mid X \leq t\} = \frac{1}{F(t)}\left(tF(t) - \int_{-\infty}^{t} F(x)dx\right) = t - \frac{1}{F(t)}\int_{-\infty}^{t} F(x)dx$, where $F(x)$ is the d.f. of X.

(b) Show that if $F(t)$ is continuous and q_γ is the γ-quantile of X, then

$$V_{\text{tail}}(X) = E\{X \,|\, X \le q_\gamma\} = q_\gamma - \frac{1}{\gamma}\int_{-\infty}^{q_\gamma} F(x)dx. \tag{6.1}$$

Show that it may be false if X is not a continuous r.v.

(c) Show that $V_{\text{tail}}(X)$ is monotone in the class of continuous r.v.'s. (*Advice:* You can either use (6.1) or observe that in the continuity case, the conditional d.f. $P(X \le x \,|\, X \le q_\gamma) = \frac{1}{\gamma}F(x)$ for $x \le q_\gamma$ and $= 1$ otherwise.)

(d) Sketch a typical graph of a continuous $F(x)$. Consider γ for which $q_\gamma < 0$ and point out the region in the graph whose area equals $\gamma\,|V_{\text{tail}}(X)|$.

8. Take real data on the daily stock prices for the stocks of two companies for one year from, say, http://finance.yahoo.com or another similar site. For different values of γ, compare the performance of the companies using the VaR and the TailVaR criteria. (The absolute values of the prices should have no effect on results. The analysis should be based on returns, that is, on the ratios of the prices on the current and the previous days. For the notion of "return" see Example 1.2.2-4.) Estimate the mean return for each company. Try to characterize and compare the performance of the companies, taking into account all three characteristics mentioned.

9.* If we interpret X as an income, then the r.v. $\widetilde{X} = -X$ may be interpreted as a loss. Considering only r.v.'s whose d.f.'s are strictly increasing, do the following.

(a) Prove that $q_\gamma(X) = -q_{1-\gamma}(\widetilde{X})$. Show it graphically.

(b) Consider, instead of (1.2.3), the function $\widetilde{V}(\widetilde{X};\, s) = E\{\widetilde{X} \,|\, \widetilde{X} \ge s\}$, the mean value of the loss given that it has exceeded a level s. Show that $\widetilde{V}(\widetilde{X};\, s) = -V(X;\, -s)$ for any s, and $\widetilde{V}(\widetilde{X};\, s) = |V(X;\, -s)|$ for all $s \ge 0$. Give a heuristic explanation.

(c) Consider the criterion preserved by the risk measure $\widetilde{V}_{\text{tail}}(\widetilde{X}) = E\{\widetilde{X} \,|\, \widetilde{X} \ge q_{1-\gamma}(\widetilde{X})\}$. Show that $X \succsim Y \Leftrightarrow \widetilde{X} \precsim \widetilde{Y} \Leftrightarrow \widetilde{V}_{\text{tail}}(\widetilde{X}) \le \widetilde{V}_{\text{tail}}(\widetilde{Y})$.

10.** For some $V(X)$ and a family $\mathcal{P}$, suppose that (1.3.2) is true. Show that the monotonicity property and Properties I-III from Section 1.3 are fulfilled. (The converse assertion is more difficult to prove, but the sufficiency of the representation (1.3.2) is understandable. Recall that $\min_x(f(x)+g(x)) \ge \min_x f(x) + \min_x g(x)$.)

Section 3

11. Show that if $u(x)$ is strictly increasing, then the rule (3.1.2) is strictly monotone. (*Advice*: Consider $E\{u(X)\} - E\{u(Y)\} = E\{u(X) - u(Y)\}$.)

12. Graph all utility functions from Section 3.1.3.

13. Consider a r.v. X such that $P(X > x) = x^{-1}$ for $x \ge 1$. This is a particular case of the Pareto distribution which we will consider in Section **2**.1.1.3 in detail. Does X have a finite expected value? Find the certainty equivalent of X for $u(x) = \sqrt{x}$.

14. Prove that the Masset criterion (3.1.4), whatever the parameter a is, positive or negative, has the following properties.

(a) (*Additivity property.*) For any two independent r.v. X_1, X_2, it is true that $C(X_1 + X_2) = C(X_1) + C(X_2)$. Interpret this, viewing X_1, X_2 as the results of two independent investments.

(b) (*An analog of Proposition 2: the independence from the initial wealth.*) If $c(X_1) \geq c(X_2)$, then $c(w+X_1) \geq c(w+X_2)$ for any w. Give an economic interpretation.

15. Write formulas for the certainty equivalents for the cases 2, 3, 5 from Section 3.1.3.

16. Let X be exponential, and $u(x) = -e^{-\beta x}$ (see Section 3.1.3). Show that for the certainty equivalent $c(X)$, we have $\frac{c(X)}{E\{X\}} \to 1$ as $E\{X\} \to 0$, that is, $c(X) \approx E\{X\}$ if $E\{X\}$ is small. Interpret it. (*Advice*: Use $\ln(1+x) = x + o(x)$.)

17. Repeat calculations of Example 3.2-1 for the case when $\xi = 0$ with probability 0.9, and is uniformly distributed on $[0,1]$ with probability 0.1.

18. An EUM customer of an insurance company has a total wealth of 100 (in some units) and is facing a random loss ξ distributed as follows: $P(\xi = 0) = 0.9, P(\xi = 50) = 0.05, P(\xi = 100) = 0.05$.

    (a) Let the utility function of the customer be $u(x) = x - 0.005x^2$ for $0 \leq x \leq 100$. Graph it. Is the customer a risk averter?

    (b) What would you say in the case $u(x) = x + 0.005x^2$?

    (c) For the case 18a, find the maximal premium the customer would be willing to pay to insure his wealth against the loss mentioned. First, set the equation clearly and explain it, then solve. Is the premium you found greater or less than $E\{X\}$? Might you predict it from the beginning?

    (d) Find the minimal premium which an insurance company would accept to cover the risk mentioned, if the company's preferences are characterized by the utility function $u_1(x) = \sqrt{x}$, and the company takes 300 as an initial wealth. Is the premium you found greater or less than $E\{\xi\}$. Might you predict it?

    (e) Solve Exercise 18d for the case when the r.v. ξ is uniformly distributed on $[0,100]$.

    (f) Solve Exercise 18c for $u(x) = 200x - x^2 + 349$. (*Advice*: Look at this function attentively before starting calculations.)

19. Find the maximal premium the customer would be willing to pay to insure half of the loss in the situations of Example 3.2-1.

20. Give an explicit example when the maximal acceptable premium for a customer does depend on the initial wealth. (*Advice*: You may take $u(x) = \sqrt{x}$ and a r.v. assuming two values.)

21. Take real data on the daily stock prices for the stocks of two companies for one year from, say, http://finance.yahoo.com or another similar site. Considering a particular utility function, for instance, $u(x) = -e^{-\beta x}$ for some β, determine which company is better for an EU maximizer with this utility function. (*Advice:* Look at the comment in Exercise 8. To estimate the expected value $E\{u(X)\}$, where X is a random return, we can use the usual estimate $\frac{1}{n}[E\{u(X_1)\} + ... + E\{u(X_n)\}]$, where $\{X_1,...,X_n\}$ is the time series based on the data. Excel is convenient for such calculations.) Add to your analysis the characteristics considered in Exercise 8. Try to describe the performance of the companies, taking into account all characteristics you computed.

22. Is Condition Z from Section 3.4.1 a requirement on (a) distribution functions, or (b) random variables, or (c) the preference order under consideration?

23. Is Condition Z from Section 3.4.1 based on the concept of expected utility?

24. Is it true that for an expected utility maximizer to be a risk averter, his/her utility function should have a negative second derivative? Give an example. (*Advice:* Look up the definition of concavity.)

25. Let $u(x) = x$ for $x \in [0,1]$, and $u(x) = \frac{1}{2} + \frac{1}{2}x$ for $x \geq 1$. Is an EU maximizer with this utility function a risk averter?

26. Check for risk aversion the criteria with utility functions from Section 3.1.3.

27. Let the utility function of a person be $u(x) = e^{ax}$, $a > 0$. Graph it. Is the person a risk averter or a risk lover? Show that in this case, the comparison of risky alternatives does not depend on the initial wealth.

28. Let X be exponential, and $u(x) = -e^{-\beta x}$ (see Section 3.1.3). Show that the certainty equivalent $c(X) \to 0$, as $\beta \to \infty$. Interpret it in terms of risk aversion.

29. In Examples 3.1.3-1 and 2, compare the expected values and the certainty equivalents for different values of α including the case $\alpha \to 0$. Interpret results.

30. Let $u_1(x)$ and $u_2(x)$ be John's and Mary's utility functions, respectively.

    (a) How do John's and Mary's preferences differ if (i) $u_1(x) = 2u_2(x) + 3$; (ii) $u_1(x) = -2u_2(x) + 3$.

    (b) Let $u_1(x) = \sqrt{x}$ and $u_2(x) = x^{1/3}$. Who is more averse to risk? In what sense?

    (c) Let $u_1(x) = -1/\sqrt{x}$ and $u_2(x) = -1/x^{1/3}$. Who is more afraid of being ruined (having a zero or negative income)? Who is more averse to risk?

31. Let Michael be an EUM with the utility function $u(x) = -\exp\{-\beta x\}$ and $\beta = 0.001$. (For this value of β, the values of expected utility in this problem will be in a natural scale.) Michael compares stocks of two mutual funds. The today price for each is \$100 per share. Michael believes that in a year the price for the first stock will be on the average either 10% higher or 10% lower with equal probabilities, while for the second stock 10% up or down should be replaced by a slightly higher figure, approximately, 11%. (a) Which mutual fund is "better" for Michael? Do we need to calculate something? (b) Now, assume that the second fund invites all people who buy 100 shares to a dinner valued at \$$k$. Which k would make difference?

32. Provide calculations to obtain (3.2.6).

33. (a) It is known that for any positive r.v. X, the function $n(s) = (E\{X^s\})^{1/s}$ is non-decreasing in s. Using this fact, prove that the r.-h.s. of (3.2.5) is non-decreasing in β. (*Advice:* Set $\xi = \ln(X)$ and write $X = e^{\xi}$.)

    (b) Using Jensen's inequality, prove that indeed $n(s)$ above is non-decreasing. (*Advice:* We should prove that if $s < t$, then $n(s) \leq n(t)$, which is equivalent to $E\{X^s\} \leq (E\{X^t\})^{s/t}$. Write $X^s = (X^t)^{s/t} = u(X^t)$, where $u(x) = x^{s/t}$. Figure out whether $u(x)$ is concave for $s < t$.)

34. Write the risk aversion function for the utility function (3.1.3) setting $u_1(x) = \ln x$.

35.* Let $R(x)$ and $R_r(x)$ be absolute and relative risk aversion functions for a utility function $u(x)$. Find the corresponding risk aversion functions for $u^*(x) = bu(x) + a$, where a, b are constants. Interpret the result in the light of the first property from Section 3.1.2.

36.* Find the absolute risk aversion for $u(x) = e^{\beta x}$, $\beta > 0$. Interpret the fact that the risk aversion characteristic is negative.

37.* Prove that, up to linear transformation, only exponential utility functions have a constant absolute risk aversion, and only power utility functions and the logarithm have a constant relative risk aversion. (*Advice:* You should consider equations $u'' = cu'$, and $xu''(x) = cu'(x)$.)

38. Consider two r.v.'s both taking values $1, 2, 3, 4$. For the first r.v. the respective probabilities are $0.1, 0.2, 0.5, 0.2$, for the second, $0.1, 0.3, 0.3, 0.3$. Which r.v. is better in the EUM-risk-aversion case? Justify the answer.

39. Consider a r.v. taking values $x_1, x_2, \ldots$ with probabilities $p_1, p_2, \ldots$, respectively. Let the x_i's be equally spaced, that is, $x_{i+1} - x_i$ equals the same number for all i. Assume that for a particular i, we replaced probabilities p_{i-1}, p_i, p_{i+1} by probabilities $p_{i-1} + \Delta$, $p_i - 2\Delta$, $p_{i+1} + \Delta$ where a positive $\Delta \leq p_i/2$. Has the new distribution become worse or better in the EUM-risk-aversion case? Justify the answer.

40. Consider two distributions, F_1 and F_2. The distribution F_1 is uniform on $[-1, 1]$. The distribution F_2 is the triangular distribution on the same interval. More precisely, the corresponding density

$$f_2(x) = \begin{cases} x+1 \text{ if } -1 \leq x \leq 0, \\ 1-x \text{ if } -0 \leq x \leq 1, \\ 0 \quad \text{otherwise.} \end{cases}$$

Let $f_1(x)$ be the density of the first distribution.

Graph the densities $f_1(x)$ and $f_2(x)$ in the same system of coordinates. Guess which distribution is better in the EUM-risk-aversion case, and give heuristic arguments in favor of your guess. Prove your statement rigourously. Point out a similarity between this exercise and Exercises 38-39.

41. John agreed with the following strange payment for a job done. A regular coin will be tossed. In the case of a head, John will be paid \$200. In the case of a tail, a die will be rolled. If the die shows one or two, John will be paid \$300, otherwise – nothing. Determine the random payment considering the mixture of the distributions of r.v.'s $X_1 = \$200$, and $X_2 = 300$ or 0 with probabilities $1/3$ and $2/3$, respectively.

42.* (a) Is the FSD rule defined in Section 3.5.2 a requirement on distributions or on the preference order?

(b) Does the FSD rule concern only the EUM criterion or all preference orders?

(c) Answer the same questions regarding the SSD.

43.* Which criteria from Section 1.2 satisfy the FSD rule?

44.* Show rigorously that in the general Δ-scheme from Section 3.5.5, (a) equivalence sets are planes or hyper-planes; (b) to determine completely the preference order, it suffices to determine $n-1$ equivalent points.

45.* Fig.'s 19abcd depict equivalence curves in the Δ-scheme for the distributions of r.v.'s taking three values. Which figures correspond to EUM, and which do not?

46.* Consider the distributions of all random variables taking only values 0, 10, 20. We identify any such a distribution F with the vector of probabilities (p_1, p_2, p_3). Let $F_1 = (0.1,\ 0.5,\ 0.4)$, $F_2 = (0.2,\ 0.2,\ 0.6)$, and $F_3 = (0.1,\ 0.8,\ 0.1)$.

(a) Mark all corresponding points in the (p_1, p_3)-plane picture.

(b) Find a distribution F_4 such that the assertion $F_1 \simeq F_2$, $F_3 \simeq F_4$ would not contradict the independence axiom.

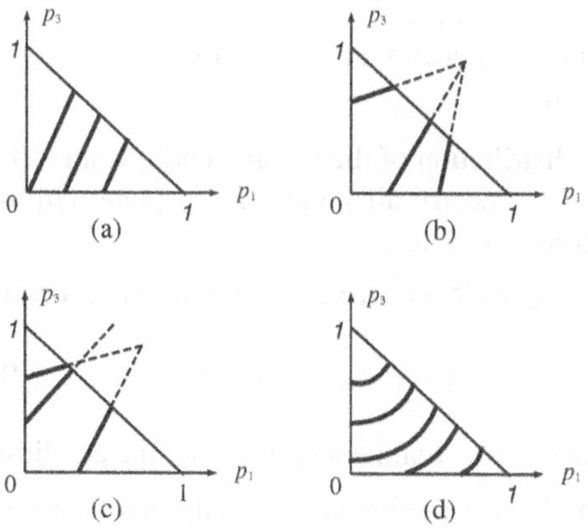

FIGURE 19.

47.* Consider the probability distributions of ALL random variables taking values $0, 10, 20$, and 40.

(a) Identify such distributions with points in $\mathbb{R}^3$. What region in $\mathbb{R}^3$ do we consider in this case? To what distributions do boundary points of this region correspond?

(b) Assume that you are an EU maximizer. Fix a distribution F_0, and consider ALL distributions $F \simeq F_0$. Where do all points corresponding to these distributions lie?

(c) Assume that the following three distributions are equivalent:
$(0.2, 0.3, 0.1, 0.4)$, $(0.3, 0, 0, 0.7)$, $(0, 0.6, 0.4, 0)$. Find one more distribution equivalent to the mentioned.

48.* Consider the distributions from Exercise 46. You are an EUM, and your utility function is increasing. What is better, F_1 or F_3 ? Find $G_1 = \frac{1}{2}(F_1 + F_2)$, $G_2 = \frac{1}{2}(F_3 + F_2)$. Mark both points in the Δ-scheme picture. Which point is better?

49.* Consider an EUM in the situation of Exercise 46. Assume that distributions $(0.1, 0.5, 0.4)$ and $(0.2, 0.3, 0.5)$ are equivalent. Find one more distribution which is equivalent to these distribution. Find all such distributions.

50.* You know that Fred is an EUM, and for him "the larger, the better". You ask Fred to compare the following two random variables: both take on values $0, 20, 40$; the first—with probabilities $(\frac{1}{4}, \frac{1}{2}, \frac{1}{4})$, respectively, and the second—with probabilities $(\frac{1}{3}, \frac{1}{6}, \frac{1}{2})$. It turns out that Fred considers these distributions equivalent for him.

(a) Is this information enough in order to predict the result of comparison by Fred of ANY two random variables with the values mentioned?

(b) Which distribution would Fred prefer: $(\frac{1}{2}, \frac{1}{4}, \frac{1}{4})$ or $(\frac{3}{5}, \frac{1}{10}, \frac{3}{10})$?

(c) Is Fred a risk lover? If yes, justify the answer. If no, does this mean that he is a risk averter?

*Section 4**

51. Consider Fig.'s 19abcd from Exercise 45 depicting equivalence curves in the Δ-scheme for the distributions of r.v.'s taking three values. Which figures correspond to axioms you know. Identify these axioms.

52. Consider distributions F_1, F_2, and F_3 from Exercise 46. Find a non-trivial and meaningful example of a distribution F_4 such that the relations $F_1 \simeq F_2$, $F_3 \simeq F_4$ would not contradict the betweenness axiom.

53. Consider probability distribution of the income taking values $0, 20$, and 40 with probabilities p_1, p_2, p_3, respectively. Specify all points in the plane (p_1, p_3), which correspond to the distributions under consideration.

    Assume that for Jane equivalence curves on the diagram mentioned are curves given by the formula

$$p_3 = c + e^{p_1}, \quad \text{where} \quad -e \le c \le 0.$$

    (a) Graph these curves, and realize why we need the condition $-e \le c \le 0$.
    (b) Is Jane a EUM? Do her preferences meet the Betweenness Axiom?
    (c) Since we consider only the narrow class of distributions concentrated on the set 0, 20, 40, we cannot figure out whether Jane is a risk lover or not. Nevertheless, reasoning heuristically, give an argument that we should not expect that Jane is a risk lover.
    (d) Jane follows the rule "the larger the better". Which distribution is better for her: $(0.1, 0.5, 0.4)$ or $(0.2, 0.2, 0.6)$?

54. In Example 4.3.1-3, find the certainty equivalent of a r.v. $X = d > 0$ or 0 with probabilities p and q, respectively. Analyze and interpret the case p close to one.

55. Let $g(s)$ defined in Section 4.3 equal $1 - e^{-s}$ for $s > 0$, and s for $s \le 0$. Graph $g(s)$. Let c be the certainty equivalent of a r.v. X. Is it true that $c \le E\{X\}$? Estimate the certainty equivalent for X equal to 1 or 0 with equal probabilities.

56. Show that all lines (4.2.8) intersect at one point. (*Advice:* Consider two lines, l_1 and l_2, corresponding to two different values d_1 and d_2. They may intersect only at a point in l_∞ where $W(F)$ is not defined since otherwise at this point $W(F)$ would have taken two different values.)

57. Show that $W(F^{(\alpha)})$ in (4.2.6) is not equal, in general, to $\alpha W(F_1) + (1-\alpha)W(F_2)$. (*Hint:* The answer does not require long calculations. Consider, for example $\alpha = 0.5$ and the case $\int_{-\infty}^{\infty} w(x)dF_2(x) = 2\int_{-\infty}^{\infty} w(x)dF_1(x)$.)

58. In the situation of Example 4.3.1-2, find the maximal premium $G_{\max}$ assuming $w = d$ and the loss to have the same distribution as X in this example.

59. Let an investor follow the RDEU criterion with $\Psi(p) = 1 - (1-p)^\beta$. In other words, Ψ behaves as a power function for p close to one, that is, for p's corresponding to large values of r.v.'s. Let X be an exponential r.v. with a distribution F. Show that in this case, the transformation of F corresponds to dividing X by β.

60. Find $G_{\max}$ in the situation of Example 4.4.2-1 in the case when $P(\xi = 0) = 0.9$, $P(\xi = 1) = 0.1$.

Section 5

61. Find the deductible d in (5.1.4) for $\lambda = m/2$ for two cases: (a) X is exponential, $E\{X\} = m$; (b) X is uniform on $[0, 2m]$. (So, the expected values are the same.) Compare results. Explain the difference from a heuristic point of view.

62.* Which regions in Fig.18 have the same area?

Chapter 2

An Individual Risk Model for a Short Period

1,2,3

This chapter and Chapter 3 are devoted to two connected topics. First, we explore the structure of the payments an insurance organization provides during a given time period. Secondly, we consider the solvency of the insurance process or, more specifically, the size of premiums sufficient for the insurance organization to meet its obligations.

We begin with one insured unit. It may be an individual or an organization; we will use the terms "client" or "insured". In this case, our goal is to explore the probability distribution of the payment of the company to a particular insured.

Next we consider the aggregate claim coming from many clients. Here, the most important problem consists in approximating the distribution of the aggregate claim in the case when the number of clients is large. "Good" approximations will allow us to estimate premiums acceptable for the insurer.

Chapter 3 concerns the portfolio of risks as a whole. In this case, the objects of study are the random number of claims coming into the insurance organization, the total payment the organization should provide, and again acceptable premiums.

In both chapters, we consider only a single and sufficiently short period of time, and all models in these chapters are static. Once we have studied how the company functions during a short period, we will proceed in Chapter 4 to insurance processes which run over extended time periods; that is, to dynamic models.

1 THE DISTRIBUTION OF AN INDIVIDUAL PAYMENT

1.1 The distribution of the loss given that it has occurred

1.1.1 Characterization of tails

Let us consider one client. If the client suffers damages potentially covered by the insurance contract, we will talk about a *loss event*. Denote by ξ the r.v. of the *loss given that the loss event occurred* and by q the probability of the loss event. Usually although not always, q is small. Regarding the r.v. ξ, the term *severity* is also frequently used.

Thus, the real loss of the insured is the r.v.

$$X = \begin{cases} \xi & \text{with probability } q, \\ 0 & \text{with probability } 1-q. \end{cases} \tag{1.1.1}$$

This may be rewritten as

$$X = I\xi, \tag{1.1.2}$$

where the *indicator* of the loss event

$$I = \begin{cases} 1 & \text{with probability } q, \\ 0 & \text{with probability } 1-q. \end{cases}$$

In this section, we do not touch on the timing of the payments issue, viewing ξ as the total amount of losses during the period under consideration. So, I is the indicator that at least one loss event has occurred.

If the insurance contract covers the whole possible loss, the amount paid by the insurance organization equals X. If the damage is not covered in full, the amount paid is a part of X.

This subsection concerns merely the distribution of ξ. Before classifying possible types of this distribution, we first recall

The exponential distribution which, in a certain sense, may be viewed as a key case. As was defined in Section **0**.3.2.2, this is the distribution of a positive r.v. $\xi = \xi_a$ with the density

$$f_a(x) = \begin{cases} 0 & \text{if } x < 0, \\ ae^{-ax} & \text{if } x \geq 0. \end{cases} \tag{1.1.3}$$

The parameter a is positive and plays the role of a scale parameter. If a r.v. ξ_1 has the density $f_1(x)$, then the r.v. $\xi_a = \xi_1/a$ has density $f_a(x)$(see Section **0**.3.2.2 and Exercise 1). The corresponding distribution function (d.f.)

$$F_a(x) = \begin{cases} 0 & \text{if } x < 0, \\ 1-e^{-ax} & \text{if } x \geq 0; \end{cases} \tag{1.1.4}$$

the tail, i.e.,

$$P(\xi_a > x) = e^{-ax}; \tag{1.1.5}$$

and

$$E\{\xi_a\} = 1/a, \; Var\{\xi_a\} = 1/a^2. \tag{1.1.6}$$

(See Section **0**.3.2.2 for detail.)

In Section **0**.3.2.2, we have also shown that the exponential distribution has the unique

Lack-of-Memory (or Memoryless) Property: for any $x, y \geq 0$,

$$P(\xi > x+y \,|\, \xi > x) = P(\xi > y), \tag{1.1.7}$$

where $\xi = \xi_a$ is defined above.

Assume that we have preliminary information that the loss has exceeded a level x. Then the l.-h.s. of (1.1.7) is the conditional probability that the real loss will be larger than x by y. If the memoryless property holds, this probability does not at all depend on the value of x.

We may interpret (1.1.7) in a similar way in the case when ξ is not the size of a loss but the duration of a process; for example, the duration of a job to be done or the time between two consecutive claims arriving at a company. Such examples will turn out to be important when we deal with dynamic models.

Assume that a process, a job for example, has already lasted x hours. Then the l.-h.s. of (1.1.7) is the conditional probability that the job will last y hours *more*. In the case of the memoryless property, this conditional probability does not depend on when the job began. One may view the situation as if at each moment the process starts over as from the very beginning—so to say, the system "does not remember" what happened before.

Certainly, the property under discussion is very special, and it is important to keep in mind that

> The exponential distribution is the only distribution with the memoryless property.

A not difficult although non-trivial proof may be found in many textbooks on Probability (see, e.g., [38], [116], [120]).

For other distributions, $P(\xi > x+y \,|\, \xi > x)$ depends on x. It is especially important to consider $P(\xi > x+y \,|\, \xi > x)$ for large x's or putting it in another way, for large deviations. Formally, we let $x \to \infty$ and set

$$Q(y) = \lim_{x\to\infty} P(\xi > x+y \,|\, \xi > x) = \lim_{x\to\infty} \frac{P(\xi > x+y,\, \xi > x)}{P(\xi > x)} = \lim_{x\to\infty} \frac{P(\xi > x+y)}{P(\xi > x)}.$$

Consider three typical situations.

a) $P(\xi > x) \sim Cx^{-\alpha}$, as $x \to \infty$, for some constants C and $\alpha > 0$.[1] Then

$$Q(y) = \lim_{x\to\infty} \frac{x^{\alpha}}{(x+y)^{\alpha}} = \lim_{x\to\infty} \frac{1}{(1+y/x)^{\alpha}} = 1 \quad \text{for } any\ y.$$

It may be interpreted as follows. If we have information that ξ has exceeded a large level x, then with large probability the surplus $\xi - x$ will be also large. For instance, in the job example, it would mean that if the job has lasted for a long time, then we should not expect that it will be over soon.

We mentioned already that the probability $P(\xi > x)$ for large x's is often called the *tail* of the distribution. The particular case above is classified as that of a *heavy tail*. Later we will consider a general definition.

b) $P(\xi > x) \sim Ce^{-ax}$, as $x \to \infty$, for some constant C and $a > 0$. In this case, the tail is asymptotically exponential for large values of x, and

$$Q(y) = \lim_{x\to\infty} \frac{e^{-a(x+y)}}{e^{-ax}} = e^{-ay}.$$

Thus, the conditional distribution, at least asymptotically, for large x's, is exponential.

---

[1]The symbol $u(x) \sim v(x)$ means that $\frac{u(x)}{v(x)} \to 1$, that is, $u(x)$ and $v(x)$ are close for large x's.

c) $P(\xi > x) \sim Ce^{-ax^{\gamma}}$, where C and a are positive and $\gamma > 1$. In this case, the tail vanishes faster than any exponential function, and

$$Q(y) = \lim_{x\to\infty} \exp\{-a[(x+y)^{\gamma} - x^{\gamma}]\} = 0 \text{ for } \textit{any} \;\; y > 0.$$

► (To show this, first note that $(1+z)^{\gamma} - 1 \sim \gamma z$ as $z \to 0$, which may be verified, for example, by L'Hôpital's rule. Then,
$(x+y)^{\gamma} - x^{\gamma} = x^{\gamma}[(1+\frac{y}{x})^{\gamma} - 1] \sim x^{\gamma}\gamma\frac{y}{x} = x^{\gamma-1}\gamma y \to \infty$, as $x \to \infty$, for any $y > 0$ and $\gamma > 1$. ◄

In this case, if we have information that ξ has exceeded a given large value of x, this means that with high probability the real value of ξ is close to x. In the job example, this means that if the job has lasted for a long time, we expect that it will end soon.

The last two cases are classified as those of *light tails*.

Let us turn to a general classification. We restrict ourselves to positive r.v.'s. For the distribution F of a r.v. ξ, set $\overline{F}(x) = 1 - F(x) = P(\xi > x)$, the *tail* of F.

A distribution F is said to be *light-tailed* if for some positive c and B,

$$\overline{F}(x) \le Be^{-cx} \tag{1.1.8}$$

for sufficiently large x's, more precisely for all $x \ge x_{cB}$, where x_{cB} is some number perhaps depending on c and B.

Thus, we are interested in the behavior of the tail $\overline{F}(x)$ for large x. The significance of the definition above is that $P(\xi > x) \to 0$, as $x \to \infty$, as an exponential tail (with the parameter c) or faster.

Note also that the constant B is involved in (1.1.8) merely to make the verification of the condition easier. As a matter of fact, without loss of generality, we can set $B = 1$.

► Indeed, if (1.1.8) holds, then $\overline{F}(x) \le Be^{-cx} = (Be^{-cx/2})e^{-cx/2}$. The function $Be^{-cx/2} \to 0$ as $x \to 0$, so for sufficiently large x's, we have $Be^{-cx/2} \le 1$ and $\overline{F}(x) \le e^{-cx/2}$; that is, (1.1.8) is true for $B = 1$ and c replaced by $c/2$. ◄

If there is no c and B for which the above property is true, we call the distribution *heavy-tailed*.

S.Asmussen cites in [10] an actuarial folklore definition of a heavy-tailed distribution F of a r.v. ξ as that for which "20% of the number of claims account for more than 80% of the total value of the claims". One may clarify it as follows. Let $q = q_{0.8}(\xi)$, the 0.8-quantile of ξ (for the definition of quantile see Section **0**.1.3.4). Then, if we consider a large number of independent claims with the same distribution F, asymptotically, 20% of them will be larger than q. The "part of the mean value" for $x > q$ is $m(q) = \int_q^{\infty} x dF(x)$. The above heuristic definition means that $[m(q)/m(0)] \ge 0.8$, where $m(0) = E\{\xi\}$.

We say that a tail $\overline{G}(x)$ is heavier than $\overline{F}(x)$ if

$$\frac{\overline{F}(x)}{\overline{G}(x)} \to 0 \text{ as } x \to \infty. \tag{1.1.9}$$

In other words, $\overline{F}(x)$ is vanishing faster than $\overline{G}(x)$.

(For the reader who prefers the big O and little o notation introduced in Section **0**.4.1, note that the definition (1.1.8) amounts to the relation $\overline{F}(x) = 0(e^{-cx})$ for some $c > 0$, and the definition (1.1.9) may be rewritten as $\overline{F}(x) = o(\overline{G}(x))$.)

EXAMPLE 1. Let $\overline{F}(x) \sim x^{-2}$ and $\overline{G}(x) \sim x^{-1}$. Clearly, $\overline{F}(x) \to 0$ faster than $\overline{G}(x)$. Formally, we have

$$\frac{\overline{F}(x)}{\overline{G}(x)} \sim \frac{x^{-2}}{x^{-1}} = \frac{1}{x} \to 0.$$

So, $\overline{G}(x)$ is heavier than $\overline{F}(x)$. □

1.1.2 Some particular light-tailed distributions

1. The *exponential distribution* itself is, obviously, light-tailed since $e^{-ax} \leq e^{-ax}$, and hence (1.1.8) is true for $c = a$ and $B = 1$.

2. *Distributions of bounded r.v.'s.* Let a r.v. ξ be less than or equal to some constant b with probability one. Then $\overline{F}(x) = P(\xi > x) = 0$ for any $x \geq b$, and (1.1.8) clearly holds for any non-negative c and B.

   Let ξ' be an unbounded r.v. with a distribution G, that is, $\overline{G}(x) > 0$ for all x. Then $\dfrac{\overline{F}(x)}{\overline{G}(x)} = 0$ for $x > b$, and $\overline{G}(x)$ is heavier than $\overline{F}(x)$.

3. *Mixtures of exponentials.* (For the definition of mixture and a discussion, see Section **0**.1.3.5.) Let the distribution $F = \sum_{j=1}^{k} w_j F_j$, where k is an integer, weights w_j are positive, $\sum_{j=1}^{k} w_j = 1$, and F_j is the exponential distribution with positive parameter a_j. For example, $F(x) = \frac{1}{3}(1 - e^{-x}) + \frac{2}{3}(1 - e^{-2x})$. We can write

$$\overline{F}(x) = \sum_{j=1}^{k} w_j \overline{F}_j(x) = \sum_{j=1}^{k} w_j e^{-a_j x}.$$

   In our particular example, $\overline{F}(x) = \frac{1}{3}e^{-x} + \frac{2}{3}e^{-2x}$. Then, setting $c = \min\{a_j\}$, we have

$$\overline{F}(x) \leq \sum_{j=1}^{k} w_j e^{-cx} = e^{-cx} \sum_{j=1}^{k} w_j = e^{-cx},$$

   and (1.1.8) is true. In the example above, $\overline{F}(x) = \frac{1}{3}e^{-x} + \frac{2}{3}e^{-2x} \leq e^{-x}$, and (1.1.8) is true for $c = B = 1$.

4. *The Γ (gamma)-distribution.* This is the distribution with the density given by

$$f_{a\nu}(x) = \frac{a^{\nu}}{\Gamma(\nu)} x^{\nu-1} e^{-ax} \quad \text{for } x \geq 0, \tag{1.1.10}$$

   and $f_{a\nu}(x) = 0$ for $x < 0$. Parameters a and ν are positive. A detailed description is given in Section **0**.3.2.3; for $\nu = 1$, (1.1.10) determines an exponential distribution.

   The Γ-distribution is light-tailed.

► To show this, let us write

$$\overline{F}(x) = P(\xi > x) = \int_x^\infty f_{a\nu}(y)dy = \int_x^\infty \frac{a^\nu}{\Gamma(\nu)} y^{\nu-1} e^{-ay} dy$$
$$= \int_x^\infty \left[\frac{a^\nu}{\Gamma(\nu)} y^{\nu-1} e^{-ay/2}\right] e^{-ay/2} dy.$$

For any $a > 0$, the function

$$K(y) = \frac{a^\nu}{\Gamma(\nu)} y^{\nu-1} e^{-ay/2} \to 0 \text{ as } y \to \infty. \tag{1.1.11}$$

Then there exists a number $d = d(a)$ such that $K(y) \leq \frac{a}{2}$ for all $y \geq d(a)$. Then, for $x \geq d(a)$,

$$\overline{F}(x) = \int_x^\infty K(y) e^{-ay/2} dy \leq \int_x^\infty \frac{a}{2} e^{-ay/2} dy = e^{-ax/2}.$$

Thus, if we set $c = a/2$, then for all $x \geq d(2c)$,

$$\overline{F}(x) \leq e^{-cx}. \blacktriangleleft$$

5. *The Weibull distribution.* Another way to generalize the exponential distribution is to consider the tail

$$\overline{F}(x) = \exp\{-ax^r\} \quad \text{for } x \geq 0, \tag{1.1.12}$$

where a and r are positive parameters. For $r = 1$, this is an exponential distribution. For $r > 1$, we deal with a light-tailed distribution. Suggestions for how to show it rigorously are given in Exercise 4.

1.1.3 Some particular heavy-tailed distributions

1. *The log-normal distribution.* This is the distribution of the r.v. $\xi = e^\eta$, where η is a normal r.v. In other words, ξ is log-normal if $\ln \xi$ is normal. This distribution appears in numerous applications in Economics, Physics, and other areas. Let $a = E\{\eta\}$ and $b^2 = Var\{\eta\}$. Then, we can represent η by $\eta = a + b\eta_0$, where η_0 is standard normal. Consequently,

$$\xi = e^{a+b\eta_0}. \tag{1.1.13}$$

Since the m.g.f. of η_0 is $M_{\eta_0}(z) = e^{z^2/2}$ (see Section **0**.4.3.6),

$$E\{\xi\} = E\{e^{a+b\eta_0}\} = e^a E\{e^{b\eta_0}\} = e^a M_{\eta_0}(b) = e^{a+b^2/2}. \tag{1.1.14}$$

Similarly,

$$E\{\xi^2\} = E\{e^{2a+2b\eta_0}\} = e^{2a} M_{\eta_0}(2b) = e^{2(a+b^2)} \tag{1.1.15}$$

and, hence,

$$Var\{\xi\} = e^{2(a+b^2)} - e^{2a+b^2} = e^{2a+b^2}(e^{b^2} - 1). \tag{1.1.16}$$

The d.f. of ξ is

$$F(x) = P(e^{a+b\eta_0} \leq x) = P(\eta_0 \leq [\ln x - a]/b) = \Phi([\ln x - a]/b),$$

where $\Phi(\cdot)$ is the standard normal d.f. Then the density

$$f(x) = \frac{d}{dx}\Phi((\ln x - a)/b)) = \frac{1}{\sqrt{2\pi}xb}\exp\{-(\ln x - a)^2)/2b^2\}.$$

The log-normal distribution is heavy-tailed, that is, $\overline{F}(x) \to 0$, as $x \to \infty$, slower than any exponential function.

► To prove this, we use the estimate (**0**.3.2.18). From the first inequality there, it follows that for $x > 2$

$$1 - \Phi(x) \geq \frac{3}{4}x^{-1}\varphi(x).$$

Since η_0 is symmetric, without loss of generality we can set $b > 0$. Note also that for $x > e^{-a}$, we have $-a < \ln x$, and $\frac{1}{b}(\ln x - a) \leq \frac{2}{b}\ln x$. The last quantity is larger than 2 for $x > e^b$. Consider $x > \max\{e^{-a}, e^b\}$ (note that a may be negative). Then

$$\overline{F}(x) = 1 - \Phi\left(\frac{1}{b}(\ln x - a)\right) \geq 1 - \Phi\left(\frac{2}{b}\ln x\right) \geq \frac{3}{4}\left(\frac{2\ln x}{b}\right)^{-1}\varphi\left(\frac{2\ln x}{b}\right)$$
$$= \frac{3b}{8\sqrt{2\pi}}(\ln x)^{-1}\exp\left\{-\frac{2}{b^2}(\ln x)^2\right\} = \frac{3b}{8\sqrt{2\pi}}\exp\left\{-\frac{2}{b^2}(\ln x)^2 - \ln\ln x\right\}.$$

Then for any $c > 0$ and $x > \max\{e^{-a}, e^b\}$,

$$\frac{\overline{F}(x)}{\exp\{-cx\}} \geq \frac{3b}{8\sqrt{2\pi}}\exp\left\{cx - \frac{2}{b^2}(\ln x)^2 - \ln\ln x\right\}.$$

It is a standard Calculus exercise to show that $cx - \frac{2}{b^2}(\ln x)^2 - \ln\ln x \to \infty$ as $x \to \infty$. Hence,

$$\frac{\overline{F}(x)}{\exp\{-cx\}} \to \infty, \quad \text{as } x \to \infty, \quad \text{for any } c > 0.$$

This implies that there is no $c, B > 0$ for which (1.1.8) is true. ◄

2. *The Pareto distribution.* Consider a r.v. ξ_1 for which $P(\xi_1 > x)$ is the function

$$\overline{F}_1(x) = \begin{cases} 1 & \text{for } x < 1, \\ x^{-\alpha} & \text{for } x \geq 1, \end{cases} \tag{1.1.17}$$

where $\alpha > 0$ is a parameter. Thus, in this case, the tail is vanishing as a power function. Since $P(\xi_1 > 1) = 1$, the r.v. ξ_1 takes on values from $[1, \infty)$ with probability one.

We call this distribution a Pareto distribution as well as the distributions of all linear transformations of ξ_1; more specifically, the distributions of r.v.'s $\xi = b\xi_1 + d$ for all d and $b > 0$. The parameter b may be viewed as a scale parameter. Since ξ_1 assumes values from $[1, \infty)$, the r.v. ξ takes on values from $[b + d, \infty)$, so $b + d$ may be called a location parameter.

Often, the term "Pareto distribution" is applied to the distribution with the tail

$$\overline{F}(x) = \left(\frac{\theta}{x + \theta}\right)^{\alpha} \quad \text{for } x \geq 0, \tag{1.1.18}$$

where the parameter $\theta > 0$. In Exercise 6f, we show that this is the distribution of the r.v. $\theta\xi_1 - \theta = \theta(\xi_1 - 1)$. In this case, θ is a scale parameter. Indeed, if a r.v. Z_1 has distribution (1.1.18) with $\theta = 1$, then the r.v. $Z_\theta = \theta Z_1$ has distribution (1.1.18).

Let $F_1(x)$ be the d.f. of ξ_1. Then its density $f_1(x)$ is given by

$$f_1(x) = F_1'(x) = -\overline{F}_1'(x) = \begin{cases} 0 & \text{for } x < 1, \\ \alpha/x^{\alpha+1} & \text{for } x \geq 1. \end{cases}$$

The pth moment of ξ_1 is

$$E\{\xi_1^p\} = \int_1^\infty x^p f_1(x)dx = \frac{\alpha}{\alpha - p} \tag{1.1.19}$$

if $p < \alpha$. For $p \geq \alpha$, the moment does not exist. The reader is encouraged to provide simple integration on her/his own. See also Exercise 6 for other questions about this distribution.

Thus, for a r.v. ξ defined as $b\xi_1 + d$, the expectation $E\{\xi\}$ exists if $\alpha > 1$, and the variance $Var\{\xi\}$ exists if $\alpha > 2$. In the last case,

$$Var\{\xi_1\} = E\{\xi_1^2\} - (E\{\xi_1\})^2 = \frac{\alpha}{(\alpha-2)(\alpha-1)^2}. \tag{1.1.20}$$

As we know from Calculus, any power function vanishes slower than any exponential function, which implies that for a Pareto distribution—whichever definition above we choose—there is no c for which (1.1.8) is true.

The tail (1.1.17) may be interpreted as "very heavy". In Exercise 6, we discuss how the degree of "heaviness" depends on α.

3. *The Weibull distribution.* For $r < 1$, the tail in (1.1.12) vanishes slower than any exponential function, so the distribution is heavy-tailed. See also Exercise 4 for detail.

Route 1 ⇒ *page 150*

2,3

1.1.4 The asymptotic behavior of tails and moments

The facts considered in this subsection may help classify tails without direct calculations. We use below the Calculus notation

$$f(x) = O(g(x)) \text{ and } f(x) = o(g(x)), \tag{1.1.21}$$

where $f(x)$ and $g(x)$ are functions. A detailed explanation of the big O and small o notation and some examples are given in Appendix, Section 4.1.

To simplify formulations, we restrict ourselves to a positive r.v. ξ. Let $\overline{F}(x) = P(\xi > x)$, and let $m_k = E\{\xi^k\}$, the kth moment of ξ.

First, note that if $m_k < \infty$ for some k, then

$$\overline{F}(x) = O(x^{-k}).$$

This immediately follows from the Markov inequality (**0**.2.5.2).

Secondly, if F is light-tailed, then $m_k < \infty$ for all k. Accordingly, if $m_k = \infty$ for some k, the distribution is heavy-tailed.

A direct way to show this is to set $u(x) = x^k$ in the general representation (**0**.2.2.1). We have

$$m_k = E\{\xi^k)\} = \int_0^\infty [1 - F(x)]d(x^k) = k\int_0^\infty \overline{F}(x)x^{k-1}dx. \qquad (1.1.22)$$

If F is light-tailed, then by definition (1.1.8) and (1.1.22), for some x_{cB},

$$m_k = k\int_0^{x_{cB}} \overline{F}(x)x^{k-1}dx + k\int_{x_{cB}}^\infty \overline{F}(x)x^{k-1}dx \leq k\int_0^{x_{cB}} x^{k-1}dx + kB\int_{x_{cB}}^\infty e^{-cx}x^{k-1}dx.$$

The first integral is finite, and the second is finite for any $c > 0$.

However, the finiteness of all moments is not sufficient for the distribution to be light-tailed. It may happen that $m_k < \infty$ for all k, but the distribution is heavy-tailed.

As an example, one may consider the Weibull distribution (1.1.12), say, with $r = 1/2$ and $a = 1$. Then $\overline{F}(x) = \exp\{-\sqrt{x}\}$. In Example 1.1.3-3, we saw that this distribution is heavy-tailed. However, it has all moments. Indeed, by (1.1.22), $m_k = k\int_0^\infty e^{-\sqrt{x}}x^{k-1}dx < \infty$.

►To prove that the last integral is finite, we recall that $e^{-x} = O(x^{-m})$ for any fixed m. In particular, $e^{-\sqrt{x}} = O((\sqrt{x})^{-2k-2}) = O(x^{-k-1})$. Hence, $e^{-\sqrt{x}}x^{k-1} = O(x^{-2})$. The function $O(x^{-2}) \leq Cx^{-2}$ for a constant C and large x's. Such a function is integrable. ◄

The following proposition gives more insight into the asymptotic behavior of tails. Let $M_\xi(z)$ be the moment generating function (m.g.f.) of ξ.

Proposition 1 *The distribution of ξ is light-tailed if and only if $M_\xi(z) < \infty$ for some $z > 0$.*

► ***Proof.*** Let $M(z) < \infty$ for some $z > 0$. We apply the inequality for deviations from Proposition **0**.4. Set $u(x) = e^{z|x|}$ in (**0**.2.5.1). Since ξ is positive, by (**0**.2.5.1), for $x > 0$,

$$P(\xi > x) = P(|\xi| > x) \leq \frac{E\{u(\xi)\}}{u(x)} = e^{-zx}E\{e^{zX}\} = e^{-zx}M_\xi(z). \qquad (1.1.23)$$

Because $M_\xi(z) < \infty$, condition (1.1.8) holds with $c = z$ and $B = M_\xi(z)$.

Conversely, let (1.1.8) be true for some $c, B > 0$, and $x > x_{cB}$. We have

$$M_\xi(z) = \int_0^\infty e^{zx}dF(x) = -\int_0^\infty e^{zx}d\overline{F}(x) = -\,e^{zx}\overline{F}(x)\Big|_0^\infty + z\int_0^\infty e^{zx}\overline{F}(x)dx.$$

Now, $\lim_{x\to\infty} e^{zx}\overline{F}(x) \leq B\lim_{x\to\infty} e^{zx}e^{-cx} = 0$ for $z < c$. Hence,

$$M_\xi(z) = \overline{F}(0) + z\int_0^\infty e^{zx}\overline{F}(x)dx \leq \overline{F}(0) + z\int_0^{x_{cB}} e^{zx}dx + zB\int_{x_{cB}}^\infty e^{zx}e^{-cx}dx.$$

The first integral is finite, and the second is finite for $z < c$. ■ ◄

An interesting discussion on possible tails of the loss distributions, especially heavy tails, may be found, e.g., in [37] and [93].

1,2,3

1.2 The distribution of the loss

Next, we consider the r.v. X from (1.1.1). We assume ξ (the loss in the case where the loss event has occurred) to be positive. Then $X = 0$ only if $I = 0$, and

$$P(X=0) = P(I=0) = 1-q. \tag{1.2.1}$$

Furthermore, for $x \geq 0$, the payment $X > x$ if, first, the loss event has occurred (the probability is q), and secondly, the loss $\xi > x$ (the probability is $\overline{F}_\xi(x) = P(\xi > x)$, the tail of the distribution of ξ). Hence, $P(X > x) = q\overline{F}_\xi(x)$, and the d.f. of X is

$$F_X(x) = P(X \leq x) = 1 - q\overline{F}_\xi(x). \tag{1.2.2}$$

This also may be rewritten as

$$F_X(x) = 1 - q + qF_\xi(x), \tag{1.2.3}$$

where $F_\xi(x) = 1 - \overline{F}_\xi(x)$ is the d.f. of ξ.

In particular, since ξ was assumed to be positive, $\overline{F}_\xi(0) = 1$ and

$$F_X(0) = 1 - q. \tag{1.2.4}$$

Because X is non-negative, $F_X(x) = 0$ for all $x < 0$.

Now, it is worthwhile to recall that if for a r.v. Z and a number c, it is true that $P(Z = c) = 0$, then the d.f. $F_Z(x)$ is continuous at the point c. If $P(Z = c) = \Delta > 0$, then $F_Z(x)$ "jumps" at the point c, and Δ is the size of the jump. For detail, see Section **0**.1.3.3; in particular, Fig.4 there.

We see that $F_X(x)$ jumps at the point 0 by $1-q$. Hence, $P(X = 0) = 1 - q$, which was already stated in (1.2.1).

EXAMPLE 1. Let ξ be exponential and $E\{\xi\} = 1/a$. Then by (1.2.2),

$$F_X(x) = \begin{cases} 1 - qe^{-ax} & \text{if } x \geq 0, \\ 0 & \text{if } x < 0. \end{cases}$$

The graph is shown in Fig.1. It is quite typical. □

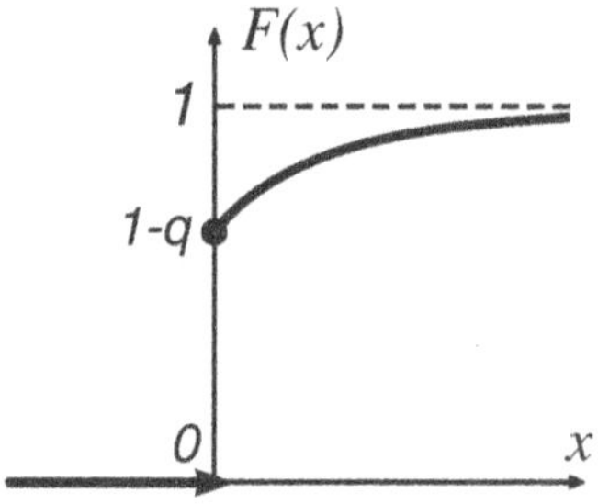

FIGURE 1.

Now consider moments of the r.v. X. From (1.1.1) it follows that

$$E\{X^k\} = qE\{\xi^k\}. \tag{1.2.5}$$

Set $\mu = E\{\xi\}$ and $v^2 = Var\{\xi\}$. Then, in view of (1.2.5),

$$E\{X\} = q\mu, \tag{1.2.6}$$

$$\begin{aligned} Var\{X\} &= E\{X^2\} - (E\{X\})^2 = qE\{\xi^2\} - q^2(E\{\xi\})^2 = q(v^2 + \mu^2) - q^2\mu^2 \\ &= qv^2 + q(1-q)\mu^2. \end{aligned} \tag{1.2.7}$$

EXAMPLE 2. In the situation of Example 1, we get

$$E\{X\} = \frac{q}{a}, \quad Var\{X\} = \frac{q}{a^2} + \frac{q(1-q)}{a^2} = \frac{2q-q^2}{a^2}. \tag{1.2.8}$$

(Look up the mean and the variance of the exponential distribution in (1.1.6).) □

1.3 The distribution of the payment and types of insurance

Let Y be the amount to be paid by the company in accordance with the insurance contract. For brevity, we will call it a *payment*. If the coverage is full, then $Y = X$. However, the insurance often pays only a part of the loss. As in Section **1.5**, we set

$$Y = r(X)$$

and call $r(x)$ a *payment function*. We assume that $r(x)$ is non-negative, non-decreasing, and $0 \le r(x) \le x$. From the last condition it follows, in particular, that $r(0) = 0$.

Consider several particular but important cases. In the first case,

$$r(x) = r_{1d}(x) = \begin{cases} 0 & \text{if } x \le d, \\ x-d & \text{if } x > d, \end{cases} \tag{1.3.1}$$

where the payment policy involves a *deductible d* (the excess-of-loss type; see Section **1.5** for detail).

Next, consider the payment function

$$r(x) = r_{2s}(x) = \begin{cases} x & \text{if } x \le s, \\ s & \text{if } x > s, \end{cases} \tag{1.3.2}$$

where s is a *maximal* or *limit payment*.

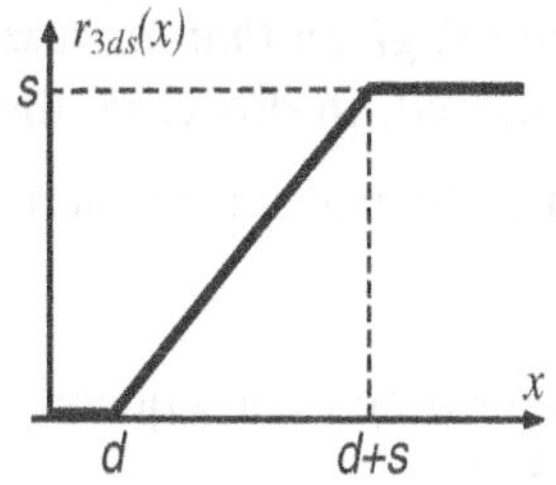

FIGURE 2. The payment function in the case of deductible and limit payment.

The combination of these types, when both restrictions—a deductible and a limit payment—are included, is given by

$$r(x) = r_{3ds}(x) = \begin{cases} 0 & \text{if } x \le d, \\ x-d & \text{if } d < x < s+d, \\ s & \text{if } x \ge s+d \end{cases} \tag{1.3.3}$$

(see the graph in Fig.2). If in the last formula $d = 0$, then we come to (1.3.2) and if $s = \infty$, we have (1.3.1).

Another type of insurance with deductible is the *franchise deductible* insurance with the payment function

$$r(x) = r_{4d}(x) = \begin{cases} 0 & \text{if } x \le d, \\ x & \text{if } x > d. \end{cases} \tag{1.3.4}$$

That is, if the loss exceeds the deductible, the loss is covered in full.

One more type is *proportional* or *quota share insurance* where

$$r(x) = r_{5k}(x) = kx \tag{1.3.5}$$

for a positive $k < 1$.

Certainly, there exist more complicated policies. For instance, in auto-insurance, limits for payments for different types of losses (car damage, medical expenses) are different.

Our goal is to write the d.f. of Y, its expectation, variance and other moments. How to compute moments is clear since, in view of (1.2.2) and the fact that $r(0) = 0$,

$$E\{Y^k\} = E\{r^k(X)\} = \int_0^\infty r^k(x)dF_X(x) = r(0)P(X=0) + q\int_0^\infty r^k(x)dF_\xi(x) = q\int_0^\infty r^k(x)dF_\xi(x). \tag{1.3.6}$$

Sometimes, it is convenient to compute the last integral by parts or use the general representation (**0**.2.2.1) with $u(x) = r^k(x)$, which leads to

$$\begin{aligned} E\{Y^k\} &= q\int_0^\infty (1 - F_\xi(x))dr^k(x) = qk\int_0^\infty r^{k-1}(x)r'(x)(1 - F_\xi(x))dx \\ &= qk\int_0^\infty r^{k-1}(x)r'(x)\overline{F}_\xi(x))dx. \end{aligned} \tag{1.3.7}$$

(See for details Section **0**.2.2. Note also that, as in Section **0**.2.2, we used the fact that $r(0) = 0$.)

Let $r(x) = r_{3ds}(x)$ defined in (1.3.3). Then $r'(x) = 1$ if $x \in (d, d+s)$, and $= 0$ for $x < d$ and $x > d+s$. At $x = d$ and $x = d+s$ we view $r'(x)$ as a function having jumps; this does not matter for integration. Consequently, by (1.3.7),

$$E\{Y^k\} = qk\int_d^{d+s} (x-d)^{k-1}\overline{F}_\xi(x)dx. \tag{1.3.8}$$

For $k = 1$,

$$E\{Y\} = q\int_d^{d+s} \overline{F}_\xi(x)dx. \tag{1.3.9}$$

Note also that if we are interested in the moments of the payment r.v. *given that the loss event has occurred,* we should just set $q = 1$ in (1.3.7)-(1.3.9). Indeed, in this case, the payment is equal to $r(\xi)$ and $E\{r^k(\xi)\} = \int_0^\infty r^k(x)dF_\xi(x)$. Integrating by parts as we did in (1.3.7), we come to the right members in (1.3.7)-(1.3.9) with $q = 1$.

EXAMPLE 1. Let ξ be exponentially distributed with parameter a and consequently, $E\{\xi\} = 1/a$. Let $r(x) = r_{3ds}(x)$ as defined in (1.3.3). Then, by (1.3.8),

$$E\{Y^k\} = qk\int_d^{d+s} (x-d)^{k-1}e^{-ax}dx.$$

The last integral is standard. In particular,

$$E\{Y\} = q\int_d^{d+s} e^{-ax}dx = \frac{q}{a}e^{-ad}(1 - e^{-as}),$$

and

$$E\{Y^2\} = 2q\int_d^{d+s} (x-d)e^{-ax}dx = \frac{2q}{a^2}e^{-ad}(1 - (1+as)e^{-as}).$$

(The reader is invited to compute the last integral by parts or look it up in a Calculus textbook, e.g., [136]). Hence,

$$\begin{aligned} Var\{Y\} &= \frac{2q}{a^2}e^{-ad}(1-(1+as)e^{-as}) - \frac{q^2}{a^2}e^{-2ad}(1-e^{-as})^2 \\ &= \frac{q}{a^2}e^{-ad}\left[2(1-(1+as)e^{-as}) - qe^{-ad}(1-e^{-as})^2\right]. \end{aligned}$$

When there is no limit payment ($s = \infty$),

$$E\{Y\} = \frac{q}{a}e^{-ad}, \quad Var\{Y\} = \frac{q}{a^2}e^{-ad}\left[2 - qe^{-ad}\right]. \tag{1.3.10}$$

If the coverage is full ($d = 0$), we come back to (1.2.8).

EXAMPLE 2 ([153, N35][2]). An insurance company offers two types of policies, Type Q and Type R. Type Q has no deductible but a policy limit of 3,000. Type R has no limit but an ordinary deductible of d. Losses follow the Pareto distribution (1.1.18) with $\theta = 2{,}000$ and $\alpha = 3$. Calculate the deductible d such that both policies have the same expected cost per loss.

The word "ordinary" distinguishes this policy from the franchise deductible policy. As we saw in Section 1.1.3, the parameter θ in (1.1.18) is a scale parameter. So, if we choose 2,000 as a unit of money, then in accordance with (1.1.18), we have $\overline{F}_\xi = 1/(1+x)^\alpha$. Since we are considering losses, we set $q = 1$ in (1.3.9). The reader may see below that if we do not do that, q will cancel anyway. Thus, for Type Q, we have $d = 0$, the policy limit $s = 1.5$, and the expected cost is $\int_0^{1.5}(1+x)^{-3}dx = 0.42$. If $s = \infty$, the expected cost equals $\int_d^\infty (1+x)^{-3}dx = \frac{1}{2(1+d)^2}$. Thus, $\frac{1}{2(1+d)^2} = 0.42$, which gives $d \approx 0.091$. Thus, the answer is $0.091 \times 2000 = 182$.

EXAMPLE 3 ([153, N4][3]). Well-Traveled Insurance company sells a travel insurance policy that reimburses travelers for any expenses incurred for a planned vacation that is canceled because of airline bankruptcies. Individual claims follow the Pareto distribution (1.1.18) with $\theta = 500$ and $\alpha = 2$. Because of financial difficulties in the airline industry, Well-Traveled imposes a limit of \$1000 on each claim. If a policyholder's planned vacation is canceled due to airline bankruptcies and he or she has incurred more than \$1000 in expenses, what is the expected non-reimbursed amount of the claim?

This problem is similar to Example 2. First, we choose \$500 as a unit of money. The main point in our problem is that we consider the distribution of the loss ξ *given that* $\xi > 2$ units. The tail of the conditional distribution is

$$P(\xi > x \,|\, \xi > 2) = \frac{P(\xi > x)}{P(\xi > 2)} = \frac{(1+x)^{-2}}{(1+2)^{-2}} = \frac{9}{(1+x)^2} \quad \text{for } x \geq 2.$$

Note that $P(\xi > x \,|\, \xi > 2) = 1$ if $x < 2$. Then, using the general formula (**0**.2.2.2), we have

$$E\{\xi \,|\, \xi > 2) = \int_0^\infty P(\xi > x \,|\, \xi > 2)dx = \int_0^2 dx + \int_2^\infty \frac{9dx}{(1+x)^2} = 5.$$

[2]Reprinted with permission of the Casualty Actuarial Society.

[3]Reprinted with permission of the Casualty Actuarial Society.

In the case $\xi > 2$, the company pays 2 units, so on the average, the expected non-reimbursed amount equals 3 units or $1,500. □

Next, we consider the d.f. $F_Y(x)$. At the end of this section, we provide a general formula, but it is worth noting that in particular cases it is often easier to proceed from particular features of these cases rather than from a general representation. We consider two such cases.

Let $r(x) = r_{3sd}(x)$ from (1.3.3). First, the payment is greater than zero if the loss event occurs (with probability q) and the loss is greater than the deductible (the probability is $\overline{F}_\xi(d)$). This leads to

$$P(Y = 0) = P(X \le d) = 1 - q\overline{F}_\xi(d). \tag{1.3.11}$$

For $0 < y < s$, the payment is greater than y if the loss event occurs and the loss $\xi > d + y$. The probability of this is $q\overline{F}_\xi(y+d)$. Furthermore, the payment is equal to s if the loss event occurs and the loss is greater than or equal to $d + s$. So,

$$P(Y = s) = P(X \ge d+s) = qP(\xi \ge d+s) = q(1 - P(\xi < d+s)). \tag{1.3.12}$$

Also, $P(Y \le s) = 1$, and $P(Y < 0) = 0$.

Thus, eventually, for the payment function (1.3.3), the d.f. of the payment Y is

$$F_Y(y) = \begin{cases} 0 & \text{if } y < 0, \\ 1 - q\overline{F}_\xi(y+d) & \text{if } 0 \le y < s, \\ 1 & \text{if } y \ge s. \end{cases} \tag{1.3.13}$$

Certainly, if $s = \infty$ (no limit payment), then we have

$$F_Y(y) = \begin{cases} 0 & \text{if } y < 0, \\ 1 - q\overline{F}_\xi(y+d) & \text{if } y \ge 0. \end{cases} \tag{1.3.14}$$

EXAMPLE 4. Again let ξ be exponentially distributed with a parameter a. Then $\overline{F}_\xi(x) = e^{-ax}$. After substitution into (1.3.13), we have

$$F_Y(y) = \begin{cases} 0 & \text{if } y < 0, \\ 1 - qe^{-a(y+d)} & \text{if } 0 \le y < s, \\ 1 & \text{if } y \ge s. \end{cases} \tag{1.3.15}$$

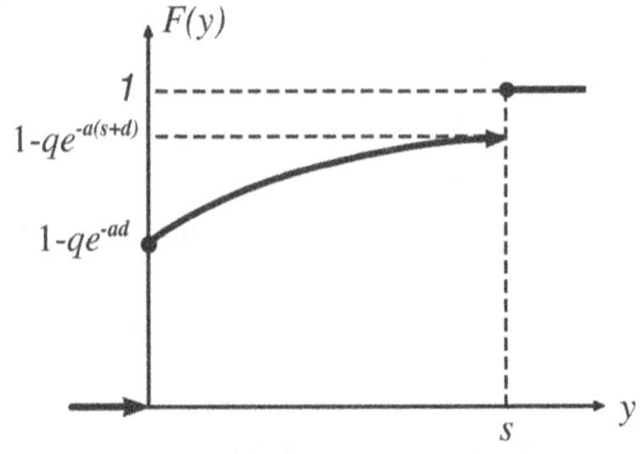

FIGURE 3.

If $s = \infty$ (no limit payment), then

$$F_Y(y) = \begin{cases} 0 & \text{if } y < 0, \\ 1 - qe^{-a(y+d)} & \text{if } y \geq 0. \end{cases} \tag{1.3.16}$$

The graph for (1.3.15) is given in Fig.3 and has two jumps. The first is the jump at 0 with a size of $1 - qe^{-ad}$.

The second jump is at the point s and has the size $qe^{-a(s+d)}$. (Certainly, if $s = \infty$, we have just one jump at zero.) To understand the significance of this instance, we should again recall that the jump of a d.f. at a point c equals the probability that the corresponding r.v. is exactly equal to c. Thus,

$$P(Y = 0) = 1 - qe^{-ad}, \;\; P(Y = s) = qe^{-a(s+d)}.$$

(Compare with Example 1.2-1. Now $P(Y = 0)$ is larger due to deductible.) □

Coming back to (1.3.13), we see that, as in the last example, the d.f. has at least two jumps: at 0 and at s. These jumps are equal to the respective probabilities that Y will assume the values 0 and s; see (1.3.11)-(1.3.12).

Next, we briefly consider the franchise deductible type policy. In this case, since $r(x) = 0$ for $0 \leq x < d$, and $r(x) = x$ for $x \geq d$, from (1.3.6) it follows that

$$E\{Y^k\} = q\int_d^\infty x^k dF_\xi(x). \tag{1.3.17}$$

Furthermore, for $0 \leq y \leq d$, we have $P(Y \leq y) = P(Y = 0) = 1 - qP(\xi > d)$. If $y > d$, then $P(Y \leq y) = 1 - qP(\xi > y)$. Thus,

$$F_Y(y) = \begin{cases} 0 & \text{if } y < 0, \\ 1 - q\overline{F}_\xi(d) & \text{if } 0 \leq y \leq d, \\ 1 - q\overline{F}_\xi(y) & \text{if } y > d. \end{cases} \tag{1.3.18}$$

In Exercise 14, we will consider particular examples.

Route 1 ⇒ page 158

2,3

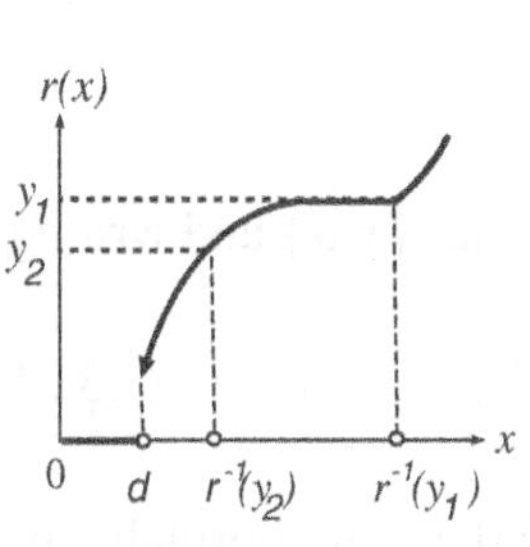

FIGURE 4.

Now, let us write a general formula for the d.f. $F_Y(y)$. Assume that $r(x)$ is non-decreasing and continuous at any x, perhaps, except one point d such that $r(x) = 0$ for $x \leq d$, and $r(x) > 0$ for $x > d$. See, for instance, Fig.4. For a payment function this restriction is fairly mild. As an example, one may consider the function $r_{4d}(x)$ in (1.3.4) for the franchise deductible insurance type.

Let $y > 0$. Obviously, the d.f.

$$F_Y(y) = P(Y \leq y) = P(r(X) \leq y).$$

We define the inverse $r^{-1}(x)$ in the same manner as we defined the inverse of a d.f. in Section **0**.3.2.1. For the type of payment functions $r(x)$ we consider, we may do it more explicitly setting $r^{-1}(y) = \max\{x : r(x) \leq y)\}$. The definition is illustrated in Fig.4. Note that the maximum above may be equal to infinity, which as we will see, cannot cause any problem. For example, if $r(x) = r_{3sd}(x)$ defined in (1.3.3), then $r^{-1}(s) = \infty$, as one can see from Fig.2.

For the inverse so defined, the event $\{r(X) \leq y\} = \{X \leq r^{-1}(y)\}$ for any $y > 0$, which may be easily seen, for example, from Fig.4. Thus, by virtue of (1.2.2),

$$F_Y(y) = P(X \leq r^{-1}(y)) = F_X(r^{-1}(y)) = 1 - q\overline{F}_{\xi}(r^{-1}(y)). \tag{1.3.19}$$

Consider, as an example, $r(x) = r_{3sd}(x)$ from (1.3.3). In this case, $r(x)$ is increasing on $(d, d+s)$ and takes values from the interval $(0,s)$; see again Fig.2. Then $r^{-1}(y) = y+d$ for $y \in [0,s)$ and for such y's,

$$F_Y(y) = 1 - q + qF_{\xi}(y+d).$$

Since, $r^{-1}(s) = \infty$,

$$F_Y(s) = 1 - q\overline{F}_{\xi}(\infty) = 1 - q \cdot 0 = 1,$$

which is natural because s is a limit payment. All of this leads again to (1.3.13).

Now, let us revisit the ordinary deductible case.

Obviously, the inclusion of a deductible into a policy decreases the payment. The ratio of the expected value of this decrease to the expected payment without deductible is called the *loss elimination ratio* (l.e.r.). Note that in the ordinary deductible case (1.3.1), the payment may be also written as $Y = \max\{0, X-d\}$, and the decrease mentioned—as $\min\{X,d\}$. Thus, the l.e.r. equals

$$\frac{E\{X\} - E\{\max\{0, X-d\}\}}{E\{X\}} = \frac{E\{\min\{X,d\}\}}{E\{X\}}.$$

If $X = 0$, which occurs with probability $1-q$, then $\min\{X,d\} = 0$. If $X = \xi$, which occurs with probability q, then $\min\{X,d\} = \min\{\xi,d\}$. Hence, $E\{\min\{X,d\}\} = qE\{\min\{\xi,d\}\}$. Since $E\{X\} = qE\{\xi\}$,

$$\text{l.e.r.} = \frac{E\{\min\{\xi,d\}\}}{E\{\xi\}} \tag{1.3.20}$$

and does not depend on q.

To compute $E\{\min\{\xi,d\}\}$, we again use (**0**.2.2.1) setting $u(x) = \min\{x,d\}$ and writing

$$E\{\min\{\xi,d\}\} = \int_0^{\infty} (1 - F_{\xi}(x))d(\min\{x,d\}). \tag{1.3.21}$$

For $u(x) = \min\{x,d\}$, the derivative $u'(x) = 1$ if $x < d$, and $= 0$ if $x > d$. Formally, the derivative $u'(x)$ does not exist at d, but this does not matter for integration: we may view $u'(x)$ as a function having a jump at d. Thus, from (1.3.21) it follows that

$$E\{\min\{\xi,d\}\} = \int_0^{d} (1 - F_{\xi}(x))dx. \tag{1.3.22}$$

EXAMPLE 5. Let ξ be exponential with parameter a. We have

$$E\{\min\{\xi,d\}\} = \int_0^d e^{-ax}dx = \frac{1}{a}(1-e^{-ad}).$$

Since $E\{\xi\} = 1/a$, from (1.3.20) we get that the l.e.r. is $1-e^{-ad}$. For $d=0$, we naturally come to zero loss elimination.

EXAMPLE 6. Let ξ have the Pareto distribution defined in (1.1.17) with parameter $\alpha > 1$. A company covers the risk without a limit payment but with a deductible d. What should d be for the company's mean payment to constitute 90% of the mean payment without deductible?

This means that the l.e.r. equals 0.1. Since $\xi > 1$, we should distinguish two cases: $d<1$ and $d \geq 1$.

If $d<1$, then $\min\{\xi,d\} = d$ because $\xi \geq 1$. Therefore, in this case, $E\{\min\{\xi,d\}\} = d$. If $d \geq 1$, using (1.3.22) and (1.1.17), we get

$$E\{\min\{\xi,d\}\} = \int_0^1 (1-0)dx + \int_1^d (1/x^\alpha)dx = 1+\frac{1}{\alpha-1}\left(1-\frac{1}{d^{\alpha-1}}\right) = \frac{1}{\alpha-1}\left(\alpha-\frac{1}{d^{\alpha-1}}\right).$$

Since $E\{\xi\} = \alpha/(\alpha-1)$ [see (1.1.19)], we obtain

$$\text{l.e.r.} = \begin{cases} \dfrac{\alpha-1}{\alpha}d & \text{if } d<1, \\ 1-\dfrac{1}{\alpha d^{\alpha-1}} & \text{if } d \geq 1. \end{cases} \tag{1.3.23}$$

We should determine values of d such that the last expression is equal to 0.1. The solution depends on α. For $d=1$, the l.e.r. equals $(\alpha-1)/\alpha$ and this quantity equals 0.1 for $\alpha = 10/9$. Hence, if $\alpha \geq 10/9$, we should consider the case $d<1$ and write $\frac{\alpha-1}{\alpha}d = 0.1$. This will lead to $d = \frac{\alpha}{10(\alpha-1)}$. [It becomes clear if we graph the function (1.3.23).] If $\alpha < 10/9$, we write $1-\frac{1}{\alpha d^{\alpha-1}} = 0.1$ and get $d = \left(\frac{10}{9\alpha}\right)^{1/(\alpha-1)}$. □

In conclusion, we will consider the effect of *inflation*. Assume that there is a gap between the moments of loss and payments, and the insurer is obligated to cover inflation losses. From a modeling point of view, this means that at the moment of payment, instead of losses ξ, the insurer should proceed from the amount $(1+v)\xi$, where v is the inflation rate during the period under consideration. The point is that the deductible is subtracted after inflation has been taken into account. In the ordinary deductible case (1.3.1), we may consider the problem as follows.

First, note that (1.3.1) may be rewritten as $r(x) = \max\{0, x-d\}$. In view of (1.1.2), without inflation, we would have $Y = I \cdot \max\{0, \xi-d\}$. (If there are no losses ($I=0$), then $Y=0$, and in the case of a loss ξ, the payment $Y = \max\{0, \xi-d\}$.)

In the case of inflation, we should replace ξ by $(1+v)\xi$, and hence the payment

$$Y = I \cdot \max\{0, (1+v)\xi - d\} = I \cdot \max\left\{0, (1+v)\left(\xi - \frac{d}{1+v}\right)\right\}$$
$$= (1+v)I \cdot \max\left\{0, \xi - \frac{d}{1+v}\right\}. \tag{1.3.24}$$

It appears as if you apply a smaller deductible, $d/(1+v)$, and after that you multiply the payment by the inflation coefficient $(1+v)$.

To illustrate this, consider the simple case when ξ is exponential with parameter a. In (1.3.10), we got that the expected payment $E\{Y\} = \frac{q}{a}\exp\{-ad\}$. To consider inflation, we should replace ξ by $(1+v)\xi$. The latter r.v. is also exponential with the parameter $a/(1+v)$. Thus, replacing a by $a/(1+v)$, we have

$$E\{Y\} = \frac{q(1+v)}{a}\exp\{-ad/(1+v)\}. \tag{1.3.25}$$

However, we can arrive at (1.3.25) in another way. Namely, we could just replace in (1.3.10) the deductible d by $d/(1+v)$, which would lead to $\frac{q}{a}\exp\{-ad/(1+v)\}$. After that, we multiply the result by $(1+v)$, which again leads to (1.3.25). Certainly, in this simple case, it does not matter which way we choose. In more complex situations, the latter technique allows us to avoid calculations when inflation is considered. The discussion is continued in Exercise 15.

---

2 THE AGGREGATE PAYMENT

In this section, we do not specify particular details of insurance contracts such as deductible or payment limits. So here, we use habitual notation and denote by X's the *payments* provided by a company to particular clients.

Consider a group consisting of a fixed number n of clients. Let X_i be the payment to the ith client. Then the cumulative payment

$$S = S_n = X_1 + \ldots + X_n.$$

In this context, the r.v.'s X_i are called sometimes *severities*. The object of our study is the distribution function and other characteristics of S. Unless stated otherwise, we assume the X_i's to be independent. If the X's are also identically distributed, we call the group *homogeneous*.

There are several approaches to this problem.

2.1 Convolutions

2.1.1 Definition and examples

Let $F_i(x)$ be the d.f. of X_i. Consider first the case when $n = 2$, so $S = X_1 + X_2$. The basic fact is that if X_1 and X_2 are independent, then the d.f. of S is

$$F_S(x) = \int_{-\infty}^{\infty} F_1(x-y)dF_2(y). \tag{2.1.1}$$

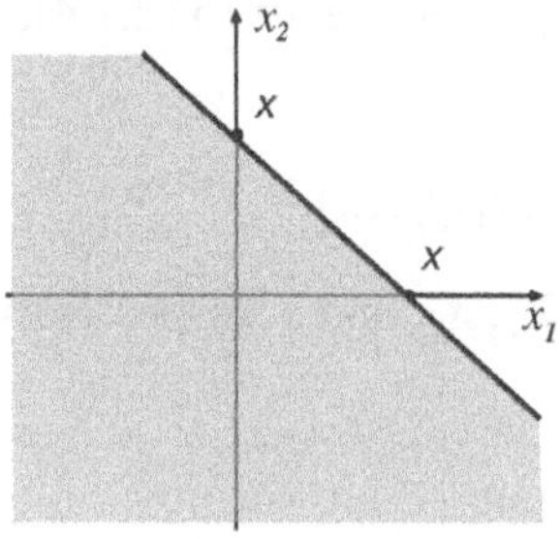

FIGURE 5.

A proof of (2.1.1) may be found in many books on Probability, e.g., in [102], [116], [120]. The idea is to consider the vector (X_1, X_2) whose values are points (x_1, x_2) in the plane. Then $P(X_1 + X_2 \le x)$ is the probability that the value of the vector (X_1, X_2) will be in the region $\{(x_1, x_2) : x_1 + x_2 \le x\}$; see Fig.5. Since X_1, X_2 are independent, this probability may be written as the double integral

$$\int\int_{x_1+x_2 \le x} dF_1(x_1)dF_2(x_2). \tag{2.1.2}$$

Direct integration will lead to (2.1.1) if we replace x_2 by y. The reader can provide details of the integration on her/his own or look at proofs, e.g., in [102], [116], [120].

The operation (2.1.1) is called *convolution* and is denoted in symbols as $F_1 * F_2$.

Once we have "convoluted" F_1 and F_2, we can continue, adding to S_2 a third r.v. X_3. In accordance with (2.1.1), in this case, the d.f. F_{S_2} must be convoluted with F_3. This leads to $F_{S_3} = F_{S_2} * F_3 = F_1 * F_2 * F_3$. Continuing this process, we get

$$F_{S_n} = F_1 * ... * F_n.$$

Examples will be considered a bit later.

Assume that the X's are continuous r.v.'s and for each i, there exists the probability density $f_i(x) = F_i'(x)$. Then, differentiating (2.1.1), for $n = 2$, we get that the density of S is

$$\begin{aligned} f_S(x) &= \frac{dF_S(x)}{dx} = \frac{d}{dx}\int_{-\infty}^{\infty} F_1(x-y)dF_2(y) = \frac{d}{dx}\int_{-\infty}^{\infty} F_1(x-y)f_2(y)dy \\ &= \int_{-\infty}^{\infty} \frac{d}{dx}F_1(x-y)f_2(y)dy = \int_{-\infty}^{\infty} f_1(x-y)f_2(y)dy. \end{aligned}$$

Thus,

$$f_S(x) = \int_{-\infty}^{\infty} f_1(x-y)f_2(y)dy. \tag{2.1.3}$$

This operation is denoted by $f_1 * f_2$ and is called the *convolution of densities*. In general, for an arbitrary integer n

$$f_{S_n} = f_1 * ... * f_n.$$

The counterpart of (2.1.3) for discrete integer-valued r.v.'s is as follows. Let X_1, X_2 take on values $0,1,2,\dots$, and $f_k^{(i)} = P(X_i = k),\ i = 1,2$. Then, setting $f_m = P(S = m)$, for $n = 2$, we have

$$f_m = \sum_{k=0}^{m} f_k^{(1)} f_{m-k}^{(2)}. \tag{2.1.4}$$

Formula (2.1.4) may be derived from (2.1.1) but the direct proof that we provide now is shorter and more illustrative. The r.v. S is equal to m if the r.v. X_1 is equal to some k and X_2 to $m-k$. The corresponding probability is $P(X_1 = k, X_2 = m-k) = P(X_1 = k)P(X_2 = m-k) = f_k^{(1)} f_{m-k}^{(2)}$, because X_1 and X_2 are independent. Summing over k leads to (2.1.4).

Consider the sequences

$$f^{(1)} = (f_1^{(1)}, f_2^{(1)}, \dots); \;\; f^{(2)} = (f_1^{(2)}, f_2^{(2)}, \dots); \;\; f = (f_1, f_2, \dots).$$

The above sequences of probabilities specify the distributions of X_1, X_2, and S, respectively. Then (2.1.4) may be written in compact form as

$$f = f^{(1)} * f^{(2)}.$$

EXAMPLE 1. Let independent r.v.'s

$$X_1 = \begin{cases} 1 \text{ with probability } p_1 = \frac{1}{3} \\ 0 \text{ with probability } q_1 = \frac{2}{3} \end{cases}, \quad X_2 = \begin{cases} 1 \text{ with probability } p_2 = \frac{1}{2} \\ 0 \text{ with probability } q_2 = \frac{1}{2} \end{cases}.$$

Clearly, S takes on values $0,1,2$. The problem is very simple and can be solved directly, but we use this example to demonstrate how (2.1.4) works. We have

$$f_0 = f_0^{(1)} f_0^{(2)} = q_1 q_2 = \frac{2}{3} \cdot \frac{1}{2} = \frac{1}{3},$$

$$f_1 = \sum_{k=0}^{1} f_k^{(1)} f_{n-k}^{(2)} = f_0^{(1)} f_1^{(2)} + f_1^{(1)} f_0^{(2)} = q_1 p_2 + p_1 q_2 = \frac{2}{3} \cdot \frac{1}{2} + \frac{1}{3} \cdot \frac{1}{2} = \frac{1}{2},$$

$$f_2 = \sum_{k=0}^{2} f_k^{(1)} f_{n-k}^{(2)} = f_0^{(1)} f_2^{(2)} + f_1^{(1)} f_1^{(2)} + f_2^{(1)} f_0^{(2)} = q_1 \cdot 0 + p_1 p_2 + 0 \cdot q_2 = p_1 p_2 = \frac{1}{3} \cdot \frac{1}{2} = \frac{1}{6}.$$

EXAMPLE 2. This is a classical example. Let X_1 and X_2 be independent and uniformly distributed on $[0,1]$. Obviously, $S_2 = X_1 + X_2$ takes on values from $[0,2]$, so it suffices to find the density $f_S(x)$ only in this interval. The densities $f_i(x) = 1$ for $x \in [0,1]$ and $= 0$ otherwise. Hence, by (2.1.3),

$$f_S(x) = \int_0^1 f_1(x-y)dy. \tag{2.1.5}$$

The integrand $f_1(x-y) = 1$ if $0 \le x-y \le 1$ which is equivalent to $x-1 \le y \le x$.

Let $0 \le x \le 1$. Then the left inequality holds automatically because $x-1 \le 0$ while $y \ge 0$. So, $f_1(x-y) = 1$ if $y \le x$, and $= 0$ otherwise. Hence, for $0 \le x \le 1$, we may integrate in (2.1.5) only over $[0,x]$, which implies that

$$f_{S_2}(x) = \int_0^x dy = x.$$

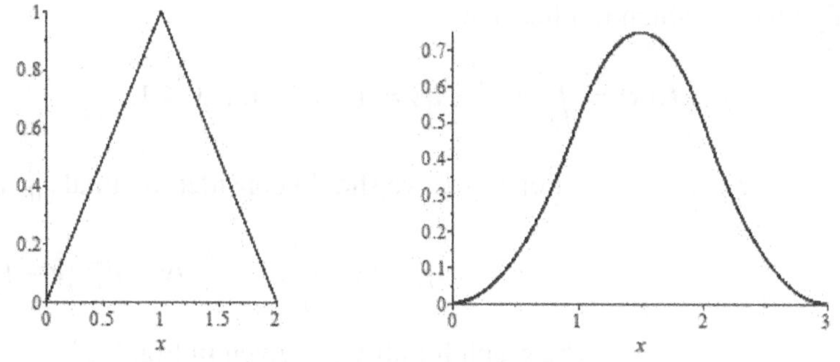

(a) (b)

FIGURE 6. For Example 2: (a) the graph of f_{S_2}; (b) the graph of f_{S_3}.

On the other hand, in view of the symmetry of the distributions of the X's, the density $f_S(x)$ should be symmetric with respect to the center of $[0,2]$; that is, the point one (see Fig.6a). So, for $1 \le x \le 2$, we should have $f_S(x) = 2 - x$. Eventually,

$$f_{S_2}(x) = \begin{cases} x & \text{if } 0 \le x \le 1, \\ 2-x & \text{if } 1 \le x \le 2, \\ 0 & \text{otherwise;} \end{cases} \tag{2.1.6}$$

see again Fig.6a. This distribution is called *triangular*. We see that while the values of X's are equally likely, the values of the sum are not. In Exercise 19, the reader is invited to give a common-sense explanation of this fact.

Now let X_3 be also uniformly distributed on $[0,1]$ and independent of X_1 and X_2. Obviously, the sum $S_3 = X_1 + X_2 + X_3$ assumes values from $[0,3]$. To find its density, we can again apply (2.1.3), replacing $f_2(y)$ by $f_3(y)$, and $f_1(x-y)$ by $f_{S_2}(x-y)$. Thus,

$$f_{S_3}(x) = \int_0^1 f_{S_2}(x-y)dy, \tag{2.1.7}$$

where f_{S_2} is given in (2.1.6). We relegate a bit tedious calculations to Exercise 19. The result is

$$f_{S_3}(x) = \begin{cases} x^2/2 & \text{if } 0 \le x \le 1, \\ (-2x^2+6x-3)/2 & \text{if } 1 \le x \le 2, \\ (x-3)^2/2 & \text{if } 2 \le x \le 3, \\ 0 & \text{otherwise.} \end{cases}$$

The graph is given in Fig.6b.

EXAMPLE 3. Let X_1 and X_2 be independent, X_1 be exponential with $E\{X_1\} = 1$, and X_2 be uniformly distributed on $[0,1]$. Then the density of the sum is given by

$$f_S(x) = \int_0^1 f_1(x-y)dy.$$

The integrand $f_1(x-y) = e^{-(x-y)}$ if $y \leq x$, and $= 0$ otherwise. Hence, for $x \leq 1$, we should integrate only up to x, which implies that

$$f_S(x) = \int_0^x e^{-(x-y)} dy = 1 - e^{-x} \text{ for } x \leq 1.$$

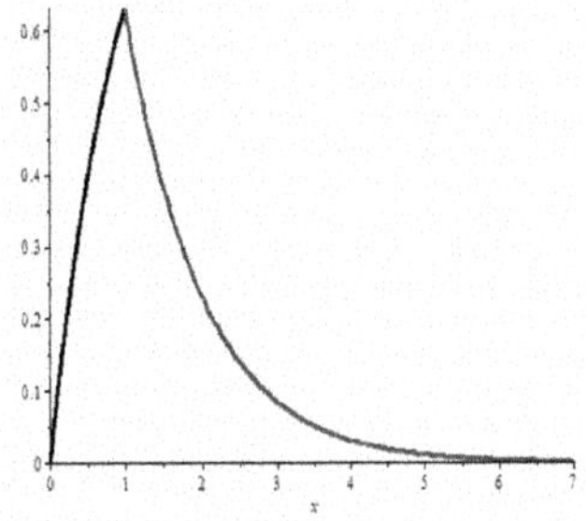

FIGURE 7.

For $x > 1$, we should consider the total $\int_0^1$, and

$$f_S(x) = \int_0^1 e^{-(x-y)} dy = e^{-x}(e-1).$$

The graph for all x's is given in Fig.7. □

In the examples above, we saw that the distribution of a sum may essentially differ from the distributions of the separate terms. Next, we consider cases when the convolution inherits properties of individual terms in the sum.

2.1.2 Some classical examples

I. Sums of normals (normal r.v.'s).

Proposition 2 *Let X_1 and X_2 be independent normals with expectations m_1 and m_2, and variances σ_1^2 and σ_2^2, respectively. Then the r.v. $S = X_1 + X_2$ is normal with expectation $m_1 + m_2$, and variance $\sigma_1^2 + \sigma_2^2$. In other words, if $\varphi_{m\sigma^2}$ is the normal density with mean m and variance σ^2, then*

$$\varphi_{m_1\sigma_1^2} * \varphi_{m_2\sigma_2^2} = \varphi_{m_1+m_2,\sigma_1^2+\sigma_2^2}.$$

We consider a short proof in the next section. Here, as one more example of convolution, we demonstrate the direct convolution for the particular case $m_1 = m_2 = 0$ and $\sigma_1 = \sigma_2 = 1$. Denoting by $\varphi(x)$ the standard normal density and applying (2.1.3), we have

$$f_S(x) = \int_{-\infty}^{\infty} \varphi(x-y)\varphi(y)dy = \int_{-\infty}^{\infty} (2\pi)^{-1/2} \exp\{-(x-y)^2/2\}(2\pi)^{-1/2} \exp\{-y^2/2\}dy$$
$$= (2\pi)^{-1} \int_{-\infty}^{\infty} \exp\{-\frac{1}{2}(x-y)^2 - \frac{1}{2}y^2\}dy = (2\pi)^{-1} \int_{-\infty}^{\infty} \exp\{xy - y^2 - \frac{1}{2}x^2\}dy.$$

Completing the square, we write $xy - y^2 - \frac{1}{2}x^2 = -(y - x/2)^2 - \frac{1}{4}x^2$. Then

$$f_S(x) = (2\pi)^{-1} \exp\{-x^2/4\} \int_{-\infty}^{\infty} \exp\{-(y-x/2)^2\}dy.$$

As is straightforward to verify, with the change of variables $y - x/2 = s/\sqrt{2}$, we may write

$$f_S(x) = \frac{1}{\sqrt{2\pi}\sqrt{2}} \exp\{-x^2/4\} \int_{-\infty}^{\infty} \frac{1}{\sqrt{2\pi}} \exp\{-s^2/2\}dy.$$

The last integral is that of the standard normal density and consequently, equals one. The expression before the integral is the normal density with zero mean and a variance of two. ■

II. Sums of Poisson r.v.'s.

Proposition 3 *Let X_1 and X_2 be independent Poisson r.v.'s with parameters λ_1 and λ_2, respectively. Then the r.v. $S = X_1 + X_2$ is a Poisson r.v. with parameter $\lambda_1 + \lambda_2$. In other words, if π_λ is the Poisson distribution with mean λ, then*

$$\pi_{\lambda_1} * \pi_{\lambda_2} = \pi_{\lambda_1+\lambda_2}.$$

Proof. Again, to demonstrate the convolution procedure, we give a direct proof. A shorter proof with use of m.g.f.'s is given in the next section. By virtue of (2.1.4), for $f_m = P(S = m)$, we have

$$f_m = \sum_{k=0}^{m} f_k^{(1)} f_{m-k}^{(2)} = \sum_{k=0}^{m} e^{-\lambda_1} \frac{\lambda_1^k}{k!} e^{-\lambda_2} \frac{\lambda_2^{m-k}}{(m-k)!}$$
$$= e^{-(\lambda_1+\lambda_2)} \frac{1}{m!} \sum_{k=0}^{m} \frac{m!}{k!(m-k)!} \lambda_1^k \lambda_2^{m-k} = e^{-(\lambda_1+\lambda_2)} \frac{1}{m!} \sum_{k=0}^{m} \binom{m}{k} \lambda_1^k \lambda_2^{m-k}.$$

The last sum above is the binomial expansion of $(\lambda_1 + \lambda_2)^m$, which leads to the Poisson formula for the probability f_m. ■

III. Sums of Γ-distributed r.v.'s. We call a r.v. Γ-r.v. if it has a Γ-distribution.

Proposition 4 *Let X_1 and X_2 be independent Γ-r.v.'s with parameters (a, ν_1) and (a, ν_2) respectively. (Notice that the scale parameter a is the same.) Then the r.v. $S = X_1 + X_2$ is a Γ-r.v. with parameters $(a, \nu_1 + \nu_2)$. In other words, if $f_{a\nu}$ denotes the Γ-density with parameters (a, ν), then*

$$f_{a\nu_1} * f_{a\nu_2} = f_{a,\nu_1+\nu_2}. \tag{2.1.8}$$

This is a very important fact having many applications in diverse areas such as physics, economics, etc., and, of course, in insurance, which we will see repeatedly.

In Exercise 21, the reader is encouraged to carry out a direct proof using (2.1.3). Exercise 21 contains some suggestions as to how to proceed. The reader also can take a look, e.g., at [122, p.192]. In the next section, we prove Proposition 4 by using m.g.f's.

EXAMPLE 1. During a day, a company received four telephone calls with claims from clients from a homogenous group. The company knows that the distribution of a particular claim given that a loss event happened, is exponential with a mean of one (unit of money). However, the real sizes of these particular claims have not yet been evaluated. What is the probability that the cumulative claim will exceed, for example, 5?

We deal with $S_4 = X_1 + X_2 + X_3 + X_4$, where the X's are exponentially distributed with the same parameter $a = 1$. The exponential distribution is the Γ-distribution with parameter $\nu = 1$. Hence, by Proposition 4, the r.v. S_4 has the Γ-distribution with parameters $a = 1$ and $\nu = 4$. So, the density

$$f_{S_4}(x) = \frac{x^3}{\Gamma(4)} e^{-x} = \frac{x^3}{3!} e^{-x} = \frac{x^3}{6} e^{-x},$$

and

$$P(S_4 > 5) = 1 - \frac{1}{6}\int_0^5 x^3 e^{-x} dx \approx 0.27.$$

EXAMPLE 2 ([153, N16][4]) includes situations different from what we considered above. Which of the following are true?

1. The sum of two independent negative binomial r.v.'s with parameters (p_1, ν_1) and (p_2, ν_2) is negative binomial if and only if $\nu_1 = \nu_2$.

2. The sum of two independent binomial r.v.'s with parameters (p_1, n_1) and (p_2, n_2) is binomial if and only if $n_1 = n_2$.

3. The sum of two independent Poisson r.v.'s with parameters λ_1 and λ_2 is Poisson if and only if $\lambda_1 = \lambda_2$.

All of the above are false.

1. First, note that the distribution (**0**.3.1.15) differs from (**0**.3.1.13) only by a shift, so the answer should be the same for both distributions. Let $p_1 = p_2 = p$, and ν_1, ν_2 be positive integers. Let us consider a sequence of independent trials each having probability p of being a success. Denote by N_{ν_1} the moment when the ν_1-th success occurs, and by N_{ν_2} the number of trials between the ν_1-th success and the success with the number $\nu_1 + \nu_2$. Then N_{ν_1} and N_{ν_2} are independent negative binomial r.v.'s having distribution (**0**.3.1.13) with the common parameter p and the respective parameters ν_1 and ν_2. Then the r.v. $N_{\nu_1} + N_{\nu_2}$ is the moment when the success with the number $\nu_1 + \nu_2$ occurs. Hence, $N_{\nu_1} + N_{\nu_2}$ have the negative binomial distribution with parameters $p, \nu_1 + \nu_2$. The condition $\nu_1 = \nu_2$ is not necessary, while the condition $p_1 = p_2$ matters.

2. Let S_1 and S_2 be independent binomial r.v.'s with parameters (p_1, n_1) and (p_2, n_2), respectively. If $p_1 = p_2 = p$, as we did above, we may interpret S_1 as the number of successes in the first n_1 trials, and S_2 as the number of successes in n_2 trials after the n_1th trial. Thus, $S_1 + S_2$ is the total number of successes in $n_1 + n_2$ trials, and this r.v. has a binomial distribution. Again, the condition $p_1 = p_2$ is essential, while the equality $n_1 = n_2$ is not a necessary condition.

3. From Proposition 4 it follows that $\lambda_1 = \lambda_2$ is not necessary. □

In conclusion, note that Propositions 3 and 4 show—above all—that the Poisson and Γ-distributions may be well approximated by normal distributions for large parameters λ and ν.

Indeed, consider, for example, independent identically distributed r.v.'s $X_1, X_2, \ldots$ having the exponential distribution with parameter a. Then, by Proposition 4, the sum $S_n = X_1 + \ldots + X_n$ has the Γ-distribution with parameters (a, n). On the other hand, by the Central Limit Theorem, S_n is asymptotically normal for large n. Consequently, the Γ-distribution with parameters (a, n) is also asymptotically normal for large n. The same reasoning applies to the Poisson case. Exercises 43 and 44 contain detailed advice on how to show the same in the general case when λ and ν may not be integers.

Route 1 ⇒ *page 166*

[4]Reprinted with permission of the Casualty Actuarial Society.

2.1.3 An additional remark regarding convolutions: Stable distributions 2,3

The following notions help to come to a deeper understanding of Propositions 2–4 above.

We say that a class of distributions is *closed with respect to convolution* if for any two independent r.v.'s, X_1 and X_2, having distributions from this class, the distribution of their sum $X_1 + X_2$ also belongs to the same class. For example, as we saw, the class of all normal distributions is closed with respect to convolution, and the same is true for the class of all Poisson distributions, or the class of all Γ-distributions with a fixed scale parameter a.

However, these classes have different structures, and to clarify this, we introduce one more notion.

Consider a r.v. X and the family of r.v.'s $Y = a + bX$ for *all* possible values of numbers a and $b > 0$. The family of the corresponding distributions is called a *type*. We refer to b as a scale factor, and a as a location constant. One may say that any two distributions from a type are the distributions of r.v.'s that may be linearly transformed to each other. In particular, this means that not only the r.v. X but any r.v. $Y = a + bX$ with $b > 0$ may serve as the "original" r.v. generating the type.

For example, since any (m, σ^2)-normal r.v. $Y = m + \sigma X$, where X is a standard normal r.v., normal distributions compose a type. The same is true for the family of all uniform distributions because the distribution uniform on $[s,t]$ may be represented as the distribution of a r.v. $Y = s + (t - s)X$, where X is uniform on $[0, 1]$.

On the other hand, Poisson distributions do *not* compose a type. Indeed, if X is a Poisson r.v., then the r.v. $a + bX$ is a r.v. assuming values a, $a + b$, $a + 2b$, ... rather than $0, 1, 2, ...$, and hence, it is not Poisson. Therefore, the class of Poisson distributions is a class of another nature than a type.

The same is true for the Γ-distribution because two Γ-distributions with different values of the parameter ν cannot be reduced to each other by a change of scale. See, for example, the graphs of Γ-densities for different ν's in Fig.**0**-13; these functions are essentially different and cannot be reduced to each other by a linear transformation of the argument.

An interesting and important question is which *types* are closed with respect to convolutions. By Proposition 2, the normal type has such a property, while—as we saw in Section 2.1.1—the uniform type is not closed with respect to convolution.

It turns out that the closedness of a type with respect to convolution is a rare property and may be considered a characterization property of the normal distribution.

Proposition 5 *The normal type is the only type of distributions with finite variances, which is closed with respect to convolution.*

A proof of this proposition is actually not difficult but beyond the scope of this book; see, e.g., [27], [38], [122, p. 255].

Thus, when the variances are finite, the sum has the same type as separate terms only in the case of normal distributions. Regarding the classes of Poisson or Γ-distributions that are closed with respect to convolution, we realize that these classes are not types, and the changes that convolution brings about are more essential than the change of scale.

Another matter is that there are distributions with infinite variances whose types have the property under discussion. These distributions are called *stable*. An example is the *Cauchy distribution* with density $f(x) = \frac{1}{\pi(1+x^2)}$. The theory of stable distributions may be found in many textbooks; see, e.g., [27], [38], [120], [122], [129].

2.1.4 The analogue of the binomial formula for convolutions

Sometimes it is useful to know that the Newton binomial formula applies to convolutions too. Namely, for any distributions F_1 and F_2, and non-negative numbers α and β such that $\alpha+\beta=1$,

$$(\alpha F_1 + \beta F_2)^{*n} = \sum_{k=0}^{n} \binom{n}{k} \alpha^k \beta^{n-k} F_1^{*k} * F_2^{*(n-k)}. \tag{2.1.1}$$

Here $F^{*k} = F * \ldots * F$, where the convolution is carried out k times. Detailed advice on how to prove (2.1.1) is given in Exercise 41.

EXAMPLE 1. Let $f_{a\nu}$ denote the Γ-density with parameters a, ν. Find the density $f = f(x) = (\alpha f_{a\nu_1}(x) + \beta f_{a\nu_1}(x))^{*3}$. The reader will see that if a is a common parameter, we can provide calculations for any $\alpha, \beta, \nu_1, \nu_2$. However, to make the calculations more transparent, let us set $\alpha = \beta = 1/2, \nu_1 = 1, \nu_2 = 2$. Then, making use of (2.1.1) and (2.1.8), we have

$$f = \frac{1}{8} f_{a1}^{*3} + 3\frac{1}{4}\cdot\frac{1}{2} f_{a1}^{*2} * f_{a2} + 3\frac{1}{2}\cdot\frac{1}{4} f_{a1} * f_{a2}^{*2} + \frac{1}{8} f_{a2}^{*3}.$$

By (2.1.8), $f_{a1}^{*3} = f_{a3}$, $f_{a1}^{*2} * f_{a2} = f_{a2} * f_{a2} = f_{a4}$ and similarly, $f_{a1} * f_{a2}^{*2} = f_{a5}$, $f_{a2}^{*3} = f_{a6}$. Thus,

$$f = \frac{1}{8}\left(f_{a3} + 3f_{a4} + 3f_{a5} + f_{a6}\right).$$

Hence,

$$\begin{aligned} f(x) &= \frac{1}{8}\left(\frac{a^3}{\Gamma(3)}x^2 e^{-ax} + \frac{3a^4}{\Gamma(4)}x^3 e^{-ax} + \frac{3a^5}{\Gamma(5)}x^4 e^{-ax} + \frac{a^6}{\Gamma(6)}x^5 e^{-ax}\right) \\ &= \frac{1}{8}\left(\frac{a^3}{2}x^2 + \frac{a^4}{2}x^3 + \frac{a^5}{8}x^4 + \frac{a^6}{120}x^5\right) e^{-ax}. \ \square \end{aligned}$$

The particular cases considered above are important and interesting but, nevertheless, are special. In general, computing convolutions is tedious and for large n, practically impossible. The method we consider next sometimes allows to avoid complicated calculations.

2.2 Moment generating functions

In this section, we use m.g.f.'s. The reader is recommended to look up this notion in Section **0**.4 and consider at least some problems from Exercises 33–40 of the current chapter.

Let again $S_n = X_1 + \ldots + X_n$, where X_i's are independent r.v.'s (for example, of payments). Let $M_i(z)$ be the m.g.f. of X_i. Then, the m.g.f. of S_n is

$$M_{S_n}(z) = M_1(z)\cdot M_2(z)\cdot \ldots \cdot M_n(z).$$

(See property (**0**.4.1.4).) Multiplication is a much easier operation than convolution. If we can compute the m.g.f.'s of individual terms, we can compute the m.g.f. of the sum. If we are able to determine which distribution the latter m.g.f. represents, due to the uniqueness property of m.g.f. (see Theorem 8 in Section **0**.4.1), the problem will be solved. Sometimes, to do this, it is enough to recognize a familiar m.g.f. In other cases, some calculations are required.

To demonstrate the power of the method of m.g.f.'s, we begin with the three classical examples corresponding to the three convolution cases considered in the previous section in Propositions 2–4. We will see that the m.g.f. method allows to essentially simplify proofs. For the m.g.f.'s of the distributions considered below, see Section **0**.4.3.

I. Sums of normals.

Let X_1 and X_2 be normal with expectations m_1 and m_2 and variances σ_1^2 and σ_2^2, respectively, and let $S = X_1 + X_2$. Since the m.g.f. of a (m, σ^2)-normal r.v. is $\exp\{mz + \sigma^2 z^2/2\}$, the m.g.f.

$$M_S(z) = \exp\{m_1 z + \sigma_1^2 z^2/2\}\exp\{m_2 z + \sigma_2^2 z^2/2\} = \exp\{(m_1 + m_2)z + (\sigma_1^2 + \sigma_2^2)z^2/2\}.$$

This is the m.g.f. of the normal distribution with expectation $m_1 + m_2$, and variance $\sigma_1^2 + \sigma_2^2$, which proves Proposition 2.

II. Sums of Poisson r.v.'s.

Now, let X_1 and X_2 be Poisson r.v.'s with respective parameters λ_1 and λ_2. The m.g.f. of a Poisson r.v. with parameter λ is $\exp\{\lambda(e^z - 1)\}$. Then the m.g.f.

$$M_S(z) = \exp\{\lambda_1(e^z - 1)\}\exp\{\lambda_2(e^z - 1)\} = \exp\{(\lambda_1 + \lambda_2)(e^z - 1)\}.$$

This is the m.g.f. of the Poisson distribution with parameter $\lambda_1 + \lambda_2$ and Proposition 3 is proved.

III. Sums of Γ-distributed r.v.'s.

Let X_1 and X_2 be Γ-r.v.'s with parameters (a, ν_1) and (a, ν_2), respectively. Since the m.g.f. of the Γ-r.v. with parameters (a, ν) is $(1 - z/a)^{-\nu}$, the m.g.f.

$$M_S(z) = (1 - z/a)^{-\nu_1}(1 - z/a)^{-\nu_2} = (1 - z/a)^{-(\nu_1 + \nu_2)}.$$

This is the m.g.f. of the Γ-distribution with parameters $(a, \nu_1 + \nu_2)$, which proves Proposition 4.

Let us consider now a typical example demonstrating how the calculations may proceed in the general case.

EXAMPLE 1. Let X_1 and X_2 be i.i.d. r.v.'s with common distribution $F = \frac{1}{3}F_1 + \frac{2}{3}F_2$, where F_1 and F_2 are exponential distributions with means 1 and 2, respectively. So, we consider a mixture of exponentials. Find the distribution of $S = X_1 + X_2$.

The m.g.f. of an exponential r.v. with mean m is $(1 - mz)^{-1}$. (Show this proceeding from results of Section **0**.4.3.) The m.g.f. of a mixture of distributions is equal to the mixture of the m.g.f. (see Section **0**.4.1). Thus, the common m.g.f. of X_1 and X_2 is $M_X(z) = \frac{1}{3}\cdot\frac{1}{1-z} + \frac{2}{3}\cdot\frac{1}{1-2z}$.

Hence, $M_S(z) = \left(\frac{1}{3}\cdot\frac{1}{1-z} + \frac{2}{3}\cdot\frac{1}{1-2z}\right)^2.$

Next, we apply the method of partial fractions writing

$$\left(\frac{1}{3}\cdot\frac{1}{1-z}+\frac{2}{3}\cdot\frac{1}{1-2z}\right)^2=\frac{a}{1-z}+\frac{b}{1-2z}+\frac{c}{(1-z)^2}+\frac{d}{(1-2z)^2}, \tag{2.2.1}$$

where a,b,c,d are coefficients. It is easy to compute (by finding the common denominator) that for the last equality to be true, one should have $a=-4/9$, $b=8/9$, $c=1/9$, $d=4/9$. Mathematical software such as *Maple* or *Mathematica* can do it automatically.

The right member of (2.2.1) may be considered as the mixture of four m.g.f.'s with weights a,b,c,d. The first two m.g.f.'s are exponential with respective means 1 and 2, and the last two correspond to the Γ-distributions with the vector-parameter (a,ν) equal to $(1,2)$ and $(1/2,2)$, respectively.

(Indeed, for example, $\dfrac{1}{1-z}$ is the m.g.f. of the standard exponential distribution. Then $\dfrac{1}{(1-z)^2}=\dfrac{1}{1-z}\cdot\dfrac{1}{1-z}$ is the m.g.f. of the convolution of two standard exponential distribution. By virtue of Proposition 4, this is the Γ-distribution with parameters $(1,2)$. The other terms in (2.2.1) are treated similarly.)

The fact that one weight in (2.2.1) is negative should not make us uncomfortable: we consider this representation as a purely mathematical construction.

If the m.g.f. of the sum is the above mixture of the mentioned m.g.f.'s, then the density of the sum is the mixture of the corresponding densities with the same weights. More specifically,

$$\begin{aligned} f_S(x) &= -\frac{4}{9}e^{-x}+\frac{8}{9}\cdot\frac{1}{2}e^{-x/2}+\frac{1}{9}xe^{-x}+\frac{4}{9}\left(\frac{1}{2}\right)^2 xe^{-x/2} \\ &= \frac{1}{9}\left[(x+4)e^{-x/2}+(x-4)e^{-x}\right] \text{ for } x\geq 0, \text{ and } =0 \text{ for } x<0. \end{aligned} \tag{2.2.2}$$

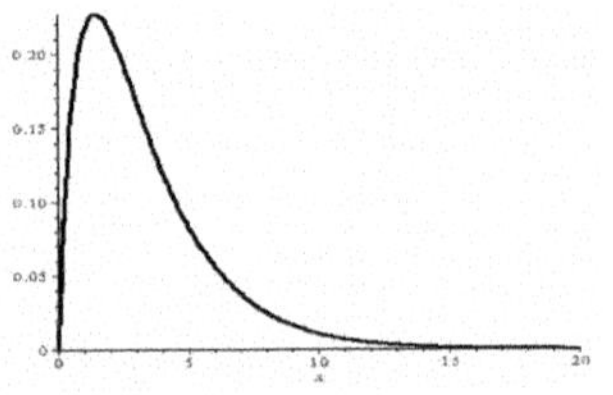

FIGURE 8.

One can double check that the last function is positive for $x>0$ with total integral of one, so it is indeed a density. The graph is given in Fig.8. □

► EXAMPLE 2. It is useful to consider another way of solving the problem of Example 1. As usual, denote by $f_{a\nu}$ the Γ-density with parameters a,ν, and by f the density of the sum. By virtue of (2.1.1),

$$f=\left(\frac{1}{3}f_{11}+\frac{2}{3}f_{1/2,1}\right)^{*2}=\frac{1}{9}f_{11}^{*2}+\frac{4}{9}f_{11}*f_{1/2,1}+\frac{4}{9}f_{1/2,1}^{*2}.$$

By (2.1.8), $f_{11}^{*2}=f_{12}$, and $f_{1/2,1}^{*2}=f_{1/2,2}$.

In order to find $f_{11}*f_{1/2,2}$, it is better to apply the m.g.f.'s method. In Exercise 40, the reader is asked to prove that $[f_{11}*f_{1/2,2}](x)=e^{-x/2}-e^{-x}$, using a method similar to what we used in Example 1 above. Then

$$f(x)=\frac{1}{9}xe^{-x}+\frac{4}{9}(e^{-x/2}-e^{-x})+\frac{4}{9}\left(\frac{1}{2}\right)^2 xe^{-x/2},$$

and we again arrive at (2.2.2). □ ◄

In general, even if we have found the m.g.f. of a sum of r.v.'s, finding the corresponding distribution may turn out to be difficult. As we will see later, the method of m.g.f.'s proves to be useful rather for qualitative theoretical analysis.

However, there is one more principal difficulty which is not relevant to analytical methods. In practice, we do not know the distributions of the addends X_i precisely. As a rule, we can only estimate some parameters of these distributions, for example, means, variances, and some moments. Accordingly, we are guaranteed at best only approximations of the distributions of sums. The most important such approximation is the one associated with the normal distribution and based upon various modifications of the CLT.

---

3 PREMIUMS AND SOLVENCY. APPROXIMATIONS FOR AGGREGATE CLAIM DISTRIBUTIONS

3.1 Premiums and normal approximation. A heuristic approach

3.1.1 Normal approximation and security loading

We already touched on the method of finding premiums with use of normal approximation in Section **1**.2.1. Now we will do it in much more detail.

Let $S_n = X_1 + ... + X_n$, where X_i's are r.v.'s. For now, we do not assume they are independent or identically distributed. Consider the normalized sum

$$S_n^* = \frac{S_n - E\{S_n\}}{\sqrt{Var\{S_n\}}}. \tag{3.1.1}$$

The goal of normalization is to consider the sum S_n in an appropriate scale; namely, after normalization, $E\{S_n^*\} = 0$ and $Var\{S_n^*\} = 1$ (see also Section **0**.2.6).

The modern probability theory establishes a wide spectrum of conditions under which the distribution of S_n^* is asymptotically normal; that is, conditions under which for any x,

$$P(S_n^* \leq x) \to \Phi(x) \text{ as } n \to \infty, \tag{3.1.2}$$

where $\Phi(x)$ is the standard normal d.f.

In the simplest case where the separate terms (or addends) X_i are independent and identically distributed (i.i.d.), and have a finite variance, (3.1.2) is always true (see Section **0**.6.2). In more general situations, some conditions are needed, but it is worth emphasizing that *these conditions are fairly mild.* Let us discuss it in more detail.

If the addends are independent but not identically distributed, the corresponding conditions require that there are no addends that in a certain sense have the same order as the whole sum. One sufficient condition of such a type will be established in Section 3.2.

Independence of addends is also not necessary. The theory of normal approximation for dependent addends is now well developed and deals with a wide variety of types of dependency. In each case, the corresponding condition of asymptotic normality means that the random addends may be dependent but "not too strong".

Probably, the simplest example for illustrative purposes is the so called m-dependence when each term in the sum depends only on the "nearest" m r.v.'s. More specifically, X_i depends only on r.v.'s X_j for which $i-m \le j \le i+m$. The last collection of r.v.'s may be viewed as a "*dependency neighborhood*".

For example, if $m = 1$, the r.v. X_1 depends only on X_2, the r.v. X_2 depends only on X_1 and X_3, the r.v. X_3 depends only on X_2 and X_4 and so on. So, the dependency neighborhood of a r.v. consists of three r.v's (counting the r.v. itself).

As a matter of fact, one may impose a much weaker condition. For example, it is sufficient to require that the dependence between two terms, X_i and X_j, becomes (in a certain sense and at a certain rate) weaker as the "distance" $|i-j|$ becomes larger. Loosely put, the addends which are far away from each other are weakly dependent. Requirements of such a type are called *mixing conditions*.

Some central limit theorems for dependent r.v.'s may be found in [120], [122]; more systematic exposition—e.g., in [63], [70]. Graph related dependency structures described in terms of dependency neighborhoods may be found, e.g., in [113]; economic models based on such structures in [87], [88]; see also references therein.

Once asymptotic normality is established, we can use it to estimate, for example, the premium for which the probability that the company will not suffer a loss is larger than a given level.

Suppose that the company specifies the lowest acceptable level β for the mentioned probability. For instance, β may be equal to 0.9 or 0.8. Often, insurance companies connect such a probability with the investment rate in the financial market and the corresponding probabilities of default. It may lead to a very high β like 0.99. As a matter of fact, β does not have to be very close to one. If, say, $\beta = 0.8$, this means that in the long run, the company will make a profit 80% of time on the average. If the single period which we consider here is short, this is not bad at all.

EXAMPLE 1. Consider k periods of time. Let the probability of avoiding a loss in a separate period be β, and let results for different periods be independent. Reasoning very roughly, assume that the company will not suffer a loss overall if the number of non-successful periods is less than the number of successful periods, which means that the number of non-successful periods is less than $k/2$. Denote the probability of this event by $p = p(k,\beta)$. This is the binomial probability that during k independent trials, the number of successes will not be less than $k/2$. More precisely, $p(k,\beta) = \sum_{j\ge k/2}\binom{k}{j}\beta^j(1-\beta)^{k-j}$. By using Excel or a calculator in statistics mode, it is easy to verify data in the following table.

TABLE 1.

	$k=3$ $\beta=0.8$	$k=3$ $\beta=0.85$	$k=3$ $\beta=0.9$	$k=12$ $\beta=0.6$	$k=12$ $\beta=0.7$	$k=12$ $\beta=0.8$
$p(k,\beta)$	0.897	0.939	0.972	0.841	0.961	0.996

Let one period be a year, and let the total period be (only) three years ($k = 3$). Then the probability of not suffering a loss at the end of the three-year period is approximately 0.9 for $\beta = 0.8$. For $\beta = 0.9$, it is close to 0.97.

If one period is a month and the total period is a year ($k = 12$), then even $\beta = 0.6$ is not bad since p in this case is about 0.84. For $\beta = 0.8$, we have $p = 0.996$, which is fairly high.

Note, however, that if a single period under consideration is a month, then the probability q of a loss event for separate clients may be small. In the numerical examples shown below, the choice of q corresponds rather to a single period of a year. □

Let us return to the general scheme and consider one period of time. Assume that the premium the company collects is proportional to the expected payment. More specifically, the total premium

$$c_n = (1+\theta)E\{S_n\}. \tag{3.1.3}$$

As was already mentioned in Section **1.5**, the coefficient θ is called a *relative security loading*. The quantity $\theta E\{S_n\}$ is called a *security loading*. Similar to Section **1.2.1**, for the least acceptable premium, we may write

$$\beta = P(S_n \le c_n) = P(S_n - E\{S_n\} \le \theta E\{S_n\}) = P\left(\frac{S_n - E\{S_n\}}{\sqrt{Var\{S_n\}}} \le \frac{\theta E\{S_n\}}{\sqrt{Var\{S_n\}}}\right)$$
$$= P\left(S_n^* \le \theta E\{S_n\} \Big/ \sqrt{Var\{S_n\}}\right), \tag{3.1.4}$$

where the normalized sum S_n^* is defined in (3.1.1).

We start with non-rigorous estimation. If we consider normal approximation acceptable, we can write that $P(S_n^* \le x) \approx \Phi(x)$ for any x, and in particular, $P\left(S_n^* \le \theta E\{S_n\} \Big/ \sqrt{Var\{S_n\}}\right) \approx \Phi\left(\theta E\{S_n\} \Big/ \sqrt{Var\{S_n\}}\right)$. Thus,

$$\Phi\left(\theta E\{S_n\} \Big/ \sqrt{Var\{S_n\}}\right) \approx \beta.$$

This, in turn, implies that

$$\theta \approx \frac{q_{\beta s}\sqrt{Var\{S_n\}}}{E\{S_n\}}, \tag{3.1.5}$$

where $q_{\beta s}$ is the β-quantile of the standard normal distribution, which means that $\Phi(q_{\beta s}) = \beta$. For instance, if $\beta = 0.9$, then $q_{\beta s} = 1.28...$.

Consider for a while, the particular case of independent and identically distributed (i.i.d) r.v.'s X_i. Let $m = E\{X_i\}$, $\sigma^2 = Var\{X_i\}$. Then, $E\{S_n\} = mn$, $Var\{S_n\} = \sigma^2 n$, and as is easy to verify, (3.1.5) may be rewritten as

$$\theta \approx \frac{q_{\beta s}\sigma}{m\sqrt{n}}. \tag{3.1.6}$$

For a distribution with mean m and standard deviation σ, the fraction σ/m is called a *coefficient of variation*. We see that θ is specified by the coefficient of variation of the addends X_i and the number of clients. Moreover, in view of (3.1.3), the total premium may be distributed between individual clients in accordance with the relative loading θ, which amounts to the premium $(1+\theta)m$ for each client.

Given the coefficient σ/m, the larger the number of clients, the less the security loading needed by the company to maintain a certain level of security. More precisely, the loading coefficient required is proportional to $1/\sqrt{n}$.

Assume that a certain number n of clients agree to buy insurance with a loading coefficient larger than or equal to the loading coefficient in (3.1.6). Then the company can function at a level of security higher than or equal to the desirable level β. This reflects the essence of the matter; namely, that insurance is efficient when the risk is redistributed among a sufficiently large number of clients. We discuss this issue in more detail in Section 3.3.

Let us come back to the general case. Set $m_i = E\{X_i\}$. Then

$$E\{S_n\} = m_1 + ... + m_n = n\overline{m}_n, \tag{3.1.7}$$

where

$$\overline{m}_n = \frac{1}{n}(m_1 + ... + m_n),$$

the average expectation.

In general, to find $Var\{S_n\}$ we should know the dependency structure between the X's. In this book, we mostly restrict ourselves to the case of independent addends. In this case,

$$Var\{S_n\} = \sigma_1^2 + ... + \sigma_n^2 = n\overline{\sigma}_n^2, \tag{3.1.8}$$

where $\sigma_i^2 = Var\{X_i\}$, and

$$\overline{\sigma}_n^2 = \frac{1}{n}(\sigma_1^2 + ... + \sigma_n^2), \tag{3.1.9}$$

the average variance.

► Note that the independence of X's is not necessary for (3.1.8) to be true: we need only the X's to be non-correlated (see Section **0**.2.4.3).

Moreover, if the r.v.'s are correlated, this does not exclude the possibility of normal approximation and the CLT may be still true under certain conditions. In this case, we should just write a correct representation for $Var\{S_n\}$. The reader familiar with the notion of covariance (see again Section **0**.2.4.3 for detail), knows that, in general,

$$Var\{S_n\} = \sigma_1^2 + ... + \sigma_n^2 + 2\sum_{i>j} Cov\{X_i, X_j\}.$$

In this case, we can use the approximation below, defining $\overline{\sigma}_n^2$ as $Var\{S_n\}/n$. ◄

From (3.1.7) and (3.1.8) it follows that we can rewrite (3.1.5) as

$$\theta \approx \frac{q_{\beta_s}\overline{\sigma}_n}{\overline{m}_n\sqrt{n}}. \tag{3.1.10}$$

This is similar to (3.1.6) and shows that when the average characteristics $\overline{\sigma}_n$ and $\overline{m}_n$ are bounded, θ again has an order of $1/\sqrt{n}$.

The choice between equivalent relations (3.1.5) and (3.1.10) is the matter of convenience. Though the latter explicitly shows the structure of the relative security loading θ, relation (3.1.5) may turn out to be more convenient in calculations.

In the general case (3.1.10), the situation is the same as in the case of i.i.d. r.v.'s, but the role of the coefficient of variation is played by $\overline{\sigma}_n/\overline{m}_n$, the fraction of the average characteristics $\overline{\sigma}_n$ and $\overline{m}_n$. Furthermore, once θ is determined, the ith client pays $(1+\theta)m_i$.

It makes sense to note also that θ in (3.1.10) should not be viewed as the real security loading coefficient to be used by the company. If the law and circumstances allow it, the company may proceed from a larger θ. The coefficient in (3.1.10) is the minimal coefficient acceptable for the company.

EXAMPLE 2. Consider a homogeneous group of $n = 2000$ clients. Assume that the probability of a loss event for each client is $q = 0.1$ and if a loss event occurs, the payment is a r.v. uniformly distributed on $[0,1]$. In accordance with (1.2.6)-(1.2.7), the expected value and the variance of the separate payment X are $m = q\frac{1}{2} = 0.05$ and $\sigma^2 = q\frac{1}{12} + q(1-q)(\frac{1}{2})^2 \approx 0.0308$. (The reader may look up the mean and the variance of the uniform distribution in Section **0**.3.2.1.) Let $\beta = 0.9$. Then the quantile $q_{0.9,s} \approx 1.281$ (see Table 2 in Appendix, Section 2).

Since the X's are identically distributed, we use (3.1.6) which gives

$$\theta \approx \frac{1.281 \cdot \sqrt{0.0308}}{0.05 \cdot \sqrt{2000}} \approx 0.1005.$$

This approximately amounts to a 10% loading. The premium for each client is $(1+\theta)m \approx (1+0.1)0.05 = 0.055$ units of money.

EXAMPLE 3. Assume that a portfolio of a company consists of two homogeneous groups of risks. For the first group, the number of clients $n_1 = 2000$ and the probability of a loss event for each client is $q_1 = 0.1$. The payment, if a loss event occurs, is a non-random amount of $z_1 = 10$ units of money. For the second group, the corresponding quantities are $n_2 = 400$, $q_2 = 0.05$, and $z_2 = 30$. In particular, $n = n_1 + n_2 = 2400$.

Assume that the loss events are independent and $\beta = 0.9$. For a particular payment X (the index is omitted), $E\{X\} = qz$ and $Var\{X\} = z^2 q(1-q)$. Hence,

$$E\{S_n\} = n_1 q_1 z_1 + n_2 q_2 z_2 = 2599.2,$$
$$Var\{S_n\} = n_1 q_1 (1-q_1) z_1^2 + n_2 q_2 (1-q_2) z_2^2 = 35100.$$

Then, by (3.1.5),

$$\theta \approx \frac{1.281 \cdot \sqrt{35100}}{\sqrt{2599.2}} \approx 0.092,$$

that is, about 9.2%. Each client from the first group should pay a premium of $(1+\theta)q_1 z_1 \approx 1.092$, while for the second group, the individual premium is $(1+\theta)q_2 z_2 \approx 1.638$. □

3.1.2 An important remark: the standard deviation principle

The representation of the premium as

$$c_n = (1+\theta)E\{S_n\} \tag{3.1.11}$$

is traditional. However, as we saw above, the loading coefficient θ depends on the distributions of X's. In the limiting case, it depends on the mean and variance of S_n. So, as a matter of fact, $E\{S_n\}$ appears in the r.-h.s. of (3.1.11) twice: as the second cofactor and implicitly, in θ.

Another form of premium representation which is in a certain sense more convenient and natural, is determined by the relation

$$c_n = E\{S_n\} + \lambda\sigma_{S_n}, \tag{3.1.12}$$

where the symbol σ_X denotes the standard deviation of a r.v. X and the coefficient λ indicates loading with respect to standard deviation. The representation (3.1.12) corresponds to the so called *standard deviation principle*.

Following the same logic as in (3.1.4), we can write

$$\beta = P(S_n \le c_n) = P(S_n - E\{S_n\} \le \lambda\sigma_{S_n}) = P\left(\frac{S_n - E\{S_n\}}{\sigma_{S_n}} \le \lambda\right) = P\left(S_n^* \le \lambda\right).$$

Then, using normal approximation, we write $P\left(S_n^* \le \lambda\right) \approx \Phi(\lambda)$. Thus, up to normal approximation,

$$\lambda = q_{\beta s}, \tag{3.1.13}$$

that is, does not depend on X's and coincides with the β-quantile of the standard normal distribution.

Assume, for simplicity, that X's are identically distributed. Set $m = E\{X_i\}$ and $\sigma^2 = Var\{X_i\}$. Then $E\{S_n\} = mn$, $Var\{S_n\} = \sigma^2 n$ and in view of (3.1.12) and (3.1.13),

$$c_n = mn + q_{\beta s}\sigma\sqrt{n}.$$

Hence, the premium per one client is

$$c = \frac{1}{n}c_n = m + q_{\beta s}\frac{\sigma}{\sqrt{n}}.$$

This is a nice result. The mean payment m may be viewed as the net premium, that is, the part of the premium which does not take the risk carried by the insurer into account. The additional part, which may be viewed as a payment for risk or security loading, is equal to $q_{\beta s}\frac{\sigma}{\sqrt{n}}$. It does not depend on m, which is natural, and is specified only by the standard deviation and the number of clients n.

In conclusion, note that the representation (3.1.12)-(3.1.13) does not contradict (3.1.11) with θ given in (3.1.10). Indeed, from (3.1.11)-(3.1.12) it follows that $(1+\theta)E\{S_n\} = E\{S_n\} + \lambda\sigma_{S_n}$. If we take the representation (3.1.5)—which is equivalent to (3.1.10)—we readily get that

$$\lambda = \theta\frac{E\{S_n\}}{\sigma_{S_n}} = \frac{q_{\beta s}\sqrt{Var\{S_n\}}}{E\{S_n\}} \cdot \frac{E\{S_n\}}{\sigma_{S_n}} = q_{\beta s}.$$

Thus, (3.1.12) and (3.1.13) do not determine a new rule of premium determination but rather represent the previous rule in another (and nicer) form. For more detail on the standard deviation principle see Section 4.

Route 1 $\Rightarrow$ *page 178*

3.2 A rigorous estimation 2,3

Our next step is to eliminate the sign $\approx$ providing rigorous estimates for θ. To this end, we should know the accuracy of normal approximation or, in other words, the rate of convergence in (3.1.2). In this section, we consider independent X_i's.

Let $Var\{X_i\} = \sigma_i^2$ and $B_n^2 = \sigma_1^2 + ... + \sigma_n^2$, the variance of the sum S_n.

Now, let $\mu_i = E\{|X_i - m_i|^3\}$, the third central absolute moment (in short, the third moment) of the r.v. X_i. Set

$$L_n = \frac{1}{B_n^3} \sum_{i=1}^{n} \mu_i. \tag{3.2.1}$$

First of all, note that in the case of i.i.d. X_i's, if all $\sigma_i = \sigma$ and all $\mu_i = \mu$, then $B_n^2 = \sigma^2 n$ and

$$L_n = \frac{1}{\sigma^3 n^{3/2}} \mu \cdot n = \frac{1}{\sqrt{n}} \frac{\mu}{\sigma^3}. \tag{3.2.2}$$

So, L_n has an order of $1/\sqrt{n}$.

The quantity L_n which is called *Lyapunov's fraction*, plays an essential role in the theory of normal approximation. First, note that L_n does not have a dimension. If we measure the X's in dollars, then the dimensions of the numerator and the denominator in (3.2.1) are the same: $(\text{dollars})^3$.

Secondly, the characteristic L_n is not sensitive to a change of scale. If we multiply all X's by a constant $c > 0$, then the denominator and the numerator in (3.2.1) will be multiplied by c^3, and L_n will not change. (If we multiply X's by a negative c, the denominator and the numerator will be multiplied by $|c|^3$.)

The same concerns shifting. If we add the same number to all X's, this does not change L_n. The reader is suggested to check it rigorously in Exercise 49.

The Lyapunov fraction may be considered a characteristic showing the extent to which the r.v.'s X_i differ. The following examples will help to clarify this.

We have already seen that in the i.i.d. case, L_n is proportional to $1/\sqrt{n}$. Consider the opposite, in a sense, case. Let X_1 be a r.v. with a positive variance, while all other $X_i \equiv 0$, $i \geq 2$. Then, $B_n = \sigma_1^2$, and

$$L_n = \frac{\mu_1}{\sigma_1^3} \not\to 0.$$

If the X_i's are not identically distributed but have "the same order", then the characteristic L_n has the same order as that in (3.2.2). More precisely, this means the following.

Let all $\mu_i \leq \mu$ and all $\sigma_i \geq \sigma$ for some positive constants μ and σ. Then $B_n^2 \geq \sigma^2 n$, and

$$L_n \leq \frac{1}{\sigma^3 n^{3/2}} \mu \cdot n = \frac{1}{\sqrt{n}} \frac{\mu}{\sigma^3}.$$

The significance of the Lyapunov fraction is demonstrated by the following celebrated *Berry-Esseen's theorem*.

Let $F_n^*(x) = P(S_n^* \leq x)$, the d.f. of the r.v. S_n^*.

Theorem 6 *There exists an absolute constant C such that for any x*

$$|F_n^*(x) - \Phi(x)| \leq CL_n. \tag{3.2.3}$$

Proofs of this theorem and comments may be found in many advanced textbooks on Probability, e.g., in [27], [38].

Regarding the constant C, the last results show that

$$C \leq 0.56. \tag{3.2.4}$$

More detailed discussion and references may be found in [122, p.259].

From (3.2.3), we immediately get that $F_n^*(x) \to \Phi(x)$ if

$$L_n \to 0.$$

This is a sufficient condition for normal convergence for the case of non-identically distributed r.v.'s. However, as a matter of fact, Theorem 6 gives us more, namely, the accuracy of normal approximation.

Note that the bound in (3.2.3) is universal, that is, it is the same for all x's and for all distributions with the same third moments and variances. This means that the bound is oriented to the worst case among all distributions mentioned and all x's. In many particular cases, the real rate may be much better, especially for large x's. Many results of this type and further discussions may be found, e.g., in [14], [100], [122].

We apply the Berry-Esseen theorem to estimate the security loading coefficient θ.

Denote by Δ_n the r.-h.s. of (3.2.3), that is, the rate. Let a premium c_n be defined as in Section 3.1.1. We want $P(S_n \leq c_n)$ to be not smaller than a given β. Similar to (3.1.4), we have

$$P(S_n \leq c_n) = P\left(S_n^* \leq \theta E\{S_n\} \Big/ \sqrt{Var\{S_n\}}\right) = F_n^*\left(\theta E\{S_n\} \Big/ \sqrt{Var\{S_n\}}\right)$$
$$\geq \Phi\left(\theta E\{S_n\} \Big/ \sqrt{Var\{S_n\}}\right) - \Delta_n, \tag{3.2.5}$$

where in the last step, we have used (3.2.3).

Let us choose θ such that

$$\Phi\left(\theta E\{S_n\} \Big/ \sqrt{Var\{S_n\}}\right) \geq \beta + \Delta_n. \tag{3.2.6}$$

Then from (3.2.5) it will follow that $P(S_n \leq c_n) \geq \beta + \Delta_n - \Delta_n = \beta$. Thus, for θ from (3.2.6),

$$P(S_n \leq c_n) \geq \beta.$$

A solution to (3.2.6) is given by the inequality

$$\frac{\theta E\{S_n\}}{\sqrt{Var\{S_n\}}} \geq q_{\beta+\Delta_n,s},$$

and

$$\theta \geq \frac{q_{\beta+\Delta_n,s}\sqrt{Var\{S_n\}}}{E\{S_n\}}. \tag{3.2.7}$$

Comparing (3.2.7) with (3.1.5), we see that the difference consists in replacing the quantile $q_{\beta s}$ by a larger quantile $q_{\beta+\Delta_n,s}$. In particular, estimate (3.2.7) is larger than the heuristic estimate (3.1.5). (It is natural that in order not to rely on heuristic calculations, we increase the functioning reliability by establishing a larger premium.)

Similar to (3.1.10), the estimate (3.2.7) may be rewritten as

$$\theta \geq \frac{q_{\beta+\Delta_n,s}\overline{\sigma}_n}{\overline{m}_n\sqrt{n}}. \tag{3.2.8}$$

EXAMPLE 1. Let a group of clients be homogeneous, and let the probability of the loss event for each client be q. The payment, if a loss event occurs, is a certain amount z.

For any particular X (the index is omitted), the mean $m = qz$, the variance $\sigma^2 = z^2q(1-q)$, and the third moment $\mu = E\{|X-qz|^3\} = |z-qz|^3q + |0-qz|^3(1-q)$. Since $z > 0$ and $q < 1$, we have $\mu = z^3(1-q)^3q + z^3q^3(1-q) = z^3q(1-q)(1-2q+2q^2)$.

Because the group is homogeneous, it suffices to consider one X_i for which

$$\frac{\mu}{\sigma^3} = \frac{z^3q(1-q)(1-2q+2q^2)}{(z^2q(1-q))^{3/2}} = \frac{1-2q+2q^2}{\sqrt{q(1-q)}}. \tag{3.2.9}$$

The last expression does not depend on z which is understandable: $\frac{\mu}{\sigma^3}$ is not sensible to change of scale.

Now, let $q = 0.1$, $\beta = 0.9$, and $n = 2000$. In this case, as is easy to compute, $\frac{\mu}{\sigma^3} \leq 2.734$, and in view of (3.2.4),

$$\Delta_n = C\frac{1}{\sqrt{n}}\frac{\mu}{\sigma^3} \leq 0.03425.$$

Then $q_{\beta+\Delta_n,s} \leq q_{0.93425,s} \leq 1.5082$. If we take the last number as an estimate for $q_{\beta+\Delta_n,s}$, then (3.2.8) will be true, and hence $P(S_n \leq (1+\theta)E\{S_n\})$ will not be less than β.

The group is homogeneous. So, $m = qz$ and $\sigma^2 = z^2q(1-q)$. Then, the coefficient of variation $(\sigma/m) = \sqrt{(1-q)/q} = 3$. Eventually, using (3.2.8), we come to the inequality

$$\theta \geq \frac{1.5082 \cdot 3}{\sqrt{2000}} \approx 0.101.$$

As is easy to compute using $q_{\beta s}$ instead of $q_{\beta+\Delta_n,s}$, the heuristic estimate (3.1.10) gives us ≈ 0.085.

It is interesting to compare results for different values of n and β.

Denote by θ_{n0} and θ_n the estimates given by (3.1.10) and (3.2.8), respectively. The reader is invited to verify the figures in the following Table 1.

TABLE 1.

	Δ_n	$q_{\beta s}$	$q_{\beta+\Delta_n,s}$	θ_{n0}	θ_n
$n = 2000,\ \beta = 0.8$	0.034	0.841	0.971	0.056	0.065
$n = 2000,\ \beta = 0.9$	0.034	1.281	1.508	0.086	0.101
$n = 2000,\ \beta = 0.95$	0.034	1.645	2.150	0.110	0.144
$n = 8000,\ \beta = 0.9$	0.017	1.281	1.386	0.042	0.046
$n = 8000,\ \beta = 0.95$	0.017	1.644	1.840	0.055	0.062

We see that the discrepancy between θ_{n0} and θ_n increases for larger β's. On the other hand, as n increases, the difference between θ_{n0} and θ_n decreases. A more detailed discussion is relegated to Exercise 50b and remarks there.

An important remark. Let us return to the case $n = 2000$ and $\beta = 0.9$, for which the heuristic estimate $\theta_{n0} \approx 0.086$, while $\theta_n \simeq 0.101$. The comparison of these two numbers should not lead us to the wrong conclusion that the *real* error of the heuristic estimate is of the order of $0.101 - 0.086 = 0.015$. This is not a real error but an estimate of the error obtained by using Theorem 6. In other words, 0.101 is the loading which guarantees the given security level due to Theorem 6. However, this does not exclude the existence of an advanced result that could give an estimate closer to the heuristic one. Such results indeed exist; some of them may be found, e.g., in monographs [14], [100].

EXAMPLE 2. Now consider the portfolio of the two groups from Example 3.1.1-3. We have already shown how to compute μ and σ for a particular X. Then, for the whole portfolio,

$$L_n = \frac{n_1 z_1^3 q_1(1-q_1)(1-2q_1+2q_1^2) + n_2 z_2^3 q_2(1-q_2)(1-2q_2+2q_2^2)}{\left(n_1 z_1^2 q_1(1-q_1) + n_2 z_2^2 q_2(1-q_2)\right)^{3/2}}.$$

It is straightforward to calculate using the particular values from Example 3.1.1-3 that $L_n \approx 0.093$.

Hence, $\Delta_n = CL_n \approx 0.560 \cdot 0.093 \approx 0.052$. Let $\beta = 0.8$. Then $q_{\beta+\Delta_n,s} \leq q_{0.852,s} \approx 1.045$. In Example 3.1.1-3, we have found that $E\{S_n\} \approx 2599$ and $VarS_n = 35100$.

Then the bound (3.2.7) gives the value $\theta_n \approx \frac{1.045 \cdot \sqrt{35100}}{2599} \approx 0.075$, while by (3.1.10), $\theta_{n0} \approx \frac{0.842 \cdot \sqrt{35100}}{2599} \approx 0.060$.

EXAMPLE 3. Let us return to Example 3.1.1-2. We have already computed that $m = 0.05$ and $\sigma^2 \approx 0.0308$. To apply (3.2.7), we need to calculate the third moment

$$\mu = E\{|X-m|^3\} = E\{|X-0.05|^3\} = |0-0.05|^3(1-q) + q\int_0^1 |x-0.05|^3 dx \approx 0.020$$

for $q = 0.1$. The above value may be obtained either by direct integration or by use of software.

Since the group is homogeneous,

$$\Delta_n = C\frac{\mu}{\sigma^3\sqrt{n}} \leq \frac{0.56 \cdot 0.02}{(0.0308)^{3/2}\sqrt{2000}} \approx 0.046.$$

Eventually,

$$\theta \geq \frac{q_{0.9+0.046,s}\sqrt{0.0308}}{0.05 \cdot \sqrt{2000}} \approx 0.126. \ \square$$

1,2,3

3.3 The number of contracts needed to maintain a given security level

Above, given a risk portfolio, we were seeking a θ ensuring that the probability of not suffering a loss is not less than a given level β. Assume now that θ is given and we are looking for a sufficient number of contracts n for which the probability mentioned is sufficiently large. This is a natural statement of the problem. A premium is a market characteristic and it is determined not only by companies but by clients also, in accordance with their attitude to risk. Certain rules and regulations also keep premiums below some level.

If we restrict ourselves to the heuristic estimate of Section 3.1.1, the problem turns out to be simple. Namely, we should find n from (3.1.10) for given β and θ, and the distribution of X's. If the group is homogeneous (X's are identically distributed), the coefficient of variation $\bar{\sigma}_n/\bar{m}_n = \sigma/m$, where σ and m are the standard deviation and the mean of a separate X. In this case, we immediately get from (3.1.10) that

$$n \approx \left(q_{\beta_s}\sigma/(m\theta)\right)^2. \tag{3.3.1}$$

EXAMPLE 1. Let us return to Example 3.1.1-2. For $\beta = 0.9$ and $n = 2000$, we got $\theta \approx 0.100$. Assume that potential buyers of the insurance product agree only on premiums with $\theta = 0.07$. How many clients should the company have in this case in order to keep the same level β ? From (3.3.1), we have

$$n \approx \left(1.281 \cdot \sqrt{0.0308}\,/(0.05 \cdot 0.07)\right)^2 \approx 4122. \tag{3.3.2}$$

There is another way to obtain the same answer. Since we have only changed θ, we could use the answer from Example 3.1.1-2 and, because n is proportional to θ^{-2}, write

$$n \approx 2000 \cdot \frac{(0.1005)^2}{(0.07)^2} \approx 4122.$$

In any case, the answer above represents only a rough approximation, and eventually we can say only that "n should be around 4000". □

Route 1 ⇒ page 199

A rigorous approximation leads to somewhat more complicated calculations since in this case, n should be a solution to the inequality (3.2.6): 2,3

$$\Phi\left(\theta E\{S_n\}\Big/\sqrt{Var\{S_n\}}\right) \geq \beta + \Delta_n. \tag{3.3.3}$$

Let the group be homogeneous, and m, σ^2, and μ be the mean, the variance, and the third moment for a separate X. Then $E\{S_n\} = mn$, $Var\{S_n\} = \sigma^2 n$, and $\Delta_n = C\mu/(\sigma^3\sqrt{n})$. Inequality (3.3.3) may be rewritten as

$$\Phi(\theta\sqrt{n}m/\sigma) \geq \beta + C\mu_3/(\sigma^3\sqrt{n}). \tag{3.3.4}$$

The l.-h.s. of (3.3.4) is an increasing function of n, while the r.-h.s. is decreasing. So, the numerical solution for the inequality (3.3.4) is easy: starting with a "non-large" n, we increase it until we find the first n for which (3.3.4) becomes true. Certainly, if we do not want the procedure to be long, we should not start from $n = 1$, but from a value of n closer to the solution we are looking for. For example, we can take, as a starting value of n, the

Adjusting calculations

	A	B	C	D	E	F	G
1	n	$g_1(n)$	$g_2(n)$	the indicator for $g_1>g_2$		the initial n	
2	4000	0.897048	0.932761	0			
3	4100	0.899837	0.932359	0		4000	
4	4200	0.902538	0.931972	0		100	
5	4300	0.905153	0.931598	0			
6	4400	0.907688	0.931237	0			
7	4500	0.910144	0.930888	0		the step in n	
8	4600	0.912525	0.93055	0			
9	4700	0.914833	0.930223	0			
10	4800	0.917072	0.929907	0			
11	4900	0.919243	0.9296	0			
12	5000	0.92135	0.929303	0			
13	5100	0.923395	0.929014	0			
14	5200	0.92538	0.928733	0			
15	5300	0.927307	0.928461	0			
16	5400	0.929178	0.928196	1			
17	5500	0.930995	0.927939	1			
18	5600	0.93276	0.927688	1			
19	5700	0.934474	0.927444	1			
20	5800	0.93614	0.927207	1			
21	5900	0.937759	0.926975	1			
22	6000	0.939332	0.926749	1			
23							

	A	B	C	D	E	F	G
1	n	$g_1(n)$	$g_2(n)$	the indicator for $g_1>g_2$		the initial n	
2	5350	0.928249	0.928328	0			
3	5351	0.928268	0.928325	0		5350	
4	5352	0.928286	0.928323	0		1	
5	5353	0.928305	0.92832	0			
6	5354	0.928324	0.928317	1			
7	5355	0.928343	0.928315	1		the step in n	
8	5356	0.928361	0.928312	1			
9	5357	0.92838	0.928309	1			
10	5358	0.928399	0.928307	1			
11	5359	0.928417	0.928304	1			
12	5360	0.928436	0.928301	1			
13	5361	0.928455	0.928299	1			
14	5362	0.928473	0.928296	1			
15	5363	0.928492	0.928293	1			
16	5364	0.92851	0.928291	1			
17	5365	0.928529	0.928288	1			
18	5366	0.928548	0.928286	1			
19	5367	0.928566	0.928283	1			
20	5368	0.928585	0.92828	1			
21	5369	0.928604	0.928278	1			
22	5370	0.928622	0.928275	1			
23							

FIGURE 9. An Excel worksheet for Example 3.3-2.

estimate for n obtained by the heuristic approach, and first choose a rough step in n. Once we find a rough estimate, we may adjust our calculations.

EXAMPLE 2. Consider the situation of Example 1. We have $m = 0.05$, $\sigma \approx \sqrt{0.0308} \approx 0.175$, $\mu_3 = 0.020$, $\theta = 0.07$, and $C = 0.56$. The reader is invited to verify that, for the particular numbers above, (3.3.4) may be written as

$$\Phi(0.02\sqrt{n}) \geq 0.9 + 2.072/\sqrt{n}. \tag{3.3.5}$$

In the Excel worksheet in Fig.9, the l.-h.s. of (3.3.5) is denoted by $g_1(n)$, the r.-h.s. —by $g_2(n)$. In the first worksheet in Fig.9a, we start with $n = 4000$, the step in n is 100, and the (rough) estimate is 5300. In the second worksheet in Fig.9b, we start with 5350, the step is 1, and the estimate is 5353. Certainly, a reasonable answer would be "around 5350 or so". □

3.4 Approximations taking into account the asymmetry of S

A r.v. X and its distribution are said to be *symmetric* about their mean m if $P(X > m+x) = P(X < m-x)$ for all $x > 0$. The r.v. S_n is not necessarily symmetric. Let, for example, X's be exponential (and hence, non-symmetric). Then, for any fixed n, the distribution of S_n is a Γ-distribution and consequently, it is also non-symmetric. Another matter is that when n is growing, the distribution of S_n approaches a normal distribution which is symmetric, so the asymmetry of S_n is diminishing. In this section, we discuss how to take the asymmetry mentioned into account.

3.4.1 The skewness coefficient

Probably the most popular, though somewhat rough, characteristic of asymmetry of a r.v. X is the so called *skewness coefficient*

$$\gamma = \gamma_X = \frac{\varkappa_3}{\sigma^3},$$

where $\sigma^2 = Var\{X\}$, $\varkappa_3 = E\{(X-m)^3\}$, the third central non-absolute moment, and $m = E\{X\}$.

(The notation above is consistent with the notation for cumulants in Section **0**.4.5.2, since the third cumulant $\varkappa_3$ coincides with the third central moment; see Section **0**.4.5.2.)

First, note that if X is symmetric about m, then $\varkappa_3 = 0$ and the skewness $\gamma = 0$. To clarify this, assume that X has a density $f(x)$. If X is symmetric about m, the density $f(x)$ is also symmetric about m. Consider the centered r.v. $\tilde{X} = X - m$. Its density is $\tilde{f}(x) = f(x+m)$ and is symmetric about 0. Then

$$\varkappa_3 = E\{(X-m)^3\} = E\{\tilde{X}^3\} = \int_{-\infty}^{\infty} x^3 \tilde{f}(x)dx = 0$$

as an integral of an odd function.

The reader is encouraged to show that the coefficient γ is dimensionless and invariant under shifting and change of scale: $\gamma_{X+c} = \gamma_X$, and $\gamma_{cX} = \gamma_X$ for any number c (in the last case, $c \neq 0$).

We say that the distribution of X is skewed to the right if $\gamma_X > 0$. If $\gamma_X < 0$, the r.v. X is said to be skewed to the left.

EXAMPLE 1. Let X have the Γ-distribution with parameters (a, ν). First, since a is a scale parameter, skewness does not depend on a and we can set $a = 1$. Now, $E\{X^2\} = \nu(\nu+1)$, $E\{X^3\} = \nu(\nu+1)(\nu+2)$ (see Section **0**.3.2.3) and hence, $E\{(X-E\{X\})^3\} = E\{(X-\nu)^3\} = E\{X^3\} - 3\nu E\{X^2\} + 3\nu^2 E\{X\} - \nu^3 = 2\nu$. Since $Var\{X\} = \nu$, we have $\gamma = 2\nu/\nu^{3/2} = 2/\sqrt{\nu}$. □

Let $S_n = X_1 + ... + X_n$, where the X's are independent r.v.'s. Let us set $m_i = E\{X_i\}$, $\sigma_i^2 = Var\{X_i\}$, and $\varkappa_{3i} = E\{(X_i - E\{X_i\})^3\}$. Then

$$E\{(S_n - E\{S_n\})^3\} = \sum_{i=1}^{n} \varkappa_{3i}.$$

Indeed, let $\tilde{X}_i = X_i - m_i$. Then $S_n - E\{S_n\} = \tilde{X}_1 + ... + \tilde{X}_n$. Clearly, $E\{\tilde{X}_i\} = 0$, and hence

$$E\{(S_n - E\{S_n\})^3\} = E\{(\tilde{X}_1 + ... + \tilde{X}_n)^3\} = \sum_i E\{\tilde{X}_i^3\},$$

because for $i \neq j \neq k$, all terms $E\{\tilde{X}_i^2 \tilde{X}_j\} = E\{\tilde{X}_i^2\}E\{\tilde{X}_j\} = 0$, and $E\{\tilde{X}_i \tilde{X}_j \tilde{X}_k\} = E\{\tilde{X}_i\}E\{\tilde{X}_j\}E\{\tilde{X}_k\} = 0$.

Since $Var\{S_n\} = \sum_{i=1}^{n} \sigma_i^2$,

$$\gamma_{S_n} = \left[\left(\sum_{i=1}^{n} \varkappa_{3i} \right) \Big/ \left(\sum_{i=1}^{n} \sigma_i^2 \right)^{3/2} \right] = \frac{1}{\sqrt{n}} \frac{\overline{\varkappa}_{3n}}{\overline{\sigma}_n^3}, \tag{3.4.1}$$

where the average third moment $\overline{\varkappa}_{3n} = \frac{1}{n}\sum_{i=1}^{n}\varkappa_{3i}$ and the average variance $\overline{\sigma}_n^2 = \frac{1}{n}\sum_{i=1}^{n}\sigma_i^2$.

Let us consider the case when the X's are identically distributed. Then σ_i^2 is equal to some σ^2, $\varkappa_{3i}$ is equal to some $\varkappa_3$, and (3.4.1) implies that

$$\gamma_{S_n} = \frac{1}{\sqrt{n}}\frac{\varkappa_3}{\sigma^3} = \frac{\gamma}{\sqrt{n}},$$

where γ is the skewness coefficient for each X. Thus, $\gamma_{S_n} \to 0$, which is not surprising because S_n approaches a symmetric r.v.

3.4.2 The Γ-approximation

Next, we discuss how to approximate the distribution of S_n taking into account possible asymmetry. We start with somewhat naive but frequently well working *Γ-approximation.*

Many sample distributions of aggregate claims have approximately the same shape as the gamma distribution; namely, they are skewed to the right and their histograms have a unique maximum. So, we can try to approximate the distribution of S_n by the distribution of a r.v. $Y_z = z + Y$, where Y has the Γ-density $f_{a\nu}(x)$ [see (1.1.10)] and z is a number.

The r.v. Y_z is called a *translated* Γ-r.v. Its density is equal to $f_{a\nu}(x-z)$ by virtue of the rule (**0**.2.6.1).

We choose z, a, and ν in such a way that

$$E\{Y_z\} = E\{S_n\},\ Var\{Y_z\} = Var\{S_n\},\ \gamma_{Y_z} = \gamma_{S_n}. \tag{3.4.2}$$

That is, we require the coincidence of the first three moments. The first two conditions from (3.4.2) give

$$z + \frac{\nu}{a} = n\overline{m}_n,\quad \frac{\nu}{a^2} = n\overline{\sigma}_n^2.$$

From Example 1, formula (3.4.1), and the fact that skewness does not depend on shifting, it follows that

$$\nu = \frac{4}{\gamma_{S_n}^2} = \frac{4n\overline{\sigma}_n^6}{\overline{\varkappa}_{3n}^2}.$$

It is now easy to determine that

$$a = \frac{2\overline{\sigma}_n^2}{\overline{\varkappa}_{3n}},\quad z = n\overline{m}_n - \frac{2n\overline{\sigma}_n^4}{\overline{\varkappa}_{3n}}.$$

In the case $m_i = m$, $\sigma_i^2 = \sigma^2$ and $\varkappa_{3i} = \varkappa_3$, we have

$$a = \frac{2\sigma^2}{\varkappa_3} = \frac{2}{\gamma\sigma},\quad z = n\left(m - \frac{2\sigma^4}{\varkappa_3}\right) = n\left(m - \frac{2\sigma}{\gamma}\right),\quad \nu = \frac{4n}{\gamma^2}, \tag{3.4.3}$$

where γ is the skewness coefficient for each X.

Examples are given in Exercises 56-57.

3.4.3 Asymptotic expansions and Normal Power (NP) approximation

The next approximation taking into account a possible asymmetry of S concerns the following refinement of the CLT.

We adopt the notation of Sections 3.1.1-3.2 and assume that the X's are independent and, for the sake of simplicity, identically distributed. For the relations below to be true for the case of non-identically distributed r.v.'s, moment characteristics should be replaced by certain average characteristics.

Let $m_i = m$, $\sigma_i = \sigma$, $\mu_i = \mu$, $\varkappa_{3i} = \varkappa$. As before, set $\gamma = \varkappa/\sigma^3$.

In our case, Theorem 6 implies that

$$|F_n^*(x) - \Phi(x)| \leq C\frac{1}{\sqrt{n}} \cdot \frac{\mu}{\sigma^3}.$$

In particular, this means that

$$F_n^*(x) = \Phi(x) + O\left(\frac{1}{\sqrt{n}}\right). \tag{3.4.4}$$

(For the notation $O(\cdot)$ see Appendix, Section 4.1.2.) In other words, the remainder, or the error of the approximation in the CLT, has the order $\frac{1}{\sqrt{n}}$. As follows from the corresponding theorems of Probability Theory, under rather mild conditions, the last relation may be replaced by the following more precise representation:

$$F_n^*(x) = \Phi(x) + \frac{\gamma}{6\sqrt{n}}(1-x^2)\varphi(x) + O\left(\frac{1}{n}\right), \tag{3.4.5}$$

where $\varphi(x) = (2\pi)^{-1/2}\exp\{-x^2/2\}$, the standard normal density. The representation (3.4.5) is called an *asymptotic (or Edgeworth's) expansion.* A sufficient condition for (3.4.5) to be true is that the X's have a bounded density and a finite moment EX_i^4. The proof is not simple; it may be found, for example, in [14], [38], [100].

The expansion (3.4.5) contains more information about the accuracy of normal approximation. The second term is given by a precise formula and has the order $\frac{1}{\sqrt{n}}$. The remainder is not written explicitly but it has a higher order than it does in (3.4.4), namely, $\frac{1}{n}$. Precise bounds for the remainder may be found, e.g., in [14] and [100].

Note also that we can continue such an expansion, getting precise terms of the orders $\frac{1}{n^{1/2}}$, $\frac{1}{n}$, $\frac{1}{n^{3/2}}$, and so on, up to a remainder $\frac{1}{n^{k/2}}$ for an arbitrary k. See, e.g., again [14], [38], [100]. Here, we restrict ourselves to the first term of the asymptotic expansion.

It is natural and important that the term $\frac{\gamma}{6\sqrt{n}}(1-x^2)\varphi(x)$ in (3.4.5) involves the skewness coefficient. This term vanishes when $\gamma = 0$, and in any case tends to zero as $n \to \infty$. As was already noted, it is not surprising because $F_n^*(x)$ tends to a symmetric, namely, normal distribution.

Thus, if we decide to neglect terms of order $\frac{1}{n}$, we adopt the approximation

$$F_n^*(x) \approx \Phi(x) + \frac{\gamma}{6\sqrt{n}}(1-x^2)\varphi(x). \tag{3.4.6}$$

As will be shown later, the expansion (3.4.5) is equivalent to the following two (equivalent) representations:

$$F_n^*(x) = \Phi\left(x + \frac{\gamma}{6\sqrt{n}}(1-x^2)\right) + O\left(\frac{1}{n}\right), \tag{3.4.7}$$

or

$$F_n^*\left(y + \frac{\gamma}{6\sqrt{n}}(y^2-1)\right) = \Phi(y) + O\left(\frac{1}{n}\right). \tag{3.4.8}$$

We will see below why it is convenient to write in (3.4.8) y rather than x.

Approximations based on (3.4.7) or (3.4.8) are called *normal power (NP) approximations*. As we will see below, the latter formula is convenient for estimating quantiles of S_n. We justify (3.4.7) and (3.4.8) in the end of this subsection, and now let us return to the situation of Section 3.1.1 and loading coefficients.

We saw in Section 3.1.1 that the insurance company does not suffer a loss with probability β if the loading coefficient θ satisfies $P\left(S_n^* \leq \theta E\{S_n\} \Big/ \sqrt{Var\{S_n\}}\right) = \beta$. In our situation, this is equivalent to

$$F_n^*\left(\frac{\theta m\sqrt{n}}{\sigma}\right) = \beta. \tag{3.4.9}$$

Applying (3.4.8), we see that for (3.4.9) to be true with an accuracy of $O(\frac{1}{n})$, it suffices to set

$$\frac{\theta m\sqrt{n}}{\sigma} = y + \frac{\gamma}{6\sqrt{n}}(y^2-1)$$

with $y = q_{\beta s}$, the β-quantile of the standard normal distribution. We get from this that

$$\theta = \frac{\sigma}{m\sqrt{n}}\left(q_{\beta s} + \frac{\gamma}{6\sqrt{n}}(q_{\beta s}^2 - 1)\right) \tag{3.4.10}$$

(compare with (3.1.6)).

EXAMPLE 1. Consider a homogeneous group of $n = 200$ clients. The probability of the loss event for each client is $q = 0.1$, and the payment if a loss event occurs is the certain amount $z = 10$.

In this case, γ does not depend on z and equals $(1-2q)/\sqrt{q(1-q)} \approx 2.66$ (see Exercise 54). For $\beta = 0.95$, formula (3.4.10) gives $\theta \approx 0.360$, while the "classical" formula (3.1.10) gives 0.348. In Exercise 59, the reader is encouraged to provide an Excel worksheet to solve this problem for various values of n and q. □

In conclusion, we show that (3.4.7) and (3.4.8) are equivalent to (3.4.5).

First, set $\varepsilon = \dfrac{\gamma}{6\sqrt{n}}(1-x^2)$ and note that such an $\varepsilon = O(\frac{1}{\sqrt{n}})$. Writing the Taylor expansion for $\Phi(x+\varepsilon)$, we have

$$\Phi\left(x + \frac{\gamma}{6\sqrt{n}}(1-x^2)\right) = \Phi(x+\varepsilon) = \Phi(x) + \Phi'(x)\varepsilon + O(\varepsilon^2) = \Phi(x) + \varphi(x)\varepsilon + O(\varepsilon^2)$$

$$= \Phi(x) + \varphi(x)\frac{\gamma}{6\sqrt{n}}(1-x^2) + O\left(\varepsilon^2\right) = \Phi(x) + \varphi(x)\frac{\gamma}{6\sqrt{n}}(1-x^2) + O\left(\frac{1}{n}\right).$$

So, we have come to the r.-h.s. of (3.4.5).

To obtain (3.4.8), let us set $q(x) = (1-x^2)\varphi(x)$ for a moment, and $x = y+\varepsilon$, where now $\varepsilon = \frac{\gamma}{6\sqrt{n}}(y^2-1)$. Making use of (3.4.5) and Taylor's expansions in ε for $\Phi(y+\varepsilon)$ and $q(y+\varepsilon)$, we have

$$\begin{aligned}
F_n^*(x) &= F_n^*(y+\varepsilon) = \Phi(y+\varepsilon) + \frac{\gamma}{6\sqrt{n}}q(y+\varepsilon) + o\left(\frac{1}{n}\right) \\
&= \Phi(y) + \varphi(y)\varepsilon + O(\varepsilon^2) + \frac{\gamma}{6\sqrt{n}}[q(y) + O(\varepsilon)] + o\left(\frac{1}{n}\right) \\
&= \Phi(y) + \varphi(y)\varepsilon + \frac{\gamma}{6\sqrt{n}}q(y) + o\left(\frac{1}{n}\right).
\end{aligned}$$

It remains to observe that ε was chosen exactly in a way that $\varphi(y)\varepsilon + \frac{\gamma}{6\sqrt{n}}q(y) = 0$, so we have arrived at (3.4.8). ■

---

4 SOME GENERAL PREMIUM PRINCIPLES

In this section, we touch on some general principles of determining *risk premiums*; that is, premiums taking into account riskiness incurred by an insurance organization. Potential clients of the organization may or may not agree with the premiums suggested. The latter case will lead to a further adjusting process, but we do not explore this issue here. Some possible preferences of individuals were discussed in Chapter 1.

Speaking of *risk* premiums, we mean that profits and expenses are not included in calculations, and we determine a pure premium for risk involved.

Two situations are usually distinguished.

In a short term insurance, a *single premium* is paid at the beginning of the period under consideration to cover the future claim (risk). In this case, the premium is a function of the random claim, and this function is what should be determined.

In the case of life insurance or, for example, pension plans, the policy is based not on a single premium but rather on a *sequence of premium payments* to be carried out at a certain rate. For example, we may talk about monthly premiums.

We call such a type of premium payments premium annuity and explore it later, in Section **10**.1. In this section, we consider the first type of premiums mentioned.

For a particular contract, denote by X the r.v. of the possible payment of a company (the company's risk), and let P be the corresponding premium. Without stating it explicitly each time, we assume that P is a function of X, which we write as $P = \pi(X)$. Certainly, this function assumes numerical values.

We list below and discuss some general *premium principles*; that is, the rules of determining the function $\pi(\cdot)$.

Note that although X is the payment corresponding to a single contract, the rule of premium determination for a particular contract may (and should) depend on the whole risk portfolio with which the company is dealing. In particular, it concerns the first principle below.

1. *The expected value principle:*

$$P = (1+\theta)E\{X\},$$

where $\theta \geq 0$ is a relative security loading coefficient.

We systematically considered this principle in this chapter and continue in Chapters 3. The choice of a value of θ in all the models of these chapters is determined by certain characteristics of the portfolio. For example, θ may depend on the number of the contracts of the portfolio; see, for instance, (3.1.10).

2. *The variance principle*:

$$P = E\{X\} + \lambda Var\{X\}, \tag{4.1}$$

where $\lambda \geq 0$ is a "weight" assigned to the variance. The sign "+" above indicates the risk aversion of the decision maker—the more the variance (riskiness), the larger the premium should be.

This is, actually, a version of the mean-variance principle that we discussed in detail in Section **1**.1.2.5. The only difference is that now we apply it to premiums, which is reflected by the fact that the weight λ is positive.

As was noted in Section **1**.1.2.5, one should use variance as a risk measure with caution: it may lead to conclusions contradicting common sense. Consider a simple example very close to Example **1**.1.2.5-1 .

EXAMPLE 1. Let $X = a > 1$ with probability one, and a random loss

$$Y = \begin{cases} 0 & \text{with probability } 1/a, \\ a & \text{with probability } 1 - 1/a. \end{cases}$$

Thus, $\pi(X) = E\{X\} + \lambda Var\{X\} = a + \lambda \cdot 0 = a$, while $\pi(Y) = E\{Y\} + \lambda Var\{Y\} = (a - 1) + \lambda a^2 \frac{1}{a}(1 - \frac{1}{a}) = (1+\lambda)a - (1+\lambda)$.

Clearly, whatever the positive λ is, we can choose a large enough for $\pi(Y) > \pi(X)$. On the other hand, $X \geq Y$ with probability one, and common sense dictates that the premium for X should not be smaller than that for Y. □

So, the rule (4.1) is not monotone in the sense that the relation $X \geq Y$ does not imply that $\pi(X) \geq \pi(Y)$. In Section **1**.1.2.5, we discussed situations when such a criterion may nevertheless be applied, and when it should not be used. We will continue this discussion below, but first consider

3. *The standard deviation principle*:

$$P = E\{X\} + \lambda\sqrt{Var\{X\}} = E\{X\} + \lambda\sigma_X, \tag{4.2}$$

where σ_X denotes the standard deviation of X and $\lambda > 0$ is a weight.

This rule has the same shortcoming as Rule 2 above: it may assign a larger premium to an obviously smaller risk. For instance, for X and Y from Example 1, $\pi(X) = a$, while $\pi(Y) = a - 1 + \lambda\sqrt{a-1} > a$ for sufficiently large a.

However, if we deal with a portfolio with a large number of separate contracts, the situation may change. In this case, the total loss may be well approximated by a normal r.v. for which the mean-variance and mean-standard-deviation criteria work well (see again Section **1**.1.2.5 and also Section 3.1.2 where we consider this principle in the case of a large portfolio). So, if we apply the rule (4.2) to the whole portfolio, the result may be quite reasonable.

This concerns both criteria 2 and 3, but as we saw in this chapter, when applying normal approximation, it is convenient to work with standard deviations. In particular, when considering the case of a homogeneous portfolio in Sections 3.1.2 and **1**.2.1, we set λ in (4.2) equal to $q_{\beta s}/\sqrt{n}$, where n is the number of contracts in the portfolio, β is a given security level, and $q_{\beta s}$ is the β-quantile of the standard normal distribution.

4. *The mean value principle.* This is a term in use, though in the utility theory framework of Section **1**.3, the term '*certainty equivalent principle*' would be more natural. The rule under consideration is specified by a given increasing function $g(\cdot)$ and is defined by the relation

$$g(P) = E\{g(X)\}, \text{ or } P = c(X) = g^{-1}(E\{g(X)\}). \tag{4.3}$$

If we view $g(x)$ as a utility function, we may view P as the certainty equivalent of X.

We chose the notation $g(x)$ instead of the traditional notation $u(x)$ in order to emphasize that $g(x)$ should not be interpreted as the utility function of the company.

If the company had expected the future *income* X, and if $g(x)$ had been the utility function of the company, then P from (4.3) would have been the amount of money equivalent to X. However, the company does not receive X but rather pays it, which is not the same. The premium P from (4.3) is what the company would agree to receive for paying (covering) the loss X in the future. The function g reflects the attitude of the company to *losses* (while a utility function deals with *income*), and consequently, if the company is a risk averter, $g(x)$ should be convex (!) assigning larger weights to large values of the loss. Later, we will also consider a principle based on the company's utility function.

EXAMPLE 2. Let $g(x) = e^{\beta x}$, where a parameter $\beta > 0$. This is a convex function. In accordance with (4.3),

$$P = \frac{1}{\beta}\ln E\{e^{\beta X}\} = \frac{1}{\beta}\ln M_X(\beta), \tag{4.4}$$

where $M_X(z)$ is the m.g.f. of X. We came to this rule in Section **1**.3.2; see (**1**.3.2.6). The particular rule (4.4) is called *the exponential principle.* By Jensen's inequality (**1**.3.4.2),

$$P = \frac{1}{\beta}\ln E\{e^{\beta X}\} \geq \frac{1}{\beta}\ln e^{E\{\beta X\}} = \frac{1}{\beta}\beta E\{X\} = E\{X\},$$

which reflects the risk aversion of the company. In Exercise 65, we continue this example and, in particular, compare (4.4) with the criterion $\frac{-1}{\beta}\ln E\{e^{-\beta X}\}$ from Section **1**.3.1.3. □

5. *The utility equivalence principle.* Now, we assume that the company's preferences correspond to the expected utility maximization with a utility function $u(x)$. In this case, P is a solution to the equation

$$E\{u(w+P-X)\} = u(w), \tag{4.5}$$

where w is the initial reserve corresponding to the insurance under consideration. We have already applied this principle for premium determination in Section **1**.3.2, considering an insurance company and a separate client, as well.

5a. *The zero utility principle.* Setting $w=0$ in (4.5), we come to P as a solution to the equation

$$E\{u(P-X)\} = u(0). \tag{4.6}$$

In this case, we compare the utility of the profit $P-X$ with the utility of zero profit.

5b. *The exponential principle.* Setting $u(x) = -e^{-\beta x}$ with $\beta > 0$ in (4.6), we come again to (4.4). We already considered this principle in Section **1**.3.2.

6. *The Escher principle* (first suggested by H. Bühlmann in [21]):

$$P = \frac{E\{Xe^{\alpha X}\}}{E\{e^{\alpha X}\}}, \tag{4.7}$$

where $\alpha \geq 0$ is a parameter.

Such premiums arise in many models—in particular, in some reinsurance schemes, or as premiums minimizing losses in some particular cases.

One can view (4.7) as follows. Assume, for simplicity, that X has a density $f(x)$. If we had chosen a premium equal to the mean value of X, we would have computed $\int_0^\infty xf(x)dx$. If we assign a weight $w(x)$ to each value x, we will deal with $\int_0^\infty xw(x)f(x)dx$.

The last integral looks like the expected value with respect to another density, namely, $w(x)f(x)$. However, for this function to be indeed a density, the integral $\int_0^\infty w(x)f(x)dx$ should be equal to one, which is not true. To fix it, we should normalize the function $w(x)f(x)$, that is, divide it by $\int_0^\infty w(x)f(x)dx$. Eventually, it leads to

$$P = \frac{\int_0^\infty xw(x)f(x)dx}{\int_0^\infty w(x)f(x)dx}. \tag{4.8}$$

The definition (4.7) corresponds to the particular case $w(x) = e^{\alpha x}$. As a matter of fact, we can use (4.8) for various weighting functions $w(x)$.

The reader who is familiar with the material of Section **1**.4.2 recognized in the above criterion a particular case of the weighted utility criterion, which we considered in the section mentioned with examples, properties, and axioms.

The Escher parameter α reflects the degree of risk aversion. For $\alpha = 0$, the premium $P = E\{X\}$, and the premium P as a function of α is increasing. (A detailed advice on how to prove it is given in Exercise 64.) Hence, $P \geq E\{X\}$ for $\alpha \geq 0$, and the larger α is, the more P differs from the mean loss.

7. *The Swiss principle* (first suggested in [23] by H. Bühlmann, B. Gagliardi, H. Gerber, and E. Straub). This principle unifies some of the rules above. Let $\lambda \in [0,1]$ and $g(x)$ be an increasing function. Then we define P as a solution to the equation

$$E\{g(X-\lambda P)\} = g((1-\lambda)P). \tag{4.9}$$

If $\lambda = 0$, then $P = g^{-1}(E\{g(X)\})$, which amounts to the mean value Principle 4.

If $\lambda = 1$, we come to the zero utility Principle 5a with $u(x) = g(-x)$ or $u(x) = -g(-x)$. Formally, both versions lead to the same result since if we insert the latter function into (4.9), the minus will cancel. However, the choice of $u(x) = -g(-x)$ is more natural. We saw that in the mean value principle, it is reasonable to choose a convex $g(x)$. Then $-g(-x)$ will be concave. (Consider the second derivatives, although it would be more illustrative to draw a typical graph of a convex $g(x)$, and then realize how the graph $u(x) = -g(-x)$ will look.)

With $\lambda = 1$ and $g(x) = xe^{\alpha x}$, we come to the Escher principle. Indeed, in this case,

$$E\{g(X-P)\} = E\{(X-P)e^{-\alpha P}e^{\alpha X}\} = e^{-\alpha P}\left\{E\{Xe^{\alpha X}\} - PE\{e^{\alpha X}\}\right\}.$$

After inserting it into (4.9), the factor $e^{-\alpha P}$ will cancel since $g(0) = 0$. Solving (4.9) as an equation in P, we will come to (4.7).

Note, however, that this is a formal argument. The function $xe^{\alpha x}$ is convex not for all x's, and the justification connected with a weighting function is more convincing.

The case $0 < \lambda < 1$ may be considered intermediate.

8. *The Orlicz principle* (first considered by J. Haezendonck and M. Goovaerts in [54]). Here we fix $\lambda \in [0,1]$ and an increasing $g(x)$, and define P by

$$E\{g(X/P^{\lambda})\} = g(P^{1-\lambda}).$$

For $\lambda = 0$, we come to the mean value rule; for $\lambda = 1$, we get the equation

$$E\{g(X/P)\} = g(1). \tag{4.10}$$

The ratio X/P may be viewed as the value of X per premium unit—a relative risk, so to say. If $g(x)$ is a utility function, then (4.10) requires the relative risk to be equivalent to one unit of money in the sense of expected utility.

A comprehensive analysis of premium determination and further references may be found in the monograph [49], and a discussion of some further properties and references, e.g., in [49], [57], [58], [78], [110], [126].

5 EXERCISES

Section 1

1. (a) Look up in Section **0**.2.6 the connection between the d.f. and the density of a r.v. X, and the d.f. and the density of the r.v. $Y = bX + c$.

   (b) Using the results of Section **0**.2.6, show that the parameter a in (1.1.3) and (1.1.10) is a scale parameter.

2. For any distribution with a non-zero mean m and standard deviation σ, the ratio σ/m is called a *coefficient of variation* (c.v.).

   (a) Will the c.v. change if we multiply the r.v. by a number?

   (b) Does the c.v. of a Γ-distribution depend on the scale parameter a?

   (c) What can you say about the Γ-distribution with c.v. equal to one?

   (d) Let ξ be log-normal and specified by (1.1.13).

      i. Does the c.v. depend on the parameter a?

      ii. Find the parameter b in the case when the c.v. equals $1/4$.

      iii. Show that, if k is the c.v., then

$$E\{\xi\} = e^a\sqrt{1+k^2}, \quad Var\{\xi\} = e^{2a}(1+k^2)k^2. \tag{5.1}$$

3. In the situation of distribution (1.1.10), find *all* c for which (1.1.8) is true.

4. Show that the tail in (1.1.12) is "light" for $r > 1$ and "heavy" for $r < 1$. (*Advice:* Write $\exp\{-ax^r\} = \exp\{-ax^{r-1}x\}$ and observe that for $r > 1$ and large x's, the quantity ax^{r-1} gets larger than any fixed number; and for $r < 1$, smaller than any fixed positive number.)

5. The distribution with the d.f. $F(x) = (x/\theta)^{\gamma}/[1+(x/\theta)^{\gamma}]$ for $x \geq 0$, where $\theta > 0$ is a parameter, is called *loglogistic*.

   (a) Show that θ is a scale parameter.

   (b) Show that, if X has such a distribution, the r.v. $1/X$ has a distribution of the same type. With which parameter?

   (c) Is this distribution heavy- or light-tailed?

6. (a) Verify (1.1.19), (1.1.20).

   (b) Consider the r.v. $\xi' = x_0\xi_1$, where a number $x_0 > 0$, and the distribution of ξ_1 is defined in (1.1.17). What values does ξ' assume? Write the density, the expected value, and the variance of ξ'.

   (c) State rigorously a fact from Calculus from which it follows that the Pareto distribution is heavy-tailed.

   (d) Show that the fact that the Pareto distribution is heavy tailed also follows from Proposition 1 of Section 1.1.4.

   (e) Which type of function $Q(y)$ from Section 1.1 corresponds to the Pareto distribution?

   (f) Let $\xi = b\xi_1 + d$, where ξ_1 is distributed in accordance with (1.1.17). Find b and d for which ξ has the distribution (1.1.18).

(g) Denote by $\overline{F}_{\alpha}(x)$ the tail in (1.1.17). Show that if $\alpha_1 > \alpha_2$, then F_{α_2} is heavier than F_{α_1} in the sense of (1.1.9).

7. Using (1.2.2), (1.2.6), and (1.2.7), write formulas for $F_Y(x)$, $E\{Y\}$, and $Var\{Y\}$ in the case of proportional insurance (1.3.5).

8. Let a random loss ξ be log-normal. Consider a proportional insurance policy where the insurer pays $k\xi$, $k < 1$. Show that given that the loss event has occurred, the payment is also log-normal. Which parameters, if any, will change in the representation (1.1.13)? Will the parameter a get smaller or larger? (*Hint*: $ke^x = e^{\ln k + x}$.)

9. The probability of a fire in a certain structure during a given period is 0.03. If a fire occurs, the damage is uniformly distributed on the interval $[0, 100]$ (say, the unit of money is \$10,000).

   (a) Assume that the insurance company covers the total damage and denote by Y the (random) amount the company will pay. Write $E\{Y\}$ and $Var\{Y\}$. Graph the distribution function of Y.

   (b) Do the same for the case when the insurance contract provides coverage above a deductible of 5.

   (c) Do the same for the case when the insurance contract provides coverage above a deductible of 5, and the maximum (limit) payment is 90 units.

10. A company insures the cost of injuries for a group of customers. The probability that a particular customer will be injured is 0.05. The cost of 40% of injuries follows the loglogistic distribution (see Exercise 5) with $\theta = 5$ units of money and $\gamma = 3$. For the remaining 60% of the injuries, the cost is loglogistic with $\theta = 3$ and $\gamma = 2$. The company establishes a deductible of 6.

    (a) Find the probability that an injury will result in a claim.

    (b) Find the probability that a particular contract (policy) will result in a claim.

11. You bought auto insurance with a deductible of \$200 and with no restriction on maximal payment. Suppose that the probability of a loss event during a year is 0.1, and the probability that two loss events will occur is negligible. Assume also that the distribution of the loss in the case of a loss event is closely approximated by the exponential distribution with a mean of \$1000.

    (a) What percent of the loss does the insurance cover on the average? (*Hint*: We consider the case where a loss event has occurred and are dealing with the payment divided by the loss.)

    (b) What is the probability that the company will pay nothing during a year?

    (c) What is the expected value and standard deviation of the payment?

    (d) Graph the d.f. of the payment.

12. Losses are modeled by the exponential distribution with a mean of 300. An insurance plan includes an ordinary deductible of 100 and pays 50% of the costs above 100 until the insured is paid 600. Then the plan pays 20% of the remaining costs. Find the expected payment in the case of loss event.

13. In the situation of Exercise 11, let the maximal payment be \$2,000. Graph the d.f. of the payment.

14. (a) Graph $r(x)$ in the case of franchise deductible (1.3.4).
    (b) Consider the problems of Examples 1.3-1,4 in this case.
    (c) Derive (1.3.18) from the general formula (1.3.19).

15. Losses follow the Pareto distribution (1.1.18) with some parameters θ and $\alpha > 1$.
    (a) Find the expected payment with deductible d.
    (b) Do the same for $\alpha = 3, \theta = 3$ in the case of inflation with a rate of $v = 4\%$. (*Advice:* Do not recalculate everything, use the previous answer and (1.3.24)).

16.* Find the loss elimination ratio as a function of deductible d, if the loss variable ξ has (a) an exponential distribution; (b) the uniform distribution on $[0,a]$; (c) the Pareto distribution in the form (1.1.18) for $\theta = 1$, $\alpha = 2$.

17.* Inflation impacts claims at a rate ν. The loss amount ξ has (a) the Pareto distribution in the form (1.1.18), (b) the Weibull distribution. Which parameters of these distributions should be changed and how?

Section 2

18. Find the distribution of the sum of independent r.v.'s

$$X_1 = \begin{cases} 0 & \text{with probability } 1/4 \\ 1 & \text{with probability } 3/4 \end{cases}, \quad X_2 = \begin{cases} 1 & \text{with probability } 2/3 \\ 2 & \text{with probability } 1/3 \end{cases}.$$

19. (a) Give a common sense explanation why values of S_2 in Example 2.1.1-2 are not equally likely.
    (b) Carry out calculations in (2.1.7).

20. Find the distribution of the sum of independent r.v.'s X_1 and X_2 if
    (a) X_1 and X_2 are exponentials with parameters $a_1 = 1$ and $a_2 = 2$, respectively;
    (b) X_1 and X_2 are uniform on $[0,1]$ and $[0,2]$, respectively.

21. Prove Proposition 4 with a direct use of (2.1.3). (*Advice:* First, since a is a common scale parameter, without loss of generality, we may set $a = 1$. Second, after integration, you will come to a constant $B(\nu_1,\nu_2) = \int_0^1 (1-t)^{\nu_1 - 1} t^{\nu_2 - 1} dt$. This is the so called B(beta)-function. It is known that it is equal to $\Gamma(\nu_1)\Gamma(\nu_2)/\Gamma(\nu_1,\nu_2)$, but we do not need to know this. Since we know that the resulting function is a density, the value of the constant may be obtained from the fact the integral of a density equals one. On the way, you will automatically obtain the value of the B(beta)-function mentioned. See also, e.g., [120, p.153].)

22. Consider Example 2.1.2-1.
    (a) What is the mean of the total claim, that is, $E\{S_4\}$?
    (b) What is the probability that S_4 will be exactly equal to $E\{S_4\}$?
    (c) Using software (for example, Excel), compute and compare $P(S_4 > E\{S_4\})$ and $P(S_4 < E\{S_4\})$.

23. Let $S_n = X_1 + ... + X_n$, where the X's are independent, and X_i has the Γ-distribution with parameters $(1, (\frac{1}{2})^i)$. Show that for large n, the distribution of S_n may be closely approximated by the standard exponential distribution.

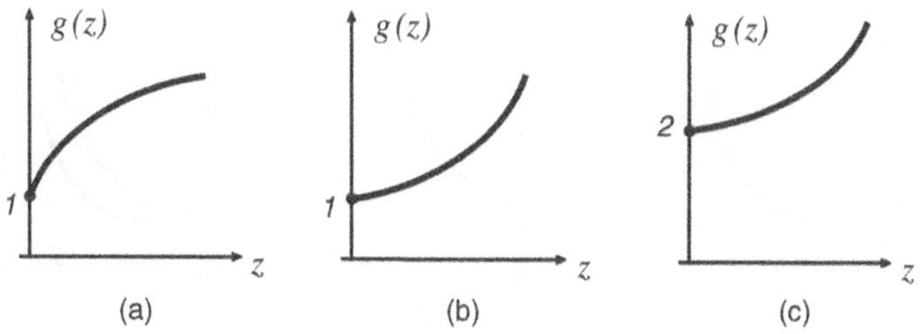

FIGURE 10.

24. (a) Can we switch F_1 and F_2 in (2.1.1), and f_1 and f_2 in (2.1.3)?
    (b) How should we change formulas (2.1.1) and (2.1.3) to obtain the d.f. and the density, respectively, of the r.v. $X_1 - X_2$?

25. Assume that the aggregate payments for two portfolios are independent and their distributions, as those of sums of many separate payments, are closely approximated by normal distributions. Let (m, σ^2)-parameters of these distributions be $(5,8)$ and $(3,7)$, respectively.
    (a) Without any calculations, find the probability that the total payment will exceed 8.
    (b) Estimate the probability that the total payment will exceed 10.

26. In a county, the daily data of traffic accidents with serious injuries are values of independent Poisson r.v.'s. The mean values are 5 and 6 for Friday and Monday, respectively, and 3 for other days. Write a formula and estimate, using software, the probability that a weekly number of such accidents will exceed 30.

27. The densities of independent r.v.'s X_1 and X_2 are $f_1(x) = C_1 x^5 e^{-4x}$ and $f_2(x) = C_2 x^8 e^{-4x}$, respectively, where C_1 and C_2 are constants.
    (a) Do we need to calculate these constants in order to find $P(X_1 + X_2 > x)$?
    (b) Estimate $P(X_1 + X_2 > 3)$.
    (c) Write C_1 and C_2.

28. Let Z_n be a Poisson r.v. with parameter λ equal to an integer n. By using Proposition 3, show that Z_n may be represented as $Y_1 + ... + Y_n$, where the Y's are independent Poisson r.v.'s with parameter $\lambda = 1$.

29. Let Z_n be a Γ-r.v. with a scale parameter of a and parameter ν equal to an integer n. By using Proposition 4, show that Z_n may be represented as $Y_1 + ... + Y_n$, where Y's are independent exponential r.v.'s with parameter a.

30. Let ξ be exponentially distributed, and $E\{\xi) = m$. Point out all z's for which the m.g.f. $M_\xi(z)$ exists.

31. Find the m.g.f. of the geometric distribution (**0**.3.1.7) by using (**0**.4.3.1) and (**0**.4.1.5).

32. Find the m.g.f. of the negative binomial distribution (**0**.3.1.13) by using (**0**.4.3.2) and (**0**.4.1.5).

33. Which function in Fig.10 looks as a m.g.f.?

34. What is the difference between the two distributions whose m.g.f.'s are graphed in Fig.11a and 11b?

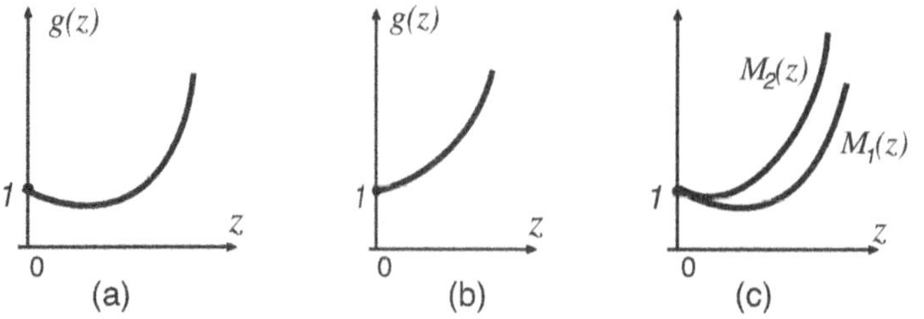

FIGURE 11.

35. Let $M(z)$ be the m.g.f. of a r.v. X, $E\{X\} = m$, $Var\{X\} = \sigma^2$. Write $M'(0)$ and $M''(0)$.

36. Compare the means and *variances* of the r.v.'s whose m.g.f.'s are graphed in Fig.11c. (The graphs of $M_1(z)$ and $M_2(z)$ are tangent at $z = 0$.)

37. (a) As we know, the m.g.f. of the standard normal distribution is $M(z) = e^{z^2/2}$. Can the function $g(z) = e^{z^4/2}$ and in general, $g(z) = e^{cz^4}$, where $c > 0$, be a m.g.f.? (*Hint*: Compute $g'(0)$ and $g''(0)$ and then jump to a conclusion.)

    (b) Can the function $1 + z^4$ be the m.g.f. of a r.v.?

    (c) * In general, show that if $g(z)$ has two continuous derivatives and $g(z) = 1 + \varepsilon(z)z^2$, where $\varepsilon(z) \neq 0$ for $z \neq 0$, and $\varepsilon(z) \to 0$, as $z \to 0$, then $g(z)$ cannot be a m.g.f.

38. Show that if $P(X \geq 3) = 0.2$, then the corresponding m.g.f. $M(z) \geq 0.2e^{3z}$. What can we say about the distribution of X if $M(z) \sim 0.2e^{3z}$ as $z \to \infty$?

39. Let X_1, X_2 be independent standard exponential r.v.'s, and let $Y = X_1 - X_2$. It is known that the r.v. Y has the density

$$f(x) = \frac{1}{2}e^{-|x|}. \qquad (5.2)$$

Such a distribution is called *two-sided exponential.*

    (a) Graph the density $f(x)$.

    (b) Prove (5.2) by the convolution method.

    (c) Find the m.g.f. of Y and prove (5.2) using the method of Example 2.2-1.

40. Let $S = X_1 + X_2$, where X_1 and X_2 are exponential r.v.'s with $E\{X_1\} = 1, E\{X_2\} = 2$. Not computing a convolution but rather making use of the method of Example 2.2-1, prove that the density

$$f_S(x) = -e^{-x} + 2\frac{1}{2}e^{-x/2} = e^{-x/2} - e^{-x} \text{ for } x \geq 0, \text{ and } = 0 \text{ for } x < 0.$$

Show that the above function is, indeed, a density.

41.* Prove that, if $\alpha, \beta \in [0, 1]$ and $\alpha + \beta = 1$, $F(x)$ and $G(x)$ are two d.f.'s, and the mixture $H(x) = \alpha F(x) + \beta G(x)$, then the m.g.f. of the convolution H^{*2} is

$$M_{H^{*2}}(z) = \alpha^2 M_F^2(z) + 2\alpha\beta M_F(z)M_G(z) + \beta^2 M_G^2(z),$$

where functions M_F, M_G are the corresponding m.g.f.'s. Show that this implies that $H^{*2}(x) = \alpha^2 F^{*2}(x) + 2\alpha\beta F(x) * G(x) + \beta^2 G^{*2}(x)$. Generalize it to the case H^{*n}, as it is presented in (2.1.1).

42.* In the fashion of Example 2.1.4-1, find the density

$$\left(\frac{1}{3}f_{a2}+\frac{2}{3}f_{a3}\right)^{*2}.$$

43.* Let Z_λ be a Poisson r.v. with parameter λ, and $Z^*_\lambda=(Z_\lambda-\lambda)/\sqrt{\lambda}$, the normalized r.v. Prove that Z^*_λ is asymptotically normal for large λ; that is, $P(Z^*_\lambda\le x)\to\Phi(x)$ as $\lambda\to\infty$.

(*Advice:* There are two ways to show this. First, using (**0**.4.3.3) and (**0**.4.1.5), we can write the m.g.f. of Z^*_λ and prove that this m.g.f. converges to the m.g.f. of the standard normal r.v. However, a more explicit and shorter way is to use the fact stated in Exercise 28. If λ is an integer, the fact we are proving follows immediately from the Central Limit Theorem (see Section **0**.6.2). If λ is not an integer, we can consider $[\lambda]$, the integer part of λ, and —using the same Proposition 3—represent the r.v. Z_λ as $Z_{[\lambda]}+R_{\lambda-[\lambda]}$, where $R_{\lambda-[\lambda]}$ is a Poisson r.v. with parameter $\lambda-[\lambda]$ and independent of $Z_{[\lambda]}$. Then

$$Z^*_\lambda=\sqrt{\frac{[\lambda]}{\lambda}}\cdot\frac{Z_{[\lambda]}-[\lambda]}{\sqrt{[\lambda]}}+\frac{R_{\lambda-[\lambda]}-(\lambda-[\lambda])}{\sqrt{\lambda}},$$

and it remains to show that (a) $\sqrt{[\lambda]/\lambda}\to 1$; (b) $(Z_{[\lambda]}-[\lambda])/\sqrt{[\lambda]}$ is asymptotically normal by Exercise 28; and (c) $\left[(R_{\lambda-[\lambda]}-(\lambda-[\lambda]))\Big/\sqrt{\lambda}\right]\to 0$ since $\lambda-[\lambda]<1$.)

44.* Let Z_ν be a Γ-r.v. with parameters a and ν, $Z^*_\nu=(Z_\nu-E\{Z_\nu\})/\sqrt{Var\{Z_\nu\}}$, the normalized r.v. Prove that Z^*_ν is asymptotically normal for large ν, that is, $P(Z^*_\nu\le x)\to\Phi(x)$ as $\nu\to\infty$. (*Advice:* Use the result of Exercise 29 and the scheme of Exercise 43.)

Section 3

45. Verify the results of Table 3.1.1-1. (In Excel, the corresponding command is 'BINOMDIST'.)

46. In the case of proportional insurance, the payment $Y=kX$. Are the coefficients of variation of Y and X the same?

47. Consider two portfolios of 2000 and 3000 cars, respectively, insured for a single period of one year with \$1000 deductible. The damage per car (per year) is distributed as follows:

The first portfolio		**The second portfolio**	
Damage (in \$1000)	probability	Damage (in \$1000)	probability
0	0.78	0	0.8
< 1 (× \$1000)	0.12	< 1 (× \$1000)	0.1
6	0.05	6	0.08
11	0.05	11	0.02

(a) Assuming all claims to be independent, compute the expectation and the variance of the total amount of claims from the two portfolios.
Use the heuristic approach to normal approximation in order to estimate the security loading coefficient θ such that the probability that the insurance company will not suffer a loss by the end of the year is 0.99.

(b) Assume that the number of cars (clients) in both portfolios became twice as large. Will θ in this case be larger or smaller? Determine how θ will change.

(c) * Estimate θ for the case (a), using rigorous calculations.

48. An insurance company has a portfolio of 2000 cars insured for a single period of one year. The probability that a particular car will be involved in an accident is 0.05. The losses for different cars are independent. The distribution of the damage, *if an accident occurs*, may be approximated by the exponential distribution with a mean of \$1000.

    (a) Graph the distribution function of a payment of the company *for a separate policy.*

    (b) Write the expected value and the variance of the payment for a separate policy, and for the total payment of the company.

    (c) Graph the distribution function of a payment of the company *for a separate car* in the case of a deductible of *\$200.*

    (d) Considering the insurance *without deductible,* assuming all claims to be independent, and using normal approximation, compute the security loading coefficient for which the probability that the company would not lose money by the end of the year is 0.95. Use the heuristic approach.

    (e) * Do the same using the rigorous estimation approach.

    (f) Not providing any calculations, write the answer for Exercise 48d for the case when the number of cars is equal to 4000.

49.* Verify that if we replace each X_i in Section 3 by r.v.'s $X_i' = a + bX_i$, where a is an arbitrary number, and $b \neq 0$, then the Lyapunov fraction will not change.

50.* (a) i. Recalculate results of Example 3.2-1 for $q = 0.1$, $n = 1000$, and $\beta = 0.7$. (*Advice:* Do not recalculate everything; try to understand what should change, and what should not.)

    ii. Compute at least some entities in Table 3.2-1.

    (b) Recalculate results of Example 3.2-2 for $q_1 = q_2 = 0.1$, $z_1 = 10$, $z_2 = 20$, $n_1 = 2000$, $n_2 = 1000$, and $\beta = 0.8$. Compare θ_{n0} and θ_n and find premiums for both portfolios.

51. (a) Find a heuristic estimate for the coefficient θ in the case when $\beta = 0.9$, $n = 4000$, $q = 0.1$, and the payment in the case of a loss event is an exponential r.v. with unit mean.

    (b) * Provide a rigorous estimate.

52. Assume that for the group from Example 3.2-1, due to a law, θ cannot be larger than 0.08.

    (a) Estimate n required for $\beta = 0.9$, making use of the heuristic approximation. (*Hint*: You can use the results in Table 1 in this example.)

    (b) * Using software (for example, providing an Excel worksheet), give a rigorous estimate for n. (The corresponding example is Example 3.3-2.)

53.* (a) Let X_1 be an exponential r.v. with $E\{X_1\} = m$, let X_2 be uniform on $[0, 2m]$, and let

$$X = \begin{cases} X_1 & \text{with probability } \frac{1}{2}, \\ -X_2 & \text{with probability } \frac{1}{2}. \end{cases}$$

    Should the skewness coefficient of X depend on m? Can you guess without calculations whether the skewness coefficient positive or negative? Find it.

    (b) Find the skewness coefficient for an arbitrary uniform distribution.

54.* Consider the risk of a certain loss of z occurring with a probability of q. Explain without calculations why the skewness coefficient does not depend on z. Show that $\gamma = (1 - 2q)/\sqrt{q(1-q)}$.

55.* Let $S_n = X_1 + ... + X_n$, where X's are independent and have the Γ-distribution with parameters a, ν. Then, by Proposition 4, S_n has the Γ-distribution with parameters $a, n\nu$. Show that the Γ-approximation of Section 3.4 leads to exactly this distribution.

56.* Provide the Γ-approximation of Section 3.4 for the distribution of $S_{10} = X_1 + ... + X_{10}$ with $m_i = 10$, $\sigma_i = 5$, $\varkappa_{3i} = 75$.

57.* Let $S_n = X_1 + ... + X_n$, where the X's are independent, identically distributed, $E\{X_i\} = m$, and $\gamma = 1$. The normal approximation for $P(S_n \leq mn)$ gives $1/2$ (show why). As a matter of fact, if the X's are not symmetric, it is not an absolutely precise approximation. Calculate what approximation the asymptotic expansion in (3.4.5)-(3.4.6) would give in this case.

58.* In Example 3.4-1, we have shown that the skewness coefficient for the Γ-distribution with main parameter ν is equal to $2/\sqrt{\nu}$. Using the result of Exercise 44 and restricting yourself, for simplicity, to integer ν's, connect this fact with the expansion (3.4.5). (*Advice*: Since we consider normalized r.v.'s, without loss of generality, we may set the scale parameter a of the Γ-distribution equal to one.)

59.* Provide an Excel worksheet for a general solution to the problem from Example 3.4.3-1. Analyze results for various n and q.

*Section 4**

60. Let X be standard exponential, and let Y be uniform on $[0,2]$. So, $E\{X\} = E\{Y\}$. Find $\pi(X)$ and $\pi(Y)$ proceeding from Principles 1-3, Principle 4 with $g(x) = x^2$, Principle 5b, and Principle 6 with $\alpha = 1/2$. Compare and interpret the results.

61. Show that for $g(x) = x^\alpha$ and $\lambda = 1$, the Orlicz principle may be reduced to the mean value (certainty equivalent) principle.

62. We say that a premium principle $\pi(.)$ satisfies the *positive homogeneity property*, if $\pi(kX) = k\pi(X)$ for any $k \geq 0$ and for all X's for which the principle is defined. We say that the *translation invariance property* holds if $\pi(X + c) = \pi(X) + c$ for any c and all X's mentioned. (a) Verify whether these properties are true for Principles 1-3 of Section 4. (b) Do the same for Principle 4. For which particular functions $g(x)$ are the properties under discussion true? (c) Explore Principle 5. Suggest particular $u(x)$ for which both properties will be true. (d) Explore Principle 6. (*Advice*: See Section **1**.1.3 for more details and comments on the properties we discuss in this and in the next Exercise 63. One should, however, keep in mind that in Section **1**.1.3 we deal with gains, while when considering premiums, we deal with possible losses. Regarding this exercise, to show that a property does not hold, it suffices to consider a particular example. Also keep in mind that some premium principles are particular cases of others.)

63. We say that a premium principle $\pi(\cdot)$ is *additive,* if for any two independent r.v.'s X_1 and X_2,

$$\pi(X_1 + X_2) = \pi(X_1) + \pi(X_2). \tag{5.3}$$

We say that $\pi(\cdot)$ is *sub-additive* if

$$\pi(X_1 + X_2) \leq \pi(X_1) + \pi(X_2). \tag{5.4}$$

Note that unlike in Section **1**.1.3, we consider here independent r.v.'s. The sub-additivity property, especially when we have the strict inequality in (5.4), is a desirable property: the total premium does not become higher when we combine independent risks in one portfolio. See also comments in Exercise 62.

(a) Verify whether (5.4) is true for Principles 1-3 of Section 4. Determine when the standard deviation principle is strictly sub-additive.

(b) Show that the exponential and Escher principles are additive.

(c) Does translation invariance follow from additivity?

(d) Let $\pi(c) = c$ for any c (which is a natural property: for a certain loss the premium should be equal to this loss). Show that in this case, translation invariance follows from sub-additivity. (*Hint*: $\pi(X+c) \leq \pi(X)+c$. On the other hand, $\pi(X) = \pi(X+c-c) \leq \pi(X+c)+\pi(-c)$.)

(e) Show that any $\pi(X)$ equal to a linear combination of cumulants of X is additive. Show that the variance, exponential, and Escher principles are presented by linear combinations of cumulants (and this is why they are additive).

64. Prove that premium (4.7) is non-decreasing as a function of α. (*Advice*: Differentiating (4.7) in α, we will have a fraction with the numerator

$$E\{X^2 e^{\alpha X}\} - (E\{Xe^{\alpha X}\})^2.$$

To prove that the last expression is not non-negative, write $E\{Xe^{\alpha X}\}=E\{Xe^{\alpha X/2}e^{\alpha X/2}\}$ and apply the Cauchy-Schwarz inequality (**0**.2.4.4)).

65. (a) Look up Exercise **1**.33 where we proved (based on the given advice) that P in (4.4) is increasing in β. So, this parameter may serve as a risk aversion characteristic, which was discussed repeatedly in Sections **1**.3.2 and **1**.3.4.

(b) Consider $\pi(X) = \frac{1}{\beta}\ln E\{e^{\beta X}\}$ and the function $c(X) = -\frac{1}{\beta}\ln E\{e^{-\beta X}\}$ from Section **1**.3.1.3. Prove that $c(X) \geq E\{X\}$ and $c(X)$ is decreasing in β. Interpret the last fact. Show that while $\pi(X)$ "pays more attention" to large values of X, the criterion $c(X)$ ignores very large values (the saturation effect). Interpret this fact in the case when X is a loss, and when it is an income. (*Advice*: First, look up general properties of certainty equivalents in Section **1**.3.4.2. Second, consider $X = 0$ and a with equal probabilities, for example, and analyze the asymptotic behavior of $\pi(X)$ and $c(X)$ for large a.)

Chapter 3

A Collective Risk Model for a Short Period

1,2,3

From a purely mathematical point of view, the model of this chapter differs from what we considered in Chapter 2 by the fact that now we will explore sums of r.v.'s where not only separate terms (addends) are random but the number of the terms is random also. In other words, our object of study is the r.v.

$$S = S_N = \sum_{j=1}^{N} X_j, \tag{0.1}$$

where N and $X_1, X_2, \ldots$ are r.v.'s., as well. If N assumes zero value, we set $S = 0$.

Such a model admits at least two interpretations.

The first concerns a *future risk* portfolio in the situation when contracts have not yet been issued, and we do not know how many clients the company may have. In this case, N may be viewed as the number of future *clients*, and the X's as future payments to the clients. In such a situation, each X may assume zero value with positive probability.

The second interpretation deals with a *settled* portfolio. However in this case, we consider this portfolio as a whole, being interested not in separate clients but rather in the total claim the company will have to pay out. In this case, N is the number of future *claims,* and the X's represent the payments *corresponding to these claims*. In such a scheme, it is natural though not necessary, to assume the X's to be positive. The term *collective* reflects the fact that we view the portfolio as one insured unit, as a whole.

We mostly follow the latter interpretation which appears much more frequently in applications.

In this chapter, when considering different characteristics of r.v.'s such as expectations, moments, m.g.f.'s, etc., we assume—not necessarily stating it each time explicitly—that in the situations under consideration, all these characteristics are well defined and finite.

Before exploring the models of this chapter in detail, it is convenient to establish some basic facts which we will use repeatedly throughout this book.

1 THREE BASIC PROPOSITIONS

In this section, (0.1) is a purely mathematical construction which will be interpreted in the following sections in different ways. Assume that the r.v.'s $X_1, X_2, \ldots$ and N are mutually independent, r.v.'s $X_1, X_2, \ldots$ are identically distributed, and N assumes values $0, 1, 2, \ldots$. Set $m = E\{X_i\}$, $\sigma^2 = Var\{X_i\}$.

Proposition 1 *The mean value of S is given by*

$$E\{S\} = mE\{N\}. \tag{1.1}$$

In particular, if N is a Poisson r.v. with parameter λ, *then*

$$E\{S\} = m\lambda. \tag{1.2}$$

Proof. By the formula for total expectation (**0**.7.2.1), $E\{S\} = E\{E\{S\,|\,N\}\}$. In the conditional expectation $E\{S\,|\,N\}$, the value of the r.v. N is given, and we deal with a sum of a fixed number of addends. Hence, $E\{S\,|\,N\} = mN$ and $E\{S\} = E\{mN\} = mE\{N\}$. ■

► In Section **5**.2.4, we will show that the condition of independence of N and the X's is not necessary, and (1.1) remains true if for each n, the event $\{N = n\}$ is specified by values of $X_1, ..., X_n$. The last condition holds, for example, if N is the first number n for which the (growing) sum $S_n = X_1 + ... + X_n$ exceeds a fixed level. See Section **5**.2.4.2 for detail. ◄

Proposition 2 *The variance of S is given by*

$$Var\{S\} = \sigma^2 E\{N\} + m^2 Var\{N\}. \tag{1.3}$$

In particular, if N is a Poisson r.v. with parameter λ, *then*

$$Var\{S\} = \lambda E\{X^2\}, \tag{1.4}$$

where X is a r.v. distributed as the X_i*'s.*

Proof. By (**0**.7.3.2), $Var\{S\} = Var\{E\{S\,|\,N\}\} + E\{Var\{S\,|\,N\}\}$. Reasoning similar to the proof of (1.1), we can write that, given N, the conditional variance $Var\{S\,|\,N\} = \sigma^2 N$. We have also shown above that $E\{S\,|\,N\} = mN$. So, $Var\{S\} = E\{\sigma^2 N\} + Var\{mN\} = \sigma^2 E\{N\} + m^2 Var\{N\}$. If N is Poisson, $E\{N\} = Var\{N\} = \lambda$, and $Var\{S\} = \lambda(\sigma^2 + m^2) = \lambda E\{X^2\}$. ■

Proposition 3 *For all z for which the m.g.f.'s below are well defined, the m.g.f. of S is*

$$M_S(z) = M_N(\ln M_X(z)), \tag{1.5}$$

where $M_N(\cdot)$ *is the m.g.f. of N, and* $M_X(z)$ *is the (common) m.g.f. of the r.v.'s* X_i.
In particular, if N is a Poisson r.v. with parameter λ, *then*

$$M_S(z) = \exp\{\lambda(M_X(z) - 1)\}. \tag{1.6}$$

Proof. We have

$$M_S(z) = E\{e^{zS}\} = E\{E\{e^{zS}\,|\,N\}\}.$$

In $E\{e^{zS}\,|\,N\}$, the value of N is given, so the conditional expectation $E\{e^{zS}\,|\,N\}$ is the m.g.f. of a sum of a *fixed* number of terms. Hence, by the main property of m.g.f.'s,

$$E\{e^{zS}\,|\,N\} = (M_X(z))^N = e^{(\ln M_X(z))N},$$

and

$$E\{e^{zS}\} = E\{e^{(\ln M_X(z))N}\}. \tag{1.7}$$

The r.-h.s. of (1.7) is the m.g.f. of N at the point $(\ln M_X(z))$, which implies (1.5).

If N is a Poisson r.v. with parameter λ, then the m.g.f. $M_N(z) = \exp\{\lambda(e^z - 1)\}$. Replacing z by $\ln M_X(z)$, we obtain (1.6). ■

Note also that in the case when the m.g.f.'s above exist, Propositions 1-2 follow from Proposition 3 (see Exercise 1) but the direct proofs above are simpler than the derivation from (1.5).

Various examples of applications of Propositions 1-3 will be given in further sections. Now, we begin to consider the collective risk model in detail starting with possible distributions of N.

---

2 COUNTING OR FREQUENCY DISTRIBUTIONS

The distribution of the r.v. N is sometimes called a *counting* or *frequency* distribution. Different types of this distribution are considered in the theory and applications. We begin with the Poisson distribution; in a certain sense, the simplest and most important distribution.

2.1 The Poisson distribution and theorem

2.1.1 A heuristic approximation

The *Poisson distribution* is that of an integer valued r.v. Z such that

$$P(Z = k) = e^{-\lambda}\lambda^k/k! \text{ for } k = 0, 1, \ldots, \tag{2.1.1}$$

where λ is a positive parameter. As is proved in almost any course in Probability,

$$E\{Z\} = \lambda, \quad Var\{Z\} = \lambda \tag{2.1.2}$$

(see also Section **0**.3.1.5 and Exercise 3).

There are at least two explanations why this distribution plays a key role in our model. First, the Poisson distribution may appear when we view the flow of claims arriving at the company as a random process in continuous time. In Chapter 4, we will consider in detail how some natural conditions on the evolution of this process lead to the Poisson distribution for the number of claims arriving during any given period of time.

Another explanation is connected with Poisson's theorem. Consider a sequence of n independent trials with the probability of success at each trial equal to p. Let N be the total number of successes. As we know, N has the *binomial distribution;* that is,

$$P(N = k) = \binom{n}{k} p^k (1-p)^{n-k}. \tag{2.1.3}$$

The Poisson theorem tells that if n is large and p is small in a way that $E\{N\} = np$ is "neither small nor large", then the distribution of N is well approximated by a Poisson distribution. To state it rigorously, we assume that the probability p depends on n, and

$$p = p_n = \frac{\lambda}{n} + o\left(\frac{1}{n}\right), \qquad (2.1.4)$$

where λ is a positive number. We again use the Calculus symbol $o(x)$ which denotes a function converging to zero, as $x \to 0$, faster than x; that is, $\frac{o(x)}{x} \to 0$ (see Appendix, Sec. 4.1 for details). In other words, the second term $o(\frac{1}{n})$ in (2.1.4) is negligible for large n with respect to the first term $\frac{\lambda}{n}$. In the first reading, the reader can even ignore the term $o\left(\frac{1}{n}\right)$.

Thus, the r.v. N in this framework depends on n, so we write $N = N_n$.

Theorem 4 *(Poisson). For any k,*

$$P(N_n = k) \to e^{-\lambda}\frac{\lambda^k}{k!} \quad as \quad n \to \infty. \qquad (2.1.5)$$

The theorem is proved practically in any textbook on probability; see, e.g., [102], [116], [122].

Consider, for example, a portfolio of n policies "functioning" independently, and suppose that for each policy, the probability of the occurrence (during a fixed period) of a claim equals the same number p. Then, we may identify policies with independent trials and in the case of "large" n and "small" p, approximate the distribution of the total number of claims by the Poisson distribution with the parameter $\lambda = pn$.[1]

It is important to note that the accuracy of the Poisson approximation may be high.

EXAMPLE 1. Assume $n = 30$, and, initially, $p = 0.5$. Then $\lambda = 15$. The Excel worksheet in Fig.1a shows the binomial [the r.-h.s. of (2.1.3)] and Poisson [the r.-h.s. of (2.1.5)] probabilities in columns B and C, respectively. The corresponding graphs are in the chart. The values of the distribution functions are given in columns E and F, and the difference between them—in column G. We see that the distributions are not close (see the chart), and the maximal difference in absolute value, between the d.f.'s is 0.0868 (for $k = 12$), which is large. This is not surprising since $p = 0.5$ is "not small".

Now, let n still be 30, but let $p = 0.1$. Then $\lambda = 3$. The result given in Fig.1b shows that now the distributions are fairly close, the chart looks just perfect, and the maximal difference between the d.f.'s is 0.0107 (for $k = 5$), which is not bad at all. It is a bit surprising that such a good approximation can appear for relatively small n.

Consider a larger n, say, $n = 60$, keeping $\lambda = 3$. Then $p = 0.05$. The result is given in Fig 1c. We see that the maximal difference between the d.f.'s is 0.0052 ($k = 5$).

In Section 2.1.2, we consider rigorous estimates of the accuracy of Poisson approximation. □

[1] We are aware that in Chapter 2, following a tradition, we denoted the probability of claim occurrence (loss event) by q. Now, when we are dealing with the binomial distribution, a stronger tradition requires to denote the probability of success by p. In this chapter, we will not work with models of Chapter 2, so such a replacement should not cause a confusion.

k	Binomial prob.'s	Poisson prob.'s	Binom.d.f.	Poisson d.f.	The difference
0	9.31323E-10	3.05902E-07	9.31323E-10	3.05902E-07	-3.04971E-07
1	2.79397E-08	4.58853E-06	2.8871E-08	4.89444E-06	-4.86557E-06
2	4.05125E-07	3.4414E-05	4.33996E-07	3.93084E-05	-3.88745E-05
3	3.78117E-06	0.00017207	4.21517E-06	0.000211379	-0.000207163
4	2.55229E-05	0.000645263	2.97381E-05	0.000856641	-0.000826903
5	0.000132719	0.001935788	0.000162457	0.002792429	-0.002629972
6	0.000552996	0.00483947	0.000715453	0.0076319	-0.006916446
7	0.001895986	0.010370294	0.00261144	0.018002193	-0.015390753
8	0.005450961	0.0194443	0.008062401	0.037446493	-0.029384093
9	0.013324572	0.032407167	0.021386973	0.069853661	-0.048466688
10	0.027981601	0.048610751	0.049368573	0.118464412	-0.069095838
11	0.050875638	0.066287387	0.100244211	0.184751799	-0.084507588
12	0.080553093	0.082859234	0.180797304	0.267611033	-0.086813729
13	0.111535052	0.095606809	0.292332356	0.363217842	-0.070885486
14	0.13543542	0.102435867	0.427767776	0.465653709	-0.037885933
15	0.144464448	0.102435867	0.572232224	0.568089576	0.004142648
16	0.13543542	0.096033625	0.707667644	0.664123201	0.043544444
17	0.111535052	0.084735551	0.819202696	0.748858752	0.070343944
18	0.080553093	0.07061296	0.899755789	0.819471712	0.080284077
19	0.050875638	0.055747073	0.950631427	0.875218785	0.075412642
20	0.027981601	0.041810305	0.978613027	0.91702909	0.061583937
21	0.013324572	0.029864504	0.991937599	0.946893594	0.045044006
22	0.005450961	0.020362162	0.99738856	0.967255755	0.030132805
23	0.001895986	0.013279671	0.999284547	0.980535426	0.018749121
24	0.000552996	0.008299794	0.999837543	0.98883522	0.011002323
25	0.000132719	0.004979876	0.999970262	0.993815096	0.006155166
26	2.55229E-05	0.002873006	0.999995785	0.996688102	0.003307683
27	3.78117E-06	0.001596114	0.999999566	0.998284216	0.00171535
28	4.05125E-07	0.000855061	0.999999971	0.999139277	0.000860694
29	2.79397E-08	0.000442273	0.999999999	0.99958155	0.000418449
30	9.31323E-10	0.000221137	1	0.999802687	0.000197313

30 — n
0.5 — p
15 — $\lambda = np$

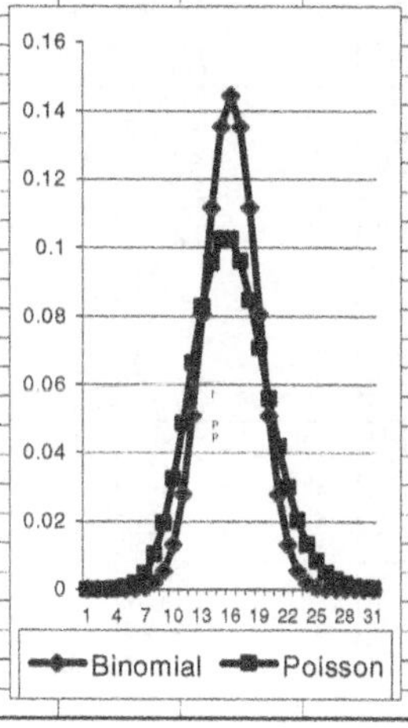

(a)

k	Binomial prob.'s	Poisson prob.'s	Binom.d.f.	Poisson d.f.	The difference
0	0.042391158	0.049787068	0.042391158	0.049787068	-0.00739591
1	0.141303861	0.149361205	0.183695019	0.199148273	-0.015453254
2	0.22765622	0.224041808	0.41135124	0.423190081	-0.011838842
3	0.236087932	0.224041808	0.647439172	0.647231889	0.000207283
4	0.177065949	0.168031356	0.824505121	0.815263245	0.009241876
5	0.102304771	0.100818813	0.926809892	0.916082058	0.010727834
6	0.04736332	0.050409407	0.974173211	0.966491465	0.007681747
7	0.018043169	0.021604031	0.992216381	0.988095496	0.004120885
8	0.00576379	0.008101512	0.997980171	0.996197008	0.001783163
9	0.001565474	0.002700504	0.999545645	0.998897512	0.000648133
10	0.000365277	0.000810151	0.999910922	0.999707663	0.000203259
11	7.37934E-05	0.00022095	0.999984716	0.999928613	5.61021E-05
12	1.29822E-05	5.52376E-05	0.999997698	0.999983851	1.38467E-05
13	1.99726E-06	1.27471E-05	0.999999695	0.999996598	3.09685E-06
14	2.69471E-07	2.73153E-06	0.999999964	0.99999933	6.34791E-07
15	3.19373E-08	5.46306E-07	0.999999996	0.999999876	1.20423E-07
16	3.3268E-09	1.02432E-07	1	0.999999978	2.13172E-08
17	3.04413E-10	1.80763E-08	1	0.999999996	3.5453E-09
18	2.44282E-11	3.01272E-09	1	0.999999999	5.57013E-10
19	1.71426E-12	4.75692E-10	1	1	8.3035E-11
20	1.0476E-13	7.13538E-11	1	1	1.1786E-11
21	5.54288E-15	1.01934E-11	1	1	1.59806E-12
22	2.51949E-16	1.39001E-12	1	1	2.08167E-13
23	9.73717E-18	1.81306E-13	1	1	2.68674E-14
24	3.15556E-19	2.26632E-14	1	1	4.21885E-15
25	8.41484E-21	2.71958E-15	1	1	0
26	1.79804E-22	3.13798E-16	1	1	0
27	2.95974E-24	3.48664E-17	1	1	0
28	3.5235E-26	3.73569E-18	1	1	0
29	2.7E-28	3.86451E-19	1	1	0
30	1E-30	3.86451E-20	1	1	0

30 — n
0.1 — p
3 — $\lambda = np$

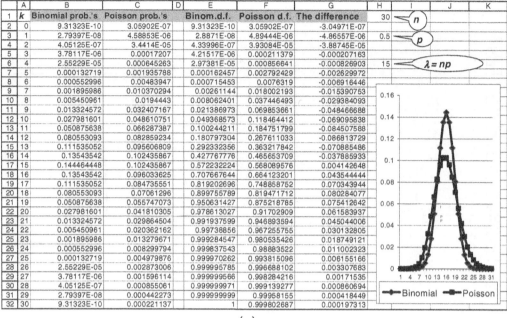

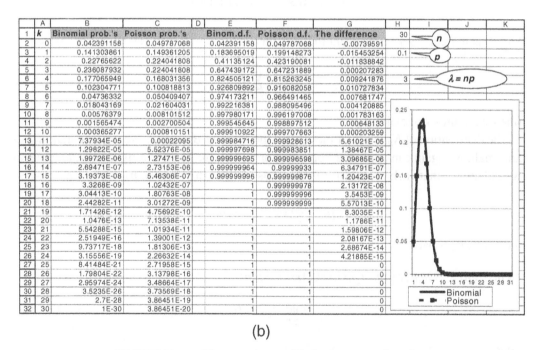

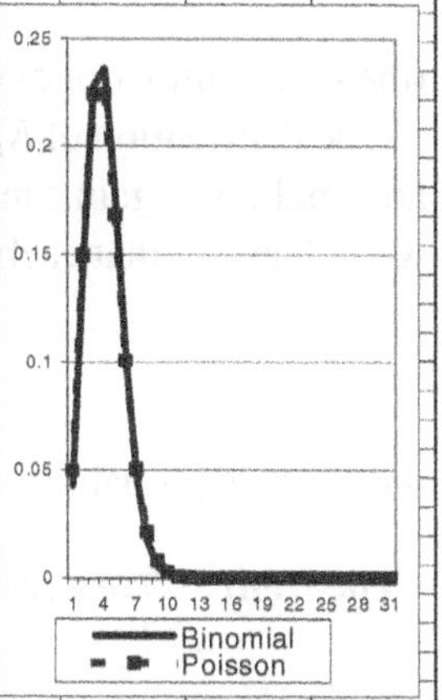

(b)

FIGURE 1. The accuracy of Poisson approximation.

Next, we consider the case of different probabilities of successes, which for the portfolio example correspond to a non-homogenous group of clients. Let

$$I_j = \begin{cases} 1 \text{ with probability } p_j, \\ 0 \text{ with probability } 1 - p_j. \end{cases}$$

	A	B	C	D	E	F	G	H	I
1	*k*	Binomial prob.'s	Poisson prob.'s		Binom.d.f.	Poisson d.f.	The difference	60	*n*
2	0	0.046069799	0.049787068		0.046069799	0.049787068	-0.003717269		
3	1	0.145483576	0.149361205		0.191553375	0.199148273	-0.007594899	0.05	*p*
4	2	0.225882394	0.224041808		0.417435769	0.423190081	-0.005754312		
5	3	0.229845243	0.224041808		0.647281012	0.647231889	4.91228E-05		
6	4	0.172383932	0.168031356		0.819664944	0.815263245	0.004401699	3	λ=*np*
7	5	0.101615792	0.100818813		0.921280735	0.916082058	0.005198677		
8	6	0.049025163	0.050409407		0.970305898	0.966491465	0.003814433		
9	7	0.019904953	0.021604031		0.990210851	0.988095496	0.002115355		
10	8	0.006940543	0.008101512		0.997151394	0.996197008	0.000954386		
11	9	0.002110574	0.002700504		0.999261969	0.998897512	0.000364457		
12	10	0.000566523	0.000810151		0.999828491	0.999707663	0.000120828		
13	11	0.000135532	0.00022095		0.999964023	0.999928613	3.54096E-05		
14	12	2.91274E-05	5.52376E-05		0.99999315	0.999983851	9.29947E-06		
15	13	5.66039E-06	1.27471E-05		0.999998811	0.999996598	2.21273E-06		
16	14	1.00014E-06	2.73153E-06		0.999999811	0.99999933	4.81346E-07		
17	15	1.61427E-07	5.46306E-07		0.999999972	0.999999876	9.64671E-08		
18	16	2.38954E-08	1.02432E-07		0.999999996	0.999999978	1.79302E-08		
19	17	3.2551E-09	1.80763E-08		1	0.999999996	3.10896E-09		
20	18	4.09267E-10	3.01272E-09		1	0.999999999	5.05513E-10		
21	19	4.76156E-11	4.75692E-10		1	1	7.74371E-11		
22	20	5.13747E-12	7.13538E-11		1	1	1.12207E-11		
23	21	5.15035E-13	1.01934E-11		1	1	1.54232E-12		
24	22	4.80535E-14	1.39001E-12		1	1	2.00284E-13		
25	23	4.17856E-15	1.81306E-13		1	1	2.32037E-14		
26	24	3.3905E-16	2.26632E-14		1	1	8.88178E-16		
27	25	2.56964E-17	2.71958E-15		1	1	-1.77636E-15		
28	26	1.8206E-18	3.13798E-16		1	1	-2.10942E-15		
29	27	1.20663E-19	3.48664E-17		1	1	-2.10942E-15		
30	28	7.48476E-21	3.73569E-18		1	1	-2.10942E-15		
31	29	4.34686E-22	3.86451E-19		1	1	-2.10942E-15		
32	30	2.36408E-23	3.86451E-20		1	1	-2.10942E-15		

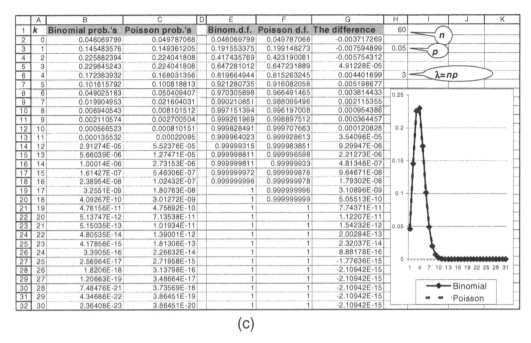

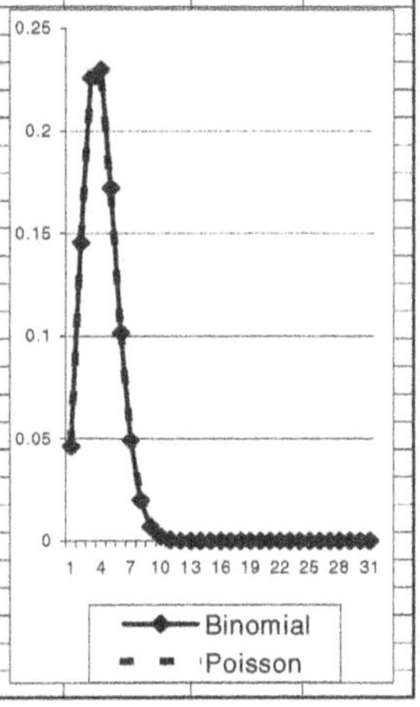

(c)

FIGURE 1. (continued).

Say, I_j is the indicator of the event that the jth customer will make a claim. Let

$$N_n = \sum_{j=1}^{n} I_j$$

(the total number of claims).

The distribution of N_n is sometimes called the *Poisson-Binomial.* We will see that, if p_j's are small, we again can apply the Poisson approximation. To state it rigorously, assume, as we did above, that each probability p_j depends on n, or in symbols, $p_j = p_{jn}$. Let

$$\bar{p}_n = \frac{1}{n}(p_{1n} + \ldots + p_{nn}),$$

the average probability.

Theorem 5 *(Generalized Poisson). Assume that*

$$\max_{j \le n} p_{jn} \to 0, \tag{2.1.6}$$

and

$$\bar{p}_n = \frac{\lambda}{n} + o\left(\frac{1}{n}\right) \tag{2.1.7}$$

for some $\lambda > 0$. Then (2.1.5) is true.

We omit the proof that may be found in many Probability textbooks, e.g., in [38], [116], [120]. Usually it is carried out with use of m.g.f.'s. In Exercise 7, we show that the "classical" Poisson theorem follows from Theorem 5.

The main condition (2.1.7) is the same as (2.1.4), we only impose it on the average probability. The significance of the additional condition (2.1.6) is that *all* probabilities p_{jn} should be small for large n. It may be shown that if, say, $p_{1n} \nrightarrow 0$, then (2.1.5) cannot be true. See Exercise 7 which contains details of how to prove it.

If (2.1.6) does not hold, we interpret it as if among all customers, there is at least one whose riskiness is comparable with the riskiness of the whole portfolio.

In applications of this scheme, we take as an approximation, the Poisson distribution with $\lambda = n\overline{p}_n = p_{1n} + ... + p_{nn}$.

Consider for example, a portfolio with d homogeneous groups. Denote by n_i and r_i the numbers of clients, and the probability of a loss event for each separate client in the ith group, respectively. In other words, $p_{jn} = r_i$ if the jth client belongs to the ith group. Set $n = n_1 + ... + n_d$, the total number of clients; $w_i = n_i/n$, the "weight" of the ith group, $\lambda_i = n_i r_i$, the expected number of claims in the ith group. Then

$$\overline{p}_n = \frac{1}{n}\sum_{j=1}^{n} p_{jn} = \frac{1}{n}\sum_{i=1}^{d} n_i r_i = \sum_{i=1}^{d} w_i r_i, \quad \lambda = \sum_{j=1}^{n} p_{jn} = \sum_{i=1}^{d} n_i r_i = \sum_{i=1}^{d} \lambda_i.$$

EXAMPLE 2. Let $d = 2$, $n_1 = 200$, $r_1 = 0.01$, $n_2 = 100$, and $r_2 = 0.02$. Then $n = 300$, $w_1 = 2/3$, $w_2 = 1/3$, $\overline{p}_n = \frac{2}{3}0.01 + \frac{1}{3}0.02 = 0.0133...$, and $\lambda = 2 + 2 = 4$. So, we can approximate $P(N_n = k)$ by the corresponding Poisson probabilities with $\lambda = 4$. We will see that the accuracy of such an approximation is relatively high. □

In conclusion, we mention a common interpretation of probabilities $P(N > n)$. When talking about the number of claims arriving, we mean a certain period of time, say a year. Let the numbers of claims in consecutive years be independent. Then, in the situation when for example $P(N > 3) = 1/5$, people sometimes say that four or more claims happen on the average once each five years. See Exercise 6.

Route 1 ⇒ *page 215*, *Route 2* ⇒ *page 207*

2.1.2 The accuracy of the Poisson approximation 3

The rate of convergence in Theorem 5 depends on what we consider: the probability for N_n to take on a particular value, the distribution function of N_n, that is, the probability for N_n to be less than or equal to a particular number, or the probability for N_n to take on a value from an *arbitrary* set.

Let n be fixed, $\lambda = p_{1n} + ... + p_{nn} > 0$, and

$$\nu_n = \frac{1}{\lambda}\sum_{j=1}^{n} p_{jn}^2. \tag{2.1.8}$$

In the classical Poisson scheme, where the r.v.'s I_j are identically distributed, for all j, probabilities $p_{jn} =$ some p, perhaps depending on n. Then $\lambda = np$, and $\nu_n = \lambda^{-1}np^2 = p$.

To understand the order of ν_n in the general case, assume that all p_{jn} have the same order; more specifically, $\frac{c_1}{n} \le p_{jn} \le \frac{c_2}{n}$ for positive constants c_1, c_2. Then $c_1 \le \lambda \le c_2$, and replacing p_{jn} in (2.1.8) by their bounds, we get that ν_n has the order of $\frac{1}{n}$. More precisely,

$$\frac{C_1}{n} \leq \nu_n \leq \frac{C_2}{n},$$

where $C_1 = c_1^2/c_2$, and $C_2 = c_2^2/c_1$.

Denote by Z_λ a Poisson r.v. with parameter λ.

Theorem 6 *For any set of integers B,*

$$|P(N_n \in B) - P(Z_\lambda \in B)| \leq 2.08\nu_n. \tag{2.1.9}$$

For $p_{jn} = p$, (2.1.9) implies

$$|P(N_n \in B) - P(Z_\lambda \in B)| \leq 2.08p.$$

Now let us consider distribution functions.

Theorem 7 *For any x,*

$$|P(N_n \leq x) - P(Z_\lambda \leq x)| \leq C\frac{\nu_n}{1-\nu_n}, \tag{2.1.10}$$

where the absolute constant $C = \frac{1}{2} + \sqrt{\frac{\pi}{8}}$.

Note that $C \approx 1.127 < 1.13$.

For $p_{jn} = p$, (2.1.10) implies

$$|P(N_n \leq x) - P(Z_\lambda \leq x)| \leq C\frac{p}{1-p}. \tag{2.1.11}$$

It is easy to calculate that if $\nu_n \leq 0.45$, then the r.-h.s. of (2.1.10) is less than the r.-h.s. of (2.1.9), and hence Theorem 7 gives a greater accuracy than Theorem 6. It is not surprising since in (2.1.10) we consider only d.f.'s. However, if we are interested in $P(N_n \in B)$ for more complicated sets B than a half-line, then we should appeal to Theorem 6.

The bound (2.1.9) is a generalization of a bound from [80], and is obtained in [104]. The bound (2.1.10) was obtained in [132]. A survey may be found in [12].

Note also that for $\nu_n \geq 0.45$, it is meaningless to talk about the Poisson approximation, and the r.-h.s. of (2.1.9) is larger than 0.93. In this case, the normal approximation would be much more adequate.

EXAMPLE 1. Consider the situation of Example 2.1.1-2. We have

$$\nu_n = \frac{1}{4}((0.01)^2 \times 200 + (0.02)^2 \times 100) = 0.015,$$

and

$$|P(N_n \leq x) - P(Z_\lambda \leq x)| \leq C\frac{0.015}{1-0.015} \approx 0.019,$$

which is not so bad. □

In conclusion, note that bounds (2.1.10) are universal, that is, serve for all possible values of p_{jn} and n. In particular cases, the real accuracy may be better than the bounds from Theorems 6 and 7.

EXAMPLE 2. Consider the situation of Example 2.1.1-1. For $p = 0.1$, the bound (2.1.11) gives the accuracy $\approx 1.13\frac{0.1}{1-0.1} \approx 0.126$, while the calculations in Example 2.1.1-1 gave the accuracy 0.01. If $p = 0.05$, then (2.1.11) results in ≈ 0.060, while particular calculations gives ≈ 0.0052. Such a discrepancy is connected also with the fact that here we consider the classical case of identically distributed r.v.'s, while (2.1.11) with the constant C came from a result for the general case of non-identically distributed addends. □

2.2 Some other "counting" distributions

The Poisson distribution plays a central role in the theory. Nevertheless, it is the simplest 2,3
distribution among the distributions considered in the literature and in practice. Let us address to other types of counting distributions. We interpret models below as those of the number of claims during a certain period and corresponding to a certain insurance portfolio.

2.2.1 The mixed Poisson distribution

Assume that the parameter of a Poisson r.v. is chosen at random in accordance with some probability distribution—that is, this parameter is also a random variable. As an example, one may consider the situation when the intensity of the flow of claims during a given period is determined at the beginning of the period by a random factor reflected in the mean value of the number of claims. (For example, we deal with road accidents, and the random factor concerns weather conditions.)

Another interpretation is connected with a division of the population of clients into classes. Assume that in each class, the distribution of the number of claims is Poisson but the intensity of the corresponding flow is different for different classes. If we do not know to which class a particular group under consideration belongs, we can view the Poisson parameter as random.

Denote the random parameter mentioned by Λ. Proceeding from the statement of the problem, we apply the Poisson formula to the conditional probability $P(N = n \,|\, \Lambda = \lambda)$, and write

$$P(N = n \,|\, \Lambda = \lambda) = e^{-\lambda}\lambda^n/n!.$$

Let $F_\Lambda(\lambda)$ be the d.f. of the r.v. Λ. By formula (**0**.7.3.8)—which may be viewed as the generalization of the formula for total probability,

$$P(N = n) = \int_0^\infty (e^{-\lambda}\lambda^n/n!)dF_\Lambda(\lambda).$$

Given Λ, the conditional expectation $E\{N \,|\, \Lambda\} = \Lambda$, and the conditional variance $Var\{N \,|\, \Lambda\}$

$= \Lambda$. Then, by (**0**.7.2.1) and (**0**.7.3.2),

$$E\{N\} = E\{E\{N \mid \Lambda\}\} = E\{\Lambda\}, \tag{2.2.1}$$

$$\begin{aligned} Var\{N\} &= E\{Var\{N \mid \Lambda\}\} + Var\{E\{N \mid \Lambda\}\} \\ &= E\{\Lambda\} + Var\{\Lambda\}. \end{aligned} \tag{2.2.2}$$

Since by (2.2.1), $E\{N\} = E\{\Lambda\}$, we have $Var\{N\} = E\{N\} + Var\{\Lambda\}$, so unlike for the Poisson distribution itself, the variance of the mixed Poisson distribution is larger than the mean value.

EXAMPLE 1 is somewhat formal. Let Λ be uniform on $[a,b]$, $a > 0$. Then

$$P(N=n) = \int_a^b \frac{e^{-\lambda}\lambda^n}{n!}\frac{1}{(b-a)}d\lambda = \frac{1}{(b-a)n!}\int_a^b e^{-\lambda}\lambda^n d\lambda.$$

The last integral is standard; we will omit the precise formula here. In particular,

$$P(N=0) = \frac{1}{(b-a)}\int_a^b e^{-\lambda}d\lambda = \frac{e^{-a}-e^{-b}}{(b-a)}.$$

By (2.2.1) and (2.2.2),

$$E\{N\} = \frac{a+b}{2}, \quad Var\{N\} = \frac{a+b}{2} + \frac{(b-a)^2}{12}. \quad \square$$

For the m.g.f. of N, we have the following nice presentation. Let us recall that for the Poisson distribution with parameter λ, the m.g.f. $M(z) = \exp\{\lambda(e^z - 1)\}$. Then the m.g.f. of N is

$$M_N(z) = E\{e^{zN}\} = E\{E\{e^{zN} \mid \Lambda\}\} = E\{\exp\{\Lambda(e^z-1)\}\}.$$

The last expression is the m.g.f. of Λ at the point $(e^z - 1)$, and hence

$$M_N(z) = M_\Lambda(e^z - 1), \tag{2.2.3}$$

where $M_\Lambda(z)$ is the m.g.f. of Λ.

EXAMPLE 2 (*The Pòlya model*). This is probably the most notable case. Let Λ have the Γ-distribution with parameters (a,ν). Then (see Section **0**.4.3.5)

$$M_\Lambda(z) = 1/(1-z/a)^\nu,$$

and by (2.2.3),

$$M_N(z) = 1/(1-(e^z-1)/a)^\nu.$$

It is straightforward to verify that $1-(e^z-1)/a$ can be rewritten as $(1-qe^z)/p$, where $p = a/(1+a)$ and $q = 1-p = 1/(1+a)$. Hence,

$$M_N(z) = \left(\frac{p}{1-qe^z}\right)^\nu. \tag{2.2.4}$$

The right member of (2.2.4) is the m.g.f. of the *negative binomial distribution*—that is, the distribution of an integer valued r.v. K_ν such that for $n = 0, 1, ...$

$$P(K_\nu = n) = \binom{\nu+n-1}{n} p^\nu q^n. \tag{2.2.5}$$

The "binomial coefficient" is defined by the general formula

$$\binom{r}{n} = \frac{r(r-1)\cdot ... \cdot(r-n+1)}{n!}, \tag{2.2.6}$$

where n is an integer, and r is an arbitrary real number, not necessarily an integer (see Sections **0**.3.1.4, **0**.4.3.2, and Exercise 8 of the current chapter for detail).

So, in general, $\binom{r}{n}$ may be negative (say, for $r = 2.5$ and $n = 4$). However, the coefficient in (2.2.5) is positive because, by (2.2.6),

$$\binom{\nu+n-1}{n} = \frac{(\nu+n-1)(\nu+n-2)\cdot ... \cdot\nu}{n!};$$

see also Exercise 8.

In the particular case $\nu = 1$, the Γ-distribution above is an exponential distribution and distribution (2.2.5) is geometric, i.e.,

$$P(K_1 = n) = pq^n; \tag{2.2.7}$$

see again Exercise 8.

The negative binomial distribution (2.2.5) serves as a good approximation in many practical situations including those that are not relevant to mixing of Poisson distributions.

EXAMPLE 3 ([153, N10][2]). Low Risk Insurance Company provides liability coverage to a population of 1000 private drivers. The number of claims during a given year from this population is Poisson distributed. If a driver is selected at random from this population, his *expected* number of claims per year is a random variable with a Gamma distribution with parameters $a = 1$ and $\nu = 2$. Calculate the probability that a driver selected at random will not have a claim during the year.

When saying "the number of claims during a *given* year", we mean the conditional distribution given the conditions of a particular year. We assume this conditional distribution is Poisson. Let N_i be the number of claims corresponding to the ith client, $i = 1, ..., n = 1000$. We adopt a model in which each N_i has the mixed Poisson distribution with a random parameter Λ_i, and each Λ_i has the Γ-distribution with $a = 1, \nu = 2$. The values of $\Lambda_1, ..., \Lambda_n$ specify conditions for the year under consideration. Assume that, *given* $\Lambda_1, ..., \Lambda_n$, the r.v.'s N_i are independent and, consequently, the (conditional) distribution of the total number of claims, $N = N_1 + ... + N_n$, is indeed Poisson. Each N_i has the negative binomial distribution with parameters $p = \frac{a}{1+a} = \frac{1}{2}$, and $\nu = 2$. In accordance with (2.2.5), $P(N_i = 0) = p^2 = 0.25$. Note that we did not assume Λ_i's are independent. The only assumption we made is the conditional independence of N_i's given $\Lambda_1, ..., \Lambda_n$.

[2]Reprinted with permission of the Casualty Actuarial Society.

EXAMPLE 4 (this example is close to the problem [151, N21]). The number of claims coming from a group of 200 clients is closely approximated by the negative binomial distribution with parameters $p = 0.2$, $\nu = 3$. What distribution would you expect for a group of 400 clients of the same type?

We adopt the model from Example 3: the number of claims, N, is Poisson with a random parameter $\Lambda = \Lambda_1 + ... + \Lambda_n$, where n is the number of clients. Each Λ_i has a Γ-distribution. Denote the parameters of this distribution by a_0, ν_0. If Λ has also a Γ-distribution with parameters (a, ν), then N has the negative binomial distribution with parameters $p = \frac{a}{1+a}$ and ν.

The mean value $E\{N\} = \frac{\nu(1-p)}{p} = \frac{\nu}{a}$. It is natural to expect that, if the number of clients is twice as large, so the expected number of claims doubles. The question is which parameter changes: ν or a? The answer depends on what we assume regarding the r.v.'s $\Lambda_1, ..., \Lambda_n$.

(a) First, suppose that claims coming from each client are independent. Then we may assume $\Lambda_1, ..., \Lambda_n$ to be independent, which implies that Λ has the Γ-distribution with parameters $a = a_0$, and $v = n\nu_0$. Thus, when n is changing, the parameter a does not change, while ν is growing in proportion to n. So, in this case, N has the negative binomial distribution with the same $p = 0.2$ and the new $\nu = 2 \cdot 3 = 6$.

(b) Assume now that values of the parameter Λ_i are specified by conditions common to all clients, say, the weather conditions for a particular time interval. Then it is natural to assume that $\Lambda_1 = ... = \Lambda_n$. In this case $\Lambda = n\Lambda_1$ and have the Γ-distribution with parameters $a = a_0/n$, and $\nu = \nu_0$. For $n = 200$, the parameter $p = 0.2$, and since $p = \frac{a}{1+a}$, the parameter $a = 1/4$. Hence, for $n = 400$, the parameter $a = 1/8$, and N is negative binomial with $p = 1/9$, and the same $\nu = 3$. □

Next, we discuss how to determine the distribution of Λ if we have known the value of N.

Let the symbol $P(\Lambda \in d\lambda)$ mean $P(\Lambda = \lambda)$ when the r.v. Λ is discrete, and $f(\lambda)d\lambda$ when Λ is continuous with density $f(\lambda)$. In both cases, $P(\Lambda \in d\lambda)$ may be viewed as the probability that Λ will take on a value from the infinitesimally small interval $[\lambda, \lambda + d\lambda]$. Then

$$P(\Lambda \in d\lambda \,|\, N = n) = \frac{P(\Lambda \in d\lambda, N = n)}{P(N = n)} = \frac{P(N = n \,|\, \Lambda = \lambda) P(\Lambda \in d\lambda)}{P(N = n)}$$
$$= \frac{e^{-\lambda}\lambda^n}{n!} \frac{P(\Lambda \in d\lambda)}{P(N = n)}.$$

If Λ takes on values $\lambda_1, \lambda_2, ...$ with probabilities $p_1, p_2 ...$, respectively, then the last relation may be written as

$$P(\Lambda = \lambda_k \,|\, N = n) = \frac{e^{-\lambda_k}\lambda_k^n}{n!} \frac{p_k}{P(N = n)}, \tag{2.2.8}$$

where

$$P(N = n) = \sum_k \frac{e^{-\lambda_k}\lambda_k^n}{n!} p_k. \tag{2.2.9}$$

If Λ has density $f(\lambda)$, then

$$P(\Lambda\in d\lambda\,|\,N=n)=\frac{e^{-\lambda}\lambda^n}{n!}\,\frac{f(\lambda)d\lambda}{P(N=n)},\tag{2.2.10}$$

where

$$P(N=n)=\int\frac{e^{-\lambda}\lambda^n}{n!}f(\lambda)d\lambda.\tag{2.2.11}$$

Formulas (2.2.8) and (2.2.10) are versions of Bayes's formula (see Section **0**.1.2) for the discrete and continuous cases, respectively. In Exercise 9, the reader is invited to verify this for the case of two values of Λ.

EXAMPLE 5. Let the number of daily claims for an auto insurance portfolio be a Poisson r.v. with a mean of three on a "good" day, and ten on a "bad" day. The probability of a good day is $3/4$, for a bad day, accordingly, $1/4$. Eight claims have arrived by the end of a day. What is the probability that the day falls into the category "bad"? By (2.2.8)-(2.2.9), $P(\Lambda=10\,|\,N=8)=e^{-10}\frac{10^8}{8!}\cdot\frac{1}{4}\Big/\left(e^{-3}\frac{3^8}{8!}\cdot\frac{3}{4}+e^{-10}\frac{10^8}{8!}\cdot\frac{1}{4}\right)\approx 0.822$; compare with the prior unconditional probability 0.25.

EXAMPLE 6. Consider the situation of Example 2. The density of the r.v. Λ is the Γ-density $f(\lambda)=\frac{1}{\Gamma(\nu)}a^\nu\lambda^{\nu-1}e^{-a\lambda}$. Then, by (2.2.10), the conditional density of Λ given $N=n$ is

$$f(\lambda\,|\,N=n)=\frac{e^{-\lambda}\lambda^n}{n!}\,\frac{f(\lambda)}{P(N=n)}=\frac{1}{P(N=n)n!}e^{-\lambda}\lambda^n\cdot\frac{a^\nu}{\Gamma(\nu)}\lambda^{\nu-1}e^{-a\lambda}$$
$$=C(n,a,\nu)\lambda^{n+\nu-1}e^{-(a+1)\lambda},$$

where $C(n,a,\nu)$ is an expression that depends on n,a,ν, and does not depend on λ. We can certainly compute $C(n,a,\nu)$, but it is not necessary. The term $\lambda^{n+\nu-1}e^{-(a+1)\lambda}$ says that we are dealing with the Γ-distribution with parameters $a+1,n+\nu$. Then $C(n,a,\nu)$ must be equal to $(a+1)^{n+\nu}/\Gamma(n+\nu)$. The reader may double-check this taking into account that $P(N=n)$ is given in (2.2.5) with $p=a/(a+1)$, but (if we have not made a mistake in our calculations) it is unnecessary.

In particular, since $E\{\Lambda\,|\,N=n\}$ is the mean value of the Γ-distribution mentioned, we have

$$E\{\Lambda\,|\,N=n\}=\frac{n+\nu}{a+1}.\tag{2.2.12}$$

Set λ_{ave}=$E\{\Lambda\}$, the unconditional mean value, and realize that in our situation, $\lambda_{ave}=\frac{\nu}{a}$, as the mean value of the Γ-distribution with parameters a,ν. Then, (2.2.12) may be rewritten as

$$E\{\Lambda\,|\,N=n\}=\lambda_{ave}\cdot\frac{n+\nu}{\lambda_{ave}+\nu}.\tag{2.2.13}$$

This is simple and nice. If the observed value of n is larger (smaller) than λ_{ave}, then the conditional expected value of Λ given the observed information, is larger (smaller) than λ_{ave}.

Let, for example, $a=2$, $\nu=4$. Then $\lambda_{ave}=E\{\Lambda\}=(\nu/a)=2$. Suppose that N took on the value 5. Then the posterior distribution of Λ is the Γ-distribution with new parameters $\tilde{a}=a+1=3$ and $\tilde{\nu}=n+\nu=5+4=9$. In particular, $E\{\Lambda\,|\,N=5\}=\frac{5+4}{2+1}=3$. □

2.2.2 Compound mixing

Let the total number of claims

$$N = Y_1 + ... + Y_K, \tag{2.2.14}$$

where integer valued r.v.'s $Y_1, Y_2, ...$ and K are independent, and r.v.'s $Y_1, Y_2, ...$ are identically distributed. For example, K may be the number of accidents corresponding to a risk portfolio and Y_i is the number of claims due to the ith accident.

The scheme (2.2.14) is called *compound mixing*. The distribution of K is called *primary*, the distribution of Y's—*secondary*.

We see that (2.2.14) is a particular instance of the scheme from Section 1: the role of N from (0.1) is played by K, and the role of S by N.

Applying (1.5) to (2.2.14), we have

$$M_N(z) = M_K(\ln M_Y(z)), \tag{2.2.15}$$

where $M_Y(z)$ is the m.g.f. of Y's. If K is Poisson with parameter λ, by (1.6),

$$M_N(z) = \exp\{\lambda(M_Y(z) - 1)\}. \tag{2.2.16}$$

EXAMPLE 1. Let K have the Poisson distribution with parameter λ, and for all i,

$$Y_i = \begin{cases} 1 \text{ with probability } p, \\ 0 \text{ with probability } 1 - p. \end{cases}$$

In the particular case of the example with accidents, it would mean that each accident causes either one claim or no claims. For example, Y_i may equal zero if the size of the damage at the ith accident does not exceed a deductible.

It is noteworthy that in this case N is also a Poisson r.v. with parameter $p\lambda$. Indeed,

$$M_Y(z) = pe^z + (1 - p) = 1 + p(e^z - 1), \tag{2.2.17}$$

and by (2.2.16),

$$M_N(z) = \exp\{p\lambda(e^z - 1)\}.$$

The last function is the Poisson m.g.f. with parameter $p\lambda$.

We will return to this in Section 3.1.2 where we give another solution to the same problem using a special feature of the Poisson distribution.

EXAMPLE 2. Let K be the same as in Example 1, and let each Y_i take on the value $0, 1, 2$ with probability p_0, p_1, p_2, respectively. Then $M_Y(z) = p_0 + p_1 e^z + p_2 e^{2z} = 1 + p_1(e^z - 1) + p_2(e^{2z} - 1)$, and by (2.2.16),

$$\begin{aligned} M_N(z) &= \exp\{\lambda(M_Y(z) - 1)\} = \exp\{\lambda[p_1(e^z - 1) + p_2(e^{2z} - 1)]\}, \\ &= \exp\{\lambda p_1(e^z - 1)\}\exp\{\lambda p_2(e^{2z} - 1)\}. \end{aligned}$$

The last expression is the product of two m.g.f.'s. The first corresponds to a r.v. Υ_1 having the Poisson distribution with parameter λp_1. A r.v. whose m.g.f. equals the second factor,

$\exp\{\lambda p_2(e^{2z}-1)\}$, may be represented as $2\Upsilon_2$, where Υ_2 has the Poisson distribution with parameter λp_2. Indeed, $E\{e^{z2\Upsilon_2}\} = E\{e^{(2z)\Upsilon_2}\}$, that is, the m.g.f. of Υ_2 at the point $2z$. (See also formula (**0**.4.1.5) for the m.g.f.'s of the linear transformations of r.v.'s.)

Thus, we can write the representation

$$N = \Upsilon_1 + 2\Upsilon_2, \tag{2.2.18}$$

where Υ_1, Υ_2 are independent Poisson r.v.'s with parameters $p_1\lambda$ and $p_2\lambda$, respectively.

In order to compute $E\{N\}$ and $Var\{N\}$, we can use (1.2) and (1.4), or proceed from the representation (2.2.18), which implies that $E\{N\} = p_1\lambda + 2p_2\lambda = (p_1+2p_2)\lambda$ and $Var\{N\} = p_1\lambda + 4p_2\lambda = (p_1+4p_2)\lambda$. This example is continued in Exercise 15.

Note also that like Example 1, this example will be also revisited in Section 3.1.2. In the current section, we use a general approach based on m.g.f.'s; in Section 3.1.2 we proceed from a special feature of the Poisson distribution.

EXAMPLE 3. Let K have the Poisson distribution with parameter λ_1, and all Y_i have the Poisson distribution with parameter λ_2. Then the m.g.f.

$$M_N(z) = \exp\{\lambda_1(e^{\lambda_2(e^z-1)}-1)\}.$$

This distribution is called *Poisson-Poisson*. From (1.2) and (1.4) it immediately follows that $E\{N\} = \lambda_1\lambda_2$ and $Var\{N\} = \lambda_1(\lambda_2+\lambda_2^2)$.

EXAMPLE 4. Let K be negative binomial with parameters p_1 and ν, and all Y_i take on values 1 or 0 with probabilities p_2 and $1-p_2$, respectively. Let $q_i = 1-p_i, i=1,2$. By (2.2.4), $M_K(z) = [p_1/(1-q_1e^z)]^\nu$. Hence, in view of (2.2.15) and (2.2.17),

$$M_N(z) = [p_1/(1-q_1\exp\{\ln(1+p_2(e^z-1))\})]^\nu = \left[\frac{p_1}{1-q_1(1+p_2(e^z-1))}\right]^\nu. \tag{2.2.19}$$

Let $\bar{p} = (1-q_1)/(1-q_1q_2)$ and $\bar{q} = 1-\bar{p}$. It is straightforward to verify that the expression in the brackets $[\cdot]$ on the r.-h.s. of (2.2.19) may be rewritten as $\bar{p}/(1-\bar{q}e^z)$, and hence N has the negative binomial distribution with parameters $\bar{p}$, ν. □

2.2.3 The $(a,b,0)$ and $(a,b,1)$ (or Katz-Panjer's) classes

Let $p_k = P(N=k)$. We say that the distribution of N belongs to the $(a,b,0)$ class if there exist a and b such that

$$\frac{p_k}{p_{k-1}} = a + \frac{b}{k} \tag{2.2.20}$$

for all $k = 1,2,\dots$. It may be proved (see, e.g., [98]) that the only possible distributions satisfying (2.2.20) are those listed in Table 1 below.

TABLE 1.

	Distributions	p_0	a	b
1	Poisson with λ	$e^{-\lambda}$	0	λ
2	Geometric with p and $q=1-p$	p	q	0
3	Negative binomial with ν, p and $q=1-p$	p^ν	q	$(\nu-1)q$
4	Binomial with n and p and $q=1-p$	q^n	$-p/q$	$(n+1)p/q$

Let us check, for example, Position 3. For the negative binomial distribution, in view of (2.2.5), $p_0 = p^\nu$, and

$$\frac{p_k}{p_{k-1}} = \binom{\nu+k-1}{k} p^\nu q^k \Big/ \binom{\nu+k-2}{k-1} p^\nu q^{k-1} = \frac{\nu+k-1}{k} q = q + \frac{(\nu-1)q}{k},$$

which leads to the result in the table. The consideration of other table entries are relegated to Exercise 19.

EXAMPLE 1 ([153, N40][3]). You are given a negative binomial distribution with $\nu = 2.5$ and $p = 1/6$. For what value of k does p_k take on its largest value?

For the negative binomial distribution, $p_k = p_{k-1}\phi(k)$ where $\phi(k) = q + \frac{(\nu-1)q}{k}$. In our example, $\phi(k) = \frac{1}{6}(5 + \frac{7.5}{k})$. We are looking for k such that $\phi(k) \geq 1$ and $\phi(k+1) \leq 1$. This is $k = 7$. □

We say that the distribution of N belongs to the $(a, b, 1)$ class if (2.2.20) is true for some a and b, and all $k = 2, 3, \ldots$. The difference is that we begin to construct probabilities recursively starting from p_1 rather than p_0, and reserving the possibility of defining p_0 as we wish, or more precisely, as it fits the data. Such a modification allows us to construct distributions having the structure close to "classical" but adjusted to particular real situations.

The distribution with $p_0 = 0$ is called *zero-truncated*, otherwise—*zero-modified.*

Consider, for example, the Poisson case when $a = 0$, $b = \lambda$, and p_0 is fixed. For other probabilities,

$$p_2 = \frac{\lambda p_1}{2} = \frac{\lambda p_1}{2!},\; p_3 = \frac{\lambda p_2}{3} = \frac{\lambda^2 p_1}{2\cdot 3} = \frac{\lambda^2 p_1}{3!},$$

and so on, which leads to

$$p_k = \frac{\lambda^{k-1} p_1}{k!}.$$

On the other hand, $1 - p_0 = \sum_{k=1}^{\infty} p_k = p_1 \sum_{k=1}^{\infty}(\lambda^{k-1}/k!) = p_1\lambda^{-1}\sum_{k=1}^{\infty}(\lambda^k/k!) = p_1\lambda^{-1}(e^\lambda - 1)$. Thus, $p_1 = (1-p_0)\lambda(e^\lambda - 1)^{-1}$, and for $k = 1, 2, \ldots$

$$p_k = \frac{1-p_0}{e^\lambda - 1}\cdot\frac{\lambda^k}{k!}. \tag{2.2.21}$$

If we set $p_0 = e^{-\lambda}$, as in the "usual" Poisson distribution, we come to Poisson probabilities for all k. If $p_0 = 0$,

$$p_k = \frac{1}{e^\lambda - 1}\frac{\lambda^k}{k!},$$

which coincides with $P(Z = k \,|\, Z > 0)$ where Z is a "usual" Poisson r.v. Other examples are considered in Exercise 20.

► A formula similar to (2.2.21) is true in the general case. Let N be an original r.v. and $p_k = P(N = k)$ be the corresponding original probabilities for which

$$p_k = p_{k-1}\phi(k) \tag{2.2.22}$$

[3]Reprinted with permission of the Casualty Actuarial Society.

for $k = 1, 2, \ldots$, where $\phi(k)$ is a given function. In the particular case above, $\phi(k) = a + b/k$.

Proposition 8 *Let $\tilde{p}_k$ be modified probabilities, that is,*

$$\tilde{p}_k = \tilde{p}_{k-1}\phi(k) \tag{2.2.23}$$

for $k = 2, 3, \ldots$. Then

$$\tilde{p}_k = (1-\tilde{p}_0)\frac{p_k}{\sum_{k=1}^{\infty} p_k} = (1-\tilde{p}_0)P(N = k \mid N > 0). \tag{2.2.24}$$

Proof. In view of (2.2.23) $\tilde{p}_k = \phi(k)\tilde{p}_{k-1} = \phi(k)\phi(k-1)\tilde{p}_{k-2} = \ldots = \phi(k)\phi(k-1) \cdot \ldots \cdot \phi(2)\tilde{p}_1$, that is, $\tilde{p}_k = \tilde{p}_1\psi(k)$, where $\psi(k) = \phi(k)\phi(k-1) \cdot \ldots \cdot \phi(2)$, and $\psi(1) = 1$. We have $1 - \tilde{p}_0 = \sum_{k=1}^{\infty} \tilde{p}_k = \tilde{p}_1 \sum_{k=1}^{\infty} \psi(k)$ and $\tilde{p}_1 = (1-\tilde{p}_0)/(\sum_{k=1}^{\infty} \psi(k))$. Thus,

$$\tilde{p}_k = (1-\tilde{p}_0)\frac{\psi(k)}{\sum_{k=1}^{\infty} \psi(k)}. \tag{2.2.25}$$

On the other hand, $p_1 = \phi(1)p_0$, and $p_k = p_1\psi(k) = p_0\phi(1)\psi(k)$. Multiplying the numerator and the denominator in (2.2.25) by $p_0\phi(1)$, we come to (2.2.24). ■ ◄

---

3 THE DISTRIBUTION OF THE AGGREGATE CLAIM 1,2,3

3.1 The case of a homogeneous group

First, we consider a *homogeneous group* of clients and claims coming from this group. Namely, we consider a fixed time period and assume that the total claim S during this period is represented by relation (0.1) where the sizes of claims, X_i, are *independent and identically distributed* (i.i.d.) r.v.'s, and the total number of claims, N, is independent of the X's. If $N = 0$, we set $S = 0$. Unless stated otherwise, we suppose $P(X_j > 0) = 1$ for all j.

Propositions 1 and 2 give a clear way to compute $E\{S\}$ and $Var\{S\}$. Examples are given in Exercise 22.

Our object of study is the distribution of S. There are several approaches to computing or estimating this distribution.

3.1.1 The convolution method

Let $g_n = P(N = n)$ and $F(x)$ be the d.f. of X_j. Since all X's are positive with probability one,

$$P(S = 0) = P(N = 0) = g_0, \tag{3.1.1}$$

and the d.f. $F_S(x)$ has a "jump" of g_0 at $x = 0$. Furthermore, for $x \geq 0$,

$$F_S(x) = P(S \leq x) = \sum_{n=0}^{\infty} P(S \leq x \mid N = n)P(N = n) = \sum_{n=0}^{\infty} P(S_n \leq x \mid N = n)P(N = n),$$

where, as usual, $S_n = X_1 + ... + X_n$, and $S_0 = 0$. The X's do not depend on N, and they are mutually independent. Hence for $n \geq 1$, we have $P(S_n \leq x \mid N = n) = P(S_n \leq x) = F^{*n}(x)$, where the symbol F^{*n} denotes $F * ... * F$, the nth convolution of F (see Section **2**.2.1.1). Thus,

$$F_S(x) = \sum_{n=0}^{\infty} g_n F^{*n}(x), \tag{3.1.2}$$

where $F^{*0}(x)$ (which corresponds to the case when $N = 0$) is defined as the d.f. of a r.v. $X \equiv 0$, that is, $F^{*0}(x) = 1$ for $x \geq 0$, and $= 0$ for $x < 0$.

If the density $f(x) = F'(x)$ exists, the density $f_{S_n}(x)$ also exists for all n except zero. Since the derivative $(F^{*0}(x))' = 0$ for $x > 0$, the d.f. $F_S(x)$ is differentiable for all $x > 0$. We call the corresponding derivative the density of S, which exists for all $x > 0$. Eventually, differentiating (3.1.2), we get that for $x > 0$,

$$f_S(x) = \sum_{n=1}^{\infty} g_n f^{*n}(x), \tag{3.1.3}$$

where f^{*n} is the nth convolution of the density f; see **2**.2.1.1. (Notice that the summation in (3.1.3) starts from $n = 1$ since $\frac{d}{dx}F^{*0}(x) = 0$ for $x > 0$.)

The same formula is true for the case when the X's are discrete r.v.'s if we understand $f_S(x)$ as $P(S = x)$, and $f^{*n}(x)$ as $P(S_n = x)$. The last probability is the result of the nth convolution of the distribution of addends.

Formulas (3.1.2)-(3.1.3) may be useful when we can write a good representation or an approximation for all convolutions F^{*n}. This is true at least for some special distributions, which will be demonstrated in the following examples.

EXAMPLE 1. Let all X_i have the Γ-distribution with parameters (a, ν). In particular, for $\nu = 1$, it is the exponential distribution with parameter a. Denote the corresponding d.f. by $\Gamma(x; a, \nu)$. By Proposition **2**.2.1.2-4, S_n has the d.f. $\Gamma(x; a, n\nu)$, and if $f(x)$ is the density of X_i, then the density of S_n is

$$f^{*n}(x) = a^{n\nu} x^{n\nu - 1} e^{-ax} / \Gamma(n\nu).$$

If ν is an integer, the function $\Gamma(x; a, \nu)$ can be written explicitly (see later Section **4**.2.1), but for now, for us it matters only that it is a "good" tractable function. By (3.1.2), we have

$$F_S(x) = \sum_{n=0}^{\infty} g_n \Gamma(x; a, n\nu). \tag{3.1.4}$$

For the density $f_S(x)$ and $x > 0$,

$$f_S(x) = \sum_{n=1}^{\infty} g_n \frac{a^{n\nu} x^{n\nu - 1} e^{-ax}}{\Gamma(n\nu)}. \ \square \tag{3.1.5}$$

If the r.v. N takes on a finite and not large number of values (and hence only a finite number of g_n's are greater than zero), then the sums in (3.1.2) and (3.1.3) are finite and may be computed numerically.

If N takes on a large or infinite number of values, we have to use truncation. Assume that we restricted ourselves to the summation in (3.1.2) over n from 0 to some k. Then the error will be equal to

$$\sum_{k+1}^{\infty} g_n F^{*n}(x) \le \sum_{k+1}^{\infty} g_n = P(N > k). \qquad (3.1.6)$$

This probability may be small for a relatively moderate k.

EXAMPLE 2. Let N have the geometric distribution in the form (**0**.3.1.9) with parameter p. Then $P(N > k) = (1-p)^{k+1}$; see (**0**.3.1.10) . Note also that $E\{N\} = (1-p)/p$. Assume that there are 50 claims on the average, i.e., $E\{N\} = 50$. Then, as is easy to calculate, $P(N > k) \le 0.05$ for $k \ge 150$. So, if 0.05 is acceptable accuracy for us, we can consider the sum $\sum_{m=0}^{150}$, which is computable even for a computer with moderate performance. For $E\{N\} = 20$, we would have $P(N > 60) \le 0.051$ and, for example, $P(N > 80) \le 0.02$. □

It makes sense to emphasize that the real error of truncation may be essentially smaller than estimate (3.1.6), since above we replaced $F^{*n}(x)$ by its roughest bound, namely, one.

EXAMPLE 3 illustrates the possibilities of concrete calculations. Let us return to (3.1.4) and assume that N has a geometric distribution with parameter p. In Fig.2, the corresponding Excel worksheet is presented. The values of p, $q = 1-p$, ν, and a, are in cells G2, G3, G6, G7, respectively. In G9 and G10, $m_S = E\{S\}$ and $\sigma_S = \sqrt{Var\{S\}}$ are computed in accordance with (1.1) and (1.3). For the distributions chosen,

$$E\{S\} = \frac{\nu}{a} \cdot \frac{q}{p}, \quad Var\{S\} = \frac{\nu}{a^2} \cdot \frac{q}{p} + \left(\frac{\nu}{a}\right)^2 \cdot \frac{q}{p^2} \qquad (3.1.7)$$

(in Exercise 23 the reader is encouraged to verify these formulas). Once we know m_S and σ_S, we can choose (cell G5) a reasonable value of x, say, in the range $(m_S - 3\sigma_S, m_S + 3\sigma_S)$. Values of n and the corresponding terms in (3.1.4) are in columns A and B. The Excel command for, say, B3 is

B3=$G$2*$G$3^A3*GAMMADIST($G$5,A3*$G$6,$G$7^(-1),TRUE)

The estimates for $P(S \le x)$ are in column D and obtained by summing over the first k terms in column B. (Not all rows are shown in Fig.2.) We see that the values of the estimates for $k = 300$ and $k = 200$ do not differ significantly, which allows us to suppose that for these k's, the estimate of 0.87 is good enough, while the estimates for $k = 150$ and lower are not. Certainly, making use of a more sophisticated program, we can easily get values of $F_S(x)$ for a large spectrum of x's. Table 2 gives five values of the d.f. for the parameters as in Fig.2.

TABLE 2.

x	20	50	75	100	150
$F_S(x)$	0.34	0.64	0.78	0.87	0.95

□

Consider two more examples.

	A	B	C	D	E	F	G	H
1	*n*	*g* $g_n P(S_n =<x)$		$P(S=<x)$				
2	0	0.01		**0.86602**	*k=300*		**0.01**	*p*
3	1	0.0099		**0.85913**	*k=200*		**0.99**	*q*
4	2	0.009801		**0.78076**	*k=150*			
5	3	0.00970299		**0.63763**	*k=100*		**100**	*x*
6	4	0.00960596		**0.40104**	*k=50*		**1**	*v*
7	5	0.0095099					**2**	*a*
8	6	0.009414801						
9	7	0.009320653					**49.5**	m_S
10	8	0.009227447					**49.9975**	σ_S
11	9	0.009135172					**-100.492**	m_S-3 σ_S
12	10	0.009043821					**199.492**	m_S+3 σ_S
13	11	0.008953383						
14	12	0.008863849						
15	13	0.00877521						
16	14	0.008687458						
17	15	0.008600584						

FIGURE 2. The worksheet for Example 3. Not all 300 numbers in Columns A and B are shown.

EXAMPLE 4. Assume that the distribution of X's may be well approximated by the (m, σ^2)-normal distribution. Then, by Proposition 2, Section **2**.2.1.2, the distribution of S_n may be approximated by the $(nm, n\sigma^2)$-normal distribution, and we can set $F^{*n}(x) = \Phi(\frac{x-nm}{\sigma\sqrt{n}})$ (verify this on your own, or see Section **0**.3.2.4). Hence,

$$F_S(x) = \sum_{n=0}^{\infty} g_n \Phi\left(\frac{x-nm}{\sigma\sqrt{n}}\right). \tag{3.1.8}$$

In Exercise 25, the reader is invited to provide an Excel worksheet and carry out calculations based on (3.1.8).

EXAMPLE 5. Let X_j take on two values, a and b, with probabilities p and q, respectively. Assume $a > b > 0$. Clearly, all values of S are combinations $ka + mb$, where k and m are integers. Then $P(S_n = x) \neq 0$ only if $x = ka + (n-k)b$ for some natural $k \leq n$. In this case, $k = k(x,n) = (x - nb)/(a-b)$, and $P(S_n = x) = \binom{n}{k} p^k q^{n-k}$.

Note also that since the biggest value of S_n is na and the smallest is nb, $P(S_n = x) = 0$ for all $x > na$ or $x < nb$, which is equivalent to $n < x/a$ or $n > x/b$, respectively. Thus,

$$P(S = x) = \sum_{x/a \leq n \leq x/b} g_n \binom{n}{k(x,n)} p^{k(x,n)} q^{n-k(x,n)} \tag{3.1.9}$$

for all x of the form $ka + mb$. Denote by $B(z; n, p)$ the binomial d.f. with parameters n and p. Then, taking into account that $P(S_n \leq x) = 1$ if $na \leq x$, and $= 0$ if $x < nb$, we have

$$P(S \leq x) = \sum_{n \leq x/a} g_n + \sum_{x/a < n \leq x/b} g_n B(k(x,n); n, p). \tag{3.1.10}$$

The above sum contains a finite number of terms, and given g_n as well as the other parameters, computer calculations need not be lengthy. In Exercise 25, the reader is encouraged to create an Excel worksheet and carry out the calculations based on (3.1.10). □

3.1.2 The case where N has a Poisson distribution

The calculations in the previous section were provided for special distributions of X's. In this section, we will see that if N has a Poisson distribution, then we can consider a sufficiently large class of the distributions of separate terms (addends) X_j. The distribution of S in the case where N is Poisson is called *compound Poisson.*

To show how we can treat S in this case and for future references, we consider some general properties of the Poisson distribution. Actually, these properties have their intrinsic value.

Consider l independent Poisson r.v.'s, $N_1,...,N_l$ with respective means $\lambda_1,...,\lambda_l$, and set $N = N_1 + ... + N_l$. Clearly, N is a Poisson r.v. with parameter $\lambda = \lambda_1 + ... + \lambda_l$.

For example, a company receives each day claims of l types; the daily numbers of claims of each type are independent and have Poisson distributions. The first question we are interested in is what is the distribution of the number of claims of a particular type given the total number of claims? We will see that this distribution is binomial, and this may be considered a distinctive feature of the Poisson distribution. In general, the joint distribution of the N_i's given N is multinomial (for the description of this distribution, see Section 0.3.1.2).

Proposition 9 *Let $p_i = \lambda_i/\lambda$, $i = 1,...,l$. (So, $p_1 + ... + p_l = 1$.) Then for any $n = 1,2,...$, and any non-negative integers $m_1,...,m_l$ such that $m_1 + ... + m_l = n$,*

$$P(N_1 = m_1,...,N_l = m_l \,|\, N = n) = \frac{n!}{m_1!\cdots m_l!}p_1^{m_1}\cdots p_l^{m_l}. \tag{3.1.11}$$

In particular, for any $i = 1,..,l$, and $k = 0,...,n$.

$$P(N_i = k \,|\, N = n) = \binom{n}{k} p_i^k (1-p_i)^{n-k}. \tag{3.1.12}$$

A proof will be given in the end of this section.

Thus, since the multinomial distribution corresponds to the case of independent trials, given $N = n$, we may identify n claims arrived with independent trials where each trial independently of the others has l possible results. The probability that a particular claim is that of type i is $p_i = \lambda_i/\lambda$. We discuss this issue in a bit more detail in Section 3.2.1.

The next fact may be considered converse to the first.

Let now N be the random number of some objects, and suppose that N is a Poisson r.v. with a mean of λ. Each object, independently of the other objects and of the number of the objects, may belong to one of l types. For each object, the probability of belonging to type i is p_i; $p_1 + ... + p_l = 1$.

For example, each day, a company deals with N claims, the size of each claim equals either \$100 or \$150 with respective probabilities p_1 and p_2.

Coming back to the general wording, denote by $N_1, ..., N_l$ the numbers of objects of types $i = 1, ..., l$. Clearly, $N_1 + ... + N_l = N$.

Proposition 10 *Let $\lambda_i = p_i\lambda$, $i = 1, ..., l$. Then the r.v.'s $N_1, ..., N_l$ are independent and have the Poisson distribution with respective parameters $\lambda_1, ..., \lambda_l$.*

The fact that the r.v.'s are Poisson is not very surprising; the fact that they are independent is less expected. If N is not a Poisson r.v., this may be not true; so the property established may be also viewed as a special property of the Poisson distribution. The r.v.'s N_i are called sometimes *marked Poisson r.v.'s*: they count only "marked" objects.

Let us come back to arriving claims, denote by N the total number of claims, and by X_j the size of the jth claim. Let N be a Poisson r.v., $E\{N\} = \lambda$. Assume that the X's are independent, and each X takes on l values $x_1, ..., x_l$ with respective probabilities $p_1, ..., p_l$.

Consider the sum $S = X_1 + ... + X_N$, and denote by N_i the number of the r.v.'s X that took on the value x_i, $i = 1, ..., l$. Then $N_1 + ... + N_l = N$ and the total aggregate claim

$$S = x_1 N_1 + ... + x_l N_l, \tag{3.1.13}$$

which may essentially simplify calculations; especially if l is not large.

EXAMPLE 1. Let us come back to the arriving claims with a size of \$100 or \$150. Assume that the number of claims during a day is a Poisson r.v. N with a mean of 40, and on the average, 75% of claims equal \$100. If we had been solving the problem in a straightforward fashion, we would have introduced the r.v.'s

$$X_j = \begin{cases} 100 \text{ with probability } 3/4, \\ 150 \text{ with probability } 1/4, \end{cases}$$

and would have considered $S = X_1 + ... + X_N$, the sum where not only the separate terms are random, but the number of terms is random also.

As we saw, this is a complex object. However, in the case under consideration, we may just write

$$S = \$100 \cdot N_1 + \$150 \cdot N_2,$$

where N_1 and N_2 are the number of claims equal to \$100 and \$150, respectively. By Proposition 10, N_1 are N_2 are independent Poisson r.v.'s with parameters $\lambda_1 = 0.75 \cdot 40 = 30$ and $\lambda_2 = 0.25 \cdot 40 = 10$, respectively.

Thus, the sum of 40 r.v.'s on the average has been reduced to the sum of only two (!) r.v.'s. Such a sum is easily tractable. The first characteristics may be written immediately:

$$E\{S\} = 100E\{N_1\} + 150E\{N_2\} = 100\lambda_1 + 150\lambda_2 = 4,500;$$
$$Var\{S\} = 100^2 Var\{N_1\} + 150^2 Var\{N_2\} = 100^2\lambda_1 + 150^2\lambda_2 = 525,000.$$

As to the probabilities that S assumed particular values, they cannot be presented by a simple formula, but may be easily computed numerically since we deal with just two variables. □

Route 1 ⇒ page 221

EXAMPLE 3. Let us come back to Examples 2.2.2-1,2. We see that the scheme of these examples differ from the scheme above only in notation. 2,3

In Example 2.2.2-1, $N = \Upsilon_1$, where Υ_1 is the number of the r.v.'s Y_i that took on the value 1. By Proposition 10, Υ_1 has the Poisson distribution with parameter $\lambda_1 = p\lambda$.

In Example 2.2.2-2, $N = \Upsilon_1 + 2\Upsilon_2$, where Υ_1 is defined as above, and Υ_2 is the number of the r.v.'s Y_i that took on the value 2. By Proposition 10, Υ_1 and Υ_2 are independent and have the Poisson distributions with respective parameter $\lambda_1 = p_1\lambda$ and $\lambda_2 = p_2\lambda$. □

Let us turn to proofs. It suffices to restrict ourselves to the case $l = 2$. The difference between this case and the general case is not essential. 1,2,3

Proof of Proposition 9. As has been already mentioned, N has the Poisson distribution with parameter $\lambda = \lambda_1 + \lambda_2$. Set also $m_1 = k$. Then $m_2 = n - k$. Since N_1, N_2 are independent,

$$P(N_1 = k, N_2 = n-k \,|\, N = n) = \frac{P(N_1 = k, N_2 = n-k,\, N_1 + N_2 = n)}{P(N = n)}$$
$$= \frac{P(N_1 = k,\, N_2 = n-k)}{P(N = n)} = \frac{P(N_1 = k)P(N_2 = n-k)}{P(N = n)}$$
$$= \frac{\exp\{-\lambda_1\}\lambda_1^k}{k!} \cdot \frac{\exp\{-\lambda_2\}\lambda_2^{n-k}}{(n-k)!} \bigg/ \frac{\exp\{-(\lambda_1+\lambda_2)\}(\lambda_1+\lambda_2)^n}{n!}.$$

The exponential terms cancel out, and we get that

$$P(N_1 = k, N_2 = n-k \,|\, N = n) = \frac{n!}{k!(n-k)!} \cdot \frac{\lambda_1^k \lambda_2^{n-k}}{(\lambda_1+\lambda_2)^n} = \binom{n}{k}\left(\frac{\lambda_1}{\lambda}\right)^k \left(\frac{\lambda_2}{\lambda}\right)^{n-k}$$
$$= \binom{n}{k} p_1^k p_2^{n-k}.$$

Let us turn to (3.1.12). In the case $l = 2$, it follows from (3.1.11) just because $P(N_1 = k, N_2 = n-k \,|\, N = n) = P(N_1 = k \,|\, N = n)$.

Consider the general case $l \geq 2$. First, it suffices to consider (3.1.12) for $i = 1$. Secondly, we may set $\widetilde{N}_2 = N_2 + \ldots + N_l$. The latter r.v. is Poisson with parameter $\widetilde{\lambda}_2 = \lambda_2 + \ldots + \lambda_l$. Since $N = N_1 + \widetilde{N}_2$, we have reduced the problem to the case $l = 2$. ■

Proof of Proposition 10. Let again $l = 2$. For $n \neq 0$, given $N = n$, the n objects may be identified with n independent trials, each of which may be successful (the object is of the first type) with probability p_1. Hence, for any n, and any $k \leq n$,

$$P(N_1 = k, N_2 = n-k \,|\, N = n) = \binom{n}{k} p_1^k p_2^{n-k}. \tag{3.1.14}$$

For $n = 0$, (3.1.14) is also true because $P(N_1 = 0,\, N_2 = 0 \,|\, N = 0) = 1$; and by convention, $\frac{0!}{0!0!} p_1^0 p_2^0 = 1$ also. By the multiplication rule,

$$P(N_1 = k,\, N_2 = m) = P(N_1 = k,\, N_2 = m,\, N = m+k) = P(N_1 = k,\, N = m+k)$$
$$= P(N_1 = k \,|\, N = m+k)P(N = m+k)$$
$$= \frac{(k+m)!}{k!m!} p_1^k p_2^m \, e^{-\lambda} \frac{\lambda^{k+m}}{(k+m)!} = e^{-p_1\lambda}\frac{(p_1\lambda)^k}{k!} e^{-p_2\lambda}\frac{(p_2\lambda)^m}{m!} = e^{-\lambda_1}\frac{\lambda_1^k}{k!} e^{-\lambda_2}\frac{\lambda_2^m}{m!}.$$

This is a *product* of Poisson probabilities, so N_1, N_2 are Poisson and independent by Proposition **0**.2. ■

Route 1 ⇒ *page 223*

3.1.3 The m.g.f. method

We are coming back to the general situation. When applying the m.g.f. method, we use the basic formula (1.5), try to find the m.g.f. of S and attempt to determine the distribution that corresponds to the mentioned m.g.f.

For example, in the compound Poisson case, in accordance with Proposition 3,

$$M_S(z) = \exp\{\lambda(M_X(z) - 1)\},$$

which may help in many situations. In particular, see Section 3.2 where we are exploring the case of several homogeneous groups.

Let us consider one more distribution of N restricting ourselves just to one but a very illustrative example which may be called classical.

EXAMPLE 1. Let N have the negative binomial distribution with parameters (p, ν). In this case, the distribution of S is called *compound negative binomial.* Assume also that the X's have the exponential distribution with parameter a.

First, consider the first version of the negative binomial distribution when $N = 1, 2, ..$ and the probabilities are specified by (**0**.3.1.13). In this case (see Section **0**.4.3.2),

$$M_X(z) = \frac{1}{1 - z/a}, \quad M_N(z) = \left(\frac{pe^z}{1 - qe^z}\right)^\nu = \left(\frac{p}{e^{-z} - q}\right)^\nu, \tag{3.1.15}$$

where $q = 1 - p$. Then, by (1.5),

$$M_S(z) = \left(\frac{p}{(M_X(z))^{-1} - q}\right)^\nu = \left(\frac{p}{1 - z/a - q}\right)^\nu = \left(\frac{p}{p - z/a}\right)^\nu = \left(\frac{1}{1 - z/(pa)}\right)^\nu. \tag{3.1.16}$$

The expression in the parentheses: $\dfrac{1}{1 - z/(pa)}$ is the m.g.f. of the exponential distribution with the parameter pa.

We see that the m.g.f. (3.1.16) coincides with the m.g.f. of the Γ-distribution with parameters pa, ν; so we have arrived to a rather tractable distribution. For $\nu = 1$, we deal with the exponential r.v. with parameter pa.

Now, consider the second version of the negative binomial distribution with the same parameters: p and ν; see (**0**.3.1.15). This case presupposes that N may take on zero value with a probability of p.

So (see again Section **0**.4.3.2), $M_N(z) = \left(\dfrac{p}{1 - qe^z}\right)^\nu$, and hence,

$$M_S(z) = \left(\frac{p}{1 - qM_x(z)}\right)^\nu = \left(\frac{p}{1 - q(1 - z/a)^{-1}}\right)^\nu. \tag{3.1.17}$$

It is straightforward to verify that the expression in the parentheses above may be rewritten as $p+\dfrac{q}{1-z/(ap)}$. Then (3.1.17) may be rewritten as

$$M_S(z) = (M(z))^{\nu}, \tag{3.1.18}$$

where

$$M(z) = p + q\frac{1}{1-z/(ap)}. \tag{3.1.19}$$

The function $\frac{1}{1-z/(ap)}$ is the m.g.f. of the exponential distribution with parameter ap. Let a (non-random) variable $Y \equiv 0$. The m.g.f. $M_Y(z) \equiv 1$. We see that $M(z)$ is the mixture of two m.g.f.'s: $M_Y(z)$ and the exponential m.g.f. with parameter ap. Denote by $F(x)$ the d.f. corresponding to $M(z)$, by $F_a(x)$ the d.f. of the exponential distribution with parameter a, and by $E(x)$ the d.f. of Y. (That is, $E(x) = 1$ for $x \geq 0$ and $= 0$ for $x < 0$.) Then from (3.1.19) it follows that

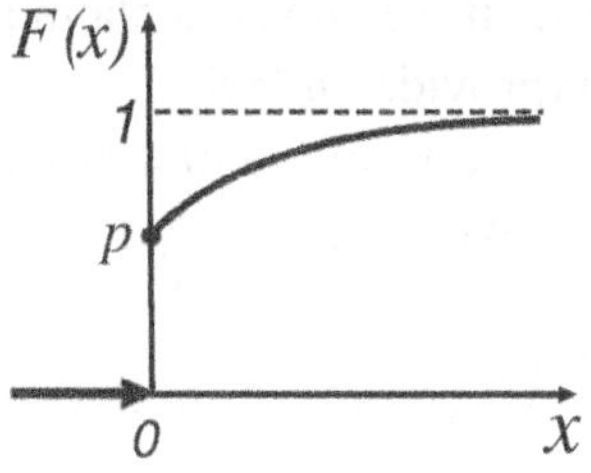

FIGURE 3.

$$F(x) = pE(x) + qF_{ap}(x), \tag{3.1.20}$$

the mixture of the corresponding distributions. The graph is given in Fig.3.

Let first $\nu = 1$. In this case, $M_S(z) = M(z)$, and the r.v. S has the distribution F. Let us denote by Z_a an exponential r.v. with parameter a. Then, in view of (3.1.20), we can write that the aggregate claim

$$S = \begin{cases} 0 & \text{with probability } p, \\ Z_{ap} & \text{with probability } q. \end{cases} \tag{3.1.21}$$

The case $S = 0$ corresponds to the case when $N = 0$, and this event indeed occurs with probability p. With probability q, the r.v. S is exponential. From (3.1.21) it follows that $E\{S\} = p \cdot 0 + q(1/pa) = q/(pa)$, which certainly coincides with what we would obtain using (1.1) or (3.1.7).

Let, say, $p = 0.01$, $a = 1$, and $\nu = 1$. Then $E\{N\} = q/p = 99$ and S is the sum of a large (random) number of exponential r.v.'s each of which has a mean of one. Nevertheless, S has a simple structure; namely, with probability 0.01 it is equal to zero, and with probability 0.99 it may be viewed as just one (!) exponential r.v. with a mean of $p^{-1} = 100$. □

If $\nu \neq 1$, the distribution of S is more complicated. However, if ν is an integer, in view of (3.1.18), S may be represented as the sum $Y_1 + ... + Y_{\nu}$, where Y's are i.i.d. r.v.'s with the above distribution F. The d.f. of this sum is the convolution $F^{*\nu}(x)$. If ν is small, calculating such a simple distribution is an easy task.

In conclusion, it is again worth noting that all constructions in Section 3.1 concern particular "good" distributions. In general, and sometimes even for these distributions, a simulation of the flow of claims—instead of direct calculations—may prove to be more efficient.

3.2 The case of several homogeneous groups 1,2,3

Now, we consider a portfolio consisting of l homogeneous groups of clients. First, we pay attention to the number of claims and the probability that a claim comes from a particular group. After that, we turn to the claim sizes and the aggregate payment.

3.2.1 The probability of coming from a particular group

Let N_i be the number of claims coming from the ith group, $i = 1,...,l$; $N = N_1 + ... + N_l$, the total number of claims. We assume N_i's to be independent and Poisson with respective parameters $\lambda_1,...,\lambda_l$.

Let $\lambda = \lambda_1 + ... + \lambda_l$, and $p_i = \lambda_i/\lambda$. In Section 3.1.2, we have proved that given N, the joint distribution of $N_1,...,N_l$ is multivariate with parameters $p_1,...,p_l$.

We mentioned also that from this it follows that the probability that a particular claim comes from the ith group is p_i. Let us clarify more carefully what it can mean.

The model is static and we count all claims arrived during a fixed period. Then we can assume that they arrive simultaneously; say, at the end of the period. Consequently, it does not matter in which order we consider these claims, and we may view this order as arbitrary.

For certainty, consider the first group. Assume we do know that N took on a particular value $n \neq 0$, and N_1 took on a value k. Then the probability that a particular claim (from the n claims arrived) came from the first group, is k/n. This is true if we choose a claim at random from n claims or consider a specific claim, say, the fifth (provided $n \geq 5$).

So, formally, if A is the event that a particular claim chosen came from the first group, then $P(A\,|\,N_1 = k, N = n) = k/n$. Then by the formula for total probability, for $n \geq 1$,

$$P(A\,|\,N=n) = \sum_{k=0}^{n} P(A\,|\,N_1 = k, N = n)P(N_1 = k\,|\,N = n) = \frac{1}{n}\sum_{k=0}^{n} kP(N_1 = k\,|\,N = n).$$

In Proposition 9 we have shown that the conditional distribution of N_1 given $N = n$ is binomial with parameters (n, p_1); see (3.1.12). The sum above is the mean of the distribution mentioned and, hence, equals np_1. Then

$$P(A\,|\,N=n) = \frac{1}{n}np_1 = p_1.$$

Without loss of generality, we can also postulate that if there are no claims, then the probability that a claim comes from the first group is also p_1. In other words, let us set by convention $P(A\,|\,N=0) = p_1$. This a formal (and non-significant) assumption. Then

$$P(A) = \sum_{n=0}^{\infty} P(A\,|\,N=n)P(N=n) = p_1\sum_{n=0}^{\infty} P(N=n) = p_1.$$

Certainly, the same concerns all other groups.

Consider also a couple of examples based on the same Proposition 9.

EXAMPLE 1. Let $l = 3$, $\lambda_1 = 2$ and $\lambda_2 = 3$, $\lambda_3 = 5$. Assume that the total number of claims, N, took on the value 6. What is the probability that $N_1 \leq 2$ and $N_2 \leq 3$? Since $p_1 = 0.2$, $p_2 = 0.3$, and $p_3 = 0.5$, in accordance with (3.1.11),

$$P(N_1 \leq 2, N_2 \leq 3\,|\,N=6) = \sum_{i=0}^{2}\sum_{j=0}^{3} \frac{6!}{i!j!(6-i-j)!}(0.2)^i(0.3)^j(0.5)^{6-i-j} = 0.83065,$$

which may be calculated even by hand.

EXAMPLE 2. Let $l = 3$, $\lambda_1 = 100$, $\lambda_2 = 200$, $\lambda_3 = 500$. Assume that the total number of claims N has assumed a particular value of 900. What is $P(N_1 \leq 120 \,|\, N = 900)$? According to (3.1.12), the conditional distribution of N_1 is binomial with parameters $p = p_1 = 1/8$ and $n = 900$. Let X be a r.v. with this distribution. Of course, computing $P(X \leq 120)$ exactly is meaningless but we can apply the normal approximation by using the central limit theorem. (In the case of the binomial distribution, it is called the *Moivre-Laplace theorem*). Since $E\{X\} = \frac{1}{8} \cdot 900 = 112.5$ and $Var\{X\} = \frac{1}{8} \cdot \frac{7}{8} \cdot 900 \approx 98.44$, we can write $P(X \leq 120) \approx \Phi((120 - 112.5)/\sqrt{98.44}) \approx \Phi(0.756) \approx 0.78$. □

3.2.2 A general scheme and reduction to one group

Now, together with N_i's and N, we adopt the following notation:

$X_{ij},\ i = 1, ..., l,\ j = 1, 2, ...,$	is the size of the jth claim coming from the ith group;
$F_i(x),\ i = 1, ..., l,$	is the *common* d.f. of X_{ij};
$M_i(z),\ i = 1, ..., l,$	is the *common* m.g.f. of X_{ij};
$S_i = \sum_{j=1}^{N_i} X_{ij},\ i = 1, ..., l,$	the total of all the claims in the ith group;
$S = \sum_{i=1}^{l} S_i,$	the total of all the claims.

It makes sense to emphasize that for each group i, the r.v.'s X_{ij} are *identically distributed.*

We assume that all r.v.'s $N_i, i = 1, ..., l$, and $X_{ij}, i = 1, ..., l,\ j = 1, 2, ...$ are mutually independent. Then the distribution of S is given by

$$F_S = F_{S_1} * ... * F_{S_l}, \tag{3.2.1}$$

the convolution of the distributions of the aggregate claims for separate groups. If we manage to find separate F_{S_i}, and if l is not large, then operation (3.2.1) may be numerically tractable.

In the case where all r.v.'s N_i have Poisson distributions, the above scheme may be simplified. For now, we will reason somewhat heuristically; at the end of the section we will provide a rigorous scheme.

Set again $\lambda_i = E\{N_i\}$, and $\lambda = \lambda_1 + ... + \lambda_l$. So, N is Poisson with parameter λ.

Let us consider the portfolio as a whole and denote by Y_k the size of the kth claim arriving, whichever group it comes from. Let B_{ik} be the event that the kth claim comes from the ith group. We know that for any k, the probability $P(B_{ik}) = p_i$, where $p_i = \lambda_i/\lambda$. Then the d.f. of Y_k does not depend on k and is equal to the function

$$F_Y(x) = P(Y_k \leq x) = \sum_{i=1}^{l} P(Y_k \leq x \,|\, B_{ik})\, P(B_{ik}) = \sum_{i=1}^{l} F_i(x)\, p_i. \tag{3.2.2}$$

So, the Y's are i.i.d. and the distribution of the Y's is a mixture of the distribution F_i.

Eventually, we may unify the groups in one homogeneous group writing

$$S = \sum_{k=0}^{N} Y_k.$$

EXAMPLE 1. Let $l = 3$, $\lambda_1 = 100$, $\lambda_2 = 200$, $\lambda_3 = 500$, and r.v.'s

$$X_{1j} = \begin{cases} 1 \text{ with probability } 1/3, \\ 2 \text{ with probability } 2/3, \end{cases} \quad X_{2j} = \begin{cases} 1 \text{ with probability } 1/4, \\ 2 \text{ with probability } 3/4, \end{cases}$$

$$X_{3j} = \begin{cases} 1 \text{ with probability } 1/6, \\ 2 \text{ with probability } 5/6. \end{cases}$$

Then $\lambda = 800$, $p_1 = \frac{\lambda_1}{\lambda} = \frac{1}{8}$, $p_2 = \frac{1}{4}$, and $p_3 = \frac{5}{8}$. The distribution of each Y_k is the mixture of the distributions above. Therefore, Y_k takes on values 1 and 2, and

$$P(Y_k = 1) = \frac{1}{8} \cdot \frac{1}{3} + \frac{1}{4} \cdot \frac{1}{4} + \frac{5}{8} \cdot \frac{1}{6} = \frac{5}{24}.$$

Hence,

$$Y_k = \begin{cases} 1 \text{ with probability } p_1 = 5/24, \\ 2 \text{ with probability } p_2 = 19/24. \end{cases}$$

Thus, $S = Y_1 + \ldots Y_N$ where N is a Poisson r.v. with parameter $\lambda = 800$. By (1.2) and (1.4),

$$E\{S\} = E\{Y_j\}E\{N\} = \frac{43}{24} \cdot 800 = \frac{4300}{3} = 1433.3\ldots,$$

$$Var\{S\} = E\{Y_j^2\}E\{N\} = \frac{81}{24} \cdot 800 = 2700.$$

The distribution of S is compound Poisson. Certainly, we cannot write this distribution in an explicit form but we can write its m.g.f. By (1.6),

$$M_S(z) = \exp\{800\,(M_Y(z) - 1)\} = \exp\left\{800\left(\frac{5}{24}e^z + \frac{19}{24}e^{2z} - 1\right)\right\}$$

$$= \exp\left\{\frac{500}{3}e^z + \frac{1900}{3}e^{2z} - 800\right\}.$$

In the case under consideration, *we can proceed further* using the construction of Section 3.1.2. In accordance with the results of this section,

$$S = N_1 + 2N_2,$$

where N_1 and N_2 are independent Poisson r.v.'s with parameters $p_1\lambda = \frac{5}{24} \cdot 800 = \frac{500}{3}$ and $p_2\lambda = \frac{19}{24} \cdot 800 = \frac{1900}{3}$, respectively. The program for calculating such a distribution may be straightforward.

EXAMPLE 2. Let $l = 2$, $\lambda_1 = 200$, $\lambda_2 = 300$. Assume the r.v.'s X_{1j} and X_{2j} are exponentially distributed with $E\{X_{1j}\} = 2$ and $E\{X_{2j}\} = 3$. Then $\lambda = 500$, $p_1 = \frac{\lambda_1}{\lambda} = 0.4$, $p_2 = 0.6$, and $S = Y_1 + \ldots Y_N$, where N is a Poisson r.v. with parameter $\lambda = 500$, and the distribution of Y's is the mixture of the exponential distributions above. The density

$$f_{Y_i}(x) = p_1 f_1(x) + p_2 f_2(x),$$

where f_1 and f_2 are the densities of r.v. X_{1j} and X_{2j}, respectively. Thus, for $x \geq 0$,

$$f_{Y_i}(x) = 0.4 \cdot \frac{1}{2} e^{-x/2} + 0.6 \cdot \frac{1}{3} e^{-x/3} = 0.2(e^{-x/2} + e^{-x/3}). \tag{3.2.3}$$

The collection of two groups above can be reduced to a homogeneous portfolio with the distribution of a particular claim given in (3.2.3). The m.g.f. of distribution (3.2.3) is the mixture of the m.g.f.'s of the above exponential distributions and equals

$$M(z) = 0.4 \cdot \frac{1}{1-2z} + 0.6 \cdot \frac{1}{1-3z} = \frac{1-2.4z}{(1-2z)(1-3z)}.$$

The m.g.f. of the r.v. S is

$$M_S(z) = \exp\{500(M(z)-1)\}.$$

Calculating $E\{S\}$and $Var\{S\}$ is easy and may be done either using (1.2) and (1.4), or directly as follows:

$$\begin{aligned} E\{S\} &= \lambda_1 E\{X_{1j}\} + \lambda_2 E\{X_{2j}\} = 200 \cdot 2 + 300 \cdot 3 = 1300, \\ Var\{S\} &= \lambda_1 E\{X_{1j}^2\} + \lambda_2 E\{X_{2j}^2\} = 200 \cdot 8 + 300 \cdot 18 = 7000. \end{aligned}$$

EXAMPLE 3 ([153, N7][4]). An insurance company pays claims at a Poisson rate of 2,000 per year. Claims are divided into three categories: "minor", "major", and "severe", with payment amounts of \$1,000, \$5,000, and \$10,000, respectively. The proportion of "minor" claims is 50%. The total expected claim payments per year is \$7,000,000. What is the proportion of "severe" claims?

Denote by λ_i, $i = 1,2,3$, the Poisson rates above. Let $\lambda = \lambda_1 + \lambda_2 + \lambda_3$ and $p_i = \lambda_i/\lambda$. The term "proportion" concerns the probabilities p_i. Let us choose \$1000 as a monetary unit. Then the total expected payment is $\lambda_1 + 5\lambda_2 + 10\lambda_3 = 7000$. Dividing it by $\lambda = 2000$, we have $p_1 + 5p_2 + 10p_3 = 3.5$. Together with $p_1 = 0.5$, and $p_1 + p_2 + p_3 = 1$, this leads to $p_3 = 0.1$. □

In conclusion, we provide a formal scheme which justifies the above construction more rigorously and gives more insight into it.

In view of (1.6), for all $i = 1, ..., l$,

$$M_{S_i}(z) = \exp\{\lambda_i(M_i(z)-1)\}. \tag{3.2.4}$$

It is worthwhile to emphasize that M_i is *not the m.g.f. of the ith claim* but the *common m.g.f. of separate claims from the ith group*. Since the S_i's are mutually independent, the m.g.f.

$$M_S(x) = \prod_{i=1}^{l} M_{S_i}(z) = \prod_{i=1}^{l} \exp\{\lambda_i(M_i(z)-1)\} = \exp\left\{\sum_{i=1}^{l} \lambda_i(M_i(z)-1)\right\}.$$

[4]Reprinted with permission of the Casualty Actuarial Society.

It is easy to verify that the last formula may be rewritten as

$$M_S(x) = \exp\left\{\lambda\left(\sum_{i=1}^{l}\frac{\lambda_i}{\lambda}(M_i(z)-1)\right)\right\} = \exp\{\lambda(M(z)-1)\}, \tag{3.2.5}$$

where

$$M(z) = \sum_{i=1}^{l} p_i M_i(z), \text{ and } p_i = \frac{\lambda_i}{\lambda}. \tag{3.2.6}$$

The m.g.f. in (3.2.6) is the mixture of the m.g.f.'s M_i with weights p_i, and hence the corresponding distribution F is the mixture of the distributions F_i. So we have arrived at the distribution in (3.2.2).

Function (3.2.5) is the m.g.f. of a compound Poisson distribution. More specifically, if W is a r.v. with the m.g.f. given by (3.2.5), then in accordance with (1.6), W may be represented (or viewed) as

$$W = Y_1 + ... + Y_N,$$

where $N, Y_1, Y_2, ...$ are mutually independent r.v.'s such that N is a Poisson r.v. with parameter λ and Y's have the common distribution F from (3.2.2) and the m.g.f. $M(z)$.

---

4 PREMIUMS AND SOLVENCY. NORMAL APPROXIMATION

4.1 Limit theorems

4.1.1 The Poisson case

In this section, we restrict ourselves to a homogeneous portfolio or more precisely, we assume the random addends X_i in (0.1) are independent and identically distributed (i.i.d.). We know that if the number of addends N is fixed and large, then the distribution of the sum may be closely approximated by a normal distribution. The question is whether it is true if N is random but takes large values with large probabilities. For example, is a normal approximation possible if N is a Poisson r.v. with a large expected value? The answer to this particular question is "yes" and reflected in Theorem 11 below. However, as we will see, in the general case, a similar result is true under certain conditions; though mild and natural. We begin with the Poisson case.

Let $N = N_\lambda$ be a r.v. having the Poisson distribution with parameter λ, and let

$$S_{(\lambda)} = X_1 + ... + X_{N_\lambda}. \tag{4.1.1}$$

We enclosed λ by parentheses in order to distinguish this notation from $S_n = X_1 + ... + X_n$. As already mentioned in Section 3.1.2, the distribution of $S_{(\lambda)}$ is called *compound Poisson*.

Set $m = E\{X_i\}$ and $\sigma^2 = Var\{X_i\}$. By (1.2) and (1.4),

$$E\{S_{(\lambda)}\} = m\lambda, \; Var\{S_{(\lambda)}\} = (\sigma^2 + m^2)\lambda.$$

Consider the normalized r.v.

$$S^*_{(\lambda)} = \frac{S_{(\lambda)} - E\{S_{(\lambda)}\}}{\sqrt{Var\{S_{(\lambda)}\}}} = \frac{S_{(\lambda)} - m\lambda}{\sqrt{(\sigma^2 + m^2)\lambda}}.$$

As we know, $E\{S^*_{(\lambda)}\} = 0$, $Var\{S^*_{(\lambda)}\} = 1$.

Theorem 11 *For any x,*

$$P(S^*_{(\lambda)} \le x) \to \Phi(x), \quad as\ \lambda \to \infty,$$

where, as usual, $\Phi(x)$ is the standard normal d.f.

Route 1 ⇒ page 233

4.1.2 The general case 2,3

This subsection is devoted to the general case and particular examples different from the Poisson scheme.

From now on, N_λ is an arbitrary integer-valued r.v. depending on a *varying general parameter* λ. We keep the same notation λ in order that the reader who decided to restrict her/himself to the compound Poisson case, could easily read the next section concerning premium evaluation.

Set $n_\lambda = E\{N_\lambda\}$ and $v_\lambda^2 = Var\{N_\lambda\}$. Most illustrations below will concern the following four examples.

1. *The case of bounded variance (BV):* in this case, we assume that n_λ is unlimitedly growing while $v_\lambda \le d$ for some number d independent of λ.

   Let, for instance, $N_\lambda = \lambda + K_\lambda$, where λ is an integer, and the "random component" K_λ is an integer valued r.v. such that $-s \le K_\lambda \le s$ for some fixed number s. Assume $E\{K_\lambda\} = 0$. Then $n_\lambda = \lambda$, $v_\lambda^2 = Var\{K_\lambda\} = E\{K_\lambda^2\} \le s^2$.

2. *The Poisson distribution (PD) case:* N_λ is Poisson with parameter λ. Then $n_\lambda = v_\lambda^2 = \lambda$.

3. *The negative binomial distribution (NBD) case:* N_λ has the negative binomial distribution (2.2.5) with fixed parameter p and (increasing) parameter $\nu = \lambda$. In this case, $n_\lambda = \lambda(1-p)/p$, $v_\lambda^2 = \lambda(1-p)/p^2$; see Section **0**.3.1.4).

4. *The geometric distribution (GD) case*: N_λ has the geometric distribution (2.2.7) with parameter $p = 1/\lambda$. In this case, $n_\lambda = (1-p)/p = \lambda - 1$ and $v_\lambda^2 = (1-p)/p^2 = \lambda(\lambda - 1)$; see Section **0**.3.1.3].

   It is worthwhile to note that the fact that the geometric distribution is a particular instance of the negative binomial distribution should not mislead us. In Case 3, the parameter p of the negative binomial distribution (see, e.g., (2.2.5)) is fixed and the parameter $\nu = \lambda$ varies while in Case 4, the varying parameter is p and $\nu = 1$.

Let us return to the general case. Let $S_{(\lambda)}$, m, and σ be defined as above. By (1.1) and (1.3),

$$E\{S_{(\lambda)}\} = mn_{\lambda}, \; Var\{S_{(\lambda)}\} = \sigma^2 n_{\lambda} + m^2 v_{\lambda}^2. \tag{4.1.2}$$

For brevity, let us set $d_{\lambda}^2 = \sigma^2 n_{\lambda} + m^2 v_{\lambda}^2$ and consider the normalized r.v.

$$S_{(\lambda)}^* = \frac{S_{(\lambda)} - E\{S_{(\lambda)}\}}{\sqrt{Var\{S_{(\lambda)}\}}} = \frac{S_{(\lambda)} - mn_{\lambda}}{d_{\lambda}}.$$

Our goal is to establish conditions of asymptotic normality of $S_{(\lambda)}^*$. As we will see, to accomplish this, we should also normalize the r.v. N_{λ} setting

$$N_{\lambda}^* = \frac{N_{\lambda} - n_{\lambda}}{v_{\lambda}}.$$

Clearly, $E\{N_{\lambda}^*\} = 0$ and $Var\{N_{\lambda}^*\} = 1$.

To make our exposition more explicit, we adopt the following notation. We say that a sequence of r.v.'s ξ_{λ} converges to a r.v. ξ *in distribution* as $\lambda \to \infty$, and write it as

$$\xi_{\lambda} \xrightarrow{d} \xi, \tag{4.1.3}$$

if the distributions of the r.v.'s ξ_{λ} converge to the distribution of ξ. In other words, the d.f.'s $F_{\xi_{\lambda}}(x) \to F_{\xi}(x)$ as $\lambda \to \infty$. (The reader may look up the notion of convergence in distribution in Section **0**.5 for more detail.)

Under notation (4.1.3), the *classical* Central Limit Theorem (CLT) may be stated as follows. For the normalized sum

$$S_n^* = \frac{S_n - mn}{\sigma\sqrt{n}},$$

where n is a *non-random* integer

$$S_n^* \xrightarrow{d} Z \quad \text{as} \quad n \to \infty, \tag{4.1.4}$$

where Z is a standard normal r.v. (See, e.g., Section **0**.6.2.)

The theorem below may be interpreted in the following way. For S_{λ} to be asymptotically normal, not only the r.v.'s S_n should be asymptotically normal for large n, but the same property should hold for N_{λ} for large λ.

For instance, this is true if N_{λ} is a Poisson r.v. since N_{λ} may be presented as a sum of independent Poisson r.v.'s (see also a remark in p.164 and Exercise **2**.43).

In the theorem stated below, all limits are those as $\lambda \to \infty$. Let, as above, Z denote a standard normal r.v.

Theorem 12 *Suppose*

$$n_\lambda \to \infty, \tag{4.1.5}$$

and either

$$\frac{v_\lambda}{\sqrt{n_\lambda}} \to 0, \tag{4.1.6}$$

or

$$\frac{v_\lambda}{\sqrt{n_\lambda}} \to c > 0 \tag{4.1.7}$$

and in addition,

$$N^*_\lambda \xrightarrow{d} Z. \tag{4.1.8}$$

Then

$$S^*_{(\lambda)} \xrightarrow{d} Z \quad as \quad \lambda \to \infty. \tag{4.1.9}$$

A proof of Theorem 12 is given in Section 4.4. Let us consider examples and discuss conditions (4.1.6)-(4.1.8).

EXAMPLE 1. In the BV case, $v_\lambda \le d$ and (4.1.6) holds if (4.1.5) is true. □

Certainly, v_λ does not have to be bounded for (4.1.6) to be true: it suffices to assume

$$v^2_\lambda = o(n_\lambda). \tag{4.1.10}$$

(for the definition of $o(\cdot)$, see Appendix, Section 4.1).

Thus, we can state

Corollary 13 *Normal convergence (4.1.9) is true if (4.1.5) and (4.1.10) are satisfied.*

EXAMPLE 2. Let

$$N_\lambda = \lambda + K_\lambda,$$

where λ is an integer, K_λ takes on integer values, and

$$-\lambda^{1/4} \le K_\lambda \le \lambda^{1/4}.$$

Then $n_\lambda \ge \lambda - \lambda^{1/4}$, $v^2_\lambda \le E\{K^2_\lambda\} \le \sqrt{\lambda}$, and $(v^2_\lambda / n_\lambda) \le [\sqrt{\lambda}/(\lambda - \lambda^{1/4})] \to 0$. □

However, the case (4.1.10) should not be considered the most important since in many applications the variance v^2_λ has *the same order as the mean* n_λ.

The classical example is, of course, the Poisson distribution for which $v^2_\lambda = n_\lambda = \lambda$. So, $(v_\lambda/\sqrt{n_\lambda}) \equiv 1$ and c in (4.1.7) equals 1. To apply Theorem 12 in this case, we need to verify condition (4.1.8).

The validity of (4.1.8) has been shown previously at the end of Section **2**.2.1.2 (p.164) as well as in Exercise **2**-43 (with a detailed advice). Thus, we have come to

Corollary 14 *In the PD case, (4.1.9) is true (which, certainly, is equivalent to Theorem 11).*

Next, we consider the NBD case; i.e., Case 3 above. Now $(v^2_\lambda / n_\lambda) = 1/p$, so $c = 1/\sqrt{p}$, and we should again verify (4.1.8). For simplicity, we restrict ourselves to the case when λ

is an integer. In view of (2.2.4), see also Exercise 8c, the r.v. N_λ in this case may be represented as the sum $Y_1 + ... + Y_\lambda$ where Y's are i.i.d. r.v.'s having the geometric distribution with parameter p. Hence, by the CLT, $N^*_\lambda \xrightarrow{d} Z$ and (4.1.8) is true.

▶ To verify (4.1.8) for the instance when λ is an arbitrary real number, we can write the m.g.f. (2.2.4) as

$$M_{N_\lambda}(z) = \left(\frac{p}{1-qe^z}\right)^\lambda = \left(\frac{p}{1-qe^z}\right)^{[\lambda]} \left(\frac{p}{1-qe^z}\right)^{\lambda-[\lambda]},$$

where $[\lambda]$ is the integer part of λ. Thus, we can represent N_λ as the sum $Y_1 + ... + Y_{[\lambda]} + U_\lambda$, where the Y's are the same as above and U_λ (the remainder, so to speak) is a r.v. independent of the Y's and having the m.g.f. $(p/(1-qe^z))^{\lambda-[\lambda]}$.

The sum $Y_1 + ... + Y_{[\lambda]}$ is asymptotically normal by the CLT, while U_λ is negligible with respect to the total sum. Indeed, $E\{Y_1 + ... + Y_{[\lambda]}\} = n_{[\lambda]} = [\lambda](1-p)/p$ and $E\{U_\lambda\} = [(\lambda - [\lambda])(1-p)/p] \leq (1-p)/p$ (see Section **0**.3.1.4). Then, $(n_{[\lambda]}/E\{U_\lambda\}) \to \infty$. The rest of the proof is carried out in accordance with the scheme of Exercises **2**-43,44. ◀

Thus, we have come to

Corollary 15 *In the NBD case, (4.1.9) is true.*

The reader should now have a grasp of the requirements of Theorem 12. The classical CLT implies that the sum S_n with a non-random large number of addends is asymptotically normal. Theorem 12 establishes conditions under which this is true for a random number of addends.

If the variance $v_\lambda^2 = o(n_\lambda)$, the condition (4.1.6) obviously holds.

If we deal with the case (4.1.7), then the normalized sum $S^*_{(\lambda)}$ is asymptotically normal if N_λ is asymptotically normal. The latter, for example, is true if we can represent N_λ as a sum of many i.i.d. addends plus perhaps a negligible remainder. As we saw, it is true, for example, in the PD and NBD cases. The instance where N_λ has a binomial distribution is explored in Exercise 37.

Consider now an example more sophisticated than the binomial case.

EXAMPLE 3. Let N_λ have the Poisson-Poisson distribution from Example 2.2.2-3 with λ_1 from this example equal to λ. For the other parameter, we will keep the same notation λ_2. Thus,

$$N_\lambda = Y_1 + ... + Y_K, \tag{4.1.11}$$

where K is a Poisson r.v. with parameter λ, and Y's are independent r.v.'s having the Poisson distribution with parameter λ_2. First of all, as has been computed in Example 2.2.2-3, $n_\lambda = E\{N_\lambda\} = \lambda\lambda_2$ and $v_\lambda^2 = Var\{N_\lambda\} = \lambda(\lambda_2 + \lambda_2^2)$. Hence, (4.1.7) is true with $c = \sqrt{1+\lambda_2}$.

To verify (4.1.8), we should just observe that scheme (4.1.11) is a particular case of the general scheme (0.1). The role of N is played by K and the role of S by N_λ. Since K is a Poisson r.v., (4.1.8) follows in this case from Corollary 14. Thus, (4.1.9) is true. □

Now we consider a situation where the conditions of Theorem 12 are not satisfied and its conclusion is false.

EXAMPLE 4. Consider the GD case. We have computed that in this case, $n_\lambda = \lambda - 1$, $v_\lambda^2 = \lambda(\lambda - 1)$, so $(v_\lambda^2/n_\lambda) = \lambda \to \infty$ and hence (4.1.7) is not true. Note that the standard deviation $v_\lambda = \sqrt{\lambda(\lambda-1)}$ and has the same order as the mean value $n_\lambda = \lambda - 1$. This means, in particular, that although λ is growing, N_λ can assume small values with a non-negligible probability and, hence, with a non-negligible probability the sum S is not large. Consequently, we should not expect S to be asymptotically normal.

As a matter of fact, the following interesting result is true.

Theorem 16 *(Rényi). If N_λ has the geometric distribution with parameter $p = 1/\lambda$, then*

$$\frac{1}{m\lambda} S_{(\lambda)} \xrightarrow{d} \xi \quad as \quad \lambda \to \infty,$$

where ξ is a standard exponential r.v.; that is, ξ is exponential and $E\{\xi\} = 1$.

We omit the proof of this theorem which may be carried out with use of m.g.f.'s (see, e.g., [12], [111]). □

4.2 Estimation of premiums

1,2,3

Formally, our model does not involve premiums since N counts claims coming from the portfolio as a whole rather than from clients who pay premiums. We can, however, talk about an amount of money $c = c_\lambda$ sufficient to cover claims with a given probability β; i.e., the amount c for which $P(S_{(\lambda)} \leq c) \geq \beta$. We may view c as an aggregate premium and define the loading coefficient θ by the relation $c = (1+\theta)E\{S_{(\lambda)}\}$. If normal approximation works in our situation, then we can apply it to the determination of the minimal acceptable θ, following the scheme of Section **2**.3.1.1. In particular, the counterpart of (**2**.3.1.5) will be

$$\theta \approx \frac{q_{\beta s}\sqrt{Var\{S_{(\lambda)}\}}}{E\{S_{(\lambda)}\}}. \tag{4.2.1}$$

The derivation is absolutely similar to what we did in Section **2**.3.1.1.

In view of (1.2)-(1.4), in the case when N is Poisson with parameter λ, the last formula may be rewritten as

$$\theta \approx \frac{q_{\beta s}\sqrt{m_2\lambda}}{m\lambda},$$

where $m_2 = E\{X_j^2\}$, the second moment of the X's.

Thus, in the compound Poisson case,

$$\theta \approx \frac{q_{\beta s}\sqrt{m_2}}{m\sqrt{\lambda}} = \frac{q_{\beta s}\sqrt{m^2+\sigma^2}}{m\sqrt{\lambda}} = \frac{q_{\beta s}}{\sqrt{\lambda}}\sqrt{1+k^2}, \tag{4.2.2}$$

and $k = \sigma/m$, the coefficient of variation of r.v.'s X. All three representations in (4.2.2) may be useful.

EXAMPLE 1. Consider the case when X's are lognormal and N is Poisson with parameter λ. We will see that in this case, to estimate θ, we should know λ, the security level β and $Var\{\ln(X_i)\}$.

Log-normality means that each $X_j = \exp\{a + b\eta_{j0}\}$, where a and b are parameters and η_{j0}'s are independent standard normal r.v.'s [see also (**2**.1.1.13)]. Because $X_j = e^a \exp\{b\eta_{j0}\}$, if we multiply all X's by e^{-a}, then we will switch to the simpler r.v.'s $\exp\{b\eta_{j0}\}$ but k and hence the representation (4.2.2) will not change. So, without loss of generality we can set $a = 0$. In this case, $m = e^{b^2/2}$, $m_2 = e^{2b^2}$ [see (**2**.1.1.14), (**2**.1.1.15)] and hence $\dfrac{\sqrt{m_2}}{m} = e^{b^2/2}$.

Now, let us observe that $\ln(X_j) = b\eta_{j0}$. Since $Var\{\eta_{j0}\} = 1$, we have $b^2 = Var\{\ln(X_i)\}$. So, the coefficient $\sqrt{m_2}/m$ depends only on $Var\{\ln(X_i)\}$.

Thus,

$$\theta \approx \frac{q_{\beta s}}{\sqrt{\lambda}} e^{b^2/2}. \tag{4.2.3}$$

For example, if $\beta = 0.9$, $b^2 = 0.2$, and $\lambda = 400$, then the loading coefficient $\theta \approx \dfrac{1.282}{20} e^{0.1} \approx 0.071$.

If we want to estimate the premium c itself, then a should be involved. The mean aggregate claim is given by $E\{S_{(\lambda)}\} = \lambda m = \lambda e^{a+b^2/2}$ and

$$c = (1+\theta)\lambda e^{a+b^2/2} = e^{a+b^2/2}\left(\lambda + q_{\beta s}\sqrt{\lambda}\sqrt{1+k^2}\right).$$

It is worth noting that the second term is much smaller than the first for large λ. □

Route 1 ⇒ page 243

2,3

EXAMPLE 2. Let X's be uniformly distributed on $[0, a]$, and let N be negative binomial with parameters $p = 1/2$ and $\nu = \lambda$. For the same reason as in Example 1, the coefficient θ does not depend on a, so we can set $a = 1$. (If $a \neq 1$, we may divide all X's by a, which will not change θ.) In this case, $E\{X\} = 1/2$ and $Var\{X^2\} = 1/12$. In view of (4.1.2),

$$E\{S_{(\lambda)}\} = \frac{n_\lambda}{2} = \frac{\lambda(1-p)}{2p},$$

$$Var\{S_{(\lambda)}\} = (1/12)n_\lambda + (1/4)v_\lambda^2 = \frac{\lambda(1-p)}{4p}\left(\frac{1}{3} + \frac{1}{p}\right).$$

Now it is easy to calculate that

$$\theta \approx \frac{q_{\beta s}}{\sqrt{\lambda}}\sqrt{\frac{3+p}{3(1-p)}}.$$

For $\beta = 0.8$, we would have $\theta \approx 1.286/\sqrt{\lambda}$. □

4.3 The accuracy of normal approximation

We restrict ourselves to the compound Poisson distribution with N being a Poisson r.v. with parameter λ.

It is remarkable that the rate of convergence to the normal distribution in this case is practically the same as in the classical case of a fixed number of addends which we considered in Theorem **2**.3.2-6.

Let

$$L_\lambda = \frac{1}{\sqrt{\lambda}} \frac{E\{|X_i|^3\}}{(E\{|X_i|^2\})^{3/2}}. \quad (4.3.1)$$

(The r.-h.s. of (4.3.1) does not depend on i since the X_i's are identically distributed.) The only difference between (4.3.1) and the Lyapunov fraction in (**2**.3.2.1) from Chapter 2 is that (4.3.1) involves non-central moments. Nevertheless, as in the case of Chapter 2, L_λ is dimensionless and not sensitive to the change of scale (see details in Chapter 2).

Theorem 17 *Let $F^*_\lambda(x)$ be the d.f. of the r.v. $S^*_{(\lambda)}$. Then there exists an absolute constant C_1 such that for any x,*

$$|F^*_\lambda(x) - \Phi(x)| \le C_1 L_\lambda. \quad (4.3.2)$$

With regard to the constant C_1, the last results show that [5]

$$C_1 \le 0.792. \quad (4.3.3)$$

Similar to the exposition in Section **2**.3.2, we can write a bound for the loading coefficient θ, that is, the counterpart of (**2**.3.2.7):

$$\theta \ge \frac{q_{\beta+\Delta_\lambda,s}\sqrt{Var\{S_{(\lambda)}\}}}{E\{S_{(\lambda)}\}}, \quad (4.3.4)$$

where Δ_λ is the r.-h.s. of (4.3.2). Let $m_2 = E\{X^2\}$. Since in our case $E\{S_{(\lambda)}\} = m\lambda$ and $Var\{S_{(\lambda)}\} = m_2\lambda$, inequality (4.3.4) may be rewritten as

$$\theta \ge \frac{q_{\beta+\Delta_\lambda,s}\sqrt{m_2}}{m\sqrt{\lambda}}. \quad (4.3.5)$$

EXAMPLE 1. Consider the situation of Example 4.2-1. Again we can set $a = 0$ because the ratio $\sqrt{m_2}/m$ does not depend on a. As we have computed before, $(\sqrt{m_2}/m) = e^{b^2/2}$.

Now, $E\{X_j^3\} = E\{e^{3b\eta_{j0}}\} = M_{\eta_{j0}}(3b) = e^{9b^2/2}$ (see also Section **2**.1.1.3) and because $m_2 = e^{2b^2}$, we have

$$L_\lambda = \frac{1}{\lambda} \frac{e^{9b^2/2}}{\left(e^{2b^2}\right)^{3/2}} = e^{3b^2/2}.$$

If as in Example 4.2-1, $b^2 = Var\{\ln(X_j)\} = 0.2$, then $\Delta_\lambda \approx 0.792 \cdot 1.350\frac{1}{\sqrt{\lambda}} \approx 1.069\frac{1}{\sqrt{\lambda}}$.

[5] The constant was obtained in [92] and also in [77]. Without computing the constant, the bound (4.3.2) was obtained earlier in [118]. A detailed discussion may be found in [12].

Set $\beta = 0.9$, $\lambda = 400$. Then $\Delta_\lambda \approx 0.053$, and $q_{\beta+\Delta_\lambda,s} \approx 1.675$, which results in the bound

$$\theta \geq \frac{1.676e^{0.1}}{20} \approx 0.093.$$

The rough approximation in Example 4.2-1 was equal to 0.071. The reader is recommended to look up the important remark in Example **2.** 3.2-1. □

4.4 Proof of Theorem 12

We prove the theorem proceeding from conditions (4.1.5), (4.1.7), and (4.1.8). The case (4.1.6) is easier, and to run a proof for this case, it suffices to set in the proof below $c = 0$.

First, we will show that under conditions of the theorem,

$$(N_\lambda/d_\lambda^2) \xrightarrow{P} k = 1/(\sigma^2 + m^2c^2). \tag{4.4.1}$$

(For the definition of convergence in probability, see Section **0.**5.) Indeed,

$$\frac{N_\lambda}{d_\lambda^2} = \frac{v_\lambda}{d_\lambda^2}\frac{N_\lambda - n_\lambda}{v_\lambda} + \frac{n_\lambda}{d_\lambda^2} = \frac{v_\lambda}{d_\lambda^2}N_\lambda^* + \frac{n_\lambda}{d_\lambda^2}. \tag{4.4.2}$$

By conditions (4.1.5)-(4.1.7),

$$\frac{v_\lambda}{d_\lambda^2} = \frac{v_\lambda}{\sigma^2 n_\lambda + m^2 v_\lambda^2} = \frac{1}{\sqrt{n_\lambda}} \cdot \frac{(v_\lambda/\sqrt{n_\lambda})}{\sigma^2 + m^2(v_\lambda^2/n_\lambda)} \to 0 \cdot \frac{c}{\sigma^2 + m^2c^2} = 0, \tag{4.4.3}$$

$$\frac{n_\lambda}{d_\lambda^2} = \frac{n_\lambda}{\sigma^2 n_\lambda + m^2 v_\lambda^2} = \frac{1}{\sigma^2 + m^2(v_\lambda^2/n_\lambda)} \to \frac{1}{\sigma^2 + m^2c^2} = k. \tag{4.4.4}$$

Thus, since $E\{N_\lambda^*\} = 0$ and $Var\{N_\lambda^*\} = 1$, relation (4.4.3) implies that $(v_\lambda/d_\lambda^2)N_\lambda^* \xrightarrow{P} 0$. Together with (4.4.4) and (4.4.2), this implies (4.4.1).

Next, we write

$$S_{(\lambda)}^* = \frac{1}{d_\lambda}(S_{(\lambda)} - mn_\lambda) = \frac{1}{d_\lambda}\left(\sum_{i=1}^{N_\lambda} X_i - mn_\lambda\right) = \frac{1}{d_\lambda}\left(\sum_{i=1}^{N_\lambda}(X_i - m) + m(N_\lambda - n_\lambda)\right)$$

$$= \frac{1}{d_\lambda}\sum_{i=1}^{N_\lambda}(X_i - m) + \frac{m}{d_\lambda}(N_\lambda - n_\lambda). \tag{4.4.5}$$

In view of (4.4.1), asymptotically N_λ is growing as kd_λ^2. The idea of the rest of the proof is to replace N_λ in the first term of (4.4.5) by the non-random number $t_\lambda = [kd_\lambda^2]$, the integer part of kd_λ^2. (We cannot replace N_λ exactly by kd_λ^2 since the sum limit is an integer.) The second term in (4.4.5) will remain unchanged. This replacement leads to the r.v.

$$Y_{\lambda 1} = \frac{1}{d_\lambda}\sum_{i=1}^{t_\lambda}(X_i - m) + \frac{m}{d_\lambda}(N_\lambda - n_\lambda),$$

and the error arising because of such a replacement is the r.v.

$$Y_{\lambda 2} = \frac{1}{d_\lambda}\left(\sum_{i=1}^{N_\lambda}(X_i - m) - \sum_{i=1}^{t_\lambda}(X_i - m)\right).$$

We will prove the theorem if we show that $Y_{\lambda 1}$ is asymptotically standard normal and the error term $Y_{\lambda 2} \xrightarrow{P} 0$. We have

$$\begin{aligned} Y_{\lambda 1} &= \frac{\sigma\sqrt{t_\lambda}}{d_\lambda} \cdot \frac{1}{\sigma\sqrt{t_\lambda}} \sum_{i=1}^{t_\lambda}(X_i - m) + \frac{m v_\lambda}{d_\lambda} \cdot \frac{N_\lambda - n_\lambda}{v_\lambda} \\ &= \frac{\sigma\sqrt{t_\lambda}}{d_\lambda} \cdot S^*_{t_\lambda} + \frac{m v_\lambda}{d_\lambda} \cdot N^*_\lambda, \end{aligned} \tag{4.4.6}$$

where as usual the symbol S^*_n denotes the normalized sum $(S_n - mn)/\sigma\sqrt{n}$.

The main point here is that, since $S^*_{t_\lambda}$ does not involve N_λ, the r.v.'s $S^*_{t_\lambda}$ and N^*_λ are independent. The r.v. $S^*_{t_\lambda}$ is asymptotically standard normal in view of (4.1.4), while $(m v_\lambda / d_\lambda) N^*_\lambda$ is asymptotically normal due to condition (4.1.8), as we will show. Then the linear combination of the terms in (4.4.6) will be asymptotically normal too.

Consider all of this in greater detail. In view of (4.1.5), $d_\lambda \to \infty$. Using the symbolism $a_\lambda \sim b_\lambda$ if $(a_\lambda / b_\lambda) \to 1$, we have

$$\frac{\sigma\sqrt{t_\lambda}}{d_\lambda} = \frac{\sigma\sqrt{[k d^2_\lambda]}}{d_\lambda} \sim \frac{\sigma\sqrt{k} d_\lambda}{d_\lambda} = \frac{\sigma}{\sqrt{\sigma^2 + m^2 c^2}}. \tag{4.4.7}$$

On the other hand,

$$\frac{m v_\lambda}{d_\lambda} \cdot N^*_\lambda = \frac{m v_\lambda}{\sqrt{\sigma^2 n_\lambda + m^2 v^2_\lambda}} \cdot N^*_\lambda = \frac{m}{\sqrt{\sigma^2 + m^2 (v^2_\lambda / n_\lambda)}} \cdot \frac{v_\lambda}{\sqrt{n_\lambda}} N^*_\lambda. \tag{4.4.8}$$

In view of (4.4.7) and (4.1.4),

$$\frac{\sigma\sqrt{t_\lambda}}{d_\lambda} \cdot S^*_{t_\lambda} \xrightarrow{d} \frac{\sigma}{\sqrt{\sigma^2 + m^2 c^2}} Z_1, \tag{4.4.9}$$

where Z_1 is a standard normal r.v.

In view of (4.4.8) and conditions (4.1.7) and (4.1.8),

$$\frac{m v_\lambda}{d_\lambda} \cdot N^*_\lambda \xrightarrow{d} \frac{m}{\sqrt{\sigma^2 + m^2 c^2}} c Z_2, \tag{4.4.10}$$

where Z_2 is a standard normal r.v. *independent* of Z_1.

Now, let us consider the sum of the r.-h. sides of (4.4.9) and (4.4.10). As is easy to compute, the variance of the sum mentioned is equal to one. So, the sum is standard normal and hence $Y_{\lambda 1} \xrightarrow{P} Z$.

Next, consider $Y_{\lambda 2}$. Since $E\{Y_{\lambda 2}\} = 0$ [check on your own using (1.1)], to show that $Y_{\lambda 2} \xrightarrow{P} 0$, it suffices to prove that $Var\{Y_{\lambda 2}\} \to 0$. Because $E\{Y_{\lambda 2}\} = 0$,

$$Var\{Y_{\lambda 2}\} = E\{Y^2_{\lambda 2}\} = \sum_{n=0}^{\infty} E\{Y^2_{\lambda 2} \mid N_\lambda = n\} P(N_\lambda = n).$$

If $n \geq t_\lambda$,

$$E\{Y_{\lambda 2}^2 \,|\, N_\lambda = n\} = \frac{1}{d_\lambda^2} E\left\{\left(\sum_{i=t_\lambda}^{n} (X_i - m)\right)^2\right\} = \frac{1}{d_\lambda^2} Var\left\{\sum_{i=t_\lambda}^{n} (X_i - m)\right\} = \frac{1}{d_\lambda^2}\sigma^2 (n - t_\lambda) \geq 0.$$

If $n < t_\lambda$,

$$E\{Y_{\lambda 2}^2 \,|\, N_\lambda = n\} = \frac{1}{d_\lambda^2} E\left\{\left(\sum_{i=n}^{t_\lambda} (X_i - m)\right)^2\right\} = \frac{1}{d_\lambda^2} Var\left\{\sum_{i=n}^{t_\lambda} (X_i - m)\right\} = \frac{1}{d_\lambda^2}\sigma^2 (t_\lambda - n) \geq 0.$$

Thus, for any n,

$$E\{Y_{\lambda 2}^2 \,|\, N_\lambda = n\} = \frac{1}{d_\lambda^2}\sigma^2 |n - t_\lambda|.$$

Hence,

$$Var\{Y_{\lambda 2}\} = \frac{\sigma^2}{d_\lambda^2}\sum_{n=0}^{\infty} |n - t_\lambda| P(N_\lambda = n) = \frac{\sigma^2}{d_\lambda^2} E\{|N_\lambda - t_\lambda|\} \leq \frac{\sigma^2}{d_\lambda^2}\sqrt{E\{(N_\lambda - t_\lambda)^2\}}$$
$$= \sigma^2 \left(E\{(N_\lambda - t_\lambda)^2 / d_\lambda^4\right)^{1/2}. \qquad (4.4.11)$$

(We used the fact that for any r.v. X, it is true that $(E\{|X|\})^2 \leq E\{X^2\}$ (for example, because $0 \leq Var\{|X|\} = E\{X^2\} - (E\{|X|\})^2$.)

Next, we will use the identity $E\{(X-a)^2\} = Var\{X\} + (a - E\{X\})^2$. (If the reader does not remember this identity, she/he is invited to prove it by direct calculations.) Thus,

$$E\{(N_\lambda - t_\lambda)^2\} = Var\{N_\lambda\} + (n_\lambda - t_\lambda)^2 = v_\lambda^2 + (n_\lambda - t_\lambda)^2.$$

Hence,

$$\frac{E\{(N_\lambda - t_\lambda)^2\}}{d_\lambda^4} = \frac{v_\lambda^2 + (n_\lambda - [kd_\lambda^2])^2}{d_\lambda^4} = \left(\frac{v_\lambda}{d_\lambda^2}\right)^2 + \left(\frac{n_\lambda}{d_\lambda^2} - \frac{[kd_\lambda^2]}{d_\lambda^2}\right)^2. \qquad (4.4.12)$$

Since $\left([kd_\lambda^2]/d_\lambda^2\right) \to k$, in view of (4.4.4) and (4.4.3), the quantity (4.4.12) vanishes and hence $Var\{Y_{\lambda 2}\} \to 0$ as $\lambda \to \infty$. ∎

---

5 EXERCISES

Section 1

1. Show that (1.1) and (1.3) follow from (1.5). (*Advice:* Look up (**0**.4.4.5).)
2. If $S_n = X_1 + ... + X_n$, where the X's are i.i.d., then as we know (see Section **0**.4.1), $M_{S_n}(z) = (M_X(z))^n$, where $M_X(z)$ is the m.g.f. of X_i. Show that Proposition 3 includes this case. (*Hint:* Write the m.g.f. of the non-random variable $N \equiv n$.)

Section 2

3. Making use of the Taylor expansion for e^x [see Appendix, (4.2.5)], show that the sum of probabilities in (2.1.1) is indeed one. Prove (2.1.2).

4. A husband and wife each can purchase insurance for which the payoff for the first claim is much higher than the others in the same year. In the husband's and the wife's cases considered separately, the numbers of claims are independent Poisson r.v.'s with the same λ. The couple has also an option to buy a joint insurance where the number of claims with priority is two. Find the distribution of the total number of the claims with priority covered for the case of the two separate insurances and for the case of the joint insurance. If the premium for the joint insurance is double the premium of the individual policy, what decision should the couple make? Show that the answer to the last question is the same for an arbitrary distribution of the number of claims (not only Poisson).

5. Provide an Excel worksheet as in Fig.1 from Section 2.1. Consider various n and p; e.g., let $\lambda = 2$ and consider $n = 10, 20, 40, 100$. Estimate the accuracy of the Poisson approximation in each case and analyze the results.

6. A portfolio consists of three homogeneous groups. The number of clients in each group, n_i, and the probabilities of loss events for each group, r_i, $i = 1,2,3$, are given in the following table:

	group 1	group 2	group 3
n_i	231	124	347
r_i	0.01	0.05	0.03

   (a) Find $\overline{p}_n$ and the parameter λ of the approximating Poisson distribution.
   (b) For the number of claims N, estimate $P(N = 15)$ and $P(N \leq 15)$. (For calculations, it makes sense to use software.)
   (c) If the data is annual data, how often, on the average, can you expect less than 16 claims a year?

7. (a) Show that the classical Poisson theorem follows from Theorem 5.
   (b) In the scheme of Section 2.1, let $I_1 = 1$ or 0 each with probability $1/2$, and for the remaining I_j, $j = 2,3,...,n$, the probability $p_j = 1/(n-1)$. Set $\overline{N}_n = \sum_{j=2}^{n} I_j$, the aggregate number of claims except for the first client. (i) Show that the distribution of $\overline{N}_n$ converges to the Poisson distribution with $\lambda = 1$. (ii) Show that condition (2.1.6) is not true. (iii) Show that the limiting distribution for the whole N_n is not a Poisson distribution. (*Advice:* Write the formula for total probability conditioning on I_1.) (iv) Find the limiting distribution for N_n; that is, $\lim_{n\to\infty} P(N_n = k)$. (*Advice:* Consider $P(N_n = k \,|\, I_1 = 1)$, and $P(N_n = k \,|\, I_1 = 0)$.) (v) What will change if $P(I_1 = 1)$ equals some fixed $p > 0$?

8.* (a) Make sure that if r in (2.2.6) is an integer, then $\binom{r}{n} = 0$ for $n = r+1, \ldots$, while for a non-integer r, the coefficient $\binom{r}{n} \neq 0$ for any n and may be negative for $n > r+1$. Make sure that, nevertheless, the coefficient in (2.2.5) is positive.
   (b) Show that for $\nu = 1$, the r.-h.s. of (2.2.4) is the m.g.f. of a geometric r.v., or more precisely, of an integer valued r.v. K_1 such that $P(K_1 = k) = pq^k$, where $p = 1 - q$, and $k = 0, 1, \ldots$. Make sure that $P(K_1 = k)$ coincides with (2.2.5) in this case. (As was noted in Section **0**.3.1.3, the geometric distribution is defined in the literature in two ways: either as the distribution of a r.v. K' assuming values $k = 1, 2, \ldots$ with probabilities pq^{k-1} or as the distribution of a r.v. K assuming values $k = 0, 1, 2, \ldots$ with probabilities

pq^k. The former distribution is that of the first success in a sequence of independent trials. Clearly, the distribution of K is that of $K' - 1$. [In Section **0**.3.1.3 we denoted K' by N but now N stands for the number of addends in (0.1).])

(c) Let now ν in (2.2.4) and (2.2.5) be an arbitrary natural number. Show that in this case, K is the sum of ν independent r.v.'s having the same distribution as K_1 above. (If, instead of the distribution of K_1, we consider the distribution of the r.v. N in the previous exercise, then K will be the distribution of the νth success in a sequence of independent trials.)

9.* In the scheme of Section 2.2.1, let Λ take on values λ_1 and λ_2, and events $A = \{N = n\}$, $B_i = \{\Lambda = \lambda_i\}, i = 1,2$. Write the Bayes formula for $P(B_i \,|\, A)$ and make sure that it is consistent with (2.2.8)-(2.2.9).

10.* In Example 2.2.1-5, find the expected value of Λ given $N = 6$. Compare it with the unconditional mean $E\{\Lambda\}$.

11.* The number N of daily claims coming from a risk portfolio is a Poisson r.v. with random parameter Λ. This parameter has the Γ-distribution with parameter $\nu = 2$ and a mean value of 3. The value of N on a particular day was 5. What can you say about the value of Λ which realized on this day. Using, for example, Excel, estimate the conditional probability that $\Lambda > 3$.

12.* Assume that the number of accidents arising in a risk portfolio is a Poisson r.v. with $\lambda = 300$, and the size of damage in each accident is an exponential r.v. with a mean of 10. The sizes of damages corresponding to different accidents are independent. Let each insurance contract involve a deductible of 2.

(a) Write the expected value and the variance for the number of claims. (b) Write the formula for the probability that the number of claims will not exceed 230. Do not compute it but tell whether you expect this probability to be very small. (c) Estimate this probability without long calculations. (*Advice:* Use the fact that the Poisson distribution with a large parameter is asymptotically normal: see Exercise **2**-43.)

13.* Assume that the number of accidents arising in a risk portfolio is negative binomial with parameters $p_1 = 1/3$ and $\nu = 4$, and the size of damage in each accident has the Pareto distribution in the form (**2**.1.1.17) with $\alpha = 2$. The sizes of damages corresponding to different accidents are independent. Let each insurance contract involve a deductible of 2. (a) Write the expected value and the variance for the number of claims. (b) Write the formula for the probability that the number of claims will not exceed 4. Do not compute it—at least it is not required—but tell whether you do or do not expect this probability to be small.

14.* Assume that the number of traffic accidents corresponding to a risk portfolio is a Poisson r.v. with $\lambda = 300$, and the probability that a separate accident causes serious injuries is $p = 0.07$. The outcomes of different accidents are independent. Find the probability that the number of accidents with serious injuries will exceed 10.

15.* Consider Example 2.2.2-2. (a) Does N assume all non-negative integer values? Find the probabilities that $N = 0,1,2,3$. (b) Verify that computing $E\{N\}$ and $Var\{N\}$ using (1.2) and (1.4) leads to the answers obtained in this example.

16.* Compute the expected value and the variance for the Poisson-Poisson distribution from Example 2.2.2-3.

17.* Give a condition under which the sum of independent negative binomial r.v.'s is negative binomial. Interpret the result considering a sequence of independent trials.

18.* Write the formulas for the expected value and the variance for the distribution from Example 2.2.2-4 making use of (1.1) and (1.3).

19.* Verify the results in Table 2.2.3-1.

20.* Calculate zero-truncated and zero-modified probabilities for the geometric distribution. With what distribution do we deal in the former case?

Section 3

21. An employer buys an insurance for his employees. The mean number of claims coming from the whole group of employees is 50, the standard deviation is 10. The individual loss has a mean of 4 (units of money) and a variance of 2. The insurance company imposes a deductible of 50 for the whole portfolio. Assuming that the distribution of the aggregated loss is closely approximated by a Γ-distribution, find the probability that the company will pay more than 200 units.

22. Compute $E\{S\}$ and $Var\{S\}$ for the following cases: (a) N has a Poisson distribution with $E\{N\} = 150$ and the X's take on values 2,3,4 with probabilities $1/3, 1/2, 1/6$, respectively. (b) N has a geometric distribution with parameter $p = 0.02$ and the d.f. of the X's is $\Gamma(x; 3, 5)$. (c) N has the negative binomial distribution with parameters $p = 0.02$, $\nu = 4$, and the X's are exponential with $E\{X\} = 3$.

23. Verify (3.1.7).

24. In the situation of Example 3.1.1-2, what should $E\{N\}$ be equal to for $P(N > 50) \leq 0.05$?

25. Following the scheme of Example 3.1.1-3, provide spreadsheets and particular calculations with parameters of your choice for the models of Examples 3.1.1-3,4,5.

26. Following the scheme of Example 3.1.1-3, provide spreadsheets and particular calculations in the case when N is a Poisson r.v. with a mean of 20.

27. In the scheme of Section 3.1.2, set $l = 3$, $x_i = i$, $p_1 = 1/2$, $p_2 = 1/3$, and $\lambda = 12$. Compute (a) $P(K_1{=}3, K_2{=}3, K_3{=}2)$, $P(K_1 \leq 3, K_2 \leq 3, K_3 \leq 2)$, $P(K_1 = 3, K_2 = 3 \,|\, N{=}8)$; (b) $P(S = 0), P(S = 3)$.

28. Let N_1 and N_2 be independent Poisson r.v.'s, $E\{N_1\} = E\{N_2\}$. Show that $N_1 - N_2 \stackrel{d}{=} Y_1 + \ldots + Y_N$, where $N = N_1 + N_2$, and the Y's are mutually independent and independent of N r.v.'s assuming values ± 1 with equal probabilities. Write the m.g.f. of $N_1 - N_2$. (*Advice:* Consider two types of claims of sizes 1 and -1, respectively. The fact that the claim is negative may be interpreted as if you are paid instead of paying. Denote by N_i, $i = 1, 2$, the number of claims of type i.)

29. Let N_1 and N_2 be independent Poisson r.v.'s, $E\{N_1\} = 100$ and $E\{N_2\} = 200$. Similar to Exercise 28, represent the r.v. $3N_1 - 5N_2$ as a sum $Y_1 + \ldots + Y_N$ where N is a Poisson r.v. and Y's are some r.v.'s that are mutually independent and independent of N.

30.* In the scheme of Section 3.1.3, for $\nu = 2$, $a = 10$, and $p = 0.1$, write $P(S = 0)$ and the density $f_S(x)$ for $x > 0$.

31.* For a portfolio, the number of loss events has the negative binomial distribution with parameters $p = 1/3$ and $\nu = 10$. The losses are independent and have the Pareto distribution (**2**.1.1.18) with $\theta = 3$, $\alpha = 2$. The insurance has an ordinary deductible of 1. Find the expected aggregate payment.

32. Let a r.v. S have the m.g.f. $M_S(z) = \exp\left\{200\left(\frac{1}{1-2z} - 1\right)\right\}$ for $z < 1/2$. (a) Which insurance model lead to this formula? (b) *Without any calculations*, write $E\{S\}$ and $Var\{S\}$.

33. An insurance portfolio consists of two homogeneous groups of clients. N_i, $i = 1, 2$, denotes the number of claims coming from the ith group during a fixed time period. Assume that r.v.'s N_i are independent and have Poisson distributions, $E\{N_i\} = 10i^2$.

    (a) If N is the total number of all claims in the portfolio, what distribution does N have? Compute $E\{N\}$, $Var\{N\}$. Write the expression for $P\{N \leq 50\}$.

    (b) Estimate $P(N_1 < 11 \,|\, N = 50)$.

    (c) Let the amount of an individual claim in the first group always be \$100, and in the second group let it be \$300. What is the distribution of S, the total amount of aggregated claims, in this case? (Include a name and describe the distribution.)

    (d) Find $E\{S\}$ and $Var\{S\}$. Write $M_S(z)$.

34. In the situation of Exercise 33, let the values of individual claims be random. More specifically, assume that an individual claim in the first group assumes values 100 and 200 with probabilities 1/3 and 2/3 respectively, while in the second group, these values are 200 and 300 with probabilities $1/2$. (a) What is the distribution of S? (b) Find $E\{S\}$ and $Var\{S\}$. Write $M_S(z)$. (c) Show that $S \stackrel{d}{=} 100N_1 + 200N_2 + 300N_3$, where N_1, N_2, and N_3 are independent Poisson r.v.'s. Find $E\{N_i\}, i = 1, 2, 3$.

35. Solve Exercises 34 and 34 for the instance where a separate claim in the first group is uniformly distributed on $[100, 200]$, while in the second group, it is uniformly distributed on $[200, 300]$. (*Advice:* Observe that the distribution of a separate "imagined" claim Y is a mixture of two uniform distributions. The intervals $[100, 200)$ and $[200, 300]$ do not intersect, so it is not difficult to write the distribution function (or the density, whatsoever) for the first group and for the second, and write or graph the mixture with appropriate weights.)

36. Solve Problem 35 for the case when the uniform distributions mentioned in this problem are those on $[100, 300]$ and on $[200, 400]$, respectively.

Section 4

37.* Apply Theorem 12 in the instance when $\lambda = 1, 2, ...,$ and N_λ is a (p, λ)-binomial r.v., that is, $P(N_\lambda = k) = \binom{\lambda}{k} p^k (1-p)^{\lambda - k}$, $k = 0, 1, ..., \lambda$. What are we dealing with in the case $p = 1$? (*Advice*: Use the CLT theorem.)

38. Assume that in the scheme of Section 4.2, the X's are exponentially distributed. (a) Does the estimate of the relative loading coefficient θ depend on the parameter of the exponential distribution mentioned? (b) Find the estimate for θ for the case when N is (i) a Poisson r.v., (ii)* a Poisson-Poisson r.v. (iii)* a negative binomial r.v.

39. Find θ and c in Examples 4.2-1 (a) for the case when the insurer pays only 80% of each claim; (b) when the distribution of X's is exponential; (c) when the distribution of the X's is uniform.

40.* Find θ and c in Examples 4.2-2 (a) for the case when the insurer pays only 80% of each claim; (b) when the distribution of X's is exponential; (c) when the distribution of the X's is lognormal.

41.* Similar to Example 4.3-1, find a precise bound for θ in the case when the X's are exponentially distributed.

Chapter 4

Random Processes and their Applications. I: Counting and Compound Processes. Markov Chains. Modeling Claim and Cash Flows

This chapter has two goals—to list some general facts from the theory of random processes, and to consider in detail various models of cash flows, including the surplus and claims processes.

We start with an overview and after that we will explore particular processes important for insurance modeling in greater detail.

1 A GENERAL FRAMEWORK AND TYPICAL SITUATIONS

1.1 Preliminaries

A random variable was defined in Chapter 0 as a function $X(\omega)$ on a space $\Omega = \{\omega\}$ of elementary outcomes ω. Usually, we omit the argument ω and write X instead of $X(\omega)$.

A *random* or *stochastic process* is defined as a collection of r.v.'s $X_t(\omega)$ where t is a running parameter. We can view $X_t(\omega)$ as a function of two arguments: ω and t.

For a *fixed* t, the function $X_t(\omega)$ as a function of ω is a random variable.

For a *fixed* ω, we have a function of t. In this case, *given* ω, the function $X_t = X_t(\omega)$ is also called a (particular) *realization, or a trajectory,* of the random process.

EXAMPLE 1. Suppose that the sample space Ω consists of only two elementary outcomes: ω_1 and ω_2. Say, we toss a coin. Assume that $X_t(\omega_1) = t$ while $X_t(\omega_2) = t^2$. So, depending on which ω occurs, we deal with either a linear function or a parabola.

In the case of a regular coin, it amounts to a random selection, with equal probabilities, of one out of two functions (rather than numbers): either t or t^2. □

As we did it when dealing with r.v.'s, usually we again omit the argument ω and just write X_t.

In practically all models of this book, t is a time parameter and X_t determines the evolution of some characteristic in time. In general, t may of any nature. For instance, t could represent the distance from the beginning of a trench dug by a gold miner to a particular place in the trench. Then X_t can represent the (random) concentration of gold at point t.

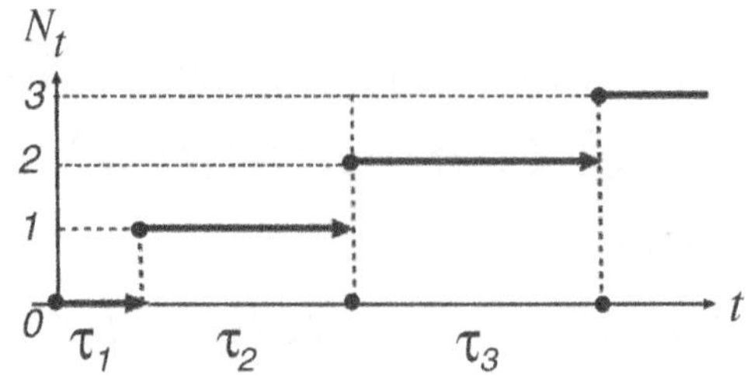

FIGURE 1.

When the quantity X_t is not a r.v. but a random vector (r.vec.), we talk about a *random vector process*. For example, X_t may be the (random) vector of stock prices for various assets in a financial market at time t.

Formally, the parameter t may take on values from an arbitrary set. When t is time, we consider two cases: the continuous time case when t takes values from an interval, and the discrete time case when $t = 0, 1, 2, \ldots$. In the latter case, $\{X_t\}$ is a sequence of r.v.'s $\{X_0, X_1, X_2, \ldots\}$.

One of the simplest processes in continuous time is a *counting process* $N_t, t \geq 0$, representing the total number of events of a certain type that have occurred prior to or at time t. For example, N_t may be the number of customers who have entered a store by time t. The same may concern the number of falling stars you observed in the sky, the number of cosmic particles registered by some device, etc. For us, the most important case is the process that counts claims being received by an insurance company.

For brevity, we call occurrence of the events above *arrivals*, and time-intervals between consecutive arrivals *interarrival times*. These are also sometimes referred to as *sojourn times*.

A typical realization of a counting process is shown in Fig.1. The symbols τ_i there denote interarrival times. For $t < \tau_1$ (no arrivals have occurred), $N_t = 0$. If $\tau_1 \leq t < \tau_1 + \tau_2$ (after the first arrival and before the second), $N_t = 1$, and so on.

Note that the realization graphed in Fig.1 is continuous from the right. This reflects the fact that, when defining N_t as the number of arrivals during the interval $(0, t]$, we include the end point t. So, if an arrival occurs at the last moment t in the interval $(0, t]$, we count it.

Another important process when considering insurance models, is a *surplus* process (sometimes called also a *reserve* process). Such a process, R_t, measures the monetary surplus of a risk portfolio at time t. We may also view it as the amount of the (monetary) reserve fund corresponding to this portfolio. For example, if u is the initial surplus, S_t is the amount paid by the company by time t, and c_t is the premium paid to the company by the same time, then $R_t = u + c_t - S_t$. The process S_t is called a *claim process*. Usually, though not always, we set $c_t = (1 + \theta)E\{S_t\}$, where θ is a loading coefficient. Later, in this chapter and Chapters 5-6, we will consider the process R_t in detail.

Next, we consider two particular but important classes of processes. For now, we do it briefly. Later we will return to these processes and explore them in more detail.

1.2 Processes with independent increments

For a half-open interval $\Delta = (t_1, t_2]$, we define the increment of a process X_t as the r.v. $X_\Delta = X_{t_2} - X_{t_1}$. For instance, for a counting process N_t, the increment N_Δ is the number of arrivals during the interval Δ.

A process $X_t, t \geq 0$, is called a process *with independent increments* if for any collection of *disjoint* intervals $\Delta_1, ..., \Delta_k$, the r.v.'s X_0, $X_{\Delta_1}, ..., X_{\Delta_k}$ are mutually independent. Note that since we included X_0 into the definition, all increments of the process do not depend on the *initial value* X_0 of the process. For example, $X_1 - X_0$ does not depend on X_0.

If time is discrete, for $t = 1, 2, ...$ we can write

$$X_t = X_0 + (X_1 - X_0) + (X_2 - X_1) + ... + (X_t - X_{t-1})$$
$$= X_0 + X_{(0,1]} + X_{(1,2]} + ... + X_{(t-1,t]}. \quad (1.2.1)$$

So, X_t is a sum of independent r.v.'s.

In the continuous time case, we can get a similar representation if for a natural n, we divide the interval $[0, t]$ into intervals $\Delta_k = \left(\frac{(k-1)t}{n}, \frac{kt}{n}\right]$, $k = 1, ..., n$, and write

$$X_t = X_0 + X_{\Delta_1} + ... + X_{\Delta_n}. \quad (1.2.2)$$

The random terms (*addends*) in the r.-h.s. of (1.2.2) are independent. Nevertheless, continuous time processes with independent increments are not as simple as one may expect proceeding from (1.2.2). This is illustrated by the following two classical models.

1.2.1 The simplest counting process

We call a counting process N_t the *simplest* if all interarrival times are independent identically distributed exponential r.v.'s. Later, when we show that this process is connected with the Poisson distribution, we will call it a *Poisson process*.

Assume that we have been watching the process defined until time t. In particular, at time t we know how much time has elapsed since the last arrival. However, by virtue of the lack of memory property of the exponential distribution (see Section **0**.3.2.2), the amount of time we must wait for the next arrival after time t does not depend on how long we have already been waiting. The process starts over as from the very beginning. So we know that the remaining time until the next arrival has an exponential distribution, and the parameter of this distribution is the same as for a separate interarrival time. This parameter is fixed since we assumed the interarrival times to be identically distributed.

Furthermore, the next interarrival time also does not depend on what happened before time t since interarrival times are independent. Thus, for any interval $\Delta = (t, t']$, the increment N_Δ does not depend on the evolution of the process before t, and hence N_t is a process with independent increments. We consider this process in detail in Section 2.

1.2.2 Brownian motion

In the case of continuous time, we call a process *continuous* if with probability one its realizations are continuous functions. A counting process (see, for example, Fig.1) is, clearly, not continuous. The next scheme, in a certain sense, is contrasting to the previous.

Let $w_t, t \geq 0$, be a continuous random process with independent increments such that $w_0 = 0$. Suppose that for any interval Δ, the increment w_Δ is a normal r.v. with zero mean and variance $|\Delta|$, where $|\Delta|$ is the length of Δ. Such a process is called the *standard Wiener process* or *Brownian motion*. Note that we do not derive but postulate independence of increments, and we skip a formal proof that such a process indeed exists, i.e., well defined. One can find such a proof in any advanced textbook on random processes, e.g., in [47] or [70]. In Example 1 below, we will discuss how to simulate trajectories of such a process.

Since $w_0 = 0$, we can write $w_t = w_t - w_0 = w_{(0,t]}$, the increment over $(0,t]$. Hence, w_t is a normal r.v. with mean zero and variance t. In particular, w_1 is a standard normal r.v., and for this reason we use in the definition of w_t the term "standard".

The process w_t may be considered a counterpart of a standard normal r.v. in the theory of processes. Originally, the term Brownian motion was used for the motion of a particle totally immersed in a liquid. The name comes from the botanist R. Brown who first considered this phenomenon. It proved that processes of the type $a + bw_t$, where a, b were real numbers, were suitable for modeling such a physical motion. The rigorous theory was built mainly by A. Einstein and N. Wiener. Nowadays, Wiener processes, or Brownian motion, are widely used for modeling various phenomena of different nature, such as diffusion in Physics, certain processes in quantum mechanics, the evolution of stock prices, surplus processes in insurance, etc.

EXAMPLE 1. To understand how realizations of w_t look, let us choose a real number $\delta > 0$ and consider points $t_k = k\delta$, where $k = 0, 1, \dots$. Let $\Delta_k = (t_{k-1}, t_k]$. Then

$$w_{t_k} = w_{\Delta_1} + \dots + w_{\Delta_k}. \tag{1.2.3}$$

The r.v.'s w_{Δ_k} are independent normal r.v.'s with zero mean and variance δ because the length of each interval is δ.

Set $\xi_k = \frac{1}{\sqrt{\delta}} w_{\Delta_k}$ for $k = 1, 2, \dots$. Since we divided w_{Δ_k} by its standard deviation, the r.v.'s ξ_k are *standard* normal (see also (**0**.2.6.3)) and independent because the intervals Δ_k do not overlap.

Because $w_{\Delta_k} = \sqrt{\delta}\xi_k$, from (1.2.3) it follows that

$$w_{t_k} = \sqrt{\delta}(\xi_1 + \dots + \xi_k). \tag{1.2.4}$$

The last representation gives a way to simulate a realization of Brownian motion. Let, say, $\delta = 0.01$ and $k = 1, 2, \dots, 100$, which means that we consider 100 equally spaced points in $[0, 1]$. The first point $t_1 = 0.01$, the last point $t_{100} = 1$.

In the Excel worksheet in Fig.2, Column A contains 100 standard normal numbers generated by Excel (not all numbers are shown in the figure). They represent ξ's. C1 contains the value of $\delta = 0.01$. In E1 the value $\sqrt{C1} \times A1$ is given, which comes from the relation $\sqrt{\delta}\xi_1 = w_{t_1}$. So, E1 corresponds to the value w_{t_1}.

The value in E2 =E1+A2*SQRT(C1), which comes from $w_{t_2} = w_{t_1} + \sqrt{\delta}\xi_2$, and so on. The cell E$k$ equals E$(k-1)$+Ak*SQRT(C1), which reflects the formula $w_{t_k} = w_{t_{k-1}} + \sqrt{\delta}\xi_k$. The graph of Column E is given in the first chart.

Cell C2 and Column G correspond to the same simulation where $\delta = 0.02$, and we consider 50 points in the same interval. It is worth emphasizing that this not a part of the

	A	B	C	D	E	F	G
1	-0.30023	δ's	0.01		-0.03002		-0.04246
2	-1.27768		0.02		-0.15779		-0.22315
3	0.244257				-0.13337		-0.18861
4	1.276474				-0.00572		-0.00809
5	1.19835				0.114117		0.161385
6	1.733133				0.28743		0.406487
7	-2.18359				0.069071		0.097681
8	-0.23418				0.045653		0.064563
9	1.095023				0.155155		0.219423
10	-1.0867	standard normals			0.046485		0.06574
11	-0.6902				-0.02254		-0.03187
12	-1.69043				-0.19158		-0.27093
13	-1.84691				-0.37627		-0.53213
14	-0.97763				-0.47403		-0.67038
15	-0.77351				-0.55138		-0.77977
16	-2.11793				-0.76318		-1.07929
17	-0.56792				-0.81997		-1.15961
18	-0.40405				-0.86037		-1.21675
19	0.134853				-0.84689		-1.19768
20	-0.36549				-0.88344		-1.24937
21	-0.32699				-0.91614		-1.29561
22	-0.37024				-0.95316		-1.34797
23	1.342642				-0.8189		-1.15809
24	-0.08528				-0.82742		-1.17016
25	-0.18616				-0.84604		-1.19648
26	-0.51321				-0.89736		-1.26906
27	1.972212				-0.70014		-0.99015
28	0.865673				-0.61357		-0.86772
29	2.375655				-0.37601		-0.53175
30	-0.65491				-0.4415		-0.62437

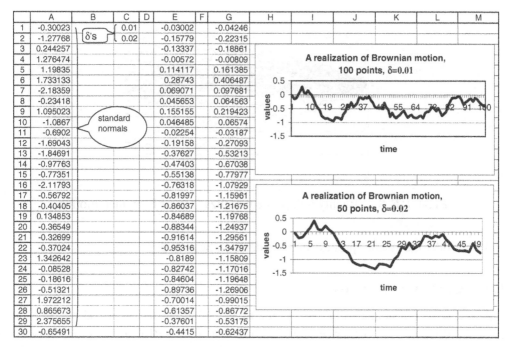

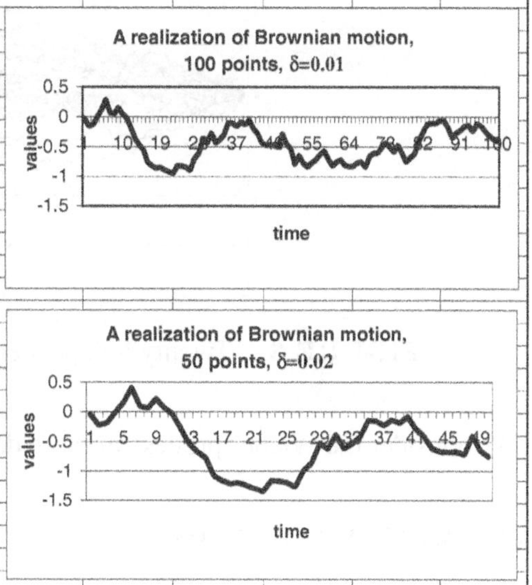

FIGURE 2. Realizations of Brownian motion on [0,1] with the step $\delta = 0.01$ (100 points) and 0.02 (50 points). In the figure, Columns A, E, and G are truncated (not all numbers are shown).

previous realization: we took the first 50 values in Column A, but assigned them to 50 equally spaced points in $[0,1]$. For example, while in the first case the number in Cell E1, -0.03, is the value of the process at the point $t = 0.01$, in the second case the number in Cell G1, -0.0425, is the value of the process at $t = 0.02$. Thus, the second graph may be viewed as a new realization.

We see that the second graph appears smoother, which is not surprising: taking only 50 points in the same interval, we eliminate 50 independent fluctuations between the chosen points.

The graphs in Fig.2 represent just two possible realizations of the process. If we regenerate in Column A other normal numbers, the realizations will be different. Fig.3 represents twenty independent realizations with the step $\delta = 0.01$.

The envelopes of the twenty curves in Fig.3 are consistent with the theory. The distribution of w_t is normal with zero mean and variance t, and hence the distribution of w_t coincides with the distribution of $\sqrt{t}Z$ where Z is standard normal. Thus, w_t grows as $\sqrt{t}$ times a random factor that does not depend on t. One may say that w_t has the order $\pm\sqrt{t}$, which is reflected in Fig.3. See also Exercise 4. □

It may be shown that in the continuous time case, *any process* with independent increments may be represented as a certain combination of the Wiener process and processes of the type bN_t, where b is a number, and N_t is the simplest (Poisson) process from the previous section; see, e.g., [45], [70]. Such combinations are called Lévy processes (after a French mathematician Paul Lévy).

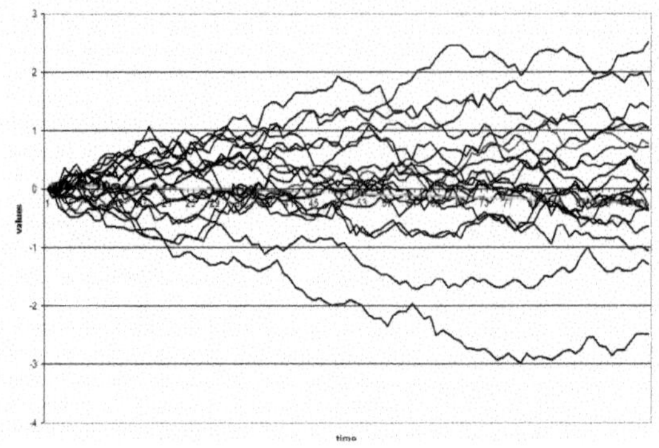

FIGURE 3. Twenty independent realizations of Brownian motion.

We consider the Wiener process in more detail in Section **5**.1.

1.3 Markov processes

Below, we use systematically the notation $P(A\,|\,X)$ that means the probability of event A given the value of a r.v. (or a r.vec.) X. In particular, $P(Y \in B\,|\,X)$ is the probability that a r.v. (or a r.vec.) Y will assume a value from a set B, given the value of X. For more detail see Section **0**.7, and in particular, p.57.

For a process $X_t, t \geq 0$, let the symbol X^t denote the whole trajectory of the process until time t. If time is discrete, X^t is the random sequence $\{X_0, X_1, ..., X_t\}$; if time is continuous, X^t is the (random) function X_u, $0 \leq u \leq t$.

We say that a process X_t is a Markov process (after A.A. Markov who first considered such processes in the beginning of the twentieth century) if for any set B on the real line, and any $t, s \geq 0$,

$$P(X_{t+s} \in B\,|\,X^t) = P(X_{t+s} \in B\,|\,X_t). \tag{1.3.1}$$

To understand this definition, suppose that t is the *present* time, and the *past trajectory* X^t *is known.* The l.-h.s. of (1.3.1) is the probability that the *future* value X_{t+s} will be in a set B, *given the whole history* of the evolution of the process by time t. The *Markov property* (1.3.1) implies that this probability, as a matter of fact, depends not on the whole past evolution but only on the last (and present for us) value X_t. This is indicated in the r.-h.s. of (1.3.1).

We may say that in the Markov case, "*given the present, the future does not depend on the past*".

For discrete time this implies, in particular, that

$$P(X_{t+1} \in B\,|\,X_0, ..., X_{t-1}, X_t) = P(X_{t+1} \in B\,|\,X_t),$$

that is, the value of the process at the next step depends only on where the process is now.

Any process with independent increments is a Markov process. Indeed, $X_{t+s} = X_t + X_{(t,t+s]}$, and X_{t+s} depends only on X_t and the increment $X_{(t,t+s]}$ which does not depend on values X_u for $u \leq t$. *The converse assertion is not true.*

EXAMPLE 1. Let $X_t = e^{bw_t}$, where b is a number, and w_t is the Wiener process. Such a process may at first glance look exotic, but as a matter fact, it may be used, for example, for modeling the evolution of stock prices. We consider it in more detail in Section **5**.1.3. We have

$$X_{t+s} = \exp\{bw_{t+s}\} = \exp\{bw_t\}\exp\{b(w_{t+s} - w_t)\} = X_t \exp\{bw_{(t,t+s]}\}. \qquad (1.3.2)$$

By the definition of w_t, the increment $w_{(t,t+s]}$ does not depend on values of the process w_u for $u \leq t$, and hence, on values X_u for $u \leq t$. Thus, for any fixed t and s, given the value of the r.v. X_t, the r.v. X_{t+s} does not depend on X_u for $u < t$. Consequently, the whole process is Markov.

Now, we show that increments of this process are dependent. For simplicity, let $b = 1$. Consider intervals $(0,1]$ and $(1,2]$. Because $w_0 = 0$, the initial value $X_0 = 1$. Then, in view of (1.3.2),

$$X_{(0,1]} = X_1 - X_0 = X_1 - 1, \quad X_{(1,2]} = X_2 - X_1 = X_1[\exp\{w_{(1,2]}\} - 1].$$

The r.v. $\exp\{w_{(1,2]}\}$ does not depend on X_1. On the other hand, both r.v., $X_{(0,1]}$ and $X_{(1,2]}$, involve X_1. Hence, the increments above are dependent.

We consider a generalization of this example in Section **5**.1.3. □

To understand the nature of Markov processes, consider the discrete time case and the following construction. Let $\xi_1, \xi_2, ...$ be independent r.v.'s. We define a process X_t, $t = 0, 1, ...$ as follows. X_0 is a given r.v.; as above, we call it an *initial value*. For $t = 1, 2, ...$, we define X_t by the recurrence relation

$$X_{t+1} = h_t(X_t, \xi_{t+1}), \qquad (1.3.3)$$

where $h_t(x,z)$, $t = 1, 2, ...$, is a function of two variables.

For example, let X_t be the capital of a person at time t, and the value of the capital at the next time, i.e., $X_{t+1} = (1 + \xi_{t+1})X_t$, where ξ_{t+1} is a random interest over the period $(t, t+1]$. If we assume ξ's to be independent of the previous history of the process, we will come to the model (1.3.3) with $h_t(x,z) = (1+z)x$.

In general, because ξ's are independent, X_{t+1} depends only on the previous value of the process X_t and the r.v. ξ_{t+1} which does not depend on the present or the past values of the process. Clearly, the process X_t so defined is Markov.

It is important that, as a matter of fact, *any discrete-time Markov process admits the representation* (1.3.3), so (1.3.3) is not an example but, in a certain sense, another definition.

The proof of this fact is constructive, and gives a way to simulate Markov processes. Let X_t be a Markov process. Since X_{t+1} depends only on X_t, the joint distribution of all r.v.'s X_t may be completely determined by the conditional distributions of each X_{t+1} given the previous value X_t. Let us consider the *conditional* distribution function of X_{t+1} given X_t, namely the function

$$F_{t+1}(z \,|\, x) = P(X_{t+1} \leq z \,|\, X_t = x).$$

We use the following fact proved in Section **0**.3.2.1. If $F(z)$ is a distribution function, $F^{-1}(z)$ is its inverse, and ξ is uniformly distributed on $[0,1]$, then the r.v. $F^{-1}(\xi)$ has the distribution function $F(z)$.

Let r.v.'s $\xi_1, \xi_2, \ldots$ be independent and uniformly distributed on $[0,1]$, and let the functions $h_t(x,z)$ from (1.3.3) be defined as follows:

$$h_t(x,z) = F_{t+1}^{-1}(z\,|\,x),$$

where $F_{t+1}^{-1}(z\,|\,x)$ is the inverse of $F_{t+1}(z\,|\,x)$ with respect to z for a fixed x. Then, by virtue of the previously stated fact, the r.v. $h_t(x, \xi_{t+1})$ has the distribution function $F_{t+1}(z\,|\,x)$.

Let us construct a process Y_t by the recurrence relation

$$Y_{t+1} = h_t(Y_t, \xi_{t+1})$$

with Y_0 having the same distribution as X_0. Note that the process Y_t is not the original process X_t, since the process Y_t has been artificially constructed. However, both processes have identical probability distributions.

It is noteworthy that we can always take $\xi_1, \xi_2, \ldots$ as particular identically distributed r.v.'s; namely, uniform r.v.'s.

The described construction gives a way for simulating Markov processes in discrete time.

EXAMPLE 2. Consider a Markov process X_t for which $X_0 = 1$, and given X_t, the distribution of X_{t+1} is the Pareto distribution (**2**.1.1.17) with $\alpha = X_t$. In this case,

$$F_{t+1}(z\,|\,x) = 1 - z^{-x} \text{ for } z \geq 1,$$

and hence

$$h(x,z) = F_{t+1}^{-1}(z\,|\,x) = (1-z)^{-1/x}.$$

So, we set $Y_0 = 1$, and

$$Y_{t+1} = (1 - \xi_{t+1})^{-1/Y_t}. \tag{1.3.4}$$

A corresponding Excel worksheet is presented in Fig.4. Column A contains twenty values of uniform ξ's; time moments t are in Column D; and twenty values of the process are in Column E. For example, in accordance with the recurrence formula (1.3.4), the command for Cell E3 is =(1-A2)^(-1/E2). □

In the following sections, we consider in detail particular types of processes important in actuarial modeling.

---

2 POISSON AND OTHER COUNTING PROCESSES

2.1 The homogeneous Poisson process

We come back to the model of Section 1.2.1. Let $\tau_1, \tau_2, \ldots$ be consecutive interarrival times. Then $T_n = \tau_1 + \ldots + \tau_n$ is the time of the nth arrival. As in Section 1.2.1, we assume that the r.v.'s τ_i are independent and exponentially distributed with the same parameter which we will denote by λ. As before, let N_t be the total number of arrivals by time t.

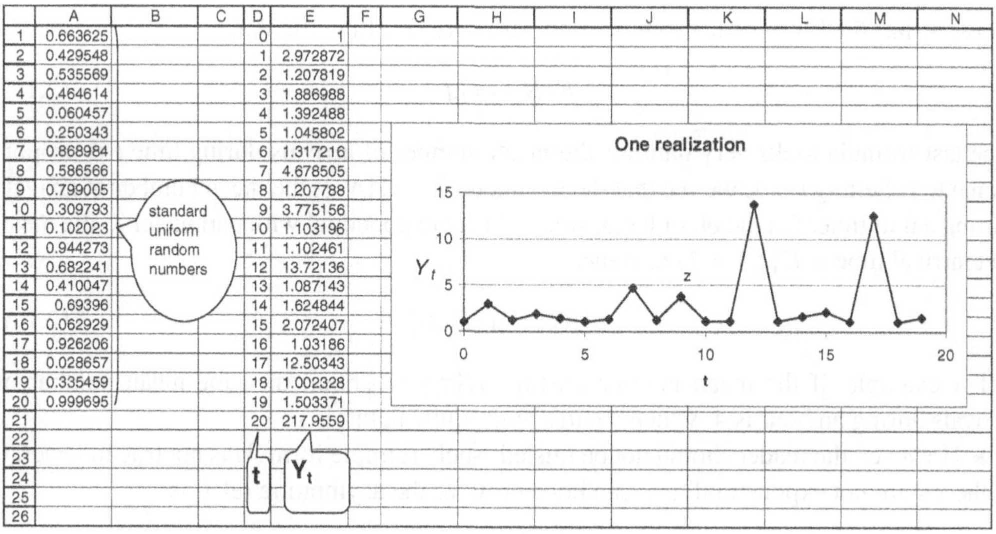

	A	D	E
1	0.663625	0	1
2	0.429548	1	2.972872
3	0.535569	2	1.207819
4	0.464614	3	1.886988
5	0.060457	4	1.392488
6	0.250343	5	1.045802
7	0.868984	6	1.317216
8	0.586566	7	4.678505
9	0.799005	8	1.207788
10	0.309793	9	3.775156
11	0.102023	10	1.103196
12	0.944273	11	1.102461
13	0.682241	12	13.72136
14	0.410047	13	1.087143
15	0.69396	14	1.624844
16	0.062929	15	2.072407
17	0.926206	16	1.03186
18	0.028657	17	12.50343
19	0.335459	18	1.002328
20	0.999695	19	1.503371
21		20	217.9559
22			
23			
24			
25			
26			

FIGURE 4. Simulation of the Markov process from Example 2. The worksheet and the graph of one realization

Now note that the number of arrivals N_t is not less than n if and only if the nth arrival occurred prior to or at time t. Consequently,

$$P(N_t \geq n) = P(T_n \leq t), \tag{2.1.1}$$

and to find the distribution of N_t, it suffices to find the distribution of T_n.

Each exponentially distributed τ has the Γ-density $f_{\lambda 1}$ in the notation of (**2**.1.1.10). By Proposition **2**.4, the density of T_n is the convolution $f_{\lambda 1} * f_{\lambda 1} * ... * f_{\lambda 1} = f_{\lambda n}$, the Γ-density with parameters λ, n. Again using (**2**.1.1.10), we arrive at the density of T_n:

$$f_{T_n}(x) = \frac{\lambda^n}{(n-1)!} x^{n-1} e^{-\lambda x} \text{ for } x \geq 0. \tag{2.1.2}$$

Hence,

$$P(T_n \leq t) = \int_0^t f_{T_n}(x)dx = \frac{\lambda^n}{(n-1)!}\int_0^t x^{n-1}e^{-\lambda x}dx = 1 - e^{-\lambda t}\left(1 + \frac{\lambda t}{1!} + ... + \frac{(\lambda t)^{n-1}}{(n-1)!}\right). \tag{2.1.3}$$

The last integral is standard and may be found in many Calculus textbooks. To check that (2.1.3) is true, one may take the derivative of its r.-h.s. and make sure that it equals $f_{T_n}(t)$, that is, the derivative of the l.-h.s. (When you differentiate, all terms will cancel except one.) At $t = 0$, the r.-h.s. and the l.-h.s. are both equal to zero. So, the r.-h.s. and the l.-h.s. are equal to each other for all t. We consider this in detail in Exercise 7.

Combining (2.1.1) and (2.1.3), we write

$$\begin{aligned} P(N_t = n) &= P(N_t \geq n) - P(N_t \geq n+1) = P(T_n \leq t) - P(T_{n+1} \leq t) \\ &= e^{-\lambda t}\frac{(\lambda t)^n}{n!}. \end{aligned} \tag{2.1.4}$$

Thus, N_t has the Poisson distribution with parameter λt. In particular,

$$E\{N_t\} = \lambda t. \tag{2.1.5}$$

The last formula looks very natural: the mean number of arrivals during time t is proportional to t. Setting $t = 1$, we see that the parameter $\lambda = E\{N_1\}$, the mean number of arrivals during a unit time. On the other hand, since τ_i's are exponential with parameter λ, the mean interarrival time is $E\{\tau_i\} = 1/\lambda$. Hence,

$$E\{N_1\} = 1/E\{\tau_i\}. \tag{2.1.6}$$

For example, if the mean interarrival time $E\{\tau_i\} = \frac{1}{4}$ hour, then the mean number of arrivals during an hour is 4, which again sounds quite natural.

▶ However, the reader should not be misled: such a simple formula is not true in general. If the τ's are not exponential, one can hope only for the asymptotic relation

$$\frac{1}{t}N_t \xrightarrow{P} 1/E\{\tau_i\} \text{ as } t \to \infty. \tag{2.1.7}$$

(See also Exercise 8 and, e.g., [10], [38], [50], [122]. For the definition of convergence in probability, $\xrightarrow{P}$, see Section **0.5**.) ◀

Consider now an interval $\Delta = (t, t+s]$ and the increment N_Δ, that is, the number of arrivals during Δ. In view of the memoryless property of the exponential distribution, we do not have to calculate the distribution of N_Δ: at any moment t, the process starts over as from the beginning, and its evolution does not depend on what happened before t. Consequently, for $P(N_\Delta = n)$ we should take the same formula (2.1.4) and replace t by $|\Delta|$, the length of the interval Δ. This gives

$$P(N_\Delta = n) = e^{-\lambda|\Delta|}\frac{(\lambda|\Delta|)^n}{n!}. \tag{2.1.8}$$

Thus, we have established the following properties of the process N_t :

P1. $N_0 = 0$ with probability one (since the time of the first arrival, τ_1, is positive with probability one, and at the time zero, we do not observe an arrival).

P2. N_t is a counting process with independent increments.

P3. For any Δ, the r.v. N_Δ has the Poisson distribution with parameter $\lambda|\Delta|$.

These properties may be considered a new definition of the process N_t since they are equivalent to the original definition in terms of the interarrival times τ_i.

Indeed, from P3 it follows that for the first arrival

$$P(\tau_1 > t) = P(N_t = 0) = e^{-\lambda t}, \tag{2.1.9}$$

that is, τ_1 is exponentially distributed. For τ_2, using P2 and (2.1.8), we have

$$P(\tau_2 > t \,|\, \tau_1 = s) = P(\text{no arrivals in } (s, s+t] \,|\, \tau_1 = s) = e^{-\lambda t}.$$

Consequently, τ_2 is also exponential and independent of τ_1. The other τ's are considered similarly.

The process described is called a *homogeneous Poisson process,* and the corresponding *flow of arrivals* (for example, a *flow of claims*)—a *Poisson flow*.

EXAMPLE 1. Assume that the flow of claims being received by the claim department of an insurance company is well approximated by a Poisson flow and that the mean time between two consecutive claims is half an hour.

(a) Find the expected value and the variance of the number of claims during the period between 2pm and 6pm. We choose an hour as a unit of time. In view of the memoryless property, we can consider any time, including 2pm, an initial time. Since $E\{\tau_i\} = (1/\lambda) = 1/2$, we have $\lambda = 2$, and $E\{N_{(2,6]}\} = E\{N_4\} = 4\lambda = 8$. In the case of the Poisson distribution, the variance $Var\{N_{(2,6]}\} = E\{N_{(2,6]}\} = 8$.

(b) Find the probability that there will be exactly 10 claims, and the probability that there will be at most 10 claims, during the same period. By (2.1.8), $P(N_{(2,6]}{=}10){=}P(N_4{=}10){=}$ $e^{-2\cdot 4}(2\cdot 4)^{10}/10! \approx 0.099$. The probability $P(N_{(2,6]} \le 10) = PoissonDist(10;\,8)$, where the symbol $PoissonDist(x;\,\lambda)$ stands for the Poisson distribution function (in x) with parameter λ. We will use this type of symbols (not looking too mathematically) when direct calculations are cumbersome, and require software to compute. Such symbols coincide with or are close to the corresponding commands in popular programs like Excel.

In our case, $PoissonDist(10;\,8) = \sum_{k=0}^{10} e^{-8}(8)^k/k!$. It hardly makes sense to estimate the last sum by hand, but using a computer, it evaluates to ≈ 0.816.

(c) Find the probability that if we start to count at 2pm, the seventh claim will come after 5pm. In view of (2.1.1), the probability in consideration is $P(T_7 > 3) = P(N_3 < 7)$. We could choose to compute either side of the last equality. If we choose the left, we can use (2.1.2) with $n = 7$, $\lambda = 2$, and write $P(T_7 > 3) = 1 - \int_0^3 f_{T_7}(x)dx = 1 - \int_0^3 \frac{2^7}{6!}x^6 e^{-2x}dx =$ $1 - GammaDist(3;2,7) \approx 0.606$, where $GammaDist(x;a,\nu)$ stands for the Γ-distribution function (in x) with parameters a,ν. There is a corresponding command in Excel.

If we prefer to compute $P(N_3 < 7)$, we write $P(N_3 < 7) = P(N_3 \le 6) = PoissonDist(6;\,6)$, since $E\{N_3\} = 2\cdot 3$. The answer is, naturally, the same.

EXAMPLE 2. Assume now that for the same problems, the information given concerns the number of claims rather than interarrival times. For example, suppose it is given that on the average the company receives 5 claims each 6 hours. This means that $E\{N_6\} = 5 = \lambda\cdot 6$, so $\lambda = 5/6$, and we may repeat all calculations with the new λ.

EXAMPLE 3. Consider the same process as in the above examples. Suppose it was noticed that, on the average, in half of the cases, half an hour or more elapses before the subsequent claim arrives. Translating into notation, it says that $P(\tau_i > 1/2) = 1/2$. Hence, $e^{-\lambda/2} = 1/2$, and $\lambda = 2\ln 2 \approx 1.38$. After that, we proceed as above. □

The model described provides a simple way to simulate Poisson variables and the Poisson process itself. We simulate a sequence of values of independent exponential variables with a parameter λ, which will represent the values of consecutive interarrival times. Then, we sum them up *until the sum exceeds one.* In accordance with what we have proved, the

	A	B	C	D	E	F	G	H	I	J	K	L	M	N
1	0.38	0.10	0.60	0.90	0.88	0.96	0.01	0.41	0.86	0.14	0.25	0.05	0.03	0.16
2	0.24	0.57	0.13	0.03	0.03	0.01	1.06	0.22	0.04	0.49	0.35	0.77	0.86	0.45
3	0.24	0.81	0.94	0.97	1.00	1.01	2.07	2.29	2.33	2.83	3.18	3.95	4.81	5.26
4														
5			5			1		4			2		1	
6														
7	4	λ												
8														

FIGURE 5. Simulation of a Poisson r.v.

number of terms in such a sum, not counting the last, has the Poisson distribution with the parameter λ.

We may simulate values of exponentially distributed r.v.'s in accordance with the inverse-distribution-function method of Section **0**.3.2.1. More precisely, consider the exponential d.f. $F(x) = 1 - e^{-\lambda x}$, and its inverse $F^{-1}(y) = -\frac{1}{\lambda}\ln(1-y)$. Then, in accordance with Proposition 7 of Section **0**.3.2.1, if Z is a r.v. uniformly distributed on $[0,1]$, then the r.v. $X = -\frac{1}{\lambda}\ln(1-Z)$ is exponential with parameter λ.

Note that $1-Z$ is also uniformly distributed on $[0,1]$; the proof is left to the reader. Therefore, we can use $\ln Z$ instead of $\ln(1-Z)$, because their distributions are identical.

Fig.5 presents an Excel worksheet. Row 1 contains random numbers—denote them by Z—uniform on $[0,1]$, generated by Excel. The parameter λ is specified in Cell A7.

Numbers in Row 2 are obtained by the formula $-\frac{1}{\lambda}\ln Z$. They are independent values of a r.v. having the exponential distribution with the parameter λ. (These values, are, certainly, pseudo independent, since we used a computer to generate them.)

Row 3 contains the successive sums of the numbers from Row 2, so the numbers in Row 2 correspond to the arrival times. We see that there were 5 arrivals in the first unit of time, 1 arrival in the second, 4 in the third, and so on. These numbers simulate a sequence of independent values of a Poisson random variable. Excel is not very convenient for such simulation, so we restrict ourselves to this simple worksheet.

2.2 The non-homogeneous Poisson process

2.2.1 A model and examples

The distribution in (2.1.8) depends on the length of the interval but not on its location, which indicates that the intensity of arrivals does not change in time. In many real situations, this may be assumed only for a short period of time. For example, the intensity of the flow of claims coming into an insurance company may vary at different moments of the day, days of the week, and seasons of the year.

To model this, we introduce a function $\lambda(t)$ which is interpreted as the instantaneous intensity of the arrival flow at time t. (The significance of the word "instantaneous" is the same as for the speed at a particular moment of a vehicle moving with a varying speed). We call the function $\lambda(t)$ an *intensity function* of the process of arrivals; or simply *intensity*.

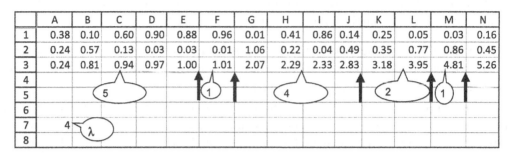

Let

$$\chi(t) = \int_0^t \lambda(s)ds. \qquad (2.2.1)$$

If $\lambda(t)$ is equal to a constant λ for all t, then $\chi(t) = \lambda t$.

For an arbitrary interval $\Delta = [t, t+u]$, set

$$\chi_\Delta = \int_\Delta \lambda(s)ds = \int_t^{t+u} \lambda(s)ds = \int_0^{t+u} \lambda(s)ds - \int_0^t \lambda(s)ds = \chi(t+u) - \chi(t). \qquad (2.2.2)$$

When $\lambda(t) = \lambda$, we have $\chi_\Delta = \lambda u = \lambda|\Delta|$, where $|\Delta|$ stands for the length of Δ.

By analogy with Properties P1–P3 from Section 2.1, we call a process N_t, $t \geq 0$, a *non-homogeneous Poisson process* if it has the following properties.

PN1. $N_0 = 0$ with probability one.

PN2. N_t is a counting process with independent increments.

PN3. There exists a non-negative function $\lambda(t)$ such that for any interval Δ, the r.v. N_Δ has the Poisson distribution with parameter χ_Δ defined in (2.2.2).

Thus, for any interval Δ,

$$P(N_\Delta = n) = e^{-\chi_\Delta} \frac{\chi_\Delta^n}{n!}. \qquad (2.2.3)$$

From (2.2.3) we obtain, in particular, that

$$E\{N_\Delta\} = \chi_\Delta.$$

Considering an interval $\Delta = [0, t]$, we have

$$P(N_t = n) = \exp\{-\chi(t)\} \frac{(\chi(t))^n}{n!}. \qquad (2.2.4)$$

The homogeneous case corresponds to a constant intensity $\lambda(t)=\lambda$: in this case $\chi(t)=\lambda t$, and we return to (2.1.4).

EXAMPLE 1. Customers arrive at a service facility according to a non-homogeneous Poisson process with a rate of 3 customers/hour in the period between 9am and 11am. After 11am, the rate is decreasing linearly from 3 at 11am to zero at 5pm. Find the probability that there will be not more than 15 customers between 10am and 4pm.

Taking 9am as an initial time, we have $\lambda(t) = 3$ for $t \in [0, 2]$, and $\lambda(t) = (8-t)/2$ for $t \in [2, 8]$. For the interval $\Delta = [1, 7]$, the expected number of customers $\chi_\Delta = \int_1^7 \lambda(s)ds =$ 11.75, and we get (for example, using Excel) that $P(N_\Delta \leq 15) = \mathit{PoissonDist}(15; 11.75) \approx$ 0.862. □

In the general case, when the rate can change in time, the number of arrivals may be small during a long period, and may be large during a short period.

EXAMPLE 2. (*Can the number of arrivals be finite during an infinite period of time?*) The question concerns the value of

$$E\{N_{[0,\infty)}\} = \chi_{[0,\infty)} = \int_0^\infty \lambda(s)ds.$$

If the integral above converges, then the mean number of arrivals is finite. Suppose, for instance, that $\lambda(s)=e^{-s}$. Then the mean $E\{N_{[0,\infty)}\}=\int_0^\infty e^{-s}ds=1$, so during the *infinite* period there will be on the average *only one* arrival. One may consider any example for which the integral above converges; say, $\lambda(s)=1/(1+s^2)$. Then the r.v. $N_{[0,\infty)}$ is again a Poisson r.v. with a finite mean value. (What is it equal to?)

EXAMPLE 3 (*An explosion*). Let under some circumstances the rate of claims start to grow very fast; from a modeling point of view, we can consider $\lambda(t)$ converging to infinity during a finite period. For example, let $\lambda(t)=1/(1-t)$. Then $\chi(t)=\int_0^t(1-s)^{-1}ds=-\ln(1-t)$ for $t<1$. Hence, $\chi(t)\to\infty$ as $t\to 1$; that is, during the period $[0,1]$ there will be an infinite number of claims with probability one. □

Consider now how interarrival times look in the non-homogeneous case. We will see that, *in general, they are no longer exponential or independent.*

Indeed, for the first arrival

$$P(\tau_1>t)=P(N_t=0)=\exp\{-\chi(t)\}. \tag{2.2.5}$$

If $\lambda(t)=\lambda$, we come to the exponential distribution (2.1.9), however if, for instance, $\lambda(t)=t$, then $P(\tau_1>t)=\exp\{-\int_0^t sds\}=\exp\{-t^2/2\}$. In this case, τ_1 has a distribution different from the exponential distribution.

Furthermore, for the nth interarrival time τ_n given the time T_{n-1} of the previous arrival, the conditional probability

$$P(\tau_n>s\,|\,T_{n-1}=t)=P(N_{(t,t+s]}=0\,|\,T_{n-1}=t)=P(N_{(t,t+s]}=0)=\exp\left\{-\int_t^{t+s}\lambda(u)du\right\}. \tag{2.2.6}$$

We see that the above probability depends on the time of the previous arrival.

However, it should not look scary because the distribution of the nth arrival is still tractable: the formula (2.1.1) is true in the general case, and

$$P(T_n\le t)=P(N_t\ge n)=1-P(N_t<n), \tag{2.2.7}$$

which may be computed as a Poisson probability.

EXAMPLE 4. Suppose that the intensity $\lambda(t)=9(8-t)^2/64$ for $0\le t\le 8$, that is, starting from 9, the intensity decreases as a parabola and equals zero at time 8. The expected number of customers during the whole period is $\chi(8)=\frac{9}{64}\int_0^8(8-t)^2dt=24$.

Is the probability that at least 20 customers will come within the first half of the period significant? (Notice that the intensity is decreasing rapidly.)

We have $E\{N_4\}=\chi(4)=\frac{9}{64}\int_0^4(8-t)^2dt=21$, so the probability should be more than 0.5. More precisely, by (2.2.7), $P(T_{20}\le 4)=P(N_4\ge 20)=1-P(N_4\le 19)=1-PoissonDist(19;21)\approx 0.615$. □

2.2.2 Another perspective: Infinitesimal approach

Conditions P3 and PN3 in the above definitions of homogenous and non-homogeneous processes, respectively, are clear; but its verification in particular cases may cause difficulties. The equivalent definition below proceeds from the behavior of the process merely in small time intervals. It is quite useful for applications and gives an additional insight into the nature of Poisson processes.

Below, the symbol $o(\delta)$ stands for a function which is "much smaller" than δ for small δ; formally, $\frac{1}{\delta}o(\delta) \to 0$ as $\delta \to 0$. It is important to emphasize that in different formulas $o(\delta)$ may denote different functions but since we do not need to specify them, we will use the same symbol in different formulas. The reader may find more comments regarding this notation in the Appendix, Section 4.1.

We define N_t from scratch. Suppose that it is a *counting process N_t* with *independent increments* such that $N_0 = 0$, and for a function $\lambda(t) \geq 0$ and any positive t and δ,

$$P(N_{(t,t+\delta]} = 1) = \lambda(t)\delta + o(\delta) \text{ as } \delta \to 0, \tag{2.2.8}$$

$$P(N_{(t,t+\delta]} > 1) = o(\delta) \text{ as } \delta \to 0. \tag{2.2.9}$$

Again, $o(\delta)$ in (2.2.8) may differ from $o(\delta)$ in (2.2.9). In (2.2.8), the term $o(\delta)$ is negligible with respect to $\lambda(t)\delta$, while relation (2.2.9) means that for small δ (i.e., for small intervals), $P(N_{(t,t+\delta]} > 1)$ is negligibly small.

This may be understood as follows. For a small time interval, the probability of an arrival is proportional to the length of the interval (up to a negligible remainder), and the probability that more than one arrival will occur is negligible. The coefficient of proportionality, $\lambda(t)$, depends on time and, as in Section 2.2.1, is interpreted as the mean number of arrivals per unit of time in a neighborhood of t. It is called the *rate* or the *intensity* at time t.

We define $\chi(t)$ and χ_Δ as in (2.2.1) and (2.2.2), respectively.

Proposition 1 *For the process N_t defined above and any interval Δ,*

$$P(N_\Delta = n) = e^{-\chi_\Delta}\frac{\chi_\Delta^n}{n!}. \tag{2.2.10}$$

A proof will be given in Section 2.2.3.

The infinitesimal approach allows us to easily solve many problems involving varying intensities.

Assume, for example, that arrivals—for instance, claims—are counted (or accepted) only with a probability p, perhaps depending on time: $p = p(t)$. In this case, the only change needed is to replace the intensity $\lambda(t)$ by $p(t)\lambda(t)$ in condition (2.2.8).

Indeed, for an arrival to be counted in a period $[t, t+\delta]$, first, the arrival must occur—the probability of this is $\lambda(t)\delta + o(\delta)$. Secondly, once an arrival has occurred, the probability that it belongs to the first type is $p(t)$. If we count only this type of arrival, we should multiply $\lambda(t)\delta + o(\delta)$ by $p(t)$. This gives $(\lambda(t)\delta + o(\delta))p(t) = p(t)\lambda(t) + p(t)o(t)$. The last term is negligible with respect to the second, and we may rewrite the whole expression as $p(t)\lambda(t) + o(t)$.

This is a particular case of the *marked Poisson process*. We have already considered this phenomenon in the static case in Section **3**.2.2.2 and, in greater detail, in Section **3**.3.1.2, and Exercises **3**-12, **3**-13.

EXAMPLE 1. After some moment which we view as initial, the flow of customers entering a service facility started to grow with intensity $\lambda(t) = 3(1+t)$. In order to keep the rate equal to the initial rate 3, the management decides to refuse some customers service depending on the character of the job to be done. Since the type of the next customer is not known in advance, the process of refusals is random. With what probability should the facility accept the claim arrived at time t? With probability $p(t) = 1/(1+t)$, since then $p(t)\lambda(t) = 3$.

EXAMPLE 2 ([153, N13][1]). During the hurricane season (August, September, October, and November), hurricanes hit the US coast with a monthly rate of 1.25, and each hurricane during this period has a 20% chance of being a "major". Outside of hurricane season (the other months), hurricanes hit at a Poisson rate of 0.25 per month, and each such a hurricane has only a 10% chance of being "major". Determine the probability that a hurricane selected at random is "major".

The problem concerns the formula

$$p_i = \lambda_i/\lambda$$

from Sections **3**.3.1.2 and **3**.3.2.1 for the probability that a particular claim comes from the ith group. In our case, claims are hurricanes, and the group is that of "major" ones. Using (2.2.1), for the *annual* intensity of hurricanes we get $\lambda = (1.25)\cdot 4 + (0.25)\cdot 8 = 7$.

To find the intensity of hurricanes from the group mentioned, we should use the same formula (2.2.1) multiplying the intensity $\lambda(s)$ by the corresponding probability of counting $p(s)$. This leads to $\lambda_1 = (1.25)\cdot 4\cdot 0.2 + (0.25)\cdot 8\cdot 0.1 = 1.2$.

Then the probability $p_1 = (1.2/7) \approx 0.1714$. □

In the last two examples, we considered two types of arrivals: counted and not counted. Assume now that claims arriving may be of l types, and the probability of the ith type, independently of what happened before, is p_i. We have already touched on this question in Section **3**.3.1.2. In the present framework, it immediately follows that the process of claims of the ith type, N_{ti}, is Poisson with the intensity $\lambda_i(t) = p_i\lambda(t)$, and, as was proved in Section **3**.3.1.2, the processes N_{ti} are independent. If the ith type implies a payment of x_i, the total payment by time t is

$$\sum_{i=1}^{l} x_i N_{ti}. \tag{2.2.11}$$

See an example in Exercise 18.

2.2.3 Proof of Proposition 1

Set $p_n(t) = P(N_t = n)$, and for a $\delta > 0$, consider two moments of time: t and $t+\delta$. Note that if $N_{t+\delta} = n$, and $N_{(t,t+\delta]}$ equals some $k \leq n$, then N_t should be equal to $n-k$. Hence,

[1]Reprinted with permission of the Casualty Actuarial Society.

for $n = 0, 1, 2, \ldots$, in view of the independence of increments,

$$p_n(t+\delta) = P(N_{t+\delta} = n) = \sum_{k=0}^{n} P(N_t = n-k, N_{(t,t+\delta]} = k)$$
$$= \sum_{k=0}^{n} P(N_t = n-k)P(N_{(t,t+\delta]} = k) = P(N_t = n)P(N_{(t,t+\delta]} = 0)$$
$$+ P(N_t = n-1)P(N_{(t,t+\delta]} = 1) + \sum_{k=2}^{n} P(N_t = n-k)P(N_{(t,t+\delta]} = k).$$

We are going to use (2.2.8)–(2.2.9). As was noted, the symbol $o(\delta)$ in these formulas may denote different functions. Nevertheless, we can combine them, and if for example, we consider the sum of remainders, then we arrive at another remainder which may be again denoted by $o(\delta)$. First,

$$P(N_{(t,t+\delta]} = 0) = 1 - P(N_{(t,t+\delta]} = 1) - P(N_{(t,t+\delta]} > 1) = 1 - (\lambda(t)\delta + o(\delta)) - o(\delta)$$
$$= 1 - \lambda(t)\delta + o(\delta),$$

and, secondly,

$$\sum_{k=2}^{n} P(N_t = n-k)P(N_{(t,t+\delta]} = k) \leq \sum_{k=2}^{n} P(N_{(t,t+\delta]} = k) \leq P(N_{(t,t+\delta]} > 1) = o(\delta).$$

Hence, $p_n(t+\delta) = p_n(t)(1 - \lambda(t)\delta + o(\delta)) + p_{n-1}(t)(\lambda(t)\delta + o(\delta)) + o(\delta) = p_n(t)(1 - \lambda(t)\delta) + p_{n-1}(t)\lambda(t)\delta + o(\delta)$. We rewrite it as

$$\frac{1}{\delta}[p_n(t+\delta) - p_n(t)] = -\lambda(t)p_n(t) + \lambda(t)p_{n-1}(t) + \frac{1}{\delta}o(\delta).$$

Letting $\delta \to 0$, and recalling that $\frac{1}{\delta}o(\delta) \to 0$, we come to the differential equation

$$p_n'(t) = -\lambda(t)p_n(t) + \lambda(t)p_{n-1}(t). \tag{2.2.12}$$

For $n = 0$, since $p_{-1}(t) = P(N_t = -1) = 0$,

$$p_0'(t) = -\lambda(t)p_0(t). \tag{2.2.13}$$

A solution to the last equation is

$$p_0(t) = \exp\{-\chi(t)\}, \tag{2.2.14}$$

where $\chi(t)$ is defined in (2.2.1). The reader who does not remember how to solve equations of this type can easily verify (2.2.14) by substitution. The reader who remembers this should take into account that $p_0(0) = 1$ (we start with no claims), and together with this initial condition, solution (2.2.14) with $\chi(t)$ given in (2.2.1) is the unique solution to the ordinary differential equation (2.2.13).

Once we know $p_0(t)$, we get from (2.2.12) that

$$p_1'(t) = -\lambda(t)p_1(t) + \lambda(t)\exp\{-\chi(t)\}, \tag{2.2.15}$$

which leads to

$$p_1(t) = \chi(t) \exp\{-\chi(t)\}.$$

The reader can again check it by substitution. Continuing the same procedure by induction, we come to solutions for any n; namely, to (2.2.4).

Consider an arbitrary interval $\Delta = [t, t+u]$. In this case, proceeding as in the previous section from the independence of increments, we can view the point t as an initial point and use the same formula (2.2.4) with $\chi(t)$ replaced by χ_Δ. ■

Route 1 ⇒ *page 261*

2.3 The Cox process

Certainly, the process of arrivals, in particular, the flow of claims arriving at an insurance company, may be more complicated than the Poisson process. An important modification is the *Cox process* which is defined as a Poisson process whose intensity $\lambda(t)$ is random. More precisely, instead of $\lambda(t)$ we consider a random process Λ_t. A similar situation for a static model was explored in Section **3**.2.2.1.

EXAMPLE 1. In the time interval $[0,2]$, two periods are distinguished. On $[0,1]$, the random intensity Λ_t equals a r.v. Z_1, and on $(1,2]$, the intensity Λ_t equals a r.v. Z_2. The value of the intensity at the switching point $t=1$ does not matter: a point is an interval of zero length, so the probability of an arrival exactly at this point is zero.

Thus, $N_{[0,1]}$ and $N_{(1,2]}$ are r.v.'s with mixed Poisson distributions described in **3**.2.2.1.

Given Z_1, the r.v. $N_{[0,1]}$ is Poisson with a mean of Z_1. A similar assertion is true for $N_{(1,2]}$. So, we can write that $E\{N_{[0,1]}\} = E\{Z_1\}, E\{N_{(1,2]}\} = E\{Z_2\}$, and for the total number of claims

$$E\{N_{[0,2]}\} = E\{N_{[0,1]}\} + E\{N_{(1,2]}\} = E\{Z_1\} + E\{Z_2\}.$$

If Z_1, Z_2 are independent, the r.v.'s $N_{[0,1]}$ and $N_{(1,2]}$ are also independent, and the distribution of $N_{[0,2]}$ is the convolution of the distributions of the r.v.'s $N_{[0,1]}$ and $N_{(1,2]}$. □

In general, the Cox process does not necessarily have independent increments. Values of the process during a particular period may contain some information about the intensity process Λ_t, which may be used for predicting the behavior of the process in the future. It is apparent in the case when the process $\Lambda_t = \Lambda$, the same r.v. not depending on time.

EXAMPLE 2. The process of claims N_t runs over the period $[0,2]$, the process $\Lambda_t = \Lambda$, and the r.v. Λ takes on values 3 and 10 with probabilities $3/4$ and $1/4$, respectively. Suppose that we have observed the process during the period $[0,1]$, and that N_1 took a value of 8. How many claims should we expect during the interval $[1,2]$, having this information? What is the probability that $N_{[1,2]}$ will exceed, say, 5?

Note that, for the sake of simplicity, we chose unit intervals, and the particular numbers are the same as in Example **3**.2.2.1-5. We have computed there that $P(\Lambda = 10 \,|\, N_1 = 8) \approx$

0.822. Hence, $E\{N_{[1,2]}\,|\,N_1=8\}\approx 3\cdot 0.178+10\cdot 0.822=8.754$, whereas the prior unconditional expectation $E\{N_{[1,2]}\}=3\cdot 0.75+10\cdot 0.25=4.75$. Next, $P(N_{[1,2]}>5\,|\,N_1=8)\approx 1-(0.822\cdot \mathit{PoissonDist}(5;\,10)+0.178\cdot \mathit{PoissonDist}(5;\,3))\approx 0.782$.

EXAMPLE 3. (*The Pòlya process*). Let Λ have the Γ-distribution with parameters a and ν as in Examples **3**.2.2.1-2 and 6.

The random parameter Λ is an intensity. So, given Λ, the r.v. N_t has the Poisson distribution with parameter $t\Lambda$. Hence, when determining the distribution of N_t, we should replace the r.v. Λ in Example **3**.2.2.1-2 by $t\Lambda$. Then the corresponding Γ-distribution in the mentioned example will be that with parameters (a/t), ν. Consequently, in accordance with the result from Section **3**.2.2.1, N_t has the negative binomial distribution with parameters $p_t=\dfrac{(a/t)}{1+(a/t)}=\dfrac{a}{t+a}$ and ν. In particular,

$$E\{N_t\}=tE\{\Lambda\}=\frac{\nu t}{a}.$$

Let, for example, $a=2$ and $\nu=4$. In this case, $\lambda_{ave}=E\{\Lambda\}=(\nu/a)=2$. Assume that we monitored the process until time $t_1=1$ and observed that $N_1=5$. In accordance with Example **3**.2.2.1-6, the conditional distribution of Λ, given $N_1=5$, is the Γ-distribution with parameters $\tilde{a}=2+1=3$, $\tilde{\nu}=5+4=9$. Hence, the number of arrivals in the next unit period, $N_{[1,2]}$, has the negative binomial distribution with parameters $p=\tilde{a}/(1+\tilde{a})=3/4$, and $\tilde{\nu}=9$ (see Section **3**.2.2.1). In particular, we should expect on the average not $\lambda_{ave}=2$ claims, but $E\{\Lambda\,|\,N=5\}=3$ claims. □

---

3 COMPOUND PROCESSES 1,2,3

A *compound process* is a process

$$S_{(t)}=\sum_{i=1}^{N_t} X_i, \tag{3.1}$$

where N_t is a counting process, and the X_i's are i.i.d. r.v.'s not depending on the evolution of the process N_t.

The interpretation for the insurance model is clear: N_t is the number of claims arrived by time t, and X_i is the size of the ith claim (in other terms, a *severity*). We write the index (t) in order not to confuse, say, $S_{(2)}$ with S_2 which usually denotes X_1+X_2 as opposed to $\sum_{i=1}^{N_2} X_i=S_{(2)}$. In the latter case, the sum contains a random number of addends, and N_2 may be larger or less than two.

We use also the habitual notation X_i from Chapter 3. It should not cause confusion with the notation X_t for a random process as we will not use these notations with different meanings in the same context.

The graph of a typical realization is given in Fig.6. At the initial time, there are no claims and therefore $S_{(0)}=0$. The first claim of a size of X_1 appears at time τ_1, and the process

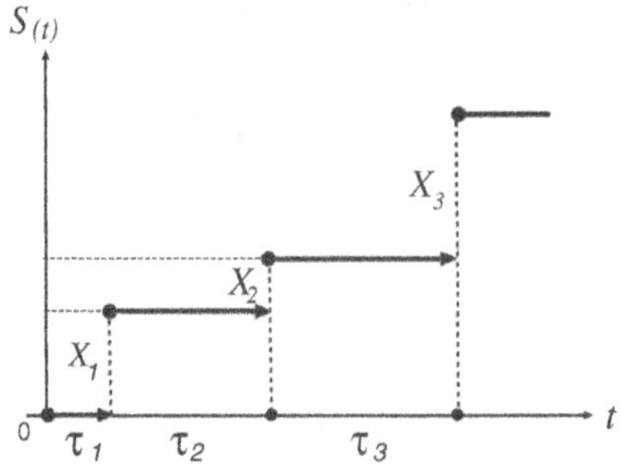

FIGURE 6. A typical realization of the compound process

jumps up by X_1. Then, during a period whose length equals the second interarrival time τ_2 nothing happens and after that, at time $\tau_1 + \tau_2$, the second claim of a size of X_2 arrives. So, the process jumps up by X_2, and the further evolution runs in the same fashion.

We call $S_{(t)}$ a *compound process*, and if N_t is a Poisson process—a *compound Poisson process*.

The scheme (3.1) is very similar to the scheme considered in detail in Chapter 3. The only difference is that now the number of addends, N_t, depends on the time parameter t. If we are interested in the distribution of $S_{(t)}$ for a particular fixed t, then all results of Chapter 3 apply to the current scheme. Dynamic problems connected with the evolution of the process $S_{(t)}$ in time will be considered later in Chapter 6.

Let us return to the distribution of $S_{(t)}$ for a particular t and consider a compound Poisson process. In this case, when using the results for the compound Poisson distribution in Chapter 3, we should replace the parameter λ of the Poisson r.v. N in Chapter 3 by $E\{N_t\}$.

First, assume the process N_t is homogeneous. Then $E\{N_t\} = \lambda t$, and in this case (see Section **3**.1),

$$E\{S_{(t)}\} = m\lambda t, \;\; Var\{S_{(t)}\} = (\sigma^2 + m^2)\lambda t,$$

where $m = E\{X_i\}$, $\sigma^2 = Var\{X_i\}$.

If t is large, we can apply the normal approximation for the distribution of $S_{(t)}$ in accordance with Theorem **3**.4.1-12. In particular, it concerns the approximation for the premium c_t such that $P(S_{(t)} \le c_t) > \beta$, where β is a given level of security. (See comments in the beginning of Section **3**.4.2.) Setting $c_t = (1+\theta)m\lambda t$, we obtain the counterpart of formula (**3**.4.2.2):

$$\theta \approx \frac{q_{\beta s}\sqrt{(\sigma^2 + m^2)\lambda t}}{m\lambda t} = \frac{q_{\beta s}}{\sqrt{\lambda t}}\sqrt{1+k^2}, \tag{3.2}$$

where the coefficient of variation $k = \sigma/m$ and β is a given security level.

It is worth emphasizing, however, that in the above result, the time moment t is fixed, and θ in (3.2) concerns this particular chosen t. The problem of finding θ for which $S_{(t)} \le c_t$ for *all* t from an interval $[0, T]$ is more complicated and will be considered in Chapter 6.

In the case when N_t is non-homogeneous, we replace the parameter λ of the Poisson r.v. N in Chapter 3 by $E\{N_t\} = \chi(t)$ defined in Section 2.2.1. In this case

$$E\{S_{(t)}\} = m\chi(t), \;\; Var\{S_{(t)}\} = (\sigma^2 + m^2)\chi(t).$$

If t is large, and $\chi(t) \to \infty$ as $t \to \infty$ (we saw in Example 2.2.1-3 that this is not always the case), then we can again apply the normal approximation for the distribution of $S_{(t)}$ in accordance with the same Theorem **3**.4.1-12. In particular, we set $c_t = (1+\theta)m\chi(t)$ and, instead of (3.2), write

$$\theta \approx \frac{q_{\beta s}}{\sqrt{\chi(t)}}\sqrt{1+k^2}. \tag{3.3}$$

Since this topic is very similar to what we did in Chapter 3, we omit direct particular examples; see also Exercises 23-24.

Route 1 ⇒ *page 339*

2,3

However, let us consider a nice

EXAMPLE 1 ([157, N10][2]). An insurance policy has aggregated losses according to compound Poisson distribution. Claim frequency follows a Poisson process. The average number of claims reported is 200. Claim severities are independent and follow an exponential distribution with a mean of 160,000.

Management considers any claim that exceeds 1 million to be a catastrophe. Calculate the median waiting time (in years) until the first catastrophe claim.

The word "severities" means the claim sizes, i.e., the X's. Let K be the number of the first catastrophe claim. For a claim, the probability that it will be a catastrophe one is $p = P(X > 10^6) = \exp\{-10^6/(16\cdot 10^4)\} \approx 0.0019305$. Since we may view an appearance of a catastrophe claim as a "success", K has the first version of the geometric distribution with the above parameter p.

Since N_t is a Poisson process, the inter-arrival times τ_i are exponentially distributed with $a = 200$.

The arrival time of the first catastrophe claim is the r.v. $T = \sum_{i=1}^{K} \tau_i$. In accordance of the result of Example **3**.3.1.3-1, the r.v. T has the exponential distribution with parameter $\bar{a} = pa \approx 0.386$.

We are looking for a t such that $P(T > t) = e^{-\bar{a}t} = 0.5$. Hence, $t = \frac{\ln 2}{\bar{a}} \approx 1.795$. □

---

4 MARKOV CHAINS. CASH FLOWS IN THE MARKOV ENVIRONMENT

4.1 Preliminaries

A *Markov chain* is a Markov process X_t in discrete time $t = 0, 1, 2, \ldots$. Often this term is applied to processes taking integer values, and we will consider this case below. Moreover, as a rule though not always, X_t will take on non-negative integer values $i = 0, 1, 2, \ldots$.

If X_t assumes a value i, the process is said to be in *state* i at time t. Usually, such a model serves for a description of the evolution of a dynamic system moving from one state

[2]Reprinted with permission of the Casualty Actuarial Society.

to another at discrete moments of time. Since the process is Markov, to determine the probabilities of its possible realizations, it suffices to specify the following characteristics.

A. ${}_tp_{ij} = P(X_{t+1} = j \,|\, X_t = i)$, the probability that the process, being in state i at time t, will make a transition to state j in the next step. In particular, ${}_tp_{ii}$ is the probability of staying in the same state i at the next time moment $t+1$.

B. $\pi_{0i} = P(X_0 = i)$, the probability of being in state i at the initial moment of time.

Probabilities ${}_tp_{ij}$ are called *transition probabilities*, and the matrix

$$
{}_t\mathcal{P} = \|{}_tp_{ij}\| = \left\| \begin{array}{cccc} {}_tp_{00} & {}_tp_{01} & {}_tp_{02} & \cdots \\ {}_tp_{10} & {}_tp_{11} & {}_tp_{12} & \cdots \\ {}_tp_{20} & {}_tp_{21} & {}_tp_{22} & \cdots \\ \vdots & \vdots & \vdots & \vdots \end{array} \right\|
$$

is called a *transition probability matrix,* or briefly, a *transition matrix.*

We call the vector of probabilities $\boldsymbol{\pi}_0 = (\pi_{00}, \pi_{01}, \pi_{02}, ...)$ an *initial probability distribution.* If the system starts from a fixed state i_0, then $\boldsymbol{\pi}_0 = (...,0,1,0,...)$, where 1 is in position i_0.

Since the process always stays in some state, the sum of all entries in the vector $\boldsymbol{\pi}_0$ is equal to one. The same is true for the sum of all probabilities in *each row* of the matrix ${}_t\mathcal{P}$ since the process moves from the state where it is in, to some state (perhaps the same).

A matrix with non-negative elements and with such a property is called *stochastic.* Any stochastic matrix may be the transition probability matrix for some chain. Indeed, we would just define such a chain setting its transition matrices ${}_t\mathcal{P}$ equal to the matrix given.

A chain is called *homogeneous* if its transition probabilities ${}_tp_{ij}$ do not depend on t, that is, ${}_tp_{ij}$ equals some p_{ij} for all t, i, and j. In this case, we write ${}_t\mathcal{P} = \mathcal{P} = \|p_{ij}\|$.

There are myriads of the examples of Markov chains. For now, we consider four.

EXAMPLE 1. For a car insurance portfolio of a company operating in a particular area, the intensity of daily claims depends on weather conditions. The company distinguishes three types: normal, rainy, and icy-road conditions, leading to three states which we label $0,1,2$. Particular transition probabilities characterize the conditions of a season in the area, and if the time period under consideration is not long, we may assume that these probabilities do not depend on time. For instance, the transition matrix

$$
\mathcal{P} = \left\| \begin{array}{ccc} 0.6 & 0.3 & 0.1 \\ 0.4 & 0.5 & 0.1 \\ 0.25 & 0.7 & 0.05 \end{array} \right\| \tag{4.1.1}
$$

may characterize a soft winter. Since among p_{ii}'s (the probabilities to stay in the same state as on the previous day) the probability $p_{00} = 0.6$ is the highest, the normal condition appears to be the most stable. The distribution $\boldsymbol{\pi}_0$ characterizes the condition at the initial time. For example, if in the beginning of the season it was raining, then $\boldsymbol{\pi}_0 = (0,1,0)$.

It is important to emphasize that the above Markov model implicitly presupposes that *given* the current condition, the information on the weather conditions on previous days is not needed for the weather forecast. In some situations, such a simplification may be acceptable, but in general it is certainly useful to know the tendency in weather conditions in the past. So, such a simple Markov model may not be as useful in providing accurate results.

However, it does not mean that we should refuse the Markov setup. We can keep using it if we extend the set of possible states of the weather on a current day. Namely, we should include into the characterization of states information about the past. For example, if we want to take into account the information about the weather on two consecutive days, we may consider, *for a current day*, such states as "rainy today, and rainy yesterday", "rainy today, and normal yesterday", "icy today, and rainy yesterday", and so on. Such a model may turn out to be adequate enough. See also Exercise 30.

EXAMPLE 2. For a group of people, let the probability that a person of age s will attain age $s+1$ be p_s. Set $q_s = 1 - p_s$, and consider a person of an initial age x from this group and two states: alive (more generally, intact) and deceased (more generally, failed). Since the person chosen is alive, $\boldsymbol{\pi}_0 = (1, 0)$. If the person lives t years more, she/he will attain the age $x+t$, and the probability that, *after that*, she/he will live at least one year more is p_{x+t}. Hence, the transition matrix

$$ {}_t\mathcal{P} = \left\| \begin{array}{cc} p_{x+t} & q_{x+t} \\ 0 & 1 \end{array} \right\|. $$

The chain is not homogeneous, and, say, for $x > 30$, it is reasonable to assume that p_{x+t} is decreasing in t. (In the range of small s, the probability p_s may actually increase. For example, p_{16} may be smaller than p_{25}; one can theorize why this might be the case. For a more detailed discussion on survival probabilities, see Chapter 7.)

Denote by h_s the probability that if a person of age s dies within a year, it happens because of an accident. Consider three states: alive, died because of an accident, died from other causes. In this case,

$$ {}_t\mathcal{P} = \left\| \begin{array}{ccc} p_{x+t} & q_{x+t}h_{x+t} & q_{x+t}(1 - h_{x+t}) \\ 0 & 1 & 0 \\ 0 & 0 & 1 \end{array} \right\|. \tag{4.1.2} $$

EXAMPLE 3 (*Rearrangement* or *shuffling*). A stack of k books lie on a desk. You take a book at random, read what you need, and put the book on the top of the stack. The states of the system may be identified with the orders in which the books can be arranged, that is, with $k!$ permutations of the numbers $1, 2, ..., k$. The reader may replace the word 'book' by 'card', and 'stack'—by 'deck'. In this case, we are talking about one of simplest ways of shuffling.

The process above cannot move from any state to any state in one step. For example, if $k = 3$, and we number books as 1,2,3, then the process can move from state (1,2,3) only to the same state and to states (2,1,3) and (3,1,2). If all books are equally likely to be chosen, the process can move from state (1,2,3) to each of the mentioned states with probability $1/3$, and with zero probability—to the other three states: (1,3,2); (2,3,1); (3,2,1). Applying the same argument to other states, we conclude that the transition matrix of our homogeneous Markov chain is

$$\begin{array}{c} \quad\quad 123\downarrow \quad 132\downarrow \quad 213\downarrow \quad 231\downarrow \quad 312\downarrow \quad 321\downarrow \\ \mathcal{P} = \left\| \begin{array}{cccccc} 1/3 & 0 & 1/3 & 0 & 1/3 & 0 \\ 0 & 1/3 & 1/3 & 0 & 1/3 & 0 \\ 1/3 & 0 & 1/3 & 0 & 0 & 1/3 \\ 1/3 & 0 & 0 & 1/3 & 0 & 1/3 \\ 0 & 1/3 & 0 & 1/3 & 1/3 & 0 \\ 0 & 1/3 & 0 & 1/3 & 0 & 1/3 \end{array} \right\| \begin{array}{l} \leftarrow 123 \\ \leftarrow 132 \\ \leftarrow 213 \\ \leftarrow 231 \\ \leftarrow 312 \\ \leftarrow 321 \end{array} \end{array} \qquad (4.1.3)$$

(verify on your own; the arrows $\leftarrow$ and $\downarrow$ show to which state (permutation) a row and a column corresponds).

If in the beginning, all arrangements are equally likely, then $\boldsymbol{\pi}_0 = (1/6, 1/6, 1/6, 1/6, 1/6, 1/6)$.

EXAMPLE 4 (*The simple random walk*). Let the initial state X_0 be equal to an integer u, and let $X_{t+1} = X_t + \xi_{t+1}$, where ξ_t's are independent r.v.'s taking values ± 1 with probabilities p and $1-p$, respectively. Then $X_t = u + \xi_1 + ... + \xi_t$.

For example, u may be the initial capital of a person who, in consecutive moments of time, either gets or loses one unit of money with the above probabilities. In this case, X_t is the total capital at time t. We assume that the capital may be negative (the person owes money). The reader may think about the simplest game of chance with tossing a coin, assuming that the coin may be non-symmetric.

The process above may be considered also the simplest model of the surplus process for a risk portfolio (see Section 1.1) in discrete time. In this case, ξ_t is interpreted as the profit of the company during the period $(t-1, t]$; that is, the premium minus the payment. In the simplest model, we can think that it assumes only values ± 1.

The term "random walk" comes from another interpretation. It concerns the motion of a particle which moves at each discrete moment of time either to the right or to the left by one unit of distance. For a particle immersed in a liquid, such a motion is connected with the bombardment by the molecules of the surrounding medium. (In fact, the particle is moving in a three-dimensional space, so we are talking about the projection of such a motion on a certain direction.)

States in the random walk may be identified with integers $0, \pm 1, \pm 2, ...$, and since the process may move from state i only to states $i+1$ or $i-1$, the transition probabilities

$$p_{i,i+1} = p, \; p_{i,i-1} = 1-p, \quad p_{ij} = 0 \text{ for all } \; j \neq i \pm 1,$$

and do not depend on t. So, in the main diagonal of the transition matrix, we have zeros; the diagonal above the main consists of p's; and in the diagonal below the main, we have $(1-p)$'s. All other elements are zeros. The chain is homogeneous, $\boldsymbol{\pi}_0 = (0, ..., 0, 1, 0, ...)$, where 1 is in the uth place.

As we will see in this and the next chapters, the simple random walk is not as simple an object as it may seem. Many results concerning this process are typical for processes of a more complicated nature. For this reason, we will repeatedly return to this model and eventually explore it in detail. □

Next, we consider transitions in many steps. Let us specify particular paths using the symbol $\to$. For example, writing $1 \to 2 \to 1 \to 3$, we will mean that the chain, starting from state 1, moved to state 2, then returned to state 1, then moved to state 3. For a homogeneous chain with a transition matrix $\mathcal{P} = \|p_{ij}\|$, the probability of such a path is $p_{12}p_{21}p_{13}$. Other simple examples are given in Exercise 33.

Consider for now a homogeneous chain and denote by $p_{ij}^{(2)}$ the probability of moving from state i to state j in two steps, *not specifying an intermediate state.* Set $\mathcal{P}^{(2)} = \left\|p_{ij}^{(2)}\right\|$.

To compute $p_{ij}^{(2)}$, we should consider all possible two-step paths $i \to k \to j$, where the intermediate state k is arbitrary. The probability of each such a path is $p_{ik}p_{kj}$, and hence $p_{ij}^{(2)} = \sum_k p_{ik}p_{kj}$. The r.-h.s. of the last formula is the (i, j)-element of the matrix product $\mathcal{P}\mathcal{P} = \mathcal{P}^2$. Thus,

$$\mathcal{P}^{(2)} = \mathcal{P}^2.$$

Skipping similar simple calculations in the general case (the reader may apply induction) and restricting ourselves to the homogeneous chain case, we state the following

Proposition 2 *Let the chain under consideration be homogeneous and $p_{ij}^{(n)} = P(X_{t+n} = j \mid X_t = i)$, the probability of moving to state j in n steps starting from state i. (Since the chain is homogeneous, this conditional probability does not depend on t.) Let $\mathcal{P}^{(n)} = \left\|p_{ij}^{(n)}\right\|$, the corresponding transition probability matrix. Then*

$$\mathcal{P}^{(n)} = \mathcal{P}^n, \tag{4.1.4}$$

where $\mathcal{P}^n$ is the nth power of the matrix $\mathcal{P}$.

► In the non-homogeneous case, we set ${}_tp_{ij}^{(n)} = P(X_{t+n} = j \mid X_t = i)$, and ${}_t\mathcal{P}^{(n)} = \left\|{}_tp_{ij}^{(n)}\right\|$, the corresponding transition matrix. Then

$${}_t\mathcal{P}^{(n)} = {}_t\mathcal{P} \cdot {}_{t+1}\mathcal{P} \cdot {}_{t+2}\mathcal{P} \cdot ... \cdot {}_{t+n-1}\mathcal{P}, \tag{4.1.5}$$

where the symbol $\cdot$ denotes the multiplication of matrices. ◄

Below, unless stated otherwise, we restrict ourselves to homogeneous chains.

EXAMPLE 5. Consider rearrangements (shuffling) of three books from Example 3. As is easy to verify, for $\mathcal{P}$ from (4.1.3),

$$\mathcal{P}^2 = \left\| \begin{array}{cccccc} 2/9 & 1/9 & 2/9 & 1/9 & 2/9 & 1/9 \\ 1/9 & 2/9 & 2/9 & 1/9 & 2/9 & 1/9 \\ 2/9 & 1/9 & 2/9 & 1/9 & 1/9 & 2/9 \\ 2/9 & 1/9 & 1/9 & 2/9 & 1/9 & 2/9 \\ 1/9 & 2/9 & 1/9 & 2/9 & 2/9 & 1/9 \\ 1/9 & 2/9 & 1/9 & 2/9 & 1/9 & 2/9 \end{array} \right\| \begin{array}{l} \leftarrow 123 \\ \leftarrow 132 \\ \leftarrow 213 \\ \leftarrow 231 \\ \leftarrow 312 \\ \leftarrow 321. \end{array} \tag{4.1.6}$$

Thus, all two-step transitions have positive probabilities. □

In Exercise 32, we consider—in the same context – Examples 1-2.

From (4.1.4), we get that

$$\mathcal{P}^{(n+m)} = \mathcal{P}^{(n)}\mathcal{P}^{(m)}. \tag{4.1.7}$$

Equation (4.1.7) is called the *Chapman-Kolmogorov equation.*

Now, let $\pi_{ti}=P(X_t=i)$, and $\boldsymbol{\pi}_t = (\pi_{t0}, \pi_{t1}, \pi_{t2}, \ldots)$, the distribution of the process at time t. By the formula for total probability,

$$\pi_{1i} = P(X_1 = i) = \sum_k P(X_1 = i \mid X_0 = k)P(X_0 = k) = \sum_k p_{ki}\pi_{0k}.$$

In the vector form, it may be written as

$$\boldsymbol{\pi}_1 = \boldsymbol{\pi}_0\mathcal{P}, \tag{4.1.8}$$

where a *row* vector is multiplied on the right by a square matrix.

Taking into account (4.1.4), we can write a similar formula for an arbitrary time t, stating it as the following proposition.

Proposition 3 *In the homogeneous case,*

$$\boldsymbol{\pi}_t = \boldsymbol{\pi}_0\mathcal{P}^t. \tag{4.1.9}$$

EXAMPLE 6. Assume that in the situation of Example 3, $k = 3$ and in the beginning, the first book is on the top, while the second book may be equally likely either in the second or third position. That is, $\boldsymbol{\pi}_0 = (1/2, 1/2, 0, 0, 0, 0)$. Using (4.1.6) and (4.1.9), it is straightforward to compute that $\boldsymbol{\pi}_2 = \boldsymbol{\pi}_0 \cdot \mathcal{P}^2 = (1/6, 1/6, 2/9, 1/9, 2/9, 1/9)$. It is interesting to compare $\boldsymbol{\pi}_0$ and $\boldsymbol{\pi}_2$ and guess the further tendency. It will be determined in Section 4.4. □

Next, we consider simulation of a Markov chain.

EXAMPLE 7. Return to Example 1. A simulation procedure is illustrated in the Excel worksheet in Fig.7. The initial state is a free input entry in Cell D1. We chose the value 2. The matrix $\mathcal{P}$ is in the array F25:H27. Random numbers generated are placed in Column A, time moments t—in Column C, the values of X_t—in Column D.

Since the initial state is 2, in the first step, we should consider the probabilities in the third row of the matrix $\mathcal{P}$, that is, 0.25, 0.7, 0.05. To simulate the motion of the chain in the first step, we generate a random number Z. If Z is less than 0.25, (which occurs with probability 0.25), then the process moves to state 0. If $0.25 \le Z < 0.95$ (which occurs with probability 0.7), it moves to state 1. Otherwise, the process stays in state 2. In the particular realization in Fig.7, the r.v. $Z \approx 0.893$ is in Cell A2. So $X_1 = 1$, and it is reflected in cell D2. The command for Cell D2 is

=IF(D1=0, IF(A2<$F$25, 0,IF(A2<$F$25+$G$25,1,2)), IF(D1=1,IF(A2<=$F$26,0, IF($F$26<A2<=$F$26+$G$26,1,2)), IF(A2<=$F$27,0, IF(A2<=$F$27+$G$27,1,2)))).

After the first step, we proceed similarly, depending on the current state. For example, since X_1 happened to be one, to simulate X_2, we should consider the second row in the matrix $\mathcal{P}$ and generate another number from $[0,1]$. It is in Cell A3, we proceed with the command in D3 similar to that in D2, and continue in the same fashion. The realization we

	A	C	D	F	G	H
1		0	2			
2	0.8927274	1	1			
3	0.6867275	2	2			
4	0.4224372	3	1			
5	0.8536943	4	2			
6	0.1696524	5	0			
7	0.3178808	6	0			
8	0.3087863	7	0			
9	0.356914	8	0			
10	0.6671957	9	1			
11	0.3213599	10	0			
12	0.2576373	11	0			
13	0.2908719	12	0			
14	0.549089	13	0			
15	0.5478683	14	0			
16	0.1898251	15	0			
17	0.9736625	16	2			
18	0.1196326	17	0			
19	0.8098697	18	1			
20	0.9927976	19	2			
21	0.2686544	20	1			
22	0.7107456	21	2			
23	0.8444166	22	1			
24	0.7566149	23	2			
25	0.5019684	24	1	0.6	0.3	0.1
26	0.0557573	25	0	0.4	0.5	0.1
27	0.8172246	26	1	0.25	0.7	0.05
28	0.3868221	27	0			
29	0.2483901	28	0			
30	0.0555742	29	0			
31	0.6490982					
32	0.7506027					

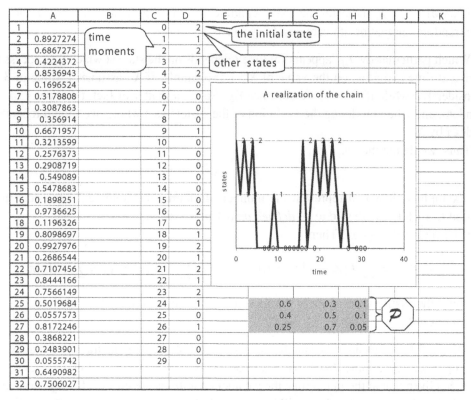

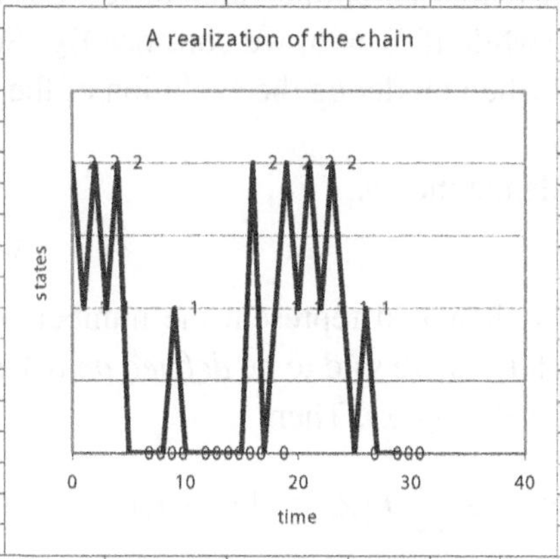

FIGURE 7. Simulation of the Markov chain from Examples 1,7 in Section 4.1

got in Column D is presented in the chart. Picking other numbers for Column A, we will get another realization.

Readers who are familiar with Excel, realize that the commands in Column D were arranged by one-moment copying, and all numbers in column A have been also generated by one command. □

4.2 Variables defined on a Markov chain. Cash flows

For the most part, this section concerns modeling cash flows. Additional examples of applications may be found, e.g., in [30]. We assume all chains under consideration to be *finite* (that is, having a finite number of states) and, unless stated otherwise, homogeneous.

4.2.1 Variables defined on states

Consider a Markov chain X_t. For each time moment t, we assign to each state i a random variable Y_{ti}. We assume that, *given* i, the r.v. Y_{ti} does not depend on other r.v.'s involved in the model.

EXAMPLE 1. As in the situation of Example 4.1-1, X_t is the indicator of the weather condition at time t, and Y_{ti} is the number of claims in period t, given that the weather conditions correspond to state i—that is, given $X_t = i$.

EXAMPLE 2. X_t indicates the health condition of a person in annual period t, and Y_{ti} is the (random) annual health care cost for year t if $X_t = i$. □

Which particular r.v. Y_{ti} "appears" at time t depends on the state at which the process will arrive at time t, that is, on the value of X_t. For example, if X_t assumes a value of 3, we consider Y_{t3}, while if $X_t = 6$, we consider Y_{t6}. So, as a matter of fact, the index i in Y_{ti} is random, and when modeling the evolution of the system as a whole, we should replace the index i by X_t.

Thus, we define the r.v.

$$Z_t = Y_{tX_t}.$$

In Example 1, it would represent the number of claims, in Example 2—the cost in the period t. Such r.v.'s are said to be *defined on a Markov chain.*

Let $F_{ti}(x) = P(Y_{ti} \leq x)$. Then

$$P(Z_t \leq x) = \sum_i P(Z_t \leq x \,|\, X_t = i)P(X_t = i) = \sum_i P(Y_{ti} \leq x)\pi_{ti} = \sum_i F_{ti}(x)\pi_{ti}.$$

Thus, the distribution of Z_t is the mixture of the distributions F_{ti} with respect to the distribution $\boldsymbol{\pi}_t$.

EXAMPLE 3. Consider the situation of Example 1 above with the transition matrix from Example 4.1-1. Assume that Y_{ti} is a Poisson r.v. with parameter λ_i. Let $\lambda_0 = 2$, $\lambda_1 = 4$, $\lambda_2 = 8$. Given that at the initial time the weather conditions are normal, find the probability that during the second day there will be at most 3 claims.

Thus, $\boldsymbol{\pi}_0 = (1,0,0)$, and by (4.1.9), $\boldsymbol{\pi}_2 = \boldsymbol{\pi}_0\mathcal{P}^2$, where $\mathcal{P}$ is given in (4.1.1). So, we easily compute, using any software we like, that $\boldsymbol{\pi}_2 = (0.505, 0.4, 0.095)$. Then, using the same symbol *PoissonDist* as in Example 2.1-1b, we have $P(Z_2 \leq 3) = \mathit{PoissonDist}(3,2) \cdot 0.505 + \mathit{PoissonDist}(3,4) \cdot 0.4 + \mathit{PoissonDist}(3,8) \cdot 0.095 \approx 0.857 \cdot 0.505 + 0.433 \cdot 0.4 + 0.042 \cdot 0.095 \approx 0.610$. □

Let

$$S_n = \sum_{t=0}^{n} Z_t = \sum_{t=0}^{n} Y_{tX_t}.$$

In the above examples, S_n represents either the total number of claims or the total payment during $n+1$ periods of time (we count the initial period).

Computing the distribution of S_n is complicated and we restrict ourselves to $E\{S_n\}$. The matrix notation allows us to do that in a nice compact form.

Let $c_{ti} = E\{Y_{ti}\}$, and vector $\mathbf{c}_t = (c_{t1}, c_{t2}, \ldots)$. In the situation of Example 2, the vector $\mathbf{c}_t$ characterizes the possible cash flow at time t. We have

$$E\{Z_t\} = \sum_i E\{Y_{ti}\}P(X_t = i) = \sum_i c_{ti}\pi_{ti} = \langle \mathbf{c}_t, \boldsymbol{\pi}_t \rangle,$$

where $\langle \cdot, \cdot \rangle$ stands for scalar (or dot) product. Then, in view of (4.1.9), $E\{Z_t\} = \langle \mathbf{c}_t, \boldsymbol{\pi}_0\mathcal{P}^t \rangle$, and

$$E\{S_n\} = \sum_{t=0}^{n} \langle \mathbf{c}_t, \boldsymbol{\pi}_0\mathcal{P}^t \rangle. \tag{4.2.1}$$

(By convention, $\mathcal{P}^0$ is the identity matrix of the corresponding size.)

First, consider the case when $\mathbf{c}_t = \mathbf{c} = (c_1, c_2, ...)$, that is, does not depend on t. Then

$$E\{S_n\} = \sum_{t=0}^{n} \langle \mathbf{c}, \boldsymbol{\pi}_0 \mathcal{P}^t \rangle = \langle \mathbf{c}, \boldsymbol{\pi}_0 \sum_{t=0}^{n} \mathcal{P}^t \rangle = \langle \mathbf{c}, \boldsymbol{\pi}_0 \mathcal{K}_n \rangle, \tag{4.2.2}$$

where the matrix

$$\mathcal{K}_n = \sum_{t=0}^{n} \mathcal{P}^t.$$

This is a geometric series of matrices, so it is tempting to use the formula for the sum of such a series.

For any number $r \neq 1$, we have $1 + r + ... + r^n = (1-r)^{-1}(1-r^{n+1}) \to (1-r)^{-1}$ as $n \to \infty$, if $|r| < 1$. For a matrix $\mathcal{B}$ and the identity matrix $\mathcal{I}$, we do have

$$\mathcal{I} + \mathcal{B} + ... + \mathcal{B}^n = (\mathcal{I} - \mathcal{B})^{-1}(\mathcal{I} - \mathcal{B}^{n+1}) \to (\mathcal{I} - \mathcal{B})^{-1}, \tag{4.2.3}$$

if the inverse matrix $(\mathcal{I} - \mathcal{B})^{-1}$ exists. The limiting relation is true provided that the absolute values of all eigenvalues of $\mathcal{B}$ are less than one (see, e.g., [61], [79]).

However, we cannot use (4.2.3) in the situation above because $\det(\mathcal{I} - \mathcal{P}) = 0$, and hence $(\mathcal{I} - \mathcal{P})^{-1}$ does not exist. Indeed, let $\mathbf{e} = (1, 1, ..., 1)$, so the transpose $\mathbf{e}^T$ is the corresponding column vector. Since the sum of the probabilities in each row of $\mathcal{P}$ is one, $\mathcal{P}\mathbf{e}^T = \mathbf{e}^T$, that is, one is an eigenvalue of $\mathcal{P}$. This means that $\det(\mathcal{I} - \mathcal{P}) = 0$.

In the next sections, we consider situations where (4.2.3) proves to be useful, but so far we have to consider $\mathcal{K}_n$ as is.

EXAMPLE 4. We return to the situation of Example 3. Now, we find $E\{S_2\}$, the expected total number of claims during three periods (we count the initial period). In our case, $\mathbf{c} = (\lambda_0, \lambda_1, \lambda_2) = (2, 4, 8)$, and by (4.2.2),

$$E\{S_2\} = \langle \mathbf{c}, \boldsymbol{\pi}_0(\mathcal{I} + \mathcal{P} + \mathcal{P}^2) \rangle,$$

where $\boldsymbol{\pi}_0 = (1, 0, 0)$, and $\mathcal{P}$ is given in (4.1.1). Calculations are easy, especially if you use software, and lead to $E\{S_2\} = 8.57$.

Calculations for larger n's are similar and tractable with a good computer program. □

4.2.2 Mean discounted payments

The notions of discount and present value we use here are introduced in Section **0**.8.3. Briefly, the discount factor v_t is the value of one unit of money to be paid at time t if the evaluation is carried out from the standpoint of the present initial time zero. In the model below, we adopt the simplest representation

$$v_t = v^t,$$

where v is a given *discount factor* for a unit of time. We assume $0 < v < 1$.

Let us return to the model of Section 4.2.1 and formula (4.2.1). As before, we suppose that the expected payment vector at time t is a constant vector $\mathbf{c}$. However, now we are evaluating the present value of this future payment from the standpoint of the initial time. The present value mentioned is $v^t\mathbf{c}$, and in our model, we should set $\mathbf{c}_t = v^t\mathbf{c}$. Inserting it into (4.2.1), we have

$$E\{S_n\} = \sum_{t=0}^{n} \langle v^t\mathbf{c}, \boldsymbol{\pi}_0 \mathcal{P}^t \rangle = \sum_{t=0}^{n} \langle \mathbf{c}, v^t \boldsymbol{\pi}_0 \mathcal{P}^t \rangle = \sum_{t=0}^{n} \langle \mathbf{c}, \boldsymbol{\pi}_0 (v\mathcal{P})^t \rangle. \tag{4.2.4}$$

The last quantity, the expected present value of the cash flow, is called in the actuarial literature an *actuarial present value.* The approach in (4.2.4) is sometimes called the "triple-product-summation approach (3π-approach)" since we multiply in $\langle v^t\mathbf{c},\, \boldsymbol{\pi}_0\mathcal{P}^t\rangle$ the discount, the amount to be paid, and the probability that a particular state occurs, and then we add all products up.

From (4.2.4), we have

$$E\{S_n\} = \langle \mathbf{c}, \boldsymbol{\pi}_0 \sum\nolimits_{t=0}^{n} (v\mathcal{P})^t \rangle. \qquad (4.2.5)$$

In the previous section, we saw that one is an eigenvalue of $\mathcal{P}$. We use now the fact that *all other eigenvalues for the stochastic matrix* $\mathcal{P}$ *are less than one*; see, e.g., [95, pp.11, 141], [61].

(This fact may be derived, for example, from the following theorem. In the complex number space, consider the circles whose centers are the diagonal elements p_{ii}, and the radiuses are the sums of the non-diagonal elements that is, $\sum_{j:j\neq i} p_{ij}$, in the corresponding rows. Then all eigenvalues belong to the union of these circles; see, e.g., [61]. For our stochastic matrix, all points in these circles are less than one in the absolute value. Another proof is connected with the Perron-Frobenius theorem; see, e.g., [95, pp.11, 141].)

The fact mentioned means that $\det(\lambda\mathcal{I} - \mathcal{P})$ may be zero only for $\lambda \leq 1$. Writing $\det(\lambda\mathcal{I} - \mathcal{P}) = \lambda\det(\mathcal{I} - \frac{1}{\lambda}\mathcal{P}) = \lambda\det(\mathcal{I} - v\mathcal{P})$, where $v = 1/\lambda$, we see that $\det(\mathcal{I} - v\mathcal{P})$ may be equal to zero only for $v \geq 1$. Hence, if $0 \leq v < 1$, we can apply formula (4.2.3) with $\mathcal{B} = v\mathcal{P}$, writing

$$E\{S_n\} = \langle \mathbf{c}, \boldsymbol{\pi}_0 (\mathcal{I} - v\mathcal{P})^{-1}(\mathcal{I} - (v\mathcal{P})^{n+1})\rangle. \qquad (4.2.6)$$

If n is not large, we should stop and use the above formula. However, if we consider the process in the long run (for large n), we may use the approximation

$$\sum\nolimits_{t=0}^{n} (v\mathcal{P})^t \to (\mathcal{I} - v\mathcal{P})^{-1} \text{ as } n \to \infty,$$

which implies that in this case, the discounted mean cash flow

$$E\{S_n\} \to \langle \mathbf{c},\, \boldsymbol{\pi}_0(\mathcal{I} - v\mathcal{P})^{-1}\rangle \text{ as } n \to \infty. \qquad (4.2.7)$$

In Section 4.3.1, we consider another way of obtaining limiting relations of this type.

In Exercise 41, we evaluate the present value of the payments in the situation of Example 4.2.1-2 for a given discount factor. Next, to cover different situations, we shall consider a somewhat different example.

EXAMPLE 1. An investor distinguishes five types of years: very bad, bad, moderate, good, and very good. The mean profit corresponding to the states mentioned is equal, in some units, to $-3, -1, 1, 2, 4$, respectively. The discount factor is $v = 0.97$.

Suppose that the change of investment conditions from year to year is well approximated by the homogeneous Markov model with the transition matrix

$$\mathcal{P} = \left\| \begin{array}{lllll} 0.1 & 0.7 & 0.2 & 0 & 0 \\ 0.2 & 0.2 & 0.6 & 0 & 0 \\ 0.05 & 0.1 & 0.7 & 0.1 & 0.05 \\ 0 & 0.1 & 0.1 & 0.6 & 0.2 \\ 0 & 0 & 0.2 & 0.5 & 0.3 \end{array} \right\|. \qquad (4.2.8)$$

The present condition of the market is moderate. Estimate the mean present value of the profit in the long run.

Thus, $\boldsymbol{\pi}_0 = (0,0,1,0,0)$, $\mathbf{c} = (-3,-1,1,2,4)$, and the estimate of $E\{S_n\}$ for large n, is $\langle \mathbf{c}, \boldsymbol{\pi}_0(\mathcal{I} - v\mathcal{P})^{-1}\rangle \approx 36.008$, which may be easily computed using the matrix-commands MMULT and MINVERSE in Excel. In Exercise 39, the reader is suggested to consider the solution for different values of v. In Exercise 40, we consider this model for a finite n. □

4.2.3 The case of absorbing states

A state i is called *absorbing* if $p_{ii} = 1$.

For instance, in Example 4.1-2, state 1 (deceased) is absorbing. If in a life insurance contract, the situations when the insured dies and when she/he ceases paying premiums (and the contract is canceled) are considered separately, then we have two absorbing states.

Consider a chain for which states $i = 0, 1, ..., k$, are non-absorbing, and the last r states are absorbing, that is, $p_{ii} = 1$ for $i = k+1, ..., k+r$. Then, the transition matrix

$$\mathcal{P} = \begin{Vmatrix} \mathcal{A} & \mathcal{B} \\ \mathcal{O} & \mathcal{I} \end{Vmatrix}, \tag{4.2.9}$$

where $\mathcal{O}$ is a matrix with zero elements, $\mathcal{I}$ is a $r \times r$ identity matrix, $\mathcal{A}$ is some $(k+1)\times(k+1)$-matrix, and $\mathcal{B}$ is a $(k+1)\times r$ matrix.

Assume that for each non-absorbing state, the probability of moving to some absorbing state is positive: for each $i = 0, 1, ..., k$, there is a state j from the set $\{k+1, ..., k+r\}$ such that $p_{ij} > 0$. This implies, in particular, that the sum of all probabilities in each row of the matrix $\mathcal{A}$ is less than one.

Since at each step the process can move to an absorbing state with a positive probability, at some random but finite time T, the process will arrive at the absorbing state and never leave it (get stuck in it). Intuitively it is clear; we will show it rigorously in the end of this subsection.

The matrix $\mathcal{I}$ from (4.2.9) will not participate in calculations below. Let now the symbol $\mathcal{I}$ denote the $(k+1)\times(k+1)$ identity matrix, that is, the matrix of the same size as $\mathcal{A}$.

An essential circumstance in what we will do below is in the fact that $\det(\mathcal{I} - v\mathcal{A}) \neq 0$ not only for $v < 1$, but for $v = 1$ too. Therefore, though the model below involves a discount factor v, we may consider the case $v = 1$ also, which will allow us to explore, for example, the time of absorption or the situations when the r.v.'s Y_{ti} are not payments but, say, the numbers of claims. So, we use below the word "payment" only for certainty.

(To show that $\det(\mathcal{I} - \mathcal{A}) \neq 0$, set b_i equal to the sum of all elements in the ith row of $\mathcal{B}$. We have assumed that $b_i < 1$. Skipping the trivial case $b_i = 0$, we divide the first row of $\mathcal{B}$ by b_1, the second row by b_2, and so on. Denote the resulting matrix by $\mathcal{K}$. Clearly, $\mathcal{K}$ is a stochastic matrix, and $\mathcal{B} = \mathcal{S}\mathcal{K}$, where $\mathcal{S}$ is the diagonal matrix with $b_1, b_2, \cdots, b_k$ in the diagonal. Since all $b_i < 1$ and all eigenvalues of $\mathcal{K}$ are less than or equal to one, all eigenvalues of $\mathcal{S}\mathcal{K}$ are less than one.)

For the payment vector $\mathbf{c} = (c_0, ..., c_{k+r})$, we assume that $c_i = 0$ for $i > k$ (no payments in absorbing states). Set $\widetilde{\mathbf{c}}=(c_0, c_1, ..., c_k)$, the vector of "real" payments, and set $\widetilde{\boldsymbol{\pi}}_0 = (\pi_{00}, \pi_{01}, ..., \pi_{0k})$. Consider the expression $\langle \mathbf{c},\, v^t \boldsymbol{\pi}_0 \mathcal{P}^t \rangle$ in (4.2.5).

Since the last r coordinates in $\mathbf{c}$ are zeros, the last r coordinates of the vector $\boldsymbol{\pi}_0\mathcal{P}^t$ do not matter. On the other hand, in view of (4.2.9), the first $k+1$ coordinates of $\boldsymbol{\pi}_0\mathcal{P}^t$ constitute the vector $\widetilde{\boldsymbol{\pi}}_0\mathcal{A}^t$. Consequently,

$$\langle \mathbf{c}, v^t\boldsymbol{\pi}_0\mathcal{P}^t\rangle = \langle \widetilde{\mathbf{c}}, v^t\widetilde{\boldsymbol{\pi}}_0\mathcal{A}^t\rangle.$$

Making use of (4.2.3), we have

$$E\{S_n\} = \sum_{t=0}^{n}\langle \widetilde{\mathbf{c}}, v^t\widetilde{\boldsymbol{\pi}}_0\mathcal{A}^t\rangle = \langle \widetilde{\mathbf{c}}, \widetilde{\boldsymbol{\pi}}_0(\sum_{t=0}^{n} v^t\mathcal{A}^t)\rangle = \langle \widetilde{\mathbf{c}}, \widetilde{\boldsymbol{\pi}}_0(\mathcal{I}-v\mathcal{A})^{-1}(\mathcal{I}-(v\mathcal{A})^{n+1})\rangle. \tag{4.2.10}$$

From this it follows that

$$E\{S_n\} \to \langle \widetilde{\mathbf{c}}, \widetilde{\boldsymbol{\pi}}_0(\mathcal{I}-v\mathcal{A})^{-1}\rangle \quad \text{as } n\to\infty. \tag{4.2.11}$$

In Section 4.3.1, we arrive at relations of this type in a different manner.

Let us discuss the significance of the last relation. As was mentioned, with probability one, absorption will happen at some random but finite time T. After absorption, there are no payments, and accordingly, S_n will not change after the absorption time; that is, $S_n = S_T$ for all $n \geq T$. Then the limiting r.v. $S_\infty = \lim_{n\to\infty} S_n$ is just S_T, that is, the total discounted cash flow until absorption. Formula (4.2.11) gives the expected total discounted cash flow, i.e., $E\{S_T\}$. (Certainly, we may reason in this way if there is no restriction on the duration of the process.)

EXAMPLE 1. Consider a medical insurance model with four states for an insured: healthy, sick, ceased paying, deceased. The transition matrix

$$\mathcal{P} = \begin{Vmatrix} 0.9 & 0.05 & 0.01 & 0.04 \\ 0.1 & 0.8 & 0.01 & 0.09 \\ 0 & 0 & 1 & 0 \\ 0 & 0 & 0 & 1 \end{Vmatrix} \tag{4.2.12}$$

gives transition probabilities corresponding annual transition periods. For example, 0.1 is the probability that an insured being sick at the beginning of a year will recover by the beginning of the next year.

Let the mean annual health care costs corresponding to the first two states—'healthy' and 'sick'—be equal to 1 and 4, respectively. That is, $\mathbf{c} = (1,4,0,0)$, and $\widetilde{\mathbf{c}} = (1,4)$. Assume also that at the initial moment, 94% of clients of the company are healthy, that is, $\boldsymbol{\pi}_0 = (0.94, 0.06, 0, 0)$, and $\widetilde{\boldsymbol{\pi}}_0 = (0.94, 0.06)$. Estimate the expected total costs for the discount $v = 0.97$.

We have

$$\mathcal{A} = \begin{Vmatrix} 0.9 & 0.05 \\ 0.1 & 0.8 \end{Vmatrix}, \quad (\mathcal{I}-v\mathcal{A})^{-1} \approx \begin{Vmatrix} 9.434 & 2.043 \\ 4.085 & 5.349 \end{Vmatrix},$$

and

$$\widetilde{\boldsymbol{\pi}}_0(\mathcal{I}-v\mathcal{A})^{-1} \approx (9.113, 2.241).$$

Thus, by (4.2.11),

$$E\{S_n\} \to \langle \widetilde{\mathbf{c}}, \widetilde{\boldsymbol{\pi}}_0(\mathcal{I}-v\mathcal{A})^{-1}\rangle \approx 1\cdot 9.113 + 4\cdot 2.241 = 18.077. \quad \square$$

It is interesting that from (4.2.11) with $v = 1$, we can immediately obtain the expected absorption time, that is, the expected number of steps until absorption.

Indeed, let us set $\widetilde{\mathbf{c}} = \mathbf{e} = (1, 1, ..., 1)$. Then, at each time until the moment of absorption, the payment equals one. Consequently, if a time moment n precedes the moment of absorption, the variable S_n is equal to the number of periods elapsed. (We count time 0.) At the moment of absorption, there is no payment, and hence at this moment, S_n is exactly equal to the moment of absorption. After absorption, there are no payments, and accordingly, S_n will not change after this point. Eventually, due to the choice of $\widetilde{\mathbf{c}}$,

$$E\{T\} = E\{S_\infty\} = \langle \mathbf{e}, \widetilde{\boldsymbol{\pi}}_0(\mathcal{I} - \mathcal{A})^{-1} \rangle.$$

To rewrite this in a more convenient form, we use two elementary facts:

(1) if a vector $\boldsymbol{\mu}$ is a row-vector, then its transpose $\boldsymbol{\mu}^T$ is the same vector presented as a column;

(2) $\langle \mathbf{a}, \mathbf{b} \rangle = \mathbf{a}\mathbf{b}^T$ for any row-vectors $\mathbf{a}$, $\mathbf{b}$ of the same dimension.

Let $\boldsymbol{\mu}$ be a vector such that $\boldsymbol{\mu}^T = (\mathcal{I} - \mathcal{A})^{-1}\mathbf{e}^T$. Then

$$E\{T\} = \widetilde{\boldsymbol{\pi}}_0(\mathcal{I} - \mathcal{A})^{-1}\mathbf{e}^T = \widetilde{\boldsymbol{\pi}}_0\boldsymbol{\mu}^T = \langle \widetilde{\boldsymbol{\pi}}_0, \boldsymbol{\mu} \rangle. \tag{4.2.13}$$

Note that, since $E\{T\}$ is a finite number, we proved along the way that T is finite with probability one.

Let $\widetilde{\boldsymbol{\pi}}_0 = (0, ..., 0, 1, 0, ...0)$, where 1 corresponds to state i. Then from (4.2.13) it follows that μ_i, the ith coordinate of $\boldsymbol{\mu}$, is $E\{T \mid X_0 = i\}$.

EXAMPLE 2. For the data from Example 1

$$(\mathcal{I} - \mathcal{A})^{-1} = \begin{Vmatrix} 40/3 & 10/3 \\ 20/3 & 20/3 \end{Vmatrix}, \quad \text{and} \quad \boldsymbol{\mu}^T = (\mathcal{I} - \mathcal{A})^{-1}\mathbf{e}^T = \begin{pmatrix} 50/3 \\ 40/3 \end{pmatrix}.$$

So, the expected lifetime for a sick person from the population under consideration constitutes 0.8 of the same time for a healthy person (the ratio of their expected lifetimes). If, as in Example 1, at the initial time, 94% of population are healthy, then $\widetilde{\boldsymbol{\pi}}_0 = (0.94, 0.06)$, and for a randomly chosen client, $E\{T\} = \langle \widetilde{\boldsymbol{\pi}}_0, \mu \rangle = 0.94 \cdot \frac{50}{3} + 0.06 \cdot \frac{40}{3} = 16.4\bar{6}$. $\square$

4.2.4 Variables defined on transitions

In some situations, variables defined on a Markov chain—say, cash flows—may not be determined by the state in which the process is at a current moment, but rather by the transition the process has made. For example, in the case of life insurance, the company pays only one time when the transition "alive"→"deceased" occurs.

For such a case, we replace the r.v.'s Y_{ti} from above by r.v.'s $Y_{t(ij)}$ assigning $Y_{t(ij)}$ to the transition $i \to j$ if it occurred during the period from time $t-1$ to time t. For brevity, we interpret $Y_{t(ij)}$ below as a payment, though other interpretations are also possible.

Assume that $E\{Y_{t(ij)}\} = v^t c_{ij}$, where v is a discount factor, and the mean payment c_{ij} does not depend on t. Let the matrix $\mathcal{C} = \|c_{ij}\|$. Given that the process at time $t-1$ is in state i, the expected payment in one step, not involving discount, is equal to $\sum_j c_{ij}p_{ij}$. Denote this number by $c_{\mathcal{P}i}$, and set the vector $\mathbf{c}_{\mathcal{P}} = (c_{\mathcal{P}1}, ..., c_{\mathcal{P}k})$, where k is the number of states of the chain.

[One may note that $c_{\mathcal{P}i}$ is the (i,i)-diagonal element of the matrix $\mathcal{C}\mathcal{P}^T$, where $\mathcal{P}^T$ is the transpose of $\mathcal{P}$, that is,

$$\mathbf{c}_{\mathcal{P}} = diagonal\,(\mathcal{C}\mathcal{P}^T).$$

This may be of help in providing a program for calculations.]

Thus, the expected value of the payment at time t is

$$\sum_i c_{\mathcal{P}i} P(X_{t-1} = i) = \sum_i c_{\mathcal{P}i} \pi_{(t-1)i} = \langle \mathbf{c}_{\mathcal{P}}, \boldsymbol{\pi}_{t-1} \rangle = \langle \mathbf{c}_{\mathcal{P}}, \boldsymbol{\pi}_0 \mathcal{P}^{t-1} \rangle.$$

Here and below $t = 1, 2, ...$, because there is no payment at time 0.

So, if S_n is the total *discounted* payment during n periods, then

$$E\{S_n\} = \sum_{t=1}^{n} v^t \langle \mathbf{c}_{\mathcal{P}}, \boldsymbol{\pi}_0 \mathcal{P}^{t-1} \rangle = v \sum_{t=1}^{n} \langle \mathbf{c}_{\mathcal{P}}, \boldsymbol{\pi}_0 v^{t-1} \mathcal{P}^{t-1} \rangle = v \langle \mathbf{c}_{\mathcal{P}}, \boldsymbol{\pi}_0 (\sum_{t=1}^{n} (v\mathcal{P})^{t-1}) \rangle. \quad (4.2.14)$$

If $v = 1$ we should stop at this stage. This is the case when we do not take into account a discount, or we view Y_{ti} not as payments but as the numbers of claims, for example.

If $v < 1$, we can continue, writing

$$E\{S_n\} = v \cdot \langle \mathbf{c}_{\mathcal{P}}, \boldsymbol{\pi}_0 (\mathcal{I} - v\mathcal{P})^{-1} (\mathcal{I} - (v\mathcal{P})^n) \rangle, \quad (4.2.15)$$

and

$$E\{S_n\} \to v \cdot \langle \mathbf{c}_{\mathcal{P}}, \boldsymbol{\pi}_0 (\mathcal{I} - v\mathcal{P})^{-1} \rangle \quad \text{as } n \to \infty. \quad (4.2.16)$$

Thus, the only difference between representations (4.2.6) and (4.2.4) is that we replaced $\mathbf{c}$ by $\mathbf{c}_{\mathcal{P}}$, and instead of $(v\mathcal{P})^{n+1}$ in (4.2.6), we wrote $(v\mathcal{P})^n$ in (4.2.15).

EXAMPLE 1. Consider the situation of Example 4.2.3-1 with the transition matrix (4.2.12). Assume that we deal with a life insurance contract paying a unit of money (say, \$100,000) upon the death of the insured. As in the example mentioned, we assume that at the initial moment, 94% of the population under consideration is healthy. So, for a randomly chosen client, $\boldsymbol{\pi}_0 = (0.94, 0.06, 0, 0)$. Let $v = 0.97$.

A payment is provided only in the case of transitions $0 \to 3$ or $1 \to 3$. Consequently, from the concrete representation (4.2.12) it follows that $c_{\mathcal{P}1} = 1 \cdot 0.04$, $c_{\mathcal{P}2} = 1 \cdot 0.09$, and $c_{\mathcal{P}3} = c_{\mathcal{P}4} = 0$. So, $\mathbf{c}_{\mathcal{P}} = (0.04, 0.09, 0, 0)$.

We can use the structure of $\mathcal{P}$ as we did in Example 4.2.3-1, but we do not need it. If we restrict ourselves to the approximation (4.2.16), we can easily compute $v\langle \mathbf{c}_{\mathcal{P}}, \boldsymbol{\pi}_0 (\mathcal{I} - v\mathcal{P})^{-1} \rangle$ directly by using, say, Excel or another software. In this case, it is convenient to keep in mind that a scalar product $\langle \mathbf{a}, \mathbf{b} \rangle$ may be written as $\mathbf{a}\mathbf{b}^T$.

The reader can verify that the answer to this problem is ≈ 0.549. □

4.2.5 What to do if the chain is not homogeneous

Non-homogeneous chains are actually handled similarly, although formulas will become a bit messy. We should replace everywhere $\mathcal{P}^t$ by ${}_0\mathcal{P}^{(t)}$ which is the product of matrices defined in (4.1.5).

In particular, it concerns (4.2.2) and the definition of the matrix $\mathcal{K}_n$ there.

In Section 4.2.3, the matrix $\mathcal{A}^t$ should be replaced by ${}_0\mathcal{A}^{(t)} = {}_0\mathcal{A} \cdot {}_1\mathcal{A} \cdot {}_2\mathcal{A} \cdot ... \cdot {}_{t-1}\mathcal{A}$, where ${}_k\mathcal{A}$ is the matrix corresponding to the transition from time k to time $k+1$. However, now we do not have a geometric series, so we can not proceed as in (4.2.11), and, hence, we are doomed to straightforward calculations.

The same concerns (4.2.5). We should replace there $(v\mathcal{P})^t$ by $v^t\,{}_0\mathcal{P}^{(t)}$ and leave (4.2.5) as it is, because now we cannot proceed as in (4.2.6).

In Section 4.2.4, we should first redefine $c_{\mathcal{P}i}$ as ${}_t c_{\mathcal{P}i} = \sum_j c_{ij} \cdot {}_{t-1}p_{ij}$, since the transition probability depends now on time. The vector of payments will then be ${}_t\mathbf{c}_{\mathcal{P}} = ({}_{t-1}c_{\mathcal{P}1}, ..., {}_{t-1}c_{\mathcal{P}k})$. In (4.2.14), we should replace $\mathbf{c}_{\mathcal{P}}$ by ${}_t\mathbf{c}_{\mathcal{P}}$ and $(v\mathcal{P})^{t-1}$ by $v^{t-1}\,{}_0\mathcal{P}^{(t-1)}$, and stop there.

In this case, calculations will be straightforward and not tractable by hand, but a good program can do it easily.

4.3 The first step analysis. An infinite horizon

Here we consider a method proven to be efficient in many problems concerning global characteristics of Markov processes in the case of an infinite time horizon. In particular, as an example, we obtain again limiting relations from Sections 4.2.2 and 4.2.3.

However, in order to illustrate the idea of the method, we start with an example that has no direct relevance to insurance.

EXAMPLE 1 ([152, N25][3]). A gambler begins with 2 chips. At each play he/she can (a) win 2 chips with probability 0.1; (b) win 1 chip with probability 0.2; (c) push (win 0 chips) with probability 0.3; (d) lose 1 chip with probability 0.3; (e) lose 2 chips with probability 0.1. Play continues as long as the gambler has exactly 2 or 3 chips. Calculate the expected number of rounds the gambler has 3 chips.

We remember that the gambler starts with 2 chips. Nevertheless, we introduce into consideration two r.v.'s: N_2 the number of rounds the gambler has 3 chips if he/she begins with 2 chips; and N_3—the number of rounds the gambler has 3 chips if he/she begins with 3 chips. Let X be the number of chips the gambler won at the *first step.*

Rounds are independent. Therefore, if for example the gambler wins 1 chip at the *first step,* we arrive at an identical situation with one exception: now the gambler has 3 chips. So, $E\{N_2 \,|\, X=1\} = E\{N_3\}$. Similarly, $E\{N_3 \,|\, X=-1\} = 1 + E\{N_2\}$ because, if the gambler starts with 3 chips, we count the first round, and if he/she loses 1 chip at the first play, he/she starts over with 2 chips.

We have

$$E\{N_2\} = E\{N_2 \,|\, X{=}-2, \text{ or } -1, \text{ or } 2\}P(X{=}-2, \text{ or } -1, \text{ or } 2) + E\{N_2 \,|\, X{=}1\}P(X{=}1)$$
$$+ E\{N_2 \,|\, X=0\}P(X=0) = 0 + E\{N_3\} \cdot 0.2 + E\{N_2\} \cdot 0.3.$$

Thus, $7E\{N_2\} = 2E\{N_3\}$.

Similarly,

$$E\{N_3\} = E\{N_3 \,|\, X{=}-2, \text{ or } 1, \text{ or } 2\}P(X{=}-2, \text{ or } 1, \text{ or } 2) + E\{N_3 \,|\, X{=}-1\}P(X{=}-1)$$
$$+ E\{N_1 \,|\, X{=}0\}P(X{=}0) = 1 \cdot 0.4 + (1 + E\{N_2\}) \cdot 0.3 + (1 + E\{N_3\}) \cdot 0.3,$$

[3] Reprinted with permission of the Casualty Actuarial Society.

which leads to $7E\{N_3\} = 10 + 3E\{N_2\}$. Together with the previous relation, it gives $E\{N_2\} = \frac{20}{43} \approx 0.465$. $\square$

4.3.1 Mean discounted payments in the case of infinite time horizon

Consider the model of Section 4.2.2. Before, we let the finite number of steps n converge to infinity in the end of calculations. Now, we assume that the time horizon n is infinite from the start. In other words, we consider the r.v.

$$S_\infty = \sum_{t=0}^{\infty} Z_t = \sum_{t=0}^{\infty} Y_{tX_t},$$

supposing that this infinite series converges. Under assumptions we make below, this will be the case.

We interpret r.v.'s Y_{ti} as random payments and, as in Section 4.2.2, set $c_{ti} = E\{Y_{ti}\}$ and $\mathbf{c}_t = (c_{1t}, c_{2t}, ...)$.

In Section 4.2.2, we obtained the representation (4.2.2) as a corollary from some more general result. Our goal here is to show that, if we are interested only in $E\{S_\infty\}$, we can calculate it by a shorter and more explicit method.

Assume that $\mathbf{c}_t = v^t\mathbf{c}$, where $\mathbf{c} = (c_1, c_2, ...)$ is a given payment vector not depending on time, and v is a discount factor. Set $\mu_i = E\{S_\infty \mid X_0 = i\}$, the expected total *discounted* payment over the infinite time interval in the case when the initial state is i.

Making use of the formula for total expectation (**0**.7.2.1), we write

$$\mu_i = E\{S_\infty \mid X_0 = i\} = \sum_j E\{S_\infty \mid X_1 = j, X_0 = i\} P(X_1 = j \mid X_0 = i), \tag{4.3.1}$$

where summation is over all possible states j at which the process can arrive *at the first step*.

Consider the case when at the first step the process moves from a state i to a state j. Since the process is Markovian, once it has arrived at state j, its future evolution depends only on where it is now and not on where it was at the initial time. So, at time $t = 1$, we can consider the process as a process starting here with the initial state j.

At the moment $t = 0$, the discount equals $v^0 = 1$. So, the total discounted payment is equal to the payment c_i made at time $t = 0$ plus the total discounted payment made at time $t = 1$ and at all time moments after $t = 1$. The only thing we should realize is that it would be a mistake to write for the second part mentioned $E\{S_\infty \mid X_0 = j\} = \mu_j$. It would have been correct if we had evaluated this part of payments from the standpoint of time $t = 1$. But we are at time $t = 0$, which means that for us the expected present value of this part is not μ_j but rather $v\mu_j$.

Eventually, $E\{S_\infty \mid X_1 = j, X_0 = i\} = c_i + v\mu_j$, and from (4.3.1) it follows that for each $i = 0, 1, ...$

$$\mu_i = \sum_j (c_i + v\mu_j) p_{ij} = c_i \sum_j p_{ij} + v \sum_j \mu_j p_{ij} = c_i + v \sum_j p_{ij}\mu_j, \tag{4.3.2}$$

because $\sum_j p_{ij} = 1$ for all i.

To find μ_i's, we should solve the system of equations in (4.3.2). It is easier to visualize it if we use the matrix notation.

Let vector $\boldsymbol{\mu} = (\mu_0, \mu_1, ...)$. As was noted repeatedly, if the symbol ^{T} stands for the transpose operation, then $\boldsymbol{\mu}^T$ is the same vector $\boldsymbol{\mu}$ viewed as a column vector. Observe that the r.-h.s. of (4.3.2) is the ith coordinate of the vector $\mathbf{c}^T + v\mathcal{P}\boldsymbol{\mu}^T$. Then the system of equations (4.3.2) may be written in the compact form

$$\boldsymbol{\mu}^T = \mathbf{c}^T + v\mathcal{P}\boldsymbol{\mu}^T. \tag{4.3.3}$$

If $0 \leq v < 1$, the matrix $(\mathcal{I} - v\mathcal{P})^{-1}$ exists, and we have

$$\boldsymbol{\mu}^T = (\mathcal{I} - v\mathcal{P})^{-1}\mathbf{c}^T. \tag{4.3.4}$$

It is worthwhile to emphasize that *the main idea of the derivation of (4.3.4) consisted in conditioning with respect to what can happen at the first step.* Note also that in our calculations, we did not suppose that the chain was finite.

Next, we show that (4.3.4) implies what we obtained in Section 4.2.2. Indeed,

$$E\{S_\infty\} = \sum_i E\{S_\infty \,|\, X_0 = i\}P(X_0 = i) = \sum_i \mu_i \pi_{0i} = \langle \boldsymbol{\pi}_0, \boldsymbol{\mu} \rangle.$$

For any row-vectors of the same dimension $\mathbf{a}, \mathbf{b}$, their scalar product $\langle \mathbf{a}, \mathbf{b} \rangle = \mathbf{a}\mathbf{b}^T$. Thus, by virtue of (4.3.4),

$$E\{S_\infty\} = \langle \boldsymbol{\pi}_0, \boldsymbol{\mu} \rangle = \boldsymbol{\pi}_0 \boldsymbol{\mu}^T = \boldsymbol{\pi}_0 (\mathcal{I} - v\mathcal{P})^{-1}\mathbf{c}^T = \langle \boldsymbol{\pi}_0 (\mathcal{I} - v\mathcal{P})^{-1}, \mathbf{c} \rangle, \tag{4.3.5}$$

which coincides with (4.2.7).

EXAMPLE 1. Let us return to Example 4.2.2-1. Substituting the data from this example into (4.3.4) and using software, it is easy to get $\boldsymbol{\mu} \approx (29.473,\ 31.854,\ 36.008, 40.620)$. The third number coincides with the answer in Example 4.2.2-1. We see also how the expected present value is increasing with the change of the initial state. The tendency is not surprising since the higher the number of a state, the "better" the state is. □

In Exercise 43, we derive (4.2.11) and (4.2.13) either from (4.3.5) or directly, making use of the first step approach.

4.3.2 The first step approach to random walk problems

4.3.2.1 The probability of returning to zero. Consider the simple random walk model (Example 4.1-4) viewing, for certainty, X_t as a surplus process. Assume $p > 1/2$. Then at each step, the expected profit $E\{\xi_t\} = 2p - 1 > 0$, so the process has a tendency to "move up" on the average. Let the initial surplus $X_0 = 0$.

Denote by T the time, if any, of the first return to zero. Formally, $T = \min\{t \geq 1; X_t = 0\}$. The probability we are computing is $P(T < \infty)$. The probability of the complement event, that is, $P(T = \infty) = 1 - P(T < \infty)$ is the probability that the process will never revisit zero. The first step approach leads to

$$P(T < \infty) = P(T < \infty \,|\, X_1 = 1)p + P(T < \infty \,|\, X_1 = -1)(1 - p).$$

Since $X_t = \sum_{k=1}^t \xi_k$ and $E\{\xi_k\} = 2p - 1 > 0$, we have $E\{X_t\} = t(2p - 1) \to \infty$ as $t \to \infty$. Then, by the law of large numbers, the process starting from (-1) will cross zero level with probability one. Hence, $P(T < \infty \,|\, X_1 = -1) = 1$, and setting $s = P(T < \infty \,|\, X_1 = 1)$, we have

$$P(T < \infty) = sp + 1 - p. \tag{4.3.6}$$

Applying the same approach to $P(T < \infty \,|\, X_1 = 1)$, we write

$$\begin{aligned} s &= P(T < \infty \,|\, X_1 = 1, \xi_2 = 1)p + P(T < \infty \,|\, X_1 = 1, \xi_2 = -1)(1-p) \\ &= P(T < \infty \,|\, X_2 = 2)p + (1-p). \end{aligned} \tag{4.3.7}$$

(When considering the first term, we have used the Markov property; when considering the second term—the fact that the condition $\{X_1 = 1, \xi_2 = -1\}$ implies that the process has already returned to zero.)

Now, $P(T < \infty \,|\, X_2 = 2)$ is the probability that the process will ever arrive at zero starting from state 2. This can be computed as the product of two probabilities: that the process will ever arrive at state 1 starting from state 2, and that the process will ever arrive at zero starting from state 1. The second probability was denoted by s, and the first equals s because the probability to visit state 1 starting from state 2 is equal to the probability to visit state 0 starting from state 1. The last assertion is true in view of the Markov property and the fact that the probabilities of moving up and down are the same at each step.

Thus, $P(T < \infty \,|\, X_2 = 2) = s^2$, and in view of (4.3.7),

$$s = s^2 p + 1 - p.$$

The last equation has two roots: $s_1 = 1$, $s_2 = (1-p)/p$. Let us recall that $E\{\xi_i\} > 0$. In this case, from a heuristic point of view, it seems very plausible that starting from state 2 the process with *some positive* probability will never return to 1, and hence will never come to zero. In other words, we conjecture that $s < 1$. A not too long proof of this fact requires, however, some additional theory, and we postpone it to Example 4.5.2-1. So, we choose $s = (1-p)/p$. Together with (4.3.6) it leads to

$$P(T < \infty) = 2(1-p).$$

Note that when deriving the last formula, we assumed that $p > 1/2$. Similarly, if $p < 1/2$, we have $P(T < \infty) = 2p$. For $p = 1/2$, both formulas lead to $P(T < \infty) = 1$, which is indeed true, as will be shown in Example 4.5.2-1.

Let again $p > 1/2$. Then

$$P(T = \infty) = 1 - P(T < \infty) = 2p - 1 > 0.$$

We saw that, if at the first step the profit becomes negative, the process will cross zero level with probability one. Consequently, $P(T = \infty)$ is the probability that at the first step the profit will be positive, and the surplus X_t will never reach zero again.

4.3.2.2 The ruin problem. Viewing again X_t as a capital (or the surplus of a risk portfolio), we fix an integer a, and set X_0 equal to some $u \in [0, a]$. Assume that the process comes to a stop when it either reaches zero level (which we call ruin) or the level a, whichever comes first. For example, an investor or a player having an initial capital of u plays until

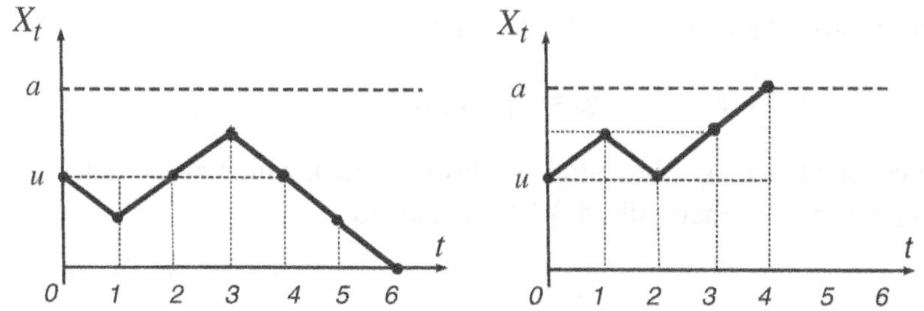

The case of ruin

The case when the process first reaches the level a

FIGURE 8.

the first moment when she/he either runs out of money or gets a planned amount a. See also Fig.8.

In the case of a risk portfolio, we can interpret a as a level after which the company stops accumulating the surplus and either invests a part of it or, for example, pays dividends.

Let A_u be the event that the process reaches zero level, and it happens before the process hits the level a. Set $q_u = P(A_u)$. The index u indicates that this probability depends on the initial level u.

Let B_u be the event that the process reaches the a-level, and it happens before the process hits zero level. Set $p_u = P(B_u)$.

One can guess that $p_u + q_u = 1$, which is true but should be proved. Theoretically, it may happen that the process will never reach either of the boundaries of the interval $[0,a]$ ("corridor" $[0,a]$ in Fig.8). We will show later that the probability of such an event is zero.

To find q_u, first note that

$$q_0 = 1, \; q_a = 0. \tag{4.3.8}$$

(If $u = 0$, the player is already ruined; if $u = a$, the player already has the amount a.)

For $u = 1, 2, ..., a-1$

$$q_u = P(A_u \,|\, \xi_1 = 1)p + P(A_u \,|\, \xi_1 = -1)(1-p). \tag{4.3.9}$$

If at the first step the process moves up, the random walk starts over but from the level $u+1$, and the probability to be ruined under this condition becomes q_{u+1}. More rigorously, $P(A_u \,|\, \xi_1 = 1) = q_{u+1}$.

Similarly, $P(A_u \,|\, \xi_1 = -1) = q_{u-1}$, and we rewrite (4.3.9) as

$$q_u = pq_{u+1} + (1-p)q_{u-1}. \tag{4.3.10}$$

This is a difference equation. Without going deep into the theory of these equations, we note that if $q_u^{(1)}$ and $q_u^{(2)}$ are two particular independent solutions, then any solution will be a linear combination of these two, that is,

$$q_u = c_1 q_u^{(1)} + c_2 q_u^{(2)}, \text{ where } c_1, c_2 \text{ are some constants.} \tag{4.3.11}$$

In the case $p = 1/2$, equation (4.3.10) may be written as

$$q_u = \frac{1}{2}(q_{u+1} + q_{u-1}),$$

and as particular solutions we can choose $q_u^{(1)} \equiv 1$ and $q_u^{(2)} = u$ (the reader can verify this by direct substitution). Then, by (4.3.11), any solution has the form

$$q_u = c_1 + c_2 u. \tag{4.3.12}$$

To specify the constants, we use the initial condition (4.3.8), which gives that $1 = q_0 = c_1$, and $0 = q_a = c_1 + c_2 a$. Together with (4.3.12), it leads to

$$q_u = \frac{a-u}{a}. \tag{4.3.13}$$

To find p_u, it suffices to observe that the distance from the initial point u to the level a, is $a-u$, and since the walk is symmetric, the probability of hitting a first, starting from u, equals the probability of hitting 0 first, starting from $a-u$. In other words, $p_u = q_{a-u} = u/a$.

We see that in the symmetric case the answer is simple: the probability of not being ruined is proportional to the initial capital. We see also that $p_u + q_u$ is indeed equal to 1, that is, the probability that the process will never reach the boundaries of the corridor mentioned is zero.

Consider the case $p \neq 1/2$. If $p = 0$, obviously $q_u = 1$, so we exclude this case. The function $q_u^{(1)} \equiv 1$ still satisfies (4.3.10), but $q_u = u$ does not. Direct substitution shows that now we can take $q_u^{(2)} = r^u$, where $r = (1-p)/p$. The general solution is given in (4.3.11). To find constants c_1 and c_2, we again use (4.3.8), which leads to $1 = q_0 = c_1 + c_2$, and $0 = q_a = c_1 + c_2 r^a$. Together with (4.3.11), it implies that

$$q_u = \frac{r^u - r^a}{1 - r^a}. \tag{4.3.14}$$

To find p_u, we follow the same logic as before, replacing in q_u the argument u by $a-u$. However, since now the walk is not symmetric, we should also replace p by $1-p$. If we do that in formula (4.3.14), the new r becomes $p/(1-p) = 1/r$. Thus,

$$p_u = \frac{(1/r)^{a-u} - (1/r)^a}{1-(1/r)^a} = \frac{r^u - 1}{r^a - 1}. \tag{4.3.15}$$

We see that again $p_u + q_u = 1$.

The analysis of the *ruin probability formula* (4.3.14) leads to some interesting corollaries.

Following for a while the game interpretation, assume that the stake at each play is reduced in half. For example, a player decides to bet not \$1 but 50¢ at each play. How does this change the ruin probability?

If we adopt the new stake as a unit of money, the initial capital in this new unit will be equal to $2u$, and the upper level—to $2a$. In the symmetric case, the ruin probability (4.3.13) will not change after such a substitution, but in the case $p \neq 1/2$, the new probability will be equal to

$$q_u^* = \frac{r^{2u} - r^{2a}}{1 - r^{2a}} = \frac{r^u + r^a}{1 + r^a} \cdot \frac{r^u - r^a}{1 - r^a} = \frac{r^u + r^a}{1 + r^a} q_u.$$

We see that if $p < 1/2$ (and hence $r > 1$), then $q_u^* > q_u$; while if $p > 1/2$ (and hence $r < 1$), then $q_u^* < q_u$.

Thus, if a game is not favorable for a player ($p < 1/2$), she/he will reach the upper threshold with a larger probability by playing higher stakes.

EXAMPLE 1. Let u=\$9, the planned level a=\$10, and p=0.4. If the stake is \$1, that is, our player is within one step of reaching the level a, then though the game is not favorable for the player, the probability to reach \$10 is $p_9 = (1-(3/2)^9)/(1-(3/2)^{10}) \approx 0.660$. (Note that this probability is larger than 0.4—the probability to "move up". The reader is invited to explain this from a heuristic point of view.)

On the other hand, if the stake had been only 10¢, the new probability of reaching \$10 would have been $(1-(3/2)^{90})/(1-(3/2)^{100}) \approx 0.0173$. □

In the case $a = \infty$, the process cannot reach the level a. So, in this case we consider only the possibility of being ruined, and hence q_u is the probability of ever being ruined during an infinite interval of time.

Letting $a \to \infty$ in (4.3.13) and (4.3.14), we have $q_u \to 1$ for $r \geq 1$, and $q_u \to r^u$ if $r < 1$. In other words, for $a = \infty$,

$$q_u = 1 \text{ if } p \leq 1/2, \text{ and } q_u = r^u, \text{ if } p > 1/2. \tag{4.3.16}$$

4.3.2.3 ► Infinitesimal increments.[4] The next approximation is useful for future modeling. It applies to the situations where the capital is changing not abruptly but in small increments during small intervals of time. For example, in the case of insurance, we may think that we observe capital each hour.

Denote the length of the time interval between consecutive steps by δ, measured in some real units of time. We will view δ as very small; at the end of constructing the model, we will let $\delta \to 0$. If δ is small, it is natural to assume that the change of the capital at each epoch t is also small. Denote this change, measured in some original units of money, by η_t. Assume that $\eta_t = \pm k$ for some k with probabilities p and $1-p$, respectively. Later we will let $k \to 0$.

It is worthwhile to emphasize that we should not confuse η_t with ξ_t above, since $\xi_t = \pm 1$ in some conditional not specified yet unit of money.

To find natural representations for k and p, assume that the mean and the variance of η_t are proportional to the length of the time period. More precisely, assume that $E\{\eta_t\} = \mu\delta$, $Var\{\eta_t\} = \sigma^2\delta$ for some μ and σ. Since, on the other hand, $E\{\eta_t\} = k(2p-1)$, and $Var\{\eta_t\} = 4k^2 p(1-p)$, we have

$$k(2p-1) = \mu\delta, \quad 4k^2 p(1-p) = \sigma^2\delta.$$

Simple algebra leads to

$$k = \sigma\sqrt{\delta} + o(\sqrt{\delta}), \quad p = \frac{1}{2}\left(1 + \frac{\mu}{\sigma}\sqrt{\delta}\right) + o(\sqrt{\delta}).$$

So, the case $\mu = 0$ corresponds to $p = 1/2$.

[4]This section may be skipped in the first reading.

Now, we fit all of this to the model of the previous subsection. The initial capital u and the level a are measured in the original units of money. We adopt k as a new unit of money, and replace η_t by ξ_t taking values ± 1. Accordingly, we replace u by $\widetilde{u} = u/k$, and a by $\widetilde{a} = a/k$. Furthermore, setting $s = \sqrt{\delta}$, we have

$$r = \frac{1-p}{p} = \frac{1-(\frac{1}{2}(1+\frac{\mu}{\sigma}s)+o(s))}{\frac{1}{2}(1+\frac{\mu}{\sigma}s)+o(s)} = \frac{1-\frac{\mu}{\sigma}s+o(s)}{1+\frac{\mu}{\sigma}s+o(s)} = 1-2\frac{\mu}{\sigma}s+o(s),$$
$$\widetilde{u} = \frac{u}{k} = \frac{u}{\sigma s+o(s)} = \frac{u}{\sigma s}(1+o(1)),\ \widetilde{a} = \frac{a}{k} = \frac{a}{\sigma s}(1+o(1)).$$

For the ruin probability $q_{\widetilde{u}}$, we again use (4.3.13) if $\mu = 0$ (which corresponds to $p = 1/2$), and use (4.3.14) if $\mu \neq 0$ (which corresponds to $p \neq 1/2$). For the former case,

$$q_{\widetilde{u}} = \frac{\widetilde{a}-\widetilde{u}}{\widetilde{a}} = \frac{a-u}{a}(1+o(1)) \to \frac{a-u}{a} \text{ as } s \to 0.$$

Thus, in the symmetric case, the result is the same formula (4.3.13).

The case $\mu \neq 0$ is much more interesting. We write

$$r^{\widetilde{u}} = (1-2\frac{\mu}{\sigma}s+o(s))^{u/k} = (1-2\frac{\mu}{\sigma}s+o(s))^{\frac{u}{\sigma s}(1+o(1))} \to \exp\{-2u\mu/\sigma^2\}$$

as $s \to 0$. To get $r^{\widetilde{a}}$, we should just replace u by a in the last formula.

Letting $s \to 0$, we eventually obtain that the ruin probability

$$q_{\widetilde{u}} \to [\exp\{-2u\mu/\sigma^2\} - \exp\{-2a\mu/\sigma^2\}]/[1-\exp\{-2a\mu/\sigma^2\}]. \tag{4.3.17}$$

For $\mu > 0$, letting $a \to \infty$, we get that the probability of ever being ruined during an infinite interval of time is

$$\lim_{s\to 0} q_{\widetilde{u}} = \exp\{-2u\mu/\sigma^2\}. \tag{4.3.18}$$

We will come to this formula again when considering the ruin problem in the case of Brownian motion; see Section **5**.2.4.4.

EXAMPLE 1. Assume that the surplus for a risk portfolio grows by $\mu = 1$ units of money per day on the average, and the standard deviation of the change of the surplus is $\sigma = 2$ per day. In this situation, one may guess that the probability that the change per day will be negative is not small. Let the initial surplus be $u = 10$ units. Then the probability that the surplus will run out completely in an infinite time period, may be estimated by $\exp\{-2u\mu/\sigma^2\} = \exp\{-5/2\} \approx 0.082$. This is relatively high probability to be ruined, so it may make sense to start with a bigger amount of u. □ ◄

4.4 Limiting probabilities and stationary distributions

This section concerns homogeneous chains. Consider, as an example a particular transition matrix of a chain with two states

$$\mathcal{P} = \begin{Vmatrix} 0.8 & 0.2 \\ 0.3 & 0.7 \end{Vmatrix},$$

and its consecutive powers

$$\mathcal{P}^2 = \begin{Vmatrix} 0.7 & 0.3 \\ 0.45 & 0.55 \end{Vmatrix}, \quad \mathcal{P}^3 = \begin{Vmatrix} 0.65 & 0.35 \\ 0.525 & 0.475 \end{Vmatrix}, \ldots,$$

$$\mathcal{P}^6 = \begin{Vmatrix} 0.60625 & 0.39375 \\ 0.590625 & 0.409375 \end{Vmatrix}, \ldots, \mathcal{P}^{10} \approx \begin{Vmatrix} 0.6004 & 0.3996 \\ 0.5994 & 0.4006 \end{Vmatrix}, \ldots . \tag{4.4.1}$$

We see an explicit convergency pattern, and what is important is that the two rows are getting closer to each other. This is not accidental. Consider an arbitrary two-dimensional transition matrix which may be written as

$$\mathcal{P} = \begin{Vmatrix} 1-\alpha & \alpha \\ \beta & 1-\beta \end{Vmatrix},$$

where α, β are non-negative and are not greater than one.

Let $\alpha+\beta > 0$; that is, either α, or β, or both are positive. It is known that in this case,

$$\mathcal{P}^t = \frac{1}{\alpha+\beta}\begin{Vmatrix} \beta & \alpha \\ \beta & \alpha \end{Vmatrix} + \frac{(1-\alpha-\beta)^t}{\alpha+\beta}\begin{Vmatrix} \alpha & -\alpha \\ -\beta & \beta \end{Vmatrix}. \tag{4.4.2}$$

The reader may prove it by induction on her/his own or look at a proof, e.g., in [61] or [79].

Assume, in addition, $\alpha+\beta < 2$; that is, at least one number is not 1. Thus, $0 < \alpha+\beta < 2$, and hence $-1 < \alpha+\beta-1 < 1$. So, $|1-\alpha-\beta| < 1$, and the second term in (4.4.2) converges to zero as $t \to \infty$. Thus,

$$\mathcal{P}^t \to \frac{1}{\alpha+\beta}\begin{Vmatrix} \beta & \alpha \\ \beta & \alpha \end{Vmatrix} = \begin{Vmatrix} \pi_0 & \pi_1 \\ \pi_0 & \pi_1 \end{Vmatrix} \quad \text{as } t \to \infty, \tag{4.4.3}$$

where

$$\pi_0 = \frac{\beta}{\alpha+\beta}, \quad \pi_1 = \frac{\alpha}{\alpha+\beta}.$$

Note that the convergence is fast: $(1-\alpha-\beta)^t \to 0$ exponentially.

In the example above, $\alpha = 0.2$, $\beta = 0.3$, and $\pi_0 = \frac{0.3}{0.2+0.3} = 0.6$, $\pi_1 = \frac{0.2}{0.2+0.3} = 0.4$, which is consistent with (4.4.1).

What is remarkable in (4.4.3) is that the rows in the limiting matrix are identical. This means that, in the long run, asymptotically, the probability that the process will be in a particular state does not depend from which state the process has started at the initial time. The process is, so to say, "gradually forgetting" the past; look again at (4.4.1).

Such a property (with some variations in definitions) is called *ergodicity*, and the chain itself—*ergodic*.

We have established it for $0 < \alpha+\beta < 2$. If this is not true, ergodicity does not take place. Let $\alpha+\beta = 0$. Then, since α and β are non-negative, both numbers are zeros, and $\mathcal{P} = \mathcal{I} = \begin{Vmatrix} 1 & 0 \\ 0 & 1 \end{Vmatrix}$, the identity matrix. In this case, $\mathcal{P}^t = \mathcal{I}$ for all t, and the process will never leave the initial state.

If $\alpha+\beta = 2$, which is possible only if both numbers, α and β, equal one, then $\mathcal{P} = \begin{Vmatrix} 0 & 1 \\ 1 & 0 \end{Vmatrix}$, and the process alternates between two states. (Starting from state 0, the chain moves to state 1, and then comes back to state 0, and keeps moving in the same fashion.) Excepting the cases

$$\mathcal{P} = \begin{Vmatrix} 1 & 0 \\ 0 & 1 \end{Vmatrix} \quad \text{or} \quad \begin{Vmatrix} 0 & 1 \\ 1 & 0 \end{Vmatrix}, \tag{4.4.4}$$

the chain is ergodic.

Certainly, the fact that we considered only two states does not matter. Consider, for instance, Examples 4.1-3 and 5, concerning a process of rearrangements. We have seen already $\mathcal{P}^2$ in this case. The reader may compute that

$$\mathcal{P}^5 \approx \begin{Vmatrix} 0.169 & 0.165 & 0.169 & 0.165 & 0.169 & 0.165 \\ 0.165 & 0.169 & 0.169 & 0.165 & 0.169 & 0.165 \\ 0.169 & 0.165 & 0.169 & 0.165 & 0.165 & 0.169 \\ 0.169 & 0.165 & 0.165 & 0.169 & 0.165 & 0.169 \\ 0.165 & 0.169 & 0.165 & 0.169 & 0.169 & 0.165 \\ 0.165 & 0.169 & 0.165 & 0.169 & 0.165 & 0.169 \end{Vmatrix}.$$

After just five steps, all numbers are very close to $1/6$. We will see soon that $1/6$ is the limiting probability. So, rather quickly, all rearrangements are getting close to be equally likely.

Consider the shuffling interpretation. We may define a shuffling to be perfect if it leads to equal probabilities of all possible permutations. We see that even a simple shuffling as in our example is asymptotically perfect.

In general, ergodicity takes place under some conditions, though as we will see, they are rather mild. For example, chains with the transition matrices

$$\mathcal{P} = \begin{Vmatrix} 1 & 0 \\ 0 & 1 \end{Vmatrix} \quad \text{or} \quad \begin{Vmatrix} 0 & 1 \\ 1 & 0 \end{Vmatrix} \tag{4.4.5}$$

do not posses the ergodicity property. In the first case, $\mathcal{P}^t = \mathcal{I}$, and the process will never leave the starting state. In the second case, the process continues to alternate between two states.

Let us consider a rigorous statement. The result below concerns finite chains and may be applied to many practical problems. Recall that for simplicity, we write p_{ij}^m instead of $p_{ij}^{(m)}$, understanding it not as a power of p_{ij} but as the corresponding element of $\mathcal{P}^m$.

Theorem 4 *Let the number of states $k < \infty$. Suppose that for some state j_0 and a natural m*

$$p_{ij_0}^m > 0 \text{ for all } i. \tag{4.4.6}$$

Then there exists a probability row-vector $\boldsymbol{\pi} = (\pi_1, ..., \pi_k)$, such that $\pi_i \geq 0$ for all i, $\pi_1 + ... + \pi_k = 1$, and

$$\mathcal{P}^t \to \begin{Vmatrix} \boldsymbol{\pi} \\ \vdots \\ \boldsymbol{\pi} \end{Vmatrix} \quad as \quad t \to \infty, \tag{4.4.7}$$

where in the limiting matrix each row is equal to the vector $\boldsymbol{\pi}$.

A proof may be found, e.g., in [35] or [120].

Condition (4.4.6) is the finite-chain version of the *Doeblin condition.* It supposes the existence of a state to which the process can move with positive probability from *any* state in a finite number of steps. One can say that this state is *accessible* from any state.

Algebraically it means that for some m, the matrix $\mathcal{P}^m$ has at least one strictly positive column. Certainly, if all elements of $\mathcal{P}$ are positive, Doeblin's condition holds automatically with $m = 1$. For instance, this is the case for the matrix (4.1.1) in the example concerning changing intensities of a claim flow.

In the rearrangements problem of Examples 1.1-3 and 5, in $\mathcal{P}$ itself there is no strictly positive column. However, all elements of $\mathcal{P}^2$ are positive, so Doeblin's condition holds for $m = 2$.

Conditions of Doeblin's type and corresponding theorems for the general case, when chains may be infinite, can be found, e.g., in [35], [36], [120].

Consider now π_t, the distribution of the process at time t.

Corollary 5 *Under the conditions of Theorem 4,*

$$\boldsymbol{\pi}_t \to \boldsymbol{\pi} \quad as \quad t \to \infty, \tag{4.4.8}$$

where $\boldsymbol{\pi}$ *is the same as in (4.4.7).*

Proof. We have

$$\boldsymbol{\pi}_t = \pi_0 \mathcal{P}^t \to \boldsymbol{\pi}_0 \cdot \begin{Vmatrix} \boldsymbol{\pi} \\ \vdots \\ \boldsymbol{\pi} \end{Vmatrix}.$$

All elements in the ith column of the last matrix equal π_i. Hence, the ith element of the product is $\pi_{00}\pi_i + \pi_{01}\pi_i + ... = \pi_i(\pi_{00} + \pi_{01} + ...) = \pi_i \cdot 1 = \pi_i$. ■

Thus, whatever the initial distribution $\boldsymbol{\pi}_0$ is, asymptotically, as $t \to \infty$, the distribution of the process at time t converges to the same distribution $\boldsymbol{\pi}$. In particular, this implies that

> In the long run, the proportion of time when the process is in state i, is equal to π_i, the ith coordinate of the vector $\boldsymbol{\pi}$.

For more detail, see also Section 4.5.4.

The next question seeks to find the limiting distribution $\boldsymbol{\pi}$. From (4.1.8) it follows that

$$\boldsymbol{\pi}_t = \boldsymbol{\pi}_{t-1} \mathcal{P} \tag{4.4.9}$$

(consider one step, viewing $t - 1$ as the initial time). Since $\boldsymbol{\pi}_t \to \boldsymbol{\pi}$, and $\boldsymbol{\pi}_{t-1} \to \boldsymbol{\pi}$, as well, letting $t \to \infty$ in (4.4.9) , we have the fundamental equation for $\boldsymbol{\pi}$:

$$\boldsymbol{\pi} = \boldsymbol{\pi}\mathcal{P}. \tag{4.4.10}$$

The probability distribution $\boldsymbol{\pi}$ satisfying (4.4.10) is called a *stationary distribution.* The choice of the term is connected with the following fact.

Let $\boldsymbol{\pi}_0 = \boldsymbol{\pi}$, that is, the process starts from the distribution $\boldsymbol{\pi}$ from the very beginning. Then $\boldsymbol{\pi}_1 = \boldsymbol{\pi}_0\mathcal{P} = \boldsymbol{\pi}\mathcal{P} = \boldsymbol{\pi}$, $\boldsymbol{\pi}_2 = \boldsymbol{\pi}_1\mathcal{P} = \boldsymbol{\pi}\mathcal{P} = \boldsymbol{\pi}$, and so on: $\boldsymbol{\pi}_t = \boldsymbol{\pi}$ for *all* t.

Thus, for an arbitrary initial distribution $\boldsymbol{\pi}_0$, the distribution $\boldsymbol{\pi}_t$ is approaching the stationary distribution $\boldsymbol{\pi}$, while if the initial distribution is $\boldsymbol{\pi}$ itself, the distribution of the process is invariant through time and corresponds to the *stationary regime* from the very beginning.

Nowadays, solving equation (4.4.10), at least numerically, is not a problem even for large dimension. We restrict ourselves to the following illustrative examples.

EXAMPLE 1. Consider the process of random rearrangements with $\mathcal{P}$ given in (4.1.3). It is straightforward to verify that for such a matrix, equation (4.4.10) is true for $\boldsymbol{\pi}=(\frac{1}{6},\frac{1}{6},\frac{1}{6},\frac{1}{6},\frac{1}{6},\frac{1}{6})$, meaning that, as $t\to\infty$, all arrangements are equally likely.

EXAMPLE 2. Consider the process of changing intensities with matrix (4.1.1). Equation (4.4.10) may be written coordinatewise as follows:

$$\begin{aligned} 0.6\pi_0 + 0.4\pi_1 + 0.25\pi_2 &= \pi_0, \\ 0.3\pi_0 + 0.5\pi_1 + \ 0.7\pi_2 &= \pi_1, \\ 0.1\pi_0 + 0.1\pi_1 + 0.05\pi_2 &= \pi_2. \end{aligned}$$

Together with $\pi_0 + \pi_1 + \pi_2 = 1$, it yields

$$\pi_0 = \frac{81}{168}, \quad \pi_1 = \frac{71}{168}, \quad \pi_2 = \frac{16}{168}. \tag{4.4.11}$$

We can interpret this as for an infinite time interval, the proportion of days, for instance, with icy conditions is $\pi_2 = 2/21$.

Let us continue our example adding the information from Example 4.2.1-3. Let Y_0, Y_1, Y_2 be independent Poisson r.v.'s with parameters $\lambda_0 = 2$, $\lambda_1 = 4$, $\lambda_2 = 8$, respectively. Since for large t, the probability to be in state i may be approximated by π_i, the r.v. Z_t, the number of claims received on day t, may be approximated by a r.v.

$$Z = \begin{cases} Y_0 \text{ with probability } \pi_0, \\ Y_1 \text{ with probability } \pi_1, \\ Y_2 \text{ with probability } \pi_2. \end{cases}$$

In particular, for large t, or more rigorously, as $t\to\infty$,

$$E\{Z_t\} \to E\{Z\} = \pi_0\lambda_0 + \pi_1\lambda_1 + \pi_2\lambda_2 = 3.4166\ldots . \tag{4.4.12}$$

We can also estimate the expected total number of claims, S_n, for n days. Let $m = \pi_0\lambda_0 + \pi_1\lambda_1 + \pi_2\lambda_2$. Since $E\{S_n\} = \sum_{t=0}^{n} E\{Z_t\}$, in view of (4.4.12),

$$E\{S_n\} \sim mn = 3.4166n.$$

For example, for a season of $n = 60$ days, the estimate is $3.4166\ldots \cdot 60 = 205$. Omitting a proof, note that the last estimate is pretty good for $n = 60$, since the rate of convergence in (4.4.12) is exponential, which is fairly rapid.

If we want to approximate $P(Z_t \le x)$, we may, as in Example 4.2.1-3, just write $P(Z_t \le x) \approx \pi_0 \cdot PoissonDist(x, 2) + \pi_1 \cdot PoissonDist(x, 4) + \pi_2 \cdot PoissonDist(x, 8)$.

Computing the distribution of S_n is more complicated. An approximation can be provided by making use of the CLT for Markov chains (see, e.g., [63], [120]), but we skip this question.

EXAMPLE 3. Let a chain have one absorbing state, say,

$$\mathcal{P} = \left\| \begin{matrix} * & * & * \\ * & * & * \\ 0 & 0 & 1 \end{matrix} \right\|, \tag{4.4.13}$$

where the stars $*$ represent positive numbers. It is easy to see that in this case the solution to (4.4.10) is $\boldsymbol{\pi} = (0, 0, 1)$, that is, the limiting distribution is concentrated at the last state. It is not surprising at all, since with probability one the process must arrive at the absorbing state. For us, it is worth noting, however, that Theorem 4 covers such cases also.

EXAMPLE 4 ([151, N16][5]). Drivers transition monthly between three states, "good", "average", "bad", according to the transition matrix

$$\mathcal{P} = \left\| \begin{matrix} 0.8 & 0.2 & 0.0 \\ 0.2 & 0.6 & 0.2 \\ 0.0 & 0.4 & 0.6 \end{matrix} \right\|.$$

What percentage of drivers will transit from "good" to "average" between month 100 and month 101?

The number 100 is "large", so we may consider the limiting probability π_0 of being in the state "good". Solving (4.4.10) similarly to what we did in Example 2, we find that $\pi_0 = 0.4$. Thus, on the average, 40% of drivers will be in the state "good" in month 100. Since $p_{01} = 0.2$, from these 40%, on the average, 20% will transit to the second state ("average"). Thus, the solution is $0.4 \cdot 0.2 = 0.08$ or 8%. □

Route 2 ⇒ page 303

4.5 The ergodicity property and classification of states

3

We turn to a more detailed analysis of possible states, which will allow us to better understand the nature of ergodicity. We consider below only homogeneous chains.

4.5.1 Classes of states

A state j is said to be *accessible* from a state i if $p_{ij}^m > 0$ for some m.

In Example 4.1-3, the state (1,3,2) is not accessible from (1,2,3) in one step, but is accessible in two steps. So, we say that the former state is accessible from the latter (and, certainly, vice versa).

[5]Reprinted with permission of the Casualty Actuarial Society.

States i and j are said to *communicate* if they are accessible from each other.

In Example 4.1-2, state 1 is accessible from state 0, but not vice versa. In the random walk from Example 4.1-4, all states communicate if $p \neq 0$ or 1. Indeed, if $j > i$, we can move from i straight up along the path $i \to i+1 \to i+2 \to \ldots \to j$, whose probability is $p^{j-i} > 0$. The case $j < i$ is considered similarly.

It may be shown that states of any chain can be partitioned into disjoint *classes* such that all states from the same class communicate, and any two states from different classes do not (since otherwise they would belong to the same class).

For example, a homogeneous chain with states labeled $0, 1, 2, 3$ and

$$\mathcal{P} = \begin{Vmatrix} 0.5 & 0.5 & 0 & 0 \\ 0.5 & 0.5 & 0 & 0 \\ 0.25 & 0.25 & 0.25 & 0.25 \\ 0 & 0 & 0 & 1 \end{Vmatrix} \tag{4.5.1}$$

has three classes: $\{0,1\}$, $\{2\}$, $\{3\}$.

The chain is said to be *irreducible* if it has only one class.

For instance, for the chains from Examples 4.1-1, 3 and 4, all states communicate, and hence these chains are irreducible. (In Example 4.1-3 for $p \neq 0, 1$.)

4.5.2 The recurrence property

Let f_i be the probability that, starting from state i, the chain will ever return to this state. State i is called *recurrent* if $f_i = 1$, and *transient* if $f_i < 1$.

For a recurrent state i, the process, starting from i, will return to i with probability one. Since this process is Markov, once it revisits i, the process will start over from i as from the beginning. After this, the process will again return to i with probability one, and so on, revisiting state i infinitely often with probability one. W. Feller, the author of one of best, if not the best, book on Probability Theory [38], called such a state *persistent* and regretted that in the first edition of his book he had called it recurrent. However, the term recurrent is widely accepted.

Denote the number of revisits to a state i given that the process has started from this state by M_i. We do not count the initial stay in i.

We saw that

> If i is recurrent, then $M_i = \infty$ with probability one.

Let $f_i < 1$. Then the probability that the process will never return to i is $P(M_i = 0) = 1 - f_i$. The probability that the process will revisit i one time and then never come back is $P(M_i = 1) = f_i(1 - f_i)$, and in general, $P(M_i = k) = f_i^k(1 - f_i)$. Thus, M_i has a geometric distribution and, in particular,

$$E\{M_i\} = f_i/(1 - f_i)$$

(see (**0**.3.1.12). Thus,

> If i is transient, then $M_i < \infty$ with probability one. Moreover, $E\{M_i\} < \infty$. (4.5.2)

From (4.5.2) it immediately follows that

> For any finite Markov chain, there exists at least one recurrent state. (4.5.3)

To show this, assume that all states are transient. Then, since all transient states may be visited only a finite number of moments of time, and the number of states is finite, after a finite number of steps no states will be visited, which is impossible.

Before turning to examples, consider two more facts.

Let $X_0 = i$ (the process starts from i), and let the indicator r.v. $I_n = 1$ or 0, depending on whether $X_n = i$ or not. Then $M_i = \sum_{n=1}^{\infty} I_n$.

On the other hand, $P(I_n = 1) = P(X_n = i) = p_{ii}^n$, and hence

$$E\{M_i\} = E\{\sum_{n=1}^{\infty} I_n\} = \sum_{n=1}^{\infty} E\{I_n\} = \sum_{n=1}^{\infty} p_{ii}^n. \tag{4.5.4}$$

Thus,

> State i is recurrent iff $\sum_{n=1}^{\infty} p_{ii}^n = \infty$. (4.5.5)

We will derive from this the following proposition.

Proposition 6 *If states i and j communicate, then i and j are either both recurrent or both transient.*

In other words,

> The recurrence and transience properties are properties of classes. (4.5.6)

Proof *of Proposition 6.* If i and j communicate, by definition, there exist m_1 and m_2 such that $p_{ji}^{m_1} \geq \alpha$ and $p_{ij}^{m_2} \geq \beta$ for some $\alpha, \beta > 0$. One of the possible paths to move from j to j in m_1+n+m_2 steps is to move in m_1 steps from j to i, to return to i in n steps, and to move from i to j in m_2 steps. Hence,

$$p_{jj}^{m_1+n+m_2} \geq p_{ji}^{m_1} p_{ii}^n p_{ij}^{m_2} = \alpha\beta p_{ii}^n. \tag{4.5.7}$$

Consequently, if i and j communicate, then $\sum_{n=1}^{\infty} p_{ii}^n$ and $\sum_{n=1}^{\infty} p_{jj}^n$ converge or diverge simultaneously.

[In more detail, since

$$\sum_{n=1}^{\infty} p_{jj}^{m_1+n+m_2} = \sum_{n=m_1+m_2+1}^{\infty} p_{jj}^n,$$

series $\sum_{n=1}^{\infty} p_{jj}^{m_1+n+m_2}$ and $\sum_{n=1}^{\infty} p_{jj}^n$ converge or diverge simultaneously. Consequently, if $\sum_{n=1}^{\infty} p_{jj}^n$ converges, then $\sum_{n=1}^{\infty} p_{ii}^n$ converges; if $\sum_{n=1}^{\infty} p_{ii}^n$ diverges, then $\sum_{n=1}^{\infty} p_{jj}^n$ diverges. Since i and j are arbitrary, we can switch i and j, and apply a similar argument.] ■

From (4.5.3) and (4.5.6) it follows that

> For any irreducible finite Markov chain, all states are recurrent.

For instance, the chains from Examples 4.1-1 and 3 are irreducible, and hence all states are recurrent.

If a finite chain has several classes, for each class we should check whether the process may leave the class and never come back with a positive probability. For finite chains, it is usually easy.

For instance, in Example 4.1-2, state 0 (alive) is transient, and other states are recurrent, moreover, absorbing.

For a chain with $\mathcal{P}$ from (4.5.1), classes $\{0,1\}$, and $\{3\}$ contain recurrent states, and class $\{2\}$—transient. A chain with

$$\mathcal{P} = \begin{Vmatrix} 0.5 & 0.5 & 0 & 0 \\ 0.5 & 0.5 & 0 & 0 \\ 0 & 0 & 0.75 & 0.25 \\ 0 & 0 & 0.3 & 0.7 \end{Vmatrix}$$

clearly has two classes but here all states are recurrent.

If a chain is infinite, the question may not be so simple.

EXAMPLE 1 is classical and concerns the simple random walk; see Example 4.1-4. Let $p \neq 0, 1$. Then, as has been shown, the chain is irreducible, and in view of (4.5.6), it suffices to check only one state, say, 0. Clearly, the process can return to 0 only in an even number of steps. Hence, $p_{00}^{2k+1} = 0$. Suppose $X_{2k} = 0$. Then the number of ξ's taking the value 1 should be equal to the number of ξ's taking the value -1. Therefore,

$$p_{00}^{2k} = \binom{2k}{k} p^k (1-p)^k = \frac{(2k)!}{k!k!} p^k (1-p)^k.$$

We apply Stirling's formula

$$k! \sim \sqrt{2\pi k} k^k e^{-k},$$

where, as usual, $a_k \sim b_k$ means $(a_k/b_k) \to 1$. (A proof may be found in many books on advanced calculus. Neat and less traditional proofs are contained, e.g., in [38, II.9] and in [116, 4.3].)

The reader can easily verify that, by Stirling's formula,

$$p_{00}^{2k} \sim \frac{1}{\sqrt{\pi k}} [4p(1-p)]^k.$$

Let $a_p = 4p(1-p)$. If $p \neq \frac{1}{2}$, then $a_p < 1$ (recall what is the maximum of $p(1-p)$). Then a_p^k is an exponential function in k. Hence,

$$\sum_{n=1}^{\infty} p_{00}^n \sim \sum_{k=1}^{\infty} p_{00}^{2k} \sim \sum_{k=1}^{\infty} \frac{1}{\sqrt{\pi k}} a_p^k < \infty.$$

Consequently, by criterion (4.5.5), state 0 is transient.

Thus, if $p \neq \frac{1}{2}$, starting from any state, the process will never come back with positive probability. This should not seem surprising. If $p > 1/2$, then $E\{\xi\} = 2p - 1 > 0$, and $E\{X_t\} = u + (2p-1)t \to \infty$ as $t \to \infty$. So, by the law of large numbers, $X_t \to +\infty$ with probability one, and may be in any state only a finite number of times. If $p < 1/2$, similarly $X_t \to -\infty$ with probability one.

If $p = \frac{1}{2}$, then $a_p = 1$, and $p_{00}^{2k} \sim \dfrac{1}{\sqrt{\pi k}}$. In this case

$$\sum_{n=1}^{\infty} p_{00}^{n} \sim \sum_{k=1}^{\infty} p_{00}^{2k} \sim \frac{1}{\sqrt{\pi}} \sum_{k=1}^{\infty} \frac{1}{\sqrt{k}} = \infty,$$

and state 0 is recurrent. Thus, in the case of the symmetric random walk, the process starting from any state will return to the same sate again and again, infinitely often. □

The above classification and partitioning of the state space into classes is useful since, for many purposes, we can restrict our attention to one class. In particular, it is worth noting the following.

If a process starts from a transient state, it can leave the class to which this state belongs. For example, in the case (4.5.1), the process starting from state 2 can move to state 0, and then it will never leave the class $\{0, 1\}$.

However,

> If the process starts from a recurrent state, it will never leave the class from which it has started to evolve.

Indeed, let i be the initial state. Then the process can move to any state which is accessible from i. If this new state had not communicated with i, it would have meant that the probability to come back to i is zero. Then the probability of returning to i would not have been one, and i would not have been recurrent. Consequently, any state to which the process can move from i communicates with i. Such a state, by definition, belongs to the same class as i.

The same argument leads to the following fact:

> In a recurrent class, the process can reach any state from any state with probability one. (4.5.8)

Indeed, if the probability to move from, say, state k to state i is less than one, then with a positive probability the process may move from i to k and not return to i, that is, the probability of returning to i, starting from i, would be less than one.

4.5.3 Recurrence and travel times

Let T_{ik} be the number of steps required for the process, starting from i, to reach state k. Then the r.v. $T_i = T_{ii}$, is a return (or recurrence) time. Set $m_{ik} = E\{T_{ik}\}$, and $m_i = m_{ii} = E\{T_i\}$.

If state i is transient, the process will never come back to state i with the positive probability $1-f_i$ defined above. So, with the mentioned positive probability, $T_i = \infty$ and, hence, $m_i = \infty$.

In general, if state i is recurrent, it also may happen that $m_i = \infty$. Such states are called *null recurrent*. Otherwise, the state is called *positive recurrent*.

A classical example of null recurrence concerns states in the symmetric random walk from Example 4.5.2-1: for $p = \frac{1}{2}$ all states, as was shown, are recurrent, but the expected return time $m_{ii} = \infty$. The direct proof is somewhat complicated; however, when we consider Markov moments in Section **5**.2.4, we will be able to prove it almost instantly.

The phenomenon mentioned is essentially connected with the fact that for the random walk, the number of states is infinite. For finite chains the situation is much simpler.

Consider a chain with a finite number of states. Let i be a recurrent state, and S be the class containing i. As was shown above, starting from i, the process may travel only inside S. We can prove also the following simple proposition.

Proposition 7 *For the chain and state i defined above,* $m_{ik} < \infty$ *for any* $k \in S$.

Proof is elementary, so we give its sketch. Since all states from S communicate, for each $j,k \in S$, there exists $m = m(j,k)$ such that $p_{jk}^{m(j,k)} > 0$. Let $M = \max_{j,k \in S} m(j,k)$ and $\alpha = \min_{j,k \in S} p_{jk}^{m(j,k)}$. Since we consider max and min over a finite set of positive numbers, $0 < \alpha \le M < \infty$.

Suppose the process starts from i. The probability that it will not enter state k during M steps is less or equal than $1-\alpha$. In notation, $P(T_{ik} > M) \le 1-\alpha$. If the process has not entered k during the first M steps, the experiment is repeated, starting from the state at which the process has arrived at the Mth step. The probability that the process will not enter k during the next M steps is again less than $1-\alpha$. Consequently, $P(T_{ik} > 2M) \le (1-\alpha)^2$.

It is clear now that $P(T_{ik} > sM) \le (1-\alpha)^s$ for any integer s. Note that $(1-\alpha) < 1$ because $\alpha > 0$. Now, by using formula (**0**.2.2.3), we can write that $E\{T_{ik}\} \le \sum_{s=0}^{\infty} M \cdot P(T_{ik} > sM) \le M\sum_{s=0}^{\infty}(1-\alpha)^s < \infty$. ■

The values that recurrence times T_i can take are also connected with the periodicity property. Assume that $p_{ii}^n = 0$ for all n not divisible by some number k. It means that, starting from i, the process can revisit i only at moments $k, 2k, \ldots$. We say that d is the *period of state i* if d is the largest number among all integers k with the property above. For example, if the chain may revisit the state i with positive probabilities only at moments $3, 6, 9, \ldots$, then $d = 3$. If $d = 1$, the state is called *aperiodic*. Clearly, if $p_{ii} > 0$, state i is aperiodic.

It may be shown, by making use of (4.5.7), that periodicity is a class property, so for an irreducible chain all states are either aperiodic or periodic with the same period. The chain with $\mathcal{P} = \begin{Vmatrix} 0 & 1 \\ 1 & 0 \end{Vmatrix}$ is obviously periodic with a period of two. The same is true for the simple random walk.

4.5.4 Recurrence and ergodicity

We are able now to state a basic ergodicity theorem.

Theorem 8 *For all states i, j of any irreducible recurrent aperiodic chain, there exists*

$$\lim_{t\to\infty} p_{ij}^t = \pi_j = \frac{1}{m_j}.$$

Analyzing this theorem, we can jump to the following conclusions.

- $\lim_{t\to\infty} p_{ij}^t = 0$ iff $m_j = \infty$, that is, only when the state j is null recurrent.
- Let $m_j = \infty$. Setting $i = j$, we get that $\lim_{t\to\infty} p_{jj}^t = 0$. On the other hand, all states in the chain under consideration communicate. Hence, in view of (4.5.7), if $p_{jj}^t \to 0$, then $p_{ii}^t \to 0$ for other states i, which means that $m_i = \infty$ for all states i. In other words, null recurrence is a class property, and for the chain under discussion (a) either all limits $\lim_{t\to\infty} p_{ij}^t$ are equal to zero, or all are positive; (b) either all expected recurrence (return) times m_j are infinite, or all are finite.
- Recall that for any chain $\sum_j p_{ij}^t = 1$. If the chain is finite, and k is the number of states, we can consider $\lim_{t\to\infty}$ of both sides writing $1 = \lim_{t\to\infty} \sum_{j=1}^k p_{ij}^t = \sum_{j=1}^k \lim_{t\to\infty} p_{ij}^t = \sum_{j=1}^k \pi_j$, and having eventually $\sum_{j=1}^k \pi_j = 1$. (For an infinite chain we cannot bring $\lim_{t\to\infty}$ inside the sum without additional conditions.) But π_j's are either all zeros, or all positive. Since the sum equals one, the former is impossible. Thus,

For any finite recurrent chain, $\pi_j > 0$, and $m_j < \infty$ for all j.

Now we state a general ergodicity theorem.

Theorem 9 *For any positive recurrent aperiodic chain and all states i, j*

$$\lim_{t\to\infty} p_{ij}^t = \pi_j = \frac{1}{m_j} > 0,$$

$$\sum_j \pi_j = 1, \tag{4.5.9}$$

and the limiting vector $\pi = (\pi_1, \pi_2, \ldots)$ is a unique solution to the equation

$$\boldsymbol{\pi} = \boldsymbol{\pi}\mathcal{P}, \tag{4.5.10}$$

satisfying (4.5.9).

We considered a number of examples for finite chains in Section 4.4. Next, we consider one example for an infinite chain.

EXAMPLE 1 (*success runs*). Assume that the process may move from state $i = 0, 1, 2, \ldots$ either to state $i+1$ with probability p, or to state 0 with probability $q = 1 - p$. In other words, we either move up by one step, or fall to the bottom and start from the state 0. Assume, for example, that we are dealing with repeated independent trials, and p is the probability of success at each trial. (For instance, for an investor each day is either lucky or not with probabilities p and q, respectively.) We are interested in the number of consecutive successes, or in other words, we consider the length of the success run. At each step, the length mentioned may either increase by one or drop to zero.

The transition matrix in this case is

$$\mathcal{P} = \begin{Vmatrix} q & p & 0 & 0 & \cdots \\ q & 0 & p & 0 & \cdots \\ q & 0 & 0 & p & \cdots \\ \vdots & \vdots & \vdots & \vdots & \vdots \end{Vmatrix}.$$

Hence, (4.5.10) implies

$$q\left(\sum_{j=0}^{\infty}\pi_j\right) = \pi_0, \;\; p\pi_k = \pi_{k+1} \text{ for all } k = 1, 2, \ldots .$$

Since $\sum_{j=0}^{\infty}\pi_j = 1$, we have $\pi_0 = q$, and eventually $\pi_i = qp^i$ for $i = 0, 1, 2, \ldots$ Thus, the limiting distribution is geometric. In Exercise 56, the reader is invited to solve a problem where the probability of success depends on the current state. □

---

5 EXERCISES

Section 1

1. Consider a counting process for which interarrival times are independent and uniform on $[0, 1]$. Show that in this case (a) increments of the process are dependent; (b) the process is not Markov. Does your answer to the second question answer the first?

2. Let the independent exponential interarrival times in the scheme of Section 1.2.1 be not identically distributed. Give an argument on whether the counting process N_t still have independent increments. Compare it with the case of identically distributed interarrival times. (*Advice*: Assume, for example, that the expected value of the kth interarrival time equals k, and observe that at any time you know how many arrivals have already occurred. Does this information matter for understanding how long we will wait for the next arrival?)

3. Provide an Excel worksheet with realizations of Brownian motion. Play a bit with it, considering different δ's and different numbers of points chosen.

4. Show that $\frac{1}{t}w_t \xrightarrow{P} 0$ as $t \to \infty$. In other words, w_t is growing slower than t, or $w_t = o(t)$ in probability.

5. What distribution does X_t from Example 1.3-1 have? Compute $E\{X_t\}$, and $Var\{X_t\}$.

6. Provide an Excel worksheet with realizations of a discrete-time Markov process X_t such that the distribution of X_{t+1} given X_t is uniform on $[0, X_t]$.

Section 2

7. Verify (2.1.3).

8.* Prove (2.1.7) proceeding from the following outline. For i.i.d. τ's, set $m = E\{\tau_i\} \neq 0$. For $\varepsilon > 0$,

$$P\left(\frac{N_t}{t} - \frac{1}{m} > \varepsilon\right) = P\left(N_t > t\left(\frac{1}{m} + \varepsilon\right)\right) \leq P(N_t \geq n_t)$$

where n_t is the integer part of

$$t\left(\frac{1}{m} + \varepsilon\right).$$

By (2.1.1),

$$P(N_t \geq n_t) = P(T_{n_t} \leq t) = P\left(\frac{T_{n_t}}{n_t} - m \leq \frac{t}{n_t} - m\right).$$

Note that

$$\frac{t}{n_t} - m \to -\frac{\varepsilon m^2}{1 + \varepsilon m}$$

as $t \to \infty$. Hence,

$$P\left(\frac{T_{n_t}}{n_t} - m \leq \frac{t}{n_t} - m\right) \leq P\left(\frac{T_{n_t}}{n_t} - m \leq -\frac{\varepsilon m^2}{2(1 + \varepsilon m)}\right)$$

for large t. Now note that T_n is the sum of i.i.d. r.v.'s, and the last probability converges to zero, as $t \to \infty$, by the LLN:

$$\frac{T_{n_t}}{n_t} \xrightarrow{P} m.$$

Thus,

$$P\left(\frac{N_t}{t} - \frac{1}{m} > \varepsilon\right) \to 0$$

for any $\varepsilon > 0$. The probability

$$P\left(\frac{N_t}{t} - \frac{1}{m} < -\varepsilon\right)$$

is considered similarly.

9. Customers arrive at a service facility according to a Poisson process with an average rate of 5 per hour. Find

   (a) the probabilities that (i) during 6 hours no customers will arrive, (ii) at most twenty five customers will arrive;

   (b) the probabilities that the waiting time between the third and the fourth customers will be (i) greater than 30 min., (ii) equal to 30 min., (iii) greater than or equal to 30 min.;

   (c) the probability that after the first customer has arrived, the waiting time for the fifth customer will be greater than an hour;

   (d) for the same waiting time—its expected value and the standard deviation.

10. Answer all questions from Exercise 9 under the condition that the mean interarrival time is 30 min.

11. For a particular flow of claims, the mean number of claims per day is 10.

    (a) Assume that the third claim has just arrived. What is the expected waiting time for the fifth claim?

    (b) What is the probability that the third claim will come in less than three hours?

(c) Assume that during two hours after the last claim, no claims arrived. Find the probability that during the next half an hour there will be no claims.

12. Let N_t be a Poisson process with rate λ, and let T_i be the time of the ith arrival. Write

    (a) $Corr\{N_2, N_4 - N_2\}$;

    (b) $E\{N_t N_{t+s}\}$, $Corr\{N_t, N_{t+s}\}$;

    (c) $E\{T_{n+m} \mid N_t = n\}$, $Var\{T_{n+m} \mid N_t = n\}$. (*Hint:* you will avoid calculations if you think about the memoryless property.)

13. Give an example of an intensity $\lambda(t)$ for which the probability that no claim will ever arrive is $1/e$.

14. Let $\lambda(t)$ be changing linearly. For an interval $\Delta = [t_1, t_2]$, we define the average intensity as $\widetilde{\lambda} = \frac{1}{2}[\lambda(t_1) + \lambda(t_2)]$. Show that N_Δ is a Poisson r.v. with the parameter $\widetilde{\lambda}|\Delta|$, where $|\Delta|$ is the length of Δ.

15. For a non-homogeneous Poisson flow of claims, the intensity $\lambda(t)$ during the first 8 hours is increasing as $10(t/8)^2$ [ending up with 10 claims/hour at the end of the period]. Find the expected value and variance of the number of claims during the whole period. Given that the fifth claim arrived at 5 h, find the probability that the sixth claim will arrive after 5h 6min.

16. A flow of arrivals N_t is a non-homogeneous Poisson process with the periodical intensity $\lambda(t) = |\sin \pi t|$. The unit of time is a day.

    (a) What is the intensity of arrivals at the end of each day? What about the beginning of each day?

    (b) When is the intensity the largest?

    (c) What is the mean and the variance of the number of arrivals during a year?

    (d) Estimate without any calculations $P(|N_t - E\{N_t\}| > \sqrt{Var\{N_t\}})$ for $t = 1$ year.

17. Let the intensity of a non-homogeneous Poisson process $\lambda(t) = 1$ for $t \in [0,1]$, and $\lambda(t) = 100$ for $t \in [1,2]$. Explain heuristically and rigorously that the interarrival times τ_1 and τ_2 are dependent. (*Advice*: Consider not a general representation but rather the conditional distribution of τ_2 given $\tau_1 = s$ for two particular values of s; for example for $s = 1/2$ and $s = 1$.)

18. Let the flow of accidents corresponding to an auto insurance portfolio be well modeled by a Poisson process with an intensity $\lambda(t)$. Each accident may cause zero, one, or two injuries with probabilities p_0, p_1 and $1 - p_0 - p_1$, respectively. (The probability of a larger number of injuries is considered negligible.)

    Can the number of injuries by time t be modeled by a Poisson process? Can it be represented as a compound Poisson process? Can it be represented as a linear combination of independent Poisson processes?

19. Assume that the occurrence of traffic accidents corresponding to a risk portfolio is well described by a Poisson process with rate $\lambda = 30$ per day. The probability that a separate accident causes serious injuries is $p = 0.1$. The outcomes of different accidents are independent. Estimate the probability that during a month the number of accidents with serious injuries will exceed 100.

20. Ann is receiving telephone calls in the claim department of an insurance company. Calls come at a Poisson rate of 1 each 15 min. Consider a time interval and the probability that there will be no more than one call during this interval. What length should the interval have for this probability to be greater than 0.8?

21. Solve Exercise 20 for the case when the intensity is decreasing as $\lambda(t) = 1/(1+t)$, and the initial time $t = 0$.

22. The intensity of a Poisson process is decreasing as $\lambda(t) = 1/(1+t)$. Compare the intensity and the probability that there will be no arrivals during the period $[0,t]$.

Section 3

23.* (a) Write formula (3.2) for the case where X's are i.i.d. and have a log-normal distribution, and the intensity $\lambda = 1$.

(b) Write formula (3.3) for the case where X's are i.i.d. and have a log-normal distribution, and the intensity $\lambda(t) = 2t$.

24.* For a risk portfolio, let us consider the probability that the total aggregate claim over the time period $[0, 10]$ will exceed the total premium paid. We need to estimate the loading coefficient θ such that the probability mentioned will be less than, say, 0.05. In which cases below would you apply approximation (3.3): (a) $\lambda(t) = t^2$; (b) $\lambda(t) = 1/(1+t)^2$; (c) $\lambda(t) = 1/(1+t)$; (d) $\lambda(t) = 100/(1+t)$?

25. For a flow of claims, the counting process N_t is the same as in Exercise 16, the size of each claim does not depend on previous claims and has the Pareto distribution in the form (**2**.1.1.17) with $\alpha = 4$. Denote by $S_{(t)}$ the aggregate claim during the period $[0,t]$. For $t =$ 1 year

(a) Find $E\{S_{(t)}\}$ and $Var\{S_{(t)}\}$.

(b) Estimate without calculations $P\left(S_{(t)} > E\{S_{(t)}\} + \sqrt{Var\{S_{(t)}\}}\right)$.

(c) Estimate the loading coefficient θ for which $P\left(S_{(t)} > (1+\theta)E\{S_{(t)}\}\right) \leq 0.05$.

26. A flow of claims arriving at an insurance company is represented by a Poisson process in continuous time. The mean time between two adjacent claims is half a day. The random value X of a particular claim is uniformly distributed on $[0, 10]$ (say, the unit of money is \$1000), the relative loading coefficient $\theta = 0.05$. Find the mean and the variance of the aggregate claim during a year. Estimate the probability that the aggregate claim at the end of a year will not exceed the premium paid during the same year.

27. For a particular group of clients, a flow of claims arriving at an insurance company may be represented as a compound Poisson process. The mean number of claims the company receives per day is 10, the amount of a particular claim is equal to either 2, 3, or 4 with probabilities 1/4, 1/2, and 1/4, respectively. Estimate the monthly premium the company should charge in order that during each separate month the probability of making profit is not less than 0.8.

28. The aggregate amount of claims arriving at an insurance company may be represented as a compound Poisson process. The mean number of claims the company receives per day equals three, the amount of a particular claim is uniformly distributed on $[0, 1]$. When charging premiums, the company proceeds from the relative loading coefficient $\theta = 0.1$.

(a) Find the expected value and the variance of the aggregate claim the company receives during 100 days. What premium does the company get during this period?

(b) Find an approximate value of the initial surplus u of the company for which the aggregate claim at the end of the period mentioned will exceed the total premium to be received plus the surplus u, with the probability less than 0.05. (*Advice:* In order not to repeat numerical calculation, carry them out at the end of the problem.)

29. John sells T-shirts at the beach. There are two types of shirts: priced \$10 and \$15. People stop at the stand according to a Poisson process with rate $\lambda = 10$ per hour. However, each customer buys a shirt with probability $p = 0.3$, and if she/he buys, a cheaper shirt is bought with probability 0.75. Nobody buys two shirts.

    (a) Does the number of people who bought shirts have a Poisson distribution? If no, give an example, if yes, find the intensity of the corresponding Poisson process.

    (b) Does the number of people who bought shirts for \$10 have a Poisson distribution?

    (c) Are the numbers of people who bought shirts for \$10 and for \$15 independent r.v.?

    (d) Find the probability that the first shirt will be purchased during the first hour, and before this purchase two customers will stop at the stand but will not buy anything. (*Advice*: This means that there will be at least three arrivals during the first hour but the first two arrivals will be unlucky for John. Look up the joint distribution of $N_1, ..., N_l$ *given* $N = n$ in Section **3**.3.2.1.)

    (e) Using normal approximation, estimate the probability that the total sales of shirts during 8 hours is greater than \$250.

*Section 4**

30. Explain why in Example 4.1-1, the probability to move from the state "rainy today, and rainy yesterday" to the state "rainy today, and normal yesterday" is zero.

31. Would you assume that h_s in Example 4.1-2 is monotone in s?

32. Compute the probabilities to move from each state to each state in two steps for Examples 4.1-1,2. Regarding Example 2, find the probability to survive two years being of age x in two ways: using multiplication of matrices and just proceeding from common sense.

33. Compute the probability of the path $0 \to 0 \to 1$ for Examples 4.1-1,2, and the probability of the path $0 \to 1 \to 2 \to 1 \to 0$ for Example 4.1-4.

34. Provide an Excel worksheet for Example 4.1-7 and play a bit, changing the transition matrix and watching what happens. Consider several realizations, generating different random numbers.

35. Compute $E\{Z_3\}$ and $Var\{Z_3\}$ in the situation of Example 4.2.1-3. Do the same for the case when r.v.'s Y_{ti} have exponential distributions, and $E\{Y_{ti}\} = i$.

36. Compute, using Excel or another software, $E\{S_5\}$ for the situation from Example 4.2.1-4.

37. Does the discount factor depend on the size of the cash flow in the models we consider here? Is it always the case in reality?

38. Mr. M. runs a business. Each year, with a probability of 0.1, Mr. M. cancels this business. If this does not happen, he faces either "bad" or "good" year with respective average incomes 1 or 2. The transition probabilities for these two states are specified by the matrix

$$\mathcal{P} = \begin{Vmatrix} 0.6 & 0.4 \\ 0.3 & 0.7 \end{Vmatrix},$$

where zero state corresponds to a "bad" year. The initial year is "good". Evaluate the expected discounted present value of the total income for the business under consideration in the long run for $v = 0.9$.

39. Using Excel or another software, provide solutions for different values of v in the situation of Example 4.2.2-1. Graph the estimate for $E\{S_n\}$ against v and explain the tendency from an economic and mathematical point of view.

40. Using software, compute the expected present value of total payments during 3 periods for the data from Example 4.2.2-1.

41. Assume that, in the situation of Example 4.2.1-2, we distinguish three health conditions: 'healthy', 'sick', and 'died', with

$$\mathcal{P} = \begin{Vmatrix} 0.85 & 0.14 & 0.01 \\ 0.6 & 0.35 & 0.05 \\ 0 & 0 & 1 \end{Vmatrix}.$$

(The example is designed merely for illustration, but it contains an attempt to reflect a possible situation on the average. Usually, for young people the numbers in the first column are larger than in the table above, and for old people—smaller. To some extent, it may make the homogeneity assumption less restrictive.)
The mean annual health care cost in the two first states—'healthy' and 'sick'—equals 1 and 4, respectively. Assume that at the initial moment, 94% of the clients of the company are healthy. Find the actuarial present value of the total cost in the long run with a discount of 0.9. Find the expected absorption time, i.e., the expected lifetime.

42. For the insurance in Example 4.2.4-1,
    (a) write the matrix C;
    (b) verify the answer obtained in this example;
    (c) under the assumption that a chosen person will die within a year with probability one, write without any calculations what S_n should be, and show that formula (4.2.15) does not contradict your answer.

43. Derive (4.2.11) and (4.2.13) from (4.3.5).

44. In the situation of Example 4.2.3-1, using the first-step-analysis approach, find the expected time of being healthy (the number of moments of being in state 0) for an insured who was healthy at the beginning, and for the insured chosen at random.

45. Show that the transition matrices in (4.4.5) do not satisfy Doeblin's condition, while the matrix in Example 4.4-4 does. Show that in the last case we do not even have to compute a power of $\mathcal{P}$.

46. Consider Example 4.4-1 for k books (or cards). Show that in the long run all $k!$ permutations asymptotically are equally likely.

47. Does Doeblin's condition of Theorem 4 hold for the transition matrix (4.4.13)? Connect it with what was said in Example 4.4-3.

48. Does Doeblin's condition of Theorem 4 hold for the transition matrix

$$\mathcal{P} = \begin{Vmatrix} 0.4 & 0.6 & 0 & 0 \\ 0.7 & 0.3 & 0 & 0 \\ 0 & 0 & 0.5 & 0.5 \\ 0 & 0 & 0.1 & 0.9 \end{Vmatrix}?$$

Argue why in this case we should not hope for ergodicity.

49. Assume that in the situation of Example 4.4-2, in the stationary regime, on a day, the number of claims occurs to be six. Find the probability that the weather on this day corresponds to the icy road condition.

50. Peter runs a small business classifying each day as good, or moderate, or bad. Transitions correspond to a Markov chain with transition matrix (4.1.1). A daily income is a random variable having the Pareto distribution in the form (**2**.1.1.18) with $\theta = 4$ and parameter $\alpha =$ 4, 5, and 6, respectively, depending on the type of the day. Find

    (a) the expected daily income in the long run;

    (b) the probability that, in the stationary regime, the income on a particular day will exceed two units of money.

51. In the situation of Example 4.2.2-1, find the stationary distribution for the transition matrix (4.2.8). Discuss the result in terms of "to what extent the investment climate is good, what proportion of the years is good".

52.** For a short period, we can assume that the probabilities in (4.1.2) do not depend on time. What classes do such a chain have?

53.** (a) Show that, if there are more than one state, and $p_{ii} = 1$ for some i, then the chain is reducible.

    (b) Give an example of a reducible chain for which *all* transition probabilities are less than one.

    (c) Show that in order to specify the classes of a chain, we do not need to know particular values of transition probabilities but only which of them are not equal to zero. Why does it not contradict the statement of Exercise 53a?

54.** Classify the states of

$$\mathcal{P} = \left\| \begin{array}{cccc} 0 & 0 & 0.5 & 0.5 \\ 1 & 0 & 0 & 0 \\ 0 & 0.75 & 0 & 0.25 \\ 0 & 1 & 0 & 0 \end{array} \right\|.$$

55.** Show that, while a chain with $\mathcal{P} = \left\| \begin{array}{cc} 0 & 1 \\ 1 & 0 \end{array} \right\|$ is clearly periodic with a period of 2, a chain with $\mathcal{P} = \left\| \begin{array}{ccc} 0 & 1/2 & 1/2 \\ 1 & 0 & 0 \\ 1/3 & 1/3 & 1/3 \end{array} \right\|$ is aperiodic.

56.** In the problem from Example 4.5.4-1, assume that from state i the process moves to state $i+1$ with probability $p_i = \frac{\lambda}{1+i}$, and to state 0 with probability $q_i = 1 - p_i$, where a number $\lambda \leq 1$, and $i = 0, 1, 2, ...$

    (a) Proceeding from (4.5.10), find the limiting distribution $\boldsymbol{\pi}$.

    (b) Let T_0 be the return time to state 0 (not counting the initial stay at 0). Find the distribution of T_0 and its mean value.

Chapter 5

Random Processes and their Applications. II: Brownian Motion and Martingales

We continue to consider different types of random processes keeping the notation from Chapter 4.

1 BROWNIAN MOTION AND ITS GENERALIZATIONS

In this section, we revisit Brownian motion, or the Wiener process, w_t defined in Section **4**.1.2.2.

1.1 More on properties of the standard Brownian motion

1.1.1 Non-differentiability of trajectories

The definition of w_t in Section **4**.1.2.2 requires the trajectories of the process to be continuous. Let us turn to differentiability.

As before, let w_Δ stand for the increment of w_t and $\Phi(x)$ denote the standard normal d.f.

To determine whether w_t has a derivative at a point t, consider a time interval $\Delta = (t, t+\delta]$ and explore the behavior of the r.v. $\eta_\Delta = w_\Delta/\delta$ as $\delta \to 0$.

By definition, the r.v. w_Δ is normal with zero mean and a standard deviation of $\sqrt{\delta}$. Then for $x \geq 0$, we have $P(|\eta_\Delta| > x) = P(|w_\Delta| > x\delta) = 2(1 - \Phi(x\delta/\sqrt{\delta})) = 2(1 - \Phi(x\sqrt{\delta})) \to 2(1 - \Phi(0)) = 1$ as $\delta \to 0$. Since the last relation is true for an arbitrary large x, this means that when δ is approaching zero, the r.v. $|\eta_\Delta|$ takes on arbitrary large values with a probability close to one. Rigorously, $|\eta_\Delta| \to \infty$ as $\delta \to 0$ in probability (for a definition of this type of convergence see Section **0**.5).

As a matter of fact, an even stronger property is true. Namely, with probability one, trajectories of Brownian motion (that is, w_t as a function of t) are nowhere differentiable; i.e., the derivative does not exist at any point. (See, e.g., [112, p.32] or the outline of a proof in [70, p.268].)

This is an amazing property. Trajectories are continuous but not smooth, and the process fluctuates infinitely frequently in any arbitrary small time interval. However, this is not an obstacle for applications. If we are interested in the increments of the process over intervals that perhaps are small but not infinitesimally small, then we are dealing with r.v.'s w_Δ which are normal in the mathematical and usual sense as well, and hence are tractable.

When in 1872, K. Weierstrass constructed a function that was continuous but non-differentiable

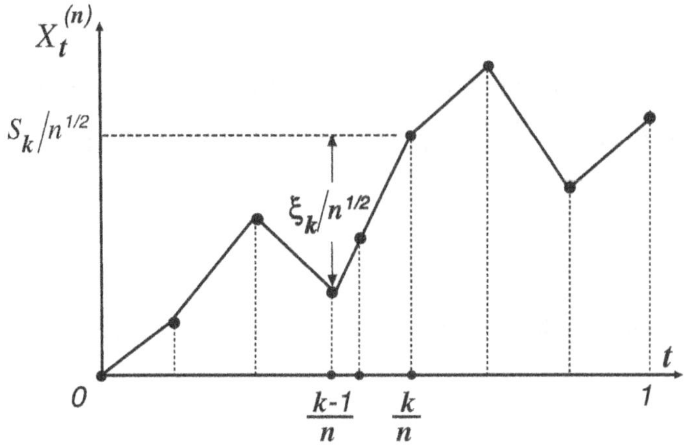

FIGURE 1.

at any point, it was a significant mathematical achievement. Some people considered this function pathological, others—a mathematical masterpiece, but regardless, this function looked exotic. Nowadays, the Wiener process whose trajectories are functions with the same property, serves as a good model for many applied problems.

1.1.2 Brownian motion as an approximation. The Donsker–Prokhorov invariance principle

Let $\xi_1, \xi_2, ...$ be i.i.d. r.v.'s having zero means and unit variances. Let $S_0 = 0$ and $S_k = \xi_1 + ... + \xi_k$.

Let us consider the time interval $[0,1]$ and for each $n = 1, 2, ...,$ construct a piecewise linear (or polygonal) random process $X_t^{(n)}$ on $[0,1]$ as follows.

We divide $[0,1]$ into n intervals $\Delta_k = (\frac{k-1}{n}, \frac{k}{n}]$, $k = 1, ..., n$, of the same length $\frac{1}{n}$. The end points of these intervals are the points $t_k = t_{kn} = k/n$. At the points t_k, we set (see also Fig.1)

$$X_{t_k}^{(n)} = \frac{1}{\sqrt{n}} S_k,$$

and for $t_{k-1} \leq t \leq t_k$, we define $X_t^{(n)}$ as a linear function whose graph connects points $(t_{k-1}, X_{t_{k-1}}^{(n)})$ and $(t_k, X_{t_k}^{(n)})$; see again Fig.1.

The process so constructed is called a *partial sum process*. We may view it as the sequence of partial sums $S_1, ..., S_n$, compressed in a way that it runs in the interval $[0,1]$. Since ξ_k's are independent, the process $X_t^{(n)}$ is that with independent increments. We will see that for large n, the fluctuations of the piecewise linear process $X_t^{(n)}$ are approaching those of Brownian motion.

Proposition 1 *For any t,*

$$X_t^{(n)} \xrightarrow{d} w_t \quad as \quad n \to \infty, \tag{1.1.1}$$

where the convergence $\xrightarrow{d}$ *means the convergence of the distributions of the corresponding r.v.'s. (see also Section* ***0.5****).*

We prove it at the end of this section, but first note that, as a matter of fact, an essentially stronger assertion is true. Namely, not only the marginal distributions (for separate t's) of the process $X_t^{(n)}$ converge to the corresponding marginal distributions of w_t, but the probability distribution of the process $X_t^{(n)}$ as a whole (that is, the joint distribution of the values of the process at different time moments t) converges to the distribution of the standard Wiener process. This fact is referred to as the *Donsker–Prokhorov invariance principle.* A rigorous statement may be found, for example, in [47], [70]. By virtue of this principle, the Wiener process may be viewed as a continuous approximation of the partial sum process.

EXAMPLE 1. Let ξ_i take on values ± 1 with equal probabilities. Then the process of partial sums corresponds to the symmetric random walk considered in Section **4**.4.3.2.2. We have shown there that, starting from a level u, the symmetric random walk will reach a level a before hitting zero level with the probability $p_u = u/a$. The corresponding ruin probability is $q_u = (a-u)/a$. Consider the process $X_t = u + w_t$. This is Brownian motion starting from level u. In view of the invariance principle, we may conjecture that the corresponding ruin probability for X_t will be the same as for the symmetric random walk. In Section 2.4.4, we will show that this is indeed true. □

Proof of Proposition 1. For a fixed $t \in (0,1]$, let $k = k(n)$ be the smallest integer which is not less than tn. Formally, $k = tn$ if tn is an integer, and $k(n) = [tn]+1$ otherwise. (As usual, $[a]$ denotes the integer part of a.) Because $t > 0$, we have $k(n) \to \infty$ as $n \to \infty$.

For k so defined, $t \in \Delta_k$. Indeed, if tn is an integer, then $t = \frac{k}{n} \in \Delta_k$. If tn is not an integer, we have $tn < k < tn+1$, which implies $\frac{k-1}{n} < t < \frac{k}{n}$.

Since the ξ's have zero means and unit variances, $E\{S_{k(n)}\} = 0$ and $Var\{S_{k(n)}\} = k(n)$. Because $X_t^{(n)}$ is a linear on $[t_{k-1}, t_k]$,

$$X_t^{(n)} = X_{t_{k(n)}}^{(n)} - \frac{t_k - t}{t_k - t_{k-1}} \cdot \frac{\xi_k}{\sqrt{n}} = \sqrt{\frac{k(n)}{n}} \cdot \frac{1}{\sqrt{k(n)}} S_{k(n)} - \frac{t_k - t}{t_k - t_{k-1}} \cdot \frac{\xi_k}{\sqrt{n}}$$

(see also Fig. 1). By construction, $\frac{k(n)}{n} \to t$ as $n \to \infty$. By the CLT,

$$\frac{1}{\sqrt{k(n)}} S_{k(n)} \xrightarrow{d} Z,$$

where Z is a standard normal r.v. Also, $\left| \frac{t_k - t}{t_k - t_{k-1}} \cdot \frac{\xi_k}{\sqrt{n}} \right| \leq 1 \cdot \frac{|\xi_k|}{\sqrt{n}} \xrightarrow{d} 0.$

Hence, $X_t^{(n)} \xrightarrow{d} \sqrt{t}Z$. The r.v. $\sqrt{t}Z$ is normal with zero mean and variance t, that is, it has the same distribution as w_t. ■

1.1.3 The distribution of w_t, hitting times, and the maximum value of Brownian motion

First, note that since by definition w_t is normal with zero mean and variance t, we can explicitly write its density and the d.f. In accordance with (**0**.3.2.16), the density of w_t is

$$f_t(x) = \frac{1}{\sqrt{2\pi t}} \exp\left\{ -\frac{x^2}{2t} \right\}, \tag{1.1.2}$$

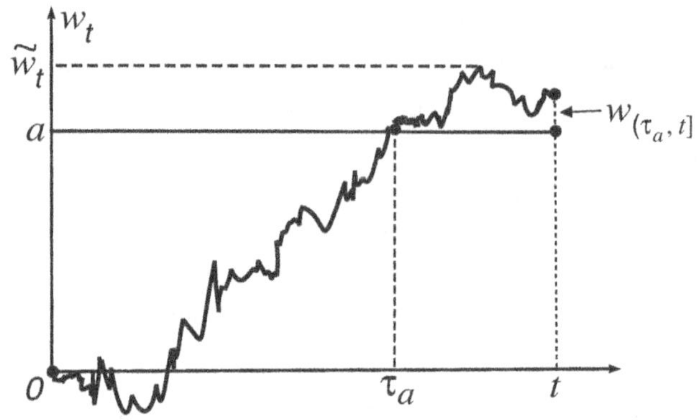

FIGURE 2.

and the d.f.

$$F_t(x) = \Phi\left(\frac{x}{\sqrt{t}}\right) \tag{1.1.3}$$

for all $t > 0$.

The next two definitions are illustrated in Fig.2. We set $\tau_a = \min\{t \geq 0;\, w_t = a\}$, the time at which the process, starting from zero, first reaches (or hits) a level a. (In Chapter 4, the symbol τ stood for an interarrival time, but it should not cause confusion here as we do not consider interarrival times in the current chapter.)

Let $\widetilde{w}_t = \max\limits_{0 \leq s \leq t} w_s$, the maximal value of the process over the interval $[0,t]$. (Since trajectories w_t are continuous with probability one, the maximum exists with probability one.)

We explore two characteristics, τ_a and $\widetilde{w}_t$, together because they are strongly connected. Namely,

$$\tau_a \leq t \quad \text{iff} \quad \widetilde{w}_t \geq a. \tag{1.1.4}$$

Indeed, if the maximum value of the process over the period $[0,t]$ was greater than or equal to a, then the process "had to cross" the level a. Since the process is continuous, it could not overshoot a and was equal to a at the moment of crossing.

By the formula for total probability,

$$P(w_t \geq a) = P(w_t \geq a \,|\, \tau_a > t)P(\tau_a > t) + P(w_t \geq a \,|\, \tau_a \leq t)P(\tau_a \leq t). \tag{1.1.5}$$

The first conditional probability clearly equals zero because if the continuous process w_t reached the a-level at the first time after time t, the process cannot be larger than a at time t.

The second conditional probability

$$P(w_t \geq a \,|\, \tau_a \leq t) = 1/2. \tag{1.1.6}$$

To show this, one may reason as follows. The random moment τ_a is less than or equal to t. The value of the process at the moment τ_a is exactly equal to a. Hence, if $t > \tau_a$, the r.v.

$$w_t = a + w_{(\tau_a,t]},$$

where $w_{(\tau_a,t]}$ is the increment of the process in the remained time interval $(\tau_a,t]$; see Fig.2. Then, w_t will be greater than or equal to a only if $w_{(\tau_a,t]} \geq 0$. In view of symmetry, $w_{(\tau_a,t]}$ is equally likely to be positive or negative, which implies (1.1.6).

► More precisely,

$$P(w_t \geq a \,|\, \tau_a \leq t) = P(a + w_{(\tau_a, t]} \geq a \,|\, \tau_a \leq t) = P(w_{(\tau_a, t]} \geq 0 \,|\, \tau_a \leq t) = P(w_{(\tau_a, t]} \geq 0). \tag{1.1.7}$$

The last step is true because w_t is the process with independent increments: once w_t has hit the level a, the future evolution of w_t does not depend on the hitting time.

The r.v. $w_{(\tau_a, t]}$ is not normal, since the length of $(\tau_a, a]$ is random, but it is a symmetric r.v. Indeed, given $\tau_a =$ some u, the r.v. $w_{(\tau_a, t]} = w_{(u,t]}$. This r.v. is normal and, in particular, symmetric for *any* u. Hence, $w_{(\tau_a, t]}$ is symmetric and, in particular, it is equally likely to be positive or negative. So, $P(w_{(\tau_a, t]} \geq 0) = \frac{1}{2}$. From this and (1.1.7), we get (1.1.6). ◄

Thus, (1.1.5) implies that $P(w_t \geq a) = \frac{1}{2}P(\tau_a \leq t)$. Combining it with (1.1.3), we eventually obtain that

$$P(\tau_a \leq t) = 2(1 - \Phi(a/\sqrt{t})). \tag{1.1.8}$$

Then, in view of (1.1.4),

$$P(\widetilde{w}_t \leq a) = 2\Phi(a/\sqrt{t}) - 1. \tag{1.1.9}$$

In (1.1.8)-(1.1.9), we consider the distribution functions of two r.v.'s: τ_a and $\widetilde{w}_t$. In (1.1.8), the argument of the d.f. is t, and a is a parameter, while in (1.1.9), the roles of these two quantities switch: a is the argument of the d.f., and t is a parameter.

EXAMPLE 1. Assume that you own a stock whose current price per share is $S_0 = 100$. The price changes in time as the process $S_t = S_0 \exp\{\sigma w_t\}$. In this context, the parameter σ is called a volatility. In Example **4**.1.3-1, we already mentioned that such a model may be sufficiently adequate in many situations; a more general model will be considered in Sections 1.3 and 2.4.6.

Let σ equal, say, 0.15. You have decided to sell your shares when the price increases by 10%. What is the probability that this will not happen within the first year?

You are going to sell your stock at the *first time* t when $S_t \geq 1.1 S_0$. This is equivalent to the inequality $\exp\{\sigma w_t\} \geq 1.1$, or $w_t \geq \frac{1}{\sigma}\ln(1.1) \approx 0.63$. (Note that the answer does not depend on S_0.) So, you will not sell your shares if w_t does not reach the level 0.63 during the time period $[0,1]$. The probability of this event is $P(\widetilde{w}_1 < 0.63)$ (why?) and equals $2\Phi(0.63) - 1 \approx 0.47$. Note that $\widetilde{w}_1$ is a continuous r.v., so it does not matter whether to write $P(\widetilde{w}_1 < x)$ or $P(\widetilde{w}_1 \leq x)$. □

Formula (1.1.9) leads to an unexpected, at first glance, conclusion. Consider the probability that during a fixed time interval $[0,T]$ the process will take on only non-positive values. (In this case, the graph of the realization will be under the t-axis.) This probability is $P(\widetilde{w}_T \leq 0)$. In accordance with (1.1.9), it is equal to zero for any arbitrary small $T > 0$. This means that starting from zero, the process cannot move down taking for a while only negative values. On the contrary, during any arbitrary small interval $[0,T]$, the process will cross zero level with probability one. It may be proved that before "leaving zero", the process fluctuates around zero infinitely often, rapidly oscillating around zero, so to speak. Since the state of the process at any time may be viewed as the initial state with respect to the future evolution, the same conclusion concerns the behavior of the process around any point. The evolution of the process is by no means smooth but consists of an infinite number of small but frequent movements up and down.

1.2 The Brownian motion with drift

In Chapter 0, we defined the normal r.v. with mean m and variance σ^2 as a r.v. $m+\sigma X$, where the r.v. X is standard normal.

We define a *Brownian motion with drift parameter* μ and *variance parameter* σ^2 as the process

$$X_t = \mu t + \sigma w_t, \quad t \geq 0, \tag{1.2.1}$$

where w_t is the (standard) Brownian motion.

Since w_t is a process with independent increments, so is X_t, and for any time interval Δ, the increment X_Δ is normal with mean $\mu|\Delta|$ and variance $\sigma^2|\Delta|$, where $|\Delta|$ stands for the length of Δ. In particular, the r.v. X_t is $(\mu t, \sigma^2 t)$-normal, and hence for any $t > 0$, the density of X_t is

$$f_t(x) = \frac{1}{\sqrt{2\pi t}\sigma} \exp\left\{-\frac{(x-\mu t)^2}{2\sigma^2 t}\right\}. \tag{1.2.2}$$

1.2.1 Modeling of the surplus process. What a Brownian motion with drift approximates in this case

For a risk portfolio, let $c_t = ct$ be the premium collected by time t. Here, c is the rate at which the premium is accumulating. Denote by $S_{(t)}$ the total claim paid during the period $[0,t]$. For simplicity, we will not consider the initial surplus. Thus, the surplus at time t is $R_t = ct - S_{(t)}$.

In some situations, a Brownian motion with drift, $\mu t + \sigma w_t$, may be a good model for the process R_t. It is worth emphasizing, however, that in such a model, μ is the expected value of the profit per unit of time, that is, the premium minus the mean total amount of claims per unit of time.

Roughly speaking, we can adopt such a model in situations when the process "looks almost continuous", and during any small period of time, the profit is small and proportional to the length of the period on the average. When $\mu = 0$, we can think about the limiting model in the framework of the invariance principle from Section 1.1.2. A simple example of an approximation with $\mu \neq 0$ is given in Section **4**.4.3.2.3.

The Brownian motion approximation may work also well in the case when $S_{(t)}$ is a compound process, but here we should be cautious in interpretations.

For example, assume that $S_{(t)}$ a compound Poisson process (see Section **4**.3). More precisely, let

$$S_{(t)} = \sum_{i=0}^{N_t} \xi_i,$$

where N_t is a homogeneous Poisson process with rate λ, and ξ_i is the size of the ith claim. (Here we chose the symbol ξ instead of X since in this context X_t stands for the random process.)

Let $m = E\{\xi_i\}$, and $\varkappa^2 = E\{\xi_i^2\}$. In our case, $E\{S_{(t)}\} = m\lambda t$ and $Var\{S_{(t)}\} = \varkappa^2\lambda t$ (see Section **4**.3). In order to approximate $R_t = ct - S_{(t)}$ by $X_t = \mu t + \sigma w_t$, we should set $E\{X_t\} = E\{R_t\}$ and $Var\{X_t\} = Var\{R_t\}$. This amounts to $\mu t = ct - m\lambda t$ and $\sigma^2 t = \varkappa^2\lambda t$, or

$$\mu = c - m\lambda, \quad \sigma^2 = \varkappa^2\lambda. \tag{1.2.3}$$

Because w_t is a continuous process, we can hope for a good approximation only if each particular claim is "small" but the number of claims during a unit period, i.e., λ is "large". Assume that this is the case, and to indicate that the ξ_i's are small, let us represent them by $\xi_i = \delta Y_i$, where the rescaling parameter δ is viewed as small, while Y's are "not small" r.v.'s and do not depend on δ. In such a setup,

$$S_{(t)} = \delta \overline{S}_{(t)}, \text{ where } \overline{S}_{(t)} = \sum_{i=0}^{N_t} Y_i.$$

Set $\overline{m} = E\{Y_i\}$, and $\overline{\varkappa}^2 = E\{Y_i^2\}$. By Theorem **3**.11, for a fixed t and large λ's, the normalized r.v. $\overline{S}^*_{(t)} = (\overline{S}_{(t)} - \overline{m}\lambda t)/\sqrt{\overline{\varkappa}\lambda t}$ is asymptotically normal, which justifies the approximation of R_t by the normally distributed r.v. X_t.

Route 2 ⇒ page 311

3

Let us consider it, however, in more detail. First, note that $m = \delta\overline{m}$ and $\varkappa^2 = \delta^2\overline{\varkappa}$. For a particular finite μ and σ, solving (1.2.3) with respect to λ and c, we have

$$\lambda = \frac{1}{\delta^2}\frac{\sigma^2}{\overline{\varkappa}^2}, \quad c = \mu + \frac{1}{\delta}\frac{\overline{m}\sigma^2}{\overline{\varkappa}^2}.$$

Then

$$E\{S_{(t)}\} = m\lambda t = \frac{1}{\delta}\frac{\overline{m}\sigma^2}{\overline{\varkappa}^2}t, \quad Var\{S_{(t)}\} = \sigma^2 t.$$

Since δ is small, the parameter λ is large, which has been expected. However, we see also that $E\{S_{(t)}\}$ and the premium are large, so the expected profit $E\{R_t\} = \mu t$ is neither large nor small only because the large claim is balanced by the large premium. Note also that in this case, $Var\{S_{(t)}\}$ is independent of δ.

This looks somewhat artificial but we should realize that we are talking about a mathematical approximation.

In any case, it is worth emphasizing that the model based on a compound process is not the only model for a surplus process. In practice, the aggregate claim during even a small time period may come from a large number of independent clients and may be closely approximated by a normal r.v. by virtue of the CLT. This circumstance itself provides hope for a good accuracy of the Brownian motion approximation.

The phenomenon of the large expectations we discussed above is not surprising from a point of view of the general theory of processes with independent increments. As has been already noted in Section **4**.1.2.2, any such process may be represented as a Lévy processes; that is, a certain combination of Brownian motion and Poisson processes. These two types of processes are essentially different. The former represents a continuous component of the process, while the latter describes possible "jumps" (as in counting processes). From the theory mentioned it follows that no combination, even infinite, of Poisson processes may lead to a Brownian motion without elimination of an infinite drift. For the corresponding theory, see, for example, [45], [46], [70].

1.2.2 A reduction to the standard Brownian motion

In this section, we provide a technical formula useful in many applications. All expectations appearing below are assumed to be finite.

First, consider a normal r.v. X with mean m and—to avoid cumbersome formulas—with unit variance. We are interested in $E\{g(X)\}$ for a function $g(\cdot)$. Denote by $E_0\{g(X)\}$ the expectation in the case when X is standard normal and observe that

$$E\{g(X)\} = E_0\left\{g(X)\exp\{mX - m^2/2\}\right\}. \tag{1.2.4}$$

Indeed,

$$\begin{aligned} E\{g(X)\} &= \int_{-\infty}^{\infty} g(x)\frac{1}{\sqrt{2\pi}}e^{-(x-m)^2/2}dx = \int_{-\infty}^{\infty} g(x)e^{xm-m^2/2}\frac{1}{\sqrt{2\pi}}e^{-x^2/2}dx \\ &= \int_{-\infty}^{\infty} g(x)e^{xm-m^2/2}\varphi(x)dx, \end{aligned}$$

where $\varphi(x)$ is the standard normal density. This leads to (1.2.4).

Next, we write a counterpart of this formula for the Brownian motion with drift.

Let $X_t = \mu t + w_t$ (for simplicity, we set $\sigma = 1$). We use the notation X^t for the (random) function X_u, $0 \le u \le t$, the whole trajectory until time t. Let $g(X^t)$ be a function of such a trajectory. Thus, $g(X^t)$ may depend on the *whole* trajectory. For example, $g(X^t)$ may equal $\max_{0\le u\le t} X_u$.

Denote by $E_0\{g(X^t)\}$ the expectation in the case $\mu = 0$, i.e., the case of the standard Brownian motion.

Proposition 2 *For any $t > 0$,*

$$E\{g(X^t)\} = E_0\left\{g(X^t)\exp\{\mu X_t - t\mu^2/2\}\right\}. \tag{1.2.5}$$

The point here is that the exponent above involves only X_t, the value of the process at the last moment. Proposition 2 is the simplest version of *Girsanov's theorem* widely used in Financial Mathematics. See, e.g., [70]; statements for the continuous and discrete time cases, as well as detailed comments, may be found in [130].

Proof. We will consider the case when $g(X^t)$ depends on the values of the process at two points: t and some $s < t$. The case of an arbitrary number of points is considered similarly. In the case where $g(X^t)$ depends on values at all points $s \in [0,t]$, one should apply the limiting argument by using the fact that the trajectory X^t is continuous.

So, let $g(X^t) = g(X_s, X_t)$ for a function $g(\cdot,\cdot)$. The r.v.'s X_s and $X_t - X_s$ are independent, have means μs and $\mu(t-s)$, and variances s and $t-s$, respectively. Then

$$
\begin{aligned}
E\{g(X^t)\} &= E\{g(X_s, X_t)\} = E\{g(X_s, X_s + X_t - X_s)\} \\
&= \int_{-\infty}^{\infty}\int_{-\infty}^{\infty} g(x, x+y)\frac{1}{\sqrt{2\pi}}\exp\{-(x-\mu s)^2/2s\} \\
&\quad \times \frac{1}{\sqrt{2\pi}}\exp\{-(y-\mu(t-s))^2/(2(t-s))\}dxdy \\
&= \int_{-\infty}^{\infty}\int_{-\infty}^{\infty} g(x, x+y)\exp\{\mu(x+y) - \frac{1}{2}\mu^2 t\} \\
&\quad \times \frac{1}{\sqrt{2\pi}}\exp\{-x^2/2s\}\frac{1}{\sqrt{2\pi}}\exp\{-y^2/(2(t-s))\}dxdy \\
&= E_0\left\{g(X_s, X_s + X_t - X_s)\exp\{\mu(X_s + X_t - X_s) - \frac{1}{2}t\mu^2\}\right\} \\
&= E_0\left\{g(X_s, X_t)\exp\{\mu X_t - \frac{1}{2}t\mu^2\}\right\}. \blacksquare
\end{aligned}
$$

The reader familiar with the notion of a Radon-Nicodim derivative realizes that we have computed the derivative of the distribution of the Brownian motion with drift with respect to that of the standard Brownian motion.

1.3 Geometric Brownian motion 2,3

Let us consider a process $Y_t = Y_0 \exp\{X_t\}$, where $Y_0 > 0$ is a certain number, and $X_t = \mu t + \sigma w_t$, a Brownian motion with drift. The process Y_t is called a *geometric Brownian motion.* Since $X_0 = 0$, the number Y_0 is the initial value of the process Y_t. Because $\ln(Y_t) = \ln(Y_0) + X_t$, and X_t is normally distributed, Y_t has a log-normal distribution.

The geometric Brownian motion is widely used for modeling investment processes, for example, the evolution of stock prices. To clarify why the future value of an asset, for instance, a future stock price, may be closely approximated by a log-normal r.v., let us consider the following simple model.

Let S_0 be the initial price of an asset. The price at the next moment of time—say, on the next day—may be written as $S_1 = S_0 R_1$, where R_1 is the growth factor during the elapsed period, or the *return* for this period per unit of money. If $R_1 > 1$, the price has grown; if $R_1 < 1$ the price has dropped. At the next moment, the price $S_2 = S_1 R_2 = S_0 R_1 R_2$, where R_2 is the return in the second period. Continuing in the same fashion, we get that at the end of the nth time period, the value of the asset is the r.v. $S_n = S_0 R_1 \cdot \ldots \cdot R_n$, where R_i is the return corresponding to the ith time period. Then $\ln S_n = \ln S_0 + \ln R_1 + \ldots + \ln R_n$, which is the sum of r.v.'s. Consequently, under mild conditions, we can use the CLT and approximate the distribution of $\ln S_n$ by a normal distribution.

The approach we will use below is based on the infinitesimal argument and shows how to treat the problem in continuous time. Let B_0 be an investment into a risk-free security with a constant interest rate r, and let B_t be the result of investment at time t. In this case (see also Section **0**.8.1), the relative growth over an infinitesimally small interval $[t, t+dt]$ is

$$\frac{dB_t}{B_t} = r\,dt, \tag{1.3.1}$$

which leads to the solution $B_t = B_0 e^{rt}$.

Now, let S_0 be an investment into a risky asset and S_t be the corresponding result at time t. By analogy, assume that

$$\frac{dS_t}{S_t} = m\,dt + \sigma dw_t, \tag{1.3.2}$$

where m, σ are parameters, and dw_t is the infinitesimal increment of the standard Brownian motion over the infinitesimal interval $[t, t+dt]$. The difference between (1.3.1) and (1.3.2) is that we have added the random component σdw_t in (1.3.2).

Since the length of the interval $[t, t+dt]$ is dt, the variance of dw_t is equal to dt. So, for the r.v. dS_t/S_t, we have

$$E\{dS_t/S_t\} = m\,dt, \;\; Var\{dS_t/S_t\} = \sigma^2 dt.$$

It is natural to call m the *expected return* (per unit of time). The quantity σ is called a *volatility* and is considered a measure of riskiness in this framework.

Solving (1.3.2) is not as easy as solving (1.3.1) since we cannot integrate (1.3.2) directly—as we know, w_t is not differentiable. The corresponding theory leads to the following solution:

$$S_t = S_0 \exp\{(m - \sigma^2/2)t + \sigma w_t\}, \tag{1.3.3}$$

that is, to the geometric Brownian motion with $\mu = m - \sigma^2/2$.

Derivations of (1.3.3) at different levels of rigor may be found in almost any textbook on Financial Mathematics (see, e.g., [29], [62], [130], [135], [145]). All derivations are based on the famous differentiation *Ito's formula* obtained first by K. Ito. We omit the proof but will use (1.3.3) later as an example.

---

2 MARTINGALES

In this section, we assume all r.v.'s under consideration to have finite expectations.

2.1 Two formulas of a general nature

Throughout this section, we systematically use the notion of conditional expectation $E\{Y \mid X\}$ introduced in Section **0**.7.1 and clarified there and in subsequent chapters. In what follows below, the symbol X in $E\{Y \mid X\}$ may stand for a random vector as well as for a random variable.

Below, we will repeatedly use the formula for total expectation (see, e.g., (**0**.7.2.1))

$$E\{E\{Y \mid X\}\} = E\{Y\}. \tag{2.1.1}$$

In addition to (2.1.1), we will need two more simple relations. In the first reading, the reader may take these relations at a heuristic level.

Consider two r.v.'s or r.vec.'s: X and $\widetilde{X}$.

> If $\widetilde{X} = g(X)$ where $g(\cdot)$ is a one-to-one function, then $E\{Y\,|\,\widetilde{X}\} = E\{Y\,|\,X\}$.

(2.1.2)

To show this, let us recall how we defined the r.v. $E\{Y\,|\,X\}$. First, we have defined the regression function $m(x) = E\{Y\,|\,X = x\}$, and after that, we set $E\{Y\,|\,X\} = m(X)$. So, when considering $\widetilde{X}$, we define the regression function $\widetilde{m}(x) = E\{Y\,|\,\widetilde{X} = x\}$, and we set $E\{Y\,|\,\widetilde{X}\} = \widetilde{m}(\widetilde{X})$.

To make it explicit, let $X, \widetilde{X}$ be r.v.'s, and $g(x) = x^3$. The general case is considered absolutely similarly, and the reader is invited to do it on her/his own. So, let $\widetilde{X} = X^3$. We have

$$\widetilde{m}(x) = E\{Y\,|\,\widetilde{X} = x\} = E\{Y\,|\,X^3 = x\} = E\{Y\,|\,X = x^{1/3}\} = m(x^{1/3}).$$

Then,

$$E\{Y\,|\,\widetilde{X}\} = \widetilde{m}(\widetilde{X}) = m(\widetilde{X}^{1/3}) = m(X) = E\{Y\,|\,X\}.$$

If $g(\cdot)$ is not a one-to-one function, then different values of X may correspond to the same value of $\widetilde{X}$, and (2.1.2) may be not true. However, we can proceed as follows.

First, let us look again at (2.1.1). In the l.-h.s., we first compute the expected value of Y given additional information (about the value of X), and after that, we compute the expected value of the conditional expectation. Such a procedure leads to the unconditional expectation in the r.-h.s.

Since conditional expectations inherit the main properties of "usual" expectations, we can write a counterpart of (2.1.1) for conditional expectations. In particular, we can replace the expectation $E\{\cdot\}$ in (2.1.1) by the conditional expectation $E\{\cdot\,|\,\widetilde{X}\}$. The only thing we need for such a generalization to be true is that the information on which the interior conditional expectation is based should be either more detailed than, or at least equal to, the information based on values of $\widetilde{X}$.

If $\widetilde{X} = g(X)$, this requirement is fulfilled because given X, we know exactly which value $\widetilde{X}$ has assumed (but perhaps not vice versa). Thus,

> If $\widetilde{X} = g(X)$ where $g(\cdot)$ is a function, then $E\{Y\,|\,\widetilde{X}\} = E\{E\{Y\,|\,X\}\,|\,\widetilde{X}\}$. (2.1.3)

(We first condition Y on X, and after that—on $\widetilde{X}$.)

If $g(\cdot)$ is a one-to-one function, then (2.1.3) is trivial: in view of (2.1.2), the l.-h.s. in (2.1.3) is equal to $E\{Y\,|\,X\}$, and the r.-h.s. equals $E\{E\{Y\,|\,X\}\,|\,X\} = E\{Y\,|\,X\}$.

We proceed to random processes.

2.2 Martingales: General properties and examples

Beginning to build a general framework, we presuppose the existence of an original basic process ξ_t on which all other processes under consideration depend. We may interpret ξ_t as a global characteristic of the "state of nature" at time t. In general, ξ_t may take on values from an arbitrary space; for example, ξ_t may be a vector process.

We will use the notation ξ^t for the whole trajectory ξ_s, $0 \leq s \leq t$ through time t, and sometimes call it the *history of the process* by time t. So, the conditional expectation of a r.v. given ξ^t is the conditional expectation given the entire history through time t. In the case of discrete time, ξ^t is just a sequence $\xi_0, \xi_1, ..., \xi_t$. The exposition below is designed in a way that the notation ξ^t will not cause confusion with the power symbol.

For *all other processes* X_t to be considered, we assume that for each t, the r.v. X_t is completely determined by values of ξ^t. In other words, X_t is a function of ξ^t. We say that X_t is *adapted* to ξ_t.

Let, for example, each ξ_t take on real values and time t is discrete. Let $X_0 = \xi_0$, $X_1 = \xi_0\xi_1, X_2 = \xi_0\xi_1\xi_2$, and so on: $X_t = \xi_0 \cdot ... \cdot \xi_t$. Clearly, the process X_t satisfies the above condition.

Now, note that since given ξ^t the value of X_t is known,

$$E\{X_t \,|\, \xi^t\} = X_t. \tag{2.2.1}$$

The process X_t is called a *martingale* with respect to the basic process ξ_t if for all $t, s \geq 0$,

$$E\{X_{t+s} \,|\, \xi^t\} = X_t. \tag{2.2.2}$$

When considering martingales, it is convenient to use the game or investment interpretation and view X_t as the total profit (perhaps negative) of a player or an investor by time t. In this case, definition (2.2.2) means that if t is the present time, then *on the average*, the future profit X_{t+s} is equal to what the player has already reached at time t.

Since the basic process is fixed, we will often omit the reference to ξ_t, just calling X_t a martingale. However, it is important to emphasize that if (2.2.2) holds, the process X_t is a martingale with respect to itself also, that is, for all $t, s \geq 0$,

$$E\{X_{t+s} \,|\, X^t\} = X_t, \tag{2.2.3}$$

where $X^t = (X_0, ..., X_t)$, the history of the process by time t. We use the same symbolism as for ξ^t and will do the same for other processes below.

To justify (2.2.3), we recall that X^t is a function of ξ^t. Hence, by general rule (2.1.3), $E\{X_{t+1} \,|\, X^t\} = E\{E\{X_{t+1} \,|\, \xi^t\} \,|\, X^t\} = E\{X_t \,|\, X^t\} = X_t$.

Therefore, if we work with a process X_t having property (2.2.3), we can take as a basic process the process X_t itself. *By convention, we will do it each time when in a particular problem the basic process is not specified.* However, in general, it is reasonable to suppose that there is one basic process on which all other processes under consideration depend.

In the discrete time case, together with the process X_t, it is also convenient to consider the process of the increments of X_t. More precisely, set $Z_0 = 0$, and $Z_{t+1} = X_{t+1} - X_t$, the profit in one step (play) after time $t = 0, 1, ...$. Then, in view of (2.2.1) and (2.2.2),

$$E\{Z_{t+1} \,|\, \xi^t\} = E\{X_{t+1} - X_t \,|\, \xi^t\} = E\{X_{t+1} \,|\, \xi^t\} - E\{X_t \,|\, \xi^t\} = X_t - X_t = 0.$$

In the game interpretation, this means that regardless of the past, the profit in one step equals zero on the average. We call such a game *fair*.

R.v.'s Z_t for which

$$E\{Z_{t+1} \mid \xi^t\} = 0 \text{ for all } t = 0, 1, ... \tag{2.2.4}$$

are called *martingale differences* with respect to ξ_t. Since all processes under consideration, including Z_t, depend on ξ_t, we will again omit the reference to ξ_t.

Like martingales, martingales-differences are martingale differences with respect to themselves too, that is,

$$E\{Z_{t+1} \mid Z^t\} = 0 \quad \text{for all } t = 0, 1, ... \tag{2.2.5}$$

The proof is similar to the proof of (2.2.3).

Note that from (2.2.4) it follows, in particular, that

$$E\{Z_t\} = 0 \quad \text{for all } t = 0, 1, ... \tag{2.2.6}$$

(Since, by the general property (2.1.1), $E\{Z_{t+1}\} = E\{E\{Z_{t+1} \mid \xi^t\}\} = E\{0\} = 0$.)

Now, if $Z_t = X_{t+1} - X_t$, then

$$X_t = X_0 + (X_1 - X_0) + (X_2 - X_1) + ... + (X_t - X_{t-1}) = X_0 + Z_1 + ... + Z_t. \tag{2.2.7}$$

The interpretation is clear: the total profit is equal to the initial capital plus the sum of all profits in the previous plays.

Thus, a martingale may be represented as the sum of martingale differences plus the initial value. The converse assertion is also true. Let X_t be equal to the very right-hand side of (2.2.7) where Z_t's are martingale differences. Then

$$E\{X_{t+1} \mid \xi^t\} = E\{X_t + Z_{t+1} \mid \xi^t\} = E\{X_t \mid \xi^t\} + E\{Z_{t+1} \mid \xi^t\} = X_t + 0 = X_t.$$

Thus, we have proved (2.2.2) for $s = 1$. In the discrete time case, this implies (2.2.2) for any natural s. Indeed, again in view of the property of conditional expectations (2.1.3),

$$E\{X_{t+s} \mid X^t\} = E\{E\{X_{t+s} \mid X^{t+s-1}\} \mid X^t\} = E\{X_{t+s-1} \mid X^t\} = ... = E\{X_{t+1} \mid X^t\} = X_t.$$

The reader is also recommended to solve Exercise 19.

Thus, in the discrete time case,

> A random sequence X_t is a martingale if and only if it may be represented by (2.2.7), where $\{Z_t\}$ is a sequence of martingale differences.

We now proceed to examples. The first two are trivial and the third is also very simple, but they shed some light on the nature of martingales.

EXAMPLE 1. Let time t be discrete and the process $S_t = \xi_1 + ... + \xi_t$, where ξ's are independent and $E\{\xi_i\} = 0$ for all i. Then, in view of the independence condition, $E\{\xi_{i+1} \mid \xi^i\} = E\{\xi_{i+1}\} = 0$. So, ξ_i's are martingale differences and, consequently, S_t is a martingale. We see that the notion of martingale in discrete time is a generalization of the sum of independent variables with zero means. In (2.2.7), we do not require the Z's to be independent, but rather to have zero conditional expectations $E\{Z_{i+1} \mid Z^i\}$.

Let in our example $E\{\xi_i\} = m \neq 0$. Then S_t is not a martingale. However,

$$\text{the process } X_t = S_t - mt \text{ is a martingale.} \tag{2.2.8}$$

Indeed, $S_t - mt = \sum_{i=1}^{t}(\xi_i - m)$, and $E\{\xi_i - m\} = 0$. This simple observation will lead to some non-trivial conclusions in Section 2.4.2. □

A generalization of (2.2.8) looks as follows.

EXAMPLE 2. Let ξ_t be a process with independent increments in continuous or discrete time. Let $m_t = E\{\xi_t\}$ and $X_t = \xi_t - m_t$. We will show that X_t is a martingale.

Since $E\{X_t\} = 0$ for any $t > 0$, we have $E\{X_{(t,t+s]}\} = E\{X_{t+s}\} - E\{X_t\} = 0$. Then $E\{X_{t+s} \,|\, \xi^t\} = E\{X_t + X_{(t,t+s]} \,|\, \xi^t\} = X_t + E\{X_{(t,t+s]} \,|\, \xi^t\}$. Since the increment $X_{(t,t+s]}$ does not depend on ξ^t, we have $E\{X_{(t,t+s]} \,|\, \xi^t\} = E\{X_{(t,t+s]}\} = 0$, and hence $E\{X_{t+s} \,|\, \xi^t\} = X_t$.

For instance, if N_t is a homogeneous Poisson process with parameter λ, then $X_t = N_t - \lambda t$ is a martingale. A further generalization of the procedure above will be considered in the end of this subsection.

EXAMPLE 3. Let time be discrete, and let $X_t = C_t \cdot \xi_1 \cdot ... \cdot \xi_t$, where ξ's are i.i.d. positive r.v.'s, and C_t is a constant. What value of C_t will make X_t a martingale?

The answer is simple. We should set $C_t = m^{-t}$, where $m = E\{\xi_i\}$. Indeed, in this case,

$$\begin{aligned}
E\{X_{t+s} \,|\, \xi^t\} &= E\{m^{-(t+s)} \cdot \xi_1 \cdot ... \cdot \xi_{t+s} \,|\, \xi^t\} = m^{-(t+s)} E\{\xi_1 \cdot ... \cdot \xi_t \cdot \xi_{t+1} \cdot ... \cdot \xi_{t+s} \,|\, \xi^t\} \\
&= m^{-s} m^{-t} \xi_1 \cdot ... \cdot \xi_t E\{\xi_{t+1} \cdot ... \cdot \xi_{t+s} \,|\, \xi^t\} = m^{-s} X_t E\{\xi_{t+1} \cdot ... \cdot \xi_{t+s}\} \\
&= m^{-s} X_t m^s = X_t.
\end{aligned}$$

EXAMPLE 4. Let time be discrete, and let $\xi_1, \xi_2, ...$ be i.i.d. r.v.'s with zero means. Consider a bilinear form of ξ's, more precisely, set

$$X_t = \sum_{1 \le i < j \le t} \xi_i \xi_j, \; t = 2, 3, ...$$

Then X_t is a martingale. To prove this, we will show that $Z_{t+1} = X_{t+1} - X_t$ is a martingale difference. We have $Z_{t+1} = \sum_{i=1}^{t} \xi_i \xi_{t+1} = \xi_{t+1} \sum_{i=1}^{t} \xi_i$. Hence,

$$E\{Z_{t+1} \,|\, \xi^t\} = \left(\sum_{i=1}^{t} \xi_i\right) E\{\xi_{t+1} \,|\, \xi^t\} = \left(\sum_{i=1}^{t} \xi_i\right) E\{\xi_{t+1}\} = 0.$$

EXAMPLE 5. Let S_t, $t = 0, 1$, or 2, be a stock price at three consecutive moments of time. All possible outcomes are shown in Fig.3a. The initial price is 10. At the end of the first period, the price may be either 14 or 8. If at time $t = 1$ the price occurs to be 14, its next value may be either 15 or 11, while if S_1 takes on a value of 8, at the end of the second time step the price may be either 10 or 5. Such models of price evolution are called *binomial tree models*. Find a probability measure P under which the process S_t is a martingale. In Financial Mathematics such a measure is also called a *risk neutral measure*. It is widely used in pricing various financial products in financial markets.

Let $p = P(S_2 = 15 \,|\, S_1 = 14)$, and $p' = P(S_2 = 10 \,|\, S_1 = 8)$ (these are probabilities to "move up" starting from 14 and 8, respectively; see also Fig.3a). In order to have $E\{S_2 \,|\, S_1\} =$

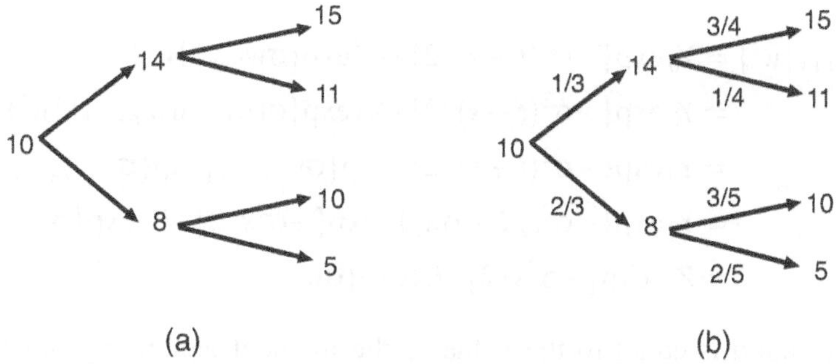

FIGURE 3.

S_1, we need $15 \cdot p + 11 \cdot (1-p) = 14$, and $10 \cdot p' + 5 \cdot (1-p') = 8$, which gives $p = 3/4$, and $p' = 3/5$; see Fig.3b.

For $E\{S_1 \,|\, S_0\} = S_0$, we should have $14 \cdot p'' + 8 \cdot (1-p'') = 10$, where $p'' = P(S_1 = 14)$, which gives $p'' = 1/3$. The probabilities p, p', p'' *completely specify the probabilities of all four possible paths in the tree:* the probability of the path $10 \to 14 \to 15$ is $\frac{1}{3} \cdot \frac{3}{4} = \frac{1}{4}$; for the path $10 \to 14 \to 11$, it is $\frac{1}{3} \cdot \frac{1}{4} = \frac{1}{12}$; the probabilities for the two paths remained are $\frac{2}{3} \cdot \frac{3}{5} = \frac{2}{5}$ and $\frac{2}{3} \cdot \frac{2}{5} = \frac{4}{15}$. See also Fig.3b.

EXAMPLE 6 is noteworthy, deep, and concerns both discrete and continuous time. Let ξ_t be any process, and V be a r.v. which may be viewed as a global characteristic of the process ξ_t. For instance, $V = \max_{0 \le t < \infty} \xi_t$ if such a maximum is finite, or V is the first moment when ξ_t will reach a particular level. Let $X_t = E\{V \,|\, \xi^t\}$, the conditional expected value of V given the information about the evolution of the process ξ_t until time t. In this setup, the process X_t is a martingale.

To prove it, it suffices to apply again property (2.1.3) from which it follows that

$$E\{X_{t+s} \,|\, \xi^t\} = E\{E\{V \,|\, \xi^{t+s}\} \,|\, \xi^t\} = E\{V \,|\, \xi^t\} = X_t. \ \square$$

The idea of the next example is similar to the idea of Example 3. Since this example is used in a number of problems below, we present it as a proposition.

Proposition 3 *Let a geometric Brownian motion $Y_t = Y_0 \exp\{\mu t + \sigma w_t\}$ where a positive Y_0 is certain, and let $\alpha = \mu + \sigma^2/2$. Then the process $Z_t = e^{-\alpha t} Y_t$ is a martingale with respect to w_t.*

Note that since there is a one-to-one correspondence between the processes Y_t, Z_t, and w_t, it does not matter whether to condition on Y^t, or Z^t, or w^t, and hence Z_t is also a martingale with respect Y_t and itself. We will see that it is more convenient to condition on w_t.

Furthermore, $Z_t = Y_0 \exp\{-\mu t - (\sigma^2 t/2) + \mu t + \sigma w_t\} = Y_0 \exp\{-\sigma^2 t/2 + \sigma w_t\}$; that is, the μt terms cancel out. We use this fact in the proof below, but for future references, it is convenient to have Proposition 3 as is stated above.

Proof. Since w_t is a process with independent increments,

$$\begin{aligned}E\{Z_{t+s}\,|\,w^t\} &= Y_0\exp\{-\sigma^2(t+s)/2\}\cdot E\{\exp\{\sigma w_{t+s}\}\,|\,w^t\}\\ &= Y_0\exp\{-\sigma^2(t+s)/2\}\cdot E\{\exp\{\sigma(w_t+w_{(t,t+s]})\}\,|\,w^t\}\\ &= Y_0\exp\{-\sigma^2(t+s)/2\}\cdot\exp\{\sigma w_t\}\cdot E\{\exp\{\sigma w_{(t,t+s]}\}\,|\,w^t\}\\ &= Y_0\exp\{-\sigma^2 t/2+\sigma w_t\}\cdot\exp\{-\sigma^2 s/2\}\cdot E\{\exp\{\sigma w_{(t,t+s]}\}\}\\ &= Z_t\cdot\exp\{-\sigma^2 s/2\}\cdot E\{\exp\{\sigma w_{(t,t+s]}\}\}.\end{aligned} \tag{2.2.9}$$

The last expectation is equal to the value of the moment generating function of the r.v. $w_{(t,t+s]}$ at point σ. The r.v. $w_{(t,t+s]}$ is normal with zero mean and variance s. Hence [see (**0**.4.3.6)], $E\{\exp\{\sigma w_{(t,t+s]}\}\} = \exp\{\sigma^2 s/2\}$, and (2.2.9) implies the assertion of the proposition. ■

EXAMPLE 7. Consider the stock price process S_t from (1.3.3). Assume that the risk free interest in the market is r and is compounded continuously. In this situation, from the standpoint of time 0, the present value of the stock price at time t is $W_t = e^{-rt}S_t$; see Section **0**.8.3. For the process S_t, the role of α from Proposition 3 is played by $(m-\frac{1}{2}\sigma^2)+\frac{1}{2}\sigma^2 = m$. Consequently, for the process W_t to be a martingale, we should set $m = r$.

In summary, for W_t to be a martingale, the expected return m should be equal to the risk free interest r. Such a situation is often referred to as a "risk neutral world" for the following reason.

Let $m = r$. Since W_t is a martingale,

$$E\{W_{t+s}\,|\,W^t\} = W_t. \tag{2.2.10}$$

Suppose that the present time is the initial time $t = 0$, and we are speculating about possible values of the future price. When comparing the possible prices at two different future moments of time, t and $t+s$, we should not compare the prices themselves (S_t and S_{t+s}) but their present values from the standpoint of the initial time (W_t and W_{t+s}). Relation (2.2.10) means that, on the average, the present value keeps the level it has already reached. In other words, whatever value W_t assumes, given this value, the conditional expected value of W_{t+s} will be equal to W_t. □

In conclusion, we establish one simple but important property of martingales.

Proposition 4 *If X_t is a martingale, then for any $t \geq 0$,*

$$E\{X_t\} = E\{X_0\}. \tag{2.2.11}$$

(For a martingale, expected values do not change in time.)

Proof. Setting $t = 0$ in (2.2.3) and taking into account that $X^0 = X_0$, we have $E\{X_s\,|\,X_0\} = X_0$. Computing the expected values of both sides, by virtue of the basic property (2.1.1), we get that $E\{X_s\} = E\{X_0\}$. It remains to replace s by t. ■

EXAMPLE 8. Let us revisit Example 7. By Proposition 4, from (2.2.10) it follows that $E\{W_t\} = E\{W_0\}$. On the other hand, $W_0 = S_0$, the initial price, and W_0 is not random.

Eventually, $E\{W_t\} = S_0$. Thus, in the situation of Example 7, the prices themselves change even on the average, but the mean present values of the future prices do not change. □

Next, we generalize the procedure from Examples 1-2 and consider how to make any process a martingale. Let ξ_t be an *arbitrary* process with finite expectations in discrete time $t = 0, 1, 2, \ldots$, and $V_t = \xi_0 + \ldots + \xi_t$. Note that *any* process V_t may be represented in this way: it suffices to set $\xi_0 = V_0$, and $\xi_t = V_t - V_{t-1}$, the increment over the period $(t-1, t]$. Let us set $A_t = E\{\xi_t \,|\, \xi^{t-1}\}$, and $B_t = A_1 + \ldots + A_t$. Then the process $M_t = V_t - B_t$ is a martingale.

Indeed, we can write $M_t = \xi_0 + \sum_{k=1}^{t} Z_t$, where $Z_t = \xi_t - A_t$. On the other hand, $E\{Z_t \,|\, \xi^{t-1}\}$ $= E\{\xi_t \,|\, \xi^{t-1}\} - E\{E\{\xi_t \,|\, \xi^{t-1}\} \,|\, \xi^{t-1}\} = E\{\xi_t \,|\, \xi^{t-1}\} - E\{\xi_t \,|\, \xi^{t-1}\} = 0$; that is, Z_t's are martingale differences.

In Example 2, to make a process with independent increments a martingale, we subtracted the corresponding expectations. We see that in general, we can do the same if we subtract conditional expectations. The process B_t below is called a *compensator*, and the representation $V_t = M_t + B_t$ is called *Doob's decomposition.*

2.3 Martingale transform

Let X_t, $t = 0, 1, 2, \ldots$, be a martingale with respect to a basic process $\{\xi_t\}$, and let $Z_{t+1} = X_{t+1} - X_t$, $t = 1, 2, \ldots$, the corresponding sequence of martingale differences. Then $X_t = X_0 + Z_1 + \ldots + Z_t$.

Consider another sequence of r.v.'s $\{Y_t\}$ such that for each t, the r.v. Y_{t+1} is a function of ξ^t. Such a sequence is called *predictable.* To clarify the significance of this definition, assume that t is the present time and ξ^t, that is, the evolution of the basic process until time t, is known. From the standpoint of time t, the future value of X_{t+1} is random, still unknown, while the value of Y_{t+1}, being a function of ξ^t, is known or "predictable". (For example, the sequence $A_{t+1} = E\{\xi_{t+1} \,|\, \xi^t\}$ in the end of the previous section depends only on ξ^t, and hence, is predictable.)

The process

$$W_t = X_0 + Y_1 Z_1 + \ldots + Y_t Z_t \tag{2.3.1}$$

is called a *martingale transform.*

EXAMPLE 1. A player participates in a sequence of independent plays (turns). At each turn, success and failure are equally likely. Let a r.v. ξ_t take the value $+1$ if the turn t is successful, and -1 otherwise. So, $\xi_t = \pm 1$ with equal probabilities. Let $X_0 = 0$ and $X_t = \xi_1 + \ldots + \xi_t$, the difference between the number of successful plays and the number of non-successful plays by time t. The process X_t is a martingale and the corresponding martingale differences $Z_t = \xi_t$.

Suppose that after the play t is over, the player makes a bet of Y_{t+1} at the next play $t+1$. The player is free to choose any Y_{t+1} depending on known results, so Y_{t+1} depends on $\xi_1, \ldots, \xi_t$.

For example, assume that the player bets \$1 if she/he lost the last game, and the player increases the stake by \$1 if she/he won. In this case, $Y_{t+1} = 1$ if $\xi_t = -1$, and $Y_{t+1} = Y_t + 1$ if $\xi_t = 1$.

As another example, assume that the player determines at what moment she/he will quit playing. Suppose that the rule of quitting is such that "to play or not to play" after a time t is completely determined by the previous history of plays, that is, by ξ^t. In this case, the moment of quitting is a r.v. τ, and the event $\{\tau \leq t\}$ is completely determined by ξ^t. We may describe such a situation setting $Y_t = 0$ for $t > \tau$. (To quit means to bet zero.) In particular, whether $Y_{t+1} = 0$ (the player quits after time t) depends on ξ^t.

Let us return to the general case of arbitrary predictable stakes Y_t. Clearly, the total profit is given by (2.3.1) with $Z_t = \xi_t$.

Generations of gamblers tried to find a sequence Y_t (a betting strategy) which would transform games from fair (or non-favorable) into favorable—that is, into games for which $E\{W_{t+1} \,|\, \xi^t\} > W_t$. $\square$

Proposition 5 *The transform W_t in (2.3.1) is a martingale.*

Proof. Since Y_{t+1} depends only on ξ^t, we have $E\{Y_{t+1}Z_{t+1} \,|\, \xi^t\} = Y_{t+1}E\{Z_{t+1} \,|\, \xi^t\} = 0$ because Z_{t+1} is a martingale difference. Hence, the sequence Y_tZ_t is that of martingale differences and, consequently, W_t is a martingale. $\blacksquare$

2.4 Optional stopping time and some applications

2.4.1 Definitions and examples

We begin with the discrete time case. Consider a process ξ_t, $t = 0, 1, \ldots$, and an integer valued r.v. τ such that

The event $\{\tau \leq t\}$ is completely determined by the values of the r.v.'s $\xi_0, \ldots, \xi_t$. (2.4.1)

The r.v. τ with property (2.4.1) is called a *Markov time*, or an *optional time*. Sometimes instead of "time" we will say "moment" meaning a time moment.

If in a particular problem, we deal with one process, say, X_t, then we can choose as a basic process the process X_t itself, and in this case, in definition (2.4.1), we may replace ξ's by X's.

The r.v. τ is interpreted as the moment of time when a certain event connected with the evolution of the process ξ_t occurs. Condition (2.4.1) means that, if t is the present time and if we know the history of the process until time t, then we do know whether the event mentioned occurred before or at time t.

A typical example is a hitting time, for example, the time when a process ξ_t first reaches a level a. In this case, $\tau = \min\{t : \xi_t \geq a\}$.

Another good example is the time when, starting from zero, ξ_t returns to zero.

As one more example, suppose that the ξ's are integer valued r.v.'s, and τ is the moment when the process takes on an odd value at the first time. Then τ is an optional time.

A typical example where a r.v. τ is not an optional time is the moment when the process attains its maximum over a certain period. More precisely, for a fixed T, consider the r.v.

$$\tau = \min\{t \leq T : \xi_t = \max_{s \leq T} \xi_s\}. \tag{2.4.2}$$

In order to determine whether a moment t is a point of maximum, we should know the *future* values of the process *after* the time t. Consequently, the event $\{\tau = t\}$ is not generally determined by ξ^t.

Let, for example, ξ_t be a stock price. Then τ in (2.4.2) is the time when the price attains its maximum. If τ were an optional moment, the stockholders would have known when to sell their shares to maximize profit. Clearly, this is not the case.

Generalizing definition (2.4.1), we also call τ an optional time in the following situations.

- The r.v. τ does not depend on ξ's at all; we could say that τ depends on the ξ's in a trivial way.

- Condition (2.4.1) holds for each finite t but with a positive probability no event $\{\tau \le t\}$ occurs. In this case, we say that $\tau = \infty$.

  For example, we know that for the random walk with $p > 1/2$ (see Section **4**.4.3.2), the probability that starting from a level u, the process will hit zero (the ruin probability) is $q_u = [(1-p)/p]^u < 1$. So, if τ is the moment of ruin, then $P(\tau = \infty) = 1 - q_u > 0$.

We will call an optional time τ for which $P(\tau < \infty) = 1$ a *stopping time*.

In Probability Theory, a r.v. X for which $P(X < \infty) < 1$, is called *improper* or *defective*. Thus, an optional but not stopping time is a defective r.v.

Before turning to results, note also that definition (2.4.1) will not change if we replace $\{\tau \le t\}$ by $\{\tau = t\}$, which is typical in the literature when only the discrete time case is considered.

Indeed, the event $\{\tau \le t\} = \{\tau = 1\} \cup ... \cup \{\tau = t\}$. For $k \le t$, if event $\{\tau = k\}$ is determined by the values of ξ^k, then it is determined by the values of ξ^t. Then the union $\cup_{k=1}^t \{\tau = k\}$ is also determined by ξ^t. Vice versa, $\{\tau = t\} = \{\tau \le t\} - \{\tau \le t-1\}$, and both events are determined by ξ^t.

The continuous time case is somewhat more complicated because for each time moment t, we should take into account the behavior of the process in an infinitesimal neighborhood of t. Without going too deeply into the theory, we just state that in this case, it is reasonable to define an *optional time* τ as a r.v. such that

The event $\{\tau < t\}$ is completely determined by the values of ξ^t, i.e., the trajectory ξ_s for all $s \in [0,t]$.	(2.4.3)

The fact that the point t is not included into the event $\{\tau < t\}$ allows us to avoid some technical difficulties. The difference between (2.4.1) and (2.4.3) is not essential because usually for a "good process" in continuous time, the probability that a certain event will happen *exactly* at a fixed time t is zero. For example, for the Poisson process, the probability that a new arrival will occur exactly at a fixed time t_0, is zero (say why).

All other definitions above are the same as in the discrete time case.

We come now to a very useful fact which allows us to solve many problems concerning global characteristics of processes in an explicit way. Cases in point are ruin probabilities, the moments of reaching certain levels, etc. The significance of corresponding theorems consists in the fact that the martingale property is preserved by optional stopping at certain random times.

In this section, we restrict ourselves to a specific property, namely, (2.2.11). We will see that under mild conditions, (2.2.11) continues to be true if we replace the certain time t by a random stopping time τ. Under such a replacement, the assertion of Proposition 4 becomes much stronger and deeper.

Thus, our next step is to establish conditions under which a martingale X_t and a stopping time τ satisfy

$$\text{The martingale stopping property: } E\{X_\tau\} = E\{X_0\}. \tag{2.4.4}$$

In Section 2.5, we consider a more general version of this property.

First, note that (2.4.4) is not always true.

EXAMPLE 1 (*the doubling strategy*) is classical. A player plays a game of chance consisting in a sequence of independent bets with probability $p > 0$ to win at each bet. Having started with a stake of one, the player plays until the first win doubling her/his bet after each loss. After the first win, the player quits. In casinos, such a strategy is also called a "negative progression".

In the scheme of Example 2.3-1, this corresponds to the bet $Y_1 = 1$, and $Y_{t+1} = 2^t I_{t+1}$ for $t = 1,2,\dots$, where $I_{t+1} = 1$ if $\xi_1 = \dots = \xi_t = -1$ (there were only losses and the player keeps playing) and $I_{t+1} = 0$ otherwise (the player won before or at time t and has quitted). Then the profit at time 0 is $W_0 = 0$, and the profit at time $t > 0$ is $W_t = Y_1\xi_1 + \dots + Y_t\xi_t$. Note that so far, $\xi_i = \pm 1$ with probabilities p and $1-p$, respectively, rather than with equal probabilities.

If the first success happens at time $k+1$, the player's profit will be $2^k - (1+2+4+\dots+2^{k-1}) = 1$. Let τ be the moment of the first win, that is, the number of plays to be played. The r.v. τ has a geometric distribution. More precisely, $P(\tau = k) = p(1-p)^{k-1}$, $E\{\tau\} = 1/p$, and $P(\tau < \infty) = 1$ if $p > 0$. Hence, τ is a stopping time, that is, the probability that a success will never happen is zero. This means that the doubling strategy allows a player to get one unit of money with probability one, i.e., without any risk to lose money. It is worth noting, however, that this presupposes that the player should, at least theoretically, have an infinite initial capital. In Exercise 26, we discuss this from a somewhat more realistic point of view, but now it is important for us to consider the case $p = 1/2$.

The sequence I_t is predictable, and the process W_t is a particular case of the process (2.3.1). Since $p = 1/2$, we have $E\{\xi_i\} = 0$ for all i's, and by Proposition 5, the process W_t is a martingale.

Now, the profit at the moment τ is equal to one, that is, $W_\tau = 1$. On the other hand, $W_0 = 0$, which means that (2.4.4) does not hold. □

Modern dictionaries (see, e.g., [143]) give three meanings of the word 'martingale'. The first concerns a strap of a horse's harness keeping the horse from rearing its head; the second—a device for keeping a sail in a certain position; the third—a system of betting in which, after a losing wager, the amount bet is doubled. Probably, the use of the word in the third definition came by analogy with the first. Non-mathematical dictionaries do not give the fourth, and nowadays widespread mathematical meaning of the word. To the author's knowledge, in the mathematical sense and by analogy with the gambling case, the term martingale was first used by J. Ville in "Étude Critique de la Notion de Collectif (1939)"[1]. Later, J. L. Doob's book [35] made the martingale an important chapter of Probability Theory. The first use of martingales in Actuarial Modeling is due to H. Gerber and F. DeVylder; see [41], [42], [34].

Next, we establish some conditions under which (2.4.4) is true.

Theorem 6 *Let X_t be a martingale and τ be a stopping time. Then the martingale stopping property (2.4.4) holds if at least one of the following conditions is true.*

1. *There exists a constant c such that the r.v. $\tau \leq c$ with probability one.*
2. *There exists a constant C such that the r.v. $|X_t| \leq C$ for all $t \leq \tau$ with probability one.*
3. *Time is discrete, $E\{\tau\} < \infty$, and there exists a constant C such that the conditional expectation $E\{|Z_{t+1}| \,|\xi^t\} \leq C$ for all $t = 0, 1, ...$ with probability one, where the martingale differences $Z_{t+1} = X_{t+1} - X_t$.*

We consider a generalization of this theorem and a proof in Section 2.5.

To comment on the conditions above, we interpret the stopping time as the time at which the process stops to run. Condition 1 means that the process will stop before or at a finite time c. Condition 2 means that before or at the stopping time τ, the process itself did not exceed a finite value C. Condition 3 concerns the discrete time case and means that the conditional expected increments of the process are bounded.

EXAMPLE 2. For illustration, we show that none of these three conditions are satisfied in the situation of Example 1. Indeed, in this case τ has the geometric distribution and assumes any positive integer value with a positive probability. So, τ is not bounded. For $t < \tau$, the profit $W_t = -(1 + ... + 2^{t-1}) = -2^t + 1$ and, hence, is not bounded. The same concerns the profit in one play. In our example, the role of the martingale differences Z_t is played by the r.v.'s $Y_t\xi_t$. If $t+1 < \tau$, then $Y_{t+1}\xi_{t+1} = -2^t$, and the r.v. $E\{|Y_{t+1}\xi_{t+1}| \,|\xi^t\}$ assumes the value 2^t. Hence, the r.v. $E\{|Y_{t+1}\xi_{t+1}| \,|\xi^t\}$ is not bounded. □

Now, we turn to examples where Theorem 6 proves to be quite useful.

2.4.2 Wald's identity

Proposition 7 *Let $\xi_1, \xi_2, ...$ be i.i.d. r.v.'s, and let $m = E\{\xi_i\}$ and be finite. Let τ be a stopping time with respect to the process ξ_t and such that $E\{\tau\} < \infty$. Set $S_\tau = \sum_{i=1}^{\tau} \xi_i$. Then*

$$E\{S_\tau\} = mE\{\tau\}. \tag{2.4.5}$$

It is recommended that the reader looks up Proposition **3**.1 where—in a slightly different notation— (2.4.5) was proved in the case when τ does not depend on the ξ's. The result

[1]The author thanks Professor Michael Sharpe for this reference.

(2.4.5) is much stronger since now the number of terms in the sum depends on the values of the terms.

Proof. Let $X_0 = 0, X_t = S_t - mt$, where $S_t = \sum_{i=1}^{t} \xi_i$. In Example 2.2-1, we showed that X_t is a martingale. The corresponding martingale differences are $Z_{t+1} = \xi_{t+1} - m$, and since the ξ's are independent, $E\{|Z_{t+1}|\,|Z^t\} = E\{|\xi_{t+1} - m|\} \leq E\{|\xi_{t+1}|\} + |m| = E\{|\xi_1|\} + |m|$. The last step is proper because the ξ's are identically distributed. So, Condition 3 of Theorem 6 holds, and we can write

$$0 = E\{X_0\} = E\{X_\tau\} = E\{S_\tau - m\tau\} = E\{S_\tau\} - mE\{\tau\},$$

which implies (2.4.5). ■

EXAMPLE 1. Consider the random walk as described in Section **4**.4.3.2. Let $X_0 = 0$ (the process starts from zero level); $X_t = \xi_1 + ... + \xi_t$; $\xi_i = \pm 1$ with probabilities p and $q = 1 - p$, respectively; $p > 1/2$; and $\tau_a = \min\{t : X_t \geq a\}$, the time of reaching a level $a > 0$ at the first time.

Since $m = E\{\xi_t\} = 2p - 1 > 0$, we have $E\{X_t\} = t(2p-1) \to \infty$ as $t \to \infty$. Then, by the law of large numbers, starting from 0, the process will cross the level a with probability one. Consequently, τ_a is a stopping time.

Formally, to find $E\{\tau_a\}$ with use of Wald's identity (2.4.5), we should first prove that $E\{\tau_a\} < \infty$. We skip this preliminary step and turn directly to the value of $E\{\tau_a\}$.[2]

By (2.4.5)—the role of S_τ is played by X_τ, we have $E\{X_{\tau_a}\} = mE\{\tau_a\} = (2p-1)E\{\tau_a\}$. Assume that a is an integer. Since in each step X_t increases or decreases by one, at time τ the process will exactly take the value a (rather than overshoot the level a). In other words, $X_{\tau_a} = a$. Then $a = (2p-1)E\{\tau_a\}$, and

$$E\{\tau_a\} = \frac{a}{2p-1}. \tag{2.4.6}$$

For example, if $p = 2/3$ and $a = 1$, then starting from zero, the process will reach the (next) level one in three steps on the average. See also Exercise 24.

EXAMPLE 2. In the previous example, let $p = 1/2$ and hence $m = 0$. Assume that $E\{\tau_a\} < \infty$. Then by Wald's identity, we would have $a = 0 \cdot E\{\tau_a\} = 0$, which contradicts the assumption $a > 0$. Consequently, $E\{\tau_a\} = \infty$, which is not trivial at all.

However, this should be understood correctly. The above assertion does not mean that the process will be moving to a infinitely long; however, the *mean* time it will take is infinite. In its turn, this may be understood as follows. Suppose we run independent replicas of the same random walk; that is, we repeat the experiment many times. Let τ_{ai} be the value of the stopping time τ_a in the ith replica. Then, by the LLN, with probability one, $\frac{1}{n}(\tau_{a1} + ... + \tau_{an}) \to \infty$ as $n \to \infty$.

[2]The finiteness of $E\{\tau_a\}$ may be proved in many ways, however it requires the knowledge of some additional facts; see, e.g., [122, p.368]. For the reader familiar with Mathematical Analysis, the easiest way could be to use Fatou's lemma (see, e.g., [70], [129]) stating that for any sequence $X_n \xrightarrow{\text{a.s.}} X$, it is true that $E\{X\} \leq \liminf E\{X_n\}$. For an integer n, consider the stopping moment $\min\{\tau_a, n\}$ whose mean is finite. By (2.4.5), $E\{S_{\min\{\tau_a,n\}}\} = mE\{\min\{\tau_a,n\}\}$, and since $E\{S_{\min\{\tau_a,n\}}\} \leq a$, we have $E\{\min\{\tau_a,n\}\} \leq a/m$. Then, by Fatou's lemma, $E\{\tau_a\} \leq \liminf E\{\min\{\tau_a,n\}\} \leq a/m < \infty$.

EXAMPLE 3. Set again $p = 1/2$ and consider the r.v. $\tau_0 = \min\{t : X_t = 0,\ t = 1,2,...\}$, the time needed to revisit 0 starting from 0. In Section **4**.4.5.2, we have proved that $P(\tau_0 < \infty) = 1$; that is, the random walk will revisit state 0 with probability one. (The reader who did not follow Route 3 may take this fact for granted.) Let us prove that though τ_0 is finite, $E\{\tau_0\} = \infty$.

(For the reader who read Section **4**.4.5, note that in other words we are going to prove the null recurrence of states of the symmetric random walk, which was stated in Section **4**.4.5.3.)

Let τ' be the time needed to reach 0 after the first step (regardless of whether the process moved up or down). Then $\tau_0 = 1 + \tau'$, and it suffices to show that $E\{\tau'\} = \infty$.

In accordance with the notation τ_a, let τ_1 be the time of reaching 1 starting from 0. Let τ'_1 be the time needed to reach 0 starting from 1, and τ'_{-1} be the time of reaching 0 starting from -1. Since we consider a symmetric random walk, the r.v.'s τ_1, τ'_1, and τ'_{-1} have the same distribution. Then $P(\tau' = k) = \frac{1}{2}P(\tau'_1 = k) + \frac{1}{2}P(\tau'_{-1} = k) = \frac{1}{2}P(\tau_1 = k) + \frac{1}{2}P(\tau_1 = k) = P(\tau_1 = k)$. Thus, τ' and τ_1 also have the same distribution. As has been proved, $E\{\tau_a\}$ is infinite for any $a \geq 1$. Consequently, $E\{\tau'\}$ is also infinite.

EXAMPLE 4. Consider the general random walk when, as in Example 1, $X_t = \xi_1 + ... + \xi_t$ but the ξ's are arbitrary i.i.d. r.v.'s with a finite mean $m > 0$. Let τ_a, $a > 0$, be defined as above. In this case, we cannot write that $X_\tau = a$ since the process may overshoot the level a. But we can write $X_{\tau_a} \geq a$, which implies that $mE\{\tau_a\} \geq a$ and hence

$$E\{\tau_a\} \geq a/m.$$

There is one case, however, when we can write a precise equality. Let the ξ's be exponentially distributed. Then, due to the memoryless property, given that the process has exceeded the level a, the overshoot has the same exponential distribution with the same parameter as the original ξ's. (See Section **2**.2.1.1.) Consequently, $E\{X_{\tau_a}\} = a + (\textit{the mean value of the overshoot}) = a + m$. This implies

$$E\{\tau_a\} = \frac{a+m}{m} = 1 + \frac{a}{m}.$$

In Exercise 25, we prove that $E\{\tau_a\} = \infty$ if $m = 0$. □

2.4.3 The ruin probability for the simple random walk

Consider again the classical random walk when $X_0 = u$, $X_t = u + \xi_1 + ... + \xi_t$ for $t = 1,2,...$, where the ξ's are independent and take on values ± 1 with probabilities p and $q = 1 - p$, respectively. We assume that u is a natural number and $0 \leq u \leq a$ for some fixed natural number $a > 0$. See also Section **4**.4.3.2 for further detail. We will see how quickly one can compute the ruin probability in this case by making use of Theorem 6.

Let $\tau = \min\{t : X_t = 0 \text{ or } a\}$, the moment of reaching the boundaries of the corridor $[0, a]$. We set $p_u = P(X_\tau = a)$, the probability that the process will first reach the level a, and $q_u = P(X_\tau = 0)$, the probability that the process will first reach the level 0, i.e., the ruin probability.

First, let $p = 1/2$. The process X_t is simultaneously a martingale and a Markov chain. The reader who skipped Section **4**.4.5.2 from Route 3 should either take the fact that in this case $P(\tau < \infty) = 1$ for granted or look at the notion of recurrence and (**4**.4.5.8) in this section. Once we know that τ is a stopping time, the rest is straightforward.

Because $0 \le X_t \le a$ for $t \le \tau$, Condition 2 of Theorem 6 holds. Consequently, by Theorem 6,

$$E\{X_\tau\} = E\{X_0\} = u.$$

On the other hand, $E\{X_\tau\} = ap_u + 0(1 - p_u) = ap_u$. Thus, $p_u = u/a$.

Let $p \neq 1/2$. We saw in Example 2.4.2-1 that for $p > 1/2$ the process will reach level a with probability one. Hence, τ is a stopping time. Since 0 is also a barrier, the same argument implies that τ is a stopping time for $p < 1/2$.

Let us set $r = q/p$ and consider the process

$$Y_t = r^{X_t}.$$

The conditional expectation

$$E\{Y_{t+1} \,|\, \xi^t\} = E\{r^{X_t + \xi_{t+1}} \,|\, \xi^t\} = r^{X_t} E\{r^{\xi_{t+1}} \,|\, X^t\} = Y_t E\{r^{\xi_{t+1}}\} = Y_t(rp + r^{-1}q).$$

By our choice of r, we have $rp + r^{-1}q = q + p = 1$, and consequently

$$E\{Y_{t+1} \,|\, \xi^t\} = Y_t.$$

Thus, Y_t is a martingale with respect to $\{\xi_t\}$, and hence with respect to itself.

For $t \le \tau$, values of Y_t lie between $r^0 = 1$ and r^a. Hence, Condition 2 of Theorem 6 again holds. (Depending on whether $r > 1$ or not, $r^a > 1$ or < 1.) Applying the theorem, we have

$$E\{Y_\tau\} = E\{Y_0\} = r^u.$$

On the other hand, $E\{Y_\tau\} = r^a p_u + 1 \cdot (1 - p_u) = 1 + p_u(r^a - 1)$. So, $p_u = (r^u - 1)/(r^a - 1)$, which coincides with (**4**.4.3.15). □

2.4.4 The ruin probability for the Brownian motion with drift

Let us consider the same exit problem for Brownian motion. Set $X_t = u + \mu t + \sigma w_t$, where w_t is the standard Brownian motion, $\sigma > 0$, the initial point $u \in [0, a]$, and $a > 0$. Here u and a do not have to be integers. The basic process in our problem is w_t, and $X_0 = u$. Let, as in the previous section, $\tau = \min\{t : X_t = 0 \text{ or } a\}$, $p_u = P(X_\tau = a)$, $q_u = P(X_\tau = 0)$.

Proposition 8 *For* $\mu = 0$,

$$p_u = u/a. \tag{2.4.7}$$

For $\mu \neq 0$,

$$p_u = \frac{1 - e^{-\gamma u}}{1 - e^{-\gamma a}}, \text{ where } \gamma = \frac{2\mu}{\sigma^2}. \tag{2.4.8}$$

In both cases, $q_u = 1 - p_u$.

In the case when there is no upper barrier, that is, when $a = \infty$, the probability p_u is the probability to never reach zero, and, accordingly, q_u is the ruin probability.

Corollary 9 *If $a \to \infty$, then $p_u \to 0$, $q_u \to 1$ for $\mu \leq 0$; and $p_u \to 1 - e^{-\gamma u}$ for $\mu > 0$. Thus, for $\mu > 0$, and $a = \infty$, the ruin probability*

$$q_u = 1 - p_u = e^{-\gamma u} = \exp\{-2u\mu/\sigma^2\}. \tag{2.4.9}$$

The reader may check that the formulas above coincide with approximations (**4**.4.3.17) and (**4**.4.3.18) and *interpret* this fact proceeding from the invariance principle of Section 1.1.2.

***Proof of Proposition** 8.* First, let $\mu = 0$. In this case, $X_t = u + \sigma w_t$, and the probability that during a time interval $[0, T]$ the process X_t will never reach the level a equals

$$P(\max_{t \leq T} X_t \leq a) = P(\max_{t \leq T} w_t \leq (a-u)/\sigma) = 2\Phi\left(\frac{a-u}{\sigma\sqrt{T}}\right) - 1$$

in accordance with (1.1.9). This probability converges to $2\Phi(0) - 1 = 0$ as $T \to \infty$, which means that the process will reach the level a with probability one. Thus, τ is a stopping time.

The rest is very similar to what we did in the previous Section 2.4.3. The process X_t is a martingale (see Example 2.2-2, and Exercise 17), and $0 \leq X_t \leq a$ for $t \leq \tau$. Hence, Condition 2 of Theorem 6 holds and $E\{X_\tau\} = E\{X_0\} = u$. On the other hand, $E\{X_\tau\} = ap_u + 0(1 - p_u) = ap_u$, where $p_u = P(X_\tau = a)$. This leads to $p_u = u/a$.

Let $\mu \neq 0$. By the LLN, with probability one, $X_t \to \infty$ if $\mu > 0$, and $X_t \to -\infty$ if $\mu < 0$. This means that with probability one the process will exit the "corridor" $[0, a]$, and hence τ is a stopping time. Another way to justify it is just to show that $P(0 \leq X_t \leq a) \to 0$ as $t \to \infty$ for $\mu \neq 0$.

Next, we will use a nice technique which we will apply in the next chapter systematically. Let $W_t = \exp\{s(\mu t + \sigma w_t)\}$, where s is a number. The exponent is the Brownian motion $\widetilde{\mu} t + \widetilde{\sigma} w_t$, where the drift $\widetilde{\mu} = s\mu$ and the parameter $\widetilde{\sigma} = s\sigma$. We apply to this process Proposition 3. In our case, the characteristic α from this proposition is equal to $\widetilde{\alpha} = \widetilde{\mu} + \widetilde{\sigma}^2/2 = s(\mu + s\sigma^2/2)$. Now we choose s for which $\widetilde{\alpha} = 0$, that is, we set $s = -\gamma$, where $\gamma = 2\mu/\sigma^2$. For the s chosen, by Proposition 3, the process W_t is a martingale.

If we multiply a martingale by a constant, the martingale property continues to hold (why?). Hence, the process $Y_t = e^{-\gamma u} W_t = \exp\{-\gamma(u + \mu t + \sigma w_t)\} = \exp\{-\gamma X_t\}$ is also a martingale. It is a counterpart of Y_t from Section 2.4.3.

Since $0 \leq X_t \leq a$ for $t \leq \tau$, the values of Y_t lie between 1 and $e^{-\gamma a}$, and Condition 2 again holds. Finally, we have

$$E\{Y_\tau\} = E\{Y_0\} = e^{-\gamma u}.$$

On the other hand, $E\{Y_\tau\} = e^{-\gamma a} p_u + 1 \cdot (1 - p_u) = 1 + p_u(e^{-\gamma a} - 1)$. This readily leads to (2.4.8). ■

Route 2 ⇒ page 339

3 2.4.5 The distribution of the ruin time in the case of Brownian motion

Next, we consider the model of the previous section in the particular case $a = \infty$. So, again $X_t = u + \mu t + \sigma w_t$, $\sigma > 0$, the initial point $u \in [0, \infty)$, and $\tau = \min\{t : X_t = 0\}$. However, now we solve a more difficult problem of finding the distribution of the r.v. τ. We have known already that for $\mu \leq 0$, this r.v. is a stopping time, i.e., $P(\tau < \infty) = 1$, while for $\mu > 0$, this is not the case because $\tau = \infty$ with a positive probability. So, τ may be an *improper* or *defective* r.v.

Set $F_u(t) = P(\tau \leq t)$. The index u indicates that this function depends on the fixed parameter u. The function $F_u(t)$ has all properties of d.f.'s with one exception: if τ is improper, then $F_u(\infty) = P(\tau < \infty) < 1$.

Set $f_u(t) = F_u'(t)$, the probability density of τ. This function has all properties of densities with an exception that, if τ is improper, then $\int_0^\infty f_u(t)dt = P(\tau < \infty) < 1$.

Let for $z > 0$,

$$M_u(z) = \int_0^\infty e^{-zx} dF_u(x).$$

The function $M_u(z)$ has all properties of m.g.f.'s except that if τ is improper, then $M_u(0) = P(\tau < \infty) < 1$.

Note that for $z > 0$ we can write that $M_u(z) = E\{\exp\{-z\tau\}\}$ if we set, by convention, $\exp\{-z\tau\} = 0$ for $\tau = \infty$.

Proposition 10 *For $\sigma > 0$, all μ, and $t > 0$,*

$$f_u(t) = \frac{u}{\sqrt{2\pi}\sigma t^{3/2}} \exp\left\{-\frac{1}{2\sigma^2}\left(2\mu u + t\mu^2 + u^2/t\right)\right\}, \tag{2.4.10}$$

$$F_u(t) = \Phi\left(-\frac{u}{\sigma\sqrt{t}} - \frac{\mu\sqrt{t}}{\sigma}\right) + \exp\left(-\frac{2\mu}{\sigma^2}u\right)\Phi\left(-\frac{u}{\sigma\sqrt{t}} + \frac{\mu\sqrt{t}}{\sigma}\right), \tag{2.4.11}$$

$$M_u(z) = \exp\left\{-\frac{u}{\sigma^2}\left(\mu + \sqrt{\mu^2 + 2z\sigma^2}\right)\right\}. \tag{2.4.12}$$

Some comments. The fact that (2.4.10) is the derivative of (2.4.11) may be verified by direct differentiation.

Since $P(\tau \leq 0) = 0$, we must have $F_u(0) = 0$. The reader is invited to check that, indeed, $F_u(t) \to 0$ as $t \to 0$. So, we have chosen a correct antiderivative of $f_u(t)$.

The reader can also verify that $F_u(\infty) = 1$ for $\mu \leq 0$, and $F_u(\infty) = \exp\left(-\frac{2\mu}{\sigma^2}u\right)$ for $\mu > 0$, which coincides with (2.4.9).

For $\mu = 0, \sigma = 1$, (2.4.11) becomes (1.1.8). (The distribution of the time when $u + w_t$ reaches zero is the same as for the time when w_t reaches u.)

It is interesting to check the value $M_u(0)$. For $\mu \leq 0$, it is indeed one, since in this case, $M_u(0) = \exp\left\{-\frac{u}{\sigma^2}\left(\mu + \sqrt{\mu^2}\right)\right\} = \exp\left\{-\frac{u}{\sigma^2}(\mu + |\mu|)\right\} = e^0 = 1$. However, for $\mu > 0$, we have $M_u(0) = \exp\left\{-\frac{u}{\sigma^2}\left(\mu + \sqrt{\mu^2}\right)\right\} = \exp\left\{-\frac{u}{\sigma^2}2\mu\right\}$, which again coincides with (2.4.9).

In Exercise 34, the reader is invited to explain why and how for $\mu = 0$ and $\sigma = 1$ (2.4.11) leads to (1.1.8).

Examples of direct applications of this theorem are given in Exercises 30-31. A less direct application concerning a hitting time is considered in the next section.

Proof of Proposition 10. We restrict ourselves below to a short but somewhat non-rigorous proof in order to demonstrate again the usefulness of the martingale stopping property (2.4.4). In Section 2.5.3, we will prove (2.4.10) in a more direct and rigorous way making use of some particular fact concerning Brownian motion.

As in Section 2.4.4, consider the process $W_t = \exp\{s(\mu t + \sigma w_t)\}$, where now s will play the role of a free parameter. We have seen in Section 2.4.4 that the characteristic α from Proposition 3 in this case is $\widetilde{\alpha} = s\mu + s^2\sigma^2/2$. So, the process $\exp\{-\widetilde{\alpha}t + s(\mu t + \sigma w_t)\}$ is a martingale. Multiplying it by e^{su}, which does not change the martingale property, we come to the process $Y_t = \exp\{-\widetilde{\alpha}t + su + s(\mu t + \sigma w_t)\} = \exp\{-\widetilde{\alpha}t + sX_t\}$ which is also a martingale.

The lack of rigor in the next step comes from the fact that we apply property (2.4.4) to the optional time τ which for $\mu > 0$ is an improper (defective) r.v., that is, not a stopping time. Justification of this step requires some work and we omit it here. (One way is to provide calculations for the barrier $a < \infty$ and let $a \to \infty$ at the very end.)

Since $X_0 = u$, making use of the martingale stopping time property, we have $E\{Y_\tau\} = E\{Y_0\} = e^{su}$. On the other hand, $E\{Y_\tau\} = E\{\exp\{-\widetilde{\alpha}\tau\}\}$, since $X_\tau = 0$ by the definition of τ. Thus, $E\{\exp\{-\widetilde{\alpha}\tau\}\} = e^{su}$, or

$$E\{\exp\{-(s\mu + s^2\sigma^2/2)\tau\}\} = \exp\{su\}. \tag{2.4.13}$$

Now, we set $s\mu + s^2\sigma^2/2 = z \geq 0$, solve for s, and choose the negative root

$$s = -\frac{1}{\sigma^2}\left(\mu + \sqrt{\mu^2 + 2z\sigma^2}\right). \tag{2.4.14}$$

(If we had chosen the positive root, the r.-h.s. of (2.4.13) would have been larger than one, while the l.-h.s. would have been less than one.)

From (2.4.13)-(2.4.14) it follows that

$$E\{\exp\{-z\tau\}\} = \exp\left\{-\frac{u}{\sigma^2}\left(\mu + \sqrt{\mu^2 + 2z\sigma^2}\right)\right\},$$

which is the m.g.f. of τ. So, we have come to (2.4.12).

The verification that (2.4.12) is indeed the m.g.f. (or Laplace transform) of (2.4.10) is lengthy but consists in pure integration, so we can turn to tables of integrals, for example, in [51]. At least, what we did completes the probabilistic part of the problem.

As was mentioned, in Section 2.5.3, we will come to the same solution in a more direct way. ■

2.4.6 The hitting time for the Brownian motion with drift

Next, we connect Proposition 10 with the problem of hitting time. Now let $X_t = \mu t + \sigma w_t$, and let $\tau_a = \min\{t : X_t = a\}$, the time of hitting a level $a > 0$. It makes sense to emphasize that now we consider only one barrier. Assume that $\mu > 0$. Then X_t is a process with a positive drift, and it will reach the level a with probability one. (Show that $P(X_t > a) \to 1$ as $t \to \infty$ for *any* a.)

The event $\{X_t = a\} = \{\mu t + \sigma w_t = a\} = \{a - \mu t - \sigma w_t = 0\}$. By virtue of symmetry, the distribution of the process $-\sigma w_t$ coincides with the distribution of σw_t. Hence, the probabilities of all events we are considering will not change if we replace the process $-\sigma w_t$ by σw_t.

Thus, we may consider the first time t when the process $a - \mu t + \sigma w_t$ will reach zero level. This is the ruin time for the process $\widetilde{X}_t = a - \mu t + \sigma w_t$. In order to use results of the previous section, in these results, we should replace μ by $-\mu$ and set $u = a$. Thus, we have arrived at

Corollary 11 *For the process $X_t = \mu t + \sigma w_t$ above,*

$$P(\tau_a \leq t) = \Phi\left(-\frac{a}{\sigma\sqrt{t}} + \frac{\mu\sqrt{t}}{\sigma}\right) + \exp\left(\frac{2\mu}{\sigma^2}a\right)\Phi\left(-\frac{a}{\sigma\sqrt{t}} - \frac{\mu\sqrt{t}}{\sigma}\right). \qquad (2.4.15)$$

EXAMPLE 1. We generalize Example 1.1.3-1. As in this example, suppose that you own a stock whose current price is $S_0 = 100$, and the price changes in time as the process $S_t = S_0 \exp\{\mu t + \sigma w_t\}$, where the expected return $\mu = 0.1$ and the volatility $\sigma = 0.15$. The difference is that now we consider a non-zero drift. You decided to sell your shares when the price has increased by 10%. Find the probability that this will not happen within the first year.

You will sell your stock at the first time t when $S_t \geq 1.1S_0$. The last inequality is equivalent to $\exp\{\mu t + \sigma w_t\} \geq 1.1$, or $\mu t + \sigma w_t \geq a = \ln(1.1)$. It remains to use (2.4.15) with $t = 1, \mu = 0.1, \sigma = 0.15$. Calculations give $P(\tau_a \leq 1) \approx 0.74$, and $P(\tau_a > 1) \approx 0.26$. □

2.5 Generalizations

2.5.1 The martingale property in the case of random stopping time

Now we turn to a proof of Theorem 6. However, we do not aim to prove it in full. Rather, our goal is to show that all its assertions are plausible and to demonstrate the main ideas of proofs. So, we will restrict ourselves to the discrete time case.

As was mentioned in Section 2.4, Theorem 6 is a corollary of the fact that, under some conditions, the martingale property is preserved by optional stopping. In other words, under some conditions, the time s in the definition (2.2.2) of a martingale may be random.

Let $t = 0, 1, \ldots$, the process X_t be a martingale, and τ be an optional stopping time with respect to a basic process ξ_t. For an event A, we denote by $\mathbf{1}_A$ the indicator of A, that is, the r.v. $\mathbf{1}_A$ taking a value of 1 if A occurs, and 0 otherwise. The symbol $\tau \wedge t$ stands for $\min\{\tau, t\}$.

Theorem 12 *Assume that*

$$|E\{X_\tau\}| < \infty, \qquad (2.5.1)$$

and

$$\lim_{t\to\infty} E\{|X_t|\mathbf{1}_{\{\tau>t\}}\} = 0. \qquad (2.5.2)$$

Then for any t,

$$E\{X_\tau \mid \xi^t\} = X_{\tau\wedge t}. \qquad (2.5.3)$$

A proof is given in Section 2.5.4. Now, we discuss the assertion of the theorem.

A. Let τ be non-random and be equal to a fixed s. Then, for $t < s$, (2.5.3) becomes the equation $E\{X_s \mid \xi^t\} = X_t$, which is the definition of a martingale. For $t \geq s$, (2.5.3) gives $E\{X_s \mid \xi^t\} = X_s$ by the definition of conditional expectation.

B. Setting $t = 0$ in (2.5.3), we have $E\{X_\tau \mid \xi_0\} = X_0$. (If we set $t = 0$ in ξ^t, we get the r.v. ξ_0.) Taking the expected values of both sides, we come to the martingale stopping property (2.4.4).

   Let us turn to conditions (2.5.1)-(2.5.2).

C. To show that (2.5.2) may not hold, consider again the classical example with the doubling strategy. We saw in Examples 2.4.1-1 and 2.4.1-2 that for $\tau > t$ the profit $W_t = -2^t + 1$, and $P(\tau > t) = 2^{-t}$. Hence, $E\{|W_t|\mathbf{1}_{\{\tau>t\}}\} = (2^t - 1)E\{\mathbf{1}_{\{\tau>t\}}\} = (2^t - 1)P(\tau > t) = 1 - 2^{-t} \not\to 0$. One may say that W_t grows too fast.

   Consider now particular cases where (2.5.1)-(2.5.2) are true.

D. Let Condition 1 of Theorem 6 hold, that is, $\tau \leq$ some c. Since we consider the discrete time case, we may assume that c is an integer. For all $t > c$, the set $\{\tau > t\}$ in (2.5.2) is empty, $\mathbf{1}_{\{\tau>t\}} = 0$, and $E\{|X_t|\mathbf{1}_{\{\tau>t\}}\} = 0$. In this case, Condition (2.5.1) holds because $|X_\tau| \leq \max_{i \leq c} |X_i| \leq |X_1| + ... + |X_c|$, and the expectation of the last r.v. is finite since $E\{X_i\}$ is finite for all i.

E. Let Condition 2 of Theorem 6 be true, that is, $|X_t| \leq$ some C for all $t \leq \tau$. Then, first, $|X_\tau| \leq C$, and (2.5.1) is obviously true. Since $\mathbf{1}_{\{\tau>t\}} = 0$ when $t \geq \tau$, we have $E\{|X_t|\mathbf{1}_{\{\tau>t\}}\} \leq CE\{\mathbf{1}_{\{\tau>t\}}\} = CP(\tau > t) \to 0$ as $t \to \infty$, because τ is a stopping time and $P(\tau < \infty) = 1$.

F. Verification of Condition 3 of Theorem 6 requires some work which we relegate to Section 2.5.5.

2.5.2 A reduction to the standard Brownian motion in the case of random time

Here we will generalize Proposition 2 from Section 1.2.2 for the case of stopping times.

Let $X_t = \mu t + w_t$ (as in Section 1.2.2, we set, for simplicity, $\sigma = 1$). We again use the notation X^t and consider a function $g(X^t)$ as a function of the whole trajectory X_u, $0 \leq u \leq t$. If we write $g(X^\tau)$, where τ is an optional time, we mean the whole trajectory until the random time τ. For example, $g(X^\tau)$ may equal $\max_{0 \leq u \leq \tau} X_u$.

Denote by $E_0\{g(X^\tau)\}$ the expectation in the case $\mu = 0$, that is, in the case of the standard Brownian motion.

We do not exclude the case when $P(\tau = \infty) > 0$, but we assume that $E\{g(X^\tau)\}$ and $E_0\{g(X^\tau)\}$ are finite.

Proposition 13 *Let τ be an optional time. Then*

$$E\{g(X^\tau)\} = E_0\left\{g(X^\tau)\exp\{\mu X_\tau - \tau\mu^2/2\}\right\}. \tag{2.5.4}$$

(Compare with (1.2.5).)

Proof. We provide it only for the case when τ is a discrete r.v. taking some values $t_1, t_2 \ldots$. In this case, we can write that

$$E\{g(X^\tau)\} = \sum_k E\{g(X^\tau)\mathbf{1}_{\{\tau=t_k\}}\} = \sum_k E\{g(X^{t_k})\mathbf{1}_{\{\tau=t_k\}}\}. \tag{2.5.5}$$

Since the event $\{\tau = t_k\}$ depends only on the trajectory of the process until time τ_k, the r.v. $g(X^{t_k})\mathbf{1}_{\{\tau=t_k\}}$ is a function of X^{t_k}. Then, by (1.2.5),

$$E\{g(X^{t_k})\mathbf{1}_{\{\tau=t_k\}}\} = E_0\left\{g(X^{t_k})\mathbf{1}_{\{\tau=t_k\}}\exp\{\mu X_{t_k} - \frac{1}{2}t_k\mu^2\}\right\}.$$

Together with (2.5.5), it implies that

$$E\{g(X^\tau)\} = E_0\left\{\left(\sum_k \mathbf{1}_{\{\tau=t_k\}}g(X^{t_k})\exp\{\mu X_{t_k} - \frac{1}{2}t_k\mu^2\}\right)\right\} = E_0\left\{g(X^\tau)\exp\{\mu X_\tau - \frac{1}{2}\tau\mu^2\}\right\}. \blacksquare$$

2.5.3 The distribution of the ruin time in the case of Brownian motion: another approach

In this section, we demonstrate how the last proposition can help to solve the problem of Section 2.4.5.

Setting, for now, the variance parameter $\sigma = 1$, consider the process $X_t = u + \mu t + w_t$ and the r.v. $\tau = \inf\{t : X_t = 0\}$. Note that τ may be defective. Let $S_t = \mu t + w_t$. Then $X_t = u + S_t$ and $\tau = \inf\{t : S_t = -u\}$. Thus, by definition, $X_\tau = 0$ and $S_\tau = -u$.

Being slightly non-rigorous, consider the event $A_t = \{\tau \in [t, t+dt]\}$, viewing dt as an infinitesimal increment of t. To justify such an approach, one should consider an interval $[t, t+\delta]$ and apply the limiting argument, letting $\delta \to 0$.

The process S_t is a Brownian motion with drift, and we can apply Proposition 13 to functions of this process. (We cannot apply this proposition directly to X_t, since it is a shifted Brownian motion.) We have

$$P(A_t) = E\{\mathbf{1}_{A_t}\} = E_0\left\{\mathbf{1}_{A_t}\exp\{\mu S_\tau - \tau\mu^2/2\}\right\} = E_0\left\{\mathbf{1}_{A_t}\exp\{\mu(-u) - \tau\mu^2/2\}\right\}. \tag{2.5.6}$$

Since the r.v. inside $E_0\{\cdot\}$ is equal to zero if $\tau \notin [t, t+dt]$, and since dt is infinitely small, we can set $\tau = t$, which implies that

$$P(A_t) = \exp\{-\mu u - t\mu^2/2\}E_0\{\mathbf{1}_{A_t}\} = \exp\{-\mu u - t\mu^2/2\}P_0\{A_t\},$$

where the probability $P_0(A_t) = P_0(\tau \in [t, t+dt])$ corresponds to the case $\mu = 0$. In the last case, $S_t = w_t$ and $\tau = \inf\{t : S_t = -u\}$, the first moment when the standard Brownian motion reaches level $(-u)$. In view of the symmetry of w_t, we can replace $-u$ by u and use (1.1.8), which gives us the d.f. of τ in this case. Thus,

$$P_0(\tau \in [t, t+dt]) = \frac{d}{dt}\left(2(1 - \Phi(u/\sqrt{t})\right)dt = \frac{u}{t^{3/2}}\frac{1}{\sqrt{2\pi}}\exp\left\{-u^2/2t\right\}dt,$$

where φ is the standard normal density.

Combining it with (2.5.6), we come to the density of τ:

$$f_u(t) = \frac{u}{\sqrt{2\pi}t^{3/2}} \exp\left\{ -\frac{1}{2}\left(2\mu u + t\mu^2 + u^2/t\right) \right\}, \tag{2.5.7}$$

which coincides with (2.4.10) for $\sigma = 1$.

To consider the case of an arbitrary σ, it suffices to notice that in the general case, $X_t = u + \mu t + \sigma w_t = \sigma(\frac{u}{\sigma} + \frac{\mu}{\sigma}t + w_t)$, and $X_t = 0$ iff $\frac{u}{\sigma} + \frac{\mu}{\sigma}t + w_t = 0$. Thus, to obtain the formula for the general case, it suffices to replace u by u/σ and μ by μ/σ in (2.5.7). This leads to (2.4.10).

We saw in Section 2.4.5 how to arrive at (2.4.11). As to the m.g.f. (2.4.13), one can derive it from (2.4.10) by direct (although complicated) integration, but the technique we used in Section 2.4.5 is certainly more simple. Comparing two methods—from this section and Section 2.4.5—we see that the former is more convenient for deriving the distribution, and the latter—for the m.g.f. ■

2.5.4 Proof of Theorem 12

If $\tau < t$, the value of the r.v. X_τ is completely determined by ξ^t, and hence the l.-h.s. of (2.5.3)

$$E\{X_\tau \,|\, \xi^t\} = X_\tau = X_{\tau \wedge t}.$$

That is, (2.5.3) is trivial in this case. Consider the case $\tau \geq t$.

Let an event $A = \{\xi^t \in B\}$, where B is an arbitrary set of values of ξ^t. (More precisely, an arbitrary set the probability of which is well defined; we skip here formalities concerning this issue.) Proceeding from the definition of conditional expectation, we should prove that

$$E\{X_\tau; A \cap \{\tau \geq t\}\} = E\{X_t; A \cap \{\tau \geq t\}\}. \tag{2.5.8}$$

Since we consider only the discrete time case, we can write

$$E\{X_t; A \cap \{\tau \geq t\}\} = E\{X_t; A \cap \{\tau = t\}\} + E\{X_t; A \cap \{\tau > t\}\}.$$

Clearly, $E\{X_t; A \cap \{\tau = t\}\} = E\{X_\tau; A \cap \{\tau = t\}\}$. The event $A \cap \{\tau > t\}$ is completely determined by values of ξ^t. By the martingale property, $X_t = E\{X_{t+1} \,|\, \xi^t\}$. Hence,

$$\begin{aligned} E\{X_t; A \cap \{\tau > t\}\} &= E\{E\{X_{t+1} \,|\, \xi^t\}; A \cap \{\tau > t\}\} = E\{E\{X_{t+1} \,|\, \xi^t\}\mathbf{1}_{A \cap \{\tau > t\}}\} \\ &= E\{E\{X_{t+1}\mathbf{1}_{A \cap \{\tau > t\}} \,|\, \xi^t\}\} = E\{X_{t+1}\mathbf{1}_{A \cap \{\tau > t\}}\} = E\{X_{t+1}\mathbf{1}_{A \cap \{\tau \geq t+1\}}\} \\ &= E\{X_{t+1}; A \cap \{\tau \geq t+1\}\}. \end{aligned}$$

Thus,

$$E\{X_t; A \cap \{\tau \geq t\}\} = E\{X_\tau; A \cap \{\tau = t\}\} + E\{X_{t+1}; A \cap \{\tau \geq t+1\}\}.$$

The second term on the right is the term on the left with t replaced by $t+1$. So, repeating the argument $m - t - 2$ times, we have

$$\begin{aligned}
&E\{X_t;A\cap\{\tau\geq t\}\}=E\{X_\tau;A\cap\{\tau=t\}\}+E\{X_\tau;A\cap\{\tau=t+1\}\}\\
&+E\{X_{t+2};A\cap\{\tau\geq t+2\}\}=E\{X_\tau;A\cap\{t\leq\tau\leq t+1\}\}+E\{X_{t+2};A\cap\{\tau\geq t+2\}\}\\
&=\ldots=E\{X_\tau;A\cap\{t\leq\tau\leq m-1\}\}+E\{X_m;A\cap\{\tau\geq m\}\}\\
&=E\{X_\tau;A\cap\{t\leq\tau\leq m-1\}\}+E\{X_m;A\cap\{\tau=m\}\}+E\{X_m;A\cap\{\tau>m\}\}\\
&=E\{X_\tau;A\cap\{t\leq\tau\leq m\}\}+E\{X_m;A\cap\{\tau>m\}\}.
\end{aligned}$$

From this it follows that

$$E\{X_\tau;A\cap\{t\leq\tau\leq m-1\}\}=E\{X_t;A\cap\{\tau\geq t\}\}-E\{X_m;A\cap\{\tau>m\}\}. \qquad (2.5.9)$$

The term on the left, $E\{X_\tau;A\cap\{t\leq\tau\leq m-1\}\}\to E\{X_\tau;A\cap\{t\leq\tau\}\}$ as $m\to\infty$, because $E\{X_\tau\}$ exists and is finite. For the second term on the right, we have

$$|E\{X_m;A\cap\{\tau>m\}\}|\leq E\{|X_m|;\tau>m\}\to 0 \text{ as } m\to\infty,$$

by (2.5.2). So, (2.5.8) has been proved. ■

2.5.5 Verification of Condition 3 of Theorem 6

We should show that if the condition mentioned holds, then the conditions of Theorem 12 are also satisfied.

For simplicity and without loss of generality, let $X_0=0$. Then $X_t=Z_1+\ldots+Z_t$. Let $Y_i=\mathbf{1}_{\{\tau\geq i\}}$. The main idea of the proof is to write

$$X_\tau=\sum\nolimits_{i=1}^{\tau}Z_i=\sum\nolimits_{i=1}^{\infty}Y_iZ_i$$

since once $i>\tau$, the terms in the sum vanish. Observe that whether $Y_i=1$ or not depends on the values of ξ^{i-1}, that is, the sequence Y_i is predictable. Then, by Proposition 5 from Section 2.3, $W_t=\sum_{i=1}^{t}Y_iZ_i$ is a martingale. Let us set $\widetilde{W}_t=\sum_{i=1}^{t}Y_i|Z_i|$ and note that the r.v.'s $\widetilde{W}_t$ are increasing in t. The proof is based on the bound

$$E\{\widetilde{W}_t\}\leq CE\{\tau\} \text{ for all } t, \qquad (2.5.10)$$

which we will prove at the end. Assume that (2.5.10) is true.

Then, there exists a proper (non-defective) r.v. $\widetilde{W}=\lim_{t\to\infty}\widetilde{W}_t=\sum_{i=1}^{\infty}Y_i|Z_i|$, and

$$E\{\widetilde{W}\}=E\left\{\lim_{t\to\infty}\widetilde{W}_t\right\}=\lim_{t\to\infty}E\left\{\widetilde{W}_t\right\}\leq CE\{\tau\}<\infty.$$

(We can pass the limit sign through the expectation by the *theorem on monotone convergence* which says that we can do that if the sequence of r.v.'s under consideration is monotone. See practically any advanced text-book on integration, or Probability Theory; say, [27, p.86], [120, p.113], [129, p.186].)

Secondly,

$$E\{|X_\tau|\}\leq E\left\{\sum\nolimits_{i=1}^{\infty}Y_i|Z_i|\right\}=E\{\widetilde{W}\}<\infty,$$

which yields (2.5.1).

Now, note that the expression on the l.-h.s. of (2.5.2)

$$E\{|X_t|\mathbf{1}_{\{\tau>t\}}\} \le E\{\{\sum_{i=1}^{t}|Z_i|\}\mathbf{1}_{\{\tau>t\}}\} \le E\{\{\sum_{i=1}^{\tau}|Z_i|\}\mathbf{1}_{\{\tau>t\}}\}$$
$$= E\{\{\sum_{i=1}^{\infty}Y_i|Z_i|\}\mathbf{1}_{\{\tau>t\}}\} = E\{\widetilde{W}\mathbf{1}_{\{\tau>t\}}\}.$$

The last expression vanishes as $t \to \infty$ by the *Lebesgue dominated convergence theorem* which says that, if r.v.'s $\eta_n \to \eta$ with probability one and $|\eta_n| \le \nu$ where ν is a r.v. with a finite mean, then $E\{\eta_n\} \to E\{\eta\}$. See, e.g., [27, p.100], [120, p.114], [129, p.187].

In our case, $|\widetilde{W}\mathbf{1}_{\{\tau>t\}}| \le \widetilde{W}$, and $\widetilde{W}\mathbf{1}_{\{\tau>t\}} \to 0$, as $t \to \infty$, with probability one since $P(\tau < \infty) = 1$. Thus, $E\{|X_t|\mathbf{1}_{\{\tau>t\}}\} \to 0$, and (2.5.2) is true.

It remains to prove (2.5.10). Let $\varkappa_{i+1} = E\{|Z_{i+1}| \; |\xi^i\}$. Then

$$E\{\widetilde{W}_t\} = E\left\{\sum_{i=1}^{t} Y_i|Z_i|\right\} = E\left\{\sum_{i=1}^{t} Y_i(|Z_i| - \varkappa_i)\right\} + E\left\{\sum_{i=1}^{t} Y_i\varkappa_i\right\}. \quad (2.5.11)$$

Recalling that Y_{i+1} is a function of ξ^i, we have $E\{Y_{i+1}(|Z_{i+1}| - \varkappa_{i+1})\,|\xi^i\} = Y_{i+1}E\{(|Z_{i+1}| - \varkappa_{i+1})\,|\xi^i\} = Y_{i+1}(E\{|Z_{i+1}|\,|\xi^i\} - \varkappa_{i+1}) = 0$. Hence, $E\{Y_{i+1}(|Z_{i+1}| - \varkappa_{i+1})\} = 0$, and the first term in (2.5.11) is zero.

To estimate the second term, note that all terms in it are non-negative and, by Condition 3, $\varkappa_i \le C$. Also, $\sum_{i=1}^{\infty} Y_i = \tau$ by the definition of the Y_i's. Consequently,

$$E\left\{\sum_{i=1}^{t} Y_i\varkappa_i\right\} \le CE\left\{\sum_{i=1}^{\infty} Y_i\right\} = CE\{\tau\}. \; \blacksquare$$

---

3 EXERCISES

Section 1

1. Provide an Excel worksheet illustrating the invariance principle of Section 1.1.2. To do this, simulate values of some independent r.v.'s, for example, exponential or uniform, construct sums S_k, and then the process $X_t^{(n)}$ (consider only points $t = k/n$). Provide charts with the graphs of particular realizations of $X_t^{(n)}$.

2. Find $P(w_1 + w_5 < 4)$, where w_t is the standard Brownian motion.

3. Let w_{t1} and w_{t2} be independent Brownian motions. Find σ for which $w_{t1} + 2w_{t2} \stackrel{d}{=} \sigma w_{t1}$. (Here, as usual, the symbol $X \stackrel{d}{=} Y$ means that the distribution of X is equal to the distribution of Y. The r.v.'s X and Y themselves may be not equal to each other.)

4. Continuing Exercise 3, consider the process $x_t = \sigma_1 w_{t1} + \sigma_2 w_{t2}$, where σ_1, σ_2 are numbers. Let $\sigma^2 = \sigma_1^2 + \sigma_2^2$. Show that the process x_t/σ is a standard Brownian motion, and hence x_t may be represented as σw_t, where w_t is a standard Brownian motion. Next, consider the case $\sigma_1 = -1, \sigma_2 = 0$. Explain why the fact that $-w_t$ is Brownian motion is almost obvious. (*Advice*: First, show that x_t is the process with independent increments. Secondly, consider the distribution of x_Δ.)

5. Let w_t be a Brownian motion, and let a be a fixed parameter. Show that the process $x_t = w_{ta}/\sqrt{a}$ has the same distribution as w_t.

6. A physical or economic process evolves as w_t. You decided to measure time in a different unit setting $t = as$, where a is a fixed scale parameter and s is time in the new unit. Argue that you cannot model the basic process by w_s. Find c for which the process cw_s has the same distribution as the basic process.

7. Assume that in the situation of Example 1.1.3-1, you decided to sell the stock not when the price increases by 10%, but when it drops by 10%. Will the answer change? If yes, find it. If no, justify the answer.

8. The prices for two stocks evolve as the processes $S_{t1} = 10\exp\{\sigma_1 w_{t1}\}$ and $S_{t2} = 11\exp\{\sigma_2 w_{t2}\}$, where w_{t1} and w_{t2} are independent Brownian motions, $\sigma_1 = 0.1$, and $\sigma_2 = 0.2$. What are the initial prices for the stocks? Find the probability that during a year the price S_{t1} will meet (catch up) the price S_{t2}. (*Advice*: Use the result of Exercise 4.)

9. Find $Corr\{w_t, w_s\}$ and show that it vanishes as $t \to \infty$ for any fixed s. (*Advice*: Use the fact that w_t is a process with independent increments.)

10. Show that $\widetilde{w}_t$ and $|w_t|$ have the same distribution.

11. Show that $\frac{1}{t}\widetilde{w}_t \xrightarrow{P} 0$ as $t \to \infty$. In other words, $\widetilde{w}_t$ is growing slower than t, or $\widetilde{w}_t = o(t)$. Show that this obviously implies that $w_t = o(t)$. In general, for which functions $g(t)$ can we say that $\widetilde{w}_t = o(g(t))$ in probability?

12. (a) Compute the probability density of the r.v. τ_a from Section 1.1.3. Show that $E\{\tau_a\} = \infty$.
    (b) Proceeding from the invariance principle, connect heuristically this fact with the null recurrence of states in the symmetric random walk (see Section **4**.4.5.3).

13. Prove that $E\{\widetilde{w}_t\} = \sqrt{2t/\pi}$. Show that to obtain this answer, it suffices to consider the case $t = 1$. Is the expected value of the maximum value of Brownian motion on the interval $[0, 2]$ twice as large as that on the interval $[0, 1]$?

14. Two types of customers are calling an insurance company. Each customer is equally likely to belong to each type, and the type of the next customer does not depend on the previous history. We are interested in the difference between the numbers of calls from customers of the first and second type. (Note that the difference so defined may be negative.) Estimate, using the Brownian motion approximation, the probability that during first $n = 100$ calls the difference mentioned will never exceed the level 10. (*Advice*: When applying the invariance principle from Section 1.1.2, set $\xi_k = \pm 1$. Also, the inequality $\max_{k \le n} S_k \le b$ may be rewritten as $\max_{k \le n}(S_k/\sqrt{n}) \le b/\sqrt{n}$, and it is reasonable to set $b = a\sqrt{n}$.)

15. Using results on the log-normal distribution from Section **2**.1.1, find $E\{Y_t\}$ and $Var\{Y_t\}$ for the geometric Brownian motion Y_t defined in Section 1.3.

Section 2

16. Explain why the r.v.'s Z_t in (2.2.7) are martingale differences with respect to ξ_t, with respect to themselves, and with respect to X^t, as well.

17. Show that any process ξ_t with independent increments such that $E\{\xi_\Delta\} = 0$ for any interval Δ is a martingale. Consider, as an example, w_t. Explain why the assertion of this exercise is formally more general than the assertion of Example 2.2-2 (though is very close).

18. Let N_t be a non-homogeneous Poisson process from Section **4**.2.2.1. Is N_t a martingale? Is the process $Z_t = N_t - E\{N_t\}$ a martingale? (*Advice*: Revisit Exercise 17. Look up (**4**.2.2.1).)

19. Show that any process X_t such that $E\{X_{(t,t+\delta]}\,|\,X^t\} = 0$ for any t and $\delta \geq 0$ is a martingale with respect to itself.

20. In the discrete time case, let $X_t = Z_1 + ... + Z_t$ and $Z_k = \xi_1 \cdot ... \cdot \xi_k$, where ξ's are independent and $E\{\xi_i\} = 0$. Are Z's independent? Show that X_t is a martingale.

21. Show that the following sequences are martingales: (a) $X_t = 2^t \exp\{-(\xi_1 + ... + \xi_t)\}$, where ξ's are independent and standard exponential; (b) $X_t = 2^t \xi_1 \cdot ... \cdot \xi_t$, where ξ's are independent and take on values 0 and 1 with equal probabilities. What is $\lim_{t\to\infty} X_t$ in both cases? (*Advice*: In the first case, use the fact that, by the LLN, $(\xi_1 + ... + \xi_t)/t \to 1$ with probability one. In the second case, realize how long there will be no zeros among ξ's.)

22. By analogy with Example 2.2-5, consider the n-step binomial tree for a price process S_t, $t = 0, 1, ..., n$, with the following property: the initial price is 100 and in each of n time steps, the price either grows by 20% or drops by 10%. Show that there are exactly 2^n paths that the stock price may follow. Specify a probability measure P for which S_t is a martingale. What is the probability of a particular path? (*Hint*: In a certain sense, this problem is simpler than that in Example 2.2-5 since the rates of growth or fall are the same for each node of the tree. However, for different paths, the numbers of steps where the price goes up may be different. Once you solve the problem, you will understand better why this model is called binomial.)

23. Let τ be an optional time with respect to a process ξ_t. Is the occurrence of event $\{\tau > t\}$ completely determined by the values of the r.v.'s $\xi_0, ..., \xi_t$?

24. In the situation of Example 2.4.2-1, explain why it takes considerably long to reach the level 1 starting from the neighbor level 0. Show that $E\{\tau_a\} > a$ if $p < 1$, not appealing to formula (2.4.6). What will happen if p gets closer to $1/2$?

25. In the situation of Example 2.4.2-4, prove that $E\{\tau_a\} = \infty$ if $m = 0$.

26. Consider the doubling strategy from Example 2.4.1-1 in the case of a game of roulette. Assume that a player always bets on red. Then at each play, the probability of success is $p = 9/19$ (there are 18 red, 18 black, and 2 green cells). Is W_t a martingale? Will the player win 1 with probability one, if there is no maximum bet and provided that the player has an infinitely large capital? Assume that in a casino, the minimal bet is \$5, the maximal bet is \$500, and the player starts with the minimal bet. What is the probability that the player will fail to run the doubling strategy? How much will she/he lose in this case? Suppose that a professor, when teaching the martingale theory, considered in his class of 100 students the doubling strategy as an example, and after the lecture, all students rushed to the casino (described above) to apply the doubling strategy. Find the probability that at least one student will lose. Proceeding from your answer, argue that the professor had to compute this probability in class. (*Advice*: A good idea is to use the Poisson approximation.)

27. Find the limit of (2.4.8) as $\mu \to 0$ and σ is fixed. Interpret the answer.

28. Suppose the surplus process for a risk portfolio is well approximated by the Brownian motion with drift; in other words, the process is $X_t = u + \mu t + \sigma dw_t$ from Section 2.4.4. The ruin probability for a certain choice of parameters occurs to be $\frac{1}{16}$. How will this probability change if (a) the parameter σ is doubled; (b) the process X_t is multiplied by 2; (c) the initial surplus is doubled?

29.** Consider (2.4.10)-(2.4.12) for $u = 0$. Interpret the answer.

30.** The level of a water reservoir changes accordingly to the process $X_t = 1 + 2t + 3w_t$, where w_t is a standard Brownian motion. Find the probability that the reservoir (a) will never be empty, (b) will not be empty during two units of time.

31.** Let us interpret the process X_t from Exercise 30 as the surplus process of a risk portfolio. Can the coefficient 2 be interpreted as a premium per unit of time?

32.** Assume that in the situation of Example 2.4.6-1, you decided to sell the stock not when the price increases by 10% but when it decreases by 10%. Will the answer change? If yes, find it. If no, justify the answer. Is the stopping time proper (non-defective) in this case? Find the probability that the price will never drop by 10%.

33.** The prices for two stocks change as the processes $S_{t1} = 10\exp\{\mu_1 t + \sigma_1 w_{t1}\}$ and $S_{t2} = 11\exp\{\mu_2 t + \sigma_2 w_{t2}\}$, where w_{t1} and w_{t2} are independent Brownian motions, $\mu_1 = 0.15$, $\mu_2 = 0.11$, $\sigma_1 = 0.15$, and $\sigma_2 = 0.1$. What are the initial prices for the stocks? Find the probability that during a year the price S_{t1} will meet the price S_{t2}.

34.** Explain why and how for $\mu = 0$ and $\sigma = 1$ (2.4.11) leads to (1.1.8).

Chapter 6

Global Characteristics of the Surplus Process. Ruin Models. Models with Paying Dividends

1,2,3

The purpose of this chapter is to present and explore some global characteristics of the *surplus* (or *reserve*) *process*. The characteristics we consider are connected, in some sense or another, either with the profitability of insurance operations or with their viability, i.e., the degree of protection against adversity.

1 A GENERAL FRAMEWORK

Given a particular portfolio, we define the surplus process R_t as the monetary fund "on hand" at time t. Merely for illustrative purposes, we may view R_t as the capital at time t if we keep in mind that it is, of course, not the whole capital of the company but rather the available reserve of high liquidity corresponding to the portfolio under consideration.

In some results below, we consider the surplus process R_t as a whole, without specifying its interior structure. However, in most cases, we assume

$$R_t = u + c_t - S_{(t)}, \tag{1.1}$$

where u is a fixed *initial surplus*, c_t is the aggregate amount of the positive cash flow by time t, and $S_{(t)}$ is the corresponding *loss process*. In this section, we mostly view c_t as the total premium collected by time t, and $S_{(t)}$ as the aggregate claim paid by the same time t. In a more general model, c_t may include results of investment, and $S_{(t)}$—some other expenses different from claim coverage.

As in previous chapters, when adopting model (1.1) in the continuous time case, we usually set

$$S_{(t)} = \sum_{i=1}^{N_t} X_i, \tag{1.2}$$

where N_t is the process that counts consecutive claims, and X_i is the amount of the ith claim.

As to premiums, in a typical model, $c_t = (1+\theta)E\{S_{(t)}\}$, where θ is a relative loading coefficient. In particular, if N_t is a Poisson process with rate λ, and $m = E\{X_i\}$, then as we saw in Chapter 4, $E\{S_{(t)}\} = m\lambda t$, and hence $c_t = (1+\theta)m\lambda t$.

In this case, a typical realization of the surplus process looks as in Fig.1. The process grows linearly, and at the random moments of claim arrivals, the process drops by the amounts of claims.

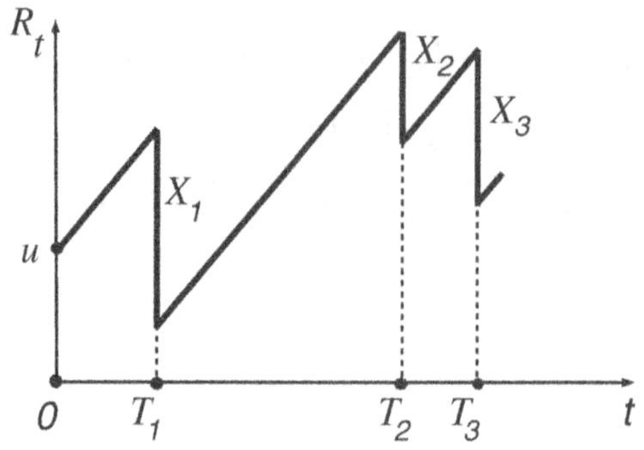

FIGURE 1. A typical realization of the surplus process; T_i is the moment of the ith claim arrival.

Subject to certain regulations and conditions, the company can choose a premium and an initial surplus that the company considers reasonable. In doing so, it proceeds from its goals, which in turn are determined by quality criteria that the company establishes for itself.

If R_t is viewed as a profit, one of possible criteria is the sum

$$\sum_{k=1}^{T} E\{g(R_k - R_{k-1})\},$$

where T is the time horizon from which the company proceeds, and $g(\cdot)$ is a utility function. (Since u stands for the initial surplus, we use a symbol different from that from Chapter 1.) Time may be discrete or continuous, and $E\{g(R_k - R_{k-1})\}$ is the expected utility of the profit during the kth period. Note that the "profit" $R_k - R_{k-1}$ may be negative.

Another criterion is the expected utility of the profit at the "final" time T, that is, $E\{g(R_T)\}$. Instead of the expected utility criterion, one may consider more flexible criteria from Section **1**.4.

A different approach is connected with paying dividends. We use here the term "dividend" for brevity and understand it in a broad sense: it may concern real dividends paid to stockholders or an amount taken from the reserve for other purposes, say, for investment.

Consider the discrete time case, and denote by d_t the dividend paid at time t. (Certainly, we do not exclude the case $d_t = 0$.) Then, instead of (1.1), for the surplus at time t we write

$$R_t = u + c_t - S_{(t)} - D_t, \tag{1.3}$$

where $D_t = d_1 + ... + d_t$, the aggregate amount of dividends paid by time t.

The choice of a dividend to be paid at time t may (and should) depend on the current situation. So, in general, d_t is a random variable depending on the realization of the process until the time t and the strategy of paying dividends. For example, if the time horizon is large and the initial surplus is low, it may prove to be reasonable to pay fewer dividends in the beginning in order to avoid ruin, to let the cash process grow and to be able to pay more

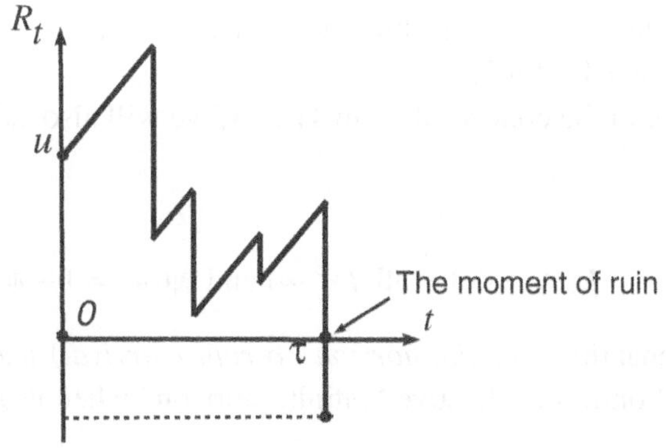

FIGURE 2. A realization of the surplus process in the case of ruin; τ is the time of ruin. For the particular realization above, the ruin has occurred at the moment of the fourth claim's arrival.

in the future. A natural criterion here is the expected discounted total payment, namely,

$$E\left\{\sum_{t=1}^{T} v^t d_t\right\}, \tag{1.4}$$

where v is a discount factor (see Section **0**.8.3). The problem consists in finding a strategy maximizing (1.4). We consider this problem in Section 3.

Criteria of another type appeal to the viability of insurance operations, which amounts to keeping the surplus at a proper level. One important example is

$$P(R_t \geq k_t \text{ for all } t \leq T), \tag{1.5}$$

where T is a time horizon, and k_t is a given level for the surplus at time t. The formula above concerns both discrete and continuous time cases.

The goal of the company in this case is either to maximize (under some natural constrains) probability (1.5), or to keep it higher than a given security level.

The simplest and most frequently considered case is that of $k_t = 0$. In this case, a traditional notation for the probability (1.5) is $\phi_T(u)$, so

$$\phi_T(u) = P(R_t \geq 0 \text{ for all } t \leq T), \tag{1.6}$$

the probability that the portfolio will be solvent during the period $[0, T]$. The initial surplus u is presented explicitly in the notation to emphasize that the probability under consideration depends on u.

As in previous chapters, we call the quantity

$$\psi_T(u) = 1 - \phi_T(u) \tag{1.7}$$

the *ruin probability* regarding the finite time horizon T. This is the probability that the surplus process R_t will assume a negative value during the period $[0, T]$; see also Fig.2.

Sometimes, (1.6) is called a *survival probability*. The term "survival" is traditional but it is important to emphasize that the same term is used in Demography and in the life-insurance modeling in another sense. Namely, in these areas, a survival probability is the probability that an individual (or a machine) will attain a certain age. We will consider the corresponding theory in Chapter 7.

To avoid confusion, in the context of Ruin Theory, we will also call survival probability a *no-ruin probability*.

For $T = \infty$, we set

$$\phi(u) = P(R_t \geq 0 \text{ for all } t < \infty) \text{ and } \psi(u) = 1 - \phi(u),$$

and call these two quantities *infinite-horizon no-ruin (survival)* and *ruin probabilities*, respectively. We will omit the adjective "infinite-horizon" when it does not cause misunderstanding.

In the actuarial literature, no-ruin and ruin probabilities are often denoted by $\tilde{\phi}(u)$ and $\tilde{\psi}(u)$ in the discrete time case, and by $\phi(u)$ and $\psi(u)$ if time is continuous. When it cannot cause misunderstanding, we will use the symbols ϕ and ψ in both cases since quite often we treat both cases simultaneously.

The probability $\psi(u)$ is one of the main objects of study in the next section. The models we consider are, to some degree, idealized and do not reflect all main features of real surplus processes. For example, we do not touch on such important issue as investment income. So, the corresponding results cannot be viewed as direct instructions for decision making. These results provide rather some useful information about the behavior of the insurance process, and may (and should) be taken into account together with other factors of the insurance business. In particular, the ruin probability should be viewed as one of the possible characteristics of the riskiness of the insurance process.

Regarding ruin models, one may also encounter the following reasoning. Ruin models do not take into account that the company invests its collected premiums and, as a result, the capital grows. On the other hand, these models do not take inflation into account. Since these two factors acting in the opposite directions may compensate each other, ruin models may occur to be more adequate to reality than they seem at first glance.

2 RUIN MODELS

We already considered ruin probabilities in the simple random walk scheme (Sections **4**.4.3.2 and **5**.2.4.3) and for the Brownian motion with drift (Section **5**.2.4.4). In this section, we consider models that are closer to real situations and to some degree more sophisticated.

Computing ruin probabilities, especially for a finite time horizon, is a difficult problem. There are skillful direct computational methods, though nowadays their significance is decreasing. Simulation of insurance processes, especially if it is carried out with the use of powerful computers, may lead to better accuracy than direct calculations do. But in this case, qualitative analysis that helps to see a general picture becomes even more important.

We start in the next section with a general theory and estimation of ruin probabilities. First, we consider the most known (and simplest) result—Lundberg's inequality, and after that we turn to a more general theorem and various applications. In Sections 2.1-2.5, we assume R_t to be a process with independent increments. In Section 2.6 belonging to Route 2, we weaken this condition assuming R_t to be a martingale. In Subsection 2.8, we will touch briefly on some computational aspects and different approaches.

Unless stated otherwise, this section concerns ruin problems with an infinite horizon.

2.1 Adjustment coefficients and ruin probabilities

2.1.1 Lundberg's inequality

As above, let R_t be a surplus process, and $u = R_0$, the initial surplus. Assume, as usual, that R_t is a process with independent increments.

It is also convenient to define the *claim surplus process* $W_t = u - R_t$. In particular, if (1.1) is true, then $W_t = S_{(t)} - c_t$, the total claim minus the total premium, and W_t does not depend on u.

For a time interval $\Delta = (t, t+s]$, we set $W_\Delta = W_{t+s} - W_t$, the increment of the process W_t over Δ. Denote by $M_\Delta(z)$ the m.g.f. of W_Δ.

To avoid superfluously complicated formulations, we assume from the very beginning that for any Δ,

$$P(W_\Delta = 0) \neq 1, \quad \text{and hence} \quad M_\Delta(z) \not\equiv 1. \tag{2.1.1}$$

(If $W_\Delta = 0$ with probability one, then with the same probability the premium exactly equals the payments, which is certainly a non-realistic and trivial case.)

We call a number $\gamma > 0$ an *adjustment coefficient* if for any Δ,

$$M_\Delta(\gamma) = 1. \tag{2.1.2}$$

Remarks:

1. First, we comment on the very definition of an adjustment coefficient. Certainly, (2.1.2) is true for $\gamma = 0$. The above definition presupposes that under some conditions there exists a positive γ for which (2.1.2) also holds. Detailed analysis will be provided later, and for now we just clarify the significance of the above definition.

   Let us consider, first, an arbitrary r.v. ξ and denote by μ and $M(z)$ its mean and m.g.f., respectively. For the sake of simplicity, suppose for now that $M(z)$ is defined for all $z \geq 0$, and assume also that $P(\xi = 0) \neq 1$. (Otherwise, $M(z) \equiv 1$, and the situation is trivial.) Since $M(z) \not\equiv 1$ and convex, $M(z)$ may equal one at most at two points, and one of them is $z = 0$. Denote by p another, if any, solution to the equation $M(z) = 1$.

   As we know, $M(z)$ is convex, $M(0) = 1$, and $M'(0) = \mu$. Hence, if $\mu \geq 0$, then the function $M(z)$ is non-decreasing, and a typical graph of $M(z)$ looks as in Fig.3a. In this case, a positive solution p does not exist.

   If $M'(0) = \mu < 0$, then starting from one at $z = 0$, the graph of $M(z)$ "goes down at least for a while"; see Fig.3bc. Since $M(z)$ is convex, two situations may take place. Either the graph looks as in Fig.3b, and a solution $p > 0$ exists and is unique; or the graph looks as in Fig.3c, and a finite positive solution does not exist. In this case, by convention, we set $p = \infty$. We will see below that it is possible only if $P(\xi \leq 0) = 1$.

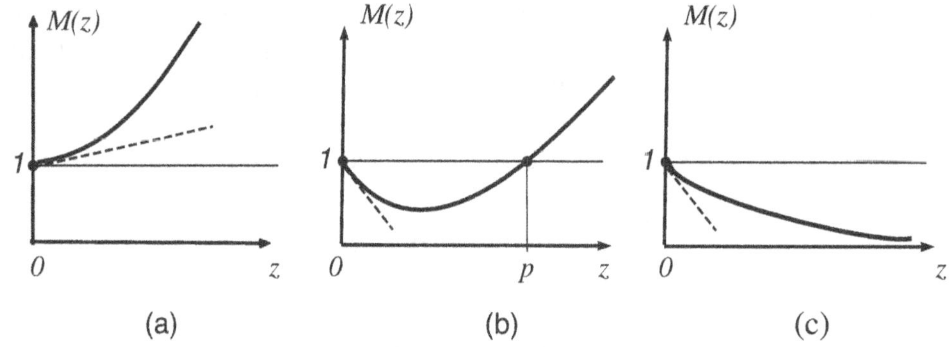

FIGURE 3. (a) The case $M'(0) = \mu \geq 0$. (b) The case $M'(0) = \mu < 0$ and a finite p. (c) The case $M'(0) = \mu < 0$ and $p = \infty$. The broken lines are tangent to $M(z)$.

In our particular case where $\xi = W_\Delta$, among the three possibilities mentioned, only the second may be viewed as realistic. Indeed, if the mean claim surplus $E\{W_\Delta\} \geq 0$ for all Δ, this may be viewed as "too bad", and there is no reason for the company to function. As we will see, in this case the ruin probability equals one. On the other hand, if $P(W_\Delta(z) \leq 0) = 1$, for the company it is "too good" to be true: in this case, clients always pay not less than the company pays them. Later, we will consider all of this in more detail.

2. It may seem non-plausible that (2.1.2) may be true with the same γ for all Δ. As a matter of fact, as we will see, the point here is that as a rule

$$M_\Delta(z) = \exp\{q_1(\Delta)q_2(z)\}, \tag{2.1.3}$$

where q_1 and q_2 are separate functions of Δ and z, respectively. So, we can try to find γ for which $q_2(\gamma) = 0$, and then (2.1.2) will be true for all Δ.

3. The reader who did not omit Section 5.2 on martingales or who is familiar with this notion, may notice that together with the independent-increments condition, (2.1.2) implies that the process $Y_t = \exp\{\gamma W_t\}$ is a martingale. We show this in detail in Section 2.6. We will also see in this section that, as a matter of fact, for subsequent results to be true, we need only Y_t to be a martingale, and the independence of increments is not necessary.

We proceed to results. Let $\psi(u)$ be a ruin probability as it was defined in Section 1.

Proposition 1 *(Lundberg's inequality). If the adjustment coefficient γ exists, then the ruin probability*

$$\psi(u) \leq \exp\{-\gamma u\}. \tag{2.1.4}$$

More remarks:

4. This famous inequality gives an estimate for the ruin probability with some leeway: we will see in examples below that, as a rule, the real ruin probability is less than the r.-h.s. of (2.1.4). But the estimate is simple and tractable and has the advantage that the total information about the process is accumulated in one parameter γ. In next sections, we consider many examples.

5. If we face the situation illustrated in Fig.3a, we can set $\gamma = 0$ in (2.1.4). The inequality will become trivial (the r.-h.s. equals one), but will be still true.

6. If for any Δ, the claim surplus $W_\Delta(z) \le 0$ with probability one, then ruin is impossible. It is also reflected in (2.1.4): in this case, we face the situation illustrated in Fig.3c (details will be provided later), so we can set $\gamma = \infty$, and the r.-h.s. of (2.1.4) equals zero for any $u > 0$.

7. This remark is *important* and concerns the initial surplus u. It is reasonable to view it not as the initial surplus at the time when the insurance process had started, but rather as the current surplus at the present time. At each time moment, we can recalculate the ruin probability, depending on the amount of the surplus at the current time.

EXAMPLE 1. Assume that the loss process $S_{(t)}$ is a homogeneous compound Poisson process with unit rate and claims having the standard exponential distribution. Let $c_t = (1+\theta)t$, where θ is the security loading coefficient. As we will compute it in Section 2.5.1, in this case, the adjustment coefficient $\gamma = \frac{\theta}{1+\theta}$ and the ruin probability itself is

$$\psi(u) = \frac{1}{1+\theta} \exp\left\{-\frac{\theta u}{1+\theta}\right\}.$$

Suppose $\theta = 0.1$ and the company has started the corresponding insurance business with an initial surplus of 30 units of money. Then the ruin probability $\psi(30) = \frac{1}{1.1}\exp\{-0.1 \cdot 30/1.1\} \approx 0.059$. Suppose that during some period the company has been lucky and the total premium collected turned out to be 5 units larger than the total payment. So, the current surplus became equal to 35 units. In this case, we may forget the previous estimate. Now, the company is in a better position, and the new ruin probability equals $\psi(35) = \frac{1}{1.1}\exp\{-0.1 \cdot 35/1.1\} \approx 0.038$. □

In this context, one more thing is noteworthy. Certainly, if the current surplus became larger than the initial, the insurer may take some money from the surplus, for example, paying dividends, and return to the initial amount. However, we should understand that in this case, the situation will have changed. Lundberg's inequality is true under the assumption that the current surplus will not be "touched". But if we suppose that in the future we may change the fund "on hand", then such a supposition should be taken into account in calculating the ruin probability. In this case, we face another situation, namely, that of paying dividends. We consider such a scheme in Section 3.

2.1.2 Proof of Lundberg's inequality

First, note that, by definition, $W_0 = 0$, and hence, $W_{[0,t]} = W_t$. Since (2.1.2) is true for any Δ, for any $T > 0$,

$$1 = M_{[0,T]}(\gamma) = E\{e^{\gamma W_T}\} = E\{e^{\gamma W_T} \,|\, \tau \le T\}P(\tau \le T) + E\{e^{\gamma W_T} \,|\, \tau > T\}P(\tau > T)$$
$$\ge E\{e^{\gamma W_T} \,|\, \tau \le T\}P(\tau \le T) = E\{e^{\gamma W_\tau} e^{\gamma(W_T - W_\tau)} \,|\, \tau \le T\}P(\tau \le T).$$

Since by definition $R_\tau < 0$, we have $W_\tau = u - R_\tau > u$. Consequently, if we replace $e^{\gamma W_\tau}$ by $e^{\gamma u}$, then the resulting expression will not get larger, and we may write that

$$1 \geq E\{e^{\gamma u}e^{\gamma(W_T - W_\tau)} \mid \tau \leq T\}P(\tau \leq T) = e^{\gamma u}E\{e^{\gamma(W_T - W_\tau)} \mid \tau \leq T\}P(\tau \leq T). \tag{2.1.5}$$

Since R_t is a process with independent increments, so is W_t. Hence, given that τ equals some $s \leq T$, the distribution of the r.v. $W_T - W_\tau$ is equal to the distribution of the r.v. $W_T - W_s$, and does not depend on the values of the process W_t for $t \leq s$. Then

$$E\{\exp\{\gamma(W_T - W_\tau)\} \mid \tau = s \leq T\} = E\{\exp\{\gamma W_{(s,T]}\} \mid \tau = s \leq T\} = E\{\exp\{\gamma W_{(s,T]}\}\} = 1$$

by the same property (2.1.2). Let $F(s) = P(\tau \leq s \mid \tau \leq T)$, the conditional d.f. of τ. Then, by the formula for total expectation,

$$E\{e^{\gamma(W_T - W_\tau)} \mid \tau \leq T\} = \int_0^T E\{e^{\gamma(W_T - W_\tau)} \mid \tau = s \leq T\}dF(s) = \int_0^T 1 \cdot dF(s) = 1.$$

From this and (2.1.5) it follows that $1 \geq e^{\gamma u}P(\tau \leq T)$, and hence $P(\tau \leq T) \leq e^{-\gamma u}$. The r.-h.s. of the last inequality does not depend on T. So, we can write that $\psi(u) = P(\tau < \infty) = \lim_{T\to\infty} P(\tau \leq T) \leq e^{-\gamma u}$. ■

2.1.3 The main theorem

Particular examples of the applications of Lundberg's inequality will be considered in Section 2.2, but first let us state a theorem that gives a precise presentation of ruin probability.

To this end, we need one more formal and very mild condition. Namely, since we consider an infinite horizon, we need an assumption concerning the behavior of the process at infinity. Loosely put, we assume that *for large time horizons the aggregate surplus should be large*. Formally, we require that

$$R_t \xrightarrow{P} +\infty \text{ as } t \to \infty \tag{2.1.6}$$

(for this type of convergence see Section **0**.5).

It is worthwhile to emphasize two things. First, (2.1.6) does *not* concern the ruin issue, and when imposing this requirement, we do not exclude that the process may take on negative values "on the way to infinity".

Secondly, this formal assumption holds in *all* reasonable models of insurance processes without paying dividends, including all particular models we consider in this book (excepting models with paying dividends in Section 3). So, in the first reading, the examples in this subsection may be even omitted.

Note also, that the condition (2.1.6) is strongly connected with the LLN. For example, let $R_t = u + c_t - S_{(t)}$, where $S_{(t)}$ is the aggregate claim, the premium $c_t = (1+\theta)E\{S_{(t)}\}$, and θ is a positive relative loading coefficient. Then

$$R_t = u + \theta E\{S_{(t)}\} - \left[S_{(t)} - E\{S_{(t)}\}\right], \tag{2.1.7}$$

and (2.1.6) will be true if $E\{S_{(t)}\} \to \infty$ (the aggregate claim is large on the average for large t), and the third term in (2.1.7) is small with respect to the second (the deviation of the payment $S_{(t)}$ from its expected value is smaller than the expected value itself). More precisely, this means that for any arbitrary small $\varepsilon > 0$,

$$P\left(|S_{(t)} - E\{S_{(t)}\}| > \varepsilon E\{S_{(t)}\}\right) \to 0, \text{ as } t \to \infty. \tag{2.1.8}$$

This is just another form of the LLN. By Chebyshev's inequality (see (**0**.2.5.3)),

$$P\left(|S_{(t)} - E\{S_{(t)}\}| > \varepsilon E\{S_{(t)}\}\right) \leq \frac{1}{\varepsilon^2} \frac{Var\{S_{(t)}\}}{(E\{S_{(t)}\})^2}.$$

We see that for (2.1.8) to be true, it suffices that

$$Var\{S_{(t)}\} = o([E\{S_{(t)}\}]^2). \tag{2.1.9}$$

(For the little o notation, see Appendix, Section 4.1.) This is a very mild condition.

EXAMPLE 1. Let time be discrete, and $S_{(t)} = S_t = X_1 + ... + X_t$, where the X's (claims at separate time moments) are i.i.d. Then $E\{S_{(t)}\} = mt$ and $Var\{S_t\} = \sigma^2 t$, where $m = E\{X_i\}$ and $\sigma^2 = Var\{X_i\}$. If $m > 0$, then $E\{S_{(t)}\} \to \infty$, and (2.1.9) is also true.

EXAMPLE 2. Let $S_{(t)}$ be a homogeneous compound Poisson process. In the notation of Section **4**.3,

$$E\{S_{(t)}\} = m\lambda t, \;\; Var\{S_{(t)}\} = (\sigma^2 + m^2)\lambda t.$$

So, again both requirements, $E\{S_{(t)}\} \to \infty$ and (2.1.9), are true.

EXAMPLE 3. If $S_{(t)}$ is a non-homogeneous compound Poisson process, using the notation of Section **4**.3, we write

$$E\{S_{(t)}\} = m\chi(t), \;\; Var\{S_{(t)}\} = (\sigma^2 + m^2)\chi(t),$$

and (2.1.6) is true if $\chi(t)$, the mean number of claims by time t, converges to infinity. □

For the reader who did not skip Chapter 5, or who is familiar with the notion of Brownian motion, consider

EXAMPLE 4. Let $S_{(t)} = \mu t + \sigma w_t$, a Brownian motion with drift. It is natural to assume $\mu > 0$. Then $E\{S_{(t)}\} = \mu t \to \infty$, $Var\{S_{(t)}\} = \sigma^2 t$, and (2.1.9) again holds. □

The reader can suggest other examples. In particular, it is not necessary for the X_i's above to be identically distributed.

Thus, we adopt the assumption (2.1.6) and turn to the main result.

Let $\tau = \tau_u$ be the moment of ruin; see also Fig.2. Formally, $\tau = \min\{t : R_t < 0\}$. If the process R_t never assumes a negative value (no ruin occurs), we indicate this writing $\tau = \infty$.

As was repeatedly noted in Chapters 4 and 5, the r.v. τ may be defective; that is, it may happen that $P(\tau_u < \infty) < 1$ and $\tau_u = \infty$ with a positive probability. Moreover, this should be the case when we are modeling real surplus processes because $P(\tau_u < \infty)$ equals the ruin probability $\psi(u)$, and we want this probability to be small.

Below we omit the index u in τ_u.

Theorem 2 *Let (2.1.6) be true, and $\gamma > 0$ be the adjustment coefficient defined above. Then*

$$\psi(u) = \frac{\exp\{-\gamma u\}}{E\{\exp\{-\gamma R_\tau\} \,|\, \tau < \infty\}}. \tag{2.1.10}$$

So, the theorem gives a precise expression for the ruin probability. The denominator in (2.1.10) looks somewhat complicated, though, as we will see in Section 2.5, there are cases when it may be easily computed.

Lundberg's inequality follows from (2.1.10) immediately. Indeed, by definition, $R_\tau \le 0$ and $\gamma > 0$. Hence, $\exp\{-\gamma R_\tau\} \ge 1$. So, the denominator in (2.1.10) is not less than one, which implies (2.1.4).

We prove Theorem 2 in Section 2.6 using a martingale technique. For the reader who does not plan yet to learn martingales, we have given above a direct proof of Lundberg's inequality.

2.2 Computing adjustment coefficients

In the first subsection below, we consider conditions under which equation (2.1.2) has a positive solution. This has a rather mathematical significance because

> If in a particular problem we manage to find an adjustment coefficient $\gamma > 0$, then we can be sure that it is unique and we can use it applying either Lundberg's inequality or Theorem 2.

This is true because, as any m.g.f., $M_\Delta(z)$ is convex. Therefore, if $M_\Delta(z) \not\equiv 1$, then—as has been already noted— $M_\Delta(z)$ may equal 1 only at two points. The first point is zero, and the other is γ. See more details below.

So, the reader who is interested rather in applications may skip Subsection 2.2.1 and move directly to Subsection 2.2.2 that deals with concrete models.

Route 1 ⇒ *page 351*

2,3

2.2.1 A general proposition

Consider a r.v. ξ whose m.g.f. $M(z)$ exists and is finite for all $z \in [0, z_0)$, where $0 < z_0 \le \infty$. We assume that z_0 is the largest number with this property; that is, $M(z) = \infty$ for $z > z_0$.

To clarify this, let us recall that if, for example, ξ is uniformly distributed on some interval, $M(z)$ exists for all z and hence $z_0 = \infty$. On the other hand, if ξ is exponential, then $M(z)$ exists only for $z < 1/\mu$, where $\mu = E\{\xi\}$. So, $z_0 = 1/\mu$. (See Section **0**.4.3.)

As for the point z_0 itself, $M(z_0)$ may or may not exist. In Example 1 below, we consider the case when $M(z_0)$ is finite. However, for the exponential distribution, $M(z)$ does not exist at $z_0 = 1/\mu$.

We do not consider $M(z)$ at negative z's.

Note also that, *in general*, we do not have to assume $\mu = E\{\xi\}$ to be finite. Since there exists a positive z for which $M(z) < \infty$, for $z > 0$ we can write $\mu = \frac{1}{z}E\{z\xi\} \le \frac{1}{z}E\{e^{z\xi}\} =$

$\frac{1}{z}M(z) < \infty$. However, we do not exclude the case where $\mu = -\infty$, that is, the negative part of X has an infinite expectation. In this case, the reasoning below remains correct.

Consider the equation

$$M(z) = 1, \ \ z \geq 0. \tag{2.2.1}$$

We assume that

$$P(\xi = 0) \neq 1, \tag{2.2.2}$$

since otherwise $M(z) \equiv 1$, and equation (2.2.1) is trivial.

Proposition 3 *If (2.2.2) holds, then the following is true.*

(a) A positive solution to (2.2.1) exists if and only if $\mu < 0$ and

$$M(z_0) \geq 1. \tag{2.2.3}$$

(b) If $\mu < 0$ but (2.2.3) does not hold, then $M(z) < 1$ for all positive z for which $M(z)$ exists.

(c) If the equation (2.2.1) has a positive solution, then this is the only positive solution, and (2.2.3) is true.

Before proving this proposition, we clarify the sense of its conditions. In Section 2.1.1, in particular in Fig.3, we have already shown the role played by μ. Consider now condition (2.2.3).

First of all, if $M(z_0) = \infty$, whatever z_0 is, finite or infinite, then (2.2.3) holds automatically. In particular, this is the case when $z_0 = \infty$ and

$$P(\xi > 0) > 0. \tag{2.2.4}$$

Indeed, (2.2.4) implies that there exists $b > 0$ such that

$$P(\xi \geq b) > 0. \tag{2.2.5}$$

(If $P(\xi \geq b) = 0$ for all $b > 0$, we can write that $P(\xi > 0) = \lim_{b \to 0, b>0} P(\xi \geq b) = 0$, which would contradict (2.2.4).) Hence, denoting by $F(x)$ the d.f. of ξ, we can write

$$M(z) = \int_{-\infty}^{\infty} e^{zx} dF(x) \geq \int_{b}^{\infty} e^{zx} dF(x) \geq e^{zb} \int_{b}^{\infty} dF(x) = e^{zb} P(\xi \geq b) \to \infty.$$

On the other hand, if $P(\xi > 0) = 0$, then in view of (2.2.2), $M(z) = E\{e^{z\xi}\} < 1$ for all $z > 0$. (Indeed, $e^{z\xi} < 1$ with a positive probability, and $P(e^{z\xi} > 1) = 0$.) So, in this case, there is no solution to (2.2.1), and condition (2.2.3) also does not hold. This situation has been illustrated in Fig.3c.

Now, let $z_0 < \infty$. Again, if $M(z_0) = \infty$, the condition (2.2.3) holds automatically. For example, let $\xi = X - c$ where X may be interpreted as a claim and is exponential, and c is viewed as a premium. Then $M(z) = e^{-cz} \frac{1}{1 - E\{X\}z} \to \infty$ as $z \to 1/E\{X\}$. The situation is illustrated in Fig.4a.

However, it may happen that $M(z_0) < \infty$, while $M(z) = \infty$ for all $z > z_0$.

EXAMPLE 1. Let $\xi = X - c$ where $c > 0$ and a positive r.v. X has the density

$$f(x) = \frac{K}{1+x^2} e^{-x}, \ x \geq 0,$$

and K is a constant for which $\int_0^\infty f(x)dx = 1$. Then

$$M(z) = e^{-cz} \int_0^\infty e^{zx} \frac{K}{1+x^2} e^{-x} dx = Ke^{-cz} \int_0^\infty e^{(z-1)x} \frac{1}{1+x^2} dx.$$

The last integral diverges (or equals infinity) for $z > 1$, and it is finite for $z < 1$. Hence, in our case, $z_0 = 1$. On the other hand,

$$M(1) = Ke^{-c} \int_0^\infty \frac{1}{1+x^2} dx < \infty. \qquad (2.2.6)$$

Thus, $M(z)$ is defined at $z \leq 1$.

Let us explore other features of this particular case. The integral in (2.2.6) equals $\arctan(\infty) = \pi/2$, so we have

$$M(1) = K_1 e^{-c} \text{ where } K_1 = K\frac{\pi}{2}.$$

Now observe that $K_1 > 1$ because $K_1 = K\int_0^\infty \frac{1}{1+x^2} > K\int_0^\infty e^{-x} \frac{1}{1+x^2} = \int_0^\infty f(x) = 1$.

Hence, $M(1) > 1$ for sufficiently small c. The situation is illustrated in Fig.4b. We see that in this case, a solution p to equation (2.2.1) exists.

On the other hand, for sufficiently large c, we have $M(1) < 1$; see Fig.4c. In this case, a solution to (2.2.1) does not exist. However, since $M(z) = \infty$ for $z > 1$ (so to say, at $z = 1$ the function $M(z)$ jumps over 1, we may view 1 as a solution. □

Not going deeply into it, note that in the situations similar to that in Example 1, z_0 may be considered a solution to (2.2.1) and used in the context of Lundberg's inequality.

We proceed to a formal

Proof of Proposition 3. First, note that since $P(\xi \neq 0) > 0$, the second derivative

$$M''(z) = E\{\xi^2 e^{z\xi}\} > 0. \qquad (2.2.7)$$

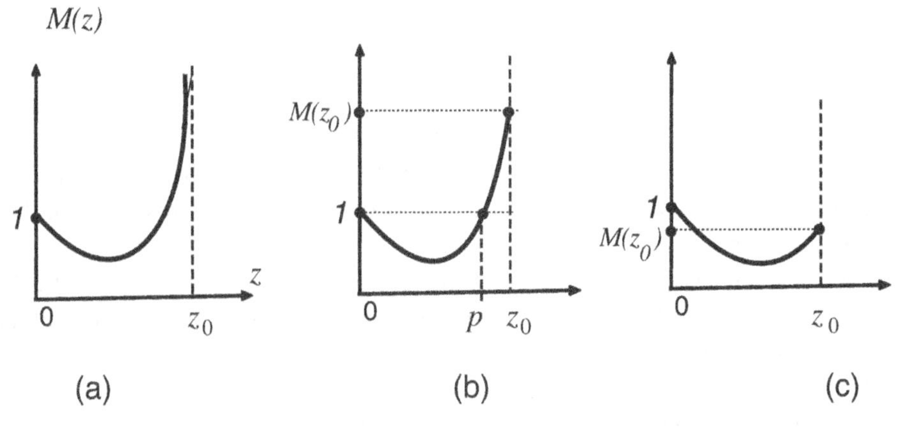

FIGURE 4.

(For the sake of rigor, note that $M''(0)$ may be infinite, which does not contradict (2.2.7), but $M''(z)$ is finite for $0 < z < z_0$. Indeed, the integral over $(-\infty, 0)$ is finite because the function $x^2 e^{zx}$ is bounded there (x's are negative, z is positive). The integral over $[0, \infty)$ is bounded because $M(z)$ is bounded for $0 < z < z_0$. See also Section **0**.4.4.)

Now, let $\mu \geq 0$. Then $M'(0) \geq 0$, and by virtue of (2.2.7), $M(z)$ is strictly increasing. Hence, (2.2.1) does not have a solution; see also Fig.3a.

Consider the case $\mu < 0$ and $M(z_0) < 1$ where z_0 is either finite or infinite. We show that in this case $M(z) < 1$ for all $z > 0$, and there is no solution to (2.2.1) (or as was told above, a solution $p = \infty$; see Fig.3c). Indeed, assume that there exists $z_1 < z_0$ such that $M(z_1) \geq 1$. Since $M'(0) < 0$, the function $M(z) < 1$ in a neighborhood of 0, and there exists z_2 such that $0 < z_2 < z_1$ and $M(z_2) < 1$. So, $M(0) = 1, M(z_2) < 1, M(z_1) \geq 1, M(z_0) < 1$ for $0 < z_2 < z_1 < z_0$, which contradicts the convexity of $M(z)$.

Now, let $\mu < 0$ and $M(z_0) \geq 1$. Let z_2 be the same as above: $M(z_2) < 1$. If $M(z_0) > 1$, then the existence of a solution follows immediately from the continuity of $M(z)$.

Let $M(z_0) = 1$. Then z_0 cannot be infinite. Indeed, if $z_0 = \infty$, then $M(0) = 1, M(z_2) < 1$, and $\lim_{z \to \infty} M(z) = 1$, which contradicts to the convexity of $M(z)$.

But if $M(z_0) = 1$ and $z_0 < \infty$, then z_0 is a solution to (2.2.1).

Now, assume that there exists a positive solution p. Because $M(z)$ is convex, and $M''(z) > 0$, the graph of $M(z)$ may intersect any line only in two points. Hence, $M(z) = 1$ at only two points: zero and p. Consequently, p is a unique positive solution. Then $M(z_0)$ is larger than 1 or equal to one, and in the latter case the solution p is z_0. ■ 1,2,3

We proceed to particular cases.

2.2.2 The discrete time case: Examples

Let $t = 0, 1, \ldots$, and $W_t = Y_1 + \ldots + Y_t$, where Y_i is the claim surplus during the ith period. We assume the Y's to be i.i.d.

As a rule, $Y_i = X_i - c$, where X_i and c are the aggregate claim and premium, respectively, corresponding to the ith period. Below, we omit the adjective "aggregate" if it cannot cause misunderstanding.

Formally, we should solve the equation (2.1.2) for all intervals $\Delta = (k, k+t]$, where k, t are integers. However, since the Y's are identically distributed, it suffices to consider $\Delta = (0, t]$, and accordingly the r.v. W_t.

Because the Y's are independent, the m.g.f. $M_\Delta(z) = M_{(0,t]}(z) = E\{\exp\{zW_t\}\} = (M_Y(z))^t$, where $M_Y(z)$ is the common m.g.f. of the r.v.'s Y_i. Thus, $M_\Delta(z) = 1$ if and only if

$$M_Y(z) = 1. \tag{2.2.8}$$

This is an equation for the adjustment coefficient γ.

We assume that

$$E\{Y\} < 0, \quad \text{and} \tag{2.2.9}$$
$$P(Y > 0) > 0 \tag{2.2.10}$$

(the index i is omitted).

So, we are in the situation of Fig.3b. To show this absolutely rigorously, one may apply Proposition 3, but as a matter of fact, we should not much worry about conditions of this proposition. As has been noted in the beginning of Section 2.2 (and as was stated in Proposition 3 itself), if we manage to find a positive solution to (2.2.8), this solution will be unique and all conditions of Proposition 3 will be true automatically.

Let now $Y_i = X_i - c$. Then $E\{Y_i\} = m - c$, where $m = E\{X_i\}$. The condition $E\{Y\} < 0$ is equivalent to the condition

$$c > m,$$

and condition (2.2.10)—to the condition $P(X_i > c) > 0$ or

$$P(X_i \leq c) < 1.$$

Both conditions are natural; regarding the latter, note that nobody will pay a premium that is larger than or equal to the future payment with probability one.

Now, note that equation (2.2.8) may be rewritten as

$$e^{-cz}M_X(z) = 1, \tag{2.2.11}$$

where $M_X(z)$ is the common m.g.f. of X's. (See (**0**.4.1.5).)

In examples below, when considering a separate claim X_i, we omit the index i.

EXAMPLE 1. Assume that the claim X has a Γ-density, say, $f(x) = xe^{-x}$. Then $E\{X\} = 2$, the m.g.f. $M_X(z) = 1/(1-z)^2$ for $z < 1$. Substituting this into (2.2.11), we come to

$$e^{-cz} = (1-z)^2. \tag{2.2.12}$$

However, we should remember that for $z \geq 1$, the m.g.f. $M_X(z)$ does not exist; so we should accept only solutions $z < 1$. In Exercise 5a, the reader is suggested to graph the r.-h.s. and the l.-h.s. of (2.2.12).

It is impossible to write a solution to (2.2.12) explicitly, but it is easy to solve it numerically, even using a graphing calculator. For example, for $c = 2.15$, we will readily find that for a solution γ, we have $0.136 < \gamma < 0.137$. Thus,

$$\psi(u) \leq \exp\{-\gamma u\} \leq \exp\{-0.136u\}.$$

(To have a correct inequality, we should take the lower (!) bound for γ.)

EXAMPLE 2. Let X be well approximated by a (m, σ^2)-normal distribution. Then $M_X(z) = \exp\{mz + \sigma^2 z^2/2\}$, and equation (2.2.11) may be rewritten as

$$-cz + mz + (\sigma^2 z^2/2) = 0.$$

The positive root is

$$\gamma = \frac{2(c-m)}{\sigma^2}. \ \square \tag{2.2.13}$$

The particular expression (2.2.13) gives an idea of the following approximation in the general case. Taking the logarithm of both sides of (2.2.11), we can rewrite it as

$$-cz + \ln M_X(z) = 0. \tag{2.2.14}$$

Assume that the root of the equation is "small". In (**0**.4.5.7), we have derived the approximation formula

$$\ln M_X(z) = mz + \frac{1}{2}\sigma^2 z^2 + o(z^2),$$

where $m = E\{X\}$, $\sigma^2 = Var\{X\}$, and $o(z^2)$ is a remainder negligible with respect to z^2 for small z's. So,

$$-cz + mz + (\sigma^2 z^2/2) + o(z^2) = 0. \tag{2.2.15}$$

If we neglect the term $o(z^2)$, the positive solution to (2.2.15) will lead us to the approximation

$$\gamma \approx \frac{2(c-m)}{\sigma^2}. \tag{2.2.16}$$

The accuracy of such an approximation can be estimated with use of bounds for the remainder in Taylor's expansion, but we restrict ourselves to an example.

EXAMPLE 3. For the data from Example 1, $\sigma^2 = 2$, and (2.2.16) gives

$$\gamma \approx \frac{2 \cdot 0.15}{2} = 0.15,$$

which is not so bad in comparison with the answer from Example 1 (≈ 0.136).

Let now $c = 2.05$. Then, as is easy to compute, $0.05 < \gamma < 0.051$, while (2.2.16) gives 0.05. □

2.2.3 The discrete time case: The adjustment coefficient for a group of insured units.

It is natural to view the loss (or claim) r.v. X_i above as the aggregate claim coming from a portfolio of insured units. Here we present it in an explicit way. Let

$$X_i = X_{i1} + \ldots + X_{in_i}, \tag{2.2.17}$$

where n_i is the number of units composing the portfolio in the ith period, and X_{ij} is the payment to the jth unit in the same period. We assume $\{X_{ij}\}$ to be i.i.d. r.v.'s. Note that in this case, X_i's are independent, but perhaps are not identically distributed because the number of terms in the sum (2.2.17) depends on i.

Denote by $\tilde{c}$ the premium for a particular unit, and set $Y_{ij} = X_{ij} - \tilde{c}$, the claim surplus of the jth unit in the ith period. Since time is discrete, for any time interval $\Delta = (t, t+k]$, where t, k are integers,

$$W_\Delta = \sum_{i=t+1}^{t+k} Y_i = \sum_{i=t+1}^{t+k} \sum_{j=1}^{n_i} Y_{ij}, \tag{2.2.18}$$

and the m.g.f. $M_\Delta(z) = (M_Y(z))^s$ where $M_Y(z)$ is the common m.g.f. of r.v.'s Y_{ij}, and $s = n_{t+1} + \ldots + n_{t+k}$, the total number of all terms in (2.2.18).

Thus, $M_{\Delta}(z) = 1$ iff $M_Y(z) = 1$, and we have come to a nice conclusion:

> The adjustment coefficient for a homogeneous portfolio is equal to the adjustment coefficient for one separate insured unit. (2.2.19)

EXAMPLE 4. When considering the above model, it is natural to assume that the loss for a particular unit is equal to zero with a positive and "substantial" probability. Let, say, $X_{ij} = 0, 10, 20$ with probabilities $0.8, 0.1, 0.1$, respectively. Then $E\{X_{ij}\} = 3$. Suppose $\widetilde{c} = 3.5$.

To compute the adjustment coefficient γ, we do not need any information about the numbers of units, and the equation for γ is the same equation (2.2.11) where X corresponds to X_{ij}, and c should be replaced by $\widetilde{c}$. Using version (2.2.14), we have

$$-3.5z + \ln(0.8 + 0.1e^{10z} + 0.1e^{20z}) = 0.$$

An analytical solution is again impossible, but it is easy to solve the equation numerically. The reader can verify that the positive solution $\gamma \approx 0.022$. □

However, the situation is different if we consider the portfolio as a whole, and set each

$$X_i = \xi_{i1} + ... + \xi_{iK_i}, \tag{2.2.20}$$

where now ξ_{ij} is the amount of the jth claim in the ith period, and the r.v. K_i is the number of claims in this period. It makes sense to emphasize that, while X_{ij}'s above were payments to separate units (and could take on zero values), ξ_{ij} are *claims* arriving at the system (and may be assumed to be positive). Suppose that all r.v.'s are independent, ξ's are identically distributed, and the same is true for K_i's.

If we know the distributions of ξ's and K_i, we can find, at least theoretically, the distribution of X, and hence the adjustment coefficient. For example, if we accept the approximation (2.2.16), we can set there $m = \widetilde{m}E\{K_i\}$ and $\sigma^2 = \widetilde{\sigma}^2 E\{K_i\} + \widetilde{m}^2 Var\{K_i\}$, where $\widetilde{m}$ and $\widetilde{\sigma}^2$ are the mean and variance of ξ's, respectively. A particular example is given in Exercise 9. The case where K_i are Poisson r.v.'s is the most interesting, but it is convenient for us to consider it later, in Section 2.2.5, after we explore the case of the Poisson process in continuous time.

2.2.4 The case of a homogeneous compound Poisson process

We turn to continuous time and consider the claim surplus process $W_t = S_{(t)} - c_t$, where

$$S_{(t)} = \sum_{i=1}^{N_t} X_i, \tag{2.2.21}$$

N_t is a Poisson process with intensity λ, and X_i are i.i.d. and do not depend on N_t. (See also Section **4**.3.) We assume X's to be positive.

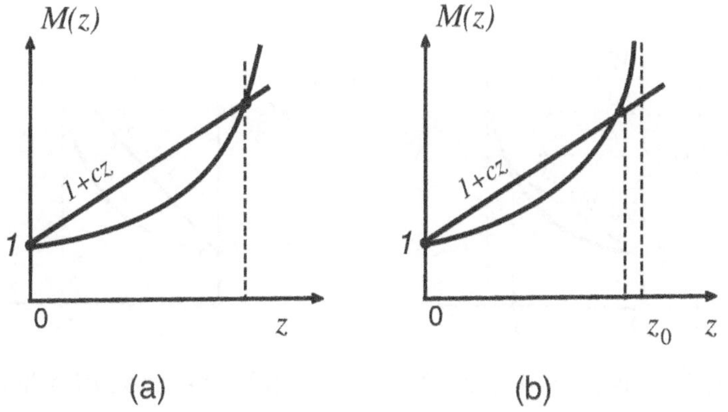

FIGURE 5.

The reader remembers that $E\{N_t\} = \lambda t$, and $E\{S_t\} = m\lambda t$, where $m = E\{X_i\}$. Denote by N_Δ, S_Δ the increments of the corresponding processes over an interval $\Delta = (t, t+\delta]$. Then $E\{N_\Delta\} = \lambda\delta$, and $E\{S_\Delta\} = m\lambda\delta$.

The point is that the r.v. S_Δ is also a compound Poisson r.v. Indeed, we can write that

$$S_\Delta = S_{(t+\delta)} - S_{(t)} = \sum_{i=N_t+1}^{N_{t+\delta}} X_i.$$

The process N_t is a process with independent increments, and at any point t "everything starts over as from the beginning". So, the r.v. S_Δ has the same distribution as the r.v.

$$S_\Delta = \sum_{i=1}^{N_\Delta} X_i.$$

In particular, the m.g.f.

$$M_{S_\Delta}(z) = \exp\{\lambda\delta(M_X(z) - 1)\},$$

where $M_X(z)$ is the m.g.f. of X's (see (**3**.1.6) in Proposition **3**.3).

Set the premium

$$c_t = (1+\theta)E\{S_{(t)}\} = (1+\theta)m\lambda t,$$

where again $\theta > 0$ is a security loading coefficient. Then the increment of the premium over the interval Δ is $c_\Delta = (1+\theta)m\lambda\delta$, and the m.g.f. of W_Δ is

$$\begin{aligned} M_{W_\Delta}(z) &= \exp\{-c_\Delta z\} M_{S_\Delta}(z) = \exp\{-(1+\theta)m\lambda\delta z\}\exp\{\lambda\delta(M_X(z)-1)\} \\ &= \exp\{\lambda\delta[M_X(z) - 1 - (1+\theta)mz]\}. \end{aligned}$$

Thus, $M_{W_\Delta}(z) = 1$ iff $M_X(z) - 1 - (1+\theta)mz = 0$, which we write as

$$M_X(z) = 1 + (1+\theta)mz, \tag{2.2.22}$$

or

$$M_X(z) = 1 + cz, \tag{2.2.23}$$

where $c = (1+\theta)m$, the premium per one claim.

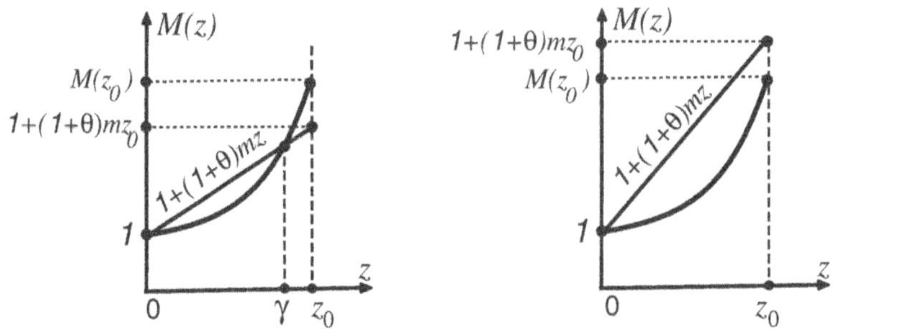

(a) The case of a "small" θ. (b) The case of a "large" θ.

FIGURE 6.

This is an equation for the adjustment coefficient.

It is noteworthy that this equation does not involve λ and δ, so the adjustment coefficient is specified only by the distribution of X (compare with (2.2.19)).

Consider (2.2.23) and/or (2.2.22) setting for simplicity $M(z) = M_X(z)$. Since X is positive, and $M(z)$ is convex, $M'(z) > M'(0) = m > 0$. Hence, $M(z)$ is strictly increasing, and if $M(z)$ is defined on $[0,\infty)$, then $M(z) \to \infty$ as $z \to \infty$. Note also that while $M'(0) = m$, the slope of the line specified by the r.-h.s. of (2.2.23) is $c > m$. See also Fig.5a. Hence, in this case a solution to (2.2.23) or (2.2.22) exists and is unique.

If $M(z)$ is defined on a finite interval $[0, z_0)$, $z_0 < \infty$, but $M(z) \to \infty$ as $z \to z_0$, we have the same; see also Fig.5b.

Consider the last case where $M(z)$ is defined on $[0, z_0]$, $z_0 < \infty$, and $M(z_0) < \infty$. Then a solution exists for all sufficiently small $\theta \geq 0$. Indeed, the closer θ is to zero, the closer the line $1 + (1+\theta)mz$ is to the line tangent to $M(z)$ at the origin; see Fig.6a. So, for small θ, the line $1 + (1+\theta)mz$ intersects the graph of $M(z)$.

► The formal proof may run as follows. Since X is positive and $e^x \geq 1 + x + \frac{1}{2}x^2$ for all $x \geq 0$, the m.g.f. $M(z) = E\{e^{zX}\} \geq E\{1 + zX + \frac{1}{2}z^2X^2\} = 1 + mz + \frac{1}{2}z^2E\{X^2\}$. Because $E\{X^2\} > 0$ and $1 + (1+\theta)mz \to 1 + mz$ as $\theta \to 0$, for sufficiently small θ, we have $M(z_0) \geq 1 + mz_0 + \frac{1}{2}z_0^2E\{X^2\} > 1 + (1+\theta)mz_0$. ◄

For large θ, a positive solution may not exist. In this case, we may set $\gamma = z_0$.

Indeed, let us first choose θ for which γ exists and equals z_0. It would be the limiting case when $M(z_0) = 1 + (1+\theta)mz_0$. For such a θ, Lundberg's bound for the ruin probability is $e^{-\gamma u} = e^{-z_0 u}$. But for a larger θ, the ruin probability will be even smaller (the positive component of the surplus process gets larger). Hence, the same bound is true for such θ's also.

EXAMPLE 1. Let X have a Γ-distribution with parameters a, ν. Then $m = \nu/a$, and (2.2.22) amounts to $(1 - z/a)^{-\nu} = 1 + z(1+\theta)\nu/a$, or

$$\left(1 + z(1+\theta)\frac{\nu}{a}\right)\left(1 - \frac{z}{a}\right)^{\nu} = 1, \tag{2.2.24}$$

where we should consider only $z < a$, since for $z \geq a$ the m.g.f. does not exist.

Solving (2.2.24) numerically does not cause any difficulty; for $\nu = 1, 2$ one can write an explicit solution. If $\nu = 1$, i.e., X is exponentially distributed, after simple algebra, (2.2.24) may be rewritten as

$$z\left(\theta - \frac{z(1+\theta)}{a}\right) = 0.$$

The positive root is

$$\gamma = \frac{a\theta}{1+\theta}. \tag{2.2.25}$$

For $\nu = 2$, since one root of (2.2.24) is $z = 0$, the equation may be reduced to a quadratic equation, and we should choose the positive root which is less than a. See also Exercise 11. □

Using expansion (**0**.4.5.3), we can get a counterpart of approximation (2.2.16), writing (2.2.23) as $1 + mz + \frac{1}{2}m_2z^2 + o(z^2) = 1 + cz$, where $m_2 = E\{X^2\}$, the second moment of X. Neglecting the term $o(z)$, we come to the approximation

$$\gamma \approx \frac{2(c-m)}{m_2}. \tag{2.2.26}$$

EXAMPLE 2. In the situation of Example 1, $m = \frac{\nu}{a}$, $m_2 = \frac{\nu}{a^2} + (\frac{\nu}{a})^2$, and since $c = (1+\theta)m$, approximation (2.2.26) gives

$$\gamma \approx \frac{2\theta a}{1+\nu}. \tag{2.2.27}$$

For $\nu = 1$, the precise γ in (2.2.25) differs from approximation (2.2.27) by the multiplier $\frac{1}{1+\theta}$, which for small θ is close to one. See another example in Exercise 14. □

2.2.5 The discrete time case revisited

We continue to consider the compound Poisson model of the previous section, but now we will refer to ruin only as the event when the current surplus becomes negative at the end of a unit interval; say, at the end of a year. Formally, it means that the no-ruin (survival) probability is defined as $\widetilde{\phi}(u) = P(R_t \geq 0,\ t = 1, 2, \ldots)$. Time is still continuous, but we check for ruin only at integer moments of time. Here, we follow a tradition and mark the no-ruin probability by a tilde to distinguish this probability from $\phi(u) = P(R_t \geq 0$ for all $t > 0)$. Let $\widetilde{\psi}(u) = 1 - \widetilde{\phi}(u)$, and as usual, $\psi(u) = 1 - \phi(u)$.

Clearly,

$$\widetilde{\psi}(u) \leq \psi(u). \tag{2.2.28}$$

(If ruin occurs at an integer time moment, then ruin has occurred at some moment.)

To find $\widetilde{\psi}(u)$, we apply the results of Section 2.2.2. Let us mark X's from this section by a tilde to distinguish them from the individual claims X in this section. More precisely, let for $k = 1, 2, \ldots$,

$$\widetilde{X}_k = \sum_{i=N_{k-1}+1}^{N_k} X_i,$$

the total claim for the unit time period $(k-1, k]$. Since N_t is a homogeneous Poisson process, at the beginning of each period, the claim process starts to run as from the very beginning, all r.v. $\widetilde{X}_k$ are i.i.d., and

$$E\{\widetilde{X}_k\} = mE\{N_k - N_{k-1}\} = m\lambda.$$

Then the premium per unit time is $c = (1+\theta)E\{\widetilde{X}_k\} = (1+\theta)m\lambda$. The r.v. $\widetilde{X}_k$ has a compound Poisson distribution, and

$$M_{\widetilde{X}_k}(z) = \exp\{\lambda(M_X(z)-1)\}.$$

Hence in our case, the equation (2.2.11) may be written as

$$\exp\{-(1+\theta)m\lambda z\}\exp\{\lambda(M_X(z)-1)\} = 1,$$

which is equivalent to $\lambda(M_X(z)-1) - (1+\theta)m\lambda z = 0$, or

$$M_X(z) = 1+(1+\theta)mz.$$

Thus, we have arrived at the same equation (2.2.22). This means that the adjustment coefficient in the discrete time case (for the scheme we consider) is the same as in the case of continuous time. So, the bounds $e^{-\gamma u}$ for $\psi(u)$ and $\widetilde{\psi}(u)$ will be the same.

Does this contradict (2.2.28)? No, since we deal with upper bounds. To compare $\psi(u)$ and $\widetilde{\psi}(u)$ we should also take into account the denominator $E\{\exp\{-\gamma R_\tau\}\,|\,\tau<\infty\}$ in (2.1.10) which is different for these two cases. The r.v. $(-R_\tau) = |R_\tau|$ is the deficit at the moment of ruin. If we consider only integer moments of time, R_t may become negative before the end of the period of ruin, and $|R_\tau|$ corresponds to the deficit accumulated during this period, whereas in the general case, $|R_\tau|$ is the deficit corresponding to the claim at the moment of ruin. So, we should expect that, on the average, $|R_\tau|$ is larger in the discrete time case.

Certainly, this is not a proof; the proof itself follows from (2.2.28) since if the numerator in (2.1.10) is the same for both cases, the denominator must take on different values in the continuous and discrete cases.

2.2.6 The case of non-homogeneous compound Poisson processes

We presuppose that the reader is familiar with the notion of a non-homogeneous Poisson process considered in Chapter 4.

Since the equation for γ in the previous section does not depend on λ, one may suppose that, as a matter of fact, the homogeneity of the process N_t is not necessary, and we can get the same result in the general case. This is indeed true.

Let $\lambda(t)$ be the intensity function for N_t. As in Section **4**.3, we set

$$\chi(t) = \int_0^t \lambda(s)ds,\ \ \chi_\Delta = \int_\Delta \lambda(s)ds$$

for any interval Δ. As was shown in Chapter 4, $E\{N_t\} = \chi(t)$, $E\{N_\Delta\} = \chi_\Delta$, and hence $E\{S_\Delta\} = m\chi_\Delta$. The r.v. S_Δ is a compound Poisson r.v., and its m.g.f.

$$M_{S_\Delta}(z) = \exp\{\chi_\Delta(M_X(z)-1)\}.$$

We set again the premium

$$c_t = (1+\theta)E\{S_{(t)}\} = (1+\theta)m\chi(t).$$

Then the increment of the premium over an interval Δ is $c_\Delta = (1+\theta)m\chi_\Delta$, and the m.g.f. of W_Δ is

$$M_{W_\Delta}(z) = \exp\{-c_\Delta z\}M_{S_\Delta}(z) = \exp\{-(1+\theta)m\chi_\Delta z\}\exp\{\chi_\Delta(M_X(z)-1)\}$$
$$= \exp\{\chi_\Delta(M_X(z)-1-(1+\theta)mz)\}.$$

Thus, $M_{W_\Delta}(z) = 1$ iff $M_X(z)-1-(1+\theta)mz = 0$, which again leads to (2.2.22). So, the equation does not involve χ_Δ, and the adjustment coefficient is again specified only by the distribution of X.

2.3 Finding an initial surplus

This section is to emphasize that having an estimate of the ruin probability, we are able to estimate the initial surplus sufficient to keep the ruin probability less than a given desirable level. Denote this level by α. Say, if we choose $\alpha = 0.05$, then we accept a 5% risk of being ruined.

We proceed from the bound

$$\psi(u) \le \exp\{-\gamma u\}, \tag{2.3.1}$$

and set $\exp\{-\gamma u\} = \alpha$. Then,

$$u = -\frac{1}{\gamma}\ln\alpha = s/\gamma, \tag{2.3.2}$$

where $s = \ln(1/\alpha)$. If $\alpha < 1$, then $s = \ln(1/\alpha) > 0$. From (2.3.1) it follows that for such a choice of the initial surplus u, the ruin probability does not exceed the security level α. The estimate s/γ is obtained with some leeway because we proceed not from the real ruin probability but from its upper bound. In any case,

> For the ruin probability to be less than α, it suffices that the initial surplus $u \ge \frac{1}{\gamma}\ln(1/\alpha) = \frac{s}{\gamma}$. (2.3.3)

EXAMPLE 1. Let us revisit Example 2.2.2-1, where we found that $0.136 < \gamma < 0.137$. If we estimate u proceeding from the inequality $u \ge s/\gamma$, then we should take the *lower* bound for γ, that is, 0.136. Indeed, if $u \ge \frac{s}{0.136}$, then we have a right to write that

$$u \ge \frac{s}{0.136} \ge \frac{s}{\gamma},$$

which implies that the ruin probability is less than α. In its turn, $\frac{1}{0.136} < 7.36$. So, we take $u \ge 7.36s$, and it will be an estimate with some leeway.

For example, for $\alpha = 0.05$, we have $s = \ln(1/.05) = \ln 20 \le 2.996 < 3$, and we come to $u \ge 3 \cdot 7.36 = 22.08$. □

Route 1 ⇒ *page 363*

However, there is another way to keep the ruin probability lower than a given level: to
2,3 increase the premium. To a certain degree, both characteristics—the initial surplus and the premium—are under the control of the insurer, and the determination of them consists in a trade-off between these characteristics. We consider it in the next section.

2.4 Trade-off between the premium and initial surplus

For the discrete time case, by (2.2.14),

$$c = \frac{1}{\gamma}\ln M_X(\gamma). \tag{2.4.1}$$

In the compound Poisson case, (2.2.23) implies

$$c = \frac{1}{\gamma}\left(M_X(\gamma) - 1\right). \tag{2.4.2}$$

(Note that these representations do not differ much for small γ. Indeed, $M_X(\gamma) - 1$ is small for γ close to 0 (say why). Since $\ln(1+x) = x + o(x)$ for small x's, we can write $\ln M_X(\gamma) = \ln[1 + (M_X(\gamma) - 1)] = M_X(\gamma) - 1 + o(M_X(\gamma) - 1)$, where the second term is small.)

Our goal is to find c and u for which the ruin probability is not greater than a given security level α. Since we have only an upper bound for the ruin probability, we have to proceed from this upper bound. For the upper bound in (2.3.1) to equal α, we should have $e^{-\gamma u} = \alpha$, or

$$\gamma = \frac{1}{u}\ln\left(\frac{1}{\alpha}\right) = \frac{s}{u}. \tag{2.4.3}$$

Now, if we replace γ in (2.4.1) and (2.4.2) by the r.-h.s. of (2.4.3), we will establish a relation between u and c, which ensures the ruin probability to be not larger than α. (As we understand, with leeway since we proceeded from an upper bound for the ruin probability.)

Thus, in the discrete time case, we have

$$c = \frac{u}{s}\ln M_X\left(\frac{s}{u}\right), \tag{2.4.4}$$

and in the compound Poisson process case,

$$c = \frac{u}{s}\left(M_X\left(\frac{s}{u}\right) - 1\right). \tag{2.4.5}$$

If, say, in the compound Poisson case, c is not equal to but larger than the r.-h.s. of (2.4.5), then the bound $e^{-\gamma u}$ will be smaller than α. From a common-sense point of view, it is understandable. If the premium is larger than required, then the ruin probability will be even smaller than we wish.

Formally, it follows from the fact that $M_X(\gamma)$ is convex, and hence the r.-h.s. of (2.4.2) is increasing in γ. (Graph $M(z)$ and consider the slope of the line connecting the point $(0,1)$ and $(z, M(z))$. We skip formalities here.)

The same is true for the discrete time case. Advice on how to show the monotonicity of the r.-h.s. of (2.4.1) is given in Exercise 12.

EXAMPLE 2. Suppose time is discrete, and X is (m, σ^2)-normal. In this case, $\ln M_X(z) = mz + \sigma^2 z^2/2$, and the reader can readily get from (2.4.4) that

$$c = m + \frac{s\sigma^2}{2} \cdot \frac{1}{u}. \tag{2.4.6}$$

The same may also be obtained by substituting γ in (2.4.3) by the explicit expression for γ in (2.2.13).

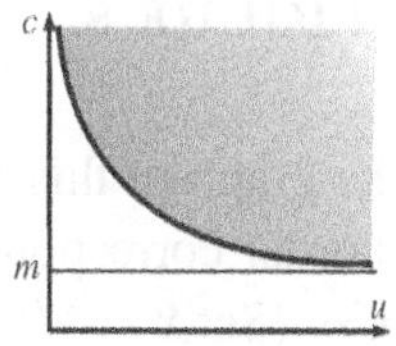

FIGURE 7.

The curve (2.4.6) is a hyperbola. Its graph is the boundary of the area depicted in Fig.7. For all points (u, c) above the curve (2.4.6) in Fig.7, the ruin probability is less than α.

We see also that $c \to m$ as $u \to \infty$. This has a natural interpretation. If the initial surplus is large, then the security loading may be small, i.e., close to zero. This means that the premium per claim may be close to the expected loss per claim. The last quantity is $E\{X\} = m$.

Let, say, $m = 10$ and $\sigma^2 = 4$. To make our formulas nicer, we choose as α not 0.05 or 0.01, but, say, $e^{-4} \approx 0.01832$. Then $s = 4$, which will make calculations simpler.

(The choice of α is rather subjective anyway. If, for example, $\alpha = 0.05$ seems proper, one may choose $\alpha = e^{-3} \approx 0.049787$, which is very close to 0.05.)

For $u = 8$, the premium c should be equal to 11, that is, the relative loading is 10%. If it is too much, but we should keep the ruin probability at the same level, then we should increase the initial surplus. For example, for the 5% loading, $c = 10.5$ and (2.4.6) leads to $u = 16$. □

Two more things are noteworthy. First, the fact that $c \to \infty$ as $u \to 0$ in (2.4.6) should not mislead us. This does not reflect the real situation but rather the circumstance that we are dealing with an estimate which is not accurate for small u. Certainly, even if the initial surplus equals zero, for any $c > m$, there will be no ruin with some positive probability. If the premium c is large (but not infinitely large), the ruin probability should be small. So, if $u = 0$, we do not need c to be infinitely large for the ruin probability to be smaller than the level α.

On the other hand, for large u, (2.4.6) gives a good approximation not only for the normal case but for *practically arbitrary* X's.

Indeed, a large u leads to a small γ; see (2.4.3). As we saw in Section 2.2.2, in this case, one can use the approximation (2.2.16). This leads to (2.4.6) as an approximation. The larger u is, the smaller γ is, and hence the better the accuracy of this approximation. Certainly, this argument is heuristic, and for rigorous estimation we should quantitatively evaluate the accuracy of the approximation.

Now, let us consider an example dealing with continuous time.

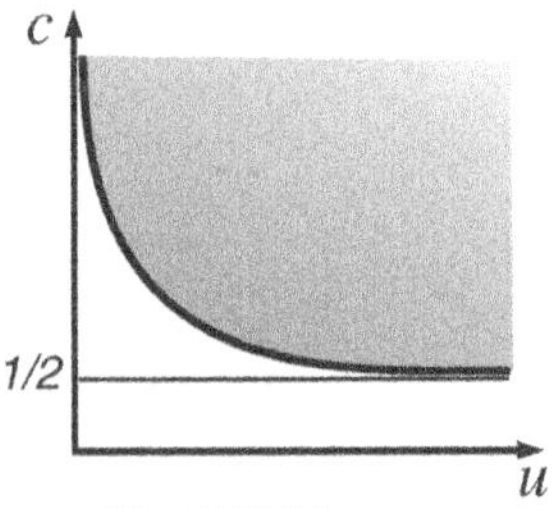

FIGURE 8.

EXAMPLE 3. Consider the compound Poisson process with X's uniformly distributed on $[0,1]$. To make the example illustrative, set $\alpha = e^{-4} \approx 0.018$ (see reasoning on this point in Example 2). Then $s = 4$. As we know, to evaluate the adjustment coefficient, we do not need any information about the intensity of the Poisson process.

In our case $M_X(z) = (e^z - 1)/z$, and (2.4.5) amounts to

$$c = \frac{u^2}{16}\left(\exp\left\{\frac{4}{u}\right\} - 1 - \frac{4}{u}\right).$$

The graph of this function is the border of the area depicted in Fig.8. All points above the border correspond to the ruin probabilities that are less than α. Note that $c \to 1/2$ as $u \to \infty$. (Set $x = 4/u$. Then $c = \frac{1}{x^2}(e^x - 1 - x) \to \frac{1}{2}$ as $x \to 0$. The last fact may be proved by L'Hôpital's rule.)

As we already noted in Example 2, the convergence mentioned is not surprising. If the initial surplus is large, the premium per claim may be close to the expected loss per claim. In our case, this is $E\{X\} = 1/2$.

As to the fact that $c \to \infty$ as $u \to 0$, the corresponding remark from Example 2 applies to this case too. □

► In conclusion, we mention an approximation concerning the case of the homogeneous compound Poisson process and a small security loading. Let $c = (1+\theta)m$ as in Section 2.2.4, and $m_2 = E\{X_j^2\}$. Then

$$\psi(u) - \frac{1}{1+\theta}\exp\left\{-\frac{2\theta m u}{(1+\theta)m_2}\right\} \to 0 \text{ as } \theta \to 0, \text{ uniformly in } u > 0. \tag{2.4.7}$$

This approximation is obtained in [69], and the accuracy of the approximation—in [69] and [12, Section 5.3]. Some refinements of (2.4.7) may also be found in [12, Section 5.3].

The main point in (2.4.7) is that the approximation is true for all u, including u depending on θ. As above, let $s = \ln(1/\alpha)$, and

$$u = s \cdot \frac{m_2}{2m} \cdot \frac{1+\theta}{\theta} \sim s \cdot \frac{m_2}{2m} \cdot \frac{1}{\theta}, \tag{2.4.8}$$

for small θ. Then, by (2.4.7), for small θ,

$$\psi(u) \approx \frac{1}{1+\theta}\alpha \approx \alpha.$$

Hence, (2.4.8) represents the trade-off between u and θ for small θ and the security level α. Certainly, since we consider small θ, the initial surplus u is large.

Note also that (2.4.8) does not contradict what we got before. Since $M_X(z) = 1 + mz + m_2(z^2/2) + o(z^2)$ for small z [see (**0**.4.5.2)], from (2.4.5) we get that for large u

$$c = (1+\theta)m = m + s \cdot \frac{m_2}{2u} + o\left(\frac{1}{u}\right).$$

From the last relation it follows that

$$\theta = s\frac{m_2}{2um} + o\left(\frac{1}{u}\right),$$

which is consistent with (2.4.8) for small θ or (which is equivalent) for large u. ◄

2.5 Three cases where the ruin probability may be computed precisely

1,2,3

As was repeatedly noted, so far we have dealt only with estimates of the ruin probability. Now we will consider cases when the denominator in the main result (2.1.10) may be computed explicitly.

2.5.1 The case with an exponentially distributed claim size

Let us consider the model $R_t = u + ct - S_{(t)}$; time may be discrete or continuous. Suppose the claims X's are exponentially distributed, $E\{X\} = m$. Denote by τ the ruin time. The value of the process R_t at time τ is the r.v. R_τ.

At the moment τ *(if it occurs)*, the process R_t makes a jump down and crosses zero level. So, $R_\tau < 0$. Certainly, the jump may occur only if at this moment a sufficiently large claim arrived. Denote by $R_{\tau-0}$ the value of the process before the jump, and by $\widetilde{X}$ the size of the jump, which is equal to the claim at the moment τ. The tilde indicates that this is not a usual claim but the claim at the moment of ruin. Then $R_\tau = R_{\tau-0} - \widetilde{X} = -(\widetilde{X} - R_{\tau-0})$, where $\widetilde{X} - R_{\tau-0}$ is the overshoot. See also Fig.9. Denote the overshoot $\widetilde{X} - R_{\tau-0}$ by D. Clearly, $D = |R_\tau|$; see again Fig.9.

Suppose the r.v. $R_{\tau-0}$ assumed a value r. Of course, the distribution of the claim $\widetilde{X}$ at the moment τ depends on r, since for ruin to occur the claim $\widetilde{X}$ must be larger than r. So, the distribution of $\widetilde{X}$ is equal to the conditional distribution of the exponential r.v. X, given that $X > r$. However, in view of the memoryless property, the overshoot $D = \widetilde{X} - r$ does not depend on r, and has the same distribution as each claim X. That is, D has the exponential distribution with the same parameter $a = 1/m$, where as usual $m = E\{X_i\}$. (See also Section 2.2.1.1.)

On the other hand, since $D = |R_\tau|$, the denominator in (2.1.10) is

$$E\{\exp\{-\gamma R_\tau\} \,|\, \tau < \infty\} = E\{\exp\{\gamma |R_\tau|\} \,|\, \tau < \infty\} = E\{\exp\{\gamma D\} \,|\, \tau < \infty\}.$$

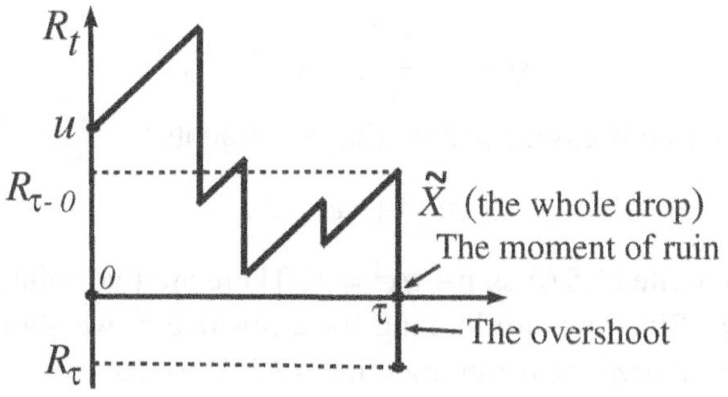

FIGURE 9.

As we saw, the r.v. D does not depend on τ, and hence

$$E\{\exp\{\gamma D\} \,|\, \tau < \infty\} = E\{\exp\{\gamma D\}\} = M_D(\gamma) = \frac{1}{1-m\gamma}.$$

Thus, $E\{\exp\{-\gamma R_\tau\} \,|\, \tau < \infty\} = 1/(1-m\gamma)$, and by (2.1.10).

$$\psi(u) = (1-m\gamma)\exp\{-\gamma u\}. \tag{2.5.1}$$

In the discrete time case, in accordance with (2.2.11), γ is the solution to the equation

$$e^{-cz} = 1 - mz. \tag{2.5.2}$$

An explicit formula for the solution does not exist, but it is easy to solve such an equation numerically (see also Exercise 17).

For the compound Poisson process, (2.2.25) implies

$$\gamma = \frac{\theta}{m(1+\theta)}, \tag{2.5.3}$$

which together with (2.5.1) gives an explicit formula:

$$\psi(u) = \frac{1}{1+\theta}\exp\left\{-\frac{\theta u}{m(1+\theta)}\right\}. \tag{2.5.4}$$

Note also that above reasoning makes sense rather for the continuous time case. In the discrete time situation, we usually view X_i as an *aggregate* claim in the ith period, that is, X_i is the sum of r.v.'s, and it would be non-realistic to assume that the distribution of X_i is exponential.

2.5.2 The case of the simple random walk

It is useful to check that Theorem 2 in this case leads to the result of Section **4**.4.3.2. In the model of Section **4**.4.3.2, we deal not with claims but with increments of the surplus process. Accordingly, the claim surplus process in our case is $W_t = Y_1 + ... + Y_t$, where the increment of the total claim surplus at the moment i is $Y_i = -1$ or 1, with probabilities p and $q = 1-p$, respectively. (Y_i indicates a loss, so when $Y_i = -1$, the surplus process moves up.) We assume $p > 1/2$.

Let u be an integer. Since the process R_t each time jumps up or down exactly by one unit, at the moment of ruin (if any), $R_\tau = -1$, and hence $E\{\exp\{-\gamma R_\tau\} \,|\, \tau < \infty\} = \exp\{\gamma\}$.

Then, by (2.1.10),

$$\psi(u) = \frac{e^{-\gamma u}}{e^{\gamma}} = e^{-\gamma(u+1)}. \tag{2.5.5}$$

On the other hand, in this case, equation (2.2.8) amounts to

$$pe^{-z} + qe^{z} = 1. \tag{2.5.6}$$

Setting $e^z = x$, we rewrite (2.5.6) as $p + qx^2 = x$. There are two solutions to this equation: $x = 1$ and $x = p/q$. Since we are looking for a positive z, we should choose the latter solution. Hence $\gamma = \ln(p/q)$. Substituting it into (2.5.5) we have

$$\psi(u) = (q/p)^{u+1} = r^{u+1}, \tag{2.5.7}$$

where, as in Section **4**.4.3.2, $r = (1-p)/p$.

The difference between (2.5.7) and the formula $\psi(u) = r^u$ in Section **4**.4.3.2 is explained by the fact that in Section **4**.4.3.2.2 we defined the ruin time as the moment when the process first reaches zero level, while here the ruin time is the moment when the process takes on a negative value. In the framework of Section **4**.4.3.2, the latter definition corresponds to the replacement of the initial capital u by $u+1$, which leads to (2.5.5).

Route 1 ⇒ page 377

2.5.3 The case of Brownian motion 2,3

As in Section **5**.2.4.4, let $R_t = u + \mu t + \sigma w_t$, where w_t is a standard Brownian motion, and $\mu > 0$. Since R_t is now a continuous process, it will not overshoot zero level, but first will hit (touch) it. In this case, it is reasonable to redefine the notion of ruin setting

$$\tau = \min\{t > 0 : R_t = 0\}. \tag{2.5.8}$$

All results above continue to be true in this case.

► As follows from the remark at the end of Section **5**.1.1.3, once the process reaches zero level at time τ, in any arbitrary small interval $(\tau, \tau+\delta]$ the process will take negative values infinitely many times, rapidly oscillating around zero for a while. So, the previous definition $\min\{t > 0 : R_t < 0\}$ is not proper: the minimum does not exist, and we should write $\inf\{t > 0 : R_t < 0\}$. In view of the continuity of the process, the latter definition is equivalent to (2.5.8). ◄

Thus, in our case, $R_\tau = 0$. Hence, the denominator $E\{\exp\{-\gamma R_\tau\} \,|\, \tau < \infty\} = 1$, and $\psi(u) = \exp\{-\gamma u\}$.

To find γ, we first realize that in our case $W_t = -\mu t - \sigma w_t$, and hence, $W_\Delta = -\mu|\Delta| - \sigma w_\Delta$, where $|\Delta|$ is the length of the interval Δ. Note that $-\sigma w_\Delta$ is a normal r.v. with zero mean and variance $\sigma^2|\Delta|$. (Multiplication by -1 does not change the distribution of a symmetric r.v.) Then $M_\Delta(z) = \exp\{-\mu|\Delta|z + \sigma^2|\Delta|z^2/2\} = \exp\{-|\Delta|z(\mu - \sigma^2 z/2)\}$, and the equation (2.1.2) is equivalent to the equation $\gamma(\mu - \sigma^2\gamma/2) = 0$. A unique positive solution is

$$\gamma = 2\mu/\sigma^2.$$

Consequently,

$$\psi(u) = \exp\{-2\mu u/\sigma^2\},$$

which coincides with (**5**.2.4.9).

The precise formula for $\psi_T(u) = P(\tau \le T)$ was obtained in Section **5**.2.4.5.

2.6 The martingale approach and a generalization of Theorem 2

We consider now the ruin problem in the martingale framework, presupposing that the reader is familiar with the basic notions of Section **5**.2. To the author's knowledge, the first use of martingales in Actuarial Modeling is due to H. Gerber (see, e.g., [41], [42]) and F. DeVylder [34].

The main goal of this section is not to prove Theorem 2, although we will do that. Rather, it is to show that this theorem is not a tricky analytical fact but a direct and almost obvious corollary from the martingale stopping property.

Assume that all processes under consideration are functions of an original process ξ_t as it was defined in Section 5.2.2. As such a process, we can take the surplus process R_t itself, but it is more convenient to define the original process separately.

As in Section 5.2.2, we denote by ξ^t the collection $\{\xi_u;\ 0 \le u \le t\}$, i.e., the whole history of the process until time t.

In the framework of this section, the surplus process R_t is a rather general process. In particular, we will not assume that it is a process with independent increments. However, let us first look at what will happen when this condition holds.

It is convenient to define the independence of increments in terms of ξ^t. More specifically, assume for a while that

$$\text{For any } t \text{ and any interval } \Delta = (t, t+s], \text{ the r.v. } R_\Delta \text{ does not depend on } \xi^t. \tag{2.6.1}$$

(Here, as in Section 5.2.2, $R_\Delta = R_{t+s} - R_t$.)

Since R_t is completely determined by ξ^t, from (2.6.1) it follows that R_Δ does not depend on R_t. Vice versa, if we take R_t as the original process (and we can do that), then property (2.6.1) will follow from the independence of increments of R_t.

As in Section 2.1.3, let $W_t = u - R_t$, the claim surplus process. Note that condition (2.1.6) is equivalent to

$$W_t \xrightarrow{P} -\infty \quad \text{as} \quad t \to \infty. \tag{2.6.2}$$

If property (2.6.1) holds for R_t, it holds for W_t also. Then, for any z, t, s and interval $\Delta = (t, t+s]$,

$$E\{e^{zW_{t+s}} \,|\, \xi^t\} = E\{e^{zW_\Delta} e^{zW_t} \,|\, \xi^t\} = e^{zW_t} E\{e^{zW_\Delta} \,|\, \xi^t\} = e^{zW_t} E\{e^{zW_\Delta}\} = e^{zW_t} M_\Delta(z), \tag{2.6.3}$$

where we denote by $M_\Delta(z)$ the m.g.f. of W_Δ.

Thus, if $M_\Delta(z) = 1$, then $E\{\exp\{zW_{t+s}\} \,|\, \xi^t\} = \exp\{zW_t\}$. As we know, there may be only one positive solution γ (if any) to the equation $M_\Delta(z) = 1$. For γ so defined, let $Y_t = e^{\gamma W_t}$. Then $E\{Y_{t+s} \,|\, \xi^t\} = Y_t$, and hence,

$$\text{The process } Y_t = e^{\gamma W_t} \text{ is a martingale.} \tag{2.6.4}$$

As a matter of fact, (2.6.4) together with (2.6.2) is the only thing we need. So, we may weaken condition (2.6.1), adopting (2.6.4) itself as the original condition. As we saw, if (2.6.1) is true, then (2.6.4) is also true, but certainly the process Y_t may be a martingale while increments of R_t are dependent.

Thus, regarding the claim surplus process W_t, we eventually assume that

(a) $W_0 = 0$;

(b) condition (2.6.2) holds;

(c) there exists a number $\gamma > 0$ for which (2.6.4) is true.

Our next step is to apply *the martingale stopping property* (**5**.2.4.4). If we had had a right to do that, we would have written

$$1=E\{e^0\}=E\{e^{\gamma W_0}\}=E\{Y_0\}=E\{Y_\tau\}=E\{\exp\{\gamma(u-R_\tau)\}\}=e^{\gamma u}E\{\exp\{-\gamma R_\tau\}\}. \quad (2.6.5)$$

Then it would have remained to recall that τ, and hence R_τ, is an improper (or defective) r.v., that is, $P(\tau<\infty)<1$. If ruin does not occur, then we can say that $\tau=\infty$. Thus, we can write—at a somewhat heuristic level – that

$$E\{\exp\{-\gamma R_\tau\}\}=E\{\exp\{-\gamma R_\tau\}\,|\,\tau<\infty\}P(\tau<\infty)+E\{\exp\{-\gamma R_\tau\}\,|\,\tau=\infty\}P(\tau=\infty). \quad (2.6.6)$$

In view of (2.6.2), $R_t=u-W_t\xrightarrow{P}+\infty$ as $t\to\infty$. Hence, if $\tau=\infty$, we can set (again reasoning a bit heuristically) that $R_\tau=\infty$, and hence $E\{\exp\{-\gamma R_\tau\}\,|\,\tau=\infty\}=0$.

Thus, from (2.6.5) and (2.6.6) it follows that

$$1=e^{\gamma u}E\{\exp\{-\gamma R_\tau\}\,|\,\tau<\infty\}P(\tau<\infty),$$

and we come to the basic result (2.1.10):

$$P(\tau<\infty)=\frac{e^{-\gamma u}}{E\{\exp\{-\gamma R_\tau\}\,|\,\tau<\infty\}}. \quad (2.6.7)$$

However, the problem is that since τ is improper and, consequently, is not a stopping time, we cannot apply the martingale stopping property directly. This obstacle, however, may be easily overcome if we apply a sort of truncation and use condition (2.6.2), which we do in the proof below.

So, we state and prove the following theorem.

Theorem 4 *Let the above conditions (a)-(c) hold. Then (2.6.7) is true.*

Proof. Let a fixed $T>0$, and $\tau_T=\min\{T,\tau\}$. Since T is fixed, τ_T is bounded, and Condition 1 of Theorem **5**.6 holds. Applying this theorem, we have

$$\begin{aligned}1&=E\{Y_0\}=E\{Y_{\tau_T}\}=E\{\exp\{\gamma W_{\tau_T}\}\}\\&=E\{\exp\{\gamma W_{\tau_T}\}\,|\,\tau\le T\}P(\tau\le T)+E\{\exp\{\gamma W_{\tau_T}\}\,|\,\tau>T\}P(\tau>T)\\&=E\{\exp\{\gamma W_\tau\}\,|\,\tau\le T\}P(\tau\le T)+E\{\exp\{\gamma W_T\}\,|\,\tau>T\}P(\tau>T)\\&=E\{\exp\{\gamma(u-R_\tau\}\,|\,\tau\le T\}P(\tau\le T)+E\{\exp\{\gamma W_T\}\,|\,\tau>T\}P(\tau>T).\end{aligned} \quad (2.6.8)$$

Let $T\to\infty$. The first term in (2.6.8) is

$$E\{\exp\{\gamma(u-R_\tau)\}\,|\,\tau\le T\}P(\tau\le T)\to e^{\gamma u}E\{\exp\{-\gamma R_\tau\}\,|\,\tau<\infty\}P(\tau<\infty). \quad (2.6.9)$$

It remains to prove that the second term in (2.6.8) vanishes as $T\to\infty$. Note that by the definition of τ, if $T<\tau$, then $R_T\ge 0$, and $W_T=u-R_T\le u$. So, fixing a number $k>0$, we get that the second term in (2.6.8) is equal to

$$\begin{aligned}&E\{\exp\{\gamma W_T\}\,|\,\tau>T\}P(\tau>T)\\&=E\{\exp\{\gamma W_T\}\,|\,\tau>T,\,W_T>-k\}P(W_T>-k\,|\,\tau>T)P(\tau>T)\\&+E\{\exp\{\gamma W_T\}\,|\,\tau>T,\,W_T\le -k\}P(W_T\le -k\,|\,\tau>T)P(\tau>T)\\&\le\exp\{\gamma u\}P(W_T>-k,\,\tau>T)+\exp\{-\gamma k\}P(W_T\le -k,\,\tau>T)\\&\le\exp\{\gamma u\}P(W_T>-k)+\exp\{-\gamma k\}.\end{aligned} \quad (2.6.10)$$

Let $T \to \infty$. Condition (2.6.2), by definition, means that $P(W_T > -k) \to 0$, as $T \to \infty$, for any fixed positive $k > 0$. Hence, from (2.6.10) it follows that

$$\lim_{T\to\infty} E\{\exp\{\gamma W_T\}\,|\,\tau > T\}P(\tau > T) \le e^{-\gamma k}$$

for any k. The term on the left does not depend on k, so we can let $k \to \infty$, which for $\gamma > 0$ implies that the limit on the left is zero. ■

2.7 The renewal approach

An essentially different approach of this subsection proves to be also quite efficient.

2.7.1 The first surplus below the initial level

In this subsection, we consider the compound Poisson process case, assuming that $R_t = u + ct - S_{(t)}$, the process $S_{(t)}$ is defined in (2.2.21), the process N_t is a homogeneous Poisson process with a constant intensity λ, and $c = (1+\theta)m\lambda$, where $m = E\{X_i\} > 0$. When considering a separate X_i, we will omit the index i.

Since it does not make sense to consider claims equal to zero, we assume also that

$$P(X > 0) = 1. \tag{2.7.1}$$

It will be convenient for us to indicate the dependence of the ruin time on u explicitly, so we set $\tau_u = \min\{t : R_t < 0\,|\,R_0 = u\}$.

Let q be the probability that the process will ever fall below the initial level u. It may happen if and only if the process $ct - S_{(t)}$ falls below zero level. Hence, q does not depend on u, and equals $\psi(0)$, the ruin probability in the case when the initial surplus equals zero.

For the same reason, the size of the drop below the level u at the moment when the process first crosses this level does not depend on u either. The distribution of the size of the drop mentioned coincides with the distribution of $|R_{\tau_0}|$, the absolute value of the deficit of the surplus at the ruin time if the process starts from zero.

It proves that q and the distribution of $|R_{\tau_0}|$ may be represented in a simple form. Let $F(x)$ be the d.f. of X's.

Theorem 5 *For any $x \ge 0$,*

$$P(\tau_0 < \infty,\ |R_{\tau_0}| \le x) = \frac{1}{(1+\theta)m}\int_0^x (1 - F(y))dy. \tag{2.7.2}$$

We prove (2.7.2) in Section 2.7.4, and now we discuss several interesting corollaries from this theorem.

First, setting $x = \infty$ in (2.7.2), we get that

$$P(\tau_0 < \infty) = \psi(0) = \frac{1}{(1+\theta)m}\int_0^\infty (1 - F(y))dy = \frac{1}{1+\theta}, \tag{2.7.3}$$

by virtue of the formula (**0**.2.2.2). Recall also that $q = \psi(0)$.

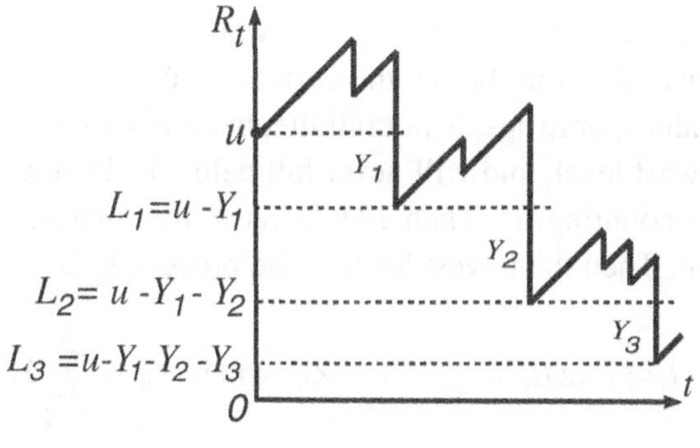

FIGURE 10. A realization of a renewal process. Y's are the drops below the corresponding levels; in particular, Y_1 is the drop below the level u.

The formula (2.7.3) is very interesting—the ruin probability for $u=0$ depends only on the security loading and does not depend on the distribution of X's at all.

Now note that, since τ_0 is an improper r.v., i.e., $P(\tau_0<\infty)=\psi(0)<1$, the overshoot $|R_{\tau_0}|$ (i.e., the deficit at the moment of ruin) is also improper: it is defined only in the case $\tau_0<\infty$. Let us consider the conditional distribution of $|R_{\tau_0}|$ given $\tau_0<\infty$, more precisely the conditional d.f. $F_1(x)=P(|R_{\tau_0}|\leq x\,|\,\tau_0<\infty)$.

From (2.7.2)-(2.7.3) it follows that

$$F_1(x)=\frac{P(\tau_0<\infty,\ |R_{\tau_0}|\leq x)}{P(\tau_0<\infty)}=\frac{1}{m}\int_0^x(1-F(y))dy.$$

The conditional density equals

$$f_1(x)=F_1'(x)=\frac{1}{m}(1-F(x)). \tag{2.7.4}$$

2.7.2 The renewal approximation

Let us return to R_t. Starting from level u, with probability $q=1/(1+\theta)$, the process at some time will drop below the initial level u. If it happens, the size of the drop below the level u will be a r.v. Y_1 having the above d.f. $F_1(x)$.

The process N_t is homogeneous, and the time between consecutive drops have the lack of memory property. Consequently, after the drop mentioned, the process will start to run as if it is at the beginning, with the exception that now the starting position is $L_1=u-Y_1$. See Fig.10.

Since the next drop cannot occur immediately after the first drop, starting from L_1, the process will be moving up for a while. Hence, L_1 is a local minimum of the process.

The process will fall below the new level L_1 with the same probability q. If it happens, the size of the new drop below the level L_1 will be a r.v. Y_2 which will not depend on Y_1, and will have the same distribution F_1. The value of the process at the moment when it falls below the level L_1, is $L_2=u-Y_1-Y_2$. See again Fig.10.

Continuing to reason in the same fashion, we define the r.v. Y_n as the size of the nth drop below the previous $(n-1)$th level, and the r.v. $L_n=u-Y_1-...-Y_n$. The r.v. L_n is the nth local minimum of R_t.

The process L_n is called a *renewal process*, and values of L_n —*record values*; see Fig.10. Since the probability of falling below the current value, that is, q, is less than one, the sequence of record values, or drops, is not infinite, but will run up to the moment when the process leaves the lowest level, and will never fall below it. Denote by K the total number of record values, not counting u. Then $P(K=n)=pq^n$, where $p=1-q$, so K has a geometric distribution. Then the lowest level of the process R_t is

$$L=\min_t R_t = L_K = u - Z_K, \text{ where } Z_K=\sum_{k=1}^{K} Y_k,$$

and the r.v.'s Y_k are independent and have the common d.f. F_1.

If $K=0$, that is, the process never falls below the initial level, then we set $Z_K=0$, and $L=u$. It occurs with probability $p=1-q$.

It is easy to understand that the no-ruin probability

$$\phi(u)=P(L\geq 0)=P(Z_K\leq u), \tag{2.7.5}$$

and we have come to a familiar object: the distribution of the sum of a random number of independent r.v.'s.

The ruin probability $\psi(u)=1-\phi(u)$.

In accordance with (**3**.3.1.2) and (2.7.3),

$$\phi(u)=p\sum_{n=0}^{\infty} q^n F_1^{*n}(u), \text{ where } p=\frac{\theta}{1+\theta} \text{ and } q=\frac{1}{1+\theta}. \tag{2.7.6}$$

We can apply to (2.7.6) methods of Section **3**.3.1.

EXAMPLE 1. Let X's take on values 1 or 2 with probabilities 0.75 and 0.25, respectively. The reader is invited to verify that in this case, $m=1.25$, and the density

$$f_1(x)=\frac{1}{m}(1-F(x))=\begin{cases} 0.8 \text{ if } x\in[0,1], \\ 0.2 \text{ if } x\in(1,2], \end{cases}$$

and equals 0 otherwise. This means that $f_1(x)=0.8g_1(x)+0.2g_2(x)$, where $g_1(x)$ and $g_2(x)$ are the densities of the uniform distributions on $[0,1]$ and $[1,2]$, respectively. In other words, f_1 is a mixture of uniform distributions; see also Exercise 18.

Let $\theta=0.2$. Then $q=\frac{1}{1+\theta}=\frac{5}{6}$. For an integer k, the part of (2.7.6) corresponding to summation $\sum_{k+1}^{\infty}$ does not exceed q^{k+1}; see Section **3**.3.1.1 for detail. For example, $q^{26}\leq 0.009$, and if we are satisfied with such an accuracy, we can restrict ourselves to $\sum_0^{25}$.

Numerical estimation of such a sum is not a very complicated problem. Denoting by G_1, G_2 the corresponding uniform d.f.'s, for the convolution F_1^{*n} we can write

$$F_1^{*n}=(0.8G_1+0.2G_2)^{*n}=\sum_{k=0}^{n}\binom{n}{k}(0.8)^k(0.2)^{n-k}G_1^{*k}*G_2^{*(n-k)};$$

see (**2**.2.1.1). There exist explicit, though cumbersome, formulas for convolutions of uniform distributions, so with good software one should not have a problem in calculations. □

In accordance with (2.7.5), $\phi(u)$ is the d.f. of Z_K. Next, we compute the m.g.f. of Z_K or, equivalently, that of its d.f. $\phi(u)$. By (**3**.1.5) and/or (**3**.3.1.16),

$$M_\phi(z) = \int_0^\infty e^{zu} d\phi(u) = \frac{p}{1 - qM_Y(z)}, \tag{2.7.7}$$

where $M_Y(z)$ is the m.g.f. of the r.v.'s Y_i. Using (2.7.4) and integrating by parts, we get that for all $z > 0$,

$$M_Y(z) = \frac{1}{m}\int_0^\infty e^{zx}(1-F(x))dx = -\frac{1}{mz}(1-F(0)) + \frac{1}{mz}\int_0^\infty e^{zx}dF(x) = \frac{1}{mz}\left[M_X(z) - 1\right],$$

because, in view of (2.7.1), $F(0) = 0$. Inserting this into (2.7.7), and substituting values for p and q, it is easy to calculate that

$$M_\phi(z) = \frac{\theta m z}{1 + (1+\theta)mz - M_X(z)}. \tag{2.7.8}$$

If for a particular X, the m.g.f. (2.7.8) is familiar for us, we can determine $\phi(u)$. For some cases, it is convenient to rewrite (2.7.8) as

$$M_\phi(z) = \frac{\theta}{1+\theta} + \frac{1}{1+\theta} \cdot \frac{\theta(M_X(z) - 1)}{1 + (1+\theta)mz - M_X(z)}. \tag{2.7.9}$$

The last formula reflects the following circumstance. With probability p the r.v. $K = 0$, and since in this case $Z_K = 0$, the d.f. of Z_K—that is, $\phi(u)$—makes a jump of p at zero. In view of (2.7.6), this may be represented in the following form:

$$\phi(u) = p + p\sum_{n=1}^{\infty} q^n F_1^{*n}(u). \tag{2.7.10}$$

The two terms in (2.7.9) correspond to the respective two terms in (2.7.10).

EXAMPLE 2. Let $F(x)$ be a mixture of exponential distributions, say, the tail

$$\overline{F}(x) = 1 - F(x) = \frac{1}{2}e^{-x} + \frac{1}{2}e^{-x/3}.$$

Then $m = \frac{1}{2}\cdot 1 + \frac{1}{2}\cdot 3 = 2$, and for $z < 1/2$,

$$M_X(z) = \frac{1}{2}\cdot\frac{1}{1-z} + \frac{1}{2}\cdot\frac{1}{1-3z}.$$

To make calculations more illustrative, set $\theta = 0.5$ in our example, realizing that it is not very realistic. Substituting it into (2.7.9), the reader can verify that in this case,

$$M_\phi(z) = \frac{1}{3} + \frac{2}{3}\cdot\frac{2-3z}{2-18z+18z^2}. \tag{2.7.11}$$

The equation $2 - 18z + 18z^2 = 0$ has two solutions: $z_1 = 0.5 + \sqrt{5}/6 \approx 0.87$, and $z_2 = 1 - z_1 \approx 0.13$.

Using the method of partial fractions we write

$$\frac{2-3z}{2-18z+18z^2} = \frac{2-3z}{18(z_1-z)(z_2-z)} = \frac{1}{18}\left(\frac{c_1}{z_1-z}+\frac{c_2}{z_2-z}\right), \tag{2.7.12}$$

where c_1, c_2 are constants that we should find. Putting the r.-h.s. of (2.7.12) into the common denominator, we get that $c_1+c_2=3$, $c_1z_2+c_2z_1=2$.

We will write all solutions up to the second digit. Solving the equations for c_1 and c_2, we readily get that $c_1=0.83$, $c_2=2.17$. Then

$$\frac{2-3z}{2-18z+18z^2} = \left(\frac{0.05}{1-z/z_1}+\frac{0.95}{1-z/z_2}\right), \tag{2.7.13}$$

where the denominators are precise but the coefficients in the numerators are computed up to the second digit. Together with (2.7.12), we have with the same accuracy that

$$M_\phi(z) = \frac{1}{3}+\frac{2}{3}\left(\frac{0.05}{1-z/z_1}+\frac{0.95}{1-z/z_2}\right).$$

The last term is a mixture of exponential m.g.f.'s. Consequently,

$$\phi(u) = \frac{1}{3}+\frac{2}{3}\left(0.05F_{z_1}(u)+0.95F_{z_2}(u)\right),$$

where F_z stands for the exponential d.f. with parameter z. Eventually,

$$\psi(u) = 1-\phi(u) = \frac{2}{3}\left(0.05\overline{F}_{z_1}(u)+0.95\overline{F}_{z_2}(u)\right)$$
$$= \frac{2}{3}\left(0.05\exp\{-z_1u\}+0.95\exp\{-z_2u\}\right) \approx 0.03\exp\{-0.87u\}+0.63\exp\{-0.13u\}. \ \square$$

Route 2 ⇒ *page 377*

2.7.3 The Cramér-Lundberg approximation

In conclusion, we present without a proof one more celebrated result of Risk Theory.

Theorem 6 *Let $\gamma>0$ be the adjustment coefficient satisfying (2.2.22). Then,*

$$\psi(u) \sim Ce^{-\gamma u} \quad \text{as } u\to\infty, \tag{2.7.14}$$

where

$$C = \frac{m\theta}{M_X'(\gamma)-m(1+\theta)}. \tag{2.7.15}$$

Proofs may be found, e.g., in [10] or [50].

To clarify the significance of the last formula, assume that X is exponential. Then $M_X(z)=1/(1-mz)$, and $M_X'(z)=m/(1-mz)^2$. By (2.5.3), $\gamma=\theta/[m(1+\theta)]$. Substituting it into (2.7.15), we readily get $C=1/(1+\theta)$, which is consistent with the precise formula (2.5.4).

Another example is given in Exercise 23.

The reader may find further interesting approximations for the ruin probability and ruin time as well as further references, e.g., in [12] and [44].

2.7.4 Proof of Theorem 5 from Section 2.7.1

Usually, this theorem is proved with the use of differential equations. Below, we mainly follow a different proof from [10] by S. Asmussen. In part, we do it for diversity, but also because the latter proof is direct and illustrative.

So, we consider the case when $R_0 = 0$. Set $\tau = \tau_0$, and for a set A from the real line denote by $I_A(x)$ the indicator A—that is, $I_A(x) = 1$ if $x \in A$ and $I_A(x) = 0$ otherwise. Let

$$\eta_A = \int_0^{\tau} I_A(R_t)dt. \tag{2.7.16}$$

Since $I_A(R_t) = 1$ or 0, depending whether R_t got into A or not, the r.v. η_A is equal to the amount of time the process R_t spent in A before the time moment τ, that is, before ruin. Our proof is based on

Lemma 7 *For any bounded set A,*

$$E\{\eta_A\} = \frac{1}{c}|A|,$$

where $|A|$ is the length of A (and $c = (1+\theta)m\lambda$, the premium rate).

We will prove this in the end of the section. From Lemma 7 and (2.7.16) it follows that

$$|A| = cE\{\eta_A\} = cE\left\{\int_0^{\tau} I_A(R_t)dt\right\}. \tag{2.7.17}$$

Next, we show that (2.7.17) implies that for any bounded function $g(y)$ defined and integrable on $[0,\infty)$,

$$\int_0^{\infty} g(v)dv = cE\left\{\int_0^{\tau} g(R_t)dt\right\}. \tag{2.7.18}$$

If $g(v) = I_A(v)$, then (2.7.18) coincides with (2.7.17). Consider a piecewise constant function

$$g(v) = \sum_k g_k I_{A_k}(v), \tag{2.7.19}$$

where $g_1, g_2, \ldots$ are numbers, and $A_1, A_2, \ldots$ are disjoint sets. The function $g(v)$ takes on the constant value g_k if $v \in A_k$. Since sets A_k are disjoint,

$$\int_0^{\infty} g(v)dv = \sum_k g_k|A_k|. \tag{2.7.20}$$

By (2.7.17),

$$\int_0^{\infty} g(v)dv = \sum_k g_k|A_k| = \sum_k g_k cE\left\{\int_0^{\tau} I_{A_k}(R_t)dt\right\}$$
$$= cE\left\{\int_0^{\tau} \sum_k g_k I_{A_k}(R_t)dt\right\} = cE\left\{\int_0^{\tau} g(R_t)dt\right\},$$

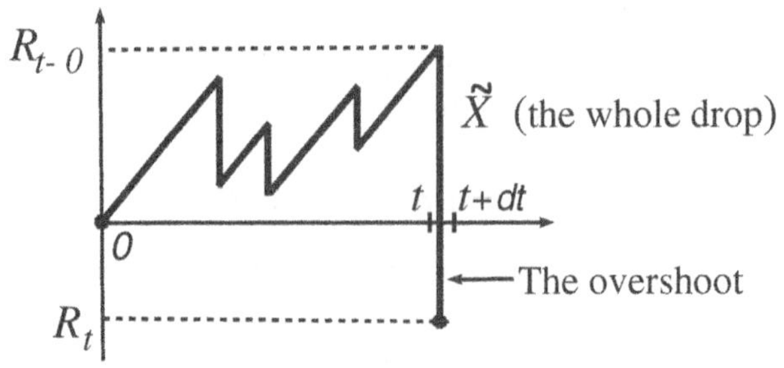

FIGURE 11.

which proves (2.7.18) for any function of the type (2.7.19). Since any bounded function may be approximated with any desired accuracy by a piecewise constant function, (2.7.18) is true for any bounded $g(y)$.

Having (2.7.18), we can turn to the direct proof of Theorem 5. Our reasoning is close to what we already did in Section 2.5.1.

The process R_t jumps down during an infinitesimally small interval $[t, t+dt]$ only if N_t jumps up (a claim arrives). In accordance with (**4**.2.2.1), the probability that this happens equals λdt. Denote by R_{t-0} the value of the process before this jump, and by $\widetilde{X}$ the size of the jump (that is, the claim). We omit an index in $\widetilde{X}$. Since dt is infinitesimally small, we may identify the value of the process after the jump with R_t, so $R_t = R_{t-0} - \widetilde{X}$; see also Fig.11.

Consider the event $\mathcal{E}_{tx}(dt)$ consisting in the following:

(i) During an interval $(t, t+dt]$ ruin occurred.

(ii) It occurred at the first time, and hence $t < \tau$ and $R_{t-0} > 0$.

(iii) The overshoot $|R_t|$ has exceeded some level $x > 0$.

Then $P(\mathcal{E}_{tx}(dt)) = P(t < \tau, \widetilde{X} > R_{t-0} + x)\lambda dt$.

Let $\mathbf{I}(\mathcal{E})$ stand for the indicator of an event $\mathcal{E}$, that is, $\mathbf{I}(\mathcal{E}) = 1$ if $\mathcal{E}$ occurs, and $= 0$ otherwise. (In the function $I_A(x)$ above, A is a set from the real line, while $\mathcal{E}$ is an event in the original space Ω of elementary outcomes.)

Using the formula for total expectation (**0**.7.2.1), we can write that $P(\mathcal{E}_{tx}(dt)) = E\{\mathbf{I}(t \leq \tau)\mathbf{I}(\widetilde{X} > R_{t-0}+x)\}\lambda dt = E\left\{\mathbf{I}(t \leq \tau)E\{\mathbf{I}(\widetilde{X} > R_{t-0}+x) \,|\, R_{t-0}, \mathbf{I}(t \leq \tau)\}\right\}\lambda dt = E\{\mathbf{I}(t \leq \tau)$ $P\left(\widetilde{X} > R_{t-0}+x \,|\, R_{t-0}, \mathbf{I}(t \leq \tau)\right)\Big\}\lambda dt$. Given R_{t-0} and $t \leq \tau$, the conditional probability $P\left(\widetilde{X} > R_{t-0}+x \,|\, R_{t-0}, \mathbf{I}(t \leq \tau)\right)$ is the probability that the amount of a claim will be larger than $R_{t-0}+x$. Then, setting $\overline{F}(x) = P(X_i > x)$, where X_i is a claim, we have

$$P(\mathcal{E}_{tx}(dt)) = E\left\{\mathbf{I}(t \leq \tau)\overline{F}(R_{t-0}+x)\right\}\lambda dt. \tag{2.7.21}$$

For a fixed t, the probability that a jump occurs at time t is zero. Consequently, for a fixed t, the distributions of the r.v.'s R_{t-0} and R_t are the same. Then we can replace R_{t-0} by R_t in the right member of (2.7.21). Thus,

$$P(\mathcal{E}_{tx}(dt)) = E\{\overline{F}(R_t+x)\mathbf{I}(t \leq \tau)\}\lambda dt.$$

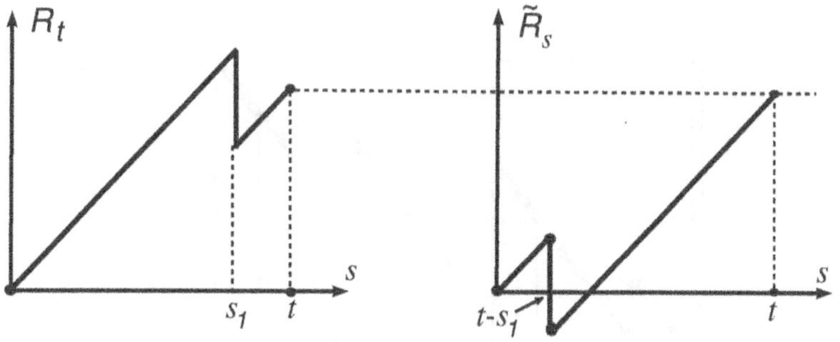

FIGURE 12.

Summing up the probabilities $P(\mathcal{E}_{tx}(dt))$, or more precisely, integrating in t, we have

$$P(\tau < \infty, |R_\tau| > x) = \int_0^\infty P(\mathcal{E}_t(dt)) = \lambda \int_0^\infty E\{\overline{F}(R_t + x)\mathbf{I}(t \le \tau)\}dt$$
$$= \lambda E\left\{\int_0^\infty \overline{F}(R_t + x)\mathbf{I}(t \le \tau)dt\right\} = \lambda E\left\{\int_0^\tau \overline{F}(R_t + x)dt\right\}.$$

Consecutively applying (2.7.18), the fact that $c = (1+\theta)m\lambda$, and the variable change $y = x + v$, we get that

$$P(\tau < \infty, |R_\tau| > x) = \frac{\lambda}{c}\int_0^\infty \overline{F}(x+v)dv = \frac{1}{(1+\theta)m}\int_0^\infty \overline{F}(x+v)dv$$
$$= \frac{1}{(1+\theta)m}\int_x^\infty \overline{F}(y)dy. \tag{2.7.22}$$

Setting $x = 0$, and recalling that $m = \displaystyle\int_0^\infty \overline{F}(y)dy$ by virtue of (**0**.2.2.2), we write

$$P(\tau < \infty) = P(\tau < \infty, |R_\tau| > 0) = \frac{1}{(1+\theta)m}\int_0^\infty \overline{F}(v)dv = \frac{1}{1+\theta}. \tag{2.7.23}$$

To get (2.7.2), it remains to subtract (2.7.22) from (2.7.23).

To complete the proof, we should provide

Proof of Lemma *7*. Let us fix, for a while, $t > 0$ and consider for $s \in [0,t]$ the process $\widetilde{R}_s = R_t - R_{t-s}$. The process $\widetilde{R}_s$ may be interpreted as R_t in reversed time. Note that $\widetilde{R}_0 = 0$, and $\widetilde{R}_t = R_t$ because $R_0 = 0$. The process $\widetilde{R}_s$ moves up linearly with the same slope c, and drops down with the same intensity λ as R_t. The distribution of jumps is also the same as for R_s, and the only difference is that, if R_s has a jump at a point s_1, the corresponding jump of $\widetilde{R}_s$ occurs at time $t - s_1$; see also Fig.12.

However, the last fact has no effect on the distribution of trajectories of $\widetilde{R}_s$, since the intensity of jumps does not depend on time, and jumps are *equally likely to occur at any time*. Thus, the distribution of the process $\widetilde{R}_s$ is the same as for R_s, that is, the probability of any collection of possible trajectories of $\widetilde{R}_s$ equals the same probability for R_s.

On the other hand, $\widetilde{R}_t = R_t = \widetilde{R}_s + R_{t-s}$. So, $\widetilde{R}_t \ge \widetilde{R}_s$ for all $s \le t$ if and only if $R_{t-s} \ge 0$ for all $s \le t$. This is the same as $R_s \ge 0$ for all $s \le t$. Thus, for any set A, the event

$$\{R_t \in A, t < \tau\} = \{R_t \in A, R_s \ge 0 \text{ for all } s \le t\} = \{\widetilde{R}_t \in A,\ \widetilde{R}_t \ge \widetilde{R}_s \text{ for all } s \le t\}.$$

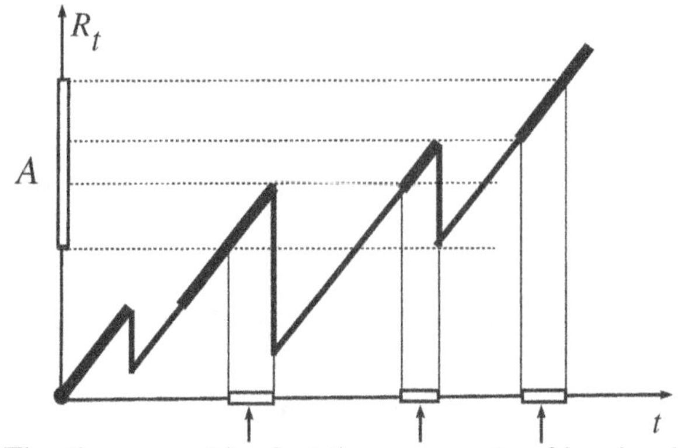
R_t
A
t
The time spent in A at the moments of leadership.

FIGURE 13. "Thick" segments indicate moments of leadership.

(The last step is true because $\widetilde{R}_t = R_t$.) Since the processes R_s and $\widetilde{R}_s$ have the same distribution, this implies, in turn, that

$$P(R_t \in A, t < \tau) = P(\widetilde{R}_t \in A, \widetilde{R}_t \geq \widetilde{R}_s \text{ for all } s \leq t) = P(R_t \in A, R_t \geq R_s \text{ for all } s \leq t). \tag{2.7.24}$$

Note also that $I_A(R_t) = \mathbf{I}(R_t \in A)$, by definition of $\mathbf{I}(\mathcal{E})$. Then, by (2.7.24), for a bounded set A,

$$E\{\eta_A\} = E\left\{\int_0^\tau I_A(R_t)dt\right\} = E\left\{\int_0^\infty \mathbf{I}(R_t \in A)\mathbf{I}(t \leq \tau)dt\right\} = E\left\{\int_0^\infty \mathbf{I}(R_t \in A, t \leq \tau)dt\right\}$$
$$= \int_0^\infty E\{\mathbf{I}(R_t \in A, t \leq \tau)\}\,dt = \int_0^\infty P(R_t \in A, t \leq \tau)dt$$
$$= \int_0^\infty P(R_t \in A, R_t \geq R_s \forall\, s \leq t)dt = E\left\{\int_0^\infty \mathbf{I}(R_t \in A, R_t \geq R_s \forall\, s \leq t)dt\right\}. \tag{2.7.25}$$

Since A is bounded, there exists M such that $A \subset [0, M]$. Denote the last integral in (2.7.25) by J. This is the total time when R_t is in A, being at the same time the largest value with respect to all previous moments $s \leq t$. We can also call such t's moments of leadership.

If $R_T > M$ at some moment T, then "in the future", for $t > T$, at a leadership moment the value of the process will be larger than M, and hence will not be in A.

Consequently, J is exactly (!) equal to $|A|/c$, the length of A divided by the slope of R_t at points of growth—see Fig.13. Note also that $J \leq (|A|/c) \leq M/c$ in any case. It remains to use the condition $R_t \xrightarrow{P} \infty$. Let $T > 0$. The last expected value in (2.7.25) equals

$$E\{J\} = E\{J \mid R_T > M\}P(R_T > M) + E\{J \mid R_T \leq M\}P(R_T \leq M)$$
$$= (|A|/c)P(R_T > M) + E\{J \mid R_T \leq M\}P(R_T \leq M). \tag{2.7.26}$$

Let $T \to \infty$. The first term in (2.7.26) converges to $|A|/c$, since $P(R_T > M) \to 1$ for any M. The second term does not exceed $(|A|/c)P(R_T \leq M) \to 0$ as $T \to \infty$. ■

1,2,3 2.8 Some recurrent relations and computational aspects

Here, we briefly discuss how to compute ruin probabilities for finite time horizons using recursive methods. The relations we consider are based on the first step analysis. We restrict ourselves to the case $R_t = u + c_t - S_{(t)}$, where u is the initial surplus, $S_{(t)}$ is the loss process, and c_t is the aggregate amount of (positive) cash collected by time t. It will be convenient for us to consider the no-ruin probability $\phi_T(u)$. The ruin probability $\psi_T(u) = 1 - \phi_T(u)$.

We start with a particular problem which requires only common sense.

EXAMPLE 1 ([153, N2][1]). BIB is a new insurer writing homeowners policies. You are given: (a) Initial surplus = \$15; (b) Number of insured homes = 3; (c) Premium per home = \$10; (d) Premiums are paid at the start of each year; (e) Size of each claim = \$40; (f) Claims are paid immediately; (g) There are no expenses; (h) There is no investment income.

Each homeowner files at most one claim per year. The probability that a given homeowner files a claim in year 1 is 20%, and in year 2, it is 10%. Claims are independent. Calculate the probability that BIB has positive surplus at the end of year 2.

The insurer will not have a positive income at the end if there is ruin in the middle of the period, so we are computing the no-ruin probability. At the beginning, the insurer has \$45, and in order to not be ruined in the first stage, there should not be more than one claim of \$40.

If there is no claim in the first period, then the insurer will have \$75 at the beginning of the second period, and to have a positive cash, the insurer must not have more than one claim. If there is one claim in the beginning, the insurer will have just \$35 at the beginning of the second period. In this case, there will be no ruin only if there is no claim in this period.

The number of claims during each period has a binomial distribution, so the no-ruin probability

$$\phi_2(15) = (0.8)^3\left[(0.9)^3 + \binom{3}{1}(0.9)^2(0.1)\right] + \binom{3}{1}(0.8)^2(0.2)\cdot(0.9)^3 = 0.7776. \;\square$$

Now, we present the same logic in a more formal way. First, let time be discrete, and $S_{(t)} = S_t = X_1 + ... + X_t$, where X_j is the size of the jth claim, and X's are i.i.d. r.v.'s. For ruin not to happen during time interval $[0, T]$, the first claim X_1 should not exceed $u + c_1$, and starting from the new level $u + c_1 - X_1$, the process should not take on negative values during time $T - 1$. We can unify both cases: $X_1 \le u + c_1$ and $X_1 > u + c_1$, setting by definition the no-ruin probability $\phi_T(u) = 0$, if $u < 0$. Then, given X_1, the conditional no-ruin probability during the last $T - 1$ periods after the first period is $\phi_{T-1}(u + c_1 - X_1)$. In view of the independence of X's, from this it follows that the no-ruin probability

$$\phi_T(u) = E\{\phi_{T-1}(u + c_1 - X_1)\}. \qquad (2.8.1)$$

[1] Reprinted with permission of the Casualty Actuarial Society.

For $T=\infty$, setting $\phi(u)=\phi_\infty(u)$, we can rewrite (2.8.1) as

$$\phi(u)=E\{\phi(u+c_1-X_1)\}, \tag{2.8.2}$$

which is an equation for $\phi(u)$.

Consider, for example, the discrete case when X_1 take on values $x_1, x_2, ...$ with probabilities $f_1, f_2, ...$, respectively. Then (2.8.1) may be written as

$$\phi_T(u)=\sum_j \phi_{T-1}(u+c_1-x_j)f_j. \tag{2.8.3}$$

It is worth emphasizing that in the last sum, as a matter of fact, terms for which $x_j > u+c_1$, vanish.

For $T=\infty$, we may write (2.8.3) as

$$\phi(u)=\sum_j \phi(u+c_1-x_j)f_j.$$

The reader is invited to make sure on her/his own that when $c_1=1$, and X_1 takes on values 0 or 1, the last equation leads to the classical equation for the ruin probability for the simple random walk; see (**4**.4.3.10).

For X's taking many values and for a finite T, calculations are not so nice as they were in Section **4**.4.3.2.2, and one should use numerical procedures. Here, we consider only simple examples in order to demonstrate the logic of calculations.

EXAMPLE 2. Let the unit of time be a year, and the premium $c=4$ be paid at the beginning of each year. Assume that the losses $X=2,4,10$ with probabilities $f_1=0.5$, $f_2=0.4$, $f_3=0.1$, respectively, are paid at the end of each year. Let $T=2$. By (2.8.3),

$$\begin{aligned}\phi_2(u) &= \phi_1(u+4-2)\frac{1}{2}+\phi_1(u+4-4)\frac{2}{5}+\phi_1(u+4-10)\frac{1}{10}\\ &= \frac{1}{10}\left(5\phi_1(u+2)+4\phi_1(u)+\phi_1(u-6)\right).\end{aligned} \tag{2.8.4}$$

Here, it makes sense to consider only integer u's. We have $\phi_1(u)=1$ for $u=6,7,....$ If $u=0,..,5$, a ruin may happen in one period only if the biggest claim occurs, so $\phi_1(u)=0.9$. Thus,

$$\phi_2(u)=\frac{1}{10}(5\cdot 0.9+4\cdot 0.9+0)=0.81 \quad \text{for } u=0,..,3;$$

$$\phi_2(u)=\frac{1}{10}(5\cdot 1+4\cdot 0.9+0)=0.86 \quad \text{for } u=4,5;$$

$$\phi_2(u)=\frac{1}{10}(5\cdot 1+4\cdot 1+0.9)=0.99 \quad \text{for } u=6,...,11;$$

$$\phi_2(u)=\frac{1}{10}(5\cdot 1+4\cdot 1+1)=1 \quad \text{for } u=12,13,... .$$

EXAMPLE 3. Consider the same problem but assume that the available surplus is invested with a risk free interest r. This means that the cash flow c_1 in (2.8.3) should include the growth of the capital, and $u+c$ above should be replaced by $(u+c)\alpha$, where $\alpha = 1+r$. Then instead of (2.8.4), we should write

$$\phi_2(u) = \frac{1}{10}\left(5\phi_1(\alpha(u+4)-2) + 4\phi_1(\alpha(u+4)-4) + \phi_1(\alpha(u+4)-10)\right),$$

and $\phi_1(u)$ should also be recomputed. To make calculations illustrative set $r = 1/9$. Now we consider all u's, not only integers. The function $g(\alpha) = \alpha(u+4)$ equals 10 for $u = 5$, so $\phi_1(u) = 1$ for $u \geq 5$. For $u < 5$ we have $\phi_1(u) = 0.9$. Note also that $\alpha(u+4)-4 \geq 5$ if $u \geq 4.1$, and $\alpha(u+4)-2 \geq 5$ if $u \geq 2.3$. Thus,

$$\phi_2(u) = \frac{1}{10}(5\cdot 0.9 + 4\cdot 0.9 + 0) = 0.81 \quad \text{for } 0 \leq u < 2.3;$$
$$\phi_2(u) = \frac{1}{10}(5\cdot 1 + 4\cdot 0.9 + 0) = 0.86 \quad \text{for } 2.3 \leq u < 4.1;$$
$$\phi_2(u) = \frac{1}{10}(5\cdot 1 + 4\cdot 1 + 0.9) = 0.99 \quad \text{for } 4.1 \leq u < 5;$$
$$\phi_2(u) = \frac{1}{10}(5\cdot 1 + 4\cdot 1 + 1) = 1 \quad \text{for } u \geq 5. \ \square$$

Nothing prevents us from continuing the recurrence procedure. Applying the same formula (2.8.3) to its interior terms we can write

$$\phi_T(u) = \sum_j \left(\sum_i \phi_{T-2}(u+c_2-x_j-x_i)f_i\right) f_j$$
$$= \sum_j \sum_i \phi_{T-2}(u+c_2-x_j-x_i)f_j f_i, \tag{2.8.5}$$

moving in the same way up to the moment when we will come to $\phi_0(u) = 1$ for $u \geq 0$, and $= 0$ for $u < 0$. Note that c_2 is the cumulative cash by time 2, and again terms inside the sum in (2.8.5) are equal to zero if $u+c_2-x_j-x_i < 0$. Calculations may be tedious even if we write a corresponding program, but the program itself should not be too complicated.

Note also that equations (2.8.3)-(2.8.5) are the so called backward equations: we condition the behavior of the process with respect to what may happen in the first period. Another approach may concern the so called forward equations. In this case, we assume that there was no ruin during the first $T-1$ periods, and consider the behavior of the process in the last period. Regarding the later methods, see, e.g., [74, Section 7.3].

The same logic may be applied to processes in continuous time. Assume that $S_{(t)}$ is a homogeneous compound Poisson process with intensity λ. Let m and $F(x)$ be the mean value and the d.f., respectively, of a separate claim. As usual, we set $c_t = (1+\theta)m\lambda$, where θ is a security loading. For simplicity, consider the case $T = \infty$.

Set $\eta = \min\{t : S_{(t)} > 0\}$, the moment of the first claim, and set $Z = S_{(\eta)}$, the value of the first claim. For the process under consideration, η is exponential with parameter λ, the d.f. of Z is F, and η and Z are independent.

Let again $\phi(u) = \phi_\infty(u)$. As before, we set $\phi(u) = 0$ if $u < 0$.

At the moment η, the conditional no-ruin probability is equal to $\phi_{T-\eta}(u+c\eta-Z)$. It is equal to zero if $Z > c\eta+u$.

In view of the memoryless property, at the moment η, the process starts over from the new level. Since $T=\infty$, the time horizon with respect to the new starting moment η is again infinite. Hence,

$$\phi(u) = E\{\phi(u+c\eta-Z)\}.$$

This is an equation for $\phi(u)$. Since we know the distributions of Z and η, we can rewrite it as

$$\phi(u) = \int_0^{\infty}\int_0^{\infty}\phi(u+ct-z)dF(z)\lambda e^{-\lambda t}dt.$$

Since $\phi(u) = 0$ for $u < 0$, it may be written as

$$\phi(u) = \lambda\int_0^{\infty}e^{-\lambda t}\left(\int_0^{u+ct}\phi(u+ct-z)dF(z)\right)dt. \tag{2.8.6}$$

The theory of solutions to equations of this type is well developed and uses various mathematical methods; see, e.g., [10], [19], [38], [50], [74]. All these methods are not very simple but give, in particular, an alternative way to obtain many results we got above by making use of the martingale or renewal approaches.

Routes 1 and 2 ⇒ *page 391*

3

3 CRITERIA CONNECTED WITH PAYING DIVIDENDS

In the situation described in the previous section, two things may happen:

- either during some finite time period, ruin will occur (for a large initial surplus and/or large premiums, the probability of this event is small), or
- the company will avoid ruin, and the surplus R_t will unlimitedly grow: $R_t \to \infty$ as $t \to \infty$. (See also condition (2.1.6).)

The last property is not realistic: no company will keep an excessive surplus of high liquidity while having an opportunity to invest a part of it or pay dividends. Moreover, a law and a general usual insurance practice requires paying some dividends if the surplus exceeds a certain level.

On the other hand, as one may guess and as we will see below, the probability that the company will remain solvent forever would be zero unless the company allows the surplus to grow. In other words, condition (2.1.6) is essential for the ruin probability not to equal one.

To resolve these issues, we should consider, as an alternative to ruin probability, other quality criteria—for example, the expected discounted amount of dividends to be paid or/and the expected life of the company. The idea to use these criteria was first aired by B. De Finetti in 1957 and was considered later by K. Borch and other scholars (see, e.g., [16], [20], [18], [138], [141]).

The goal of this section is to illustrate some ideas and results in this area. We restrict ourselves to the discrete time case and consider the model (1.3)-(1.4) from Section 1.

3.1 A general model

Denote by d_t the dividend paid at time $t = 1,2,\ldots$. The surplus process is governed by the relation

$$R_t = u + ct - S_t - D_t, \tag{3.1.1}$$

where $S_t = X_1 + \ldots + X_t$, the claims X_i are i.i.d. r.v.'s, $D_t = d_1 + \ldots + d_t$, and c is a premium per unit interval of time. As before, when it does not cause misunderstanding, we omit the index i in X_i.

Assume that at the moment, if any, when $R_t < 0$, the company stops operating.

Let $v < 1$ be a discount factor. We consider an infinite time horizon, and the criterion

$$E\left\{\sum_{t=1}^{\infty} v^t d_t\right\}, \tag{3.1.2}$$

the expected total amount of discounted dividends to be paid.

The variables d_t represent a strategy of paying dividends. In general, since d_t may depend on the history of the process until time t, it is a r.v.

Since $v < 1$, dividends to be payed in the future are less valuable than payments now. However, it does not mean that the company should pay large dividends in the beginning. If the company pays too much in earlier stages, this will reduce the current surplus and will make possible an earlier ruin. In this case, the total amount of dividends may be small because the time of functioning will be small.

Thus, an optimal strategy should reflect a trade-off between two issues: the desire to pay dividends in earlier stages, and the necessity to keep the company functioning during a sufficiently long period.

We will show in Section 3.3 that under some mild conditions the optimal strategy maximizing (3.1.2) has the following threshold structure.

- If at the end of an underwriting period the surplus R_t exceeds an optimal threshold level z^*, then the amount $R_t - z^*$ is paid out as the dividend payment during that period.
- If the surplus R_t is less than z^*, then no dividends are paid, and the company keeps the surplus R_t until the next underwriting period.

In other words,

$$d_t = \max\{R_t - z^*, 0\}.$$

We will see in Section 3.3 that the optimal level z^* does not depend on the initial surplus u.

To find the level z^*, we consider the threshold strategy for all z's and the function

$$V(u,z) = E\left\{\sum_{t=1}^{\infty} v^t d_{tz}\right\}, \tag{3.1.3}$$

where

$$d_{tz} = \max\{R_t - z, 0\}.$$

Once we know $V(u,z)$, we can try to find its maximizer in z, that is, the optimal level z^*. Since this level does not depend on u, we can do that for $u = 0$.

Certainly, this does not mean that the amount of dividends itself does not depend on u. Let us consider this in more detail.

If the initial level $u > z$, then by definition of the strategy d_{tz}, the company should immediately pay off the surplus $u - z$, that is,

$$V(u,z) = V(z,z) \quad \text{for } u > z.$$

Let $u \leq z$. Since the goal of the company is to maximize the amount of dividends, the initial surplus u that the company keeps for functioning may be viewed as an investment for getting dividends in the future. Then the variable $V(u,z) - u$ may be viewed as the profit of the company. We will prove in Section 3.3 that in the case of the optimal level z^*,

$$\text{the function } V(u,z^*) - u \text{ is increasing in } u \text{ when } u < z^*. \tag{3.1.4}$$

This means that the optimal behavior is to start with the initial surplus $u = z^*$ and proceed following the optimal threshold strategy.

The next question concerns the ruin probability. We will see that, under the above threshold strategy, it is equal to one, provided that with a positive probability the claim may exceed the premium. Such a condition is natural since otherwise nobody will pay such a premium.

Let $P(X > c + a) \geq \delta > 0$ for some $a > 0$. Let $k = [z^*/a] + 1$, where as usual $[x]$ is the integer part of x. Then with the probability δ^k, all claim surpluses $X_t - c$, $t = 1, ..., k$, will be larger than a, and R_k will be negative. If it does not happen during the first k steps, then it will happen with the same positive probability during the next k steps, and so on. So, the probability that ruin will ever happen is one.

(More rigorously, let

$$A_i = \bigcap_{t=ik+1}^{(i+1)k} \{X_t > c + a\}.$$

For ruin to occur, it suffices that at least one of the events A_i occurs. The probability of this is one, since A_i's are independent and $P(A_i) > \delta^k > 0$.)

The fact that in the case of a threshold strategy the ruin probability equals one is not a reason to refuse the approach above: we nevertheless deal with the maximal amount of dividends. Moreover, if the time before the ruin is sufficiently large, say, larger than the time horizon for the company, the fact mentioned is not essential.

Nevertheless, we can apply a more cautious approach by introducing into consideration the expected time of ruin. Let $D(u,z)$ be the mentioned expected time for the initial surplus u and the threshold level z. Having at its disposal both characteristics, $V(u,z)$ and $D(u,z)$, the insurer can establish a more flexible criterion. One example consists in maximizing $V(u,z)$ under the restriction

$$D(u,z) \geq D_0,$$

where D_0 is a given level determined by the preferences of the insurer.

Analytical solutions of the problems above are complicated even in simple cases (see, e.g., [16], [141]). So, we restrict ourselves to one example, namely, to the simple random walk model. Results for this model illustrate well what we can expect in more general cases. As to the general situation, it is worth emphasizing that numerical solutions based on simulation of the process R_t are quite tractable, and with use of modern software do not present essential difficulties.

3.2 The case of the simple random walk

Let $c = 1$, and the size of the claim at each period is the r.v.

$$X = \begin{cases} 0 \text{ with probability } p, \\ 2 \text{ with probability } q, \end{cases}$$

where $q < p$. In this case, $m = E\{X\} < 1$, and hence $c > m$.

Thus, for each period, the profit of the company is $c - X = \pm 1$ with probabilities p and q, respectively.

Consider the threshold strategy with a level z. Assume that u and z are integers, and let $w_n(u,z)$ be the probability that the first dividend will be paid at the moment n. By the definition of the strategy we use,

$$\begin{aligned} &w_0(u,z) = 0 \text{ for } u \leq z; \ w_0(u,z) = 1 \text{ for } u > z; \\ &w_n(u,z) = 0 \text{ for } u < 0, \text{ since in this case the insurer is ruined in the very beginning;} \\ &w_n(u,z) = 0 \text{ for } u > z \text{ and } n > 0, \text{ since in this case the first payment occurred} \\ &\qquad \text{at the initial time.} \end{aligned} \tag{3.2.1}$$

We apply the first step approach in a way similar to what we did in Section **4**.4.3.2. With probability p the process moves up, the surplus becomes $u+1$, the random walk starts over, and the probability that a dividend will be paid at time n "becomes" $w_{n-1}(u+1,z)$. The same concerns the case when the process in the first step moves down. Thus,

$$w_n(u,z) = pw_{n-1}(u+1,z) + qw_{n-1}(u-1,z). \tag{3.2.2}$$

Let

$$\hat{w}(u,z) = \sum_{n=0}^{\infty} v^n w_n(u,z), \tag{3.2.3}$$

the generating function of the sequence of probabilities $\{w_n\}$. (See also Section **0**.4.1.) We chose the same letter $v \in (0,1)$ as for discount on purpose.

Applying (3.2.2) for $u \le z$, and taking into account conditions (3.2.1), we have

$$\hat{w}(u,z) = \sum_{n=1}^{\infty} \nu^n (pw_{n-1}(u,z) + qw_{n-1}(u-1,z)) = p\nu\hat{w}(u+1,z) + q\nu\hat{w}(u-1,z).$$

So, for the generating function we have the equation

$$\hat{w}(u,z) = p\nu\hat{w}(u+1,z) + q\nu\hat{w}(u-1,z) \tag{3.2.4}$$

for $u \le z$. Similarly one can get that

$$\hat{w}(u,z) = 0 \text{ for } u < 0, \ \ \hat{w}(u,z) = 1 \text{ for } u > z. \tag{3.2.5}$$

Setting $\hat{w}(u,z) = r^{u+1}$, where r is a number, and inserting it into (3.2.4), we see that such a function satisfies (3.2.4) if

$$r = p\nu r^2 + q\nu. \tag{3.2.6}$$

Thus, if r_1 and r_2 are the roots of the quadratic equation (3.2.6), the functions r_1^{u+1} and r_2^{u+1} are solutions to (3.2.4). Without going too deeply into the theory, note that then any solution $\hat{w}(u,z) = c_1 r_1^{u+1} + c_2 r_2^{u+1}$, where c_1, c_2 are constants. To find constants c_1, c_2, we use (3.2.5), writing $\hat{w}(z+1,z) = 1$, $\hat{w}(-1,z) = 0$.

Eventually it leads to the solution

$$\hat{w}(u,z) = \frac{r_1^{u+1} - r_2^{u+1}}{r_1^{z+2} - r_2^{z+2}}. \tag{3.2.7}$$

Next, we consider a connection between $V(u,z)$ and $\hat{w}(u,z)$. Since u and z are integers, each dividend paid is equal to one. A dividend is paid if the surplus equals $z+1$, and once a dividend is paid, the surplus becomes equal to z. The probability that paying dividends starts from a moment n is w_n, and after that "everything starts over" from the level z. Hence, for $u \le z$,

$$V(u,z) = \sum_{n=0}^{\infty} w_n(u,z)\,[v^n \cdot 1 + v^n V(z,z)] = [1 + V(z,z)]\,\hat{w}(u,z). \tag{3.2.8}$$

Setting $u = z$, we have $V(z,z) = [1 + V(z,z)]\,\hat{w}(z,z)$, from which it follows that

$$V(z,z) = \frac{\hat{w}(z,z)}{1 - \hat{w}(z,z)}. \tag{3.2.9}$$

Combining (3.2.8) and (3.2.9), we have

$$V(u,z) = \frac{\hat{w}(u,z)}{1 - \hat{w}(z,z)}.$$

Substituting (3.2.7), we get eventually that for $u \le z$,

$$V(u,z) = \frac{r_1^{u+1} - r_2^{u+1}}{(r_1^{z+2} - r_2^{z+2}) - (r_1^{z+1} - r_2^{z+1})}. \tag{3.2.10}$$

The denominator depends only on z, while the numerator—only on u. So, as was expected, the optimal level z^* does not depend on u.

We skip detailed calculations leading to an optimal z. To find it, one should take the derivative of the denominator in (3.2.10), set it equal to zero, and divide the whole equation by r_2^z. Then the unknown z will be contained only in the expression $(r_1/r_2)^z$. Solving the equation with respect to this expression, one can readily get that the optimal level

$$z^* = \frac{1}{\ln(r_1) + |\ln(r_2)|} \ln\left[\frac{\ln(r_2)}{\ln(r_1)} \cdot \frac{r_2(1-r_2)}{r_1(r_1-1)}\right] \tag{3.2.11}$$

where $r_1 > 1$ is the larger and $r_2 < 1$ is the smaller root of equation (3.2.6). (The values in (3.2.11) may be negative; in this case one should set $z^* = 0$.)

Table 1 shows the values of z^* for different values of p, $q = 1-p$, and ν. These calculations, as well as simulation of the process with X's having different distributions, and a proof of (3.1.4) were provided by Sarah Borg in her master's thesis [18]. Note that, though we have assumed u, z to be integers, for a more complete picture, the case of arbitrary u, z was considered.

It can be seen that for each value of p, the value of z^* increases as ν increases. The higher ν is, the more the company is concerned about the future payments. So, the company increases the level of the surplus in order to increase the time before ruin.

We see that z^* initially increases in p, and then decreases as p gets closer to 1. It is also understandable. Consider the extreme case $p = 1$. Then with probability one there will be no claim, and therefore the whole surplus could be paid out as dividends. So, z^* in this case should be zero. Then, if p is close to one, we should expect z^* to be small.

TABLE 1: Values of z^* for different p and ν.

	ν				
p	**0.90**	**0.92**	**0.94**	**0.96**	**0.98**
0.60	0.15	0.54	1.14	2.21	4.65
0.65	0.69	1.14	1.79	2.86	4.99
0.70	1.02	1.46	2.07	3.01	4.73
0.75	1.17	1.57	2.10	2.89	4.27
0.80	1.19	1.54	1.99	2.63	3.73
0.85	1.11	1.40	1.78	2.30	3.17
0.90	0.95	1.18	1.48	1.89	2.58
0.95	0.68	0.85	1.07	1.38	1.88
0.98	0.007	0.09	0.19	0.034	0.58

Next, we briefly consider the expected life $D(u,z)$ for the random walk model. Assume, as before, that u, z are integers, and $p > q$.

The same first step approach leads to the equation

$$D(u,z) = 1 + pD(u+1,z) + qD(u-1,z), \quad 1 \le u \le z.$$

As can be verified by direct substitution, the solution to this equation is

$$D(u,z) = \frac{p}{(p-q)^2}\left[\left(\frac{p}{q}\right)^{z+1} - \left(\frac{p}{q}\right)^{z-u}\right] - \frac{u+1}{p-q}.$$

A general expression for $D(u,z)$ in terms of some series and other examples may be found in [16].

3.3 Finding an optimal strategy

In this section, we prove that the optimal strategy has properties described in Section 3.1. Assume that the optimal strategy exists and set

$$V(u) = \max E\left\{\sum_{t=1}^{\infty} v^t d_t\right\}, \tag{3.3.1}$$

where max is over all possible strategies $\{d_1, d_2, ...\}$ of paying dividends, and u is an initial surplus. So, $V(u)$ is the expected discounted amount of dividends under the optimal strategy. It is convenient to consider the function $V(u)$ for all u, setting $V(u) = 0$ for all $u < 0$. We assume also that $V(u)$ is continuous at all points u except perhaps $u = 0$.

Consider a time moment t. If the company is still functioning, it has a surplus $R = R_t \geq 0$, and should specify its policy for the next period. The process we consider is a Markov process, which means, in particular, that the policy may depend on the current surplus but does not depend on what strategy the company chose before time t. The company again faces the infinite time period, and should solve the optimization problem as if it is at the very beginning.

The company receives the next premium c, and pays out the claim $X = X_{t+1}$ and a dividend d (which perhaps equals zero). So, the surplus $R_{t+1} = R_t + c - d - X$. If, at the next time, the company applies the optimal strategy (which we do not know yet but assume that it exists), then given X, the expected discounted amount of dividends after time $t+1$ will be $V(R_t + c - d - X)$.

From the standpoint of the present time t, the total amount of dividends is $d + vE\{V(R + c - d - X)\}$, where $R = R_t$ and d cannot exceed the current surplus R. To find the optimal behavior at the period $[t, t+1]$, we should maximize the last expression in d, which leads to the equation

$$V(R) = \max_{0 \leq d \leq R} [d + vE\{V(R + c - d - X)\}]. \tag{3.3.2}$$

The reader familiar with the optimization theory recognizes in the above reasoning the so called optimality principle, and realizes that we have derived the Bellman equation.

Let the function

$$w(y) = vE\{V(y - X)\} - y.$$

Then

$$V(R) = \max_{0 \leq d \leq R} [R + c + w(R + c - d)] = R + c + \max_{0 \leq d \leq R} w(R + c - d) = R + c + \max_{c \leq y \leq c+R} w(y), \tag{3.3.3}$$

where we changed variables, setting $y = R + c - d$.

Assume now that the function $w(y)$ has a unique maximum at a point y_0. This is an implicit condition we impose to find the optimal solution. Consider three cases.

(i) $y_0 \leq c$. In this case, $\max_{c \leq y \leq c+R} w(y)$ is attained at the point $y = c$ (graph $w(y)$ with a unique maximum at y_0, and place c on the right of y_0). Then the optimal $d = R + c - y = R$.

(ii) $c < y_0 \leq c + R$. Then $\max_{c \leq y \leq c+R} w(y)$ is attained at the point $y = y_0$, and the optimal payment $d = R + c - y_0 = R - \tilde{z}$, where $\tilde{z} = y_0 - c > 0$.

(iii) $y_0 > c + R$. Then $\max_{c \leq y \leq c+R} w(y)$ is attained at the point $y = c + R$, and the optimal payment $d = 0$.

Setting $z^* = \max(0, y_0 - c) \geq 0$, we see that in all three cases above, the optimal payment

$$d = \begin{cases} R - z^* & \text{if } R > z^*, \\ 0 & \text{if } R \leq z^*. \end{cases} \tag{3.3.4}$$

So, the fact that the optimal strategy has the threshold structure is proved.
To prove (3.1.4), we write (3.3.3) as

$$V(u) - u = c + \max_{c \leq y \leq c+u} w(y). \tag{3.3.5}$$

If $y_0 \leq c$, then $V(u) - u = c + w(c)$ and, hence, does not depend on u. It is natural—in this case the optimal dividend payment at the initial moment would be $d = u$, and the company will start from zero level.

Let $y_0 > c$. If $u > z^* = y_0 - c$, then as was proved, the optimal behavior consists in immediate payment of the surplus $u - z^*$ as a dividend. After that the process starts from the level z^*.

If $u < z^*$, there should be no dividend at the initial moment, and $V(u) - u = c + w(c + u)$. In this case, $c + u < y_0$, and hence $w(c + u)$ is increasing in u up to the moment when $c + u = y_0$. This is equivalent to $u = z^*$. ■

---

4 EXERCISES

Sections 1 and 2

1. Is $\psi_T(u)$, as a function of T, increasing?
2. Consider the process R_t in continuous time, and set $\tilde{\phi}_n(u) = P(R_t \geq 0 \text{ for all } t = 1, 2, ..., n)$, that is, we count only integer moments of time. Which is larger: $\tilde{\phi}_n(u)$ or $\phi_n(u)$?
3. Show that in general, not assuming that $R_t = u + c_t - S_{(t)}$, for (2.1.6) to be true it suffices to require
$$E\{R_t\} \to \infty \quad \text{and} \quad Var\{R_t\} = o([E\{R_t\}]^2).$$
4. Look over Exercises 30-36 from Chapter 2.
5. Problems below concern Example 2.2.2-1.

(a) Graph the r.-h. and l.-h. sides of (2.2.12). Show that for (2.2.12) to have a positive solution $\gamma < 1$, the premium c should be indeed greater than 2. Check the numerical answers in Example 2.2.2-1.

(b) Write a program (it suffices to provide a spreadsheet) which would allow to compute γ for different c's. Compare the results with what the approximation (2.2.16) gives.

(c) Show that $\gamma \to 1$ as $c \to \infty$. Explain why in the case of large c, (2.1.4) is not a good estimate for the ruin probability. (*Advice*: Show that the ruin probability should vanish when $c \to \infty$.)

6. Assume that for some $c > m$ there exists a positive solution γ to equation (2.2.11).

   (a) Proceeding from the results of Section 2.2.1 and using Figures 4ab, show at a heuristic level that $\gamma \to 0$ as $c \to m$ from the right (being greater than m), and in this case, the ruin probability converges to one. Explain why it is not surprising from an economic point of view.

   (b) Prove that $\gamma \to 0$ as $c \to m$ rigorously.

   (c) Consider the case $c \to \infty$. Explain at a heuristic level that in this case, the ruin probability should converge to zero. Show that it follows from (2.1.4). (*Hint*: If c is very large, with probability close to one, the surplus at the next step will be very large.)

7. Assume that $M_X(z)$ in (2.2.11) is defined for all z.

   (a) Assume also that a positive solution γ to equation (2.2.11) exists for all $c > m$. Proceeding from the results of Section 2.2.1 and using Figures 3-5, show at a heuristic level that $\gamma \to \infty$ as $c \to \infty$, and the ruin probability converges to zero. Explain why it is not surprising from an economic point of view.

   (b) Prove the result of Exercise 7a rigorously.

   (c) Let the r.v. X above be bounded by a number b. Explain why in this case the instance $c > b$ has no economic sense. Show that in this case a positive solution to (2.2.11) does not exist, and that $\gamma \to \infty$ as $c \to b$ if X is not degenerate.

8. If you should solve (2.2.24) for two different a's and the same ν, would you solve (2.2.24) twice or just one time?

9. Using (2.2.16), estimate the ruin probability in the situation (2.2.20) for $u = 100$, $\theta = 0.1$, ξ's having the standard exponential distribution, and K_i's having the geometric distribution with parameter $p = 0.1$.

10. Estimate the adjustment coefficient in the situation of Example 2.2.4-1 for $\nu = 3$, $\theta = 0.1$, and $a = 1$ and 2. (*Advice*: First solve Exercise 8.)

11. Find the adjustment coefficient in the situation of Example 2.2.4-1 for $\nu = 2$. Show that for small θ the answer does not contradict approximation (2.2.27). (*Advice*: First solve Exercise 8, and think for what a solutions to (2.2.24) are simpler.)

12.** Making use of the result of Exercise **1**.33b, show that the r.-h.s. of (2.4.1) is non-decreasing in γ.

13. A flow of claims is represented by a compound Poisson process S_{N_t} in continuous time. The mean time between adjacent claims is half an hour. The random value X of a *particular* claim is uniformly distributed on $[0, 10]$ (say, the unit of money is \$1000). The initial surplus (capital) is 100, the relative loading coefficient $\theta = 0.2$.

    (a) Estimate the ruin probability.

(b) For which initial surplus is the ruin probability less than 0.05?

(c) * Let $u = 50$. Find θ for which the ruin probability is less than 0.05.

14. For a particular group of clients, a flow of claims arriving at an insurance company may be represented as a compound Poisson process. Let the amount of a particular claim be equal to either 2, 3, or 4, with probabilities $1/4$, $1/2$, and $1/4$, respectively. Let the mean number of claims the company receives per day be 10. Assume that the company chooses for its activity a relative loading coefficient $\theta = 0.1$.

    (a) Write an equation for the adjustment coefficient γ.

    (b) Does this equation involve λ ?

    (c) Find an approximate solution using software. Compare it with approximation (2.2.26).

    (d) Find an approximate value of the initial capital for which the ruin probability for the company will be less than 0.03.

    (e) Think how to answer all questions above, if a particular claim is, say, uniformly distributed on $[2,4]$. Do you expect the ruin probability to be smaller?

15. Think how to answer all questions in Exercise 14 if the number of claims during a day is exactly equal to 10, that is, we consider the discrete time scheme, and day is a unit of time.

16.* In the discrete time case, for the claim X in one period, we have $E\{X\} = 3$, $Var\{X\} = 2.5$. The required level $\alpha = 0.05$. Estimate proper combinations of the initial surplus and the loading coefficient for "large" u.

17. Provide a graph illustrating a solution to (2.5.2).

18.* (a) Without calculating anything, show that if X is a discrete r.v., the density (2.7.4) is a mixture of uniform distributions.

    (b) Find the density (2.7.4) for the case when X takes on only one value.

    (c) Find the density (2.7.4) for the case when X takes on values 1,2,3 with probabilities $1/5, 2/5, 2/5$, respectively.

19.* Show that, if X is exponential with a parameter a, then the density (2.7.4) is exponential with the same parameter. Explain that this fact is non-surprising in light of what was discussed in Section 2.5.1.

20.* Clarify from a heuristic point of view the significance of the fact that the density (2.7.4) is decreasing.

21.* In the framework of Section 2.7, let the r.v.'s X_i be continuous. Does it mean that the r.v. Z_K is continuous? (*Advice:* Think about $\phi(0)$ and how it is connected with Z_K.)

22.** In the case of the compound Poisson process, for some data the ruin probability $\psi(u) = 0.3e^{-2u} + 0.4e^{-u/2}$. Find θ and γ. (Advice: Use the Cramér-Lundberg approximation (2.7.14) and (2.7.3).)

23.** (a) Making use of (2.2.26), show that the constant C in (2.7.15) is close to one for small θ.

    (b) Making use of (2.7.14), estimate the ruin probability for large u in the case where X has the Γ-distribution with parameters $a = 1$, $\nu = 2$, and $\theta = 0.1$.

24. Realize why the values of $\phi_2(u)$ in Examples 2.8-2 and 3 are the same, while ranges for u are different.

25. Assume that in the situations of Example 2.8-2, you are asked only to find u for which $\phi(u) \geq 0.99$. Realize that in this case you can avoid most of the calculations, coming to the answers very quickly. Find this answer.

26. For $X = 2, 10$ with probabilities 0.6, 0.4, respectively, solve the problem of Example 2.8-2 and the problem of Example 2.8-3 for $r = 0.2$.

27. Show that for $T < \infty$, the counterpart of (2.8.6) will be

$$\phi_T(u) = \lambda \int_0^T e^{-\lambda t} \left(\int_0^{u+ct} \phi_{T-t}(u+ct-z)dF(z) \right) dt + e^{-\lambda T}.$$

(*Hint*: The conditional no-ruin probability given $\eta \geq T$, is certainly one (no claim arrived within the period $[0, T]$).)

*Section 3***

28. Regarding the model of Section 3.1, give a common sense explanation to why one should expect a higher optimal level z^* for higher values of discount v.

29. Can the optimal level z^* be zero? (*Advice*: Think about the case of small v.)

30. In the model of Section 3.2, let $p = 0.7$, $v = 0.9$. Using Excel or other software, provide the graph of $V(0, z)$. Interpret it. Why are too small and too large values of z not optimal? Estimate the optimal z^*. Explain why for any $u < z^*$, the maximizer of $V(u, z)$ will be the same.

31. Using Excel or other software, graph z^* against p for $v = 0.9$.

Chapter 7

Survival Distributions

1,2,3

We begin to consider situations when the obligations assumed by the insurance company are connected, in one way or another, with the lifetimes of insured units. For the most part, we address life insurance and annuities.

There are two main features of such insurance mechanisms.

The first is the same as in non-life insurance and consists in the redistribution of risk between clients of the insurance organization. The second feature concerns the time lag between the moments when the company pays benefits and the time of policy issue, that is, the time when the first premium is paid. In this case, the random nature of the insurance process is specified by the probability distributions of the lifetimes of insured units.

In the current chapter, we consider various types of such distributions and their characteristics. Chapters 8-11 concern insurance models themselves.

1 THE PROBABILITY DISTRIBUTION OF LIFETIME

1.1 Survival functions and force of mortality

Most of the results below concern the lifetimes of objects of rather general nature; for certainty and because this is the most important for us, as a rule we will talk about people. For other objects (say, machines), we will use the term "failure" rather than "death".

Let X be the (random) *lifetime*, or in another terminology, the *age-at-death* of a particular individual. Set $F(x) = P(X \leq x)$, the d.f. of X. We assume that $F(0) = 0$, which indicates the fact that once an individual has been born, her/his lifetime is not equal to zero.

When talking about an individual, we view her/him as a typical representative of some group of people. In many applications, such a group is homogeneous; that is, the lifetimes of all its members have the same distribution F. However, the homogeneity property is not necessary. For example, the life tables for the total population of a country (see Section 1.5 for details) contain information about the duration of life for all citizens of the country under consideration who constitute, of course, a non-homogeneous group. This information concerns the average lifetime, and we may view it of interest when a person is chosen at random.

Thus, in general, the distribution F is the average distribution $\frac{1}{n}(F_1 + ... + F_n)$, where n is the number of the members of the group, and F_i is the lifetime distribution of its ith member.

Let us come back to one individual and her/his lifetime X. In the theory under consideration, the tail of this distribution, $P(X > x)$, is usually denoted by $s(x)$ and is called a *survival function*. Clearly,

$$s(x) = P(X > x) = 1 - F(x). \tag{1.1.1}$$

Note that $s(x)$ is a non-increasing function, and $s(0) = 1$ because $F(0) = 0$.

EXAMPLE 1. In three different countries, typical survival functions

$$s(x) = [1 - x/100]^{\alpha} \ \text{ for } \ 0 \leq x \leq 100, \tag{1.1.2}$$

where $\alpha = 0.5$, 1, and 2, respectively, and time x is measured in years. In which country do people live the longest?

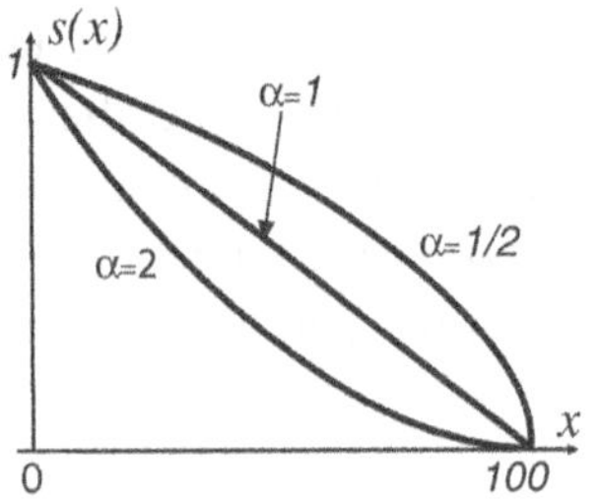

FIGURE 1.

First of all, since in all cases $s(100) = 0$, people in these countries do not live more than 100 years, and consequently, we must set $s(x) = 0$ for $x > 100$. The graphs are given in Fig.1. Note that for $\alpha = 1$, the distribution is uniform on $[0, 100]$.

Certainly, for *any* x, the probability to survive x years is larger for $\alpha = 1/2$ than for other α's. For example, $s(90)$ equals 0.32, 0.1, and 0.01 for $\alpha = 0.5$, 1, and 2, respectively, so the first country is "much better" in terms of longevity. □

Henceforth, we assume the d.f. $F(x)$ to be smooth, so the distribution has density $f(x) = F'(x)$. For an infinitesimal interval dx we, as usual, write that

$$P(x < X \leq x + dx) = f(x)dx. \tag{1.1.3}$$

Consider $P(x < X \leq x + dx \,|\, X > x)$, the probability that the individual under consideration will die within the interval $[x, x + dx]$, given that she/he has survived x years. In other words, this is the probability that a person of age x will die within the small interval of the length dx after time x. In view of (1.1.3),

$$P(x < X \leq x + dx | X > x) = \frac{P(x < X \leq x + dx, X > x)}{P(X > x)} = \frac{P(x < X \leq x + dx)}{P(X > x)} = \frac{f(x)dx}{s(x)}, \tag{1.1.4}$$

provided that $s(x) \neq 0$. Set

$$\mu(x) = \frac{f(x)}{s(x)}, \tag{1.1.5}$$

again assuming that $s(x) \neq 0$. If $s(x) = 0$, we set $\mu(x) = \infty$ by definition.

From (1.1.4) it follows that for $s(x) \neq 0$

$$P(x < X \leq x + dx | X > x) = \mu(x)dx. \tag{1.1.6}$$

In the general probability theory, the function $\mu(x)$ is called a *hazard rate*. In the case where the r.v. X is a lifetime, $\mu(x)$ is called the *force of mortality* of X.

The larger $\mu(x)$, the larger the probability that a person of age x will die "soon"; i.e., within a small time interval.

Since $f(x) = F'(x)$ and $F(x) = 1 - s(x)$, we can also write that

$$\mu(x) = -\frac{s'(x)}{s(x)}. \tag{1.1.7}$$

Sometimes it is convenient to present (1.1.7) in the form

$$\mu(x) = -\frac{d}{dx}\ln s(x). \tag{1.1.8}$$

Three examples below illustrate possible situations and are relevant to the classification of tails considered in Section **2**.1.1.

EXAMPLE 2. Let X be exponential with parameter a. Then $s(x) = e^{-ax}$, $f(x) = ae^{-ax}$ and by (1.1.5),

$$\mu(x) = a.$$

Thus,

> In the lack-of-memory case, the force of mortality is constant.

EXAMPLE 3. Now consider the case when $s(x)$ is decreasing slower than any exponential function; for example $s(x)$ is a power function. For instance, let X have the Pareto distribution (**2**.1.1.18) with $\theta = 1$. Then $s(x) = 1/(1+x)^{\alpha}$ for some $\alpha > 0$ and all $x \geq 0$, and by (1.1.7),

$$\mu(x) = \frac{\alpha/(1+x)^{\alpha+1}}{1/(1+x)^{\alpha}} = \frac{\alpha}{1+x}. \tag{1.1.9}$$

This is not a realistic model not only for a human being but practically for any object's life time since in the instance (1.1.9), the older the object, the less its chances are of dying.

EXAMPLE 4. Let now $s(x)$ be decreasing faster than any exponential function, for instance, $s(x) = e^{-x^2}$. Then, by (1.1.5),

$$\mu(x) = \frac{2xe^{-x^2}}{e^{-x^2}} = 2x \to \infty$$

as $x \to \infty$, which is much more realistic. □

Later, we will consider other examples of the force of mortality. One of most important cases concerns the Gompertz-Makeham law when the force of mortality grows exponentially as $\mu(x) = Be^{\alpha x} + A$, where A, B, and α are parameters. We consider it in more detail in Section 1.6.

Now assume that the force of mortality, or in general the hazard rate, $\mu(x)$ is given, and we want to find the survival function $s(x)$. In this case, (1.1.7) may be considered an equation for $s(x)$. Because $s(0) = 1$, the solution to this equation is

$$s(x) = \exp\left\{-\int_0^x \mu(z)dz\right\}. \tag{1.1.10}$$

(See, e.g., the mathematically similar case in Section **0**.8.1 and how we obtained formula (**0**.8.1.7) from (**0**.8.1.3). The difference between (1.1.7) and (**0**.8.1.3) is only in notation and interpretation. In Exercise 1, the reader is suggested to verify (1.1.10) by differentiation.)

The representation (1.1.10) is illustrative and convenient. Above all, it may be viewed as a generalization of the exponential distribution. If the force of mortality is constant, say, $\mu(x)$ equals some $a > 0$, then (1.1.10) implies that

$$s(x) = \exp\left\{-\int_0^x a dz\right\} = e^{-ax},$$

that is, we are dealing with an exponential distribution.

In general, the integral in the exponent in (1.1.10) is a non-linear (and non-decreasing since $\mu(x) \geq 0$) function.

It is worth emphasizing that the "survival terminology" is used only for interpretation. As a matter of fact, (1.1.10) is true for the distribution of any continuous positive random variable with hazard rate $\mu(x)$.

EXAMPLE 5 ([153, N30][1]). Acme Products will offer a warranty on their products for x years, where x is the largest integer for which there is not more than a 1% probability of product failure. Acme introduces a product with a hazard function for failure at time t of $0.002t$. Calculate the length of warranty that Acme will offer on this new product.

The tail probability $s(x) = \exp\left\{-\int_0^x 0.002t dt\right\} = \exp\left\{-0.001x^2\right\}$. The solution to the inequality $s(x) \geq 0.99$ is $x \leq \left(\frac{-\ln(0.99)}{0.001}\right)^{1/2} \approx 3.17$. So, $s(3) > 0.99$, while $s(4) < 0.99$. The warranty should cover three years. □

If X is bounded, i.e., $X \leq c$ for some $c > 0$, then $s(c) = P(X > c) = 0$, and in this case, by convention, we set $\mu(x) = \infty$ for $x > c$.

EXAMPLE 6. Let X be uniform on $[0,1]$. Then $s(x) = 1 - x$ for $x \in [0,1]$, and $= 0$ for $x > 1$. Using (1.1.7), it is easy to check that in this case, $\mu(x) = 1/(1-x)$ for $x < 1$. Since $\mu(x) \to \infty$ as $x \to 1$, it is natural to set $\mu(x) = \infty$ for all $x \geq 1$.

However, it is noteworthy that if we proceed from $\mu(x) = 1/(1-x)$ so to speak not knowing with which distribution we are dealing, then we get $s(x) = 0$ for $x > 1$, and this will not depend on how we define $\mu(x)$ for $x > 1$. Indeed, if $x > 1$, the r.-h.s. of (1.1.10) is equal to

$$\exp\left\{-\int_0^x \mu(z)dz\right\} = \exp\left\{-\int_0^1 \mu(z)dz - \int_1^x \mu(z)dz\right\} = \exp\left\{-\int_0^1 \frac{1}{1-z}dz - \int_1^x \mu(z)dz\right\} = 0,$$

since $\int_0^1 \frac{1}{1-z}dz = \infty$. □

Next, consider the representation for the density $f(x)$ given $\mu(x)$. From (1.1.10) it follows that

$$f(x) = -s'(x) = \mu(x)\exp\left\{-\int_0^x \mu(z)dz\right\}. \tag{1.1.11}$$

[1]Reprinted with permission of the Casualty Actuarial Society.

Compare with the formula $\mu e^{-\mu x}$ for the exponential density with parameter μ.

Now consider representation (1.1.10) for $x = \infty$, defining $s(\infty)$ as $\lim_{x\to\infty} P(X > x)$. We interpret $s(\infty)$ as the probability of "living forever". Then $P(X < \infty)$, the probability of ever dying, equals $1 - s(\infty)$. If X is the lifetime of an individual, in order to be realistic, we should set $s(\infty) = 0$. It follows from (1.1.10) that for this to be true, we should have $\lim_{x\to\infty} \int_0^x \mu(z)dz = \infty$, or in another notation,

$$\int_0^\infty \mu(z)dz = \infty. \tag{1.1.12}$$

Thus, for $s(\infty) = 0$, which is equivalent to $P(X < \infty) = 1$, the integral in (1.1.12) should diverge.

EXAMPLE 7. Can a force of mortality $\mu(x)$ equal $1/(1+x)^2$? Certainly not. If it were true, we would have had

$$s(\infty) = \exp\left\{-\int_0^\infty \frac{1}{(1+z)^2}dz\right\} = \exp\{-1\} = \frac{1}{e}. \;\square \tag{1.1.13}$$

In a more general setting, condition (1.1.12) may be non-necessary. For example, in the case of an insurance policy covering accidental death, the company is interested only in the random time X of death if it is a result of an accident. In this case, such an accident may not happen at all, and there is nothing unnatural in the assumption $P(X < \infty) < 1$. (In previous chapters we called such a r.v. and its distribution *defective* or *improper*.)

Next, we show that

> If a death or, in general, a failure may happen from several *independent* causes, then the hazard rate is the sum of the hazard rates corresponding to the separate causes.

(1.1.14)

To show what this means precisely, it suffices to consider the case of two causes. For example, we may distinguish the cases where death comes from natural reasons and where it is a result of an accident.

Denote by X_1 the lifetime under the assumption that the failure results *only* from the first cause. In the example above, it would mean that when considering X_1 we do not take into account the possibility of an accident. Denote by X_2 the corresponding r.v. with regard to the second cause. Say, in the same example, X_2 will be the moment of the accident, if any, with a lethal outcome.

Then the actual lifetime $X = \min\{X_1, X_2\}$. Assume X_1, X_2 to be independent and denote by $\mu_i(x)$ the hazard rate of X_i. Then, by (1.1.10) and by virtue of independence,

$$\begin{aligned} s(x) &= P(X > x) = P(\min\{X_1,X_2\} > x) = P(X_1 > x, X_2 > x) = P(X_1 > x)P(X_2 > x) \\ &= \exp\left\{-\int_0^x \mu_1(z)dz\right\}\exp\left\{-\int_0^x \mu_2(z)dz\right\} = \exp\left\{-\int_0^x (\mu_1(z)+\mu_2(z))\,dz\right\}. \end{aligned} \tag{1.1.15}$$

Comparing this with (1.1.10), we see that

$$\mu(z) = \mu_1(z) + \mu_2(z). \tag{1.1.16}$$

Clearly, this may be generalized to the case of three and more causes. We consider this instance in greater detail in Section 2.

Now, let us focus our attention in the rightmost member of (1.1.15). We see that for $s(\infty) = 0$, it suffices that (1.1.12) holds for only one of the hazard rates, $\mu_1(t)$ or $\mu_2(t)$. It is natural since now we can suppose that with positive probability one cause may not act (happen) at all.

EXAMPLE 8. Let us come back to the above example with two death causes. Let $\mu_1(z) = 1$; that is, X_1 is standard exponential, and $\mu_2(z) = 1/(1+z)^2$. Then, as was shown in Example 7, $\lim_{x\to\infty} P(X_2 > x) = e^{-1}$. This may be interpreted as if with probability $1/e$ there will be no accident. In accordance with (1.1.15), the total survival function

$$s(x) = \exp\left\{-\int_0^x \left(1+(1+z)^{-2}\right) dz\right\} = \exp\left\{-x - \frac{x}{1+x}\right\}, \quad \text{and} \quad s(\infty) = 0. \ \square$$

1.2 The time-until-death for a person of a given age

Consider a person age x. It is customary to use, for brevity, the term "*life-age-x*" or the symbol (x). The future (remaining) lifetime, or the *time-until-death* after x is denoted by $T(x)$. In other words, $T(x) = X - x$ *given* $X > x$. Hence, the distribution of $T(x)$ is the *conditional* distribution of $X - x$ given $X > x$. In particular, the survival probability

$$P(T(x) > t) = P(X > x+t \,|\, X > x).$$

This is the probability that the person of age x will live at least t years more. The traditional notation for this *conditional* survival function is ${}_tp_x$.

The corresponding d.f.

$$P(T(x) \le t) = P(X \le x+t \,|\, X > x) = 1 - P(X > x+t \,|\, X > x) = 1 - {}_tp_x.$$

In keeping with tradition, we denote $P(T(x) \le t)$ by ${}_tq_x$. So,

$${}_tq_x = 1 - {}_tp_x.$$

Note again that ${}_tq_x$ is the d.f. of $T(x)$.

Given a survival function $s(x)$,

$${}_tp_x = P(X > x+t \,|\, X > x) = \frac{P(X > x+t)}{P(X > x)} = \frac{s(x+t)}{s(x)}, \tag{1.2.1}$$

provided $s(x) \ne 0$.

Clearly, for a new-born $(x = 0)$

$${}_tp_0 = s(t).$$

From (1.2.1) and (1.1.10) it follows that

$${}_tp_x = \frac{s(x+t)}{s(x)} = \frac{\exp\{-\int_0^{x+t}\mu(z)dz\}}{\exp\{-\int_0^x \mu(z)dz\}} = \exp\left\{-\int_x^{x+t} \mu(z)dz\right\}. \tag{1.2.2}$$

EXAMPLE 1. What will happen if the force of mortality is doubled? From (1.2.2) it follows that if the new force of mortality, say, $\mu^*(x) = 2\mu(x)$, then the new probability ${}_tp_x^* = ({}_tp_x)^2$. A traditional example here concerns non-smokers and smokers. Assume that in a given country, the force of mortality for non-smokers is half that of smokers for all x's. Assume that for 20-year-old non-smokers the probability of attaining age 70, that is, ${}_{50}p_{20} = 0.95$. Then for smokers it is $(0.95)^2 = 0.9025$, so the difference is not dramatic. However, if for a 65-year non smoker the probability to live at least 15 years more, i.e., ${}_{15}p_{65} = 0.4$, then for smokers, this probability will be much less: $0.4^2 = 0.16$. □

The following two facts follow practically immediately from (1.2.1)-(1.2.2). Let $\mu_x(t)$ be the hazard rate for the r.v. $T(x)$, and let $f_{T(x)}(t)$ be the density of $T(x)$. As above, let $\mu(x)$ be the hazard rate for X. Then

$$\mu_x(t) = \mu(x+t), \tag{1.2.3}$$

and

$$f_{T(x)}(t) = \mu_x(t)\,{}_tp_x = \mu(x+t)\,{}_tp_x. \tag{1.2.4}$$

Indeed, since ${}_tq_x$ is the d.f. of $T(x)$, the density

$$f_{T(x)}(t) = \frac{\partial}{\partial t}\,{}_tq_x = \frac{\partial}{\partial t}(1 - {}_tp_x) = -\frac{\partial}{\partial t}\,{}_tp_x = \mu(x+t)\exp\left\{-\int_x^{x+t}\mu(z)dz\right\}, \tag{1.2.5}$$

where in the last step we differentiated (1.2.2) with respect to t. Since the exponent above equals ${}_tp_x$, we have $f_{T(x)}(t) = \mu(x+t)\,{}_tp_x$. On the other hand, by the definition of a hazard rate,

$$\mu_x(t) = \frac{f_{T(x)}(t)}{P(T(x) > t)} = \frac{f_{T(x)}(t)}{{}_tp_x} = \frac{\mu(x+t)\,{}_tp_x}{{}_tp_x} = \mu(x+t),$$

which implies (1.2.3).

Other frequently used probability characteristics are defined and denoted as follows:

$q_x = {}_1q_x = P(T(x) \leq 1)$, the probability that a life-age-x will die within one year;
$p_x = {}_1p_x = P(T(x) > 1)$, the probability that a life-age-x will live at least one year more;
${}_{t|u}q_x = P(t < T(x) \leq t+u) = P(x+t < X \leq x+t+u \,|\, X > x)$.

Following traditional notation, we omit the 1 in ${}_{t|1}q_x$ writing ${}_{t|}q_x$.

Clearly,

$${}_{t|u}q_x = P(T(x) \leq t+u) - P(T(x) \leq t) = {}_{t+u}q_x - {}_tq_x = {}_tp_x - {}_{t+u}p_x, \tag{1.2.6}$$

since ${}_tp_x = 1 - {}_tq_x$.

EXAMPLE 2. (a) Prove that for any $t \geq 1$,

$${}_tp_x = p_x \cdot {}_{t-1}p_{x+1}. \tag{1.2.7}$$

From a heuristic point of view, this is almost obvious. To attain age $x+t$, the person of age x should survive the first year (the probability of this is p_x) and *after that*, being $x+1$ years old, the person should live at least $t-1$ years. The formal proof is as follows:

$$
\begin{aligned}
{}_tp_x &= P(X>x+t\,|\,X>x) = \frac{P(X>x+t)}{P(X>x)} \\
&= \frac{P(X>x+1)}{P(X>x)}\cdot\frac{P(X>x+t)}{P(X>x+1)} = P(X>x+1\,|\,X>x)P(X>x+t\,|\,X>x+1) \\
&= P(X>x+1\,|\,X>x)P(X>x+1+t-1\,|\,X>x+1) = p_x\cdot{}_{t-1}p_{x+1}.
\end{aligned}
$$

It does not make sense, however, to carry out such formal calculations each time. In many problems below, it is sufficient to reason as we did in the beginning of this example.

(b) In Exercise 13, in a similar way, we prove that for integer $t=1,2,\dots$,

$$ {}_tp_x = p_x\cdot p_{x+1}\cdot\ldots\cdot p_{x+t-1} \tag{1.2.8} $$

(where $p_x = {}_1p_x$).

(c) Prove that

$$ {}_{t|u}q_x = {}_tp_x\cdot{}_uq_{x+t}. \tag{1.2.9} $$

We could apply, for example, (1.2.6) and (1.2.1) but a more illustrative approach is to use the same logic as above. For the event $\{t<T(x)\le t+u\}$ to occur, the person of age x should first attain the age $x+t$ (the probability of this is ${}_tp_x$) and *after that*, being $x+t$ years old, the person will die within u years. The probability of the latter event is ${}_uq_{x+t}$, and the two probabilities mentioned should be multiplied. A formal proof runs similar to what we did above and we skip it.

EXAMPLE 3. Let us return to the situation of Example 1.1-1.

(a) Find ${}_{60}p_{20}$. By (1.2.1), ${}_{60}p_{20}=s(80)/s(20)$. For $\alpha=2$, we have $\dfrac{s(80)}{s(20)}=\dfrac{(0.2)^2}{(0.8)^2}=$ 0.0625, while for $\alpha=1/2$ the fraction $\dfrac{s(80)}{s(20)}=\dfrac{\sqrt{0.2}}{\sqrt{0.8}}=0.5$. In the latter case, on the average, half of the population of 20-year-old people will attain the age of 80, while in the former case the corresponding share is less than 7%.

(b) Find ${}_{60|10}q_{20}$ for $\alpha=2$. By (1.2.6) ${}_{60|10}q_{20}={}_{60}p_{20}-{}_{70}p_{20}=\dfrac{s(80)}{s(20)}-\dfrac{s(90)}{s(20)}=$ $\dfrac{(0.2)^2}{(0.8)^2}-\dfrac{(0.1)^2}{(0.8)^2}=\dfrac{3}{64}$.

EXAMPLE 4. Find ${}_5p_{30}$ if the force of mortality $\mu(x)=1/70$ for all x's. We saw that a constant force of mortality corresponded to the exponential distribution with the parameter equal to the (single) value of $\mu(x)$. Thus, X is exponential with parameter $a=1/70$. In view of the lack-of-memory property, ${}_tp_x=P(X>x+t\,|\,X>x)=P(X>t)=s(t)=e^{-at}$. Thus, ${}_5p_{30}=\exp\{-\frac{1}{70}\cdot 5\}\approx 0.931$.

To what extent is the exponential distribution model realistic? Certainly, we cannot assume that the lack of memory property is true for the total lifetime: how long a person will live does depend on her/his age. However, for a young person (say, age 30 as above), the probability that she/he will not die within a fixed and relatively short period of time (as 5 years), the assumption of the constancy of the mortality rate is not artificial since the causes of death in this case are weakly related to age. The reader may look at the graph of a real force of mortality in Section 1.5.1. We continue this discussion in Exercise 21.

EXAMPLE 5 ([158, N1][2]). For an individual who is currently age 25, $\mu(x) = \frac{1}{110-x}$ for $0 \le x \le 110$. Calculate the expected number of years lived between ages 30 and 70 for that individual.

We proceed from the following two simple facts we formally prove in Exercises 7a-7b. First, the above type of mortality force corresponds to a uniform distribution; in our case, on $[0,110]$. Secondly, for a uniform X, the remaining life time $T(x)$ is also a uniform r.v.; in our case, $T = T(25)$ is uniform on $[0,85]$. The expected number of years mentioned is the r.v. $S = S(T)$ equal 0 if $T < 5$; equal to $T-5$ if $5 \le T \le 45$; and equal to 40 if $T > 45$. Hence,

$$E\{S\} = \int_5^{45} (t-5)\frac{1}{85}dt + 40 \cdot \frac{40}{85} \approx 28.24. \ \square$$

Next, we consider the mean future lifetime $E\{T(x)\}$. It is called a *complete-expectation-of-life* and in actuarial calculations is denoted by $\overset{\circ}{e}_x$.

Since ${}_tq_x$, as a function of t, is the d.f. of $T(x)$, formally we can write

$$\overset{\circ}{e}_x = E\{T(x)\} = \int_0^\infty t d\,{}_tq_x,$$

where d denotes differentiating with respect to t. One can calculate the last integral directly by using the formulas ${}_tq_x = 1 - {}_tp_x$ and (1.2.2). However, it is more convenient to use formula (**0**.2.2.2) and write

$$E\{T(x)\} = \int_0^\infty P(T(x) > t)dt = \int_0^\infty {}_tp_x dt.$$

From this and (1.2.1), using the variable change $y = x+t$, we get

$$E\{T(x)\} = \int_0^\infty \frac{s(x+t)}{s(x)}dt = \frac{1}{s(x)}\int_0^\infty s(x+t)dt = \frac{1}{s(x)}\int_x^\infty s(y)dy. \tag{1.2.10}$$

EXAMPLE 6. Find $\overset{\circ}{e}_{50}$ if $s(x) = (1-x/100)^2$ for $x \le 100$. By (1.2.10), $E\{T(50)\} = \frac{1}{s(50)}\int_{50}^\infty s(y)dy$. Since $s(y) = 0$ for $y > 100$ (see Example 1.1-1), we have

$$E\{T(50)\} = \frac{1}{s(50)}\int_{50}^{100} s(y)dy = \frac{1}{(1/2)^2}\int_{50}^{100} (1-y/100)^2 dy.$$

The change of variable $u = 1 - y/100$ will lead to $E\{T(50)\} = 50/3$. So, in the case under consideration, 50-year-old people live on the average only $16\frac{2}{3}$ years more. (See comments in Example 1.1-1.)

EXAMPLE 7. Find $\overset{\circ}{e}_x$ in the situation of Example 4. As was shown, we have the exponential distribution with parameter $1/70$. Hence, $E\{X\} = 70$. By virtue of the lack-of-memory property, $E\{T(x)\} = E\{X\}$ for any x. Hence, $\overset{\circ}{e}_x = 70$. Certainly, now the example

[2]Reprinted with permission of the Casualty Actuarial Society.

is artificial even for small x, since we are computing the whole life duration; see also Exercise 21. □

To compute $Var\{T(x)\}$, we first compute $E\{T^2(x)\}$. Again, one can do it directly, but it is more convenient to use (**0**.2.2.2) in the following way:

$$E\{T^2(x)\} = \int_0^\infty P(T^2(x) > t)dt = \int_0^\infty P(T(x) > \sqrt{t})dt = \int_0^\infty {}_{\sqrt{t}}p_x dt. \qquad (1.2.11)$$

With the change of variable $u = \sqrt{t}$, we get

$$E\{T^2(x)\} = \int_0^\infty 2u\,{}_u p_x du = \frac{1}{s(x)}\int_0^\infty 2u \cdot s(x+u)du. \qquad (1.2.12)$$

For the variance, we write $Var\{T(x)\} = E\{T^2(x)\} - (E\{T(x)\})^2$. In Exercise 29, we consider a particular example.

1.3 Curtate-future-lifetime

Often people count only the number of complete years survived, that is, the integer part of $T(x)$. This characteristic is called a *curtate-future-life-time* and is denoted by $K(x)$. By virtue of (1.2.9),

$$P(K(x) = k) = P(k \leq T(x) < k+1) = {}_k p_x \cdot {}_1 q_{x+k} = {}_k p_x \cdot q_{x+k}, \qquad (1.3.1)$$

(we omit the prefix 1 in ${}_1 q_{x+k}$).

The mean $E\{K(x)\}$ is denoted by e_x. By (1.3.1),

$$e_x = E\{K(x)\} = \sum_{k=0}^{\infty} kP(K(x) = k) = \sum_{k=0}^{\infty} k\,{}_k p_x q_{x+k}.$$

Sometimes it is more convenient to use formula (**0**.2.2.3), which gives

$$E\{K(x)\} = \sum_{k=0}^{\infty} P(K(x) > k) = \sum_{k=0}^{\infty} P(T(x) \geq k+1).$$

Since $T(x)$ is a continuous r.v., $P(T(x) = k+1) = 0$, and hence $P(T(x) \geq k+1) = P(T(x) > k+1) = {}_{k+1}p_x$. Eventually, by the variable change $n = k+1$,

$$E\{K(x)\} = \sum_{k=0}^{\infty} {}_{k+1}p_x = \sum_{n=1}^{\infty} {}_n p_x. \qquad (1.3.2)$$

EXAMPLE 1. Let X be exponential with $E\{X\} = r$. We know that such an example is artificial, and we are only considering it for illustrative purposes. As shown in Example 1.2-6, by the memoryless property, $\overset{\circ}{e}_x = E\{T(x)\} = E\{X\} = r$. By the same property, ${}_n p_x = P(T(x) > n) = P(X > n) = e^{-n/r}$. In view of (1.3.2),

$$e_x = \sum_{n=1}^{\infty} {}_n p_x = \sum_{n=1}^{\infty} e^{-n/r} = \sum_{n=1}^{\infty} (e^{-1/r})^n = \frac{e^{-1/r}}{1 - e^{-1/r}}, \qquad (1.3.3)$$

as a geometric series. Thus, the rounding procedure changes the expectation but not much. For example, if $r = 70$, then $\overset{\circ}{e}_x = 70$, while (1.3.3) gives $e_x \approx 69.501$. □

1.4 Survivorship groups

Consider now a group of l_0 newborns. In practice, demographers usually set $l_0 = 100,000$. Assume all lifetimes to be mutually independent with the same survival function $s(x)$. Let $\mathcal{L}(x)$ be the number of survivors to age x, and let $l_x = E\{\mathcal{L}(x)\}$.

We can view the survival to age x of a particular newborn as a success which happens with probability $s(x)$, and $\mathcal{L}(x)$ as the number of successes in l_0 independent trials. Hence, $\mathcal{L}(x)$ has the binomial distribution with parameters $s(x)$ and l_0. Then

$$\begin{aligned} l_x &= l_0 s(x), \qquad (1.4.1)\\ Var\{\mathcal{L}(x)\} &= l_0 s(x)(1-s(x)),\\ P(\mathcal{L}(x)=k) &= \binom{l_0}{k} s^k(x)(1-s(x))^{l_0-k}. \end{aligned}$$

From (1.4.1) we have

$$s(x) = l_x/l_0, \qquad (1.4.2)$$

which provides a way of estimating the survival function. For instance, if in a particular *homogeneous* group of 100,000 newborns, 96,381 persons survived 30 years, we may suppose that $s(30) \approx 0.96$. We omit here more precise statistical calculations such as those of confidence intervals, etc.

Now, denote by ${}_n\mathcal{D}(x)$ the number of deaths that occurred in the time interval $(x, x+n]$, and set ${}_nd_x = E\{{}_n\mathcal{D}(x)\}$. The probability that a particular person will die within the interval mentioned is $P(x < X \le x+n) = s(x) - s(x+n)$. So, similar to the argument above, ${}_n\mathcal{D}(x)$ has the binomial distribution with parameters $s(x) - s(x+n)$ and l_0. In particular, in view of (1.4.1),

$${}_nd_x = l_0[s(x) - s(x+n)] = l_x - l_{x+n}.$$

Set $d_x = {}_1d_x$ and observe that due to (1.4.2), the force of mortality

$$\mu(x) = -\frac{s'(x)}{s(x)} = -\frac{l'_x}{l_x},$$

where l'_x is the derivative of l_x in x. For a small interval $[x, x+\delta]$, the quantity $(l_{x+\delta} - l_x)/\delta$ may be considered an estimate of l'_x. If in a particular situation, we view $\delta = 1$ as small, the increment $l_{x+1} - l_x = -d_x$ may be viewed as an estimate of l'_x. Thus, we have arrived at the estimate

$$\mu(x) \approx \frac{d_x}{l_x} = \frac{l_x - l_{x+1}}{l_x}. \qquad (1.4.3)$$

We see that such an estimate coincides with q_x.

For instance, if as in the example above, from 100,000 newborns, 96,381 survived 30 years, and if 120 individuals died between 30 and 31 years, then we may write that

$$\mu(x) \approx \frac{120}{96381} \approx 0.0012.$$

In Section 1.5.2, we consider the estimation of $\mu(x)$ and interpolation of $\mu(x)$ within integer years in greater detail.

Next, note that from (1.4.2) and (1.2.1) it follows that

$$ {}_np_x = l_{x+n}/l_x. \tag{1.4.4}$$

Assume, for instance, that in the situation of the example above, in a group of 100,000 newborns, 95301 survived 35 years. Then the probability that a 30-year-old person will live at least 5 years more, that is, ${}_5p_{30}$, may be estimated as $(95301/96381) \approx 0.988$.

1.5 Life tables and interpolation

1.5.1 Life tables

Below we consider "Life table for the total population: United States, 2002".[3] It represents estimates of survival probabilities and other characteristics, based on the data on the entire population in the years around 2002. Traditionally, these estimates are given in terms of characteristics l_x and d_x, but it does not mean that just some particular 100,000 newborns were observed until the last died.

The table is reproduced from the "National Vital Statistics Reports" [3] of the National Center for Health Statistics; see also references there. One may find many interesting facts in such reports.

The definitions of the positions in the table are also cited from [3]. The values in the table concern only entire years. In the next section, we consider some possible interpolations for fractional years.

Column 1 shows the age interval between the two ages indicated.

Column 2 shows the probability (more precisely an estimate of this probability) of dying within the interval indicated in Column 1. The figures in this column are a result of the analysis of some data and form the basis of the life table. All subsequent columns are derived from them.

Column 3 contains the numbers l_x of persons from the original synthetic (virtual, imagined) cohort of 100,000 live births, who survive to the beginning of each interval. The calculation of l_x is based on (1.4.2) and (1.2.8). Having q_x's, we know $p_x = 1 - q_x$ and we can compute $l_x = l_0 p_0 \cdot \ldots \cdot p_{x-1}$. For example, $l_2 = 10^5(1-0.006971)(1-0.000472) = 99256.02903 \approx 99256$. As a matter of fact, we compute an estimate of the expected value of the number of survivors using only its integer part by tradition.

Column 4 shows the number d_x of persons dying in each successive age interval out of the original 100,000 births. Formally, we compute the corresponding expected value (more precisely, its estimate). Clearly, $d_x = l_x - l_{x+1}$, although the reader may notice that, for example, in the rows 6-7 and 7-8 we see that $99,163 - 99,148 = 15$ while in the next column this difference is estimated as 14. This is a result of round-off of the original products of probabilities.

We just quite briefly touch on the characteristics in Columns 5-6.

[3]There are fresher tables but they do not differ much from this table used also in the first edition. For the convenience of people who used in their study the first edition, we keep this table here too.)

TABLE 1. Life table for the total population: United States, 2002. The table is reproduced from "National Vital Statistics Reports" [3] of the National Center for Health Statistics.

1	2	3	4	5	6	7	8
Age	Probability of dying between ages x to $x+1$ q_x	Number surviving to age x l_x	Number dying between ages x to $x+1$ d_x	Person-years lived between ages x to $x+1$ L_x	Total number of person-years lived above age x T_x	Expectation of life at age x e_x	Force of Mortality (exponential interpolation) $\mu(x)$
0–1	0.006971	100 000	697	99 389	7 725 787	77.3	0.006995
1–2	0.000472	99 303	47	99 279	7 626 399	76.8	0.000472
2–3	0.000324	99 256	32	99 240	7 527 119	75.8	0.000324
3–4	0.000239	99 224	24	99 212	7 427 879	74.9	0.000239
4–5	0.000203	99 200	20	99 190	7 328 667	73.9	0.000203
5–6	0.000176	99 180	17	99 171	7 229 477	72.9	0.000176
6–7	0.000144	99 163	14	99 155	7 130 306	71.9	0.000144
7–8	0.000142	99 148	14	99 141	7 031 151	70.9	0.000142
8–9	0.000152	99 134	15	99 127	6 932 009	69.9	0.000152
9–10	0.000145	99 119	14	99 112	6 832 883	68.9	0.000145
10–11	0.000151	99 105	15	99 097	6 733 771	67.9	0.000151
11–12	0.000153	99 090	15	99 082	6 634 674	67.0	0.000153
12–13	0.000186	99 075	18	99 065	6 535 592	66.0	0.000186
13–14	0.000225	99 056	22	99 045	6 436 526	65.0	0.000225
14–15	0.000266	99 034	26	99 021	6 337 481	64.0	0.000266
15–16	0.000346	99 008	34	98 990	6 238 460	63.0	0.000346
16–17	0.000573	98 973	57	98 945	6 139 470	62.0	0.000573
17–18	0.000680	98 917	67	98 883	6 040 525	61.1	0.000680
18–19	0.000849	98 849	84	98 807	5 941 642	60.1	0.000849
19–20	0.000942	98 765	93	98 719	5 842 835	59.2	0.000942
20–21	0.000934	98 672	92	98 626	5 744 116	58.2	0.000934
21–22	0.000985	98 580	97	98 532	5 645 490	57.3	0.000985
22–23	0.000939	98 483	93	98 437	5 546 958	56.3	0.000939
23–24	0.000949	98 391	93	98 344	5 448 521	55.4	0.000949
24–25	0.000948	98 297	93	98 251	5 350 177	54.4	0.000948
25–26	0.000930	98 204	91	98 158	5 251 927	53.5	0.000930
26–27	0.000953	98 113	94	98 066	5 153 768	52.5	0.000953
27–28	0.000913	98 019	90	97 974	5 055 703	51.6	0.000913
28–29	0.000940	97 930	92	97 884	4 957 728	50.6	0.000940
29–30	0.000994	97 838	97	97 789	4 859 845	49.7	0.000994
30–31	0.001024	97 740	100	97 690	4 762 056	48.7	0.001025
31–32	0.001063	97 640	104	97 588	4 664 365	47.8	0.001064
32–33	0.001061	97 536	104	97 485	4 566 777	46.8	0.001062
33–34	0.001185	97 433	115	97 375	4 469 293	45.9	0.001186
34–35	0.001251	97 317	122	97 257	4 371 917	44.9	0.001252
35–36	0.001369	97 196	133	97 129	4 274 661	44.0	0.001370
36–37	0.001454	97 063	141	96 992	4 177 532	43.0	0.001455
37–38	0.001568	96 922	152	96 846	4 080 540	42.1	0.001569
38–39	0.001718	96 770	166	96 686	3 983 694	41.2	0.001719
39–40	0.001913	96 603	185	96 511	3 887 008	40.2	0.001915
40–41	0.002072	96 419	200	96 319	3 790 497	39.3	0.002074
41–42	0.002236	96 219	215	96 111	3 694 178	38.4	0.002239
42–43	0.002357	96 004	226	95 890	3 598 067	37.5	0.002360
43–44	0.002634	95 777	252	95 651	3 502 177	36.6	0.002637
44–45	0.002826	95 525	270	95 390	3 406 525	35.7	0.002830
45–46	0.003061	95 255	292	95 109	3 311 135	34.8	0.003066
46–47	0.003301	94 964	313	94 807	3 216 026	33.9	0.003306
47–48	0.003509	94 650	332	94 484	3 121 219	33.0	0.003515
48–49	0.003888	94 318	367	94 135	3 026 735	32.1	0.003896
49–50	0.004134	93 951	388	93 757	2 932 600	31.2	0.004143

TABLE 1 (continued).

1	2	3	4	5	6	7	8
	Probability of dying between ages x to $x+1$	Number surviving to age x	Number dying between ages x to $x+1$	Person-years lived between ages x to $x+1$	Total number of person-years lived above age x	Expectation of life at age x	Force of Mortality (exponential interpolation)
Age	q_x	l_x	d_x	L_x	T_x	e_x	$\mu(x)$
50–51	0.004422	93 563	414	93 356	2 838 843	30.3	0.004432
51–52	0.004822	93 149	449	92 925	2 745 487	29.5	0.004834
52–53	0.005003	92 700	464	92 468	2 652 563	28.6	0.005016
53–54	0.005549	92 236	512	91 980	2 560 094	27.8	0.005564
54–55	0.005845	91 724	536	91 456	2 468 114	26.9	0.005862
55–56	0.006719	91 188	613	90 882	2 376 658	26.1	0.006742
56–57	0.006616	90 576	599	90 276	2 285 776	25.2	0.006638
57–58	0.007621	89 976	686	89 634	2 195 500	24.4	0.007650
58–59	0.008344	89 291	745	88 918	2 105 866	23.6	0.008379
59–60	0.009429	88 546	835	88 128	2 016 948	22.8	0.009474
60–61	0.009747	87 711	855	87 283	1 928 820	22.0	0.009795
61–62	0.010877	86 856	945	86 384	1 841 536	21.2	0.010937
62–63	0.011905	85 911	1 023	85 400	1 755 153	20.4	0.011976
63–64	0.012956	84 888	1 100	84 338	1 669 753	19.7	0.013041
64–65	0.014099	83 789	1 181	83 198	1 585 414	18.9	0.014199
65–66	0.015308	82 607	1 265	81 975	1 502 217	18.2	0.015426
66–67	0.016474	81 343	1 340	80 673	1 420 242	17.5	0.016611
67–68	0.018214	80 003	1 457	79 274	1 339 569	16.7	0.018382
68–69	0.019623	78 545	1 541	77 775	1 260 295	16.0	0.019818
69–70	0.021672	77 004	1 669	76 170	1 182 520	15.4	0.021910
70–71	0.023635	75 335	1 781	74 445	1 106 350	14.7	0.023919
71–72	0.025641	73 555	1 886	72 612	1 031 905	14.0	0.025975
72–73	0.027663	71 669	1 983	70 678	959 294	13.4	0.028053
73–74	0.030539	69 686	2 128	68 622	888 616	12.8	0.031015
74–75	0.033276	67 558	2 248	66 434	819 994	12.1	0.033842
75–76	0.036582	65 310	2 389	64 115	753 560	11.5	0.037268
76–77	0.039775	62 921	2 503	61 670	689 444	11.0	0.040588
77–78	0.043338	60 418	2 618	59 109	627 775	10.4	0.044305
78–79	0.047219	57 800	2 729	56 435	568 666	9.8	0.048370
79–80	0.052518	55 071	2 892	53 624	512 230	9.3	0.053947
80–81	0.057603	52 178	3 006	50 676	458 606	8.8	0.059329
81–82	0.062260	49 173	3 061	47 642	407 930	8.3	0.064283
82–83	0.071461	46 111	3 295	44 464	360 288	7.8	0.074143
83–84	0.073437	42 816	3 144	41 244	315 825	7.4	0.076273
84–85	0.084888	39 672	3 368	37 988	274 581	6.9	0.088709
85–86	0.093123	36 304	3 381	34 614	236 593	6.5	0.097748
86–87	0.101914	32 923	3 355	31 246	201 979	6.1	0.107489
87–88	0.111270	29 568	3 290	27 923	170 733	5.8	0.117962
88–89	0.121196	26 278	3 185	24 686	142 810	5.4	0.129193
89–90	0.131694	23 093	3 041	21 573	118 125	5.1	0.141211
90–91	0.142761	20 052	2 863	18 621	96 552	4.8	0.154039
91–92	0.154390	17 189	2 654	15 862	77 931	4.5	0.167697
92–93	0.166569	14 535	2 421	13 325	62 069	4.3	0.182204
93–94	0.179282	12 114	2 172	11 028	48 744	4.0	0.197576
94–95	0.192507	9 942	1 914	8 985	37 716	3.8	0.213821
95–96	0.206215	8 028	1 656	7 201	28 730	3.6	0.230943
96–97	0.220375	6 373	1 404	5 671	21 530	3.4	0.248942
97–98	0.234947	4 968	1 167	4 385	15 859	3.2	0.267810
98–99	0.249887	3 801	950	3 326	11 474	3.0	0.287531
99–100	0.265146	2 851	756	2 473	8 148	2.9	0.308083
100+	1.00000	2 095	2 095	5 675	5 675	2.7	1.000000

Column 5 shows the number of person-years lived by the synthetic cohort within the corresponding interval. More precisely,

$$L_x = l_0 \int_0^1 t dF(x+t) + l_{x+1} = \int_0^1 t d(-l_{x+t}) + l_{x+1} = \int_0^1 l_{x+t} dt.$$

(The integration above is over t with the last step consisting in integration by parts.) The figures given in Column 5 are a result of approximate computations of the integral based on values of l_x at integer points.

Column 6 shows the number of person-years that would be lived after the beginning of the corresponding age interval. More precisely,

$$T_x = l_0 \int_0^\infty t dF(x+t) = \int_0^\infty t d(-l_{x+t}) = \int_0^\infty l_{x+t} dt.$$

(Again, integration is over t, and the last step consists in integrating by parts.) The figures given are a result of an approximation for the above integral.

Column 7 shows the life expectation at the beginning of the corresponding interval. The figures are based on (1.3.2).

Column 8 has been added and is not contained in the official table from [3]. We saw in Section 1.4 that one of the possible estimates for $\mu(x)$ was given by (1.4.3) and coincided with the estimate for q_x in Column 2. It corresponds to the so called linear interpolation which we discuss in the next Section 1.5.2.

For comparison, we give in Column 8 an alternative estimate given by the formula $\ln(l_x/l_{x+1})$. Such an estimate corresponds to the so called exponential interpolation which we also discuss in detail in Section 1.5.2. Since Column 3 contains rounded numbers, when calculating figures in Column 8, we proceeded directly from the data in Column 2. More precisely, since $l_{x+1} = l_x \cdot p_x = l_x \cdot (1 - q_x)$, we can write that

$$\ln(l_x/l_{x+1}) = -\ln(l_{x+1}/l_x) = -\ln(1 - q_x). \tag{1.5.1}$$

This is the formula we use for figures from Column 8. See again Section 1.5.2 for detail. The reader can observe that for small and moderate ages, both estimates in Column 2 and in Column 8, are close, but for large ages they somewhat differ. See also Example 1.5.2-1.

In Fig.2-4, we graph the empirical survival function, density, and force of mortality, respectively. These empirical functions are the *estimates* of the respective theoretical functions $s(x)$, $f(x)$, and $\mu(x)$. The estimates are based on the data presented in Table 1. More precisely, in Fig.2 we graph l_x/l_0, and in Fig.3 we graph d_x/l_0. The latter empirical curve has been slightly smoothed in the region of ages 82-84. The graph in Fig.4 represents the figures from Column 8. All graphs look quite natural, the survival function is concave in practically the whole range of ages, and the force of mortality is low for moderate ages. All of this attests to good living conditions in this country at this time.

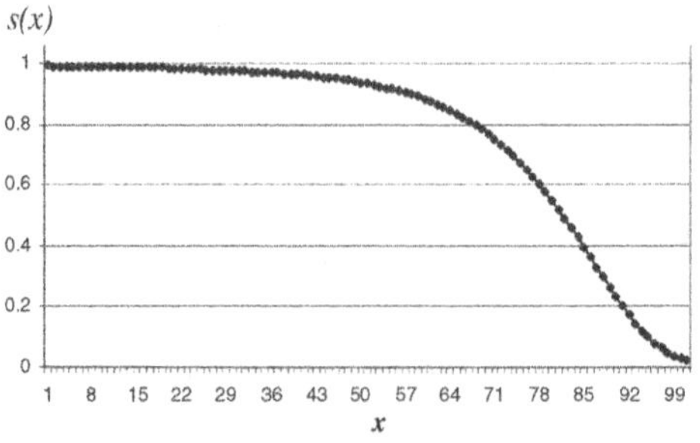

FIGURE 2. An estimate of the survival function $s(x)$ based on Table 1.

EXAMPLE 1. On the basis of Table 1, we estimate the values of the following characteristics.

(a) The density value $f(11)$. A rough estimate is $\frac{d_{11}}{l_0} = \frac{15}{100000} = 0.00015.$

(b) The probability $s(70)$. We may estimate it by $\frac{l_{70}}{l_0} = \frac{75335}{100000} = 0.75335.$

(c) The probability that a person of age 30 will live at least 50 years more. We have ${}_{50}p_{30} = \frac{s(80)}{s(30)} = \frac{l_{80}}{l_{30}} \approx \frac{52178}{97740} \approx 0.534.$

(d) The probability that a person of age 20 will die between the ages 70 and 80. In this

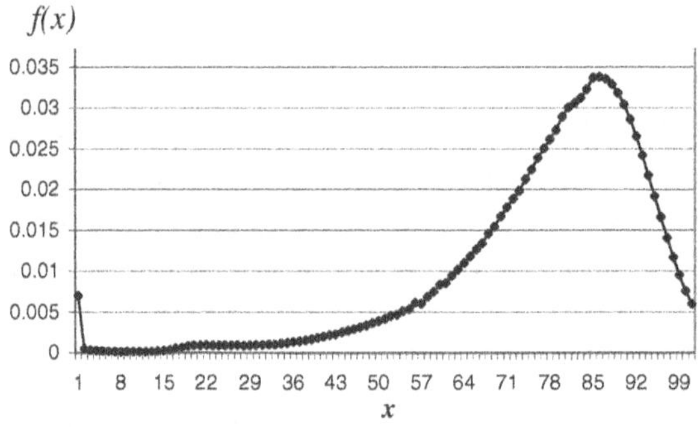

FIGURE 3. An estimate of the density $f(x)$ based on Table 1.

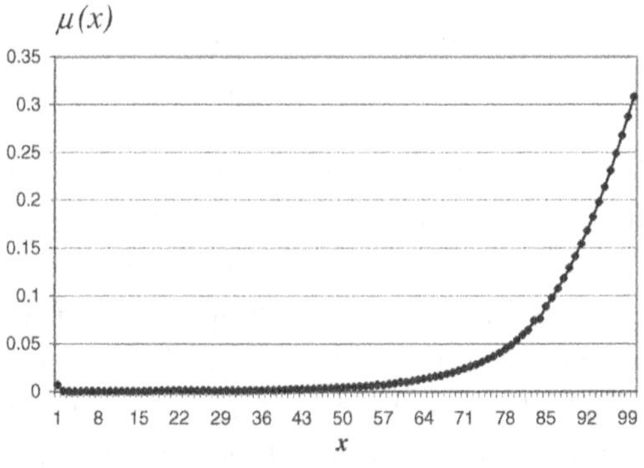

FIGURE 4. An estimate of the force of mortality $\mu(x)$ based on Table 1.

case, we deal with ${}_{50|10}q_{20} = \dfrac{l_{70} - l_{80}}{l_{20}} \approx \dfrac{75335 - 52178}{98672} \approx 0.235.$

(e) The mode of T, that is, the "most probable" value, and local extrema of $f(x)$. Analyzing the table, we see that the maximum value of d_x is attained at $x = 85$.

There are also local maxima at $x = 0, 19, 21, 26$. The fact that we observe a local maximum at $x = 0$ is understandable since it concerns the mortality of newborns. As to the three other ages, we should remember that we are dealing with estimates, and we must not think that, for example, the age of 20 is less dangerous than that of 21. However, we may conclude that the region of $19 - 26$ is characterized by having a local mortality maximum (see also Fig.3). The reader is invited to give her/his explanation of this fact.

The minimum corresponds to the region of 6 to 10.

(f) The force of mortality attains its minimum in the same region of $6 - 10$ and is increasing except $x = 0$. In Exercise 34, the reader is encouraged to interpret all these results in more detail. □

1.5.2 Interpolation for fractional ages

Except the first year of life, life tables give usually the survival probabilities for integer years, that is, for the curtate lifetime K. To find the probabilities ${}_tq_x$ for all t's and x's, we should establish interpolation rules for the distribution of the lifetime between integers. We consider here the following two types of interpolation.

Linear interpolation, or the uniform distribution over each year of age. Assume that for any interval with integer endpoints, if death occurs in this interval, all possible time moments are equally likely to be the time of death. This means that the density $f_X(x)$ is constant in any interval $(k, k+1)$ where k is an integer. Since $f_X(x) = -s'(x)$, this implies that $s'(x)$ is constant in each $(k, k+1)$, and hence $s(x)$ is linear in each $(k, k+1)$. Consequently, taking an intermediate point $k+t$, where $0 < t < 1$, we may write that

$$s(k+t) = (1-t)s(k) + ts(k+1). \tag{1.5.2}$$

In case (1.5.2), we may readily write a formula for the force of mortality. By (1.1.7), $\mu(k+t) = -s'(k+t)/s(k+t)$, and differentiating (1.5.2) in t, we have $s'(k+t) = s(k+1) - s(k)$. This leads to

$$\mu(k+t) = \frac{s(k) - s(k+1)}{s(k+t)} \quad \text{for } t \in [0,1]. \tag{1.5.3}$$

For $t = 0$, it coincides with q_k.

Exponential interpolation. Now let us assume that within any interval $(k, k+1)$, the force of mortality is a constant μ_k. Then by (1.1.10), for $t \in (0,1)$,

$$\ln s(k+t) = -\int_0^{k+t} \mu(z)dz = -\int_0^k \mu(z)dz - \int_k^{k+t} \mu(z)dz = \ln s(k) - \int_k^{k+t} \mu_k dz = \ln s(k) - \mu_k t. \tag{1.5.4}$$

Setting $t = 1$, we have $\ln s(k+1) = \ln s(k) - \mu_k$, which gives

$$\mu_k = \ln s(k) - \ln s(k+1) = \ln[s(k)/s(k+1)]. \tag{1.5.5}$$

We used this interpolation in Table 1.5.1-1. Substitution into (1.5.4) leads to

$$\ln s(k+t) = (1-t)\ln s(k) + t\ln s(k+1). \tag{1.5.6}$$

Comparing it with (1.5.2), we see that, in this case, linear interpolation is being carried for $\ln s(x)$ rather than for $s(x)$.

Some straightforward examples are considered in Exercise 37. In Exercise 38, we show that linear interpolation leads to precise formulas if X is uniformly distributed, and the same is true for exponential interpolation if X is an exponential r.v.

Now, let us compare the two above estimates of $\mu(k)$ for an integer k. As has been noted repeatedly, $q_k = \frac{s(k)-s(k+1)}{s(k)} = \frac{l_k - l_{k+1}}{l_k}$, and the estimate based on (1.5.3) equals q_k. The estimate based on (1.5.5) is equal to $-\ln(1-q_k)$ as shown in (1.5.1).

As we know, $-\ln(1-z) \approx z$ for small z, or more precisely, $-\ln(1-z) = z + \frac{1}{2}z^2 + o(z^2)$; see (**0**.4.2.8). Thus, for small q_k, both estimates should be close; for larger q_k, they will slightly differ. The latter case corresponds to large ages.

EXAMPLE 1. Consider estimates of $\mu(k)$ in Table 1.5.1-1. The estimates based on (1.5.3) coincide with q_k, and Column 2 contains estimates based on linear interpolation. Column 8 corresponds to exponential interpolation. We see that for ages from one to relatively moderate numbers, the estimates either coincide up to the 6th digit or differ at most at the 5th digit. For larger ages, the difference is larger. For k equal to 70 for example, the two respective estimates are 0.023635 and 0.023919. However, the estimates for $\mu(90)$ are 0.142761 and 0.154039. For $k = 99$, the corresponding numbers are 0.265146 and 0.308083.

Note again that we consider such precise numbers merely for illustration. Of course, it does not mean that the real estimates are accurate up to the sixth digit. On the contrary, the accuracy of the original estimates of q_x from which we proceed may be much less, and in this case, the difference between the two interpolations may not have essential significance.

EXAMPLE 2 ([153, N31][4]). You are given the following: (a) $q_{70} = 0.04$; (b) $q_{71} = 0.05$; (c) deaths are uniformly distributed within each year of age. Calculate the probability that (70) will die between ages 70.5 and 71.5. We may write

$$\begin{aligned} P(0.5 \le T(70) \le 1.5) &= P(0.5 \le T(70) \le 1) + P(1 < T(70) \le 1.5) \\ &= P(0.5 \le T(70) \le 1) + P(T(70) \le 1.5 \,|\, T(70) > 1)P(T(70) > 1) \\ &= P(0.5 \le T(70) \le 1) + P(T(71) \le 0.5)P(T(70) > 1). \end{aligned}$$

Using the uniformity assumption, we have $P(0.5 \le T(70) \le 1) = \frac{1}{2}P(0 \le T(70) \le 1) = \frac{1}{2}q_{70}$, and similarly, $P(T(71) \le 0.5) = \frac{1}{2}q_{71}$. Thus,

$$P(0.5 \le T(70) \le 1.5) = \frac{1}{2}q_{70} + (1-q_{70})\frac{1}{2}q_{71} = 0.044. \ \square$$

[4]Reprinted with permission of the Casualty Actuarial Society.

Certainly, the assumptions above are merely used for approximation purposes. For actual survival functions, neither the densities nor the forces of mortality are constant within integer years. However, such assumptions may be close to reality (see, for example, a remark on uniformity at the end of the next Section 1.6), and in any case, they lead, as a rule, to a good interpolation accuracy. (Not saying that, as we saw, for moderate ages different interpolations lead to close results.)

1.6 Analytical laws of mortality

Over many years, there has been a great deal of interest in finding analytical representations for survival functions. All such attempts were based on the belief that the duration of human life is subject to relatively simple universal laws. These laws, if any, have not been found yet, and the formulas below should be viewed merely as possible approximations of "real" survival functions. Such approximations may be helpful in calculations and in a theoretical analysis.

De Moivre in 1729 suggested the simplest approximation: the uniform distribution on an interval $[0,\omega]$. In this case, for $x\in[0,\omega]$,

$$s(x)=1-\frac{x}{\omega}, \text{ and } \mu(x)=1/(\omega-x), \tag{1.6.1}$$

which may be easily obtained using (1.1.7). In the actuarial literature, one may often see references to this case as to De Moivre's law of mortality.

Certainly, if we consider the future lifetime of a newborn, this is a very rough approximation. However, as we will see below, for some particular age intervals, such an approximation may turn out to be not too bad.

Gompertz (1825) considered the exponential force of mortality $\mu(x)=Be^{\alpha x}$ where B and α are positive parameters. Substituting it into (1.1.10), after integration we get that

$$s(x)=\exp\{-B(e^{\alpha x}-1)/\alpha\}.$$

Makeham (1860) suggested to model $\mu(x)$ by

$$\mu(x)=Be^{\alpha x}+A, \tag{1.6.2}$$

where an additional parameter $A\geq -B$. We may interpret (1.6.2) in the spirit of the scheme at the end of Section 1.1, where the duration of life is influenced by two independent factors, and the corresponding hazard rates are summed. In the case (1.6.2), $\mu_2(z)$ from (1.1.16) is the constant A, and hence the moment of the lethal accident, X_2, is an exponential r.v. with parameter A. In our interpretation, the first term in (1.6.2) corresponds to death from other causes.

The interpretation above is, however, rather speculative and should be considered optional. Moreover, the estimate of parameter A based on real data may occur to be negative. If it is larger than $-B$, the function $\mu(x)$ based on such an estimate will be positive, and we can use it. So, at least in this instance, we should perceive the representation (1.6.2) as a whole, not interpreting its separate terms.

Again applying (1.1.10), we obtain that for the case (1.6.2),

$$s(x)=\exp\{-Ax-B(e^{\alpha x}-1)/\alpha\}. \tag{1.6.3}$$

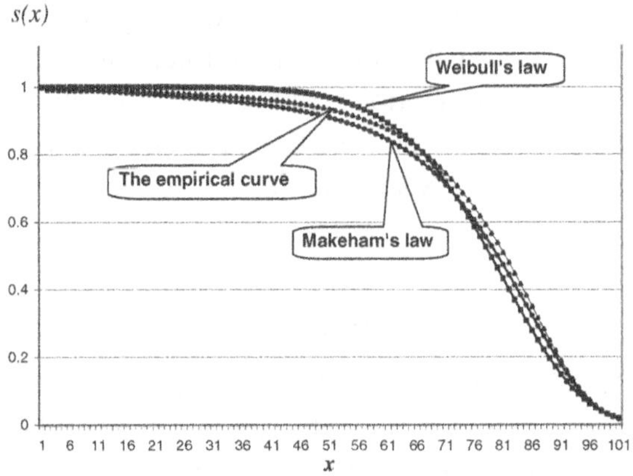

FIGURE 5. The empirical survival function based on the data from Table 1 and the Makeham and Weibull approximations.

In 1889, Makeham generalized his law using three terms and representing the force of mortality by

$$\mu(x) = A + Cx + Be^{\alpha x}, \tag{1.6.4}$$

where C is another parameter. See also Exercise 32.

In 1939, Weibull suggested to approximate the force of mortality by the power function $\mu(x) = Ax^n$, where A, n are parameters. In this case, the integration in (1.1.10) leads to

$$s(x) = \exp\{-Ax^{n+1}/(n+1)\}.$$

It is interesting to check how these laws fit the data from Table 1.

In Fig.5, the empirical survival function is graphed along with approximations concerning the Makeham and Weibull laws. The curves in Fig.5 correspond to the functions

$$s_{\text{Makeham}}(x) = \exp\{-0.001x - 0.0005(e^{0.09x} - 1)\},$$
$$s_{\text{Weibull}} = \exp\{-4(x/100)^7\}.$$

We see that visually both curves fit reasonably well. The parameters chosen are not optimal and serve merely for illustrative purposes.

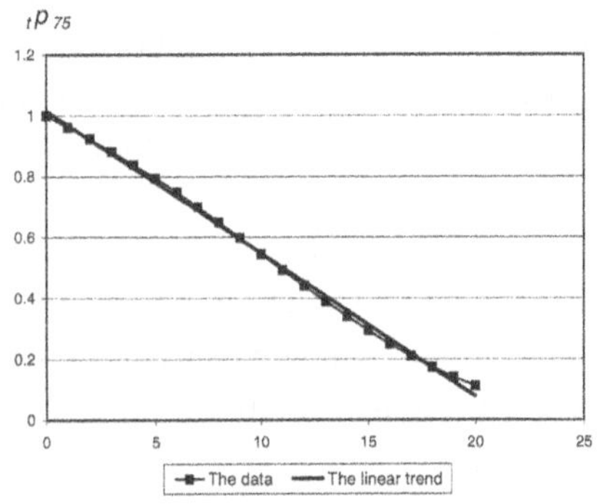

FIGURE 6.

One more interesting fact. In Fig.6, we graph the empirical survival function for $T(75)$ in the age interval $[75, 95]$; that is, the function l_{75+t}/l_{75} for $0 \le t < 20$. The straight line in Fig.6 indicates the linear trend. We know that the linear survival function corresponds to the uniform distribution. So, we watch, actually, an amazing fact: the distribution of the remaining lifetime $T(75)$ of people aged 75 is very close to the distribution uniform on $[0, 20]$. For example, it is almost equally likely that a chosen at random 75-year-old person will die at age 76 or at age 94.

Route 1 ⇒ page 437

2 A MULTIPLE DECREMENT MODEL

2,3

First, note that the theory above may serve for modeling not only lifetimes of real individuals but the durations of various insurance contracts which may terminate for reasons other than death of the client. For example, a medical insurance may terminate if the client ceases paying premiums, or moves to another area. So, in general, it makes sense to talk about the *time-until-termination* rather than the time-until-death.

Whether we talk about a real lifetime or the duration of a contract in general, it is often important to distinguish the causes of termination. For example, the benefit payment in a life insurance contract may depend on whether death occurs because of natural causes or as a result of an accident. The same contract may include also a one-time benefit payment in the case of disability, after which the contract terminates. In the last case, we distinguish the following three instances: "natural" death, a lethal accident, and disability.

We begin with a general framework.

2.1 A single life

Assume that there are m possible causes of the termination of a contract. The model below involves two r.v.'s: the time-until-termination T and the number J of the cause which will become the reason for termination. The r.v. T is a continuous positive r.v., and the r.v. J takes on values $1, ..., m$.

In this context, termination itself is called a *decrement*, and the corresponding model—the *multiple decrement model*.

For certainty, we will talk about the future lifetime of a person of age x (a life-age-x), assuming that $T = T(x)$ and $J = J(x)$. The symbols for all characteristics below will involve the corresponding index x. In situations that do not involve age, this index should be omitted.

Let

$$ {}_tq_x^{(j)} = P(T(x) \le t, J(x) = j), $$

the probability that the decrement will occur before or at time t *and* it will happen due to cause j.

As we know, the probability density of a one-dimensional r.v. is the derivative of the corresponding d.f. Then, with regard to the continuous component T of the r.vec. (T, J), it is natural to define the function

$$ f_{TJ}(t, j) = \frac{\partial}{\partial t}\, {}_tq_x^{(j)}. \tag{2.1.1} $$

We omitted the index x in the l.-h.s. When it does not cause confusion, we will omit the index TJ also, writing simply $f(t, j)$.

The function $f(t,j)$ may be considered the joint density of the vector (T,J), but we should keep in mind that since the r.v. J is discrete, while for a fixed j the function $f(t,j)$ plays the role of a density, for a fixed t this function is a mass function (that is, it deals with the probability that J will be *equal* to j). More precisely, for any set B in the real line,

$$P(T \in B, J = j) = \int_B f(t,j)dt,$$

while for any set B and any set K of integers,

$$P(T \in B, J \in K) = \sum_{j \in K} \int_B f(t,j)dt. \tag{2.1.2}$$

Let ${}_tq_x = P(T(x) \leq t)$, the marginal d.f. of the r.v. $T(x)$. In actuarial notation, this characteristic is often denoted by ${}_tq_x^{(\tau)}$ in order to emphasize that the function refers to all causes. We will omit the superscript (τ).

Since $P(T(x) \leq t) = \sum_{j=1}^m P(T \leq t, J = j)$,

$${}_tq_x = \sum\nolimits_{j=1}^m {}_tq_x^{(j)}. \tag{2.1.3}$$

As usual, we set ${}_tp_x^{(j)} = 1 - {}_tq_x^{(j)}$, ${}_tp_x = 1 - {}_tq_x$.

Denote by $f_T(t)$ the marginal density of T. Similarly to (2.1.3),

$$f_T(t) = \sum\nolimits_{j=1}^m f(t,j). \tag{2.1.4}$$

One can derive (2.1.4) either proceeding from the usual logic of finding marginal distributions (see Section **0**.1.3.2), or differentiating both sides of (2.1.3). In the latter case, we recall that $f(t,j) = \frac{\partial}{\partial t}\,{}_tq_x^{(j)}$ by definition, and $f_T(t) = \frac{\partial}{\partial t}\,{}_tq_x$ as is true for any density.

Following the general definition (1.1.5) of hazard rate, we define the *force of decrement* (mortality if we talk about a lifetime) corresponding to $T(x)$ as

$$\mu_x(t) = \frac{f_{T(x)}(t)}{P(T(x) > t)} = \frac{f_{T(x)}(t)}{{}_tp_x}. \tag{2.1.5}$$

The *force of decrement due to cause* j is defined as

$$\mu_x^{(j)}(t) = \frac{f(t,j)}{P(T(x) > t)} = \frac{f(t,j)}{{}_tp_x}. \tag{2.1.6}$$

Note that, while the numerator involves the number of only one cause, the denominator in (2.1.6) is the same as in (2.1.5) and concerns all causes. The significance of $\mu_x^{(j)}(t)$ may be clarified as follows. For an infinitesimally small dt,

$$\mu_x^{(j)}(t)dt = P(t < T(x) \leq t + dt,\ J = j \,|\, T(x) > t). \tag{2.1.7}$$

The r.-h.s. of (2.1.7) is the probability that a t-year-old person will die within the (infinitesimally small) interval $[t, t+dt]$, and it will happen from cause j.

From (2.1.4)-(2.1.6) it immediately follows that

$$\mu_x(t) = \sum_{j=1}^{m} \mu_x^{(j)}(t). \tag{2.1.8}$$

Using the general formula (1.1.10) that connects the hazard rate and the tail of the distribution, we can write

$$_tp_x = \exp\left\{-\int_0^t \mu_x(s)ds\right\}. \tag{2.1.9}$$

Two things are noteworthy. First, similar formulas for $_tp_x^{(j)}$ may be not true since the denominator in (2.1.6) contains a probability concerning all causes. This is discussed in detail in Section 2.2.

Secondly, as seen from (2.1.8)-(2.1.9), in order that $_tp_x \to 0$ as $t \to \infty$, it suffices to require $\int_0^\infty \mu_x^{(j)}(s)ds$ to diverge only for some j rather than for all j. This has a clear interpretation. The contract will ever terminate iff at least one cause acts. We discussed it in detail at the end of Section 1.1.

From (2.1.6) it follows that

$$f(t,j) = {}_tp_x \cdot \mu_x^{(j)}(t). \tag{2.1.10}$$

This allows us, in particular, to write a nice formula for $P(J=j\,|\,T=t)$. Following the usual logic of calculating conditional probabilities (see, e.g., Section **0**.7.1 and, in particular, (**0**.7.1.5)), we write

$$P(J=j\,|\,T=t) = \frac{f(t,j)}{f_T(t)}.$$

In view of (2.1.10),

$$P(J=j\,|\,T=t) = \mu_x^{(j)}(t)\frac{_tp_x}{f_T(t)}. \tag{2.1.11}$$

It is crucial that the second factor does not depend on j, while the total sum of these probabilities, $\sum_{j=1}^{m} P(J=j\,|\,T=t)$, naturally equals one. Thus, since $P(J=j\,|\,T=t)$ is proportional to $\mu_x^{(j)}(t)$,

$$P(J=j\,|\,T=t) = \frac{\mu_x^{(j)}(t)}{\sum_{j=1}^{m}\mu_x^{(j)}(t)} = \frac{\mu_x^{(j)}(t)}{\mu_x(t)}. \tag{2.1.12}$$

(Summing up both sides of (2.1.11), we have $\dfrac{_tp_x}{f_T(t)} = \sum\limits_{j=1}^{m}\mu_x^{(j)}(t)$. The last equality in (2.1.12) follows from (2.1.8).)

The marginal probability is given by

$$P(J=j) = \int_0^\infty f(t,j)dt. \tag{2.1.13}$$

EXAMPLE 1[5]. Let $m=2$, $\mu_x^{(1)}(t) = 2t$, and $\mu_x^{(2)}(t) = 3t^2$ for a fixed x. The constants 2 and 3 are chosen to make calculations below simpler. Actually, the choice of constants is

[5]The idea of this example is the same as in examples from Section 10.2 of [19].

connected with the choice of a unit of time, so for a certain choice of units, the constants above may be realistic. We discuss this at the end of the example.

Our goal is to demonstrate how we can find the joint, marginal, and conditional distributions for the random vector $(T(x), J(x))$, and also $E\{T(x)\}$.

By (2.1.8), $\mu_x(t) = 2t + 3t^2$. Since $\mu_x, \mu_x^{(1)}$, and $\mu_x^{(2)}$ do not depend on x, we omit the index x in calculations. By (2.1.9),

$$ {}_tp_x = P(T(x) > t) = \exp\left\{-\int_0^t (2s + 3s^2)ds\right\} = \exp\left\{-t^2 - t^3\right\} \text{ for } t \geq 0. $$

By (2.1.10),

$$ f(t,1) = 2t\exp\left\{-t^2 - t^3\right\},\ f(t,2) = 3t^2\exp\left\{-t^2 - t^3\right\}. $$

By (2.1.4),

$$ f_T(t) = (2t + 3t^2)\exp\left\{-t^2 - t^3\right\}. $$

In view of (2.1.13), $P(J = 1) = \int_0^\infty f(t,1)dt = \int_0^\infty 2t\exp\left\{-t^2 - t^3\right\}dt$. The last integral is not analytically computable, but using appropriate software, it is easy to obtain that $P(J = 1) \approx 0.527$, and accordingly, $P(J = 2) = 1 - P(J = 1) \approx 0.483$.

By (2.1.12),

$$ P(J = 1\,|\,T = t) = \frac{2t}{2t + 3t^2},\ P(J = 2\,|\,T = t) = \frac{3t^2}{2t + 3t^2}. $$

Now,

$$ E\{T\} = \int_0^\infty t f_T(t)dt = \int_0^\infty t(2t + 3t^2)\exp\left\{-t^2 - t^3\right\}dt \approx 0.669 $$

(to obtain this number we again use software).

The conditional density of T given $J = j$ is $f(t\,|\,j) = f(t,j)/P(J = j)$. Hence,

$$ E\{T\,|\,J = 1\} = \int_0^\infty t\frac{f(t,1)}{P(J = 1)}dt = \frac{1}{P(J = 1)}\int_0^\infty 2t^2\exp\left\{-t^2 - t^3\right\}dt \approx \frac{0.315}{0.527} \approx 0.598. $$

We may find $E\{T\,|\,J = 2\}$ in the same way, but it is easier to write that

$$ E\{T\} = E\{T\,|\,J = 1\}P(J = 1) + E\{T\,|\,J = 2\}P(J = 2). $$

The only unknown here is $E\{T\,|\,J = 2\}$. Calculations lead to $E\{T\,|\,J = 2\} \approx 0.733$.

If the unit of time is one year, the answers do not appear realistic, but if we assume that the unit is 100 years, and, for example, $x = 10$ years, then $E\{T\} = 66.9$ years, $E\{T\,|\,J = 1\} = 59.8$ years, and $E\{T\,|\,J = 2\} = 73.2$ years, which does not look improbable.

EXAMPLE 2. Let $q_{50}^{(1)} = 0.05$, $q_{51}^{(1)} = 0.06$, $q_{50}^{(2)} = 0.01$, $q_{51}^{(2)} = 0.02$, and $q_{52}^{(2)} = 0.03$. Find ${}_{2|}q_{50}^{(2)}$, i.e., the probability that a life (50) will terminate between ages 52 and 53, and it will happen from cause 2. (We omit 1 in ${}_{t|1}q_x$.)

We have ${}_{2|}q_{50}^{(2)} = {}_2p_{50} \cdot q_{52}^{(2)}$ (the probability that the person will survive age 52, and subsequently, being 52 years old, will die within a year from cause 2). Next, ${}_2p_{50} = p_{50}p_{51}$. By (2.1.3), $q_{50} = q_{50}^{(1)} + q_{50}^{(2)} = 0.06$. Similarly, $q_{51} = 0.08$, and consequently, $p_{50} = 1 - q_{50} = 0.94$, and $p_{51} = 0.92$. Then ${}_2p_{50} = 0.94 \cdot 0.92 = 0.8648$, and ${}_{2|}q_{50}^{(2)} = 0.8648 \cdot 0.03 = 0.0259$. □

2.2 Another view: net probabilities of decrement

For the sake of simplicity, let us consider only two causes. Suppose that we can represent the termination time $T(x)$ by the relation

$$T(x) = \min\{T_1(x), T_2(x)\}, \tag{2.2.1}$$

where the r.v. $T_1 = T_1(x)$ may be interpreted as the time of termination due to the first cause if the second cause does not occur (or, more precisely, does not act), and $T_2 = T_2(x)$ is the corresponding characteristic for the second cause. Since termination may occur from either cause, whichever occurs first, we have written above $\min\{T_1(x), T_2(x)\}$.

We do not claim here that r.v.'s $T_1(x)$ and $T_2(x)$ with such properties can always be determined. We simply assume for a while that it is possible to do.

Below, we omit the argument x.

For us, it is important to keep in mind that the r.v.'s T_1, T_2 may be dependent. For example, healthier people—at least for some groups—are less prone to accidents.

In the above model, the event $\{J = 1\}$ (termination comes from the first cause) coincides with the event $\{T_1 < T_2\}$, and the event $\{J = 2\}$ with $\{T_1 \geq T_2\}$. (We suppose T_1, T_2 are continuous r.v.'s, so where we write the non-strict inequality $\geq$ is not important.)

In such a setup,

$$ {}_tq^{(1)} = P(T \leq t, J = 1) = P(T \leq t, T_1 < T_2) = P(T_1 \leq t, T_1 < T_2), \tag{2.2.2}$$

$$ {}_tq^{(2)} = P(T \leq t, J = 2) = P(T \leq t, T_1 \geq T_2) = P(T_2 \leq t, T_1 \geq T_2). \tag{2.2.3}$$

If we know the joint distribution of (T_1, T_2), we know the functions ${}_tq^{(j)}$ and can define and calculate the density $f(t, j)$ and the rates $\mu^{(j)}(t)$, as we did it in (2.1.1) and (2.1.6). (We again omit x.)

Denote by $\tilde{\mu}^{(j)}(t)$ the marginal hazard rate for T_j. The tilde is used to distinguish this characteristic from $\mu^{(j)}(t)$, the force of decrement due cause j.

The question is whether $\tilde{\mu}^{(j)}(t)$ coincides with $\mu^{(j)}(t)$.

In general, it certainly does not since, for instance, $\tilde{\mu}^{(1)}(t)$ is a characteristic of the marginal distribution of T_1, while $\mu^{(1)}(t)$ is specified by the joint distribution of (T_1, T_2). (As has been previously noted, the denominator in the definition of $\mu^{(1)}(t)$ in (2.1.6) or the conditional probability in (2.1.7) involve the r.v. $T = \min\{T_1, T_2\}$.)

We illustrate this below by an example, but it is worthwhile to emphasize that it is not necessary: for the reason mentioned above, we should not expect that $\tilde{\mu}^{(j)}(t) = \mu^{(j)}(t)$.

EXAMPLE 1. Let T_1, T_2 be uniformly distributed on the triangular $\{(t_1,t_2) : t_1 \geq 0,\, t_2 \geq 0,\, t_1+t_2 \leq 1\}$; see Fig.7.

It is easy to see from this figure that $P(T_1 > t) = (1-t)^2$. (This probability equals the area of the middle triangle divided by the total area with the latter being equal to $1/2$.)

Hence, by (1.1.7), the marginal hazard rate $\tilde{\mu}^{(1)}(t) = \dfrac{2(1-t)}{(1-t)^2} = \dfrac{2}{1-t}$. On the other hand, $P(\min\{T_1,T_2\} > t) = (1-2t)^2$. (We should calculate the area of the smallest designated triangle and divide it by the total area.)

The hazard rate corresponding to the last tail probability is $\mu(t) = \dfrac{4(1-2t)}{(1-2t)^2} = \dfrac{4}{1-2t}$.

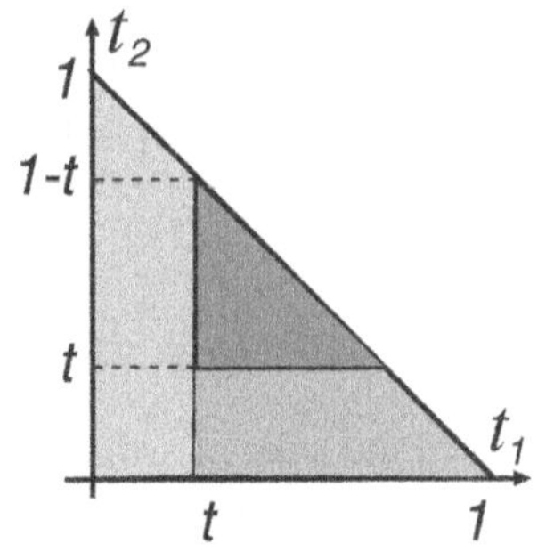

FIGURE 7.

We know that $\mu(t) = \mu^{(1)}(t) + \mu^{(2)}(t)$. On the other hand, by the symmetry of the joint distribution, $\mu^{(1)}(t) = \mu^{(2)}(t)$. Hence, $\mu^{(1)}(t) = \frac{1}{2}\mu(t) = \dfrac{2}{1-2t} \neq \dfrac{2}{1-t}$. Certainly, the point of this example is that T_1, T_2 are dependent. (See, for instance, Example **0**.1.3.2-2.) □

However, if the causes are independent, the situation is simpler.

Proposition 1 *If T_1 and T_2 are independent, then*

$$\tilde{\mu}^{(j)}(t) = \mu^{(j)}(t). \tag{2.2.4}$$

A short proof will be given in Section 2.4. In Exercise 42, we discuss the assertion of this proposition from a heuristic point of view.

Thus, if the causes act independently, the model is transparent, and the marginal hazard rates coincide with the forces of decrement.

Before considering an example, note again that if T_1, T_2 are independent, then

$$P(\min\{T_1,T_2\} > t) = P(T_1 > t,\, T_2 > t) = P(T_1 > t)P(T_2 > t). \tag{2.2.5}$$

EXAMPLE 2. Let T_1 be the future lifetime of (50) in the case when we do not take into account the possibility of accidents, and let T_2 be the time of a lethal accident, if any. So, the real lifetime $T = \min\{T_1,T_2\}$. Assume that T_1 and T_2 are independent, T_1 is uniformly distributed on $[0,50]$ (which corresponds to De Moivre's law with $\omega = 100$), and T_2 is exponential with a hazard rate of $\mu = 0.01$.

Thus, the marginal force of mortality for T_1 is $\tilde{\mu}^{(1)}(t) = 1/(50-t)$; see also Exercise 7a. Clearly, $\tilde{\mu}^{(2)}(t) = \mu$.

By Proposition 1, $\mu_{50}^{(1)}(t) = 1/(50-t)$, $\mu_{50}^{(2)}(t) = \mu$, and $\mu_{50}(t) = \mu_{50}^{(1)}(t) + \mu_{50}^{(2)}(t) = \mu + 1/(50-t)$. The last relation also follows directly from (1.1.16).

To compute ${}_tp = P(T > t)$, we can apply (2.1.9), but it is much easier to make use of (2.2.5) which implies that

$$P(T > t) = P(T_1 > t)P(T_2 > t) = (1 - t/50)e^{-\mu t} = (1 - t/50)e^{-0.01t}.$$

For example, ${}_{20}p_{50} = \frac{30}{50}e^{-0.01\cdot 20} \approx 0.491$. □

Let us return to the general case. Assume that the density $f(t, j)$, and hence the forces of decrement $\mu^{(j)}(t)$, are given. If we even can determine T_1, T_2 as above (which is not always true), these r.v.'s may turn out to be dependent. The question is whether it is possible to construct a model with some independent r.v.'s T_1', T_2', distinct from T_1, T_2 and with the following properties.

(i) The marginal hazard rates of T_1', T_2' coincide with the respective forces of decrement.

(ii) The r.v. $T' = \min\{T_1', T_2'\}$ has the *same distribution as the original time-until-termination* T.

In this case, to find the distribution of T, we could consider the model with T_1', T_2', which is simpler.

The positive answer to the above question follows almost immediately from Proposition 1. Let ${}_tp'^{(j)}$ be the tail probability function (or the survival function) corresponding to $\mu^{(j)}(t)$. (We again omit the index x even if formally it should be involved.) More precisely, let

$${}_tp'^{(j)} = \exp\left\{-\int_0^t \mu^{(j)}(s)ds\right\} \tag{2.2.6}$$

[compare with (2.1.9)]. Set

$${}_tq'^{(j)} = 1 - {}_tp'^{(j)}.$$

The function ${}_tq'^{(j)}$ is called a *net probability of decrement*, or an *independent rate of decrement*, or an *absolute rate of decrement*.

Consider independent r.v.'s T_1', T_2' having the marginal d.f.'s ${}_tq'^{(1)}, {}_tq'^{(2)}$, respectively. More precisely, $P(T_j' \le t) = {}_tq'^{(j)}$ and, consequently, $P(T_j' > t) = {}_tp'^{(j)}$.

We can always define such r.v.'s on some space of elementary outcomes and work with them. For example, we can simulate values of these r.v.'s. Certainly, these r.v.'s are artificially constructed. However, if the model based on such r.v.'s allows us to compute the distributions of interest, this is just what is needed. We can view T_1', T_2' as r.v.'s representing some "modified causes" or as functions of the original r.v.'s T_1, T_2, but this is only an optional interpretation. The model we are building serves for computational purposes.

So, let us consider model (2.2.1)-(2.2.3) replacing T_1, T_2 by T_1', T_2'. Proposition 1 asserts that in this new model the forces of decrement coincide with the original forces $\mu^{(j)}(t)$. Moreover, let us recall that, by (2.1.8)-(2.1.10), $f(t, j)$ is specified by $\mu^{(1)}(t), \mu^{(2)}(t)$ in a unique way. Consequently, the analog of the density $f(t, j)$ in our new model is the same as the original density $f(t, j)$. (In particular, we use the same symbol $f(t, j)$ in both models.)

Then, of course, *all* other probabilities [such as $P(J = j \mid T = t)$ or $P(J = j)$] and all expectations [such as $E\{T\}$ or $E\{T \mid J = j\}$] will be the same in the new model as they were in the original model.

To clarify the last assertion, consider one instance.

Let $T' = \min\{T_1', T_2'\}$. Since T_1' and T_2' are independent, in accordance with (2.2.5),

$$P(T' > t) = P(T_1' > t)P(T_2' > t) = {}_tp'^{(1)}\,{}_tp'^{(2)}. \tag{2.2.7}$$

Substituting (2.2.6), we have

$$P(T'>t)=\exp\left\{-\int_0^t \mu^{(1)}(s)ds\right\}\exp\left\{-\int_0^t \mu^{(2)}(s)ds\right\}=\exp\left\{-\int_0^t (\mu^{(1)}(s)+\mu^{(2)}(s))ds\right\}, \tag{2.2.8}$$

which coincides with ${}_tp$ in (2.1.9)-(2.1.8). In particular, in view of (2.2.7), we can write

$${}_tp = {}_tp'^{(1)}\,{}_tp'^{(2)}. \tag{2.2.9}$$

EXAMPLE 3. In a certain sense, in this example the beginning and end of Example 2 are switched. Let us put aside for a while the computations in Example 2 and start from scratch. We restrict ourselves to two causes of death of (50): natural and causes resulting from an accident. Assume that somehow we have determined that the respective forces of decrement are $\mu^{(1)}(t) = 1/(50-t)$ and $\mu^{(2)}(t) = \mu = 0.01$. Suppose, for example, that we want to compute ${}_{20}p_{50}$. Formally, we can substitute the above formulas for $\mu^{(1)}(t)$ and $\mu^{(2)}(t)$ into (2.1.9). However, it is much easier to use (2.2.9) adding the now necessary index $x = 50$. Because $\mu^{(1)}(t)$ is the force of mortality of the distribution uniform on $[0,50]$, the "net probability" ${}_tp_x'^{(1)}$ corresponds to the distribution mentioned and, hence, equals $1-t/50$. Similarly, ${}_tp'^{(2)} = e^{-\mu t}$. Then, by (2.2.9), ${}_tp_{50} = {}_tp'^{(1)}\,{}_tp'^{(2)} = (1-t/50)e^{-\mu t}$ for $t \le 50$, and for ${}_{20}p_{50}$, we will get the same answer as in Example 2.

It makes sense to emphasize again that we did not assume in our calculations that the real causes acted independently. We just proceeded from an artificial model with (other) independent causes. However, this model was constructed in a way that led to a correct result. □

Some additional remarks.

1. As mentioned before, T_1', T_2' are artificially constructed, and the corresponding model is designed merely to simplify reasoning and calculations. In particular, T_1' is not equal to T_1, i.e., to the termination time in the hypothetical situation when only the first cause is in effect. If we know the hazard rates $\mu^{(1)}(t)$ and $\mu^{(2)}(t)$, we can find the joint distribution of (T,J) but we cannot find, proceeding only from the rates $\mu^{(1)}(t)$ and $\mu^{(2)}(t)$, the distributions of T_1 and T_2. It may be shown that different joint distributions of (T_1,T_2) can lead to the same joint distribution of (T,J).

This reflects the real world situation. We may observe the moment of termination and determine the cause, but proceeding from this information, we cannot say what would have happened if the cause had not occurred.

Assume, for example, that the termination of a contract may occur either if the client dies or moves to another area. The insurance organization has data on when terminations occurred and, in each case, the reason of termination. However, the organization cannot estimate from this data the distribution of lifetimes.

2. The reader may readily generalize the above model to the case of $m \geq 2$ causes. In this case, definition (2.2.6) should be considered for $j = 1,...,m$. We may define in a similar manner r.v.'s $T'_1,...,T'_m$, and replace (2.2.9) by

$$ {}_tp = \prod_{j=1}^{m} {}_tp'^{(j)}. $$

3. Next, we will write down an approximate relation between the quantities $q'^{(j)}_x$ (which equals ${}_1q'^{(j)}_x$) and $q^{(j)}_x$ (which equals ${}_1q^{(j)}_x$). We will do so under the interpolation assumption that all $\mu^{(j)}_x(t)$ are constant in the interval $(x, x+1)$. Then the same is true for $\mu_x(t) = \sum_j \mu^{(j)}_x(t)$. The values of the functions at the endpoints of $(x, x+1)$ do not matter if we are interested in the probabilities ${}_tq^{(j)}$ and ${}_tq'^{(j)}$ which are specified by integrals of μ's. (See, for example, (2.2.6) and recall that ${}_tq'^{(j)} = 1 - {}_tp'^{(j)}$.) It may be shown that under the assumption we have made,

$$ \frac{q^{(j)}_x}{q_x} = \frac{\ln p'^{(j)}_x}{\ln p_x}, \tag{2.2.10} $$

which together with $q'^{(j)}_x = 1 - p'^{(j)}_x$ gives the desired relation.

A particular example is given in Exercise 50.

► To prove (2.2.10), denote by c_{jx} and c_x the (constant) values of $\mu^{(j)}_x(t)$ and $\mu_x(t)$ on $(x, x+1)$, respectively. By (2.1.10), $q^{(j)}_x = \int_0^1 f(t,j)dt = \int_0^1 {}_tp_x\mu^{(j)}_x(t)dt = c_{jx}\int_0^1 {}_tp_xdt$. Similarly, $q_x = \int_0^1 f_T(t)dt = \int_0^1 {}_tp_x\mu_x(t)dt = c_x\int_0^1 {}_tp_xdt$. Hence, $\frac{q^{(j)}_x}{q_x} = \frac{c_{jx}}{c_x}$. On the other hand, by (2.2.6), for $p'^{(j)}$, we have $\ln({}_tp'^{(j)}) = -\int_0^1 \mu^{(j)}_x(s)ds = -c_{jx}$. Similarly, $\ln(p_x) = -\int_0^1 \mu_x(s)ds = -c_x$. Combining all of this, we come to (2.2.10). ◄

4. As was already mentioned, among the r.v.'s $T'_1,...,T'_m$ only one, T'_i say, must be proper; that is, such that $P(T'_i > x) \to 0$ as $x \to \infty$. It may happen that for some T'_j it is not true, which means that this cause with positive probability will never act. See also the scheme at the end of Section 1.1.

2.3 Survivorship group[6]

Consider a group of l_a people of the same age a. All lifetimes are independent and correspond to the model of Section 2.1. Let ${}_n\mathcal{D}^{(j)}_x$ be the number of people who will die within the age interval $[x, x+n]$ from cause j, and let $\mathcal{L}(x)$ be the number of people who will survive to age x. Similar to what was shown in Section 1.4, both random variables are binomial with probabilities of success $P(x-a \leq T(a) \leq x-a+n, J=j)$ and $P(T(a) > x-a)$, respectively. The latter probability is ${}_{x-a}p_a$, and the former is ${}_{x-a}p_a \cdot {}_nq^{(j)}_x$ (the individual

[6]In the exposition of this section and the example below we follow the logic of Section 10.2 from [19].

lives until age x, and *after that, as an x-year old*, dies within n years from cause j). Hence, setting ${}_nd_x^{(j)} = E\{{}_n\mathcal{D}_x^{(j)}\}$ and $l_x = E\{\mathcal{L}(x)\}$, similar to Section 2.1, we have

$$ {}_nd_x^{(j)} = E\{{}_n\mathcal{D}_x^{(j)}\} = l_a \cdot {}_{x-a}p_a \cdot {}_nq_x^{(j)}, \tag{2.3.1} $$

$$ l_x = l_a \cdot {}_{x-a}p_a. \tag{2.3.2} $$

Using the second equality, we replace $l_a \cdot {}_{x-a}p_a$ by l_x in the first. This leads to

$$ {}_nd_x^{(j)} = l_x \cdot {}_nq_x^{(j)}. $$

In (2.3.2), we set $a = x-1$, which implies

$$ l_x = l_{x-1} \cdot {}_1p_{x-1} = l_{x-1}p_{x-1}. \tag{2.3.3} $$

(We omit the prefix 1 in ${}_1p_{x-1}$ and similar characteristics.) Next, we set $n = 1$ in (2.3.1), and omitting the prefix one as usual, we come to

$$ d_x^{(j)} = l_x \cdot q_x^{(j)}. \tag{2.3.4} $$

EXAMPLE 1. Let $l_{70} = 100$ and the number of causes $m = 3$. The corresponding decrement probabilities are given below:

x	$q_x^{(1)}$	$q_x^{(2)}$	$q_x^{(3)}$
70	0.02	0.01	0.05
71	0.03	0.02	0.06
72	0.04	0.03	0.07

Find l_x and $d_x^{(j)}$. In view of (2.1.3), $q_x = q_x^{(1)} + q_x^{(2)} + q_x^{(3)}$ and $p_x = 1 - q_x$. Using the recurrence formula (2.3.3) and formula (2.3.4), we have

x	$q_x^{(1)}$	$q_x^{(2)}$	$q_x^{(3)}$	q_x	p_x	$l_x = l_{x-1}p_{x-1}$	$d_x^{(1)} = l_xq_x^{(1)}$	$d_x^{(2)} = l_xq_x^{(2)}$	$d_x^{(3)} = l_xq_x^{(3)}$
70	0.02	0.01	0.05	0.08	0.92	100	2	1	5
71	0.03	0.02	0.06	0.11	0.89	92	2.76	1.84	5.52
72	0.04	0.03	0.07	0.14	0.86	81.88	3.2752	2.4564	5.7316

□

2.4 Proof of Proposition 1

The index x is everywhere omitted. Let $f_j(t)$ and $F_j(t)$ be the marginal density and d.f. of T_j, respectively. The event $\{J = 1\}$ is equivalent to $\{T_1 < T_2\}$. Then, following the general conditioning rule (**0**.7.3.9) and taking into account the independence of T_1 and T_2, we have

$$
\begin{aligned}
{}_tp^{(1)} &= P(T > t, J = 1) = P(\min\{T_1, T_2\} > t,\ T_2 > T_1) = P(T_2 > T_1 > t) \\
&= \int_t^{\infty} P(T_2 > T_1 > t \mid T_1 = s) f_1(s)ds = \int_t^{\infty} P(T_2 > s \mid T_1 = s) f_1(s)ds \\
&= \int_t^{\infty} P(T_2 > s) f_1(s)ds = \int_t^{\infty} (1 - F_2(s)) f_1(s)ds.
\end{aligned}
$$

Then

$$
f(t,1) = \frac{\partial}{\partial t}\,{}_tq^{(1)} = \frac{\partial}{\partial t}(1 - {}_tp^{(1)}) = -\frac{\partial}{\partial t}\int_t^{\infty}(1 - F_2(s))f_1(s)ds = (1 - F_2(t))f_1(t).
$$

On the other hand, by (2.2.5),

$$
{}_tp = P(T > t) = (1 - F_1(t))(1 - F_2(t)).
$$

Hence, by definition,

$$
\mu^{(1)}(t) = \frac{f(t,1)}{{}_tp} = \frac{(1 - F_2(t))f_1(t)}{(1 - F_1(t))(1 - F_2(t))} = \frac{f_1(t)}{(1 - F_1(t))}.
$$

Again by definition, the last quantity is the marginal hazard rate of T_1. ■

---

3 MULTIPLE LIFE MODELS

In this section, we study survival characteristics of a set (or cohort) of lives when this set is considered as a whole, as "one client". For example, a joint family pension plan terminates only when both spouses die. Another example is a family life insurance plan with benefits payable at the time of the first death. In such situations, we refer to a set of lives as a "*status*".

If we know how to define the future lifetime of a status, we will talk about a *survival status*. Certainly, how long the status will exist depends on the type of the status (actually, on the type of the insurance contract under consideration). In any case, the future lifetime (or the time-until-failure, or the termination time) for a status is a function of the lifetimes of the persons (lives) involved.

Consider a status of two lives, (x) and (y), with respective future lifetimes $T(x)$ and $T(y)$. In general, these r.v.'s are dependent. The following two types are probably the most important.

1. The *joint-life status*. The status is intact until the first death occurs, and hence the lifetime of the status is $\min\{T(x), T(y)\}$. Traditionally, the status itself is denoted by the symbol $x:y$, and the status lifetime by $T(x:y)$ or $T(xy)$.

2. The *last-survivor status*. In this case, termination occurs upon the last death, and hence the lifetime of the status is $\max\{T(x), T(y)\}$. The status is denoted by $\overline{x:y}$, and the status lifetime by $T(\overline{x:y})$ or $T(\overline{xy})$.

Certainly, these two models do not exhaust all possible situations. For example, a joint life insurance contract may involve benefits payable upon each death, and the amount of benefits paid may depend on which death occurs first. We will consider such examples later.

In any case, to describe the situation, we should know the joint distribution of the r.v.'s $T(x)$ and $T(y)$.

3.1 The joint distribution

It may be characterized either by the joint d.f.

$$F_{T(x)T(y)}(u,v) = P(T(x) \le u,\, T(y) \le v),$$

or by the joint survival function

$$s_{T(x)T(y)}(u,v) = P(T(x) > u,\, T(y) > v).$$

To make our exposition simpler, we will sometimes omit the index $T(x)T(y)$, and since x and y are fixed, we write T_1 and T_2 instead of $T(x)$ and $T(y)$. Later we will come back to actuarial notation.

All calculations below are not specific and are based on general probability theory formulas from Section **0**.1.3.2. First of all, the marginal d.f.

$$F_{T_1}(u) = P(T_1 \le u) = P(T_1 \le u,\, T_2 \le \infty) = F_{T_1T_2}(u,\infty) = F(u,\infty), \tag{3.1.1}$$

if we omit the index T_1T_2. Similarly, since the distributions of T's are continuous, the marginal survival function

$$s_{T_1}(u) = P(T_1 > u) = P(T_1 > u,\, T_2 \ge 0) = P(T_1 > u,\, T_2 > 0) = s_{T_1T_2}(u,0).$$

Analogously,

$$F_{T_2}(v) = P(T_2 \le v) = F_{T_1T_2}(\infty, v), \text{ and } s_{T_2}(v) = P(T_2 > v) = s_{T_1T_2}(0,v).$$

Next, we consider a connection between the joint d.f. and the survival function. In the one-dimensional case, for any r.v. T, we write $P(T \le x) = 1 - P(T > x)$. The counterpart of this identity in the two-dimensional case is the identity

$$P(T_1 \le u,\, T_2 \le v) = 1 - P(T_1 > u) - P(T_2 > v) + P(T_1 > u, T_2 > v).$$

(Since when subtracting $P(T_1 > u)$ and $P(T_2 > v)$, we subtract the probability of the set $\{T_1 > u, T_2 > v\}$ twice.) Thus,

$$F_{T_1T_2}(u,v) = 1 - s_{T_1}(u) - s_{T_2}(v) + s_{T_1T_2}(u,v). \tag{3.1.2}$$

In the same way, we get that

$$s_{T_1T_2}(u,v) = 1 - F_{T_1}(u) - F_{T_2}(v) + F_{T_1T_2}(u,v). \tag{3.1.3}$$

EXAMPLE 1. Let each r.v. T_i ($i = 1, 2$) take on values from $[0, 1]$, and let the joint density

$$f_{T_1T_2}(u, v) = C_k[1 - k(u - v)^2] \text{ for } 0 \le u \le 1,\ 0 \le v \le 1, \text{ and } = 0 \text{ otherwise.} \quad (3.1.4)$$

Here k is a parameter taking values from $[0, 1]$, and the constant $C_k = 6/(6 - k)$. Such a constant is chosen in order that the total integral of the density equals one. The reader can readily verify that this is true by direct integration.

The unit intervals in (3.1.4) are chosen for simplicity. Since we interpret T's as lifetimes, the unit of time is not one year but an appropriate number consistent with the ages x and y.

The parameter k characterizes the dependency between the r.v.'s T_1 and T_2. If $k = 0$, then $f_{T_1T_2}(u, v) = 1$ for all $u \in [0, 1]$ and $v \in [0, 1]$, and T_1 and T_2 are independent and uniformly distributed on $[0, 1]$. (See Example **0**.1.3.2-2 which differs from what we consider now only by scale.)

If $k > 0$, then the r.v.'s are dependent and, in a certain sense, the larger k, the stronger this dependence is. In particular, note that density (3.1.4) reaches its maximum in the line $u = v$, which means that the r.v.'s T_1 and T_2 "have tendency to be close to each other rather than to differ significantly". We may interpret this as if the persons we consider are connected with each other; for instance, they have common living conditions.

We consider $k \le 1$ because otherwise the density would be negative for some u, v.

In accordance with (**0**.1.3.11), the d.f.

$$F_{T_1T_2}(u, v) = \int_0^u \int_0^v f_{T_1T_2}(t, s)dtds = C_k \int_0^u \int_0^v (1 - k(t - s)^2)dtds$$

$$= C_k \left[uv - k\left(\frac{uv(u^2+v^2)}{3} - \frac{u^2v^2}{2}\right)\right] \quad \text{for } 0 \le u \le 1,\ 0 \le v \le 1, \quad (3.1.5)$$

which may be obtained by straightforward integration. Since $P(T_2 \le 1) = 1$,

$$F_{T_1}(u) = F_{T_1T_2}(u, 1) = C_k\left[u - k\left(\frac{u(u^2 + 1)}{3} - \frac{u^2}{2}\right)\right] \quad \text{for } 0 \le u \le 1. \quad (3.1.6)$$

Since the distribution is symmetric, the marginal distribution of T_2 is the same:

$$F_{T_2}(v) = C_k\left[v - k\left(\frac{v(v^2 + 1)}{3} - \frac{v^2}{2}\right)\right], \quad \text{for } 0 \le v \le 1.$$

Inserting all of this into (3.1.3), after some algebra, we get that the joint survival function

$$s_{T_1T_2}(u, v) = 1 - C_k\left[u + v - uv - k\left(\frac{u(u^2+1) + v(v^2+1)}{3} - \frac{u^2+v^2}{2} - \frac{uv(u^2+v^2)}{3} + \frac{u^2v^2}{2}\right)\right]. \quad (3.1.7)$$

One can also compute

$$E\{T_1\} = \int_0^1 u dF_{T_1}(u).$$

Here, straightforward calculations lead to a simple answer, namely, $E\{T_1\} = 1/2$. Thus, the answer does not depend on k.

We will revisit this example several times later, using the formulas above for various illustrations. In Exercise 53, the reader is invited to carry out the calculations in more detail. □

3.2 The lifetime of statuses

Now we consider the lifetime for the joint-life and the last-survivor status, which corresponds to the r.v.'s $T = \min\{T_1, T_2\}$ and $\overline{T} = \max\{T_1, T_2\}$, respectively.

In the first case, it is more convenient to deal with the joint survival function since

$$P(T > t) = P(\min\{T_1, T_2\} > t) = P(T_1 > t, T_2 > t) = s_{T_1 T_2}(t,t). \tag{3.2.1}$$

From (3.2.1) it follows that

$$\text{If } T_1 \text{ and } T_2 \text{ are independent, then } P(T > t) = P(T_1 > t)P(T_2 > t) = s_{T_1}(t)s_{T_2}(t). \tag{3.2.2}$$

In the case of the last-survivor status, it is easier to work with the joint d.f., writing

$$P(\overline{T} \le t) = P(\max\{T_1, T_2\} \le t) = P(T_1 \le t, T_2 \le t) = F_{T_1 T_2}(t,t). \tag{3.2.3}$$

From (3.2.3) we readily get that

$$\text{If } T_1 \text{ and } T_2 \text{ are independent, then } P(\overline{T} \le t) = P(T_1 \le t)P(T_2 \le t) = F_{T_1}(t)F_{T_2}(t). \tag{3.2.4}$$

Now, we will write the same in actuarial notation. Let ${}_tp_{x:y}$ stand for $P(T(x:y) > t) = P(T > t)$, and ${}_tp_{\overline{x:y}}$ for $P(T(\overline{x:y}) > t) = P(\overline{T} > t)$. Following the same logic in notation as above, let us set ${}_tq_{x:y} = 1 - {}_tp_{x:y}$, ${}_tq_{\overline{x:y}} = 1 - {}_tp_{\overline{x:y}}$. Note also that $T_1 = T(x)$, and $T_2 = T(y)$. Hence, $F_{T_1}(t) = {}_tq_x$, $F_{T_2}(t) = {}_tq_y$, $s_{T_1}(t) = {}_tp_x$, and $s_{T_2}(t) = {}_tp_y$. Then (3.2.1)-(3.2.4) may be rewritten as

$${}_tp_{x:y} = s_{T_1 T_2}(t,t), \text{ and in the independence case, } {}_tp_{x:y} = {}_tp_x \cdot {}_tp_y; \tag{3.2.5}$$

$${}_tq_{\overline{x:y}} = F_{T_1 T_2}(t,t), \text{ and in the independence case, } {}_tq_{\overline{x:y}} = {}_tq_x \cdot {}_tq_y. \tag{3.2.6}$$

These formulas must be understood correctly. The factor ${}_tp_x$ in (3.2.5) concerns the first life, and the term ${}_tp_y$ concerns the second. These two lives may have different distributions, so ${}_tp_x$ may not equal ${}_tp_y$ even if $x = y$. Thus, as a matter of fact, the quantities ${}_tp_x$ and ${}_yp_y$ should be supplied by additional indices which are omitted for the sake of simplicity.

Unlike (3.2.5), in the corresponding formula (3.2.2), the distributions of the two lives are denoted by different symbols, so this formula explicitly reflects the circumstance just mentioned. The same concerns the factors ${}_tq_x$ and ${}_tq_y$ in (3.2.6).

We can also connect characteristics for the two statuses by making use of the connection between the joint distribution and corresponding survival functions. Setting $u = v = t$ in (3.1.3), we obtain from (3.2.5)-(3.2.6) that

$$
\begin{aligned}
{}_tq_{x:y} &= 1 - {}_tp_{x:y} = 1 - s_{T_1T_2}(t,t) = 1 - [1 - F_{T_1}(t) - F_{T_2}(t) + F_{T_1T_2}(t,t)] \\
&= F_{T_1}(t) + F_{T_2}(t) - F_{T_1T_2}(t,t) = {}_tq_x + {}_tq_y - {}_tq_{\overline{x:y}}.
\end{aligned}
$$

We rewrite this in the following nice form:

$$
{}_tq_{x:y} + {}_tq_{\overline{x:y}} = {}_tq_x + {}_tq_y. \qquad (3.2.7)
$$

Replacing each q above by $1 - p$ and canceling all ones, we come to the equation

$$
{}_tp_{x:y} + {}_tp_{\overline{x:y}} = {}_tp_x + {}_tp_y. \qquad (3.2.8)
$$

For a straight proof of (3.2.7) and (3.2.8) see Exercise 64.

EXAMPLE 1. Let us revisit Example 3.1-1. In view of (3.1.5) and (3.2.6),

$$
{}_tq_{\overline{x:y}} = F_{T_1T_2}(t,t) = C_k\left[t^2 - k\left(\frac{2t^4}{3} - \frac{t^4}{2}\right)\right] = C_k\left(t^2 - k\frac{t^4}{6}\right). \qquad (3.2.9)
$$

Since ${}_tq_{\overline{x:y}}$ is the d.f. of the r.v. $\overline{T} = T(\overline{x:y})$, its density is

$$
f_{T(\overline{x:y})}(t) = \frac{d}{dt}\,{}_tq_{\overline{x:y}} = C_k\left(2t - k\frac{2t^3}{3}\right).
$$

Consequently, we can compute

$$
E\{T(\overline{x:y})\} = \int_0^1 t f_{T(\overline{x:y})}(t)dt = \int_0^1 tC_k\left(2t - k\frac{2t^3}{3}\right)dt = C_k\left(\frac{2}{3} - k\frac{2}{15}\right) = \frac{6}{6-k}\left(\frac{2}{3} - k\frac{2}{15}\right).
$$

For instance, for $k = 0$, when the r.v.'s T_1 and T_2 are independent, $E\{T(\overline{x:y})\} = \frac{2}{3}$, which certainly can be obtained directly. Indeed, in the independence case, T_1 and T_2 are uniform on $[0,1]$, and in view of (3.2.4), $F_{\overline{T}}(t) = F_{T_1}(t)F_{T_2}(t) = t\cdot t = t^2$. Then $E\{T(\overline{x:y})\} = \int_0^1 t\,dt^2 = \frac{2}{3}$.

In the case of dependence, the situation is more complicated. For example, for $k = 1$, we have $E\{T(\overline{x:y})\} = \frac{6}{5}\left(\frac{2}{3} - 1\cdot\frac{2}{15}\right) = \frac{16}{25}$, which is a bit larger than $2/3$.

Consider the joint-life status and, accordingly, $T(x:y)$. In this case, by (3.1.7) and (3.2.5),

$$
\begin{aligned}
{}_tq_{x:y} &= 1 - {}_tp_{x:y} = 1 - s_{T_1T_2}(t,t) = C_k\left[2t - t^2 - k\left(\frac{2t(t^2+1)}{3} - t^2 - \frac{2t^4}{3} + \frac{t^4}{2}\right)\right] \\
&= C_k\left[2\left(1 - \frac{k}{3}\right)t - (1-k)t^2 - \frac{2k}{3}t^3 + \frac{k}{6}t^4\right].
\end{aligned}
$$

For example, for $k = 1$,

$$
{}_tq_{x:y} = \frac{6}{5}\left[\frac{4}{3}t - \frac{2}{3}t^3 + \frac{1}{6}t^4\right],
$$

and the density

$$f_{T(x:y)}(t) = \frac{d}{dt}\,{}_tq_{x:y} = \frac{6}{5}\left[\frac{4}{3} - 2t^2 + \frac{2}{3}t^3\right].$$

This leads to

$$E\{T(x:y)\} = \int_0^1 t f_{T(x:y)}(t)dt = \int_0^1 t\frac{6}{5}\left[\frac{4}{3} - 2t^2 + \frac{2}{3}t^3\right]dt = \frac{1}{5}.$$

In Exercise 54, we compare it with the independence case.

EXAMPLE 2. Consider a married couple aged 50 and 60. Assume that their future lifetimes are independent, the mortality for the first life follows De Moivre's law with $\omega = 100$, and for the second life, the force of mortality of the future life is close to $\mu = 0.05$.

(a) Find ${}_{20}q_{50:60}$.This is the probability that at least one of spouses will die within 20 years. De Moivre's law corresponds to a uniform distribution. In Exercise 7b, we establish the simple fact that in this case, the conditional distributions are also uniform. Hence, the distribution of $T_1 = T(50)$ is uniform on $[0, 50]$. The distribution of $T_2 = T(60)$ is exponential with $E\{T_2\} = \frac{1}{\mu} = 20$. By (3.2.5),

$${}_{20}p_{50:60} = {}_{20}p_{50} \cdot {}_{20}p_{60} = P(T_1 > 20)P(T_2 > 20) = \frac{30}{50} \cdot e^{-\mu \cdot 20} = 0.6 \cdot e^{-1} \approx 0.22.$$

Then ${}_{20}q_{50:60} = 1 - {}_{20}p_{50:60} \approx 0.78$.

(b) Find ${}_{20}p_{\overline{50:60}}$, the probability that at least one spouse will survive 20 years. Here we use (3.2.6), which implies that

$${}_{20}q_{\overline{50:60}} = {}_{20}q_{50} \cdot {}_{20}q_{60} = P(T_1 \le 20)P(T_2 \le 20) = \frac{20}{50} \cdot (1 - e^{-\mu \cdot 20}) = 0.4 \cdot (1 - e^{-1}) \approx 0.25.$$

Then ${}_{20}p_{\overline{50:60}} = 1 - {}_{20}q_{\overline{50:60}} \approx 0.75$.

(c) Find the probability that the sixty year old person will die first. It is more convenient to find the complementary probability $P(T(60) \ge T(50))$. Using the formula for total probability in its integral form [see (**0**.7.3.9)], we have

$$P(T_2 \ge T_1) = \int_0^\infty P(T_2 > T_1 \,|\, T_1 = t) f_{T_1}(t)dt = \int_0^{50} P(T_2 > T_1 \,|\, T_1 = t)\frac{1}{50}dt.$$

Because T_1 and T_2 are independent, $P(T_2 > T_1 \,|\, T_1 = t) = P(T_2 > t)$. Consequently,

$$P(T_2 \ge T_1) = \frac{1}{50}\int_0^{50} P(T_2 > t)dt = \frac{1}{50}\int_0^{50} e^{-\mu \cdot t}dt = \frac{1}{50}20(1 - e^{-0.05 \cdot 50}) \approx 0.37.$$

Thus, $P(T_2 < T_1) \approx 1 - 0.37 = 0.63$.

EXAMPLE 3. In the case of independent lives, compute ${}_{4|2}q_{60:63}$ in terms of characteristics p_x. We have ${}_{4|2}q_{60:63} = {}_4q_{60:63} - {}_2q_{60:63}$. In view of (3.2.6), ${}_4q_{60:63} = 1 - {}_4p_{60:63} = 1 - {}_4p_{60} \cdot {}_4p_{63}$, and ${}_2q_{60:63} = 1 - {}_2p_{60:63} = 1 - {}_2p_{60} \cdot {}_2p_{63}$.

Next we use that ${}_tp_x = p_x p_{x+1} \cdot \ldots \cdot p_{x+t-1}$ [see (1.2.8)]. First, we compute ${}_2p_{60} = p_{60}p_{61}$, and ${}_2p_{63} = p_{63}p_{64}$. After that, we write ${}_4p_{60} = {}_2p_{60}p_{62}p_{63}$, and ${}_4p_{63} = {}_2p_{63}p_{65}p_{66}$.

EXAMPLE 4. In the situation of Example 1, let the unit of time be 50 years and $k = 1$.

(a) Compute the probability that both persons will not survive 30 years, and the first person will die first. The probability that both lives will not survive 30 years is ${}_tq_{\overline{x:y}}$ for $t = 3/5$. By (3.2.9), ${}_tq_{\overline{x:y}} = C_k\left(t^2 - k\frac{t^4}{6}\right) = 1.2\left(0.36 - \frac{0.1296}{6}\right) = 0.40608$ if $t = 0.6$. Since the joint distribution of $T(x)$ and $T(y)$ is symmetric, it is equally likely either the first or the second person will die first. So, the answer is $\frac{1}{2} \cdot {}_tq_{\overline{x:y}} = 0.20304$.

(b) Find the probability that the first person will not survive 40 years, the second 30 years, and the second person will die first. We adopt the notation of Example 3.1-1 and denote by u and v possible values of the first and the second lifetimes, respectively. The event whose probability we are computing corresponds to the set depicted in Fig.8.

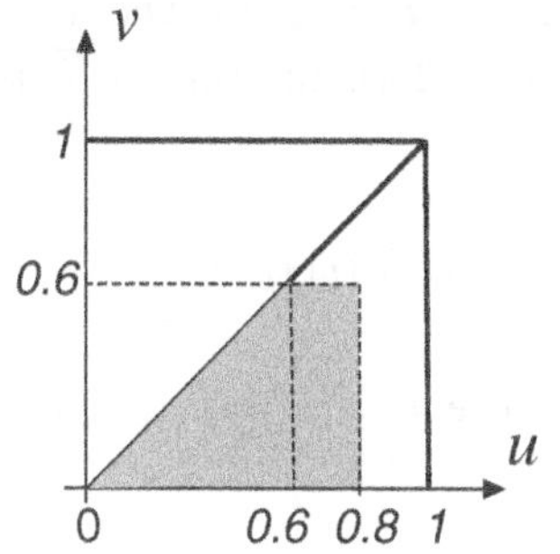

FIGURE 8.

In this figure, we see that the probability mentioned is equal to $P(T_1 \le 0.8, T_2 \le 0.6) - P(T_1 \le 0.6, T_2 \le 0.6, T_2 > T_1)$. As above, in view of symmetry, the latter probability is equal to $\frac{1}{2}P(T_1 \le 0.6, T_2 \le 0.6) = 0.20304$, as we have already computed. To compute the former probability $P(T_1 \le 0.8, T_2 \le 0.6)$, we should set in (3.1.5) $k = 1$, $u = 0.8$, and $v = 0.6$, which gives $F_{T_1T_2}(0.8, 0.6) = 0.52224$. So, the answer is $0.52224 - 0.20304 = 0.3192$.

EXAMPLE 5 ([153, N32][7]). John, age 40, and Mary, age 50, are independent lives following the same mortality as follows:

Age (x)	40	50	60
${}_{10}q_x$	0.039	0.085	0.192.

Calculate the probability that John and Mary both live at least 10 years and then both die during the following 10 years.

Following the same logic as above, we have

$$\begin{aligned} P(T > 10, \overline{T} \le 20) &= P(\overline{T} \le 20 \,|\, T > 10)P(T > 10) \\ &= P(T(\overline{40:50}) \le 20 \,|\, T(40:50) > 10)P(T(40:50) > 10) \\ &= P(T(\overline{50:60}) \le 10)P(T(40:50) > 10) \\ &= {}_{10}q_{50} \cdot {}_{10}q_{60} \cdot (1 - {}_{10}q_{40}) \cdot (1 - {}_{10}q_{50}) \approx 0.01435. \end{aligned}$$

Actually, it is enough to write the last line which explicitly shows the logic of calculations. □

Next we discuss forces of mortality. In Section 1.1, we have already seen that if we consider the minimum of two independent r.v.'s, the corresponding hazard rates are summed. Consequently, for a joint-life status in the independency case, we can write that the force of mortality of $\min\{T(x), T(y)\}$ is

$$\mu_{x:y}(t) = \mu_x^{(1)}(t) + \mu_y^{(2)}(t) = \mu^{(1)}(x+t) + \mu^{(2)}(y+t). \tag{3.2.10}$$

[7] Reprinted with permission of the Casualty Actuarial Society.

We used above (1.2.3) and denoted by $\mu_x^{(i)}(t)$ the force of mortality for the ith life. See also the remark right after (3.2.6).

EXAMPLE 6. Let two lifetimes be independent, $\mu_x^{(1)}(t) = 0.02$ and $\mu_x^{(2)}(t) = 0.03$. Then for $T(x:y) = \min\{T(x), T(y)\}$, by (1.1.16), we have $\mu_{x:y}(t) = 0.05$. and hence $T(x:y)$ is an exponential r.v. with $E\{T(x:y)\} = 20$. This is a well known fact in Probability Theory: the minimum of independent exponential r.v.'s is also exponential.

Certainly, we can compute any probabilities connected with $T(x:y)$. For example, by using the result of Example **0**.7.3-1, we get that the probability that the first person will die first is $[\mu_x^{(1)}/(\mu_x^{(1)} + \mu_x^{(2)})] = 0.4$. □

For the last-survivor status, formulas are not as nice as (3.2.10). Nevertheless, probabilities are computable. For brevity, let us set again $T_1 = T(x)$, $T_2 = T(y)$, and $\overline{T} = \max\{T_1, T_2\}$. Assume the lives to be independent. Then, by (3.2.4), the d.f. of $\overline{T}$ is $F_{\overline{T}}(t) = F_{T_1}(t)F_{T_2}(t)$, and the density

$$f_{\overline{T}}(t) = F'_{\overline{T}}(t) = F'_{T_1}(t)F_{T_2}(t) + F_{T_1}(t)F'_{T_2}(t) = f_{T_1}(t)F_{T_2}(t) + F_{T_1}(t)f_{T_2}(t).$$

If we come back to actuarial notation and use (1.2.4), we can write $f_{T(x)}(t) = \mu_x(t)\,{}_tp_x = \mu(x+t)\,{}_tp_x$ and $f_{T(y)}(t) = \mu_y(t)\,{}_tp_y = \mu(x+t)\,{}_tp_y$. Hence,

$$f_{\overline{T}}(t) = \mu_x(t)\,{}_tp_x \cdot {}_tq_y + \mu_y(t)\,{}_tp_y \cdot {}_tq_x.$$

Note also that $P(\overline{T} > t) = 1 - {}_tq_{\overline{x:y}} = 1 - {}_tq_x \cdot {}_tq_y$. Eventually,

$$\mu_{\overline{x:y}}(t) = \frac{f_{\overline{T}}(t)}{P(\overline{T} > t)} = \frac{\mu_x(t)\,{}_tp_x \cdot {}_tq_y + \mu_y(t)\,{}_tp_y \cdot {}_tq_x}{1 - {}_tq_x \cdot {}_tq_y}. \tag{3.2.11}$$

By virtue of (1.2.3), we can replace $\mu_x(t)$ and $\mu_y(t)$ by $\mu(x+t)$ and $\mu(y+t)$, respectively.

EXAMPLE 7. Estimate $\mu_{\overline{70:70}}(2)$ proceeding from the data from Table 1.5.1-1, and assuming the lifetimes are independent. By (1.4.4), ${}_2p_{70} = (l_{72}/l_{70}) \approx (71669/75335) \approx 0.95134$, and hence ${}_2q_{70} \approx 0.048663$. In Table 1.5.1-1, the two types of estimates for $\mu_{70}(2) = \mu(72)$ are 0.027663 and 0.028053, corresponding to the linear and exponential interpolation, respectively. These numbers are close, and we may take their average ≈ 0.027858. Then, (3.2.11) leads to

$$\mu_{\overline{70:70}}(2) \approx \frac{2 \cdot 0.027858 \cdot 0.95134 \cdot 0.048663}{1-(0.048663)^2} \approx 0.002585. \ \square$$

3.3 A model of dependency: conditional independence

Next, we consider a simple model of dependency between two lives involved in one insurance contract. A survey of some dependency structures used in Actuarial Modeling and references may be found, e.g., in the book [33] by M. Denuit, J. Dhaene, M. Goovaerts, and P. Kaas.

3.3.1 A definition and the first example

Assume that the durations of two lives are affected by some common factor. For example, for a married couple it may be identical living conditions. In this case, the lifetimes are dependent. Suppose, however, that once the influence of the factor mentioned has been specified (for example, the conditions under which the spouses live are known), the lifetimes may be considered independent.

To model this situation, we consider the r.v.'s $T(x)$ and $T(y)$ as above, and a r.v. or a r.vec. ζ which is identified with the common factor on which $T(x)$ and $T(y)$ depend. We assume that *given* ζ, the r.v.'s $T(x)$ and $T(y)$ are independent, which amounts to the following assumption on the conditional distributions. For any sets B_1 and B_2 from the real line,

$$P(T(x) \in B_1, T(y) \in B_2 \,|\, \zeta) = P(T(x) \in B_1 \,|\, \zeta) \cdot P(T(y) \in B_2 \,|\, \zeta). \tag{3.3.1}$$

Random variables having this property are called *conditionally independent*. To find the joint distribution of $T(x)$ and $T(y)$, we should take the expectation of both sides of (3.3.1) with respect to ζ. By virtue of the general formula of total expectation [see, e.g., (**0**.7.2.1)], this leads to the relation

$$P(T(x) \in B_1, T(y) \in B_2) = E\{P(T(x) \in B_1 \,|\, \zeta)P(T(y) \in B_2 \,|\, \zeta)\}. \tag{3.3.2}$$

EXAMPLE 1. Let $T(x)$ and $T(y)$ be exponentially distributed with a common parameter ζ which we assume to be random. Suppose that ζ is uniformly distributed on an interval $[a,b]$, and given ζ, the r.v.'s $T(x)$ and $T(y)$ are independent. (In other words, if we know that ζ took a particular value μ, we consider $T(x)$ and $T(y)$ independent and exponential with the parameter μ.) Let $f_\zeta(\mu)$ be the density of ζ. Then, by the general formula (**0**.7.3.9), the joint survival function

$$\begin{aligned} s_{T(x)T(y)}(u,v) &= P(T(x) > u,\, T(y) > v) = \int P(T(x) > u,\, T(y) > v \,|\, \zeta = \mu) f_\zeta(\mu)d\mu \\ &= \int_a^b P(T(x) > u)P(T(y) > v)\frac{1}{b-a}d\mu = \int_a^b e^{-\mu u}e^{-\mu v}\frac{1}{b-a}d\mu \\ &= \frac{1}{b-a}\int_a^b e^{-\mu(u+v)}d\mu = \frac{e^{-a(u+v)} - e^{-b(u+v)}}{(b-a)(u+v)}. \end{aligned}$$

Considering the joint-lifetime status, for $T(x:y) = \min\{T(x), T(y)\}$, we have

$$_tp_{x:y} = P(T(x:y) > t) = P(T(x) > t,\, T(y) > t) = \frac{e^{-2at} - e^{-2bt}}{2t(b-a)}. \tag{3.3.3}$$

For example, let $a = 0.02$ and $b = 0.08$, so the conditional expected values, $E\{T(x) \,|\, \zeta = \mu\}$, varies from $\frac{1}{b} = 12.5$ to $\frac{1}{a} = 50$. If ζ had assumed just one value equal to its mean 0.05, then we would have had

$$_tp_{x:y} = P(T(x) > t)P(T(y) > t) = \exp\{-0.05t\}\exp\{-0.05t\} = \exp\{-0.1t\}. \tag{3.3.4}$$

However, in case (3.3.3), we have

$$ {}_tp_{x:y} = \frac{1}{2t\cdot 0.06}\left(e^{-0.04t} - e^{-0.16t}\right) = \frac{25}{3t}\left(e^{-0.04t} - e^{-0.16t}\right). \tag{3.3.5} $$

For large t, the last probability becomes essentially larger than (3.3.4). The reader can see this in the table below where we give values of ${}_tp_{x:y}$ for both cases.

t	20	30	40	50
the independence case	0.135	0.049	0.018	0.007
the dependency case	0.170	0.081	0.041	0.022

□

3.3.2 The common shock model

This model is another particular example of conditional independence. It was first discussed in [89], [90] together with other interesting bivariate distributions.

Assume that the lifetimes of two persons are independent unless a common shock causes the death of both. For example, it may be a lethal traffic accident in which two spouses are involved. Following traditional notation, denote by Z the moment of the shock, and by $T^*(x)$, $T^*(y)$ the durations of the lives in the absence of the shock. The three r.v.'s defined are assumed to be mutually independent.

Clearly, the lifetime of the first person is the r.v. $T(x) = \min\{T^*(x), Z\}$, and the lifetime of the second is $T(y) = \min\{T^*(y), Z\}$. *Given* Z, the r.v. $T(x)$ and $T(y)$ are independent. The joint survival function

$$ \begin{aligned} s_{T(x)T(y)}(u,v) &= P(T(x) > u,\, T(y) > v) = P(\min\{T^*(x),Z\} > u,\, \min\{T^*(y),Z\} > v) \\ &= P(T^*(x) > u,\, Z > u,\, T^*(y) > v,\, Z > v) = P(T^*(x) > u,\, T^*(y) > v,\, Z > \max\{u,v\}) \\ &= P(T^*(x) > u)P(T^*(y) > v)P(Z > \max\{u,v\}), \end{aligned} $$

in view of the mutual independence of $T^*(x)$, $T^*(y)$, and Z. Thus,

$$ s_{T(x)T(y)}(u,v) = s_{T^*(x)}(u)\cdot s_{T^*(y)}(v)\cdot s_Z(\max\{u,v\}). \tag{3.3.6} $$

Then for the marginal survival functions, we have

$$ s_{T(x)}(t) = s_{T(x)T(y)}(t,0) = s_{T^*(x)}(t)\cdot s_{T^*(y)}(0)\cdot s_Z(\max\{t,0\}) = s_{T^*(x)}(t)\cdot s_Z(t). \tag{3.3.7} $$

Similarly,

$$ s_{T(y)}(t) = s_{T^*(y)}(t)\cdot s_Z(t). \tag{3.3.8} $$

Note right away that from (3.3.7) and (1.1.8) it follows that the hazard rate for $T(x)$ is

$$ \begin{aligned} \mu_{T(x)}(t) = \mu_x(t) &= -\frac{\partial}{\partial t}\ln(s_{T^*(x)}(t)\cdot s_Z(t)) = -\frac{\partial}{\partial t}\ln(s_{T^*(x)}(t)) - \frac{\partial}{\partial t}\ln(s_Z(t)) = \\ &= \mu_1^*(t) + \mu_Z(t), \end{aligned} \tag{3.3.9} $$

where, for brevity, $\mu_1^*(t)$ denotes the force of mortality for $T^*(x)$, and $\mu_Z(t)$ is the hazard rate for Z. Similarly,

$$\mu_{T(y)}(t) = \mu_2^*(t) + \mu_Z(t), \tag{3.3.10}$$

where $\mu_2^*(t)$ stands for the force of mortality for $T^*(y)$.

The main point here is that the r.v.'s $T^*(x)$ and $T^*(y)$ are not observable. We can observe when people die and whether it happens due to a common shock, but we do not know how long a person would have lived if the shock had not happened. In other words, we know (more precisely, may know) the marginal distributions of $T(x)$ and $T(y)$, and the distribution of Z, but we do not know the distributions of $T^*(x)$ and $T^*(y)$. However in this particular model, we can find these distributions solving (3.3.7) and (3.3.8) with respect to $s_{T^*(x)}(t)$ and $s_{T^*(y)}(t)$.

EXAMPLE 1. Assume that Z is exponential with parameter λ. *Often, the term common-shock-model is applied to this particular case.* From (3.3.7) it follows that ${}_tp_x = s_{T(x)}(t) = s_{T^*(x)}(t)e^{-\lambda t}$, and

$$s_{T^*(x)}(t) = e^{\lambda t} s_{T(x)}(t). \tag{3.3.11}$$

In particular, $T^*(x)$ and $T(x)$ may be exponential (that is, $s_{T^*(x)}(t)$ and $e^{\lambda t}s_{T(x)}(t)$ are exponential functions) only simultaneously.

If $T(x)$ and $T^*(x)$ are exponential and μ_1 and μ_1^* are the corresponding (constant) hazard rates, then (3.3.11) may be rewritten as $\exp\{-\mu_1^* t\} = \exp\{\lambda t\}\exp\{-\mu_1 t\}$. Hence,

$$\mu_1^* = \mu_1 - \lambda. \ \square$$

Consider now the joint-life status and, accordingly, the r.v. $T = T(x:y)$. From (3.3.6), we have

$$s_T(t) = P(T > t) = s_{T(x)T(y)}(t,t) = s_{T^*(x)}(t) \cdot s_{T^*(y)}(t) \cdot s_Z(t). \tag{3.3.12}$$

Similar to (3.3.9),

$$\mu_T(t) = \mu_1^*(t) + \mu_2^*(t) + \mu_Z(t). \tag{3.3.13}$$

We may obtain $s_{T^*(x)}(t)$ from (3.3.7) and $s_{T^*(y)}(t)$ from (3.3.8). Inserting the corresponding expressions into (3.3.13), we readily obtain

$$s_T(t) = s_{T(x)}(t) \cdot s_{T(y)}(t)/s_Z(t), \tag{3.3.14}$$

or in actuarial notation

$${}_tp_{x:y} = {}_tp_x \cdot {}_tp_y / s_Z(t). \tag{3.3.15}$$

As previously mentioned [see the remark right after (3.2.6)], the factor ${}_tp_x$ corresponds to the first life, while ${}_tp_y$ corresponds to the second, so these factors may not coincide even if $x = y$.

EXAMPLE 2. Let two future lifetimes with the same initial age of 70 belong to different groups (for example, specified by gender), and life tables give estimates $l_{70} = 69000$ and $l_{72} = 64000$ for the first group, and $\tilde{l}_{70} = 65000$ and $\tilde{l}_{72} = 61000$ for the second. Find

${}_2p_{70:70}$ if the distribution of Z is exponential with $\lambda = 0.05$. To avoid ambiguities, denote by ${}_t\tilde{p}_y$ the characteristic corresponding to the second life. Then, by (3.3.15),

$${}_2p_{70:70} = \left[{}_2p_{70} \cdot {}_2\tilde{p}_{70}/s_T(2)\right] = \left[{}_2p_{70} \cdot {}_2\tilde{p}_{70}/e^{-\lambda \cdot 2}\right] = e^{2\lambda}(l_{72}/l_{70})(\tilde{l}_{72}/\tilde{l}_{70}) \approx 0.962,$$

as is easy to compute. See also Exercise 62. □

---

4 EXERCISES

Section 1

1. Show by differentiation that (1.1.10) is a solution to (1.1.7).
2. Suppose that for some population, for people who survived 30 years, the probability of dying before 40 years is negligible. How does the survival function look in this case?
3. Find the force of mortality function for the case of Example 1.1-1. Write $s(120)$.
4. Can the function $|\cos x|e^{-x}$ be a survival function?
5. How can the mortality force be changed for the share of newborns who will survive the first year will become 10% larger? Find the answer for the case of the first k years. When does the problem have a solution? Is a solution unique?
6. Let $\mu(x) = (1+x)^{-1}$. Find the survival function. What is noteworthy?
7. Let a life time X be uniform on $[0, \omega]$ (De Moivre's law).
   (a) Show that the force of mortality is $\mu(x) = 1/(\omega - x)$.
   (b) Prove that $T(x)$ is uniformly distributed on $[0, \omega - x]$.
8. Determine the distribution associated with $\mu(x) = 1/(\omega - x)^{\alpha}$ for $x < \omega$ and parameters ω and α. For what values of α does the problem make sense?
9. Determine the distribution associated with $\mu(x) = \alpha/(\omega - x)$ for $x \leq \omega$ and parameters ω and α. For what values of α does the problem make sense?
10. Which function, if any, can serve as a force of mortality at least theoretically: (a) $\mu(x) = xe^{-x}$, (b) $\mu(x) = 0$ if $x \leq 1$, and $\mu(x) = x^{-1/2}$ otherwise.
11. In a country, for a typical person of age 50, the probability ${}_{30}p_{50} = 0.5$.
    (a) Find the same probability for a country where the force of mortality for people of age fifty and older is three times higher.
    (b) Find the same probability for a country where the force of mortality for people of age fifty and older is 0.01 less.
12. Specify t, u, and x for which ${}_{t|u}q_x = q_x$, ${}_{t|u}q_x = {}_{20}p_{30} - {}_{45}p_{30}$.
13. Prove (1.2.8) (a) arguing heuristically, and (b) rigorously, applying induction.
14. Let X be exponential and $E\{X\} = 75$. Find ${}_{20|10}q_{50}$.

15. For $s(x) = (1-x/100)^{1/2}$, find the median (that is, 0.5-quantile), ${}_{10|10}q_{75}$, and $\overset{\circ}{e}_{75}$.

16. (a) In a country, the survival function for women is closely approximated by $s_f(x) = (1-x/100)^{1/2}$, while for men, $s_m(x) = (1-x/90)^{1/2}$. We assume that the probability of the birth of a boy is $1/2$. (i) Considering just the ratio of mean values, estimate the average proportion between men and women of age 50. Comment on a way of getting a more precise solution. (ii) Find ${}_{20|10}q_{50}$ for a person aged 50 taken at random.

    (b) In general, let a population consist of two groups with conditional survival functions ${}_tp_x^{(1)}$ and ${}_tp_x^{(2)}$ and the probabilities (weights) that a newborn belongs to these groups be w_1 and w_2, respectively. Write a general formula for ${}_tp_x$ for a representative of the population, and the weights $w_1(x)$ and $w_2(x)$ of the groups among people of age x.

17. The lifetimes of 100 people are independent with the same survival function $s(x) = (1-x/100)^{1/2}$. Find the distribution of $\mathcal{L}(60)$, as well as its mean and variance.

18. Following the same logic as in Example 1.2-2, prove that ${}_np_x = {}_mp_x \cdot {}_{n-m}p_{x+m}$ for $m \leq n$.

19. If ${}_{20}p_{50} = 0.8$, and ${}_{15}p_{55} = 0.85$, what is ${}_5p_{50}$?

20. Which is larger: the probability to survive to age 70 for a 60-year-old person or for a 65-year-old person (provided that they are chosen at random from a homogeneous population)? State the question in general, using letters rather than numbers.

21. (a) For young people, causes of death are related mostly to accidents. Proceeding from this, explain why the assumption of the approximate constancy of $\mu(x)$ may look reasonable for x's between 30 and 40 years.

    (b) For what x's do we need to know values of $\mu(x)$ to compute ${}_{10}p_{30}$? Which formula shows it explicitly? Relate this question to part (a).

    (c) Why is the assumption that $\mu(x)$ is constant not unreasonable in Example 1.2-4 but rather artificial in Example 1.2-6?

22. Assume that the lifetime of 30% of newborns has a constant mortality rate of 1/50 year^{-1}, while for 70%, the rate is 1/80 year^{-1}. In other words, the distribution of X is a mixture of exponential distributions. Is it true for $T(x)$? Find the distribution of $T(20)$. Interpret your results from a common sense point of view.

23. Estimate $\overset{\circ}{e}_{50}$ in the case of $s(x) = \exp\{-(x/70)^2\}$. (*Advice*: Use (1.2.10), and observe that the resulting integrals are relevant to the standard normal distribution function.)

24. Let $\mu(x) = 1/[2(100-x)]$ for $x < 100$. Find $\overset{\circ}{e}_{75}$.

25. Prove that

$$e_x = p_x(1 + e_{x+1}).$$

26. By using the results of Exercises 13 and 25, find ${}_2p_{60}$ given e_{60}, e_{61}, and e_{62}.

27. Show that ${}_np_x = \dfrac{e_x e_{x+1} \cdot \ldots \cdot e_{x+n-1}}{(1+e_{x+1})(1+e_{x+2}) \cdot \ldots \cdot (1+e_{x+n})}$.

28. Show that if X is exponential with $E\{X\} = r$, then $K(x)$ has the geometric distribution (**0**.3.1.9) with parameter $p = 1 - e^{-\mu}$, where the (constant) force of mortality $\mu = 1/r$. (*Advice*: First, recall that $T(x)$ is distributed as X (why?). Secondly, look up Example 1.3-1.)

29. Find $Var\{T(10)\}$ in the situation of Example 1.2-4.

30. Show that $e_x \leq \overset{\circ}{e}_x$ for any distribution of T. (*Hint*: The problem is very simple.)

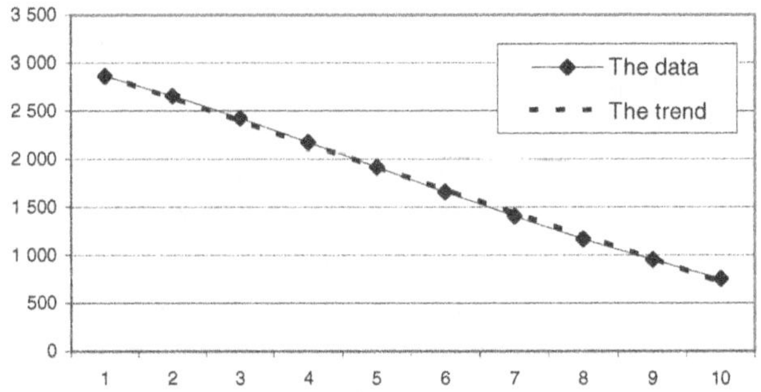

FIGURE 9. The broken line corresponds to a trend line.

31. In Example 1.3-1, show that $e_x \sim r$ as $r \to \infty$. Can we say in general that if $\overset{\circ}{e}_x$ is large, then $\overset{\circ}{e}_x \sim e_x$?

32. Show that the second term in (1.6.4) corresponds to the survival function of the same type as in Example 1.1-4.

33. By using Table 1.5.1-1, estimate: ${}_{40}p_{40}$, $\mu(20)$, the probability for a person of an age of 40 to die between 50 and 60. Compare ${}_{2|}q_0$ and ${}_{2|}q_{20}$.

34. Give a common sense explanation of the facts mentioned in Example 1.5.1-1e.

35. Graph values d_x from Table 1.5.1-1 for $x = 90, ..., 99$. Do not use the data for ages 100+ since they concern more than one year. Add a trend line. The corresponding graph obtained by Excel is shown in Fig.9. We see that the graph is close to linear. Explain that in this case, we can approximate the density of $T(90)$ by a linear decreasing function. Extrapolate (extend) the graph as a linear function, and find the age a at which the corresponding line crosses the x-axis. Show that, if we do not take into account people who survive the age of a, then the density $f_{T(90)}(t)$ may be closely approximated by the linear function $2(a-t)/a^2$. What percent of people were not taken into account? Show that if we do take into account people who survive a, we should approximate $f_{T(90)}(t)$ by another function for $x \geq 100$. For example, for a rough approximation, it may be a linear function with another slope than the one for $90 \leq x \leq 99$.

36. Graph values l_x from Table 1.5.1-1 for $x = 75, ..., 99$. Do not use the data for ages 100+ since they concern more than one year. Add a trend line. Discuss the possibility of approximating the distribution of $T(75)$ by a uniform distribution.

37. Let $s(60) = 0.830$, $s(61) = 0.825$, $s(62) = 0.820$. Using the linear and exponential interpolations described in Section 1.5.2, find $s(60.5)$ and $s(61.3)$. Using both methods, estimate $\mu(60.5)$ and the probability that a person whose age is 60 years 3 months will live at least one year more.

38. Show that (a) if the distribution of X is uniform on an interval $[0, \omega]$, then the linear interpolation (1.5.2) leads to precise values of $s(x+t)$; (b) the exponential interpolation leads to precise values of $s(x+t)$ if X has an exponential distribution. Are these distributions the only distributions with such properties?

39. In a country, for a person chosen at random, the survival probabilities $s(20) \approx 0.97$, $s(50) \approx 0.91$, and $s(60) \approx 0.65$. Find the parameters of the Makeham law (1.6.3). (*Advice*: Compute first the logarithms of $s(\cdot)$.)

Section 2*

40. In the double decrement scheme, $\mu_x^{(1)}(t) = t^2$ and $\mu_x^{(2)}(t) = 2t+1$. For which t's, given that death occurred at time t, is the probability that it occurred from cause 1 larger than the corresponding probability for cause 2?

41. Let $q_{50}^{(1)} = 0.05$, $q_{51}^{(1)} = 0.06$, $q_{52}^{(1)} = 0.07$, $q_{50}^{(2)} = 0.01$, $q_{51}^{(2)} = 0.02$, and $q_{52}^{(2)} = 0.03$. Find ${}_{2|}q_{50}^{(1)}$ and $P(J = 1 \mid 2 < T \le 3)$.

42. Give a heuristic proof of Proposition 1 writing $T \approx t$ instead of $t < T(x) \le t + dt$, and writing the r.-h.s. of (2.1.7) for $j = 1$ as $P(T \approx t,\, J = 1 \mid T \ge t) = P(\min\{T_1, T_2\} \approx t,\, T_1 < T_2 \mid T \ge t)$.

43. In the situation of Example 2.3-1, find ${}_{2|}q_{70}$, $Var\{\mathcal{D}_{71}^{(2)}\}$ (where $\mathcal{D}_{71}^{(2)} = {}_1\mathcal{D}_{71}^{(2)}$), and $P(\mathcal{D}_{71}^{(2)} = 3)$ and compare the last result with its Poisson approximation.

44. What data should be added to Table 1.5.1-1 in order that we will be able to estimate $\mu_x^{(j)}$?

45. Show that if the functions $\mu_x^{(j)}(t)$ are constant, then the r.v.'s $T(x)$ and J are independent.

46. In the case $\mu_x^{(j)}(t) = j/20$ for $j = 1, 2$, find $f_{TJ}(t, j)$, the marginal distribution of T, and $P(J = j)$.

47. In the case $\mu_x^{(j)}(t) = jt^{j-1}/20$ for $j = 1, 2$, find $f_{TJ}(t, j)$, the marginal distribution of T, and $P(J = j)$.

48. The force of mortality for accidental death is $\mu_x^{(1)}(t) = \frac{1}{30-t}$; for death from other causes, it is $\mu_x^{(2)}(t) = \frac{1}{2(30-t)}$ (for all t for which it makes sense). Find ${}_{10}p_x$ and the probability that death will be a result of an accident.

49. There are three causes for the termination of an insurance on (x). The corresponding forces of decrement are $\mu_x^{(1)}(t) = \frac{1}{30-t}$, $\mu_x^{(2)}(t) = \frac{1}{2(30-t)}$, and $\mu_x^{(3)}(t) = 0.04e^{0.1t}$. Find ${}_{10}p_x$, and estimate the probability that death will be a result of the first cause. Compare also the mortality rates above with those in Section 1.6.

50. Probabilities $q_x'^{(i)}$, $i = 1, 2$, are given in the following table.

t	0	1	2
$q_{x+t}'^{(1)}$	0.02	0.03	0.04
$q_{x+t}'^{(2)}$	0.03	0.04	0.05

Provide an Excel worksheet for estimating $q_t^{(i)}$ and ${}_tp_x$.

Section 3*

51. As was stated, for example, in (**0**.1.3.12), if $F(u,v)$ is a joint d.f., the density $f(u,v) = \dfrac{\partial^2 F(u,v)}{\partial u \partial v}$. Write $f(u,v)$ in terms of the survival function $s(u,v)$.

52. Let $h(t)$ be a decreasing function such that $h(0) = 1$ and $h(\infty) = 0$. For example, $h(t) = e^{-t^2}$. Any such a function may be the tail (survival) function for a positive r.v. T. That is, there exists a r.v. T such that $P(T > t) = h(t)$. Is it true that the function $s(u,v) = h(u+v)$ may be the joint survival function for a r.vec. (T_1, T_2) ? For example, is the function $e^{-(u+v)^2}$ a survival function?

53. Carry out all calculations in Example 3.1-1.

54. In Example 3.2-1, compare $E\{T(x:y)\}$ for $k=1$ and $k=0$.

55. For two independent lives, ${}_{30}p_{40:50}=0.9$. Assume that life conditions for both lives changed in a way that the new forces of mortality for each life decreased by 10%. Find the new value for the probability mentioned.

56. For two independent lives with the future lifetimes T and $\tilde{T}$, the characteristics ${}_tp=P(T>t)$, and ${}_t\tilde{p}=P(\tilde{T}>t)$ are given. Find the probabilities that (a) both will survive n years; (b) exactly one will survive n years; (c) no one will survive n years; (d) at least one will survive n years; (e) both will die in the nth year; (f) no one will die in the nth year; (g) exactly one will die in the nth year; (h) one will die in the nth year, and one in the next year.

57. For a husband and wife aged 50 and 40, respectively, the survival functions are the same as in Exercise 16. Under the independence assumption, find
${}_{20}p_{50:40}$, ${}_{50}p_{50:40}$,
${}_{20}p_{\overline{50:40}}$, ${}_{50}p_{\overline{50:40}}$.

58. Consider a husband and a wife aged 50 and 40, respectively. Assume that the lives are independent, for the husband the mortality is 10% higher than that for the average distribution from Table 1.5.1-1, and for the wife 10% lower. Find ${}_{20}p_{50:40}$ and ${}_{20}p_{\overline{50:40}}$.

59. Show that for a status of three independent lives (x), (y), (z), similarly to (3.2.5) and (3.2.6),

$$ {}_tp_{x:y:z}={}_tp_x\cdot{}_tp_y\cdot{}_zp_z,\ {}_tq_{\overline{x:y:z}}={}_tq_x\cdot{}_tq_y\cdot{}_zq_z. $$

Does the remark after equations (3.2.5) and (3.2.6) apply to this case?

60. Show that (3.3.15) implies (3.2.5).

61. Similar to Example 3.3.1-1, consider a random parameter ζ uniformly distributed on $[1,2]$ (in appropriate units). Given ζ, the random lifetimes T_1 and T_2 are independent and exponential with the forces of mortality ζ and $\zeta+0.5$, respectively. Find the probability that the first person will die before the second.

62. Consider the situation of Example 3.3.2-2 under the condition that $T(x),T(y)$ are exponential.

    (a) Show that in order to find ${}_kp_{70:70}$ for any integer k, it suffices to know l_{70},l_{71}, $\tilde{l}_{70}$, $\tilde{l}_{71}$, and λ.

    (b) Show that, in general, to estimate ${}_kp_{x:y}$ for any k, it suffices, to know p_{x_1}, $\tilde{p}_{y_1}$, and $p_{x_2:y_2}$ for some particular x_1,y_1,x_2, and y_2.

63. In a country, for a husband and wife of ages 25 and 20, the distributions of the lifetimes have the same mortality rate $\mu(x)\approx 0.001$ for $x\in[20,40]$. Considering the common shock model and assuming that the common shock time Z has the exponential distribution with $E\{Z\}=2000$, find ${}_{15}q_{25:20}$ and compare the answer with the value of the same characteristic in the case of the absence of common shock. Find ${}_{15}q_{\overline{25:20}}$.

64. Let as usual $\mathbf{1}_A$ be the indicator of an event A. Prove (3.2.7) proceeding from the relation $\phi(\min\{T_1,T_2\})+\phi(\max\{T_1,T_2\})=\phi(T_1)+\phi(T_2)$, and setting the function $\phi(T)=\mathbf{1}_{\{T\le t\}}$ for a fixed t. Which function $\phi(T)$ should we choose to prove directly (3.2.8)?

Chapter 8

Life Insurance Models

1,2,3

In this chapter, we consider an insurance which provides for payment of a single benefit (a *sum insured*) at some random time in the future. As a rule, we will talk about life insurance, although the same model may be applied in other situations (for example, for describing contracts offering warranties for machines).

1 A GENERAL MODEL

1.1 The present value of a future payment

In all models below, the initial time $t = 0$ is the time of policy issue, and the symbol Ψ stands for the time of benefit payment[1].

In the case of the life insurance for a person of age x, the r.v. Ψ may coincide with the moment of death $T = T(x)$ (insurances payable at the moment of death), or it may differ from T. For example, if the benefit is paid at the end of the year of death, then the payment time Ψ is $K+1$, where the curtate time $K = [T]$, the integer part of T. In the case of the so called n-year term life insurance contract, where the insurance provides a payment only if the insured dies within n years, the moment of payment $\Psi = T$ if $T \leq n$. If $T > n$, then no payment is provided, which we will indicate by setting $\Psi = \infty$. We will see that such a way of writing leads to correct results.

We will consider various types of life insurance in Section 2.

The main feature of any life insurance contract consists in the time lag between the moment of payment and the time of policy issue. If Ψ assumes a value t, then the present value of the payment of a unit of money (from the standpoint of the initial time) is equal to a discount factor v_t; see Section **0**.8.3 for detail.

In this and in the next chapters, for discrete time $t = 0, 1, 2, \ldots$, we adopt the model

$$v_t = v^t, \tag{1.1.1}$$

where the discount $v = v_1$ is the *discount for a unit time interval.* (See also Section **0**.8.)

[1]The letter Ψ is rarely used for time moments in notation, but we have a shortage of symbols since most of them are used traditionally for fixed purposes. Also, not any letter is good for denoting a time moment. For example, N would be good for an integer r.v., but traditionally it is not used for continuous variables. So, the reader is suggested to adopt Ψ as the notation for the moment of payment.

In the case of continuous time, we assume that interest is compounded continuously and write

$$v = e^{-\delta} \text{ and } v_t = e^{-\delta t} \text{ for } t \geq 0, \tag{1.1.2}$$

where δ is the unit-time-interval *interest rate*, or the *force of interest.* (See again Section **0**.8.3.) In financial or insurance practice, time is usually measured in years, and in this case, δ is an annual rate.

To make our exposition unified, we set $v = e^{-\delta}$ in the discrete time case too, viewing δ as a positive parameter. Under this assumption, in both discrete and continuous time cases, (1.1.1) and (1.1.2) are true, and we can use whichever is more convenient.

The relations (1.1.1) and (1.1.2) presuppose, in particular, that the interest rate is certain (non-random) and does not change in time. This is serious simplification. In reality, interest rate or investment earning are neither certain nor constant. So, in general, $\delta = \delta(t)$ is a random process. We slightly touched on this question in Section **0**.8.1, but modeling of insurance processes with a random or varying interest is beyond the scope of this book.

If the moment of payment Ψ is random, then the present value of the future payment of a *unit of money* is also random and, by (1.1.2), is equal to the r.v.

$$Z = e^{-\delta\Psi} = v^{\Psi}. \tag{1.1.3}$$

The expected value of Z is traditionally denoted by A, frequently with indices and other signs, depending on the type of contract under consideration. Thus,

$$A = E\{Z\} = E\{e^{-\delta\Psi}\}.$$

The quantity A is called the *actuarial present value* (APV) or the *net single premium* of the contract. The term '*single*' means that we are talking about a premium paid one time at the moment of policy issue. The term '*net*' is related to the fact that such a premium does not reflect the riskiness of the contract. As we know, and as we will repeatedly see again, for the company to fulfill its obligation with a sufficiently large probability, the real premium should be larger than the net premium; in other words, the premium should include a security loading. We will consider this issue in Chapter 10.

An important—and nice from a mathematical point of view—fact is that, as a function of δ, the APV A is the moment generating function (m.g.f.) of Ψ. More precisely,

$$A = E\{e^{-\delta\Psi}\} = M_{\Psi}(-\delta), \tag{1.1.4}$$

where $M_{\Psi}(\cdot)$ is the m.g.f. of Ψ.

By (1.1.3), the lth moment

$$E\{Z^l\} = E\{e^{-l\delta\Psi}\} = M_{\Psi}(-l\delta).$$

So, if we know the m.g.f. $M_{\Psi}(\cdot)$, then we know all moments of Z. In particular, $E\{Z^2\} = M_{\Psi}(-2\delta)$, and

$$Var\{Z\} = E\{Z^2\} - (E\{Z\})^2 = M_{\Psi}(-2\delta) - (M_{\Psi}(-\delta))^2. \tag{1.1.5}$$

We will refer to this as the *rule of double rate*.

The first example below concerns the *whole life insurance* of a life-age-x. This type of insurance provides for payment of a unit of money at the moment of death. In this case, $\Psi = T(x)$, and the traditional notation for the APV is $\bar{A}_x$.

EXAMPLE 1. Consider the *whole life insurance* of a life-age-x in the case when mortality follows the De Moivre law (**7**.1.6.1) with $\omega = 100$. Then $T(x)$ is uniformly distributed on $[0, \omega - x]$ (see Exercise **7**-7b), and the m.g.f. $M_\Psi(z) = M_{T(x)}(z) = (e^{(\omega-x)z} - 1)/[(\omega - x)z]$ in view of (**0**.4.3.4.) In accordance with (1.1.4),

$$\bar{A}_x = \frac{e^{-(\omega-x)\delta} - 1}{(\omega - x)(-\delta)} = \frac{1 - e^{-(\omega-x)\delta}}{(\omega - x)\delta} = \frac{1 - e^{-s}}{s}, \tag{1.1.6}$$

where $s = (\omega - x)\delta$. By (1.1.5),

$$Var\{Z\} = \frac{1 - e^{-2(\omega-x)\delta}}{2\delta(\omega - x)} - \left(\frac{1 - e^{-(\omega-x)\delta}}{\delta(\omega - x)}\right)^2 = \frac{1 - e^{-2s}}{2s} - \left(\frac{1 - e^{-s}}{s}\right)^2. \tag{1.1.7}$$

Let $x = 60$ and $\omega = 100$, and let us adopt $\delta = 0.04$ as the average annual force of interest for the remaining 40 years. In this case, $s = 1.6$, and $\bar{A}_x = \frac{1-e^{-1.6}}{1.6} \approx 0.499$. As may be easily computed using (1.1.7), $Var\{Z\} \approx 0.050$, and hence the standard deviation $\sigma_Z \approx 0.225$. Thus, the present value of the obligation to pay \$1 at the moment of death is, on the average, about 50¢ with a standard deviation of 22.5¢. □

Next, we consider a whole life insurance with benefits payable at the end of the year of death. In this case, $\Psi = K(x) + 1$. The APV for this type of insurance is denoted by A_x (without a bar).

EXAMPLE 2. Consider the situation of the previous example but with benefits payable at the end of the year of death. Assume that x and ω are integers. Since T is uniformly distributed on $[0, \omega - x]$, the r.v. $\Psi = K(x) + 1$ assumes values $1, ..., \omega - x$ with the same probability $q = 1/(\omega - x)$. For example, if $\omega = 100$ and $x = 60$, then Ψ equals $1, ..., 40$ with the same probability $1/40$.

The m.g.f.

$$M_\Psi(z) = E\{e^{z\Psi}\} = \sum_{k=1}^{\omega-x} e^{zk} q = q \sum_{k=1}^{\omega-x} (e^z)^k = qe^z \cdot \frac{1 - e^{z(\omega-x)}}{1 - e^z},$$

as a geometric series. Setting $v = e^{-\delta}$, we get that

$$A_x = M_\Psi(-\delta) = qe^{-\delta} \cdot \frac{1 - e^{-\delta(\omega-x)}}{1 - e^{-\delta}} = \frac{v(1 - v^{\omega-x})}{(\omega - x)(1 - v)}.$$

For the values of ω, x, and δ in Example 1, $A_x \approx 0.4889$. In Exercise 2, the reader is suggested to explain why $A_x < \bar{A}_x$ in this case, and what we can expect in the general case. In Exercise 3, we compute $Var\{Z\}$.

EXAMPLE 3. Let $\mu(x) = \mu$. Then X is exponential, and by the lack-of-memory property, $T(x)$ is also exponential with the same parameter μ. Then, by (**0**.4.3.5), $M_{T(x)}(z) = \mu/(\mu - z)$, and by (1.1.4) and (1.1.5),

$$\bar{A}_x = \frac{\mu}{\mu+\delta} \quad \text{and} \quad Var\{Z\} = \frac{\mu}{\mu+2\delta} - \left(\frac{\mu}{\mu+\delta}\right)^2. \tag{1.1.8}$$

Both characteristics do not depend on x. □

If we know the distribution of Ψ, then we can find not only moments but the distribution of Z itself. If $F_\Psi(t)$ is the d.f. of Ψ, then from (1.1.3) it follows that the d.f.

$$F_Z(x) = P(Z \le x) = P(e^{-\delta\Psi} \le x) = P(\Psi \ge -(\ln x)/\delta) = 1 - F_\Psi(-(\ln x)/\delta). \tag{1.1.9}$$

By (1.1.3), $0 \le Z \le 1$. Therefore it makes sense to consider above only $x \in [0,1]$, and in this case, $(-\ln x)$ is non-negative.

EXAMPLE 4. Assume that the survival function $s(x) = (1-x/100)^{1/2}$; see also Example **7.1.1-1**. Let us consider the distribution of Z for the whole life insurance with benefits payable upon death for a life-age-50.

In this case, $\Psi = T(50)$, and [see also (**7.1.2.1**)] ${}_tp_{50} = P(T(50) > t) = \dfrac{s(t+50)}{s(50)} = \dfrac{\sqrt{1-(50+t)/100}}{1/\sqrt{2}} = \sqrt{1-t/50}$ for $t \in [0,50]$.

Then $F_\Psi(t) = 1 - \sqrt{1-t/50}$, and by (1.1.9),

$$F_Z(x) = \sqrt{1+(\ln x)/50\delta}.$$

For instance, if $\delta = 0.04$, then the probability that the present value of \$1 to be paid at the moment of death is less than 50¢ is $\sqrt{1+(\ln(1/2))/50\cdot 0.04} = \sqrt{1-(\ln 2)/2} \approx 0.808$. □

1.2 The present value of payments for a portfolio of many policies

Consider a portfolio of n contracts with a benefit of one unit of money for each contract, and with the payment times $\Psi_1, ..., \Psi_n$, respectively. Let $Z_i = e^{-\delta\Psi_i}$, the present value of the payment corresponding to the ith contract. Then the present value for the whole portfolio is the r.v.

$$Z = Z_1 + ... + Z_n. \tag{1.2.1}$$

This follows from the very concept of present value. Though the payments are provided at different moments of time, we evaluate these payments with respect to one moment, namely, the initial moment of time. So, all Z's correspond to the same moment of time, and therefore we can sum them up.

While all of this is true, it is still not very transparent. Consider it in more detail, observing how the process of payments runs in time. Assume that payments are to be withdrawn from an investment fund growing at a rate δ. Let G be the initial amount of this fund.

The variables $\Psi_1, ..., \Psi_n$ may be arbitrary and in general random, so the fact that Ψ_1 corresponds to the first contract does not mean that the payment for this contract comes first. Denote by $\Psi_{(1)}, ..., \Psi_{(n)}$ the same time moments $\Psi_1, ..., \Psi_n$ arranged in the ascending order. In particular, $\Psi_{(1)} = \min\{\Psi_1, ..., \Psi_n\}$, the moment of the first payment, and $\Psi_{(n)} = \max\{\Psi_1, ..., \Psi_n\}$, the moment of the last payment. In statistics, such r.v.'s are called *order statistics*.

Since the fund is growing at the rate δ, at the moment of the first payment, $\Psi_{(1)}$, the fund will have amount $Ge^{\delta\Psi_{(1)}}$. For the company to fulfill its obligation, this amount should not be less than the unit of money that the company must pay at this moment. In other words, we should have $Ge^{\delta\Psi_{(1)}} \geq 1$. Let us rewrite it as

$$G \geq e^{-\delta\Psi_{(1)}}. \tag{1.2.2}$$

At the moment $\Psi_{(1)}$, the company will pay 1, and after that the fund will become equal to $Ge^{\delta\Psi_{(1)}} - 1$.

At the moment of the second payment, $\Psi_{(2)}$, the fund will grow to the amount

$$(Ge^{\delta\Psi_{(1)}} - 1)e^{\delta(\Psi_{(2)} - \Psi_{(1)})} = Ge^{\delta\Psi_{(2)}} - e^{\delta(\Psi_{(2)} - \Psi_{(1)})}.$$

The fund will be solvent at this moment if $Ge^{\delta\Psi_{(2)}} - e^{\delta(\Psi_{(2)} - \Psi_{(1)})} \geq 1$. Dividing this inequality by $e^{\delta\Psi_{(2)}}$, we represent it as

$$G \geq e^{-\delta\Psi_{(1)}} + e^{-\delta\Psi_{(2)}}. \tag{1.2.3}$$

We see that (1.2.2) follows from (1.2.3), so it suffices to consider only the latter condition. Similarly, for the time of the third payment, we come to the condition

$$G \geq e^{-\delta\Psi_{(1)}} + e^{-\delta\Psi_{(2)}} + e^{-\delta\Psi_{(3)}},$$

and for the kth payment, we get

$$G \geq e^{-\delta\Psi_{(1)}} + ... + e^{-\delta\Psi_{(k)}}.$$

Each time, the solvency condition includes the corresponding conditions for the previous payments. For the last (nth) payment we have

$$G \geq e^{-\delta\Psi_{(1)}} + ... + e^{-\delta\Psi_{(n)}}. \tag{1.2.4}$$

This condition ensures the possibility of the nth payment, and of all payments before.

The next step is nice. Since $\Psi_{(1)}, ..., \Psi_{(n)}$ are the same variables $\Psi_1, ..., \Psi_n$, simply rearranged in ascending order, we can write that

$$e^{-\delta\Psi_{(1)}} + ... + e^{-\delta\Psi_{(n)}} = e^{-\delta\Psi_1} + ... + e^{-\delta\Psi_n}.$$

Consequently, the condition (1.2.4) may be written as

$$G \geq e^{-\delta\Psi_1} + ... + e^{-\delta\Psi_n} = Z_1 + ... + Z_n.$$

Thus, for the fund to be solvent, it is necessary and sufficient that the initial fund amount G is equal to or greater than the sum $Z_1 + ... + Z_n$. This precisely means that the present value of all future payments equals $Z_1 + ... + Z_n$. If the Z's are random, then the present value is also random.

In conclusion, it makes sense to emphasize that the above argument is not necessary for proving (1.2.1); this relation does follow from the very definition of present value. We just gave more insight into the nature of this notion.

Let us come back to (1.2.1). Assume that the contracts are signed at the same moment of time, and the r.v.'s $\Psi_1, \dots, \Psi_n$ are independent and have the same distribution. We can interpret this as if the contracts are of the same type and act independently.

Let $A = E\{Z_i\}$, the actuarial value of each contract, and $\sigma^2 = Var\{Z_i\}$. Then, in view of (1.2.1), $E\{Z\} = nA$ and $Var\{Z\} = n\sigma^2$.

Denote by h the single premium per one contract, paid at the initial time. Then $H = hn$ is the total amount of premiums paid.

As we did before repeatedly, we can estimate the value of H that is sufficient for the company to fulfill its obligations with a given probability β. Calculations are very similar to what we did, for example, in Section **2**.3.1.1. The probability under consideration is $P(Z \leq H)$. In order for it to be equal to β, we should have

$$\beta = P(Z \leq H) = P\left(\frac{Z - E\{Z\}}{\sqrt{Var\{Z\}}} \leq \frac{H - E\{Z\}}{\sqrt{Var\{Z\}}}\right) = P\left(\frac{Z - nA}{\sigma\sqrt{n}} \leq \frac{H - nA}{\sigma\sqrt{n}}\right).$$

Since Z is the sum of i.i.d. r.v.'s, for large n, we can use normal approximation, writing

$$\beta = P\left(\frac{Z - nA}{\sigma\sqrt{n}} \leq \frac{H - nA}{\sigma\sqrt{n}}\right) \approx \Phi\left(\frac{H - nA}{\sigma\sqrt{n}}\right).$$

We obtain from this that $\dfrac{H - nA}{\sigma\sqrt{n}} \approx q_{\beta s}$, where $q_{\beta s}$ is the β-quantile of the standard normal distribution. From the last relation, we get the estimate

$$H \approx nA + q_{\beta s}\sigma\sqrt{n}, \text{ and } h \approx A + \frac{q_{\beta s}\sigma}{\sqrt{n}}. \tag{1.2.5}$$

More precise estimates may be obtained similarly to what we did in **2**.3.2.

EXAMPLE 1 ([151, N4][2]). A fund will pay death benefits of \$10,000 on each of 900 independent lives of age 30. You are given: $\delta = 0.04$, $\mu = 0.01$, and the death benefits are payable at the moment of death.

The initial amount of the fund is established so that the probability that sufficient funds will be in hand to withdraw the benefit payment at the death of each individual is 0.95. Calculate the initial fund amount.

We are considering the continuous time case, and the lifetimes are exponentially distributed. In accordance with (1.1.8), $A = \dfrac{\mu}{\mu + \delta} = 0.2$, $E\{Z^2\} = \dfrac{\mu}{\mu + 2\delta} = \dfrac{1}{9}$, and $Var\{Z\} = \frac{1}{9} - (\frac{1}{5})^2 \approx 0.0711$. Then, in accordance of (1.2.5), for a unit payment,

$$H \approx 0.2 \cdot 900 + 1.64 \cdot \sqrt{0.0711}\sqrt{900} \approx 193.12.$$

So, for the fund to be solvent with 95% probability, it is enough to have at the beginning \$10,000 × 193.12 = 1.9312 million. Note that the total amount to be paid is \$10,000 × 900 = 9 million. □

[2] Reprinted with permission of the Casualty Actuarial Society.

2 SOME PARTICULAR TYPES OF CONTRACTS

In this section, we consider some important particular types of life insurance. In most types, we distinguish two cases: when benefits are payable at the moment of death, and when they are provided at the end of the year of death. For brevity, we use slightly informal language and refer to the former case as that of continuous time, and to the latter case as the discrete time case.

In practice, most contracts presuppose payments at the moment of death. (Certainly, there is a time gap between the death and the moment of payment, but companies add the earned interest corresponding to this period.) However, the available information comes from discrete time tables, which leads us to models involving only complete years. We establish relations between the actuarial characteristics corresponding to both cases in Sections 7.1.5.2 and 2.1.3.

2.1 Whole life insurance

2.1.1 The continuous time case (benefits payable at the moment of death)

We already considered this type. The payment follows the death of the insured whenever the death occurs. In this case, $\Psi = T = T(x)$ for a life-age-x.

Then the present value of the payment is $Z = e^{-\delta T(x)}$, and the APV $\bar{A}_x = E\{e^{-\delta T(x)}\} = M_{T(x)}(-\delta)$, where $M_{T(x)}(z)$ is the m.g.f. of $T(x)$. In terms of the characteristics $\mu_x(t) = \mu(x+t)$, and ${}_tp_x$ from Chapter 7, the density of $T(x)$ is

$$f_{T(x)}(t) = \mu_x(t)\,{}_tp_x = \mu(x+t)\,{}_tp_x \tag{2.1.1}$$

[see (7.1.2.4)]. Then, in general,

$$\bar{A}_x = \int_0^\infty e^{-\delta t}\mu_x(t)\,{}_tp_x dt = \int_0^\infty v^t \mu_x(t)\,{}_tp_x dt. \tag{2.1.2}$$

As we saw, $E\{Z^2\} = E\{e^{-2\delta T(x)}\}$, which corresponds to the APV for the doubled interest rate 2δ. The traditional notation for this is ${}^2\bar{A}_x$. The superscript on the left means that the force of interest is doubled.

A similar notation is applied to all other types of insurance discussed below.

Making use of this notation, we can write that

$$Var\{Z\} = {}^2\bar{A}_x - (\bar{A}_x)^2. \tag{2.1.3}$$

Concrete examples were considered in Section 1.1.

2.1.2 The discrete time case (benefits payable at the end of the year of death)

The title above presupposes that we have chosen a year as a unit of time. As a matter of fact, the formulas below are true for any choice of the time unit.

We already saw that in this case, $\Psi = K(x)+1$ and takes on values $1,2,\dots$. In accordance with (**7**.1.3.1), $P(K(x)=k) = {}_kp_x \cdot q_{x+k}$. Setting, as usual, $v = e^{-\delta}$, we obtain that the APV

$$A_x = E\{e^{-\delta\Psi}\} = E\{v^{\Psi}\} = \sum_{k=0}^{\infty} v^{k+1} P(K(x)=k) = \sum_{k=0}^{\infty} v^{k+1}\,{}_kp_x \cdot q_{x+k}. \tag{2.1.4}$$

Certainly, the infinite series in (2.1.4) is a mathematical abstraction. Terms for large k's are small due to both factors, $P(K(x)=k)$ and v^{k+1}, and the contribution of these terms may be much less significant than that of terms for moderate k's.

EXAMPLE 1. Let us estimate the net premium A_{50} for $\delta = 0.04$, proceeding from the data in Table **7**.1.5.1-1 for the total population of the USA in 2002. The table estimates $P(K(50)=k)$ by d_{50+k}/l_{50} for $k \le 49$, and gives an estimate for $P(K(50) \ge 50)$ as d_{100+}/l_{50}. Thus, our estimate is

$$\sum_{k=0}^{49} \exp\{-\delta(k+1)\}\frac{d_{50+k}}{l_{50}} + \exp\{-\delta\cdot 51\}\frac{d_{100+}}{l_{50}}. \tag{2.1.5}$$

Straight calculations lead to the estimate 0.326.

We present all such estimates in the Illustrative Table in Appendix, Section 3. All figures for A_x in this table are computed in the way mentioned. The data for l_x were slightly smoothed in the region of ages 82-84, and rounded. The same concerns probabilities q_x which are now computed as l_{x+1}/l_x with $l_{x+1} = l_x - d_x$.

The column in the Illustrative table for ${}^2A_x$ corresponds to the double rate, and all values in it are obtained in the same way as above.

Since in the table under consideration, we use empirical data concerning a particular year, applying rather straightforward estimates, rounding some numbers, and not taking into account mortality for ages greater than 100, *we view this table as illustrative and will use it for illustrative purposes.*

Next, we come back to A_{50} and estimate the error of the above approximation that concerns a remainder of the series (2.1.4); namely, the quantity $\sum_{50}^{\infty} = \sum_{k=50}^{\infty} e^{-\delta(k+1)} P(K=k)$. For ages greater than 100, the table contains only one aggregate estimate. If we replace the factors $e^{-\delta(k+1)}$ for $k \ge 50$ by their upper bound $e^{-\delta(51)}$, then we will get that

$$\sum_{k=50}^{\infty} e^{-\delta(k+1)} P(K=k) \le \sum_{k=50}^{\infty} e^{-\delta(50+1)} P(K=k) \le e^{-\delta\cdot 51} P(K \ge 50).$$

This upper bound leads to the last term in (2.1.5). The value of this term is ≈ 0.0029.

Now, let us estimate the minimum of the remainder $\sum_{k=50}^{\infty}$. Assume that the probabilities $P(K=k)$ are not increasing for $k \ge 50$ (which is quite plausible) and are negligible for $k \ge 60$. Then

$$\sum_{k=50}^{\infty} e^{-\delta(k+1)} P(K=k) \approx \sum_{k=50}^{59} e^{-\delta(k+1)} P(K=k).$$

Since $e^{-\delta(k+1)}$ is decreasing, the minimum of this sum corresponds to the case when all probabilities $P(K=k)$ for $50 \le k \le 59$, are the same. In this case, $P(K=k) \approx$

$\frac{1}{10}P(K\geq 50) = \frac{1}{10}\frac{2097}{93566} \approx 0.00224$, and

$$\sum_{k=50}^{59} e^{-\delta(k+1)}P(K=k) \approx 0.00224 \sum_{k=50}^{59} e^{-\delta(k+1)} = 0.00224 \cdot e^{-51\delta}\frac{1-e^{-10\delta}}{1-e^{-\delta}} \approx 0.0025.$$

We see that the difference between the last number and 0.0029 is not large. Let us also recall that the total estimate is about 0.326, so the remainder under discussion constitutes less than 1% of the total estimate. One may conjecture that this is much smaller than the accuracy of estimation in the life table Table **7**.1.5.1-1 itself, and the error which is due to the fact that we are considering a constant δ.

EXAMPLE 2. Assume that the density of the r.v. $T = T(90)$, the future lifetime of a life-age-90, may be well approximated by the linear decreasing function $f_T(t) = \frac{2}{169}(13-t)$ for $t \in [0,13]$. In Exercise **7**.35, we saw that such a model relatively well fit the data from Table **7**.1.5.1-1. Set $\delta = 0.04$, and observe that

$$P(K=k) = P(k \leq T < k+1) = \int_k^{k+1} f(t)dt = \frac{1}{2}(f(k)+f(k+1)) = \frac{2}{169}(13-k) - \frac{1}{169}.$$

(In the third step we took into account that the density is a linear function.) Then

$$A_{90} = \sum_{k=0}^{12} v^{k+1}P(K=k) = \sum_{k=0}^{12} e^{-0.04\cdot(k+1)}\left(\frac{2}{169}(13-k) - \frac{1}{169}\right) \approx 0.830.$$

(There are formulas for such sums [see, e.g., (**9**.4.4)], but it is better to use software.)

Like Example 1, this example also serves for illustration. There are more sophisticated statistical methods for evaluating the distribution of X, and hence for estimating A_x. □

Another way of computing A_x is based on the following backward recursion relation:

$$A_x = vq_x + p_x vA_{x+1} = v(q_x + p_x A_{x+1}). \tag{2.1.6}$$

Above, $q_x = {}_1q_x = P(T(x) \leq 1)$, and $p_x = 1 - q_x$. We will prove (2.1.6) a bit later.

If we know the value of A_n for some n and the probabilities p_x, we can make use of (2.1.6), moving backward and computing A_x for all integers $x < n$.

Consider, for instance, the situation of Example 2. Suppose we accept the estimate of A_{90} that we obtained in this example. The survival function for ages less than 90 is not as simple as that for $x > 90$. But we can estimate A_x for x's less than 90 by using (2.1.6). In Exercise 8, the reader is invited to provide the corresponding Excel worksheet.

To prove (2.1.6), we again apply the first step analysis. Let us start with a heuristic reasoning.

The insured will either die within the first year or will survive one year. The former event occurs with probability q_x, the payment in this case will be made at the end of the first year, and its discounted value is equal to v.

The latter event occurs with probability p_x. In this case, the insured at the end of the first year will be $x+1$ years old, and the insurance process will start over, as from the very beginning. The present value of the future payment in this case is A_{x+1} from the standpoint of the time $t = 1$, the end of the first year. To evaluate the present value of such payments from the standpoint of the initial time, we should multiply A_{x+1} by the discount v.

Thus, the present value Z assumes the value v with probability q_x, and, *on the average*, the value vA_{x+1} with probability p_x. This is reflected in (2.1.6).

Rigorously, we may write it as follows:

$$\begin{aligned} E\{Z\} &= E\{Z \mid T(x) \leq 1\}P(T(x) \leq 1) + E\{Z \mid T(x) > 1\}P(T(x) > 1) \\ &= E\{Z \mid T(x) \leq 1\}q_x + E\{Z \mid T(x) > 1\}p_x. \end{aligned}$$

As was explained above, $E\{Z \mid T(x) \leq 1\} = v$, and $E\{Z \mid T(x) > 1\} = vA_{x+1}$, which leads to (2.1.6).

► Absolutely rigorously speaking, the very last relation also should be proved, and one can do it by writing

$$\begin{aligned} E\{Z \mid T(x) > 1\} &= E\{v^{K(x)+1} \mid T(x) > 1\} = vE\{v^{K(x)} \mid T(x) > 1\} \\ &= vE\{v^{1+K(x+1)} \mid T(x) > 1\} = vE\{v^{1+K(x+1)}\} = vA_{x+1}. \end{aligned}$$

(We used the fact that, once the insured has survived one year, her/his curtate lifetime is equal to one year plus how long she/he will live after attaining age $x+1$.) ◄

From this point onward, when considering relations similar to (2.1.6), we will not always repeat the same standard argument each time but will rather restrict ourselves to heuristic proofs.

2.1.3 A relation between A_x and $\bar{A}_x$

The relations we discuss here are based on the linear interpolation procedure from Section **7**.1.5.2, where we assumed that the lifetime is uniformly distributed within each year of age. We will show that under this assumption,

$$\bar{A}_x = \frac{i}{\delta} A_x, \tag{2.1.7}$$

where

$$i = e^{\delta} - 1. \tag{2.1.8}$$

Since we usually possess data on survival probabilities only for complete years, we can compute, proceeding from this data, only A_x. The formula (2.1.7) gives a way to estimate $\bar{A}_x$.

The quantity i is the profit that one gets at the end of a unit time interval after investing one unit of money at the beginning of this interval. If time is measured in years, this is an *effective annual interest rate* (shortly *interest*), or in other terminology an *annual yield*, provided that interest is compounded continuously. See Section **0**.8.1 for more detail.

It is worth noting that for small δ, the "correction" coefficient $\frac{i}{\delta}$ in (2.1.7) is close to one. Say, if $\delta = 0.04$, then $\dfrac{i}{\delta} = \dfrac{e^{0.04} - 1}{0.04} \approx 1.0202$. This fact becomes understandable if we recall that $e^x = 1 + x + \frac{x^2}{2} + o(x^2)$. (See Appendix– (4.2.6). For the notation $o(x)$, see Appendix, Section 4.1.) Making use of this expansion, we have

$$1 \leq \frac{i}{\delta} = \frac{e^{\delta} - 1}{\delta} = 1 + \frac{\delta}{2} + o(\delta),$$

so we can expect that $\frac{i}{\delta}$ differs from 1 approximately by $\delta/2$. In the numerical example above, this is indeed the case.

▶ Using Appendix–(4.2.2), one can show that

$$\frac{\delta}{2}+\frac{\delta^2}{6}\le\frac{i}{\delta}-1\le\frac{\delta}{2}+e^{\delta}\frac{\delta^2}{6}. \blacktriangleleft \tag{2.1.9}$$

To prove (2.1.7), consider a life time $T=T(x)$ and the corresponding curtate time $K=K(x)=[T(x)]$; we omit x if it does not cause misunderstanding. Let $T_{\text{frac}}=T_{\text{frac}}(x)$ be the fractional part of $T(x)$, that is, $T_{\text{frac}}=T-K$. So, $T=K+T_{\text{frac}}$. The uniformity assumption is equivalent to the assumption that T_{frac} is uniformly distributed on $[0,1]$, regardless the value the r.v. K has taken. In other words, K and T_{frac} are independent, and T_{frac} is uniform on $[0,1]$. Then

$$\begin{aligned}\bar{A}_x&=E\{e^{-\delta T}\}=E\{e^{-\delta(K+T_{\text{frac}})}\}=E\{e^{-\delta K}e^{-\delta T_{\text{frac}}}\}=E\{e^{-\delta K}\}E\{e^{-\delta T_{\text{frac}}}\}\\&=e^{\delta}E\{e^{-\delta(K+1)}\}E\{e^{-\delta T_{\text{frac}}}\}=e^{\delta}A_xE\{e^{-\delta T_{\text{frac}}}\}.\end{aligned} \tag{2.1.10}$$

The last factor $E\{e^{-\delta T_{\text{frac}}}\}=M_{T_{\text{frac}}}(-\delta)$, where $M_{T_{\text{frac}}}(z)$ is the m.g.f. of T_{frac}. Hence, by (**0**.4.3.4), $E\{e^{-\delta T_{\text{frac}}}\}=(1-e^{-\delta})/\delta$, and by virtue of (2.1.10), $\bar{A}_x=e^{\delta}\frac{1-e^{-\delta}}{\delta}A_x=\frac{e^{\delta}-1}{\delta}A_x$. ■

2.1.4 The case of benefits payable at the end of the m-thly period

Next, we divide each year into m subintervals—for example, as months ($m=12$) or quarters ($m=4$)—and consider the insurance with a benefit payable at the end of the m-thly period in which death occurs.

Denote by $A_x^{(m)}$ the corresponding APV. Formally, we can apply the above model to this case since the unit of time in this model was not specified. However, because available information usually concerns complete years, it is useful to specify a connection between $A_x^{(m)}$ and A_x.

As in the previous section, we will provide an approximation formula assuming that the lifetime is uniformly distributed within each year. Then, as we will prove below,

$$A_x^{(m)}=\frac{i}{i^{(m)}}A_x, \tag{2.1.11}$$

where

$$i^{(m)}=m[(1+i)^{1/m}-1]. \tag{2.1.12}$$

The characteristic $i^{(m)}$ is a *nominal annual interest rate* corresponding to the annual interest rate i. More precisely, $i^{(m)}$ is an annual rate such that if interest is compounded at this rate m times in a year, then the total annual interest will be equal to i. We considered this notion in more detail in Section **0**.8.2, and showed there that the characteristic with the mentioned property should satisfy (2.1.12).

In the same section, we proved that $\lim_{m\to\infty}i^{(m)}=\delta=\ln(1+i)$, so the result (2.1.7) from the previous section follows from (2.1.11) as a limiting case.

Note also that since $i^{(m)}$ is decreasing in m (see Section **0**.8.2), the "correction coefficient" $i/i^{(m)}$ is increasing from 1 to i/δ, and hence for any m

$$1 \le \frac{i}{i^{(m)}} < \frac{i}{\delta} = 1 + \frac{\delta}{2} + o(\delta). \tag{2.1.13}$$

▶ In particular, from this and (2.1.9) it follows that

$$0 \le \frac{i}{i^{(m)}} - 1 \le \frac{\delta}{2} + e^{\delta}\frac{\delta^2}{6}.$$

Proof of (2.1.11). Denote by $K^{(m)}$ the number of complete periods of the length $1/m$ that the insured survived. We will call these periods *m-ths*. Set $R^{(m)} = K^{(m)} - mK$. This is the number of complete m-ths lived in the year of death. Then $K^{(m)} = mK + R^{(m)}$.

Let, for example, $m = 12$. We will then call m-ths months, though as a matter of fact months have different lengths. Let, say, $T = 25.34$. Then the insured lived $[25.34 \cdot 12] = [304.08] = 304$ complete months, so $K^{(m)} = 304$. The insured lived 25 complete years, so $mK = 12 \cdot 25 = 300$, and $R^{(m)} = 4$—that is, the insured lived 4 complete months in the last year.

Let us come back to the general case. Under the assumption made, the r.v. K and $R^{(m)}$ are independent, and $R^{(m)}$ takes on values $0, 1, ..., m-1$ with the same probability $1/m$.

Now we should recall that δ is an annual rate, and we measure time in years. So, the payment time is equal not to $K^{(m)} + 1$ (as it would have been, if we had chosen an m-th as a unit of time), but $\frac{1}{m}(K^{(m)} + 1)$. Then

$$\begin{aligned}
A_x^{(m)} &= E\{\exp\{-\delta(K^{(m)}+1)/m\}\} = E\{\exp\{-\delta(mK + R^{(m)}+1)/m\}\} \\
&= E\{\exp\{-\delta K - \delta(R^{(m)}+1)/m\}\} = E\{\exp\{-\delta K\}\}E\{\exp\{-\delta(R^{(m)}+1)/m\}\} \\
&= E\{e^{-\delta K}\}E\{\exp\{-\frac{\delta}{m}(R^{(m)}+1)\}\} = e^{\delta}E\{e^{-\delta(K+1)}\}E\{\exp\{-\frac{\delta}{m}(R^{(m)}+1)\}\} \\
&= A_x \cdot e^{\delta}E\{\exp\{-\frac{\delta}{m}(R^{(m)}+1)\}\}.
\end{aligned} \tag{2.1.14}$$

The last expectation equals $M_{R^{(m)}+1}(-\delta/m)$, where $M_{R^{(m)}+1}(z)$ is the m.g.f. of $R^{(m)} + 1$. Since $R^{(m)} + 1$ assumes values $1, ..., m$ with equal probabilities,

$$M_{R^{(m)}+1}(z) = \sum_{k=1}^{m} e^{zk}\frac{1}{m} = \frac{1}{m}e^{z}\frac{1-e^{mz}}{1-e^{z}}.$$

Taking into account that $e^{-\delta} = v$ and $v = \frac{1}{1+i}$ [see (2.1.8)], we have

$$\begin{aligned}
M_{R^{(m)}+1}(-\delta/m) &= \frac{1}{m}v^{1/m}\frac{1-v}{1-v^{1/m}} = \frac{1}{m}\frac{1-v}{v^{-1/m}-1} = \frac{1}{m}\frac{1-1/(1+i)}{(1+i)^{1/m}-1} \\
&= \frac{1}{1+i}\cdot\frac{i}{m[(1+i)^{1/m}-1]} = \frac{1}{1+i}\cdot\frac{i}{i^{(m)}}.
\end{aligned}$$

Substituting it into (2.1.14) and noticing that $e^{\delta} = 1 + i$, we arrive at (2.1.11). ■ ◀

2.2 Deferred whole life insurance

2.2.1 The continuous time case

An *m-year deferred whole life insurance* provides for a benefit only if the insured survives m years. In accordance with the convention made in Section 1.1, in this case, we set the payment time $\Psi = \infty$ (the payment will never happen) if $T(x) \le m$, and $\Psi = T(x)$ if $T(x) > m$. Whether we include the event $\{T(x) = m\}$ into the former case or into the latter does not matter because $P(T(x) = m) = 0$. Because $Z = e^{-\delta\Psi}$,

$$Z = \begin{cases} 0 & \text{if } T(x) \le m, \\ e^{-\delta T(x)} = v^{T(x)} & \text{if } T(x) > m. \end{cases} \tag{2.2.1}$$

(If $\delta = 0$, then by convention, we set $0 \cdot \infty = \infty$, and $Z = e^{-\delta\Psi} = e^{-0 \cdot \infty} = e^{-\infty} = 0$.) Certainly, we could write the representation (2.2.1) without involving Ψ into consideration: the present value equals zero if no payments are provided, and equals $e^{-\delta T(x)}$ if payments are provided at time $T(x) > m$.

In this case, the APV is denoted by ${}_{m|}\bar{A}_x$. If $f_{T(x)}(t)$ is the density of $T(x)$, then by (2.2.1),

$${}_{m|}\bar{A}_x = E\{Z\} = \int_m^\infty e^{-\delta t} f_{T(x)}(t)dt. \tag{2.2.2}$$

The following formula may simplify calculations:

$${}_{m|}\bar{A}_x = {}_mp_x v^m \bar{A}_{x+m}. \tag{2.2.3}$$

Derivation of (2.2.3) is based on an argument similar to what we used in establishing (2.1.6). If the insured attains age $x+m$, the insurance becomes a "usual" whole life insurance whose actuarial present value is $\bar{A}_{x+m}$. However, it happens with the probability $P(T(x) > m) = {}_mp_x$. Also, to make evaluation proper from the standpoint of the initial time, we should multiply A_{x+m} by the discount factor v^m.

► Formally, it may be written as follows:

$$E\{Z\}=0 \cdot P(T(x) \le m) + E\{Z \mid T(x) > m\}P(T(x) > m) = E\{e^{-\delta T(x)} \mid T(x) > m\}\, {}_mp_x$$
$$= {}_mp_x E\{e^{-\delta(m+T(x+m))}\} = {}_mp_x e^{-\delta m} E\{e^{-\delta T(x+m)}\} = {}_mp_x v^m A_{x+m}. \blacktriangleleft$$

EXAMPLE 1 ([151, N2][3]). For a 5-year deferred whole life insurance of 1, payable at the moment of death of (x), you are given: Z is the present value r.v. of this insurance; $\delta = 0.1$; $\mu = 0.04$. Calculate $Var\{Z\}$.

In our case, $m = 5$, and the distribution of T is exponential. In accordance with (2.2.3) and (1.1.8), $E\{Z\}=e^{-\mu m}e^{-m\delta}\bar{A}_{x+m}=e^{-m(\mu+\delta)}\dfrac{\mu}{\mu+\delta} \approx 0.1419$; $E\{Z^2\}=e^{-\mu m}e^{-m2\delta} \cdot {}^2\bar{A}_{x+m}=$ $e^{-m(\mu+2\delta)}\dfrac{\mu}{\mu+2\delta} \approx 0.0502$; and $Var\{Z\} = E\{Z^2\} - (E\{Z\})^2 \approx 0.0502 - (0.1419)^2 \approx 0.0301$.

□

[3]Reprinted with permission of the Casualty Actuarial Society.

2.2.2 The discrete time case

In this case, an m-year deferred whole life insurance is the same type of insurance as above with the exception that a payment, if any, is provided at the end of the year of death. Under such an assumption, $\Psi = \infty$ if $T(x) \leq m$, and $\Psi = K(x)+1$ for $T(x) > m$. Accordingly, $Z = 0$ if $T(x) \leq m$, and $Z = e^{-\delta(K(x)+1)} = v^{K(x)+1}$ for $T(x) > m$.

The symbol for the APV is ${}_{m|}A_x$ (the bar is removed), and the counterparts of the formulas (2.2.2) and (2.2.3) are

$$ {}_{m|}A_x = \sum_{k=m}^{\infty} v^{k+1} P(K(x)=k), \tag{2.2.4}$$

and

$$ {}_{m|}A_x = {}_m p_x v^m A_{x+m}, \tag{2.2.5}$$

respectively. The reader is invited to verify (2.2.4) and (2.2.5) on her/his own applying the same heuristic argument as above.

To establish a connection between ${}_{m|}A_x$ and ${}_{m|}\bar{A}_x$, we can combine (2.2.3) and (2.2.5) with (2.1.7), keeping in mind that this is an approximation formula. (When is it precise? See also Exercise 13.)

Variances, as is easily verified, may be computed by the same rule of double rate: $E\{Z^2\}$ is equal to the APV with replacement of δ by 2δ. The corresponding symbols for this operation are ${}^{2}_{m|}A_x$ and ${}^{2}_{m|}\bar{A}_x$.

2.3 Term insurance

2.3.1 Continuous time

An *n-year term insurance* provides for a payment only if death occurs within n years. In this case, $\Psi = T(x)$ if $T(x) \leq n$, and $\Psi = \infty$ if $T(x) > n$. Accordingly,

$$ Z = \begin{cases} e^{-\delta T(x)} = v^{T(x)} & \text{if } T(x) \leq n, \\ 0 & \text{if } T(x) > n. \end{cases} \tag{2.3.1}$$

Setting $m = n$ in (2.2.1), and comparing it with (2.3.1), we see that n-term insurance is, in a sense, the opposite of n-deferred insurance. We will use this circumstance below.

The traditional notation for the APV in this case is $\bar{A}^{1}_{x:\overline{n|}}$, where the bar means that we deal with "continuous time", and the superscript 1 marks this particular type of insurance. In Section 2.4.2, we consider a somewhat different type, where in the corresponding symbol there will be no superscript 1.

From (2.3.1) it follows that

$$ \bar{A}^{1}_{x:\overline{n|}} = \int_0^n e^{-\delta t} f_{T(x)}(t)dt, \tag{2.3.2}$$

which may be used for direct calculations.

It is worthwhile to emphasize that, as follows from (2.3.2), the value of $\bar{A}^{1}_{x:\overline{n|}}$ is determined by the distribution of $T(x)$ in the interval $[0,n]$, and does not depend on the mortality rate after n years. The same is true for other moments, for example, the variance.

EXAMPLE 1. Let us provide a quick rough estimate of $\bar{A}^{\,1}_{20:\overline{10}|}$ proceeding from the data in Table **7**.1.5.1-1 and $\delta = 0.03$. As was told repeatedly, it is not realistic to assume that the distribution of $T(x)$ is exponential. However, if we look at the table mentioned, we will see that the variation of the estimates for $\mu(x)$ in the interval $[20, 30]$ is rather due to random fluctuations, and is close to 0.0095. We know that the density $f_{T(x)}(t)$ for $t \in [0, n]$ is completely determined by the values of $\mu(x)$ on $[x, x+n]$; see (**7**.1.2.4)-(**7**.1.2.5). So, when using (2.3.2), we can assume $f_{T(20)}(t)$ to be exponential with $\mu = 0.0095$.

It makes sense to present right away a general formula for the exponential distribution. We have

$$\bar{A}^{\,1}_{x:\overline{n}|} = \int_0^n e^{-\delta t}\mu e^{-\mu t}dt = \mu\int_0^n e^{-(\mu+\delta)t}dt = \frac{\mu}{\mu+\delta}\left(1 - e^{-(\mu+\delta)n}\right)$$

[compare with (1.1.8)]. In our case, $\bar{A}^{\,1}_{20:\overline{10}|} \approx 0.0785$. □

It is important that if we know the values of $\bar{A}_x$ and l_x for integer x's, then the computation of $\bar{A}^{\,1}_{x:\overline{n}|}$ is straightforward for any n.

Indeed, first recall that if we know l_x's, then we know probabilities ${}_np_x = l_{x+n}/l_x$; see Section **7**.1.4.

Secondly, we can use the formula

$$\bar{A}_x = \bar{A}^{\,1}_{x:\overline{n}|} + v^n\,{}_np_x\bar{A}_{x+n}, \tag{2.3.3}$$

which implies

$$\bar{A}^{\,1}_{x:\overline{n}|} = \bar{A}_x - v^n\,{}_np_x\bar{A}_{x+n}\,.$$

The proof of (2.3.3) almost immediately follows from (2.2.1) and (2.3.1). Denote by Z_1 the r.v. in (2.2.1) with $m = n$, and let Z be defined as in (2.3.1). For clarity, let us write them down together:

$$Z = \begin{cases} e^{-\delta T(x)} & \text{if } T(x) \le n, \\ 0 & \text{if } T(x) > n. \end{cases}, \quad Z_1 = \begin{cases} 0 & \text{if } T(x) \le n, \\ e^{-\delta T(x)} & \text{if } T(x) > n. \end{cases} \tag{2.3.4}$$

Now it is easy to see that $Z + Z_1 = e^{-\delta T(x)}$, which corresponds to the whole life insurance. Taking the expected values of both sides, using the symbols $\bar{A}^{\,1}_{x:\overline{n}|}$, ${}_{n|}\bar{A}^{\,1}_x$, and $\bar{A}_x$, respectively, and applying (2.2.5), we come to (2.3.3). In Exercise 17, we consider a simple heuristic proof of (2.3.3).

A corresponding example will be considered for the discrete time case.

► The next observation concerns the correlation between Z and Z_1 in (2.3.4). Since $Z \cdot Z_1 = 0$, the covariance

$$Cov\{Z, Z_1\} = -E\{Z\}E\{Z_1\};$$

see (**0**.2.4.7).

EXAMPLE 2. Assume that for some x, n and δ, we have $\bar{A}^{\,1}_{x:\overline{n}|} = 0.01$, ${}^2\bar{A}^{\,1}_{x:\overline{n}|} = 0.0005$, ${}_{n|}\bar{A}^{\,1}_x = 0.1$, and ${}^2_{n|}\bar{A}^{\,1}_x = 0.0136$. It worth emphasizing that these numbers are marginal characteristics of Z and Z_1. Nevertheless, we can find the correlation between the two types of insurance.

Indeed, $Var\{Z\} = {}^2\bar{A}^{\,1}_{x:\overline{n}|} - (\bar{A}^{\,1}_{x:\overline{n}|})^2 = 0.0004$; $Var\{Z_1\} = {}^2_{n|}\bar{A}^{\,1}_x - ({}_{n|}\bar{A}^{\,1}_x)^2 = 0.0036$. Thus, $\sigma_Z = 0.02$, and $\sigma_{Z_1} = 0.06$. Next, $Cov\{Z, Z_1\} = -\bar{A}^{\,1}_{x:\overline{n}|} \cdot {}_{n|}\bar{A}^{\,1}_x = -0.001$. Eventually

$$Corr\{Z, Z_1\} = \frac{-E\{Z\}E\{Z_1\}}{\sigma_Z \sigma_{Z_1}} = \frac{-0.001}{0.02 \cdot 0.06} = -\frac{5}{6}. \;\square\blacktriangleleft$$

2.3.2 Discrete time

For an n-year term life insurance, in the case of "discrete time",

$$Z = \begin{cases} e^{-\delta(K(x)+1)} = v^{K(x)+1} & \text{if } T(x) \le n, \\ 0 & \text{if } T(x) > n. \end{cases}$$

We write $K(x)+1$ because if $K(x) = k$, then the payment is provided at the end of the year of death, that is, at the moment $k+1$.

The traditional notation for the APV is $A^{\,1}_{x:\overline{n}|}$. The counterparts of (2.3.2) and (2.3.3) look as follows:

$$A^{\,1}_{x:\overline{n}|} = \sum_{k=0}^{n-1} v^{k+1} P(K(x) = k) = \sum_{k=0}^{n-1} v^{k+1}\,{}_k p_x q_{x+k}, \tag{2.3.5}$$

[compare with (2.1.4)], and

$$A_x = A^{\,1}_{x:\overline{n}|} + v^n\,{}_n p_x A_{x+n}, \text{ or } A^{\,1}_{x:\overline{n}|} = A_x - v^n\,{}_n p_x A_{x+n}\,. \tag{2.3.6}$$

EXAMPLE 1. Using the Illustrative Life Table at the end of the book, compute $A^{\,1}_{30:\overline{35}|}$ for the interest rate $\delta = 0.04$. First of all, ${}_{35}p_{30} = (l_{65}/l_{30}) = (82609/97743) \approx 0.8452$. Next, from the Illustrative table, we have $A_{30} \approx 0.1666$, and $A_{65} \approx 0.5055$. Then $A^{\,1}_{30:\overline{35}|} = A_{30} - v^{35}\,{}_{35}p_{30}A_{65} \approx 0.1666 - e^{-0.04\cdot 35}0.8452 \cdot 0.5055 \approx 0.06124$. $\square$

The next formula suggests a straight recursive way of computing $A^{\,1}_{x:\overline{n}|}$. Namely,

$$A^{\,1}_{x:\overline{n}|} = vq_x + vp_x A^{\,1}_{x+1:\overline{n-1}|}. \tag{2.3.7}$$

The logic of the derivation is the same. With probability q_x the insured will die within the first year. In this case a unit of money will be paid at the end of the year, and the present value of this payment is v. With the complement probability p_x, the insured will attain age $x+1$, and the contract will become equivalent to the $(n-1)$-term contract for a life-age-$(x+1)$. The formal proof uses the formula for total expectation, and is very similar to what we did above.

To use (2.3.7), first we replace n in this formula by the letter k, and x by $x+n-k$. This leads to

$$A^{\,1}_{x+n-k:\overline{k}|} = v\left[q_{x+n-k} + p_{x+n-k} A^{\,1}_{x+n-(k-1):\overline{k-1}|}\right]. \tag{2.3.8}$$

Denote $A^{\,1}_{x+n-k:\overline{k}|}$ by $g(k)$. Then (2.3.8) may be rewritten as

$$g(k) = v\left[q_{x+n-k} + p_{x+n-k}\, g(k-1)\right]. \tag{2.3.9}$$

	A	B	C	D	E	F	G
1	k	$l_{65\text{-}k}$	$q_{65\text{-}k}$	$g(k)$	$g(k)$ for v^2	v	
2	0	84518		0	0		0.96
3	1	86165	0.019114	0.01835	0.017615914		
4	2	87698	0.01748	0.034089	0.032061011	v^2	
5	3	89142	0.016199	0.047746	0.043997675		0.9216
6	4	90491	0.014908	0.059465	0.053682589		
7	5	91748	0.013701	0.069456	0.0614225		
8							
9							
10		Variance		0.056598			

FIGURE 1.

This is a recursion formula. Note that $g(0) = A^{1}_{x+n:\overline{0}|} = 0$, since $A^{1}_{x:\overline{0}|} = 0$ for any x. (The time period has zero length, and nothing will be paid.) Then from (2.3.9) it follows that $g(1) = vq_{x+n-1}$. Setting $k = 2, ..., n$ consecutively in (2.3.9), we can get $g(n) = A^{1}_{x:\overline{n}|}$ in n steps.

EXAMPLE 2. Let $x = 60$, $n = 5$, and for the corresponding part of the population, $l_{60} = 91748$, $l_{61} = 90491$, $l_{62} = 89142$, $l_{63} = 87698$, $l_{64} = 86165$, and $l_{65} = 84518$. Estimate $A^{1}_{60:\overline{5}|}$ for $v = 0.96$. We can readily provide the recursive procedure. The corresponding Excel worksheet is given in Fig.1. Years are considered there in descending order, so l_{60} is in Cell B7, and l_{65} in B2. The probabilities q_x are estimated by $(l_x - l_{x+1})/l_x$ and are given in Column C. The command for C3 is =(B3-B2)/B3. The quantities $g(k)$ are computed recursively in Column D by (2.3.9), where $p_x = 1 - q_x$. The corresponding command for D3 is =$G$2*(C3+(1-C3)*D2). The result is $g(5)$ given in Cell D7. Thus, $A^{1}_{60:\overline{5}|} \approx 0.0694$.

Now it will not take long to compute the variance. We should repeat the same procedure for the double rate 2δ, which corresponds to the squared discount v^2. To this end, we should just copy the Column D, keeping the same commands and replacing v by v^2. It is done in Column E with the command for E3 equal to =$G$5*(C3+(1-C3)*E2). Thus, in Cell E7 we have ${}^{2}A^{1}_{60:\overline{n}|} \approx 0.0614$. Then, we compute the variance as ${}^{2}A^{1}_{60:\overline{5}|} - (A^{1}_{60:\overline{5}|})^2 \approx 0.0566$ in Cell D10=E7-D7^2. □

Note that the arrival of powerful computers reduces the demand for such procedures as above. In our particular case, what we really need to know are the probabilities ${}_kp_x$. If we know them, then we may immediately compute the probabilities $P(K = k) = {}_kp_x \cdot q_{x+k}$, and then compute APVs by direct (straightforward) formulas like (2.3.9). Nevertheless, procedures as above may make programs more flexible and less time consuming.

2.4 Endowments

2.4.1 Pure endowment

In an *n-year pure endowment insurance*, benefits are payable at the end of the nth year, provided that the insured survives this term. In this case $\Psi = n$ if $T(x) > n$, and $\Psi = \infty$ if

$T(x) \leq n$. Accordingly,

$$Z = \begin{cases} 0 & \text{if } T(x) \leq n, \\ e^{-\delta n} = v^n & \text{if } T(x) > n. \end{cases} \tag{2.4.1}$$

The traditional notation for the APV is $A_{\substack{1\\x:\overline{n}|}}$ or ${}_nE_x$. The latter notation is used more frequently in the annuity context (see Chapter 9) but, for the understandable reason, we will prefer to use it in this chapter also.

Clearly,

$${}_nE_x = E\{Z\} = v^n P(T(x) > n) = v^n\, {}_np_x.$$

To compute the variance, recall that for a r.v. $X = c$ or 0 with probabilities p and q, respectively, $Var\{X\} = c^2 pq$. Hence, since $P(T(x) > n) = {}_np_x$,

$$Var\{Z\} = v^{2n}\, {}_np_x(1 - {}_np_x). \tag{2.4.2}$$

In Exercise 25, we prove the same making use of the rule of double rate (1.1.5).

EXAMPLE 1. As in Example 2.3.2-2, let $l_{60} = 91748$, and $l_{65} = 84518$. Then ${}_5p_{60} = (l_{65}/l_{60}) \approx 0.9211$, and for $v = 0.96$, we have ${}_5E_{65} \approx (0.96)^5 \cdot 0.9211 \approx 0.7510$. The fact that the number we got is much higher than the answer in Example 2.3.2-2 is not surprising. For the insured, there is much more chance of surviving five years than of dying before the end of the term. So, the pure endowment should cost more than the 5-year-term insurance. By (2.4.2), the standard deviation $\sigma_Z = v^n \sqrt{{}_np_x(1 - {}_np_x)} \approx (0.96)^5 \cdot \sqrt{0.9211 \cdot 0.0789} \approx 0.2198$. □

2.4.2 Endowment

In an *n-year endowment insurance,* the benefit is paid upon death if the insured dies within the interval $[0, n]$, and it is paid at the end of the period mentioned if the insured survives n years.

More precisely, in the continuous time case, $\Psi = \min\{T(x), n\}$ and

$$Z = \begin{cases} e^{-\delta T(x)} = v^{T(x)} & \text{if } T(x) \leq n, \\ e^{-\delta n} = v^n & \text{if } T(x) > n. \end{cases} \tag{2.4.3}$$

The APV is denoted by $\bar{A}_{x:\overline{n}|}$ (there is no superscript 1). Denote by Z_1 the r.v. in (2.3.1), and by Z_2 the r.v. in (2.4.1). Let us write these r.v.'s down as follows:

$$Z_1 = \begin{cases} e^{-\delta T(x)} & \text{if } T(x) \leq n, \\ 0 & \text{if } T(x) > n. \end{cases}, \qquad Z_2 = \begin{cases} 0 & \text{if } T(x) \leq n, \\ e^{-\delta n} & \text{if } T(x) > n. \end{cases} \tag{2.4.4}$$

We see that $Z = Z_1 + Z_2$, and hence

$$\bar{A}_{x:\overline{n}|} = E\{Z\} = E\{Z_1\} + E\{Z_2\} = \bar{A}^{1}_{x:\overline{n}|} + {}_nE_x = \bar{A}^{1}_{x:\overline{n}|} + v^n\, {}_np_x. \tag{2.4.5}$$

In Exercise 32, the reader is encouraged to provide heuristic and rigorous proofs of this relation.

For the discrete time case, we define Ψ as $\min\{K(x)+1,n\}$. Then

$$Z=\begin{cases} e^{-\delta(K(x)+1)}=v^{K(x)+1} & \text{if } K(x)<n,\\ e^{-\delta n}=v^n & \text{if } K(x)\geq n.\end{cases}$$

To clarify why we wrote $K+1$ above, consider the situation when the payment is made at the moment n. It may happen in the following two cases.

- The insured dies within the interval $[n-1,n)$. Then the payment will be provided at the end of the nth year, that is, at the moment $t=n$. In this case, $K=n-1$, and $\min\{K(x)+1,n\}$ is indeed equal to n.
- The insured attains the age of $x+n$. Then the payment is again provided at the moment n. On the other hand, in this case $K(x)\geq n$, and $\min\{K(x)+1,n\}=n$.

The APV is denoted by $A_{x:\overline{n|}}$.

The counterpart of (2.4.4) for the discrete case is

$$Z_1=\begin{cases} e^{-\delta(K(x)+1)} & \text{if } K(x)<n,\\ 0 & \text{if } K(x)\geq n.\end{cases},\quad Z_2=\begin{cases} 0 & \text{if } K(x)<n,\\ e^{-\delta n} & \text{if } TK(x)\geq n.\end{cases} \tag{2.4.6}$$

We again have $Z=Z_1+Z_2$, and

$$A_{x:\overline{n|}}=A^{\ 1}_{x:\overline{n|}}+{}_nE_x=A^{\ 1}_{x:\overline{n|}}+v^n\,{}_np_x. \tag{2.4.7}$$

Since the last relation is true for any rate δ, it is true for the doubled rate 2δ also, so we can write

$${}^2A_{x:\overline{n|}}={}^2A^{\ 1}_{x:\overline{n|}}+{}^2_nE_x={}^2A^{\ 1}_{x:\overline{n|}}+v^{2n}\,{}_np_x. \tag{2.4.8}$$

This gives a good way of computing variances.

EXAMPLE 1. We combine the results from Example 2.3.2-2 and Example 2.4.1-1, which deal with the same data. From the former example, we have $A^{\ 1}_{60:\overline{5|}}\approx 0.069$, and from the latter, ${}_5E_{65}\approx 0.751$. Hence,

$$A_{60:\overline{5|}}=A^{\ 1}_{60:\overline{5|}}+{}_5E_{60}\approx 0.069+0.751=0.820.$$

Let us consider variances. In Example 2.3.2-2, we got ${}^2A^{\ 1}_{60:\overline{5|}}\approx 0.0614$. Next we compute ${}^2_5E_{60}=E\{Z_2^2\}$, where Z_2 is the present value of the corresponding pure endowment. In Example 2.4.1-1, we computed ${}_5p_{60}\approx 0.9211$. Then $E\{Z_2^2\}=v^{2n}\,{}_np_x\approx(0.96)^{2\cdot 5}\cdot 0.9211\approx 0.6124$. Now, by (2.4.8), ${}^2A_{60:\overline{5|}}=0.0614+0.6124=0.6738$, and $Var\{Z\}={}^2A_{60:\overline{5|}}-(A_{60:\overline{5|}})^2\approx 0.6738-(0.820)^2=0.0014$.

► It is interesting to observe that, while $Var\{Z_1\}\approx 0.057$ and $Var\{Z_2\}\approx 0.048$ (see Examples 2.3.2-2 and 2.4.1-1), the variance of the sum $Z=Z_1+Z_2$ is about only 0.0015! This is connected with the fact that the r.v.'s Z_1 and Z_2 are negatively correlated. We considered a similar fact in Section 2.3.1, and will come back to this in Exercise 18. □ ◄

Route 1 ⇒ page 467

3 VARYING BENEFITS

3.1 Certain payments

Fixed benefits that do not depend on the payment time and other circumstances are called *level benefits*. In this case, we can choose this fixed amount of benefits to be the unit of money, which we did in all models above. In general, the size of the benefits to be paid can vary in time and/or may depend on causes of death. The latter case is considered in Section 4; this section concerns the dependence on time.

Let c_t be the benefit to be paid if the payment time occurs to be equal to t. The present value of such a payment is $c_t e^{-\delta t}$. Since the time of payment is the r.v. Ψ, the benefit is the r.v. c_Ψ. Then the present value of the benefit to be paid is the r.v.

$$Z = c_\Psi e^{-\delta\Psi}. \tag{3.1.1}$$

The APV

$$A = E\{c_\Psi e^{-\delta\Psi}\}. \tag{3.1.2}$$

The rule of double rate does *not* work here, since in our case,

$$E\{Z^2\} = E\{c_\Psi^2 e^{-2\delta\Psi}\}.$$

If we can compute the last quantity, we can find the variance writing

$$Var\{Z\} = E\{c_\Psi^2 e^{-2\delta\Psi}\} - (E\{c_\Psi e^{-\delta\Psi}\})^2.$$

Consider whole life insurance. For APVs, we will keep the same notations A_x and $\bar{A}_x$. In accordance with (3.1.2),

$$\bar{A}_x = \int_0^\infty c_t e^{-\delta t} f_{T(x)}(t)dt. \tag{3.1.3}$$

In the discrete time case, we have

$$A_x = \sum_{k=0}^{\infty} c_{k+1} e^{-\delta(k+1)} P(K(x) = k).$$

EXAMPLE 1. A special life insurance on a life-age-40 provides for a payment of 3 times the annual salary paid to the family if and only if the insured dies before the retirement age of 65. Assume that all possible moments of death are equally likely within the interval $[40, 65]$, and the probability of attaining the age of 65 is 0.85. The initial salary is \$60,000, and it is growing at an annual rate of 4%. The investment rate $\delta = 6\%$. Find the net single premium, that is, the APV.

(a) We take, as a unit of money, the tripled initial salary \$180,000. First, consider the model when the salary is growing continuously. This means that $c_t = \exp\{0.04t\}$ if $t \le 25$, and $c_t = 0$ otherwise.

Under the assumption we made, the density $f(t) = f_{T(40)}(t)$ is *constant* on $[0,25]$. On the other hand, $P(T(40) < 25) = \int_0^{25} f(t)dt$. Since this probability equals 0.15, the density $f(t) = 0.15 \cdot \frac{1}{25} = 0.006$ on $[0,25]$. What $f(t)$ equals on $[25,\infty)$ does not matter, since $c_t = 0$ for $t > 25$ and integration in (3.1.3) should be carried out over $[0,25]$. Thus, by (3.1.3),

$$\bar{A}_x = \int_0^{\infty} c_t e^{-\delta t} f_{T(x)}(t)dt = \int_0^{25} c_t e^{-\delta t} f_{T(x)}(t)dt = \int_0^{25} e^{0.04t} e^{-0.06t} \cdot 0.006dt$$

$$= 0.006 \frac{1 - e^{-0.02 \cdot 25}}{0.02} \approx 0.118 \text{ units.}$$

Note that the expected value of the real payment is $\int_0^{25} e^{0.04t} \cdot f(t)dt = \int_0^{25} e^{0.04t} \cdot 0.006dt \approx 0.258$ units, so the net single premium is less than half the expected benefit. In dollars, the premium is $0.118 \cdot \$180,000 = \$21,240$.

(b) In reality, the increase of salaries is carried out in a discrete way. Assume that the salary is increased by 4% at the beginning of each year. Then $c_t = 1.04^{[t]}$, where $[t]$ is the integer part of t. The benefits are still paid at the time of death. In this case,

$$\bar{A}_x = \int_0^{25} 1.04^{[t]} e^{-\delta t} \cdot 0.006dt = \sum_{k=0}^{24} \int_k^{k+1} 1.04^{[t]} e^{-\delta t} \cdot 0.006dt = 0.006 \sum_{k=0}^{24} 1.04^k \int_k^{k+1} e^{-\delta t} dt.$$

The last integral is equal to $\frac{i}{\delta} e^{-\delta(k+1)}$, where $i = e^{\delta} - 1$; see also Section 2.1.3. Thus,

$$\bar{A}_x = 0.006 \frac{e^{0.04} - 1}{0.04} \sum_{k=0}^{24} 1.04^k e^{-0.06(k+1)} = 0.006 \frac{e^{0.04} - 1}{0.04} e^{-0.06} \sum_{k=0}^{24} (1.04e^{-0.06})^k$$

$$= 0.006 \frac{e^{0.04} - 1}{0.04} e^{-0.06} \frac{1 - (1.04e^{-0.06})^{25}}{1 - 1.04e^{-0.06}} \approx 0.1135.$$

Certainly, we had to expect that the answer would be less than that in the case of continuous salary growth (why?). Eventually, the net single premium is $0.1135 \cdot \$180,000 \approx \$20,445$. □

Now consider a special case when the benefit is either growing linearly in time, or increases each year by a fixed amount at the beginning of the year. More precisely, we consider either

$$c_t = ct,$$

where c is the rate of growth, or

$$c_t = c \cdot [t+1],$$

where c is the absolute increase per year. Without loss of generality, one can set $c = 1$.

The former insurance is called *continuously increasing* (whole life) insurance; the APV in this case is denoted by $(\bar{I}\bar{A})_x$ (I comes from "increasing", the bar indicates that "time is continuous"). In the latter case, we deal with an *annually increasing* insurance. (It would be more precise to simply call it a discretely increasing annuity, since formally we do not specify the unit of time here.) The notation for the APV in this case is $(I\bar{A})_x$.

EXAMPLE 2. Find $(\bar{I}\bar{A})_{50}$ for $\delta = 0.04$, if X follows De Moivre's law with $\omega = 100$. Thus, $T(50)$ is uniform on $[0,50]$, and $(\bar{I}\bar{A})_{50} = \int_0^{50} te^{-0.04t}\frac{1}{50}dt$. Integration by parts or use of software lead to $(\bar{I}\bar{A})_{50} \approx 7.52$.

EXAMPLE 3. For the same δ and the same distribution of X as in Example 2, find $(I\bar{A})_{50}$. Similarly to what we did in Example 1b,

$$(I\bar{A})_x = \int_0^{50} [t+1]e^{-\delta t}\frac{1}{50}dt = \sum_{k=0}^{49}\int_k^{k+1}(k+1)e^{-\delta t}\frac{1}{50}dt = \frac{1}{50}\frac{i}{\delta}\sum_{k=0}^{49}(k+1)e^{-\delta(k+1)}.$$

There are formulas for such sums [see, e.g., (**9**.4.4)], but it is easier to compute the last quantity using software. For $\delta = 0.04$, the answer is ≈ 7.64.

The fact that the answer is larger than that in Example 2 does not contradict common sense. Although the benefit in Example 2 is growing continuously, it starts from zero, while in Example 3, the initial value is one. □

In the discrete time case, the annually increasing whole life insurance corresponds to $\Psi = K+1$ and $c_k = k+1$. Then the present value $Z = (K+1)v^{K+1}$. The standard notation for the APV is $(IA)_x$.

EXAMPLE 4. Find $(IA)_x$ if $\mu(x) = \mu$. We know (see Exercise **7**-28) that K has a geometric distribution (**0**.3.1.9) with parameter $p = 1-e^{-\mu}$. Then, setting $v = e^{-\delta}$, we have

$$(IA)_x = E\{(K+1)v^{K+1}\} = \sum_{k=0}^{\infty}(k+1)e^{-\delta(k+1)}(1-e^{-\mu})e^{-\mu k} = e^{-\delta}(1-e^{-\mu})\sum_{k=0}^{\infty}(k+1)e^{-(\delta+\mu)k}.$$

Making use of the general formula

$$\sum_{k=0}^{\infty}(k+1)q^k = 1/(1-q)^2, \tag{3.1.4}$$

we get

$$(IA)_x = e^{-\delta}(1-e^{-\mu})\Big/(1-e^{-(\delta+\mu)})^2.$$

(The formula (3.1.4) may be found in practically any algebra or calculus textbook, and in any probability textbook, since with its help we derive the formula (**0**.3.1.12) for the mean of the geometric distribution (e.g., [116, p.174], [122, p.67]). See also (**9**.4.4).)

One more simple example is considered in Exercise 44. □

Another special case is an n-year *term decreasing insurance* where the payments are decreasing from n to zero either linearly (*continuously decreasing*) in accordance with the formula $c_t = n - t$, or in the discrete way when $c_t = n - [t]$ (*annually decreasing*). The symbols for the APV in these cases are $(\bar{D}\bar{A})_{x:\overline{n}|}$ and $(D\bar{A})_{x:\overline{n}|}$, respectively. In Exercise 40b, we compute these quantities when X is uniform.

The notation for the discrete case is, naturally, $(DA)_{x:\overline{n}|}$.

It is important to keep in mind that for certain varying benefits, we can compute the APVs combining term insurances with level benefits.

EXAMPLE 5. Consider a discrete-time whole life insurance on a life-age-50, providing for payment of \$100K during the first 10 years; \$110K in the 11th year after policy issue; and in each year after, an increase of \$10K until the insured reaches the age of 70. If the insured attains this age, the benefit will be level (will not change) at \$200 K. Write the APV in terms of the characteristic $\bar{A}_x$ and survival probabilities.

We take \$10,000 as a unit of money. The company will pay 10 units for sure, and this is equivalent to the whole life insurance with a benefit of 10. If the insured attains the age of 60, an additional one unit will be added to the benefit. This is equivalent to an additional 10-year deferred whole life insurance with unit benefit. If the insured attains the age of 61, again an additional unit will be added to the benefit, and this is equivalent to an additional 11-year deferred whole life insurance with unit benefit. We can continue to reason in this fashion until age 69 when the last additional unit will be added to the benefit, and the whole benefit will be equal to 20 units. Thus,

$$APV = 10\bar{A}_{50} + v^{10}\,{}_{10}p_{50}\bar{A}_{60} + v^{11}\,{}_{11}p_{50}\bar{A}_{61} + \ldots + v^{19}\,{}_{19}p_{50}\bar{A}_{69}.$$

EXAMPLE 6 ([153, N1][4]) demonstrates straight calculations. XYZ Bank has issued a 5-year interest free loan, collecting annual payments of \$10,000 at the end of each year. To protect itself from loan defaults, XYZ has purchased default insurance that pays the balance of the loan at the time of default. The probabilities of default at each payment due date are given in the table below.

Payment number	1	2	3	4	5
Probability of default (given no prior default)	0.04	0.08	0.10	0.10	0.06

The annual interest for the default insurance is 6%. Calculate the expected present value of the insurance benefit.

It is convenient to present calculations in the following table.

k	q_k	${}_kp$	$P(K=k)$	$v^k(6-k)P(K=k)$
1	0.04	0.9600	0.0400	0.1887
2	0.08	0.8832	0.0768	0.2734
3	0.10	0.7949	0.0883	0.2225
4	0.10	0.7154	0.0795	0.1259
5	0.06	0.6725	0.0429	0.0320

Here, k stands for the number of a payment, and q_k is the given probability of default. In the third column, we compute the probability that no default happens before or at the moment of the kth payment. Thus, ${}_kp = {}_{k-1}p \cdot (1-q_k)$. For example, the number in the second row equals $(1-0.04)(1-0.08)$. The fourth column contains $P(K=k)$, where K is the moment of default, if any. Clearly, $P(K=k) = {}_{k-1}p \cdot q_k$.

[4]Reprinted with permission of the Casualty Actuarial Society.

Let us choose \$10,000 as a unit of money. If default happens at the moment of the kth payment, then the insurance pays $(6-k)$ units, which should be multiplied by the discount v^k, where $v = \frac{1}{1+0.06} \approx 0.9434$. The fifth column gives the values of $v^k(6-k)P(K=k)$.

The APV is the sum of all values in the fifth column, which is equal to 0.8426. So, the APV is $\$10,000 \cdot 0.8426 = \$8,426$. □

3.2 Random payments

In general, the size of a benefit may depend not only on the time of "failure" (say, death or product-failure), but on other circumstances as well. In this case, the size of the payment may be random, even if the time of failure is known. Formally, this means that in the general model, we should consider r.v.'s rather than certain payment functions. We restrict ourselves to one example.

EXAMPLE 1. Let us consider an insurance against unemployment and call the loss of job "failure". The insurance covers a period of 10 years. Assume the probability that failure occurs during this period to be 0.1; that is, 0.9 is the probability that the insured will either survive 10 years not losing the job or will die within the period $[0,10]$ having the job. Assume also that all moments in the interval $[0,10]$ are equally likely to be a failure moment.

Denote by ξ the time, *starting from the failure moment*, at which the insured will either find a new job or die. We assume that the company reimburses the loss of the salary by a single payment at the time mentioned. Such an assumption is somewhat artificial, but makes the example simpler. In Chapter 9 where we study annuities, in Example **9**.3.1-4, we consider the same problem assuming that the company pays to the insured her/his current salary from the moment of failure until the moment when the insured gets a job or die. In Exercise 42, we consider the problem under the additional assumption that if the insured does not find a job during a year, then the company pays the annual salary, and the contract terminates. This also makes the problem more realistic.

Now, suppose ξ is exponentially distributed with parameter $b=2$ and independent of the time of failure. The salary is growing exponentially at a rate of 4%; the initial annual salary rate equals 1. The interest rate $\delta = 6\%$. If the insured loses the job within the interval $[0,10]$, the company fulfills its obligations in full even if the insured finds a job (or die) after the ten year period. For simplicity, we will provide all calculations in the scheme of continuous time.

We deal with a 10-year-term insurance with the exception that, at the moment of failure, the real size of the future payment is unknown. If the failure occurs at moment t, the lost annual salary is $e^{0.04t}$. The company will reimburse this loss at the random moment $t+\xi$, and the amount of reimbursement will be equal to $e^{0.04t}\xi$. (We measure time in years, and deal with annual salaries. The company pays the salary that was lost, and does not take into account that the salary would have been growing if the insured had not lost the job.)

The payment will be made at the moment $t+\xi$. Consequently, the present value of such a payment is the r.v.

$$R_t = e^{-\delta(t+\xi)}e^{0.04t}\xi = e^{-0.06\cdot(t+\xi)}e^{0.04t}\xi = e^{-0.02t}e^{-0.06\xi}\xi.$$

Denote by X a r.v. uniformly distributed on $[0,10]$. We identify X with the moment of failure if it occurs. Then, for the present value of the total benefit, we can write the representation

$$Z = \begin{cases} 0 & \text{with probability } 0.9, \\ e^{-0.02X}e^{-0.06\xi}\xi & \text{with probability } 0.1. \end{cases}$$

The actuarial present value

$$A = E\{Z\} = 0.1 \cdot E\{e^{-0.02X}e^{-0.06\xi}\xi\} = 0.1 \cdot E\{e^{-0.02X}\}E\{e^{-0.06\xi}\xi\},$$

because X and ξ are independent. Since X is uniform on $[0,10]$, the expectation $E\{e^{-0.02X}\} = M_X(-0.02) = [(1-e^{-10\cdot 0.02})/(0.02\cdot 10)] \approx 0.906$. For the second expectation, we have

$$E\{e^{-0.06\xi}\xi\} = \int_0^\infty e^{-0.06t}t \cdot f_\xi(t)dt = \int_0^\infty e^{-0.06t}t \cdot 2e^{-2t}dt \approx 0.471.$$

Thus, $A \approx 0.1 \cdot 0.906 \cdot 0.471 \approx 0.0427$. □

---

4 MULTIPLE DECREMENT AND MULTIPLE LIFE MODELS

4.1 Multiple decrements

We use the general framework and the notation from Section 7.2.1. When it does not cause misunderstanding, we omit the symbol x denoting the age of the insured.

Assume that the benefit to be paid depends on the moment of payment and the cause of failure. Denote by c_{jt} the amount of the benefit to be paid if the failure occurs at moment t from cause j. We denote the time of payment by Ψ, and the number of the cause by J. Both quantities are random. The present value of the insurance is the r.v.

$$Z = c_{J\Psi}e^{-\delta\Psi}, \tag{4.1.1}$$

and the APV

$$A = E\{c_{J\Psi}e^{-\delta\Psi}\}. \tag{4.1.2}$$

[Compare with (3.1.1) and (3.1.2). In this subsection, we will use the symbol A for all types of insurance.]

For example, for the whole life insurance $\Psi = T = T(x)$, and we can compute the expectation (4.1.2) in the standard way in terms of the joint density $f(t,j)$ defined in Section 7.2.1. Keeping in mind that the r.v. T is continuous and J is discrete, we write

$$A = E\{c_{JT}e^{-\delta T}\} = \sum_{j=1}^{m}\int_0^\infty c_{jt}e^{-\delta t}f(t,j)dt,$$

where m is the number of possible causes.

If the benefits are payable at the end of the year of death, then $\Psi = K+1 = K(x)+1$, and

$$A = \sum_{j=1}^{m}\sum_{k=1}^{\infty} c_{j,k+1}v^{k+1}P(K(x)=k, J=j).$$

To compute the above probability, one can write $P(K(x)=k, J=j)=\int_k^{k+1} f(t,j)dt$.

It is convenient to "condition" the expectation in (4.1.2) with respect to Ψ. Consider again the case $\Psi=T$. In accordance with the general rules (**0**.7.2.1) and (**0**.7.2.3),

$$A = E\{c_{JT}e^{-\delta T}\} = E\{E\{c_{JT}e^{-\delta T}\,|\,T\}\} = E\{E\{c_{JT}\,|\,T\}e^{-\delta T}\}. \tag{4.1.3}$$

Set

$$c(T) = E\{c_{JT}\,|\,T\}.$$

This is the mean payment *given* the failure moment T. When T assumes a value t, the r.v. $c(T)$ takes on the value $c(t)$. Naturally, $c(t)=E\{c_{JT}\,|\,T=t\}$.

By (4.1.3),

$$A = E\{c(T)e^{-\delta T}\} = \int_0^\infty c(t)e^{-\delta t} f_T(t)dt, \tag{4.1.4}$$

where $f_T(t)$ is the density of T; see also Section **7**.2.1.

Our first goal is to compute $c(t)$. As in Section **7**.2.1, let $\mu(t)$ and $\mu^{(j)}(t)$ denote the total force of decrement and the force of decrement due to cause j, respectively. (We omit the index x.) Recall that $\mu(t)=\sum_{j=1}^m \mu^{(j)}(t)$, and

$$P(J=j\,|\,T=t) = \mu^{(j)}(t)\,/\mu(t)\,. \tag{4.1.5}$$

Then

$$c(t) = \sum_{j=1}^m c_{jt}P(J=j\,|\,T=t) = \sum_{j=1}^m c_{jt}\cdot\frac{\mu^{(j)}(t)}{\mu(t)}. \tag{4.1.6}$$

EXAMPLE 1. A whole life insurance for a particular type of client pays 20 units if death comes from natural causes and 10 units if it is a result of an accident. The respective hazard rates are $\mu^{(1)}(t)=0.03$ and $\mu^{(2)}(t)=0.01$. (The age x is suppressed in the notation.) Find the APV for $\delta=0.05$.

Thus, c_{1t}=20 and c_{2t}=10. The total mortality force is $\mu(t)$=$\mu^{(1)}(t)$+$\mu^{(2)}(t)$=0.04, so T is exponentially distributed with parameter μ=0.04. By (4.1.5), $P(J$=$1\,|\,T$=$t)$=0.75, and by (4.1.6), $c(t)=20\cdot 0.75+10\cdot 0.25=17.5$. In view of (4.1.4), $A=17.5\int_0^\infty e^{-\delta t}f_T(t)dt=$ $17.5M_T(-\delta)=17.5\frac{\mu}{\mu+\delta}=17.5\cdot\frac{4}{9}=\frac{70}{9}$. □

To come to another convenient representation, recall that $f_T(t)=\mu(t)\,{}_tp$, where ${}_tp=P(T>t)$. (See (**7**.1.2.4); we have omitted x.) Then, by virtue of (4.1.6),

$$c(t)f_T(t) = \sum_{j=1}^m c_{jt}\cdot\frac{\mu^{(j)}(t)}{\mu(t)}\mu(t)\,{}_tp = {}_tp\sum_{j=1}^m c_{jt}\mu^{(j)}(t).$$

Substituting it into (4.1.4), we have

$$A = \int_0^\infty\left(\sum_{j=1}^m c_{jt}\mu^{(j)}(t)\right)e^{-\delta t}\,{}_tpdt. \tag{4.1.7}$$

EXAMPLE 2. A special whole life insurance on (50) pays 2 units if death is a result of an accident, and the tripled current annual income if death comes from other causes. The force of mortality in the latter case is $\mu^{(1)}(t) = \frac{1}{50-t}$. The hazard rate for the time of the accident is $\mu^{(2)}(t) = \mu = 0.01$. Suppose that the income grows as $e^{0.02t}$ until the time $t = 20$, and after this time it remains constant. The rate $\delta = 0.05$. Find the APV.

To find the distribution of $T(50)$, we can use general formulas from **7.2.1**, but it is more convenient to appeal to the scheme of Section **7.2.2**.

In Example **7.2.2-3**, we have found that for the life time $T(x)$ under consideration, ${}_tp_{50} = (1-t/50)e^{-\mu t}$ for $t \le 50$.

The payoff function $c_{1t} = 3\min\{e^{0.02t}, e^{0.02 \cdot 20}\} = 3\min\{e^{0.02t}, e^{0.4}\}$, while $c_{2t} = 2$. Hence, by (4.1.7),

$$A = \int_0^{50} \left(3\min\{e^{0.02t}, e^{0.4}\}\frac{1}{50-t} + 2\mu\right) e^{-\delta t}(1-t/50)e^{-\mu t}\,dt$$

$$= \int_0^{50} 3\min\{e^{0.02t}, e^{0.4}\}\frac{1}{50}e^{-\delta t}e^{-\mu t}\,dt + \int_0^{50} 2\mu e^{-\delta t}(1-t/50)e^{-\mu t}\,dt.$$

Both integrals may be computed directly, but nowadays one may use software. In any case, $A \approx 1.428$.

EXAMPLE 3. The company where Mr. T. is working pays a special death benefit of \$5000 times the number of years in service. The benefit is paid at the end of the year of death, provided it occurs while Mr. T. is still in service. A non-complete year is counted. Mr. T. is now exactly 61. He has been in service for 10 years prior to the present time. During the period under consideration, Mr. T. can withdraw from this insurance plan (say, changing his job or retiring). The information on the forces of decrement for the initial group of $l_{61} = 100$ people of age 61 is given in the first four columns of the table below; $v = 0.97$. We consider death as the cause 1, and withdrawal as cause 2; k is the age at the beginning of the corresponding period.

k	l_k	$d_k^{(1)}$	$d_k^{(2)}$	$c_{1,k+1}$	r_k	$v^{k-60}c_{1,k+1}r_k$
61	100	1	4	11	0.01	0.1056
62	95	2	3	12	0.02	0.221184
63	90	3	2	13	0.03	0.34504704
64	85	0	85	0	0	0

The third and fourth columns contain the number of people who left the group in the period k for the first and the second cause, respectively. In particular, from the table it follows that once Mr. T. attains the age of 64, he will immediately retire. Let, for a moment, \$5000 be a unit of money. If Mr. T. survives k complete years ($K = k$), then the payment will be $c_{1,k+1} = 10 + k - 60 = k - 50$, $c_{2,k} = 0$. Set $r_k = P(K = k, J = 1)$. Then $r_k = \frac{l_k}{l_{61}}\frac{d_k^{(1)}}{l_k} = \frac{d_k^{(1)}}{l_{61}}$, which is reflected in the sixth column. If the payment is provided in the first year, the discount is v; for the year k, the discount is v^{k-60}. Thus, the APV per \$1 is $\sum_{k=61}^{63} v^{k-60} c_{1,k+1} r_k$; see the seventh column. The sum of the numbers in this column is 0.67183104, so the APV is $\$5000 \cdot 0.67183104 \approx \$3,359.16$. □

4.2 Multiple life insurance

Consider, for example, the last-survivor status of two lives, $\overline{x:y}$, and an insurance that provides for payment of some benefits at the moment of the last death. Following the logic of the traditional notation, it is natural to denote the APV in this case by $\bar{A}_{\overline{x:y}}$. It would have been very nice if we had been able to reduce the calculation of $\bar{A}_{\overline{x:y}}$ to the calculation of the "separate" APVs, $\bar{A}_x$ and $\bar{A}_y$. However, as we will see, as a rule it is impossible. So, the straightforward approach may turn out to be optimal: find the distribution of the lifetime $T_{\overline{x:y}}$ of the status, treat it as one life, and compute the corresponding APV.

EXAMPLE 1. Let us revisit Example **7**.3.2-1. We found there that the density $f_{T(\overline{x:y})}(t) = C_k(2t - k\frac{2t^3}{3})$, where $C_k = 6/(6-k)$, and the parameter $k \leq 1$ characterizes the dependence between the two lifetimes, T_1 and T_2. For $k = 0$ the lifetimes are independent, and the larger k is, the stronger the dependence between T_1 and T_2. To make it more interesting, consider growing benefits; for example, assume $c_t = t$. Then the corresponding notation should be $(\bar{I}\bar{A})_{\overline{x:y}}$, but to make it simpler, we will write just $\bar{A}_{\overline{x:y}}$.

Let $\delta = 0.1$. (Since the whole time period under consideration is one, it is natural to assume that the unit of time is not one year, and hence it is natural to take a larger interest rate than a usual annual interest rate.) We have

$$\bar{A}_{\overline{x:y}} = \int_0^1 te^{-\delta t} f_{T(\overline{x:y})}(t)dt = C_k \int_0^1 te^{-0.1t}\left(2t - k\frac{2t^3}{3}\right)dt.$$

One may compute this integral directly (by parts) or use software to obtain that, taking constants up to the second digit,

$$\bar{A}_{\overline{x:y}} = \frac{(3.72 - 0.72k)}{6-k}.$$

The last function is slowly decreasing in k from 0.62 to 0.5968, so to the larger extent the lives are dependent, the less the APV. □

As in Section **7**.3, denote by T_1 and T_2 the separate lifetimes in a status. Note that for *any* function $\phi(t)$,

$$\phi(\min\{T_1, T_2\}) + \phi(\max\{T_1, T_2\}) = \phi(T_1) + \phi(T_2). \tag{4.2.1}$$

This simple observation leads to a nice connection between the insurances on the last-survivor and the joint-life status. Assume that the present value of an insurance depends—perhaps in a complicated way—only on the lifetime T of the status and is represented by a function $\phi(T)$. For example, if we consider an insurance paying c_t at time t, then $\phi(t) = c_t e^{-\delta t}$.

The APV of an insurance that is specified by $\phi(t)$ is $E\{\phi(T)\}$. Then, by virtue of (4.2.1),

$$\boxed{\bar{A}_{x:y} + \bar{A}_{\overline{x:y}} = \bar{A}_x + \bar{A}_y,} \tag{4.2.2}$$

where $\bar{A}_{x:y}$, $\bar{A}_{\overline{x:y}}$, $\bar{A}_x$, and $\bar{A}_y$ are the APVs of the *same* insurance applied to the statuses $x:y$, $\overline{x:y}$, and separate lives (x), (y), respectively.

Certainly, a similar formula is true for the discrete time case. Because the function $\phi(t)$ is arbitrary, the relation (4.2.2) concerns all types of insurances we considered earlier (temporal, deferred, etc.).

EXAMPLE 2 ([158, N13][5]. You are given the following information about two policyholders who are age x and age y, respectively: (a) The future lifetimes of (x) and (y) are independent; (b) The force of mortality is constant, with $\mu_x = 0.03$ and $\mu_y = 0.08$; (c) $\delta = 0.05$; (d) A fully continuous last survivor insurance on (x) and (y) pays a benefit of 100,000.

Calculate the actuarial present value of this insurance benefit.

First, we recollect that for the continuous-time exponential case, $\bar{A}_x = \frac{\mu}{\mu+\delta}$; see (1.1.8).

Furthermore, in the independency case, the mortality force of $T(x:y)$ is the sum of the mortality forces for $T(x)$ and $T(y)$. Thus, $T(x:y)$ is an exponential r.v. with the mortality force $\mu_x + \mu_y$. (For more detail, see, for instance, Example **7**.3.2-6.)

Keeping all of this in mind and proceeding from (4.2.2), we have

$$\bar{A}_{\overline{x:y}} = \bar{A}_x + \bar{A}_y - \bar{A}_{x:y} = \frac{\mu_x}{\mu_x + \delta} + \frac{\mu_y}{\mu_y + \delta} - \frac{\mu_x + \mu_y}{\mu_x + \mu_y + \delta}$$
$$= \frac{0.03}{0.03 + 0.05} + \frac{0.08}{0.08 + 0.05} - \frac{0.11}{0.011 + 0.05} \approx 0.30288.$$

Multiplying this by 100,000, we get \$30,288.

EXAMPLE 3. Return to Example **7**.3.2-2. Consider a whole life insurance with the payment $c_t = t$. Set $\delta = 0.04$. We should again use the symbol $(\bar{I}\bar{A})_{50:60}$, but for simplicity we will write just $\bar{A}$.

The r.v. T_1 is uniform on $[0, 50]$, and T_2 is exponential with $\mu = 0.05$. Since the lifetimes are independent, by virtue of (**7**.3.2.2) or (**7**.3.2.5), the survival function for the $\min\{T_1, T_2\}$ is

$${}_tp_{x:y} = {}_tp_x \cdot {}_tp_y = (1 - t/50)e^{-0.05t} \text{ for } t \leq 50, \text{ and } = 0, \text{ otherwise.}$$

The d.f. ${}_tq_{x:y} = 1 - {}_tp_{x:y}$, and the density of $T_{x:y}$ is

$$f(t) = \frac{d}{dt}\, {}_tq_{x:y} = -\frac{d}{dt}\, {}_tp_{x:y} = (0.07 - 0.001t)e^{-0.05t} \text{ for } t \leq 50, \text{ and } = 0, \text{ otherwise.}$$

Then for the joint-life status,

$$\bar{A}_{50:60} = \int_0^\infty te^{-\delta t} f(t)dt = \int_0^{50} te^{-0.04t}(0.07 - 0.001t)e^{-0.05t}dt$$
$$= \int_0^{50} t(0.07 - 0.001t)e^{-0.09t}dt \approx 5.847.$$

[5]Reprinted with permission of the Casualty Actuarial Society.

(Again, we can integrate by parts or use software.)

Next, since T_1 is uniformly distributed on $[0,50]$,

$$\bar{A}_{50} = \int_0^{50} te^{-\delta t}\frac{1}{50}dt = \int_0^{50} te^{-0.04t}\frac{1}{50}dt \approx 7.425.$$

For the exponentially distributed r.v. T_2, we have

$$\bar{A}_{60} = \int_0^{\infty} te^{-\delta t}\mu e^{-\mu t}dt = 0.05\int_0^{\infty} te^{-0.09t}dt \approx 6.173.$$

Then, by (4.2.2),

$$\bar{A}_{\overline{50:60}} = \bar{A}_{50} + \bar{A}_{60} - \bar{A}_{50:60} \approx 7.751.$$

The reader can check the calculations above, but at least we got $\bar{A}_{\overline{50:60}}$ greater than $\bar{A}_{50:60}$. Would it be true if the benefits had been level? See Exercise 52.

EXAMPLE 4. (a) The lifetimes T_1 and T_2 of a husband and wife are independent and uniformly distributed on $[0,50]$; the discount factor is 0.96. A special insurance pays one unit upon the death of the husband, provided that he dies first. Find the APV and the variance of the present value.

First, to make formulas nicer, we consider 50 years as a unit of time, replacing $[0,50]$ by $[0,1]$. Denote by δ the interest rate corresponding to this new time scale. The present value Z takes on a non-zero value $e^{-\delta t}$ if T_1 gets into the interval $[t, t+dt]$ and $T_2 > t$. The probability of the product of these events is $1\cdot dt \cdot P(T_2 > t) = (1-t)dt$. Then

$$E\{Z\} = \int_0^1 e^{-\delta t}(1-t)dt = \frac{1}{\delta^2}\left(e^{-\delta} - 1 + \delta\right), \tag{4.2.3}$$

which may be obtained, for example, by integration by parts.

The discount factor over 50 years is $v = (0.96)^{50}$, so $\delta = -\ln v = -50\ln(0.96) \approx 2.0411$. Inserting it into (4.2.3), we obtain the APV $\bar{A} = E\{Z\} \approx 0.2811$.

By the double rate rule,

$$E\{Z^2\} = \frac{1}{4\delta^2}\left(e^{-2\delta} - 1 + 2\delta\right) \approx 0.1850,$$

and $Var\{Z\} \approx 0.1850 - (0.2811)^2 \approx 0.1050$.

(b) Let an insurance pay c units at the moment of the husband's death if he dies first, and $\tilde{c}$ units if he dies after his wife. Say, $c = 1$ and $\tilde{c} = 1/2$.

Denote by Z the present value of the insurance in the previous case (a), and by Z_1 the present value of an individual life insurance for the husband. Then the present value of the insurance under consideration is the r.v. $Z_2 = \tilde{c}Z_1 + (c-\tilde{c})Z$ (since $Z = 0$ if $T_1 > T_2$). Next, $E\{Z_1\} = \int_0^1 e^{-\delta t}dt = \frac{1}{\delta}\left(1 - e^{-\delta}\right) \approx 0.4263$, and

$$E\{Z_2\} = \tilde{c}E\{Z_1\} + (c-\tilde{c})E\{Z\} \approx \frac{1}{2}\cdot 0.2811 + \frac{1}{2}\cdot 0.4263 = 0.3537.$$

We will continue to consider this problem in Exercises 56-57. □

5 ON THE ACTUARIAL NOTATION 1,2,3

The reader certainly has grasped the logic of the *actuarial notation* we used above. It is traditional and guided and revised by the International Actuarial Association's Permanent Committee on Notation. Some general features of the notation system may be seen in Fig.2.

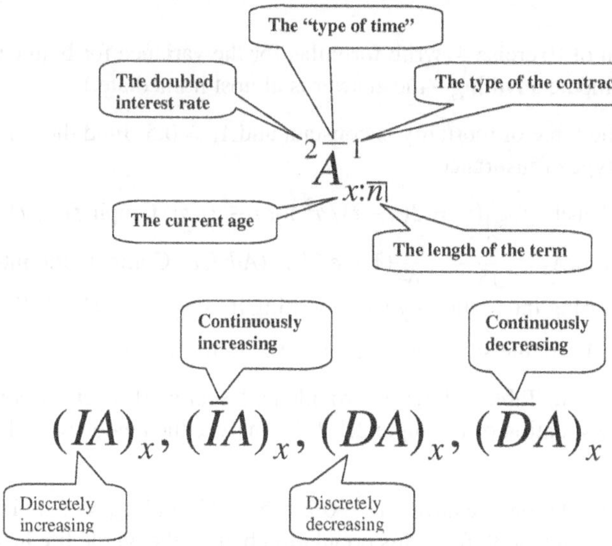

FIGURE 2.

6 EXERCISES

The use of software for computing integrals or sums in problems below is recommended.

Sections 1 and 2

1. (a) Show that obviously $\bar{A}_x \le 1$ and $A_x \le 1$. Can we write for one of these characteristics a more precise bound?

   (b) Usually, the life insurance for a younger person costs less than for an older person. Why? Is it true for $\delta = 0$? Let $\delta > 0$. Is it always true that $A_{x+t} \ge A_x$ and $\bar{A}_{x+t} \ge \bar{A}_x$? Give, first, a heuristic argument, and then justify your answer mathematically. (*Hint*: Think about "dangerous ages". Is $\mu(x)$ monotone?)

2. (a) Can we *always* expect that $A_x \le \bar{A}_x$? (b) Can $A_x = \bar{A}_x$? (c) When is A_x close to $\bar{A}_x$?

3. Compute $Var\{Z\}$ in Example 1.1-2.

4. Under the assumption $\mu(x) = \mu$,

$$\bar{A}_x = \frac{\mu}{\mu+\delta}, \quad A_x = e^{-\delta}\frac{1-e^{-\mu}}{1-e^{-(\mu+\delta)}}. \tag{6.1}$$

The first formula was obtained in (1.1.8). Prove the second. Recalling that $e^x = 1+x+o(x)$ for small x, compare the two formulas in (6.1) for small μ and δ. (*Advice*: First, look up Exercise **7**-28. Then either derive (6.1) directly, or write the m.g.f.'s of $K(x)$ following (**0**.4.3.1).)

5. In the situation of Exercise 4, write formulas for the variance for both cases. (*Hint*: You do not have to compute anything—the answer is almost immediate.)

6. Assume that the force of mortality is constant and $\bar{A}_x = 0.5$. Find the variance of the present value for this type of insurance.

7. (a) Let the density $f_{T(x)}(t) = 2(a-t)/a^2$ for $t \in [0,a]$. Graph $f_{T(x)}(t)$ for all t. Show that in this case, $\bar{A}_x = \frac{2}{a\delta} - \frac{2}{a^2\delta^2}(1-e^{-a\delta})$. (*Advice*: Compute the integral $\int_0^a e^{-\delta t}[2(a-t)/a^2]dt$ using integration by parts, writing it as $\int_0^a [2(a-t)/a^2]d(-e^{-\delta t}/\delta)$.)

   (b) Estimate $\bar{A}_{90}$ in the situation of Exercise **7**-35 for $\delta = 0.04$.

8. Using the data from Table **7**-1.5.1-1, provide an Excel worksheet for computing A_x for $x = 60, ..., 89$ in the situation of Example 2.1.2-2. Analyze the tendency of the change of A_x as x is varying.

9. For a certain population, we have $A_{30} = 0.2$ if $\delta = 3\%$, and $A_{30} = 0.09$ if $\delta = 6\%$. One hundred people at an age of 30 from this population bought the whole life insurance. Proceeding from $\delta = 3\%$, find a single premium sufficient for the company to fulfill its obligation with a probability of 0.99.

10. Consider $n = 100$ whole life insurance contracts with a benefit of \$100,000. Let all lives be independent and distributed as in Example 1.1-1, and $\delta = 0.04$. Find the total single premium sufficient for the company to carry out all payments with probability 95%.

11. Given a rate δ, find the distribution of the present value Z in the case of the whole life insurance ($\Psi = T$) under the assumption that the force of mortality is a constant μ. When is Z distributed uniformly on $[0,1]$? (*Hint*: $P(Z \le x) = P(e^{-\delta T} \le x) = P(T \ge -(\ln x)/\delta) = 1 - F_T(-(\ln x)/\delta)$, where $F_T(x)$ is the d.f. of T.)

12. Solve the problem from Example 2.2.1-1 by using (2.2.2).

13. Give an example of a particular distribution for which the approximation formula (2.1.7) is precise. Show it directly. Is this the only possible example?

14. (a) Find ${}_{10|}\bar{A}_{50}$ if X is uniformly distributed on $[0,100]$ and $\delta = 0.04$. (*Advice*: You can use the results of the calculations in Example 1.1-1 and (2.2.3).)

   (b) Find ${}_{10|}A_{50}$ in the same situation. (*Hint*: The distribution is uniform everywhere, including the intervals between complete years.)

15. Following just from the corresponding definitions, give a heuristic explanation of the formulas $\bar{A}_{x:\overline{n}|} = \bar{A}^{\,1}_{x:\overline{n}|} + {}_nE_x$ and $A_{x:\overline{n}|} = A^{\,1}_{x:\overline{n}|} + {}_nE_x$.

16. Which is larger: A_x or $A_{x:\overline{n}|}$?

17. Reasoning similarly to what we did when proving (2.1.6), show that (2.3.3) is almost obvious from a heuristic point of view.

18. (a) Consider two types of insurances, the n-year-term and the n-year-deferred, under the same conditions (the size of benefits, the discount, and the survival function). Are the corresponding present values positively or negatively correlated, or neither?

    (b) Answer the same question for the n-year-term and the n-year-pure-endowment insurances.

    (c) In the situation of Problem 18b, assume that you know the actuarial values and the variances for the insurances mentioned. How would you calculate the actuarial value and the variance for the n-year-endowment insurance?

19. Let $A_x = u, A_{x+n} = s, A_{x:\overline{n}|} = w$. Find $A^{\;1}_{x:\overline{n}|}$.

20. For a certain population we have ${}_{35}p_{30} = 0.95$, and for the rate

    $\delta = 3\%\ :\ A_{30} = 0.2,\quad A_{65} = 0.44$, while for

    $\delta = 6\%\ :\ A_{30} = 0.09,\quad A_{65} = 0.235$.

    (a) Proceeding from the 3% rate, find $A_{30:\overline{35}|}, A^{\;1}_{30:\overline{35}|}, {}_{35|}A_{30}$.

    (b) In the case of 1000 clients and the 35-year endowment insurance, find the single premiums sufficient for the company to fulfill its obligations with a probability of 0.95.

21. An actuary uses a demographic model where the lifetime of 30% of newborns has a constant mortality rate of 1/50 year^{-1}, while for 70%, the rate is 1/80 year^{-1}. The interest rate equals 5%. Under the above assumption, compute the expectation and the variance of the present value of the whole life insurance for a life-age-20 payable at the moment of death. (*Advice*: Use the result of Exercise **7**-22.)

22. You figured out somehow that an insurance company estimates the net single premium (APV) for the whole life insurance which pays \$100,000 at the end of the year of death for 30-year-old people as \$25,000; the same for 50-year-old people is \$40,000, and the 20-years-endowment insurance for 30-year-old people with the same benefit is \$55,000. You also know that the probability for a 30-year-old person to live 20 years more is about 0.95. From what average rate of interest did the actuary of the company proceed?

23. Take the values of l_x for $x = 60, ..., 70$ from Table **7**-1.5.1-1, and provide an Excel worksheet to compute $A^{\;1}_{60:\overline{10}|}$ similarly to what we did in Example 2.3.2-2. Compute the variance using the same worksheet. For 100 analogous contracts, find a single premium per contract for the security level $\beta = 0.95$.

24. For the case of De Moivre's law with $\omega = 100$ and discount $\delta = 0.04$, find $\bar{A}_{50:\overline{20}|}$ and $\bar{A}^{\;1}_{50:\overline{20}|}$.

25. Prove (2.4.2) by using the rule of double rate (1.1.5).

26. Using Table **7**-1.5.1-1, find ${}_{30}E_{30}$ for $\delta = 0.05$.

27. In each of the following, tell which quantity is larger: (i) $\bar{A}_x$ or A_x; (ii) A_x or $A^{\;1}_{x:\overline{n}|}$; (iii) A_x or $A_{x:\overline{n}|}$; (iv) A_x or ${}_{m|}A_x$.

28. Find $\lim_{m\to\infty} {}_{m|}A_x$ and $\lim_{m\to\infty} {}_{m|}\bar{A}_x$. Explain why the answer remains true even if $\delta = 0$.

29. Consider two groups of clients of the same age x. In each group, the distribution of $T(x)$ is the same for all clients. However, if $T^{(1)}(x)$ and $T^{(2)}(x)$ are the future lifetimes for clients from the first and from the second group, respectively, then $P(T^{(1)}(x) \geq t) > P(T^{(2)}(x) \geq t)$ for all t. Which group is healthier? Give a heuristic argument and then show rigorously that the APV for one group is larger than for the other, for any insurance we considered. Which

group is it? The reader who read Section **1**.3.5.1 recognizes that we are talking about the first stochastic dominance, but we do not need to know this notion in order to answer the above question. The reader who did not skip Section **1**.3.5.2 is recommended to connect this problem with the notion of the FSD. (*Hint*: Heuristically it is almost obvious; for a rigorous proof one may use (**0**.2.2.1).)

30. Prove that ${}_{m|}\bar{A}_x = \bar{A}_x - \bar{A}^{\,1}_{x:\overline{m}|}$, and ${}_{m|}A_x = A_x - A^{\,1}_{x:\overline{m}|}$.

31. Which of the following formulas are correct?

$$\text{(a)}\ A_x = A_{x:\overline{n}|} + v^n\,{}_np_xA_{x+n}, \qquad \text{(b)}\ A_x = A^{\,1}_{x:\overline{n}|} + {}_np_xA_{x+n}$$

32. Give a heuristic and rigorous proofs of (2.4.5) and (2.4.7).

33. The characteristics A_x, $\bar{A}_x$, $A_{x:\overline{n}|}$, $\bar{A}_{x:\overline{n}|}$, $A^{\,1}_{x:\overline{n}|}$, $\bar{A}^{\,1}_{x:\overline{n}|}$, ${}_{m|}A_x$, ${}_{m|}\bar{A}_x$, and ${}_nE_x$ all depend on δ.

    (a) Which of them—under a very mild condition—are decreasing in δ ?

    (b) For each APV, write without calculations the limit as $\delta \to 0$. (*Hint*: If $\delta = 0$, the present value of \$1 to be paid in the future is \$1.)

    (c) Make sure that the same limits follow from the corresponding mathematical formulas for the expectations.

34. Do we underestimate or overestimate the APV $A = E\{e^{-\delta\Psi}\}$ if, instead of computing A, we replace the r.v. Ψ by its expected value $E\{\Psi\}$ in $Z = e^{-\delta\Psi}$? In particular, state which is larger: A_x or $\exp\{-\delta(e_x+1)\}$; and $\bar{A}_x$ or $\exp\{-\delta\,\overset{\circ}{e}_x\}$. (*Advice*: Appeal to Jensen's inequality from Section **1**.3.4.2.)

35.* Denote by $A(\delta)$ the APV for an insurance with a unit benefit paid at some random time Ψ and for a rate of δ. Using the result of Exercise **1**-33, prove that, if $\delta_1 \le \delta \le \delta_2$, then

$$[A(\delta_1)]^{\delta/\delta_1} \le A(\delta) \le [A(\delta_2)]^{\delta/\delta_2}. \tag{6.2}$$

In the situation of Example 1.1-1 for $\omega = 100$ and $x = 60$, compute the exact values of the APVs for $\delta_1 = 0.04, \delta = 0.045$, and $\delta_2 = 0.05$, and analyze how much $A(\delta)$ differs from the left and right members of (6.2).

*Section 3**

36. Does the rule of double rate work in the case of varying payments?

37. Find the APV for the whole life insurance with $c_t = e^{kt}$, and $\mu(x) = \mu$. When does the problem have a solution? When is the insurance in this case equivalent to the whole life insurance with a level benefit but with another interest rate?

38. An insurance provides for payment of $c_t = e^{0.01t}$ for clients with the constant force of mortality $\mu_1 = 0.02$. Another insurance provides for payment of one unit for another type of clients with a constant force of mortality μ_2. It has turned out that the net single premiums for both insurances are the same. For which δ is it possible? Given δ, find μ_2.

39. Let the future life time $T(x)$ be uniformly distributed on $[0,n]$. Figure out without any calculations, which is larger: $(\bar{D}\bar{A})_{x:\overline{n}|}$ or $(\bar{I}\bar{A})_{x:\overline{n}|}$. Explain why your answer is not true for all distributions of $T(x)$. (*Hint*: (a) When the benefit is increasing in time, larger values of the benefit correspond to the moments of time at which the present value of a unit to be paid is getting smaller. (b) Regarding the second question, it suffices to give an example, and this example may concern a r.v. which is "practically non-random".)

40. (a) Given δ, find $(\bar{I}\bar{A})_{60}$ (i) in the case when X is uniform on $[0,100]$; (ii) when the force of mortality $\mu(x) = \mu$. (iii) In both cases, find the variances for $\delta = 0.04$, and $\mu = 0.05$.

(b) For the cases (i), (ii) above, find $(\bar{D}\bar{A})_{60:\overline{20}|}$.

41. Compute the APV for the 10-year-term and the 10-year-endowment insurances for the age of 50. The benefit is payable at the end of the year of death and equals $50 + 5m$ if the insured dies within the mth year. In the endowment case, 100 units are paid if the insured attains the age of 60. Survival probabilities may be taken from Table **7**-1.5.1-1. Certainly, the best way is to use software.

42. Solve the problem of Example 3.2-1 under the additional assumption that if the insured does not find a job during a year, then the company pays the annual salary at the end of the year period after the failure, and the contract terminates.

43. A special whole life insurance policy on a life-age of 30 has a benefit that starts at 10 and increases by 2 per year until age 50. Starting with the age of 50, the benefit is level at 50. At the age of 60, the contract terminates. Find the APV in the case of discrete time and discount $v = 0.97$ if the density of $T(30)$ is constant on $[0,30]$, and ${}_{30}p_{30} = 0.96$. (*Advice*: There are formulas for the sum at which you will arrive—see, e.g., (**9**.4.4), but it is reasonable just to use software.)

44. Compute $(IA)_{70}$ if T is uniformly distributed on $[0,30]$ and $v = 0.96$. (*Advice*: There are formulas for the sum at which you will arrive—see, e.g., (**9**.4.4), but it is reasonable just to use software.)

45. Explain the significance of the notation below and the formula itself:

$$(IA)^{1}_{x:\overline{n}|} = \sum_{k=0}^{n-1}(k+1)v^{k+1}P(K=k) = \sum_{k=0}^{n-1}(k+1)v^{k+1}\,{}_{k}p_x \cdot q_{x+k}.$$

46. A 4-year endowment insurance with benefits payable at the end of the year of death of (x) is characterized by the table below; k stands for values of the curtate lifetime. Find the APV for $v = 0.95$ and the probability that the present value will exceed the APV.

k	${}_{k+1}p_x$	benefit c_{k+1}
0	0.99	5
1	0.97	4
2	0.94	2
3	0.90	1

*Section 4**

47. In the situation of Example 4.1-1, in the case of non-accidental death, the company returns the net single premium with a rate of 2%. In the case of an accidental death, the benefit is equal to 10. Find the APV. (*Hint*: You will come to an equation for A.)

48.[6] A multiple decrement model has two causes with $\mu^{(1)}(t) = 0.2t$, $\mu^{(2)}(t) = 0.8t$ in some units. The benefit $c_{jt} = 10j$, $j = 1,2$. (a) Write a precise formula for ${}_tp = P(T > t)$. (b) Write a precise value for $\bar{A}$.

[6] The nice idea of this exercise is borrowed from [43, C7-5].

49. In the situation of Example 7.2.1-1, let $c_{1t} = t$, $c_{2t} = t^2$, and $\delta = 0.06$. Write $\overline{A}_x$ in integral form. Using software, estimate the value.

50. In the situation of Exercise 7-49, let $c_{1t} = 1$, $c_{2t} = 2, c_{3t} = 25e^{-0.1t}, \delta = 0.06$. Write $\overline{A}_x$ in integral form. Using software, estimate the value.

51. This problem is close to that of Example 4.1-3. A company has a special plan which pays (a) a death benefit of $5000 times the number of years in service, paid at the end of the year of death, provided it occurs while the employee is still in service; (b) a one-time payment (above the retirement pension) of $2000 times the number of years in service, paid at the end of the year of retirement either to the employee or to her/his beneficiary. Mr. T. is now exactly 61 years old. His salary increases by 5% every year on his birthday, and his annual salary now is $60,000. He has been in service for 10 years prior to the present time. During the period under consideration, Mr. T. can withdraw from this insurance plan (say, by changing his job). We consider death as cause 1, withdrawal as cause 2, and retirement as cause 3. The information on the forces of decrement for the initial group of $l_{61} = 100$ people of age 61 is given in the table below; $v = 0.97$. Provide a spreadsheet for computing the APV. (*Advice*: Look up, first, Section 7.2.3.)

x	$q_x^{(1)}$	$q_x^{(2)}$	$q_x^{(3)}$
61	0.02	0.01	0.05
62	0.03	0.00	0.06
63	0.04	0.00	0.07
64	0.00	0.00	1.00

52. In the case of level benefit, which is larger: $\overline{A}_{x:y}$ or $\overline{A}_{\overline{x:y}}$?

53. In the situation of Example 7.3.1-1, for $k = 1$, $\delta = 0.1$, and $c_t = 1$, find $\overline{A}_{x:y}$ and the corresponding variance. (*Advice*: You can do it directly, or you may use Example 4.2-1 and (4.2.2). All integrals that appear in this problem are tractable, but the use of software is more expeditious and is highly recommended.)

54. In the situation of Example 4.2-1, find $(\overline{I}\overline{A})_{x:y}$ for $k = 1$ and the corresponding variance.

55. In the situation of Example 7.3.1-1, find the APV for the insurance providing for payment of 2 units on the first death and one unit on the second. Let $\delta = 0.1$ and, for simplicity, $k = 1$. Integrals which will arise are tractable but tedious, so either just write them down or use software.

56. Consider the problem of Example 4.2-4a when (i) T_1 and T_2 are exponentially distributed with the same parameter a; (ii) T_1 is uniform on $[0, 40]$, and T_2 —on $[0.50]$.

57. In the situation of Example 4.2-4a, find the APV and the variance of the present value for the following insurances.

    (a) An insurance providing payments of one unit upon each death. (*Hint*: You may reduce the problem to a very simple one.)

    (b) An insurance providing payment of two units upon the first death and one unit upon the second.

    (c) If the husband dies first, the insurance pays c_1 units upon the first death, and c_2 units upon the second. If the wife dies first, the insurance pays c_3 units upon the first death, and c_4 units upon the second.

58. Consider a status of two lives. Write a simple formula for the APV for the insurance paying one unit upon both deaths.

Chapter 9

Annuity Models

1,2,3

An *annuity* is a series of payments made at certain intervals (as months or years) during some period which is, as a rule, random. Typical examples are pensions which are life annuities paid while the retired person lives, or an alimony which is paid until one of spouses dies.

As a good (and often convenient) approximation, actuaries use also models where annuity payments are carried out in a continuous-time fashion.

Regular payment of premiums by an insured, say, in the case of life insurance, is also an annuity. In this case, the *annuitant*—that is, the party receiving the annuity—is the insurance company, while the single payment of benefits constitutes the losses of the company.

Accumulated values, though they are strongly connected with annuities, will be considered in Section **10**.1.3 after we consider the notion of a net premium rate.

Below, we systematically explore two models: continuous-time and those where payments are provided at the beginning of certain periods, that is, in a discrete way.

1 TWO APPROACHES TO THE EVALUATION OF ANNUITIES

1.1 Continuous annuities

For certainty, we adopt a year as a time unit.

Consider an annuity that is *payable continuously* at a rate c_t depending, in general, on time. More precisely, we assume that the payment during an infinitesimally small interval $[t, t+dt]$ is equal to $c_t dt$. One may compare it with a water flow pouring into a basin with an instant speed of c_t at time t.

The present value of such a payment is equal to $v^t c_t dt = e^{-\delta t} c_t dt$, where $v = e^{-\delta}$ and δ is an annual rate of interest. Hence, if the payments are made during a time interval $[0, \Psi]$, then the present value of the total payment is equal to

$$Y = \int_0^{\Psi} e^{-\delta t} c_t dt. \tag{1.1.1}$$

Certainly, this is an abstraction, but it may serve as a good approximation if payments are carried out sufficiently frequently, say, monthly. (We still keep a year as a time unit.)

Usually, Ψ is random. In the case of life annuities on (x), the r.v. Ψ may coincide with the future lifetime $T(x)$ or may differ from it.

The notation for the APV (or the *net single premium*) $E\{Y\}$ is $\bar{a}$, with indices and signs when it is needed.

EXAMPLE 1. Consider an annuity on a life-age-x with $\Psi = T = T(x)$. Assume for simplicity that the force of mortality is a constant μ.

(a) First, let the payment rate be constant; say, $c_t = 1$. Then

$$Y = \int_0^T e^{-\delta t} dt = \frac{1}{\delta}(1 - e^{-\delta T}). \tag{1.1.2}$$

Since T is an exponential r.v. with parameter μ, the expectation $E\{e^{-\delta T}\} = \mu/(\mu+\delta)$ (see, e.g., (**8**.1.1.8)), and

$$\bar{a} = E\{Y\} = \frac{1}{\delta}\left(1 - E\{e^{-\delta T}\}\right) = \frac{1}{\delta}\left(1 - \frac{\mu}{\mu+\delta}\right) = \frac{1}{\mu+\delta}.$$

(b) Now consider a linearly growing payment rate; for example, $c_t = t$. Then

$$Y = \int_0^T te^{-\delta t} dt = -\frac{1}{\delta}Te^{-\delta T} + \frac{1}{\delta^2}\left(1 - e^{-\delta T}\right),$$

which may be obtained by integration by parts. For the net single premium, we have

$$\bar{a} = -\frac{1}{\delta}E\{Te^{-\delta T}\} + \frac{1}{\delta^2}\left(1 - E\{e^{-\delta T}\}\right). \tag{1.1.3}$$

The first term $E\{Te^{-\delta T}\} = \int_0^\infty te^{-\delta t}\mu e^{-\mu t} dt = \mu/(\mu+\delta)^2$, as is easy to calculate integrating again by parts. Substituting it into (1.1.3), replacing $E\{e^{-\delta T}\}$ by $\mu/(\mu+\delta)$, and doing some algebra, we get that

$$\bar{a} = 1/(\mu+\delta)^2 . \ \square \tag{1.1.4}$$

We see that if c_t is not constant, calculations turn out to be somewhat tedious even in such a simple case as in Example 1b. The approach we consider next may be helpful in computing APVs, and it is based on the following representation.

As usual, denote by $\mathbf{1}_A$ the indicator of an event A; more precisely, $\mathbf{1}_A = 1$ if A occurs, and $\mathbf{1}_A = 0$ otherwise. Note that $E\{\mathbf{1}_A\} = P(A)$. Then we can rewrite (1.1.1) as

$$Y = \int_0^\infty e^{-\delta t} c_t \mathbf{1}_{\{\Psi \geq t\}} dt. \tag{1.1.5}$$

(Indeed, for $t \leq \Psi$ the integrand is the same as in (1.1.1), and for $t > \Psi$ the indicator equals zero, so as a matter of fact, we integrate over $[0, \Psi]$.)

Taking the expectation of both sides and passing the expectation operation through the integral, we come to

$$\bar{a} = E\{Y\} = \int_0^\infty e^{-\delta t} c_t E\{\mathbf{1}_{\{\Psi \geq t\}}\} dt = \int_0^\infty e^{-\delta t} c_t P(\Psi \geq t) dt. \tag{1.1.6}$$

If $\Psi = T(x)$, then $P(\Psi \geq t) = P(T(x) \geq t) = {}_tp_x$. (Since T is a continuous r.v., it does not matter whether we write $T \geq t$ or $T > t$.) Denoting by $\bar{a}_x$ the APV of annuities in this case, we have

$$\bar{a}_x = \int_0^\infty e^{-\delta t} c_t \cdot {}_tp_x dt. \tag{1.1.7}$$

(Usually, the symbol $\bar{a}_x$ stands for the APV in the case of a constant payment rate; see Section 3.1. We keep the same symbol here to avoid complicated notation.)

EXAMPLE 2. Let us revisit Example 1b. In this case, ${}_tp_x = e^{-\mu t}$, and by (1.1.7),

$$\bar{a}_x = \int_0^\infty e^{-\delta t} t e^{-\mu t} dt = \int_0^\infty t e^{-(\mu+\delta)t} dt. \tag{1.1.8}$$

The last integral is standard and equals $1/(\mu+\delta)^2$, which coincides with (1.1.4). (The variable change $s = (\mu+\delta)t$ leads to $(\mu+\delta)^{-2}\int_0^\infty se^{-s}ds$. The last integral equals one.) We see that calculations turned out to be a bit easier than what we did in Example 1b. □

Nevertheless, the advantages of the second approach should not be overestimated. Formula (1.1.6) is indeed useful in computing expectations. But, for example, in calculating variances, representation (1.1.5) is not so helpful. [See, however, how one can apply this technique in computing second moments in (7.1.2.11)].

The approach we used in Example 1 is called the *aggregate payment technique.* The alternative approach is referred to as the *current payment technique.* We will use both—whichever turns out to be more convenient.

1.2 Discrete annuities

The model is very similar to what we considered above. Let Ψ be an integer-valued r.v. Consider Ψ time intervals of a unit length, say, Ψ complete years. Denote by c_t the payment at an integer time t. If the first payment is made at the initial time $t = 0$, the second payment is made at the beginning of the second period (that is, at time $t = 1$), and so on, then the present value of the total payment during Ψ periods is equal to

$$Y = c_0 + c_1 v + c_2 v^2 + ... + c_{\Psi-1} v^{\Psi-1}. \tag{1.2.1}$$

Such annuities are called *annuities-due.*

Now consider $\widetilde{\Psi}$ intervals, and assume that payments are provided at the end of each interval. Then the present value of the total payment is

$$\widetilde{Y} = vc_1 + c_2 v^2 + ... + c_n v^{\widetilde{\Psi}}.$$

This type is called *annuities-immediate* or *payable in arrears.*

In both cases, the r.v.'s Ψ and $\widetilde{\Psi}$ coincides with the number of payments. We use different symbols to emphasize that, for the same lifetime and for the same period of payments, Ψ and $\widetilde{\Psi}$ are different.

For example, consider, a life annuity on (x) providing for payments until the death of the annuitant. For the annuities-due, the last payment is made at the beginning of the year of death—that is, at time $t = K$, where $K = K(x)$ is the curtate lifetime. Then the number of payments $\Psi = K + 1$, and the present value

$$Y = c_0 + c_1 v + c_2 v^2 + ... + c_K v^K. \tag{1.2.2}$$

In the case of an annuity-immediate, the last payment is provided at the same time $t = K$ since at the end of the year of death, the company will not pay. In this case, the number of intervals (or, equivalently, the number of payments) equals K. So, $\widetilde{\Psi}=K$, and the present value

$$\widetilde{Y}=c_1v+c_2v^2+...+c_Kv^K \text{ if } K>0, \text{ and } \widetilde{Y}=0 \text{ if } K=0.$$

We see that there is a simple relation between Y and $\widetilde{Y}$:

$$Y=c_0+\widetilde{Y}.$$

For this reason, it suffices, at least theoretically, to study just one type – for example, annuities-due.

If we somehow manage to write a good expression for the sum in (1.2.1), then we can find the distribution of Y and its moments. In the following sections, we will demonstrate that this is quite easy if c_t is constant.

However, in general, such a summation may turn out to be difficult or even analytically impossible. In this case, at least for computing APVs, the current payment technique may help.

Indeed, similar to what we did in Section 1.1, we can rewrite (1.2.1) as

$$Y=\sum_{t=0}^{\infty}c_tv^t\mathbf{1}_{\{\Psi-1\geq t\}}. \tag{1.2.3}$$

The traditional notation for the APV of annuities-due is $\ddot{a}$ (with indices when needed). From (1.2.3) it follows that

$$\ddot{a}=E\{Y\}=\sum_{t=0}^{\infty}c_tv^tP(\Psi\geq t+1). \tag{1.2.4}$$

Consider a life annuity-due for (x), and denote the corresponding APV by $\ddot{a}_x$. *Like in the case of continuous time, $\ddot{a}_x$ stands usually for the APV in the case of a level (constant) payment rate (see Section 3.1). We keep the same symbol here to make the notation simpler.*

We saw that for a life annuity-due, the r.v. $\Psi=K(x)+1$. Replacing t by k for further convenience, we get from (1.2.4) that

$$\ddot{a}_x=\sum_{k=0}^{\infty}c_kv^kP(K\geq k)=\sum_{k=0}^{\infty}c_kv^k\,{}_kp_x. \tag{1.2.5}$$

EXAMPLE 1. (a) An organization provides annual payments to a 20 year old person for six years (for example, until the person will study in a university). Payments are made at the beginning of each year; 20 units in the each of the first two years, 25 in the third year, 30 in the fourth year, and 35 in each of the remaining years. The values of l_{20+k} are given in the spreadsheets in Fig.1ab; $v=0.96$. Find the net single premium and the corresponding variance.

In our problem, $c_k=0$ for $k>5$, and, by (1.2.5), to compute the APV we should compute $\sum_{k=0}^{5}c_kv^k\,{}_kp_{20}$. It is convenient to do so using a worksheet.

In Column C in Fig.1a, we compute ${}_kp_{20}=l_{20+k}/l_{20}$. Column D contains values of payments. In Column E, we compute the products $c_kv^k\,{}_kp_{20}$. Then the APV $\ddot{a}_{20}$ equals the sum of the values in Column E. We see that this is a very easy procedure.

	A	B	C	D	E	F
1	k	l_{20+k}	$_kp_{20}$	c_k		
2	0	98345	1	20	20	
3	1	98299	0.99953226	20	19.1910194	
4	2	98150	0.99801718	25	22.9943159	
5	3	97998	0.99647161	30	26.4484291	
6	4	97645	0.9928822	35	29.5155378	v
7	5	97280	0.98917078	35	28.2289996	
8						
9	APV		146.378302			0.96
10				$c_k \cdot {}_kp_{20} \cdot v^k$		
11						

(a) A worksheet for Example 1a; the current payment technique.

	A	B	C	D	E	F	G	H	I
1	k	l_{20+k}	$P(K=k)$	c_k	$c_k v^k$	PV	$(PV)^2$		
2	0	98345	0.0004677	20	20	20	400	0.009355	0.187096
3	1	98299	0.0015151	20	19.2	39.2	1536.64	0.059391	2.328124
4	2	98150	0.0015456	25	23.04	62.24	3873.818	0.096197	5.987292
5	3	97998	0.0035894	30	26.54208	88.78208	7882.258	0.318675	28.29261
6	4	97645	0.0037114	35	29.72713	118.5092	14044.43	0.439838	52.12485
7	5	97280	0.9891708	35	28.53804	147.0473	21622.89	145.4548	21388.74
8									
9	APV								
10				Variance		(PV)· P(K=k)			
11			146.3783						
12							$(PV)^2\cdot P(K=k)$		
13			51.048494		v		0.96		
14									

(b) A worksheet for Example 1b; the aggregate payment technique. PV stands for 'present value'.

FIGURE 1.

(b) In the worksheet in Fig.1b, we apply the aggregate payment technique. In Column C, we compute $P(K=k) = (l_{20+k} - l_{20+k+1})/l_{20}$ with one exception: in Cell C7 we compute $P(K \geq 5) = l_{25}/l_{20}$. This is the probability that the annuitant will attain the age of 25 and will receive the last payment. In column E, we compute $c_k v^k$, and in Column F – the values of the present value in the cases $K = k$ for $k = 0, ..., 4$. Cell F7 corresponds to the case $K \geq 5$. The command, say, for Cell F5 is '=SUM($E$2:E5)', which corresponds to the expression in (1.2.2) for $K = k$.

Column G contains the squares of the values in Column F. It prepares us for computing the variance. In Column H, we multiply each value of the present value by its probability. For example, Cell H5=F5*C5. In Column I, we do the same with the squares. For example, I5=G5*C5.

Now, to compute the APV, it suffices to add up all numbers in Column H, which is done in Cell C11. The sum of all numbers in Column I is $E\{Y^2\}$, so to compute the variance in C13 we should subtract from this sum the square of the APV.

We see that the procedure is longer, but it allows us to compute the variance, and in the same manner, all other moments.

EXAMPLE 2 ([158, N12][1]). An insurance company has agreed to make payments to a worker who is age x and was injured at work. The payments are 120,000 per year, paid annually, starting immediately and continuing for the remainder of the worker's life. After the first 500,000 is paid by the insurance company, the remainder will be paid by a reinsurance company. The survival function ${}_tp_x = (0.6)^t$ for $0 \le t \le 6.5$, and ${}_tp_x = 0$ for $t > 6.5$. The annual effective interest $i = 0.05$. Calculate the APV of the payment to be made by the *reinsurer*.

We use the general presentation (1.2.5). Let \$1000 be a monetary unit. Since $120 \times 4 < 500$ while $120 \times 5 > 500$, the first 500 will be paid at the beginning of the fifth year; that is at the time moment $k = 4$. (The payments starts at $k = 0$.)

The surplus over 500,000 at this moment is $600 - 500 = 100$. Hence, *for the reinsurer*, $c_k = 0$ for $k < 4$, $c_4 = 100$, and $c_k = 120$ for $k > 4$.

Also, ${}_kp_x = 0$ for $k > 6$, and the discount $v = \frac{1}{1+i} = \frac{1}{1.05}$. Hence, by virtue of (1.2.5),

$$\ddot{a}_x = \sum_{k=4}^{6} c_k v^k {}_kp_x = 100 \cdot \left(\frac{0.6}{1.05}\right)^4 + 120 \cdot \sum_{k=5}^{6} \left(\frac{0.6}{1.05}\right)^k \approx 22,151. \ \square$$

2 LEVEL ANNUITIES. A CONNECTION WITH INSURANCE

In the particular case when payments are constant (or level) during the payment period, calculations turn out to be simpler, and the aggregate payment technique leads to nice results. Once payments are constant, we can assume, without loss of generality, that they are made at a unit rate. So, we set $c_t = 1$ for both the continuous and discrete cases.

2.1 Certain annuities

Consider an annuity that is payable continuously at a constant rate of one per unit of time during a period $[0, \tau]$. As we saw already, the present value of the total payment is equal to

$$\int_0^\tau e^{-\delta t} dt = \frac{1}{\delta}(1 - e^{-\delta\tau}). \qquad (2.1.1)$$

The traditional notation of the present value in the l.-h.s. of (2.1.1) is $\bar{a}_{\overline{\tau}|}$. The bar indicates, as usual, that we deal with a continuous time model. The time τ is put into the angle above in order to indicate that this is the duration of the period under consideration, This distinguishes this notation from the notation $\bar{a}_x$ which we will use for the expected present value of the whole life annuity on an age-life-x. In the latter case, the index is an initial age rather than the length of the total payment period. Thus,

$$\bar{a}_{\overline{\tau}|} = \frac{1}{\delta}(1 - e^{-\delta\tau}). \qquad (2.1.2)$$

[1]Reprinted with permission of the Casualty Actuarial Society.

For the discrete annuities-due with unit payments at the beginning of each period, the present value of the total payment during n periods is equal to

$$1+v+v^2+...+v^{n-1}=\frac{1-v^n}{1-v}. \tag{2.1.3}$$

The quantity in the l.-h.s. of (2.1.3) is denoted by $\ddot{a}_{\overline{n}|}$. Setting $d=1-v$, we write

$$\ddot{a}_{\overline{n}|}=\frac{1}{d}(1-v^n). \tag{2.1.4}$$

For an annuity-immediate, where during the same n intervals payments are provided at the end of each interval, the present value of the total payment is denoted by $a_{\overline{n}|}$ and is equal to

$$v+v^2+...+v^n=v\frac{1-v^n}{1-v}=v\ddot{a}_{\overline{n}|}. \tag{2.1.5}$$

Thus, $a_{\overline{n}|}=v\ddot{a}_{\overline{n}|}$.

2.2 Random annuities

Now, let the payment interval be random. To emphasize this, we replace the symbol τ by Ψ which again stands for a r.v. Then in the case of continuous payment, the r.v.

$$Y=\bar{a}_{\overline{\Psi}|}$$

is the (random) present value of annuities, and in accordance with (2.1.2),

$$Y=\frac{1}{\delta}(1-Z), \quad \text{where } Z=e^{-\delta\Psi}. \tag{2.2.1}$$

The main point here is that Z is the present value of an insurance (!) which provides for payment of a unit of money at time Ψ.

By virtue of (2.2.1), if we know the distribution of Z, then we can readily compute the distribution of Y, and vice versa. See also Exercise 6 and the comments included there.

Let us consider the APVs (or net single premiums). Setting $\bar{a}=E\{Y\}$ and$\bar{A}=E\{Z\}$, we get from (2.2.1) that

$$\bar{a}=\frac{1}{\delta}(1-\bar{A}), \tag{2.2.2}$$

or

$$\bar{A}+\delta\bar{a}=1. \tag{2.2.3}$$

For the variance, we immediately obtain from (2.2.1) that

$$Var\{Y\}=\frac{1}{\delta^2}Var\{Z\}. \tag{2.2.4}$$

In the discrete case, we have practically the same representation. Let discrete payments be provided during a period $[0,\Psi]$ where Ψ is an integer-valued r.v. Then Ψ is the number of payments. Setting

$$Y=\ddot{a}_{\overline{\Psi}|},$$

the present value of the annuity-due, we derive from (2.1.4) that

$$Y = \frac{1}{d}(1-Z), \text{ where } Z = v^{\Psi} = e^{-\delta\Psi}. \tag{2.2.5}$$

The r.v. Z is again the present value of an insurance providing a single payment of one unit at time Ψ.

The only difference between (2.2.5) and (2.2.1) is in the denominator. In the discrete case, it equals $d = 1-v$; in the continuous case, it is δ. Note that this difference is not very significant since $d = 1-v = 1-e^{-\delta} = \delta - \frac{\delta^2}{2} + o(\delta)$ for small δ; see (**0**.4.2.6). For example, if $\delta = 0.05$, then $d = 1-e^{-\delta} \approx 0.04877$.

▶ Making use of (**0**.4.2.2), one can show that

$$\frac{\delta^2}{2} - \frac{\delta^3}{6} \le \delta - d \le \frac{\delta^2}{2} - e^{-\delta}\frac{\delta^3}{6}. \blacktriangleleft$$

Setting again $\ddot{a} = E\{Y\}$ and $A = E\{Z\}$, we get from (2.2.5) the counterparts of (2.2.2)-(2.2.4) for the discrete case:

$$\ddot{a} = \frac{1}{d}(1-A), \quad \text{or } A + d\ddot{a} = 1, \tag{2.2.6}$$

and

$$Var\{Y\} = \frac{1}{d^2}Var\{Z\}. \tag{2.2.7}$$

---

3 SOME PARTICULAR TYPES OF LEVEL ANNUITIES

3.1 Whole life annuities

First, consider the *whole life annuity* providing continuous payment at a rate of one to a life-age-x until death. This means that $\Psi = T = T(x)$. The APV of the annuity is denoted by $\bar{a}_x$. From (2.2.2), (2.2.4), and (**8**.2.1.3) it follows that

$$\bar{a}_x = \frac{1}{\delta}(1-\bar{A}_x), \tag{3.1.1}$$

$$Var\{Y\} = \frac{{}^2\bar{A}_x - (\bar{A}_x)^2}{\delta^2}. \tag{3.1.2}$$

EXAMPLE 1. Consider the situation of Example **8**.1.1-1, where T follows De Moivre's law. In this example, we found that $\bar{A}_x = \frac{1-e^{-s}}{s}$, where $s = (\omega - x)\delta$. Then, by (3.1.1),

$$\bar{a}_x = \frac{e^{-s}-1+s}{\delta s}. \tag{3.1.3}$$

For $x = 60$, $\omega = 100$, and $\delta = 0.04$, we have computed in Example **8**.1.1-1 that $\bar{A}_x \approx 0.499$. Then, using (3.1.1) directly, we have

$$\bar{a}_x \approx \frac{1}{0.04}(1-0.499) = 12.525.$$

Since the payments are carried out at the unit rate per unit of time, the total amount to be paid *without discounting* is equal to $\int_0^{T(60)} 1 \cdot dt = T(60)$. The r.v. $T(60)$ is uniformly distributed on $[0,40]$, so $E\{T\} = 20$. Thus, on the average, 20 units will be paid. In order to provide it, the initial fund should be equal to 12.525 on the average.

For the variance, we should just divide the expression (**8**.1.1.7) by δ^2. □

In the case of the whole life annuity-due on (x), we denote the APV by $\ddot{a}_x$. In this case, $\Psi = K+1$, and in accordance with (2.2.6) and (2.2.7),

$$\ddot{a}_x = \frac{1}{d}(1-A_x), \tag{3.1.4}$$

$$Var\{Y\} = \frac{{}^2A_x - (A_x)^2}{d^2}. \tag{3.1.5}$$

EXAMPLE 2. Consider a whole life annuity-due for the same case as in Example 1. To this end, we use results from Example **8**.1.1-2, where we have calculated that

$$A_x = \frac{v(1-v^r)}{rd},$$

with $r = \omega - x$. Substituting it into (3.1.4), we immediately get that

$$\ddot{a}_x = \frac{rd - v + v^{r+1}}{rd^2}.$$

In Example **8**.1.1-2, for $\omega = 100, x = 60$, and $\delta = 0.04$, we computed that $A_x \approx 0.4889$. Then, by (3.1.4), $a_x \approx [(1-0.4889)/(1-e^{-0.04})] = 13.035$. This is larger than the answer in Example 1. (Why? See also Exercise 15.)

EXAMPLE 3. Let $\mu(x) = \mu$. Then, as was obtained in (**8**.1.1.8), $\bar{A}_x = \frac{\mu}{\mu+\delta}$, and $Var\{Z\} = \frac{\mu}{\mu+2\delta} - \left(\frac{\mu}{\mu+\delta}\right)^2$. Hence, by (3.1.1),

$$\bar{a}_x = \frac{1}{\mu+\delta}, \quad Var\{Y\} = \frac{1}{\delta^2}\left(\frac{\mu}{\mu+2\delta} - \left(\frac{\mu}{\mu+\delta}\right)^2\right) = \frac{\mu}{(\mu+2\delta)(\mu+\delta)^2}. \tag{3.1.6}$$

Certainly, the case of the exponential distribution is simple, and we could readily compute the same directly, without using (3.1.1).

EXAMPLE 4. In the Illustrative life table in Appendix, Section 3, using (3.1.4), we provide the values of $\ddot{a}_x$ based on the original data from Table **7**.1.5.1-1. Let, for example, $x = 70$. Then $A_{70} = 0.5729602$, and

$$\ddot{a}_{70} = \frac{1}{1-e^{-0.04}}(1-0.5729602) = 10.89093833.$$

Thus, the value of the annuity of one dollar per year paid at the beginning of each year starting from the age 70, is \$10.89. This is certainly smaller than the expected amount to be paid: $e_{70} = 14.7$ (see Table **7**.1.5.1-1), so on the average, the annuitant will get \$14.7.

The reader can easily recalculate all of this for an annuity of, say, \$50,000 per year (for, example, for such a pension annuity). □

2,3

Route 1 ⇒ page 482

EXAMPLE 5. We revisit Example **8**.3.2-1. As was promised there, consider the problem under the assumption that the company pays to the insured her/his current salary from the moment of the loss of the job until the moment when the insured finds a new job or dies. Denote by ξ the duration of the payment period, and assume ξ to be exponential with parameter $b = 2$, and to be independent of the time of failure. For simplicity, let us consider the case when the company pays the salary continuously.

If the job is lost at a moment t, the lost annual salary is $e^{0.04t}$. The payment of the salary is a continuous annuity over the random period $[0, \xi]$. If the company had paid one unit of money per year, from the standpoint of the time t the present value of this payment would have been $\frac{1}{\delta}(1 - e^{-\delta\xi})$ by virtue of (2.2.1). Since the annual salary is not one but $e^{0.04t}$, the present value mentioned is $e^{0.04t}\frac{1}{\delta}(1 - e^{-\delta\xi})$. From the standpoint of the initial time 0, the present value of the same payment is $R_t = e^{-\delta t}e^{0.04t}\dfrac{1}{\delta}(1 - e^{-\delta\xi}) = e^{-0.02t}\dfrac{1}{\delta}(1 - e^{-\delta\xi})$, since $\delta = 0.06$. Consequently,

$$R_t = e^{-0.02t}Y, \text{ where } Y = \frac{1}{\delta}(1 - e^{-\delta\xi}).$$

The r.v. Y is the present value of the whole life continuous annuity with the lifetime ξ. The rest can be handled as in Example **8**.3.2-1. The present value of the insurance is the r.v.

$$Z = \begin{cases} 0 & \text{with probability } 0.9, \\ e^{-0.02X}Y & \text{with probability } 0.1, \end{cases}$$

where X is uniformly distributed on $[0, 10]$. The net single premium

$$A = E\{Z\} = 0.1 \cdot E\{e^{-0.02X}Y\} = 0.1 \cdot E\{e^{-0.02X}\}E\{Y\}$$

because X and Y are independent. We have computed in Example **8**.3.2-1 that $E\{e^{-0.02X}\} \approx 0.906$. Because ξ is exponential, (3.1.6) implies that

$$E\{Y\} = \frac{1}{b+\delta} = \frac{1}{2.06} \approx 0.485,$$

1,2,3

where b is the parameter of ξ; see Example **8**.3.2-1. Thus, $A \approx 0.1 \cdot 0.906 \cdot 0.485 \approx 0.0439$. □

Certainly, the aggregate payment technique above is not the only way to compute the APV. We can use the general formulas (1.1.7) and (1.2.5) where, in the case of whole life insurance, we set $c_t = 1$ (or $c_k = 1$ in the discrete case.) This leads to

$$\bar{a}_x = \int_0^\infty e^{-\delta t}\,{}_tp_x dt. \tag{3.1.7}$$

$$\ddot{a}_x = \sum_{k=0}^{\infty} v^k {}_kp_x. \tag{3.1.8}$$

We will use these formulas repeatedly.

Next, we consider the recurrence formula

$$\ddot{a}_x = 1 + vp_x\ddot{a}_{x+1}. \tag{3.1.9}$$

We prove (3.1.9) below, although it is almost obvious heuristically. For an annuity-due, the first unit payment is made at the very beginning, and this corresponds to the term 1 in (3.1.9). If the annuitant survives the first year (the probability of this is p_x), then the annuity to be paid will be equivalent to the whole life annuity for a life-age-$(x+1)$. The APV of this annuity is $\ddot{a}_{x+1}$. Since we evaluate it from the standpoint of the time $t = 0$, we should multiply $\ddot{a}_{x+1}$ by the discount factor v. All of this is again relevant to the first step analysis considered in Section **4**.4.3.

A formal proof of (3.1.9) may run as follows. The term for $k = 0$ in (3.1.8) equals one because ${}_0p_x = 1$. Making the variable change $m = k-1$ and using (**7**.1.2.7), we can write that

$$\ddot{a}_x = 1 + \sum_{k=1}^{\infty} v^k {}_kp_x = 1 + v\sum_{k=1}^{\infty} v^{k-1} p_x \cdot {}_{k-1}p_{x+1}$$

$$= 1 + vp_x \sum_{k=1}^{\infty} v^{k-1} {}_{k-1}p_x = 1 + vp_x \sum_{m=0}^{\infty} v^m {}_mp_{x+1} = 1 + vp_x\ddot{a}_{x+1}. \tag{3.1.10}$$

The recursion relation (3.1.9) may be applied in practical calculations. If we manage to evaluate $\ddot{a}_x$ for large x's, say, for $x = 100$, then we can move backward and calculate $\ddot{a}_x$ for all other x's. On the other hand, for very old people, the probability of dying within a short period is high, so we can assume that, say, $\ddot{a}_{100}$ is close to one. See also Example **8**.2.1.2-2.

We consider a counterpart of (3.1.9) for continuous payments in Exercise 13 and do particular calculations in Exercise 11.

3.2 Temporary annuities

Let $T = T(x)$ be the lifetime of (x), and let $K = K(x) = [T(x)]$, the curtate lifetime. In the discrete case, we deal with annuities-due, not stating it explicitly each time.

An n-year *temporary life annuity* provides for regular payments either until the moment of death or by the moment when the annuitant attains the age $x+n$ (whichever comes first). So, at the moment $x+n$ and on, the insurance organization does not pay. More precisely,

$$\begin{aligned} &\Psi = \min\{n, T\} && \text{in the continuous payment case, and} \\ &\Psi = \min\{n, K+1\} && \text{in the case of annuities-due.} \end{aligned} \tag{3.2.1}$$

Note again that in the discrete case, Ψ is the number of the intervals in which payments are provided. To clarify why we wrote $K+1$ above, consider three cases.

- If $K \geq n$ (the annuitant has survived n years), then $\min\{n, K+1\} = n$; that is, there were exactly n payments starting from the initial zero time.

- If the annuitant died within the last interval $[n-1,n]$, then $K=n-1$, and $\min\{n,K+1\}$ is still equal to n. So again, there will be n payments. The annuitant received the last nth payment at time $t=n-1$, at the beginning of the year of death.

- If $K<n-1$, then $\min\{n,K+1\}=K+1$, which is exactly equal to the number of payments in this case: starting at the time $t=0$, and until the moment $t=K$.

The insurance with a single unit payment at the time Ψ defined in (3.2.1) is an n-year endowment insurance (see Section **8**.2.4.2). So, we can again use (2.2.1) in the continuous case and (2.2.5) in the discrete case.

For the APV, the logic of the traditional notation leads to the respective notations $\bar{a}_{x:\overline{n}|}$ and $\ddot{a}_{x:\overline{n}|}$. In accordance with (2.2.2),

$$\bar{a}_{x:\overline{n}|}=\frac{1}{\delta}(1-\bar{A}_{x:\overline{n}|}), \tag{3.2.2}$$

and due to (2.2.6),

$$\ddot{a}_{x:\overline{n}|}=\frac{1}{d}(1-A_{x:\overline{n}|}), \tag{3.2.3}$$

where $d=1-v=1-e^{-\delta}$.

EXAMPLE 1. Let $x=60$, $n=5$, $v=0.96$, and l_x for $x=60,...,65$ are the same as in Example **8**.2.4.2-1. (The reader may remember that we purposely kept the same data in the example mentioned and in Examples **8**.2.3.2-2 and **8**.2.4.1-1.) In Example **8**.2.4.2-1, we calculated that $A_{60:\overline{5}|}\approx 0.820$, and $Var\{Z\}\approx 0.0014$. Then, by (3.2.3),

$$\ddot{a}_{60:\overline{5}|}\approx\frac{1}{0.04}(1-0.82)=4.50.$$

Making use of (2.2.7) and what we computed in Example **8**.2.4.2-1, we can immediately compute the variance:

$$Var\{Y\}=\frac{1}{d^2}Var\{Z\}=\frac{0.0014}{(0.04)^2}=0.875. \square$$

Another and direct way to compute APVs is to again use (1.1.7) and (1.2.5). First, consider the continuous payment case, and observe that an n-year temporal annuity may be viewed as a whole life annuity with varying payments $c_t=1$ for $t\le n$ and $c_t=0$, otherwise. (To cease paying means to start paying nothing.) The present value may be presented as in (1.1.5) with $\Psi=T(x)$ and c_t as we have defined. Then (1.1.7) will lead to

$$\bar{a}_{x:\overline{n}|}=\int_0^n e^{-\delta t}\,{}_tp_x\,dt. \tag{3.2.4}$$

In the discrete case, we should set $c_k=1$ for $k=0,...,n-1$, and $c_k=0$ for $k\ge n$, which implies

$$\ddot{a}_{x:\overline{n}|}=\sum_{k=0}^{n-1}v^k\,{}_kp_x. \tag{3.2.5}$$

The next relation allows us to compute $\ddot{a}_{x:\overline{n}|}$ in terms of $\ddot{a}_x$. Namely, we can write that

$$\ddot{a}_x = \ddot{a}_{x:\overline{n}|} + v^n\,{}_np_x\ddot{a}_{x+n}, \tag{3.2.6}$$

and hence

$$\ddot{a}_{x:\overline{n}|} = \ddot{a}_x - v^n\,{}_np_x\ddot{a}_{x+n}. \tag{3.2.7}$$

As are many similar relations above, (3.2.6) is quite understandable from a heuristic point of view. The whole life annuity consists of the annuity paid during the time interval $[0,n]$, and the annuity after the annuitant attains the age $x+n$. The last event has the probability ${}_np_x$, and the APV of the annuity after the time moment $x+n$ should be discounted by v^n.

We give a formal proof at the end of this subsection. Note also that (3.1.9) follows from (3.2.6) if we set $n=1$, since for annuities-due, where payments are made at the beginning of each period, $\ddot{a}_{x:\overline{1}|} = 1$ (why?).

A counterpart for the case of continuous payment is considered in Exercise 23.

EXAMPLE 2 ([152, N12][2]). The probability that a newborn lives to be 25 is 70%. The probability that a newborn lives to be 35 is 50%. The following annuities-due have APV equal to 60,000: a life annuity of 7,500 on (25), a life annuity of 12,300 on (35), and a life annuity of 9,400 on (25) that makes at most ten payments. What is the interest rate?

Taking 1000 as a unit of money, we have: $7.5\ddot{a}_{25} = 60$, $12.3\ddot{a}_{35} = 60$, $9.4\ddot{a}_{25:\overline{10}|} = 60$, ${}_{25}p_0 = s(25) = 0.7$, and ${}_{35}p_0 = s(35) = 0.5$. From this, we get ${}_{10}p_{25} = \frac{s(35)}{s(25)} \approx 0.71$, $\ddot{a}_{25} = 8, \ddot{a}_{35} \approx 4.88$, and $\ddot{a}_{25:\overline{10}|} = 6.4$. By (3.2.6),

$$\ddot{a}_{25} = \ddot{a}_{25:\overline{10}|} + v^{10}\,{}_{10}p_{25}\ddot{a}_{35},$$

from which we obtain that

$$v^{10} \approx \frac{8-6.4}{0.71\cdot 4.88} \approx 0.464.$$

Then $v = 0.926$, and $\delta = -\ln v = 0.076 = 7.6\%$. □

Now consider the following generalization of (3.2.6):

$$\ddot{a}_{x:\overline{m}|} = \ddot{a}_{x:\overline{n}|} + v^n\,{}_np_x\ddot{a}_{x+n:\overline{m-n}|} \quad \text{for all } m = n, n+1, \dots. \tag{3.2.8}$$

The logic is the same: we break the period $[0,m]$ into the periods $[0,n]$ and $[n,m]$. The proof is given at the end of the current subsection. The relation (3.2.6) follows from (3.2.8) if we set $m = \infty$. (Show that $\ddot{a}_{x:\overline{\infty}|} = \ddot{a}_x$.)

EXAMPLE 3 ([152, N2][3]). Given $v = 0.95$, ${}_{10}p_{25} = 0.87$, $\ddot{a}_{25:\overline{15}|} = 9.868$, and $\ddot{a}_{35:\overline{5}|} = 4.392$, calculate $A^{\;1}_{25:\overline{10}|}$.

We follow the following logic. If we calculate $A_{25:\overline{10}|}$, then knowing v and ${}_{10}p_{25}$, we will be able to calculate $A^{\;1}_{25:\overline{10}|}$. In view of (3.2.3), in order to calculate $A_{25:\overline{10}|}$, it suffices to know $\ddot{a}_{25:\overline{10}|}$. The last characteristic may be found with the use of (3.2.8).

[2]Reprinted with permission of the Casualty Actuarial Society.
[3]Reprinted with permission of the Casualty Actuarial Society.

So, first we compute ${}_{10}E_{25} = v^{10}{}_{10}p_{25} = (0.95)^{10}0.87 \approx 0.5209$. Then, by (3.2.8), $\ddot{a}_{25:\overline{10}|} = \ddot{a}_{25:\overline{15}|} - v^{10}{}_{10}p_{25}\ddot{a}_{35:\overline{5}|} \approx 9.868 - 0.521 \cdot 4.392 \approx 7.580$. Then, by (3.2.3), $A_{25:\overline{10}|} = 1 - d\ddot{a}_{25:\overline{10}|} \approx 1 - 0.05 \cdot 7.580 \approx 0.621$. At last, by (**8**.2.4.7), $A^{1}_{25:\overline{10}|} = A_{25:\overline{10}|} - {}_{10}E_{25} \approx 0.621 - 0.521 = 0.100$. □

► The formula (3.2.8) gives a recursion procedure for computing $\ddot{a}_{x:\overline{n}|}$. As was already noted, $\ddot{a}_{x:\overline{1}|} = 1$. Then, setting $n = 1$ in (3.2.8), we have $\ddot{a}_{x:\overline{m}|} = 1 + vp_x\ddot{a}_{x+1:\overline{m-1}|}$. Let us replace the letter m by the letter k and x by $x+n-k$ in the last formula. This leads to

$$\ddot{a}_{x+n-k:\overline{k}|} = 1 + v\,p_{x+n-k}\ddot{a}_{x+n-(k-1):\overline{k-1}|} \quad \text{for all } k = 1, 2, ..., n. \tag{3.2.9}$$

Set $h(k) = \ddot{a}_{x+n-k:\overline{k}|}$. From (3.2.9) it follows that

$$h(k) = 1 + v\,p_{x+n-k}\,h(k-1) \quad \text{for all } k = 1, 2, ..., n. \tag{3.2.10}$$

(Compare with (**8**.2.3.9). We could derive (3.2.10) from (**8**.2.3.9) and (3.2.3), but such a derivation will not be any shorter than the direct proof.)

Observe that $h(n) = \ddot{a}_{x:\overline{n}|}$, while $h(1) = \ddot{a}_{x+n-1:\overline{1}|} = 1$, so we can provide a backward recursion. An example is considered in Exercise 14. ◄

We proceed to a formal proof of (3.2.8). Similar to what we did when proving (3.1.10), and using (**7**.1.2.7), we have

$$\ddot{a}_{x:\overline{m}|} = \sum_{k=0}^{m-1} v^k{}_kp_x = \sum_{k=0}^{n-1} v^k{}_kp_x + \sum_{k=n}^{m-1} v^k{}_kp_x = \ddot{a}_{x:\overline{n}|} + \sum_{k=n}^{m-1} v^k{}_kp_x$$
$$= \ddot{a}_{x:\overline{n}|} + v^n{}_np_x \sum_{k=n}^{m-1} v^{k-n}{}_{k-n}p_{x+n}.$$

Under the variable change $s = k - n$, this implies that

$$\ddot{a}_{x:\overline{m}|} = \ddot{a}_{x:\overline{n}|} + v^n{}_np_x \sum_{s=0}^{m-n-1} v^s{}_sp_{x+n} = \ddot{a}_{x:\overline{n}|} + \ddot{a}_{x+n:\overline{m-n}|}. \blacksquare$$

3.3 Deferred annuities

In an m-year *deferred whole life annuity* on a life-age-x, the process of payments starts at the time moment $x+m$ (that is, after m years after the time of policy issue), provided that the annuitant attains the age $x+m$. A typical example is a pension plan.

It is probably most convenient to define it formally as the whole life annuity with payments

$$\begin{array}{ll} c_t = 0 \ \text{ if } t < m, \text{ and } c_t = 1 \text{ for } t \geq m & \text{in the case of continuous payments,} \\ c_k{=}0 \ \text{ if } k{=}0, ..., m{-}1, \text{ and } c_k{=}1 \text{ if } k{=}m, m{+}1, ... & \text{in the case of discrete payments.} \end{array} \tag{3.3.1}$$

The APVs (or net single premiums) are denoted by ${}_{m|}\bar{a}_x$ and ${}_{m|}\ddot{a}_x$, respectively. Proceeding from (3.3.1) and the general formulas (1.1.7) and (1.2.5), we readily write that

$${}_{m|}\bar{a}_x = \int_m^\infty e^{-\delta t}{}_tp_x\,dt, \tag{3.3.2}$$

$${}_{m|}\ddot{a}_x = \sum_{k=m}^{\infty} v^k{}_kp_x. \tag{3.3.3}$$

The present value itself may be presented in different ways. We prefer here the following. For a moment, let Y_x denote the present value of the whole life annuity on (x). Then in both cases—continuous and discrete—for the present value Y of the m-year deferred whole life annuity on a life-age-x,

$$\begin{aligned} Y &= 0 \qquad\qquad \text{if } T(x) < m, \\ Y &= v^m Y_{x+m} \ \text{ if } T \geq m. \end{aligned} \tag{3.3.4}$$

The logic is the same as that which we have applied repeatedly. If $T(x) < m$, then the annuitant gets nothing. If $T(x) \geq m$, then the annuitant attains the age $x+m$, and the annuity becomes a usual whole life annuity whose present value is Y_{x+m} from the standpoint of the time $t = m$.

Another form of representing Y is discussed in Exercise 17.

Since $P(T(x) \geq m) = {}_mp_x$, from (3.3.4) it follows that

$$E\{Y\} = v^m\, {}_mp_x E\{Y_{x+m}\},\ \ E\{Y^2\} = v^{2m}\, {}_mp_x E\{Y^2_{x+m}\}. \tag{3.3.5}$$

From the first relation it follows that

$${}_{m|}\bar{a}_x = v^m\, {}_mp_x\, \bar{a}_{x+m}, \tag{3.3.6}$$

$${}_{m|}\ddot{a}_x = v^m\, {}_mp_x\, \ddot{a}_{x+m} \tag{3.3.7}$$

[compare with (**8**.2.2.3), (**8**.2.2.5)].

Note that we could also derive (3.3.6)-(3.3.7) directly from (3.3.2)-(3.3.3). Detailed advice on how to do that is given in Exercise 18.

EXAMPLE 1. Mr. Doubt, a participant in a pension plan, is exactly 60 years old, and he has been in service in his current job 10 years prior to the present time. The retirement benefit of the plan pays annually (at the beginning of each year) the kth part of the salary at the time of retirement, multiplied by the number of years in service at the time of retirement.

Mr. D.'s salary increases by $100m\%$ each year on his birthday which, for simplicity, is January 1. (That is, if the increase is, say, 3%, then $m = 0.03$.) Mr. D. thinks about two opportunities: to retire at age 61, or to wait until the age of 68, when (in accordance with some rules) he will have to retire. Find m and k for which the latter opportunity is more profitable on the average, and compute the APV as a function of m and k. The interest rate during the period under consideration is about 4%. For the population to which Mr. D. belongs, $l_{60} = 84723$, $l_{61} = 82545$, $l_{68} = 75344$, $\ddot{a}_{61} = 14.12$, and $\ddot{a}_{68} = 11.13$.

Denote by B the current salary. The APV concerning the former opportunity is

$$C_1 = v \cdot p_{60} \cdot B(1+m)k \cdot 11 \cdot \ddot{a}_{61},$$

while for the alternative opportunity, the APV is

$$C_2 = v^8 \cdot {}_8p_{60} \cdot B(1+m)^8 k \cdot 18 \cdot \ddot{a}_{68} = Bv^8 \cdot p_{60} \cdot {}_7p_{61}(1+m)^8 k \cdot 18 \cdot \ddot{a}_{68}.$$

[We used (**7**.1.2.7).] Canceling out common factors, we see that $C_2 \geq C_1$ iff

$$v^7 \cdot {}_7p_{61} \cdot (1+m)^7 \cdot 18 \cdot \ddot{a}_{68} \geq 11\ddot{a}_{61}. \tag{3.3.8}$$

So, the decision should not depend on k nor on the size of the salary, and Mr. D. can proceed in his calculations at once from the age of 61. The probability ${}_7p_{61} = (l_{68}/l_{61}) \approx 0.9127$, the discount $v = e^{-.04} \approx 0.9608$, and (3.3.8) is equivalent to

$$1+m \geq \frac{1}{v}\left(\frac{11\ddot{a}_{61}}{18\ddot{a}_{68}\cdot {}_7p_{61}}\right)^{1/7} \approx \frac{1}{0.9608}\left(\frac{11\cdot 14.12}{18\cdot 11.13\cdot 0.913}\right)^{1/7} \approx 1.0168.$$

Thus, the retirement at 68 is more profitable if $m \geq 0.017$, i.e., the annual increase is not smaller than 1.7%.

Since ${}_8p_{60} = (l_{68}/l_{60}) \approx 0.913$, the APV in this case is $C_2 = Bv^8 \cdot {}_8p_{60}(1+m)^8 k\cdot 18\cdot \ddot{a}_{68} \approx (0.9608)^8\cdot 0.913\cdot 18\cdot 11.3(1+m)^8 kB = 134.860(1+m)^8 kB$.

For example, if $m = 0.02$ and $k = 0.025$, then $C_2 \approx 3.95B$; that is, the present value amounts approximately to four annual salaries. (Certainly, Mr. D. will get more than this (why?). We should also take into account that he has only 10 years of service in this job.)

A natural question is why we did not include in our calculations the salary that Mr. D. would receive if he decides to retire at the age of 68. We could do that. But then we would have to take into consideration the additional income which Mr. D. could have, being retired since the age of 61 (say, finding another job). Moreover, even if Mr. D. does not have an additional income after the earlier retirement, he would enjoy himself being free and, for example, traveling. This also should be taken into account. So, we compared only the APVs of *retirement* benefits, leaving Mr. D. to think about other issues. □

► Consider variances. From both relations in (3.3.5), we get that

$$\begin{aligned} Var\{Y\} &= E\{Y^2\} - (E\{Y\})^2 = v^{2m}\,{}_mp_x E\{Y_{x+m}^2\} - (v^m\,{}_mp_x E\{Y_{x+m}\})^2 \\ &= v^{2m}\,{}_mp_x\left[E\{Y_{x+m}^2\} - {}_mp_x(E\{Y_{x+m}\})^2\right]. \end{aligned}$$

Replacing $E\{Y_{x+m}^2\}$ by $Var\{Y_{x+m}\} + (E\{Y_{x+m}\})^2$, we eventually get that

$$\begin{aligned} Var\{Y\} &= v^{2m}\,{}_mp_x\left[Var\{Y_{x+m}\} + (1 - {}_mp_x)(E\{Y_{x+m}\})^2\right] \\ &= v^{2m}\,{}_mp_x\left[Var\{Y_{x+m}\} + {}_mq_x(E\{Y_{x+m}\})^2\right], \end{aligned}$$

where, as usual, ${}_mq_x = 1 - {}_mp_x$.

Thus, in view of (3.1.2) and (3.1.5), for the continuous case,

$$Var\{Y\} = v^{2m}\,{}_mp_x\left[\frac{1}{\delta^2}({}^2\bar{A}_{x+m} - (\bar{A}_{x+m})^2) + {}_mq_x(\bar{a}_{x+m})^2\right], \tag{3.3.9}$$

and for the discrete case,

$$Var\{Y\} = v^{2m}\,{}_mp_x\left[\frac{1}{d^2}({}^2A_{x+m} - (A_{x+m})^2) + {}_mq_x(\ddot{a}_{x+m})^2\right]. \tag{3.3.10}$$

If we wish to represent it in annuity terms only, we may write, by virtue of (3.1.1), that $\bar{A}_{x+m} = 1 - \delta\bar{a}_{x+m}$. For the characteristics for the double rate, we have ${}^2\bar{A}_{x+m} = 1 - 2\delta\cdot {}^2\bar{a}_{x+m}$, where ${}^2\bar{a}_x$ stands for the APV of the whole life annuity with the rate 2δ. Substituting it into (3.3.9), after some algebra one can get that

$$Var\{Y\} = v^{2m}\,{}_mp_x\left[\frac{2}{\delta}\left(\bar{a}_{x+m} - {}^2\bar{a}_{x+m}\right) - {}_mp_x(\bar{a}_{x+m})^2\right]. \tag{3.3.11}$$

We certainly can do the same for the discrete case, but we should be cautious when considering the double rate. Of course, by virtue of (3.1.4), we can write, that $A_{x+m} = 1 - d\ddot{a}_{x+m}$, but is it correct to write ${}^2A_{x+m} = 1 - 2d \cdot {}^2\ddot{a}_{x+m}$? Certainly not, since we double δ not $d = 1 - v$. When doubling δ, we square $v = e^{-\delta}$. Then $d = 1 - v$ should be replaced by $1 - v^2 = (1-v)(1+v) = d(2-d) = 2d - d^2$. Eventually, ${}^2A_{x+m} = 1 - (2d - d^2) \cdot {}^2\ddot{a}_{x+m}$. Substituting it into (3.3.10) and doing some algebra, we obtain that

$$Var\{Y\} = v^{2m}{}_mp_x \left[\frac{2}{\delta}\left(\ddot{a}_{x+m} - {}^2\ddot{a}_{x+m}\right) - {}_mp_x(\ddot{a}_{x+m})^2 + {}_m\ddot{a}_{x+m}\right] \tag{3.3.12}$$

[compare with (3.3.11)]. ◄

3.4 Certain and life annuities

An n-year *certain and life annuity* on (x) guarantees payments during the first n years, and if the annuitant attains the age $x+n$, the annuity proceeds as a whole life annuity. (So, if the annuitant dies within the first n years, the payment will be made to the beneficiary until the nth year.) In other words, the r.v.

$$\begin{aligned} \Psi &= \max\{n, T\} \quad \text{in the continuous payment case,} \\ \Psi &= \max\{n, K+1\} \text{ in the case of annuities-due.} \end{aligned} \tag{3.4.1}$$

[Compare with (3.2.1).]

Denote by ${}_{n|}Y_x$ the present value of the n-year deferred annuity for the same person, keeping this notation for both cases: continuous and discrete. Also recall the notations $\bar{a}_{\overline{n}|}$ and $\ddot{a}_{\overline{n}|}$ for *certain* annuities during n periods in the continuous and discrete cases, respectively; see representations in (2.1.2) and (2.1.4). Then, by definition, for the present value Y of the annuity under discussion, we can write

$$\begin{aligned} Y &= \bar{a}_{\overline{n}|} + {}_{n|}Y_x \text{ in the case of continuous payments,} \\ Y &= \ddot{a}_{\overline{n}|} + {}_{n|}Y_x \text{ in the case of discrete payments.} \end{aligned} \tag{3.4.2}$$

The traditional notation for the APV in this case (the reader should prepare her/himself for what will happen now) is $\bar{a}_{\overline{x:\overline{n}|}}$ and $\ddot{a}_{\overline{x:\overline{n}|}}$, respectively. The appearance of the "big bar" is logical. We already indicated the maximum of two variables by this symbol; see the beginning of Section **7**.3. The bar here indicates that we consider the maximum in (3.4.1).

It remains to combine (3.4.2), (2.1.2), (2.1.4), (3.3.6), and (3.3.7), which immediately implies that

$$\bar{a}_{\overline{x:\overline{n}|}} = E\{Y\} = \bar{a}_{\overline{n}|} + {}_{n|}\bar{a}_x = \frac{1 - e^{-\delta n}}{\delta} + v^n{}_np_x\bar{a}_{x+n}, \tag{3.4.3}$$

$$\ddot{a}_{\overline{x:\overline{n}|}} = E\{Y\} = \ddot{a}_{\overline{n}|} + {}_{n|}\ddot{a}_x = \frac{1 - v^n}{1 - v} + v^n{}_np_x\ddot{a}_{x+n}. \tag{3.4.4}$$

In view of the same representation (3.4.2),

$$Var\{Y\} = Var\{{}_{n|}Y_x\},$$

that is, $Var\{Y\}$ coincides with the expressions (3.3.11) and (3.3.12) for the variances of deferred annuities.

We will come to another useful representation if for ${}_{n|}\bar{a}_x$ and ${}_{n|}\ddot{a}_x$ we use (3.3.2) and (3.3.3), respectively. It will lead to formulas

$$\bar{a}_{x:\overline{n|}} = \frac{1-e^{-\delta n}}{\delta} + \int_n^\infty e^{-\delta t}\,{}_tp_x dt, \tag{3.4.5}$$

$$\ddot{a}_{x:\overline{n|}} = \frac{1-v^n}{1-v} + \sum_{k=n}^{\infty} v^k\,{}_kp_x. \tag{3.4.6}$$

At least in the latter case, it makes sense to recall that the first term is $1+v+v^2+...+v^{n-1}$, so we can rewrite (3.4.6) in the following nice form:

$$\ddot{a}_{x:\overline{n|}} = 1+v+...+v^{n-1}+v^n\cdot{}_np_x+v^{n+1}\cdot{}_{n+1}p_x+... = \sum_{k=0}^{n-1} v^k + \sum_{k=n}^{\infty} v^k\cdot{}_kp_x. \tag{3.4.7}$$

EXAMPLE 1. Given $\ddot{a}_x = 10.1$, $v = 0.96$, $p_x = 0.99$, and $p_{x+1} = 0.98$, find $\ddot{a}_{x:\overline{3|}}$.

While $\ddot{a}_{x:\overline{3|}} = 1+v+v^2+{}_{3|}\ddot{a}_x$, the whole life APV $\ddot{a}_x = 1+vp_x+v^2\,{}_2p_x+{}_{3|}\ddot{a}_x$. Then the difference $\ddot{a}_{x:\overline{3|}}-\ddot{a}_x = v(1-p_x)+v^2(1-{}_2p_x)$. Also recall that ${}_2p_x = p_x\,p_{x+1}$ [see (7.1.2.8)]. Hence, $\ddot{a}_{x:\overline{3|}}-\ddot{a}_x = 0.96\cdot 0.01+(0.96)^2(1-0.99\cdot 0.98) = 0.037$, and $\ddot{a}_{x:\overline{3|}} = \ddot{a}_x+0.037 = 10.137$.

EXAMPLE 2 ([159, N12][4]). For a special annuity product on a life aged 50 with a single benefit premium of \$50,000 paid immediately, you are given the following information: an annual benefit of K at the beginning of each year is guaranteed for the first five years; after five years, an annual benefit of K at the beginning of each year will be given until death; $i = 0.06$; $a_{50} = 12.267$; $A^{\;1}_{50:\overline{5|}} = 0.029$; $A_{55} = 0.305$. Calculate the annual benefit K.

This exam problem is on the knowledge of a number of formulas.

The APV of the product with a unit benefit premium (rather than K) is

$$a = \sum_{k=0}^{4} v^k + v^5\cdot{}_5p_{50}\cdot\ddot{a}_{55}. \tag{3.4.8}$$

We should find all entities in (3.4.8). First, $v = \frac{1}{1+i} = \frac{1}{1.06} \approx 0.943$.

Secondly, $\ddot{a}_{55} = \frac{1-A_{55}}{1-v} \approx \frac{1-0.305}{1-0.943} \approx 12.279$.

Now, note that a_{50} is an annuity immediate. We need to know $\ddot{a}_{50} = 1+a_{50} = 13.267$. Then $A_{50} = 1-(1-v)\ddot{a}_{50} \approx 1-(1-0.943)\cdot 13.267 \approx 0.249$.

Furthermore, $A_{50} = A^{\;1}_{50:\overline{5|}} + {}_5p_{50}\,v^5 A_{55}$. All entities in this relation are known except ${}_5p_{50}$. After simple calculations, we get ${}_5p_{50} \approx 0.966$.

Now we know everything to compute a in (3.4.8). Substitution leads to $a \approx 13.323$.

The real annual benefit is K, so $K\cdot a = 50,000$. This implies $K \approx \frac{50,000}{13.323} \approx 3,752$. □

Route 1 ⇒ page 505

[4]Reprinted with permission of the Casualty Actuarial Society.

2,3

4 MORE ON VARYING PAYMENTS

In the general case of payments varying in time, we are doomed to direct calculations. For example, for computing APVs, we should appeal to the general formulas (1.1.7) and/or (1.2.5). Powerful computers and good software make such calculations tractable.

In special cases, we can, nevertheless, write nice representations.

First, consider a standard increasing life annuity on (x) that is payable continuously. Without loss of generality, we can assume the *rate of the growth of payment* to be the unit rate, and set $c_t = t$. Then the present value

$$Y = \int_0^T c_t e^{-\delta t} dt = \int_0^T t e^{-\delta t} dt = \frac{1 - e^{-\delta T}}{\delta^2} - \frac{1}{\delta} T e^{-\delta T}. \tag{4.1}$$

The term $Te^{-\delta T}$ is the present value of a continuously increasing whole life insurance (see Section **8**.3.1). The notation for $E\{Y\}$ is $(\bar{I}\bar{a})_x$, and from (4.1) it follows that

$$(\bar{I}\bar{a})_x = \frac{1 - \bar{A}_x}{\delta^2} - \frac{1}{\delta}(\bar{I}\bar{A})_x. \tag{4.2}$$

Recalling that $1 - \bar{A}_x = \delta \bar{a}_x$, we rewrite it as

$$(\bar{I}\bar{a})_x = \frac{\bar{a}_x - (\bar{I}\bar{A})_x}{\delta}. \tag{4.3}$$

Let us consider the discrete case in the same manner, setting $c_k = k + 1$. The present value

$$Y = \sum_{k=0}^{K} c_k v^k = \sum_{k=0}^{K} (k+1) v^k.$$

We use the general formula

$$\sum_{k=0}^{m} (k+1) q^k = \frac{1}{1-q} \left(\sum_{k=0}^{m} q^k - (m+1) q^{m+1} \right) = \frac{1 - q^{m+1}}{(1-q)^2} - \frac{(m+1) q^{m+1}}{1-q}. \tag{4.4}$$

[It may be derived in many ways—in particular, as follows:

$$\sum_{k=0}^{m} (k+1) q^k = \frac{d}{dq} \sum_{k=0}^{m} q^{k+1} = \frac{d}{dq} \left(\frac{q - q^{m+2}}{1-q} \right). \tag{4.5}$$

One can readily verify that the last derivative equals the right member of (4.4).]

Thus,

$$Y = \frac{1 - v^{K+1}}{(1-v)^2} - \frac{(K+1) v^{K+1}}{1-v}. \tag{4.6}$$

From (4.6) it follows that the APV

$$(I\ddot{a})_x = E\{Y\} = \frac{1 - A_x}{d^2} - \frac{1}{d}(IA)_x, \tag{4.7}$$

where $d = 1 - v$. Since $1 - A_x = d\ddot{a}_x$, we have

$$(I\ddot{a})_x = E\{Y\} = \frac{\ddot{a}_x - (IA)_x}{d}. \tag{4.8}$$

The last formulas are nice and may be of help, but it makes sense also to note that for such a simple type of varying payment, the direct calculation of the APV—especially if we have good software—may turn out to be just as expeditious.

EXAMPLE 1. (a) Consider an annually decreasing annuity-due on (50), assuming the lifetime X to be uniform on $[0,100]$. Let $v = 0.96$. Since $T(x)$ is uniform, $P(K = k) = 1/50$ for $k = 0, ..., 49$. Then,

$$A_{50} = E\{v^{K+1}\} = \sum_{k=0}^{49} v^{k+1}\frac{1}{50} = \frac{1}{50}v\frac{1-v^{50}}{1-v} \approx 0.418,$$

$$(IA)_{50} = E\{(K+1)v^{K+1}\} = \sum_{k=0}^{49}(k+1)v^{k+1}\frac{1}{50} = \frac{1}{50}v\sum_{k=0}^{49}(k+1)v^k$$

$$= \frac{1}{50}v\left(\frac{1-v^{59}}{(1-v)^2} - \frac{50v^{50}}{1-v}\right) \approx 7.324,$$

where in the step before the last, we used (4.4). Hence, by (4.7),

$$(I\ddot{a})_{50} = \frac{1-0.418}{(0.04)^2} - \frac{7.324}{0.04} \approx 180.863. \tag{4.9}$$

(b) Let us compute the same directly. Since T is uniformly distributed, ${}_kp_x = (50-k)/50$ for $k \leq 50$. Then, by (1.2.5),

$$(I\ddot{a})_x = \sum_{k=0}^{\infty} c_k v^k {}_kp_x = \sum_{k=0}^{49}(k+1)v^k\left(1 - \frac{k}{50}\right) = \sum_{k=0}^{49}(k+1)v^k - \frac{1}{50}\sum_{k=0}^{49}(k+1)kv^k. \tag{4.10}$$

The first sum may be computed with the use of (4.4). For the second, we need a formula which certainly exists and may be derived by computing the second derivative in (4.5). We skip lengthy calculations here. Note also that the second sum in (4.10) is tractable even for a good calculator. The reader can verify that the answer is the same as in (4.9). □

> In many problems, practical or just for study purposes, varying payments may be represented as a combination of level annuities.

EXAMPLE 2 ([151, N1][5]). A special 30-year annuity-due on a person of age 30 pays 10 for the first 10 years, 20 for the next 10 years and 30 for the last 10 years. Given ${}_{20}E_{30} = m$, $\ddot{a}_{30:\overline{10}|} = u$, $\ddot{a}_{30:\overline{30}|} = v$, and $\ddot{a}_{50:\overline{10}|} = w$, find the APV.

[5]Reprinted with permission of the Casualty Actuarial Society.

Consider the annuity paying 20 during the whole 30-year period. Its APV is $20\ddot{a}_{30:\overline{30|}} = 20v$. If we subtract from this $10\ddot{a}_{30:\overline{10|}} = 10u$, we will get the APV of the annuity paying 10 the first 10 years and 20 during the remaining time. (We can imagine that the annuitant pays 10 units back each year during the first 10 years.) The new annuity has an APV of $20v - 10u$. Next, we add payments of 10 during the last 10 years. From the standpoint of the initial moment, the APV of this additional annuity is $10v^{20}\,{}_{20}p_{30}\ddot{a}_{50:\overline{10|}} = 10\,{}_{20}E_{30}\ddot{a}_{50:\overline{10|}} = 10mw$. So, the answer is $(-10u + 20v + 10mw)$. □

In the next example, we consider temporary increasing annuities which are defined as increasing annuities acting in a given period.

EXAMPLE 3 ([150, N9][6]). A person aged 20 buys a special five-year temporary life annuity-due with payments 1, 3, 5, 7, 9. Given: $\ddot{a}_{20:\overline{4|}} = 3.41$, $a_{20:\overline{4|}} = 3.04$, $(I\ddot{a})_{20:\overline{4|}} = 8.05$, $(Ia)_{20:\overline{4|}} = 7.17$ (where the symbols $a_{20:\overline{4|}}$ and $(Ia)_{20:\overline{4|}}$ correspond to the annuities-immediate, that is, to the annuities payable at the end of each time interval). Calculate the net single premium.

The point here is that, while the question concerns a 5-year annuity, the data correspond to a 4-year period. The payments 1, 3, 5, 7, 9 may be represented as $1, 1+2\cdot 1, 1+2\cdot 2, 1+2\cdot 3, 1+2\cdot 4$. Then the total payment is equivalent to one unit paid at the beginning + the level annuity-immediate with unit rate + the increasing annuity-immediate with a starting payment of 2. So, the APV$= 1 + a_{20:\overline{4|}} + 2(Ia)_{20:\overline{4|}} = 1 + 3.04 + 2\cdot 7.17 = 18.38$. □

---

5 ANNUITIES WITH m-thly PAYMENTS

Here we consider the case when payments are made m times a year. It is especially important when one deals with pension plans.

Let i be a given effective annual interest rate, and $d = 1 - v = \frac{i}{1+i}$, the corresponding annual discount rate; for definitions see Section **0**.8.3. We revisit the scheme of Section **8**.2.1.4, keeping the same notation and terms. In particular, we call m-ths the periods of length $1/m$ from this scheme. (Say, if $m = 12$, an m-th is a month.) Consider a whole life annuity with payments of $1/m$ made m times a year at moments $0, 1/m, ..., (m-1)/m$. So, for each complete year, the annual payment is one. Denote the present value of this annuity by $Y^{(m)}$, and the APV by $\ddot{a}_x^{(m)}$.

Let us choose for a while an m-th as a unit of time, and denote by $v^{(m)}$ the discount factor corresponding to this new unit of time. In view of the general formula $v_t = v^t$, the new discount $v^{(m)} = v^{1/m}$, where v is the annual discount. Set $\widetilde{d}^{(m)} = 1 - v^{(m)}$, the counterpart of $d = 1 - v$ in this case. We will soon see why it makes sense to use a tilde here.

Consider, first, the annuity which pays not $1/m$ but one unit of money at the beginning of each m-th, and denote the APV for this annuity by $\widetilde{Y}^{(m)}$. By virtue of the general representation (2.2.5),

[6]Reprinted with permission of the Casualty Actuarial Society.

$$\widetilde{Y}^{(m)} = \frac{1}{\widetilde{d}^{(m)}}\left(1 - Z^{(m)}\right), \tag{5.1}$$

where $Z^{(m)}$ is the present value of the whole life insurance providing for payment of a unit of money at the end of the m-th of death. This is exactly the insurance we considered in Section **8**.2.1.4. In particular, $E\{Z^{(m)}\} = A_x^{(m)}$.

It is important that, although (5.1) corresponds to the new unit of time, the dimension of the left and the right members of (5.1) is money, and we keep the unit of money as it was.

Now, we come back to the original annuity with payments $\frac{1}{m}$. Clearly, $Y^{(m)} = \frac{1}{m}\widetilde{Y}^{(m)}$, and hence

$$Y^{(m)} = \frac{1}{m} \cdot \frac{1}{\widetilde{d}^{(m)}}\left(1 - Z^{(m)}\right) = \frac{1}{m(1 - v^{1/m})}\left(1 - Z^{(m)}\right) = \frac{1}{d^{(m)}}\left(1 - Z^{(m)}\right), \tag{5.2}$$

where

$$d^{(m)} = m(1 - v^{1/m}).$$

The quantity $d^{(m)}$ is a *nominal annual rate of discount for the case when interest is compounded m*thly. We introduced and discussed it in Section **0**.8.5. Briefly, this is an annual discount rate which leads to the effective annual rate of discount d, if interest is compounded m times a year.

In particular, as was shown in Section **0**.8.5,

$$d^{(m)} \to \delta \text{ as } m \to \infty. \tag{5.3}$$

Now we return to (5.2) which implies that

$$\ddot{a}_x^{(m)} = \frac{1}{d^{(m)}}\left(1 - A_x^{(m)}\right). \tag{5.4}$$

Note that we could come to (5.4) in another way by reasoning—slightly heuristically—as follows. The general representation (2.2.5) implies that (5.4) must be true for an appropriate discount characteristic (a new d) in the denominator. Since one unit of money is paid during a year, this characteristic should be annual. Since payments are provided $m$thly, we should consider the case when interest is compounded m times a year. Then the discount characteristic should be the annual rate leading to the annual discount rate d. In Section **0**.8.5, we have shown that $d^{(m)}$ is exactly the characteristic with this property.

Nevertheless, the detailed derivation above makes the picture more transparent.

Now, we can either leave (5.4) as it is, or apply the approximation (**8**.2.1.11) from Section **8**.2.1.4. Assuming that the lifetime is uniformly distributed within each year, we write that $A_x^{(m)} = (i/i^{(m)})A_x$, where $i^{(m)} = m[(1+i)^{1/m} - 1]$, the nominal annual interest rate. (See Sections **8**.2.1.4 and **0**.8.2.)

On the other hand, $A_x = 1 - d\ddot{a}_x$. Substituting it into (5.4), after simple algebra we get that

$$\ddot{a}_x^{(m)} = \alpha(m)\ddot{a}_x - \beta(m), \tag{5.5}$$

where

$$\alpha(m) = \frac{di}{d^{(m)}i^{(m)}}, \quad \beta(m) = \frac{i - i^{(m)}}{d^{(m)}i^{(m)}}. \tag{5.6}$$

The coefficients $\alpha(m)$ and $\beta(m)$ depend only on m and the interest rate i. They are a bit cumbersome but calculable. We omit simple but tedious calculations showing that

$$\alpha(m) \approx 1, \ \beta(m) \approx \frac{m-1}{2m}, \tag{5.7}$$

which is used in practice.

In the table below, we present the values of $\alpha(m)$ and $\beta(m)$ for $m = 4, 6, 12$, and ∞, and for $i = 0.03, 0.04, 0.05$, and 0.06. In particular, it shows the degree of accuracy of the approximation (5.7).

$m\downarrow\ i\rightarrow$	0.03	0.04	0.05	0.06	approx. (5.7)
			$\alpha(m)$		
4	1.0000683	1.0001202	1.0001850	1.0002653	1
6	1.0000708	1.0001246	1.0001929	1.0002751	1
12	1.0000723	1.0002731	1.0001970	1.0002810	1
∞	1.0000728	1.0001282	1.0001984	1.0002820	1
			$\beta(m)$		
4	0.3796529	0.3811888	0.3827173	0.3842386	0.375
6	0.4214919	0.4230847	0.4246698	0.4262475	0.416...
12	0.4632610	0.4648889	0.4665080	0.4681195	0.4583...
∞	0.5049631	0.5066014	0.5082319	0.5098546	0.5

Once we know a connection between $\ddot{a}_x^{(m)}$ and $\ddot{a}_x$, we can write the relation between the respective temporary annuities. Proceeding from (3.2.7), which is clearly correct for any unit of time, we have

$$\begin{aligned}\ddot{a}^{(m)}_{x:\overline{n}|} &= \ddot{a}^{(m)}_x - v^n\,{}_np_x\ddot{a}^{(m)}_{x+n} = \alpha(m)\ddot{a}_x - \beta(m) - v^n\,{}_np_x(\alpha(m)\ddot{a}_{x+n} - \beta(m))\\ &= \alpha(m)(\ddot{a}_x - v^n\,{}_np_x\ddot{a}_{x+n}) - \beta(m)(1 - v^n\,{}_np_x) = \alpha(m)\ddot{a}_{x:\overline{n}|} - \beta(m)(1 - v^n\,{}_np_x).\end{aligned} \tag{5.8}$$

In the last step, we used (3.2.7) again. An example is considered in Exercise 39.

6 MULTIPLE DECREMENT AND MULTIPLE LIFE MODELS

6.1 Multiple decrement

If the present value of an annuity is specified only by the value of the failure time T, then we can find the APV and other characteristics directly, considering the standard annuity scheme for the lifetime T. If there is a factor that may cause a change in the payment rate but not a complete cessation, the situation is more sophisticated. We restrict ourselves to several examples.

EXAMPLE 1. ([151, N12][7]). Harold has been disabled and will begin receiving disability premiums. You are given: the discount factor $v = 0.95$; the hazard function for recovery is $\mu_{62+t}^{\text{recovery}} = 0.1(3-t)$; the hazard function for death is $\mu_{62+t}^{\text{death}} = 0.1t$; payments of \$10,000 begin today, his 62nd birthday, and he will receive \$10,000 as long as he has not recovered or died; there will be no payments made beyond his 65th birthday. Calculate the APV of Harold's disability payments.

The problem is easy since in accordance with (**7.2.1.8**), the total force of decrement is

$$\mu_{62}(t) = \mu_{62+t}^{\text{recovery}} + \mu_{62+t}^{\text{death}} = 0.1(3-t) + 0.1t = 0.3,$$

that is, a constant. In this case, it is easier to compute the APV directly by (3.2.5). Let $p = P(T(62) > 1) = e^{-\mu} = e^{-0.3} \approx 0.741$. Then ${}_kp_{62} = e^{-\mu k} = p^k$. Since there will be at most four payments, we deal with

$$\ddot{a}_{62:\overline{4}|} = 10000(1 + vp + v^2p^2 + v^3p^3) \approx 25483.35.$$

EXAMPLE 2. Let all data be the same as above, excepting $\mu_{62+t}^{\text{recovery}} = 0.2(3-t)$. Then

$$\mu_{62}(t) = \mu_{62+t}^{\text{recovery}} + \mu_{62+t}^{\text{death}} = 0.2(3-t) + 0.1t = 0.6 - 0.1t,$$

and, in accordance with (**7.2.1.9**),

$${}_tp_{62} = \exp\left\{-\int_0^t (0.6 - 0.1s)ds\right\} = \exp\left\{-0.6t + 0.05t^2\right\}$$

for $t \leq 3$. Then

$$\ddot{a}_{62:\overline{4}|} = 10000\sum_{k=0}^{3} v^k\, {}_kp_{62} = 10,000\sum_{k=0}^{3}(0.95)^k \exp\left\{-0.6k + 0.05k^2\right\} \approx 21023.7963,$$

as one may calculate using even a calculator.

EXAMPLE 3. Let us return to Example 1 and assume, for variety, that payments are provided continuously. Suppose that, if Harold is alive but is not recovered, the company pays at an annual rate of $c = 10,000$. If Harold recovers, the company continues to pay up to Harold's death at a smaller rate of $\widetilde{c} = 5000$. Find the APV under the assumption that the causes of decrement mentioned are acting independently.

The hazard rates for recovery and for death are given by the functions

$$\mu^{(1)}(t) = 0.1(3-t) \text{ if } t \leq 3, \text{ and } = 0, \text{ otherwise, and}$$
$$\mu^{(2)}(t) = 0.1t,$$

respectively.

[7]Reprinted with permission of the Casualty Actuarial Society.

We can apply the scheme of Section **7.2.2** from which we know that if the causes are independent, then the hazard rate $\mu^{(j)}(t)$ coincides with the marginal hazard rate corresponding to the cause j; see Section **7.2.2** for detail. For us, this means that the hazard rate for the pure whole life annuity acting independently equals $\mu^{(2)}(t)$.

Let us now observe that the annuity under consideration may be represented as the sum of the following two annuities:

- the whole life annuity with the payment rate $\widetilde{c}$ and, as we know, with the hazard rate $\mu^{(2)}(t)$;
- the multiple decrement whole life annuity with the payment rate $c-\widetilde{c}$ and the hazard rates $\mu^{(1)}(t)$ and $\mu^{(2)}(t)$.

Indeed, denote by $\mathcal{E}_t$ the event that Harold is alive at time t, and let $\mathcal{C}_t$ denote the event that Harold is still disabled at time t. Consider an infinitesimally small interval $[t,t+dt]$. The company pays cdt if Harold is alive but not recovered, and pays $\widetilde{c}dt$ if Harold is alive and healthy. We can represent it as

$$[\widetilde{c}\mathbf{1}_{\mathcal{E}_t}+(c-\widetilde{c})\mathbf{1}_{\mathcal{E}_t\mathcal{C}_t}]\,dt,$$

where, as usual, $\mathbf{1}_{\mathcal{E}}$ is the indicator of an event $\mathcal{E}$.

The last relation corresponds to the sum of the two annuities mentioned.

Now note that for the first annuity, ${}_tp_x=\exp\left\{-\int_0^t\mu^{(2)}(s)ds\right\}$, and the similar formula is true for the second annuity. Then, in accordance with (1.1.7), the APV of the first annuity is

$$\bar{a}'=\widetilde{c}\int_0^{\infty}e^{-\delta t}\exp\left\{-\int_0^t\mu^{(2)}(s)ds\right\}dt=\widetilde{c}\int_0^{\infty}e^{-\delta t}\exp\{-0.05t^2\}dt=\widetilde{c}\int_0^{\infty}\exp\{-\delta t-0.05t^2\}dt.$$

Up to a constant multiplier, the integrand is a normal density, so completing the square, we can calculate the integral directly. Another way is to use software. In any case, recalling that $\delta=-\ln(0.95)$, one can easily get that the integral is approximately equal to 3.50, and hence

$$\bar{a}'\approx 3.50\widetilde{c}.$$

The second annuity is the whole life annuity with the hazard rate $\mu^1(s)+\mu^2(s)$. First, we calculate

$$\int_0^t(\mu^{(1)}(s)+\mu^{(2)}(s))ds=\begin{cases}0.3t & \text{if } t\le 3,\\ 0.9+\int_3^t 0.1s\,ds=0.45+0.05t^2 & \text{if } t>3.\end{cases}$$

Then the APV for the second annuity is equal to

$$\bar{a}''=(c-\widetilde{c})\left(\int_0^3+\int_3^{\infty}\right)=$$

$$=(c-\widetilde{c})\left(\int_0^3e^{-\delta t}\exp\{-0.3t\}dt+\int_3^{\infty}e^{-\delta t}\exp\{-0.45-0.05t^2\}dt\right).$$

The last integrals are also analytically calculable (especially the first!). So, either directly or using software, we will come to ≈ 1.85 for the first integral, and ≈ 0.82 for the second. Eventually,

$$\bar{a}'' \approx (c - \tilde{c})2.67,$$

and the total APV

$$\bar{a} = \bar{a}' + \bar{a}'' \approx 5000(3.50 + 2.67) = 30850. \quad \square$$

6.2 Multiple life annuities

First, we establish the counterpart of (**8**.4.2.2) for annuities. Consider an annuity whose present value is completely specified by a function $\phi(T)$, where T is the lifetime of a particular status (single or multiple life) under consideration. Consider a status of two lives, (x) and (y). Then, absolutely similarly to what we did in Section 4.2, one may derive that

$$\bar{a}_{x:y} + \bar{a}_{\overline{x:y}} = \bar{a}_x + \bar{a}_y. \tag{6.1}$$

Here the $\bar{a}$'s denote the APVs of the annuity specified by the same function $\phi(\cdot)$ for the statuses $x:y$, $\overline{x:y}$, and separate lives (x), (y), respectively.

The same relation is certainly true for the discrete time case. Since the function $\phi(t)$ is arbitrary, it concerns all various types of annuities we considered earlier.

EXAMPLE 1. (a) Consider two persons of ages x and y, whose future lifetimes T_1 and T_2 are independent and exponentially distributed with parameters $\mu_1 = 0.04$ and $\mu_2 = 0.05$, respectively. The persons buy the joint annuity paying continuously at unit rate until the first death.

As we know (e.g., see Example **7**.3.2-6), the r.v. $T = \min\{T_1, T_2\}$ is exponential with parameter $\mu = \mu_1 + \mu_2$. Then in accordance with Example 1.1-1a,

$$\bar{a}_x = \frac{1}{\mu_1 + \delta}, \quad \bar{a}_y = \frac{1}{\mu_2 + \delta}, \quad \bar{a}_{x:y} = \frac{1}{\mu_1 + \mu_2 + \delta}. \tag{6.2}$$

(b) Consider the annuity paying at the same unit rate until the second death, so the lifetime of the status is $\overline{T} = \max\{T_1, T_2\}$. We can compute $\bar{a}_{\overline{x:y}}$ directly, but it is better to apply (6.1), which immediately leads to

$$\bar{a}_{\overline{x:y}} = \frac{1}{\mu_1 + \delta} + \frac{1}{\mu_2 + \delta} - \frac{1}{\mu_1 + \mu_2 + \delta}. \quad \square \tag{6.3}$$

Now consider the whole life annuity paying at a rate c_1 until the first death, and at a rate c_2 after the first death and until the second. This may be considered as the sum of two annuities: paying at the rate c_2 until the second death, and paying at the rate $c_1 - c_2$ until the first death. By virtue of (6.1), the APV in this case is

$$\bar{a} = (c_1 - c_2)\bar{a}_{x:y} + c_2\bar{a}_{\overline{x:y}} = (c_1 - c_2)\bar{a}_{x:y} + c_2(\bar{a}_x + \bar{a}_y - \bar{a}_{x:y}) = (c_1 - 2c_2)\bar{a}_{x:y} + c_2(\bar{a}_x + \bar{a}_y). \tag{6.4}$$

The same formula is true for the discrete case.

It is noteworthy that in the case $c_1 - c_2 < 0$, (6.4) keeps working: if $c_1 < c_2$, we may view this as if the insurer pays c_2 until the second death but simultaneously subtract $|c_1 - c_2|$ until the first death.

EXAMPLE 2. In the case of Example 1, formula (6.4) gives

$$\bar{a} = \frac{c_1 - 2c_2}{\mu_1 + \mu_2 + \delta} + \frac{c_2}{\mu_1 + \delta} + \frac{c_2}{\mu_2 + \delta}.$$

EXAMPLE 3 ([152, N5][8]). A pair of twins age (30) purchases a fully continuous joint life contract involving an annuity along with life insurance. Namely, the contract pays: the annuity of 1000 per year while both are alive; 1000 at the moment of the first death; the annuity of 600 per year after the first death until the second death; 800 at the moment of the second death. The future lifetimes of the twins are i.i.d.; $\delta = 0.05$; $\mu_x(t) = 0.04$ for all x and t. Find the APV.

Let the unit of money be \$100. The pure annuity part corresponds to the case of Example 2 with $\mu_1 = \mu_2 = 0.04$, $c_1 = 10$, $c_2 = 6$, so the APV of this part is

$$\bar{a} = \frac{10 - 2 \cdot 6}{0.08 + 0.05} + 2 \cdot \frac{6}{0.04 + 0.05} \approx 117.9487.$$

Now, let us consider the life insurance part. It may be viewed as the sum of two insurances: for the joint and for the last-survivor statuses.

If T_1 and T_2 are the separate lifetimes, the r.v. $T = \min\{T_1, T_2\}$ is exponentially distributed with parameter $\widetilde{\mu} = 2\mu_x(t) = 0.08$ (see, e.g., Example **7**.3.2-6).

In the continuous-time exponential case, $\bar{A}_x = \frac{\mu}{\mu+\delta}$; see (**8**.1.1.8).

Keeping this in mind, we have $\bar{A}_{30:30} = [\widetilde{\mu}/(\widetilde{\mu} + \delta)] = \frac{0.08}{0.08+0.05} \approx 0.6154$ and $\bar{A}_{30} = \frac{0.04}{0.04+0.05} \approx 0.4444$. By (**8**.4.2.2), $\bar{A}_{\overline{30:30}} \approx 2 \cdot 0.4444 - 0.6154 = 0.2734$.

Consequently, the total APV for the insurance part equals $\bar{A} = 10\bar{A}_{30:30} + 8\bar{A}_{\overline{30:30}} \approx 10 \cdot 0.6154 + 8 \cdot 0.2734 = 8.3412$.

Eventually, the APV of the whole product $\approx 100 \cdot (117.948 + 8.3412) = 12,628.92$.

EXAMPLE 4. The lifetimes T_1 and T_2 of a husband and wife are independent and uniformly distributed on $[0, 50]$; the discount $v = 0.96$. A special annuity pays continuously at a rate of c until the first death. If it is the husband's death, the annuity continues at the same rate. If the wife dies first, the rate changes to a rate of $\widetilde{c}$. Find the APV.

This example is close to Example **8**.4.2-4, and as we did there, we will consider 50 years as a unit of time. The annuity under consideration may be viewed as the sum of two annuities:

- the whole life annuity for the wife with the rate c, and
- the annuity payable in the period $[T_2, T_1]$ at rate $\widetilde{c}$, provided that $T_1 > T_2$.

[8]Reprinted with permission of the Casualty Actuarial Society.

By (3.1.3) from Example 3.1-1, the present value per unit payment rate for the first annuity is

$$\bar{a}' = \frac{e^{-\delta} - 1 + \delta}{\delta^2}, \tag{6.5}$$

where δ is the rate corresponding to the 50-years period. As we computed in Example **8**.4.2-4, $\delta \approx 2.0411$, and inserting it into (6.5), we get $\bar{a}' \approx 0.2811$.

Consider the second annuity, for a while assuming the payment rate to be unit rate. The present value of this annuity is the r.v.

$$Y_1 = \begin{cases} 0 & \text{if } T_1 < T_2, \\ \int_{T_2}^{T_1} e^{-\delta u} du = \frac{1}{\delta}(e^{-\delta T_2} - e^{-\delta T_1}) & \text{if } T_1 \geq T_2. \end{cases} \tag{6.6}$$

To compute the expected value of this r.v., we should take the integral of the function $\frac{1}{\delta}(e^{-\delta s} - e^{-\delta t})$ multiplied by the joint density of the r.v.'s T_1, T_2 over the region $\{(t,s) : t \geq s\}$.

The joint density of two independent (!) r.v.'s uniformly distributed on $[0,1]$ is the function $f(t,s) = 1 \cdot 1 = 1$. So,

$$\bar{a}'' = E\{Y_1\} = \int_0^1 \int_0^t \frac{1}{\delta}(e^{-\delta s} - e^{-\delta t}) ds dt = \frac{e^{-\delta}(2+\delta) + \delta - 2}{\delta^3}, \tag{6.7}$$

which can be easily verified by double integration. Inserting the value of δ, we get $\bar{a}'' \approx 0.0513$. Eventually, the APV for the total annuity is $\bar{a} = c\bar{a}' + \tilde{c}\bar{a}'' \approx 0.2811c + 0.0513\tilde{c}$. □

---

7 EXERCISES

The use of software for computing integrals or sums in problems below is recommended.

Sections 1-3

1. A special 3-year temporary annuity-due is specified by the following table.

k	0	1	2
payment	10	20	15
q_{x+k}	0.02	0.03	0.04

   Find the mean and variance of the present value in the case of $v = 0.96$. (*Hint*: It may happen that you do not need all the data given.)

2. Consider the annuity-due payable to (30) during 10 years. The payment in the kth year is equal to $\sqrt{k+1}$. Survival probabilities are evaluated in accordance with Table **7**-1.5.1-1. Using spreadsheet software, compute (a) the APV applying the current payment technique; (b) the APV and variance applying the aggregate payment technique.

3. What is the difference between $\bar{a}_{20}$ and $\bar{a}_{\overline{20}|}$?

4. Explain from a heuristic point of view that typically $\bar{a}_{x+1} \le \bar{a}_x$ and $\ddot{a}_{x+1} \le \ddot{a}_x$. Is it always true? Give a heuristic argument and a mathematical explanation. Proceeding from (2.2.2) and (2.2.6), show that your reasoning is consistent with what we did in Exercise **8**.1b.

5. The problem concerns level annuities with unit rate.

   (a) Show that if $\delta > 0$, then the r.v. Y is a bounded r.v. What is its largest value in the continuous and discrete time cases? Explain why the answer should be expected.

   (b) What is Y equal to if $\delta = 0$? Give also an heuristic explanation. Is Y bounded in this case? (Certainly, in the real life all r.v.'s are bounded, so we are talking about the Y from our model.) What is Y equal to if $\delta = 0$ in the case of the whole life annuity with continuous payment? Illustrate your conclusion considering the case (3.1.6) including the expression for the variance.

6. Proceeding from relation (2.2.1), solve the following problems. (a) How is Y distributed if Z is uniform on $[0,1]$? (b) Given an interest rate δ and a constant force of mortality μ, find the distribution of Y for the whole life continuous annuity. Consider the case $\delta = \mu$ separately. (*Advice*: Look up Exercise **8**.11.)

7. Given an interest rate δ, find the distribution of Y for the whole life continuous annuity if $T(x)$ is uniform on $[0,c]$. Analyze the case $\delta \to 0$.

8. (a) Certainly, $\bar{a}_x$ cannot be negative. Show that the numerator in (3.1.3) is indeed non-negative. (*Advice*: Consider the numerator at $s = 0$, and take its derivative in s.)

   (b) Consider the limit of (3.1.3) as $\delta \to 0$ (for example, applying L'Hôpital's rule). Explain the answer from an economic and probabilistic point of view.

9. Is it correct that the larger the value of an annuity to be paid, the less the value of the corresponding insurance plan?

10. Is it true that $\bar{a}_{x:\overline{n}|} = \frac{1}{\delta}(1 - \bar{A}^{\,1}_{x:\overline{n}|})$? If not, what was not taken into account?

11. (a) Assuming that $\ddot{a}_{100} = 1$ and using (3.1.9) and Table **7**-1.5.1-1, provide a worksheet for computing $\ddot{a}_x$ for $x = 0, ..., 99$ and $\delta = 0.04$.

    (b) Proceeding from the assumption in Example **8**.2.1.2-2, calculate $\ddot{a}_{90}$. After that, provide a worksheet for computing $\ddot{a}_x$ for smaller integer x's.

12. Proceeding from (2.2.4), write a formula for $Var\{Y\}$ in terms of the APVs of annuities, using the symbol ${}^2\bar{a}$ for annuities corresponding to the double rate.

13. Consider the recursion formula

$$\bar{a}_x = \bar{a}_{x:\overline{1}|} + v p_x \bar{a}_{x+1}. \tag{7.1}$$

    (a) Show that it is almost obvious from a heuristic point of view. Compare it with (3.1.9).

    (b) Prove it rigorously. (*Advice*: Split the integral in (3.1.7) into two parts, and make the variable change $s = t - 1$ in the second integral.)

14. Provide a worksheet for the recursion procedure (3.2.10). Using data from Table **7**-1.5.1-1, compute, for instance, $\ddot{a}_{30:\overline{15}|}$.

15. In each pair, which is large: (i) $\bar{a}_x$ or $\ddot{a}_x$; (ii) $\ddot{a}_x$ or $\ddot{a}_{x:\overline{n}|}$; (iii) $\ddot{a}_x$ or A_x ?

16. Give examples of when $\bar{a}_x > \bar{A}_x$, and when $\bar{a}_x < \bar{A}_x$. (*Advice*: Consider the cases when $T(x)$ "is large and when it is small." A good particular example may concern the exponential case.)

17. Show that for an m-deferred whole life annuity, in the continuous payment case, the present value

$$Y = 0 \qquad \text{if } T(x) < m$$
$$Y = v^m \bar{a}_{\overline{T-m}|} \text{ if } T \geq m$$

[the definition of $\bar{a}_{\overline{\tau}|}$ is given in (2.1.2)]; and in the case of discrete payments,

$$Y = 0 \qquad \text{if } K = 0, ..., m-1,$$
$$Y = v^m \ddot{a}_{\overline{K+1-m}|} \text{ if } K = m, m+1, ...$$

[for $\ddot{a}_{\overline{n}|}$, see (2.1.4)].

18. Derive (3.3.6)-(3.3.7) directly from (3.3.2)-(3.3.3). (*Advice*: Replace ${}_tp_x$ by ${}_mp_x \cdot {}_{t-m}p_{t+m}$ and $e^{-\delta t}$ by $e^{-\delta m}e^{-\delta(t-m)}$.)

19. Show that

$$\ddot{a}_{x:\overline{n}|} = \sum_{k=0}^{n-1} {}_kE_x.$$

From which formula does it immediately follow? By what technique did we get this formula?

20. (a) Which of the quantities $\bar{a}_x$, $\bar{a}_{x:\overline{n}|}$, ${}_{m|}\bar{a}_x$, $\ddot{a}_x$, $\ddot{a}_{x:\overline{n}|}$, ${}_{m|}\ddot{a}_x$ are non-increasing in δ ?
    (b) Write the limits of the quantities above for $\delta \to 0$ and $\delta \to \infty$ without calculations, proceeding from common sense.
    (c) Considering, as an example, for instance, ${}_{m|}\bar{a}$, show how your answers follow from the corresponding mathematical representations.

21. Which is larger: $\ddot{a}_{\overline{x:\overline{n}|}}$ or $\ddot{a}_x$? When $\ddot{a}_{\overline{x:\overline{n}|}} = \ddot{a}_x$? When $\ddot{a}_{\overline{x:\overline{n}|}}$ is close to $\ddot{a}_x$? Find the limit of $\ddot{a}_{\overline{x:\overline{n}|}}$ as $\delta \to 0$.

22. Write formula (3.2.6) for the case $\delta = 0$ and prove it directly.

23. Write the counterparts of (3.2.6) and (3.2.8) for the case of continuous payment.

24. Check whether you can write without calculations, not memorizing but proceeding from common sense, the relations between (i) $\bar{a}_x$ and $\bar{a}_{x:\overline{n}|}$; (ii) $\ddot{a}_x$ and $\ddot{a}_{x:\overline{n}|}$; (iii) $\ddot{a}_x$ and ${}_{m|}\ddot{a}_x$; (iv) $\bar{A}_x$ and $\bar{a}_x$; (v) A_x and $\ddot{a}_x$; (vi) $A_{x:\overline{n}|}$ and $\ddot{a}_{x:\overline{n}|}$; (vii) $A^{\,1}_{x:\overline{n}|}$ and $\ddot{a}_{x:\overline{n}|}$. Write other relations of the above types that you consider valuable. (*Advice*: Certainly, you may include other characteristics in relations. A relation, say, between $\ddot{a}_x$ and ${}_{m|}\ddot{a}_x$, as a matter of fact, should be a relation between ${}_{m|}\ddot{a}_x$ and $\ddot{a}_y$, with some y different from x.)

25. Given $\bar{a}_x = 10.1, \bar{a}_{x+20} = 8.5$, $\bar{a}_{\overline{x:\overline{20}|}} = 6.2$, and $\delta = 0.03$, find $\bar{A}^{\,1}_{x:\overline{20}|}$ and ${}_{20}p_x$.

26. Given $\ddot{a}_x = 10.1$, $\ddot{a}_{\overline{x:\overline{2}|}} = 10.1145$, and $p_x = 0.985$, find v.

27. Ann is 20 years old now. Her kind and wealthy uncle wants to make a single deposit into a special account for Ann to receive the annuity-due of \$20,000 per year during all her life. For some reasons, Ann prefers to start receiving money later but in larger amounts, for example, starting from the age of thirty. Setting $\delta = 0.04$ and using the Illustrative Table, calculate how much Ann should receive per year in the latter case under the condition that the value of the uncle's gift will not change.

28. Mr. A usually spends a part of his vacation in a resort "Not too bad" and comes back the next year with probability 0.9. The amounts of money spent each year are independent r.v.'s uniformly distributed on $[6, 10]$ (in some units of money, say \$100). The annual discount is 0.96. Compute (a) the expected present value and (b*) the variance of the present value of the money spent by Mr. A until the first time when he does not show up in the next year.

29. For a certain population, we have ${}_{35}p_{30} = 0.95$, and for the rate
$\delta = 3\%$, the APVs : $A_{30} = 0.2$, $A_{65} = 0.44$; while for
$\delta = 6\%$, the APVs : $A_{30} = 0.09$, $A_{65} = 0.235$.
    (a) For $\delta = 3\%$, compute the actuarial present value of annuities of \$100,000 paid at the beginning of each year to a person starting from her/his 30-years, until death. Find the mean and variance of the present value.
    (b) Do the same if payments are made until death but not longer than 35 years.

30. Using the current payment technique, compute $\bar{a}_{70}$ as a function of δ under the condition that $T(70)$ has the density $f_T(x) = 0.005(20-x)$ on $[0,20]$.

31. Michael is 20 years old. His parents (or a bank—in this case, it would be a loan) have agreed to pay him \$20,000 a year during 5 years, so he would be able to get his education. For simplicity, assume that the payments are made at the beginning of each year. Using the Illustrative Table, compute the net single premium and the corresponding variance for the annuities mentioned.

32. Using the Illustrative Table and linear interpolation, estimate $\bar{a}_{20}$.

*Section 4**

33. Let $T(50)$ be uniformly distributed on $[0,50]$, $v = 0.96$. Find the APV of the 15-year deferred continuous annuity paying at an annual rate of 2 for the first 10 years after the beginning of payment and at unit rate thereafter but totally not longer than 20 years.

34. (a) Explain from an heuristic point of view and show rigorously that

$$(I\ddot{a})_x = \sum_{k=0}^{\infty} {}_{k|}\ddot{a}_x, \text{ and } (I\bar{a})_x = \sum_{k=0}^{\infty} {}_{k|}\bar{a}_x. \tag{7.2}$$

    (b)[9] Write the series $\sum_{k=0}^{\infty} {}_{k|}\bar{a}_x$ for the case when $T(x)$ is uniformly distributed on $[0,M]$, for an integer M and $\delta = 0$. Compute this series.

35. Consider a continuous annuity on (30) paying at unit rate in the first year, at a rate of 1.2 during the second year (if the annuitant survives the first year), at a rate of 1.4 during the third year, and so on, up to a rate of 5 units. Then the rate is level at 5. The interest rate $\delta = 0.04$. Find the net single premium (a) if the lifetime X is uniform on $[0,100]$; (b) if $T(30)$ is exponential and $\overset{\circ}{e}_{30} = 40$. (c) How would you compute the APV for the corresponding annuity-due having data from a life table similar to what we have in the Illustrative Table?

36. (This problem is close to [151, N1]). A special 45-year annuity-due on (20) pays 30 for the first 5 years, 20 for the next 15 years, and 10 for the last 25 years. Given ${}_{20}p_{20}$, the discount v, $\ddot{a}_{20:\overline{45}|}$, $\ddot{a}_{20:\overline{5}|}$, and $\ddot{a}_{40:\overline{25}|}$, find the APV.

37. Write the counterpart of (7.2) for $(\bar{I}\bar{a})_x$.

*Sections 5-6**

38. Under the assumption that the lifetime is uniformly distributed within each year of age, write the relation between $\bar{a}_x$ and $\ddot{a}_x$. Make sure that your answer does not contradict (5.5) if we let $m \to \infty$. How will the relation look for $\delta = 0$? Explain why the answer should be expected.

[9] The nice idea of this exercise is taken from [43, Ex.C.4-5.].

39. In the same spreadsheet you provided in Exercise 11a, compute the APVs for the cases of monthly and quarterly payments.

40. A participant K. of a pension plan is now exactly 61. His salary increases by 3% every year on his birthday, and his annual salary for the age year starting now is \$60,000. He has been in service for 10 years prior to the present time. For simplicity, we assume that retirement takes place on a birthday. There are two causes for the termination of the plan before retirement: death and withdrawal. The corresponding probabilities may be estimated from the table below. In particular, one can see that the earliest retirement time is K.'s 63rd birthday, and the latest is his 65th.

Let $\delta = 0.04$. For any further information refer to the Illustrative Table.

Age x	l_x	$d_x^{(d)}$	$d_x^{(w)}$	$d_x^{(r)}$
61	100	1	9	0
62	90	2	8	0
63	80	3	7	10
64	60	4	0	6
65	50	0	0	50

(a) The retirement benefit (which is a part of the plan) pays a monthly annuity-due at an *annual* rate of 2% of the last salary times the number years of service at the time of retirement. Calculate the APV of the retirement part of the plan.

(b) The plan includes a death benefit of \$5,000 times the number of years in service paid at death, provided it occurs while K. is still in service. Calculate the APV of this part of the plan and the total APV.

(c) K. has an offer to move to another company. If he withdraws from the plan *now*, he will receive his own past contributions to the plan + interest, amounting to \$12,700. K. wants the new employer to compensate him for the loss of value in the pension plan. What should this compensation be? Assume for simplicity that K.'s contributions during the last 10 years were level and made at the beginning of each year.

41. We revisit Example 6.1-3. Assume that the annuity terminates if Harold dies or recovers, whichever comes first. However, if Harold attains the age of 65, he will be paid a continuous life annuity at rate $\tilde{c}$. Explain why in this case we do not need the assumption of the independence of the causes. Find the APV.

42. Let us look at (6.2). If $\mu_2 = 0$, then $\bar{a}_{x:y} = 1/(\mu_1 + \delta)$. Can it be predicted? What does it mean? If $\mu_1 = \mu_2 = 0$, then $\bar{a}_{x:y} = 1/\delta$. Show that it is obvious and could be obtained without considering the special exponential case. Analyze (6.3) in the same manner.

43. (a) Let $T_1(50)$ and $T_2(50)$ be independent and uniformly distributed on $[0, 50]$, and $\delta = 0.04$. Find $\bar{a}_{50:50}$ and $\bar{a}_{\overline{50:50}}$.

(b) Solve the problem if $T_1(50)$ is uniform on $[0, 40]$, and the other conditions are the same.

44. Write a general formula for the deferred continuous annuity paying at unit rate after the first death and until the second.

45. The lifetimes T_1 and T_2 of a husband and wife are independent and exponentially distributed with the same parameter μ. A special annuity pays continuously at unit rate during the period $[T_1, T_2]$, provided that $T_1 < T_2$. Find the APV for a given δ. (*Hint*: If you use the result of Exercise 44, the problem will not require lengthy calculations.)

46. Solve Exercise 45 in the case when the forces of mortality for T_1 and T_2 are different: μ_1 and μ_2. Make sure that the answer does not contradict the previous.

Chapter 10

Premiums and Reserves

1,2,3

In Chapters 2 and 3, we already considered premium determination; more precisely, premiums ensuring the solvency of insurance mechanisms. The models of those chapters concerned short-term insurance and *single premiums*; that is, premiums paid just one time at the beginning of the period under consideration.

As has been already mentioned in Section **2**.4, in the case of life insurance or future annuities, for example, pension plans, the policies are based not on single premiums but rather on *sequences* of premium payments to be carried out at certain rates. We will call such sequences of payments *premium annuities*. In this case, when determining a premium rate, the company usually proceeds from the total future random loss, which is the difference between the payments the company will provide in accordance with the contract and the total amount of premiums the company will receive until the contract is terminated. The corresponding models are explored in Section 1.

After we have considered problems concerning premiums, we will proceed—in Section 2—to the notion of reserve, which may be defined as the value of the future liability of the insurer.

1 PREMIUM ANNUITIES

1.1 General principles

Consider a risk portfolio. Denote by Z the present value of the future payments of the company, and by Y_P the present value of the total premium to be paid. The subscript P indicates that the latter present value depends on the premium rate or the premium rates if we deal with many different contracts with different premiums. So far, these rates are involved in Y_P implicitly, but soon we will write explicit expressions.

The present value of the total loss of the company is the r.v.

$$L_P = Z - Y_P. \tag{1.1.1}$$

For the reader who took Route 2, note that the ideology of premium principles in this case is close to what we discussed in Section **2**.4 but we apply these principles to the total loss L_P. However, to understand the material below, one does not need to know the material of the section mentioned.

We distinguish the three following approaches.

A. *The equivalence principle.* In this case, we require the losses to be equal to zero on the average, i.e.,

$$E\{L_P\} = 0.$$

Clearly, it implies

$$E\{Z\} = E\{Y_P\}, \tag{1.1.2}$$

which is an equation for P, so far given in an implicit way. Premiums based on this principle are called *benefit premiums* or *net premiums* (more precisely, *net premium rates*).

B. *The percentile principle.* We fix a small probability level γ, and require

$$P(L_P > 0) \leq \gamma. \tag{1.1.3}$$

Let $I_P = -L_P = Y_P - Z$, the profit of the portfolio. Then (1.1.3) is equivalent to the relation $P(I_P < 0) \leq \gamma$, or $q_\gamma(I_P) \geq 0$, where $q_\gamma(I_P)$ is the γ-quantile of I_P. Thus, in essence, we deal with the VaR criterion; see Section **1**.1.2.2.

Premiums based on (1.1.3) are called *percentile premiums.* (The words '*percentile*' and 'quantile' may be considered synonyms; the only difference is that in the former case probability is measured in percents.)

C. *The utility equivalence principle.* Given a utility function $u(x)$ and an initial surplus (or wealth) w corresponding to the portfolio, we compare the expected utility in the case of insurance (i.e., $E\{u(w-L_P)\}$) and the expected utility in the case where the insurance is not carried out (i.e., just $u(w)$ because in this case, the utility is not random). We require the former expected utility to be larger than the latter, or in the boundary case,

$$E\{u(w-L_P)\} = u(w). \tag{1.1.4}$$

In the case $u(x) = -e^{-\beta x}$, this is equivalent to

$$E\{\exp\{\beta L_P\}\} = 1. \tag{1.1.5}$$

(See similar calculations in **1**.3.2, where we had observed that in the exponential case the solution did not depend on w.) In this case, premiums are called *exponential.*

Certainly, if in (1.1.5) we set $\beta = 0$, then we come to an identity. However, if $\beta \neq 0$ but is close to zero, the exponential premium will be close to the net premium. Heuristically, it is seen immediately because for small β, the function $e^{\beta x} \approx 1 + \beta x$, and substituting this in (1.1.5), we would come to $E\{L_P\} \approx 0$.

To make it more accurate, let us appeal to (1.1.4), and instead of $u(x) = -e^{-\beta x}$, consider $u_1(x) = \frac{1}{\beta}(1 - e^{-\beta x})$, which people often do. This is just a linear transformation of the previous utility function, so the solution should be the same (see the first property of the EUM criterion in Section **1**.3.1.2). On the other hand, $u_1(x) \to x$ as $\beta \to 0$. So, in the limit, we deal with the equation $E\{w - L_P\} = w$, which again leads to $E\{L_P\} = 0$. In Section 1.5, we consider a particular example.

1.2 Benefit premiums: The case of a single risk

In this section, we follow the equivalence principle and consider *level premiums*, that is, premiums paid at a constant rate.

1.2.1 Net rate

Let us consider one contract and denote by Z the present value of the benefit payment. Let Y be the present value of the premium annuity (the total premium) *in the case where the premiums are paid at unit rate*. Then, if P is the level premium rate of the contract, the present value of the total premium to be paid is $Y_P = PY$.

Thus,

$$L_P = Z - PY, \tag{1.2.1}$$

and the requirement $E\{L_P\} = 0$ amounts to the relation $0 = E\{Z\} - PE\{Y\}$, or

$$P = \frac{E\{Z\}}{E\{Y\}}. \tag{1.2.2}$$

The numerator is the APV of a benefit payment, which we explored in Chapter 8. For example, for the whole life insurance on (x) with a unit benefit, the APV $E\{Z\} = \bar{A}_x$ or A_x, depending on whether the benefit is payable upon death or at the end of the year of death. If the insurance pays not one but, say, \$100,000 upon death, then in the right member of (1.2.2) we should set $E\{Z\} = 100{,}000\bar{A}_x$.

As another example, we may consider a deferred annuity on (x), say, a pension plan paying one unit of money at the beginning of each year starting with the age of $x+n$, provided that the annuitant attains this age. Then the numerator in (1.2.2) is $E\{Z\} = {}_{n|}\ddot{a}_x = {}_np_x \cdot v^n \ddot{a}_{x+n}$; see Section **9**.3.3.

The denominator in (1.2.2) depends on the period of premium payments, which in turn depends on the type of the contract. We distinguish the cases of continuous and discrete time premium annuities. In the former case, premiums are paid continuously, in the latter—at the beginning of each time interval.

In the case of the deferred annuity above and continuous time, the premiums are paid until death or the age $x+n$, whichever comes first. So, $E\{Y\} = \bar{a}_{x:\overline{n|}}$. In the discrete time model, $E\{Y\} = \ddot{a}_{x:\overline{n|}}$.

In the case of whole life insurance, the premiums are paid during the whole life, and hence $E\{Y\} = \bar{a}_x$ in the case of continuous payment, and $E\{Y\} = \ddot{a}_x$ if the premiums are paid at the beginning of each year.

Consider an n-year term life insurance contract. In this case, the premiums are paid only the first n years provided that the insured is alive. So, $E\{Y\} = \bar{a}_{x:\overline{n|}}$ or $\ddot{a}_{x:\overline{n|}}$ in the case of continuous and discrete payments, respectively.

The period of premium payments may be shorter than the term of the contract. For example, if for an n-year term insurance the premium annuity is provided during the first $m < n$ years, the premium will be equal to the ratio $A^{1}_{x:\overline{n|}}/\,\ddot{a}_{x:\overline{m|}}$.

Let us consider some particular cases.

EXAMPLE 1. Consider a whole life insurance on (x) with unit benefit payable at the moment of death and the continuous premium annuity. In this case,

$$P = \frac{\overline{A}_x}{\bar{a}_x}. \tag{1.2.3}$$

(For now we keep using P without additional indices and signs as a general notation for premiums. Later, for this particular type of benefit premiums, we will use the notation $\overline{P}_x$.)

Now, assume that the force of mortality $\mu_x(t) = \mu$ and the interest rate is δ. Then $\overline{A}_x = \mu/(\mu+\delta)$, and $\bar{a}_x = 1/(\mu+\delta)$; see Examples **8**.1.1-3, **9**.1.1-1a. Then (1.2.3) implies $P = \mu$, the mortality rate. It is noteworthy that in this case the premium does not depend on δ.

EXAMPLE 2. Let $v = 0.96$ and $x = 60$, and suppose the insured comes from the group for which $l_{60} = 91748$, $l_{61} = 90491$, $l_{62} = 89142$, $l_{63} = 87698$, $l_{64} = 86165$, and $l_{65} = 84518$. Find the premium rate for a 5-year term life insurance with a benefit of \$100,000. The premiums are paid at the beginning of each year until the contract termination. Thus, the rate is

$$100,000 \times \frac{A^{1}_{60:\overline{5}|}}{\ddot{a}_{60:\overline{5}|}}.$$

For the above data, in Example **9**.3.2-1, we have obtained $\ddot{a}_{60:\overline{5}|} \approx 4.50$ (by providing fairly lengthy calculations).

In Example **8**.2.3.2-2, we computed (applying a recursive procedure) $A^{1}_{60:\overline{5}|} \approx 0.069$.

Hence, for a unit benefit, $P \approx (0.069/4.50) \approx 0.0153$, and the rate is $0.0153 \times 100,000 = 1,530$. Thus, for a death benefit of $100,000$, the insured will pay at most $1,530 \times 5 = 7,650$. This is not surprising: the term is short and the probability that the company will pay nothing is $(l_{65}/l_{60}) \approx 0.92$.

EXAMPLE 3 ([151, N5][1]). For a special decreasing 15-year term life insurance on a person aged 30, you are given: mortality follows De Moivre's law with $\omega = 100$; the benefit payment is 2000 for the first 10 years and 1000 for the last 5 years; the death benefit is payable at the end of the year of death; $v = 0.95$. Calculate the level annual premium.

The remaining lifetime T is uniform on $[0,70]$, and the curtate lifetime K takes on values $0,...,69$ with equal probabilities $1/70$. The insurance may be represented as the sum of the following two insurances: a 10-year term insurance with a benefit of 1000, and a 15-year term insurance with the same benefit.

The APV for the first insurance is

$$1000 \cdot \sum_{k=0}^{9} v^{k+1}\frac{1}{70} = \frac{1000}{70} v \frac{1-v^{10}}{1-v} \approx 108.914.$$

For the second, it is

$$1000 \cdot \sum_{k=0}^{14} v^{k+1}\frac{1}{70} = \frac{1000}{70} v \frac{1-v^{15}}{1-v} \approx 145.678.$$

[1]Reprinted with permission of the Casualty Actuarial Society.

The APV for the premium annuity-due may be computed by the formula $\ddot{a}_{x:\overline{n}|} = \sum_{k=0}^{n-1} v^k \,{}_kp_x$ [see (**9**.3.2.5)]. In our case, this amounts to

$$\sum_{k=0}^{14}(1-k/70)v^k \approx 9.806.$$

(One may use formula (**9**.4.4) or just software.) Hence,

$$P \approx \frac{108.914+145.678}{9.806} \approx 25.963. \ \square$$

Route 1 ⇒ page 509

2,3

EXAMPLE 4. Consider two independent lifetimes $T_1 = T(x)$ and $T_2 = T(y)$ with constant forces of mortality $\mu_1 = 0.02$ and $\mu_2 = 0.04$, respectively. Let $\delta = 0.06$. In the continuous time case, find the premium rate for the following insurances with unit benefit.

(a) The benefit is payable at the moment of the first death. The force of mortality for $T = \min\{T_1, T_2\}$ is $\mu = \mu_1 + \mu_2 = 0.06$, and in accordance with what we have obtained in Example 1, the premium $P = \mu = 0.06$.

(b) The benefit is payable at the moment of the second death. Using again the formula from Example **8**.1.1-3, we have $\overline{A}_x = \mu_1/(\mu_1+\delta) = 0.25$. Similarly, $\overline{A}_y = 0.4$. Now, $\overline{A}_{x:y} = \mu/(\mu+\delta)$; so, $\overline{A}_{x:y} = 0.5$. By (**8**.4.2.2), $\overline{A}_{\overline{x:y}} = \overline{A}_x + \overline{A}_y - \overline{A}_{x:y} = 0.15$. Then the premium annuity $\bar{a}_{\overline{x:y}} = (1-\overline{A}_{\overline{x:y}})/\delta = 14.16\ldots$. Thus,

$$P = \frac{\overline{A}_{\overline{x:y}}}{\bar{a}_{\overline{x:y}}} \approx 0.0106.$$

(c) The benefit is payable at the moment of the second death, while the premiums are paid until the first. In this case,

$$P = \frac{\overline{A}_{\overline{x:y}}}{\bar{a}_{x:y}}.$$

Using now the result of Example **9**.1.1-1a, we have $\bar{a}_{x:y} = 1/(\mu+\delta) = 25/3$. Consequently, $P = \frac{0.15}{25/3} = 0.018$. $\square$

1,2,3

On notation. The traditional notation for premiums may be clarified by the following examples, where the expression in the parentheses indicates the type of insurance.

(a) $\overline{P}(A_x)$. A whole life insurance with unit benefit payable at the end of the year of death and premiums paid continuously.

(b) $P(\overline{A}_x)$. A whole life insurance with unit benefit payable at the time of death and premiums paid at the beginning of each year.

(c) ${}_mP(\overline{A}_x)$. An m-year payment whole life insurance with unit benefit payable at the time of death and premiums paid at the beginning of each year but not more than m times.

(d) $P({}_{m|}\ddot{a}_x)$. An m-year deferred annuity with premiums paid at the beginning of each year but not more than m times.

(Certainly, in all examples above—and all below—premiums are not paid after the insured dies.)

In the case when an insurance and the corresponding premium annuity are either both considered in continuous time, or both in discrete, we can simplify notation by denoting premiums in the following manner.

(a) P_x. A whole life insurance with unit benefit payable at the end of the year of death and premiums paid at the beginning of each year.

(b) $\overline{P}_x$. A whole life insurance with unit benefit payable at the time of death and premiums paid continuously.

(c) ${}_mP_x$. A whole life insurance with unit benefit payable at the end of the year of death and premiums paid at the beginning of each year, but not more than m times.

(d) $P^{1}_{x:\overline{n}|}$. An n-year term insurance with unit benefit payable at the end of the year of death with premiums paid at the beginning of each year.

In conclusion, we systematize *some* typical cases in the table below. For any insurance, we consider the cases of continuous and discrete payments in the same row. The bar over P indicates that the premiums are paid continuously. For the whole life insurance, we demonstrate the difference between benefits payable upon death and at the end of the year of death. For other cases, we consider, for brevity, one type. For the table to look simpler, we suppress the notations for premiums mentioned above.

TABLE 1.

Whole life insurance with a benefit payable at the moment of death	$\overline{P}=\overline{A}_x/\overline{a}_x$ and $P=\overline{A}_x/\ddot{a}_x$
Whole life insurance with a benefit payable at the end of the year of death	$\overline{P}=A_x/\overline{a}_x$ and $P=A_x/\ddot{a}_x$
n-Year term life insurance with a benefit payable at the moment of death	$\overline{P}=\overline{A}^{1}_{x:\overline{n}\rvert}/\overline{a}_{x:\overline{n}\rvert}$ and $P=\overline{A}^{1}_{x:\overline{n}\rvert}/\ddot{a}_{x:\overline{n}\rvert}$
n-Year endowment life insurance with a benefit payable at the moment of death	$\overline{P}=\overline{A}_{x:\overline{n}\rvert}/\overline{a}_{x:\overline{n}\rvert}$ and $P=\overline{A}_{x:\overline{n}\rvert}/\ddot{a}_{x:\overline{n}\rvert}$
m-Year payment whole life insurance with a benefit payable at the moment of death	$\overline{P}=\overline{A}_x/\overline{a}_{x:\overline{m}\rvert}$ and $P=\overline{A}_x/\ddot{a}_{x:\overline{m}\rvert}$
m-Year payment n-year term life insurance with a benefit payable at the moment of death	$\overline{P}=\overline{A}^{1}_{x:\overline{n}\rvert}/\overline{a}_{x:\overline{m}\rvert}$ and $P=\overline{A}^{1}_{x:\overline{n}\rvert}/\ddot{a}_{x:\overline{m}\rvert}$
m-Year payment n-year term endowment life insurance with a benefit payable at the moment of death	$\overline{P}=\overline{A}_{x:\overline{n}\rvert}/\overline{a}_{x:\overline{m}\rvert}$ and $P=\overline{A}_{x:\overline{n}\rvert}/\ddot{a}_{x:\overline{m}\rvert}$
n-Year pure endowment life insurance	$\overline{P}={}_nE_x/\overline{a}_{x:\overline{n}\rvert}$ and $P={}_nE_x/\ddot{a}_{x:\overline{n}\rvert}$
n-Year deferred whole life annuity-due	$\overline{P}={}_nE_x\cdot\ddot{a}_{x+n}/\overline{a}_{x:\overline{n}\rvert}$ and $P={}_nE_x\cdot\ddot{a}_{x+n}/\ddot{a}_{x:\overline{n}\rvert}$

1.2.2 The case where "Y is consistent with Z"

We saw in Section **9**.2.2 that each type of level annuity corresponds to a certain type of insurance, which is reflected by the formulas

$$Y = \frac{1 - Z_Y}{\delta} \text{ and } Y = \frac{1 - Z_Y}{d}$$

for the cases of continuous and discrete annuities, respectively. Unlike in the formulas (**9**.2.2.1) and (**9**.2.2.5), here we mark Z by the index Y to avoid confusion: the insurance that corresponds to Y may not coincide with the insurance whose present value is denoted by Z in the formulas (1.2.1) and (1.2.2) above. In other words, in general, $Z_Y \neq Z$.

For example, for an n-year term insurance, the premium annuity paid during the term $[0,n]$ corresponds to an n-year *endowment* insurance rather than to an n-year *term* insurance; see Section **9**.3.2. In particular, in the continuous case, $E\{Z\} = \overline{A}^{\,1}_{x:\overline{n}|}$, while $E\{Z_Y\} = \overline{A}_{x:\overline{n}|}$.

Also, for Z to be equal to Z_Y, the premium annuity and the insurance should both correspond to the same model with regard to time: either the continuous or discrete time model. We will call such cases *fully continuous time* and *fully discrete time*, respectively.

So, we will talk about *consistency* between Z and Y if $Z = Z_Y$. Below, we consider four such cases.

First, this are two types of whole life insurance: fully continuous and fully discrete. In these cases, $E\{Z\} = E\{Z_Y\} = \overline{A}_x$ and $E\{Z\} = E\{Z_Y\} = A_x$, respectively.

The next two cases are two types of n-year endowment insurance: fully continuous and fully discrete. In these cases, the respective relations are $E\{Z\} = E\{Z_Y\} = \overline{A}_{x:\overline{n}|}$ and $E\{Z\} = E\{Z_Y\} = A_{x:\overline{n}|}$.

When dealing with a consistency model, we may simplify some representations.

Let the random time moment Ψ play the same role as it played in Chapters 8 and 9. For example, in the continuous-time scheme, for a whole life insurance $\Psi = T(x)$. For an n-year endowment insurance, $\Psi = \min\{T, n\}$. Then, in the case of consistency,

$$Z = e^{-\delta\Psi} \text{ and } Y = \frac{1 - Z_Y}{\delta} = \frac{1 - Z}{\delta}.$$

(See Sections **8**.1.1, **9**.2.2.)

Then the loss

$$L_P = Z - PY = e^{-\delta\Psi} - P\frac{1 - e^{-\delta\Psi}}{\delta} = e^{-\delta\Psi}\left(1 + \frac{P}{\delta}\right) - \frac{P}{\delta}.$$

Writing $\overline{P}$ instead of P to emphasize that we are considering the *fully continuous time case*, we have

$$L_P = e^{-\delta\Psi}\left(1 + \frac{\overline{P}}{\delta}\right) - \frac{\overline{P}}{\delta} = Z\left(1 + \frac{\overline{P}}{\delta}\right) - \frac{\overline{P}}{\delta}. \tag{1.2.4}$$

We will use both representations in (1.2.4), whichever is more convenient.

Similarly, in the *fully discrete time case*,

$$L_P = v^{\Psi}\left(1 + \frac{P}{d}\right) - \frac{P}{d} = Z\left(1 + \frac{P}{d}\right) - \frac{P}{d}. \tag{1.2.5}$$

Above, the premium rates P or $\overline{P}$ were arbitrary. Let us consider net rates. In the consistency case, we can make use of the relations

$$\ddot{a} = \frac{1-A}{d} \text{ or } \bar{a} = \frac{1-\bar{A}}{\delta}. \tag{1.2.6}$$

(The indices are skipped since it may concern different types of insurance.) The general representations for the premium in this case are the following:

$$P = \frac{A}{\ddot{a}} = \frac{A}{(1-A)/d} = \frac{dA}{1-A}, \text{ and } \overline{P} = \frac{\bar{A}}{\bar{a}} = \frac{\bar{A}}{(1-\bar{A})/\delta} = \frac{\delta\bar{A}}{1-\bar{A}}. \tag{1.2.7}$$

Cases in point are presented in the table below.

TABLE 2.

Whole life insurance with a benefit payable at the moment of death	$\overline{P} = \dfrac{\delta\bar{A}_x}{1-\bar{A}_x}$
Whole life insurance with a benefit payable at the end of the year of death	$P = \dfrac{dA_x}{1-A_x}$
n-Year endowment life insurance with a benefit payable at the moment of death	$\overline{P} = \dfrac{\delta\bar{A}_{x:\overline{n}\rceil}}{1-\bar{A}_{x:\overline{n}\rceil}}$
n-Year endowment life insurance with a benefit payable at the end of the year of death	$P = \dfrac{dA_{x:\overline{n}\rceil}}{1-A_{x:\overline{n}\rceil}}$

EXAMPLE 1. Consider the 5-year endowment insurance from Example **8**.2.4.2-1. For the data from this example, $A_{60:\overline{5}\rceil} = 0.820$. Then we can immediately write that

$$P = P_{60:\overline{5}\rceil} = \frac{dA_{60:\overline{5}\rceil}}{1-A_{60:\overline{5}\rceil}} = \frac{(1-0.96)\cdot 0.820}{1-0.820} \approx 0.182. \ \square$$

1.2.3 Variances

If Y and Z are consistent in the sense defined above, we can use (1.2.4) and (1.2.5), which implies that

$$Var\{L_P\} = \left(1+\frac{\overline{P}}{\delta}\right)^2 Var\{Z\} \text{ and } Var\{L_P\} = \left(1+\frac{P}{d}\right)^2 Var\{Z\} \tag{1.2.8}$$

for the cases of continuous and discrete time, respectively. The variance $Var\{Z\}$ may be computed by the double rate rule; see Chapter 8. It is worth emphasizing that formulas (1.2.8) are true for any premium, not only for the net one.

In the case of the net premium, combining (1.2.6) and (1.2.7), we can write that

$$1+\frac{\overline{P}}{\delta} = \frac{1}{1-\bar{A}} = \frac{1}{\delta\bar{a}} \text{ and } 1+\frac{P}{d} = \frac{1}{1-A} = \frac{1}{d\ddot{a}}.$$

Thus, in the case of the benefit (net) premium, we can rewrite (1.2.8) as

$$Var\{L_P\} = \frac{1}{(\delta\bar{a})^2}Var\{Z\} = \frac{1}{(1-\bar{A})^2}Var\{Z\} \text{ in the continuous case;} \qquad (1.2.9)$$

$$Var\{L_P\} = \frac{1}{(d\ddot{a})^2}Var\{Z\} = \frac{1}{(1-A)^2}Var\{Z\} \text{ in the discrete case.} \qquad (1.2.10)$$

EXAMPLE 1. Let us continue the previous Example 1.2.2-1, where we have computed (using certain data) that the net premium $P = P_{60:\overline{5}|} = 0.182\ldots$. In Example **8**.2.4.2-1, we got that for the insurance under consideration, $A_{60:\overline{5}|} = 0.820$ and $Var\{Z\} = {}^2A_{60:\overline{5}|} - (A_{60:\overline{5}|})^2 \approx 0.0014$. Hence, by (1.2.10), $Var\{L_P\} \approx \frac{1}{(1-0.82)^2}0.0014 \approx 0.043$. So, we deal with the loss L with zero mean and the standard deviation $\sigma_L \approx 0.208$.

EXAMPLE 2. Consider the whole life insurance in the case of continuous time and a constant force of mortality μ. We saw in Example 1.2.1-1 that, for such an insurance, the net premium $P = \mu$. In Example **8**.1.1-3, we came to $Var\{Z\} = \frac{\mu}{\mu+2\delta} - \left(\frac{\mu}{\mu+\delta}\right)^2 = \frac{\mu\delta^2}{(\mu+2\delta)(\mu+\delta)^2}$. Then, by (1.2.8),

$$Var\{L_P\} = \left(1+\frac{\mu}{\delta}\right)^2 \cdot \frac{\mu\delta^2}{(\mu+2\delta)(\mu+\delta)^2} = \frac{\mu}{\mu+2\delta}.$$

Let, say, $\mu = 0.06$, and $\delta = 0.06$. Then $Var\{L_P\} = 1/3$, and the standard deviation $\sigma_L \approx 0.57$. So, if the benefit equals \$1000, then the loss deviates from zero on the average approximately by \$600. It is certainly too large but we should remember that we consider a net premium. The situation may change dramatically if we add a sufficient security loading, which we will see in Section 1.4.2. □

Route 1 ⇒ page 517

2,3

In the general case, computation of variances is more difficult.

EXAMPLE 3. Consider an n-year term insurance on (x) with unit benefit payable at the moment of death. For simplicity, assume that the premium is paid continuously. Set $T = T(x)$. Since the insurance that corresponds to the premium annuity is the n-year endowment, the present values

$$Z = \begin{cases} e^{-\delta T} & \text{if } T \le n, \\ 0 & \text{if } T > n, \end{cases} \quad \text{and} \quad Z_Y = \begin{cases} e^{-\delta T} & \text{if } T \le n, \\ e^{-\delta n} & \text{if } T > n, \end{cases}$$

and they are not the same. Set

$$Z_1 = \begin{cases} 0 & \text{if } T \le n, \\ e^{-\delta n} & \text{if } T > n. \end{cases}$$

Then $Z_Y = Z + Z_1$, and

$$L_P = Z - P\frac{1-Z_Y}{\delta} = Z - P\frac{1-Z-Z_1}{\delta} = Z\left(1+\frac{P}{\delta}\right) + \frac{P}{\delta}Z_1 - \frac{P}{\delta}.$$

Consequently,

$$Var\{L_P\} = \left(1+\frac{P}{\delta}\right)^2 Var\{Z\} + \left(\frac{P}{\delta}\right)^2 Var\{Z_1\} + 2\left(1+\frac{P}{\delta}\right)\frac{P}{\delta}\,Cov\{Z,Z_1\}. \quad (1.2.11)$$

Since $Z \cdot Z_1 = 0$,

$$Cov\{Z,Z_1\} = -E\{Z\}E\{Z_1\} = -A^{\;1}_{x:\overline{n}|} \cdot {}_nE_x, \quad (1.2.12)$$

while

$$Var\{Z\} = {}^2A^{\;1}_{x:\overline{n}|} - (A^{\;1}_{x:\overline{n}|})^2, \; Var\{Z_1\} = v^{2n}\,{}_np_x(1-{}_np_x). \quad (1.2.13)$$

Combining the last three relations, we can compute the variance for *any* P. □

1.2.4 Premiums paid m times a year

The premiums we now consider are called *true fractional* and concern the case when they are paid by m installments of equal size at the beginning of each m-thly period. The *annual* premium rate is denoted by $P^{(m)}$ (with other indices if needed). In other words, the premium paid at the beginning of each "m-th" is equal to $P^{(m)}/m$. We do not suppose, however, any adjustment in paying benefits.

In this case, in the general formula (1.2.2), we should hold the same numerator, while $E\{Y\}$ is the corresponding APV of the type $a^{(m)}$ that we considered in Section **9**.5.

EXAMPLE 1. Consider a whole life insurance on (x) with benefits payable upon death, while premiums are paid m-thly. Then the benefit *annual* true fractional premium is $P^{(m)} = \bar{A}_x \big/ \ddot{a}_x^{(m)}$. [The complete traditional notation for such a premium is $P^{(m)}(\bar{A}_x)$.]

Let us compare it with the *annual* rate when payments are provided *one time a year*, that is, with $P = \bar{A}_x/\ddot{a}_x$. [The complete notation for such a premium would be $P(\bar{A}_x)$.]

In Section **9**.5, we showed that $\ddot{a}_x^{(m)} = \alpha(m)\ddot{a}_x - \beta(m)$, where $\alpha(m)$ and $\beta(m)$ are defined in (**9**.5.6), and

$$\alpha(m) \approx 1, \; \beta(m) \approx \frac{m-1}{2m}. \quad (1.2.14)$$

Hence,

$$P^{(m)} = \frac{\bar{A}_x}{\ddot{a}_x^{(m)}} = \frac{\bar{A}_x}{\alpha(m)\ddot{a}_x - \beta(m)} = \frac{\bar{A}_x}{\ddot{a}_x} \cdot \frac{1}{\alpha(m) - \beta(m)/\ddot{a}_x} = P \cdot \frac{1}{\alpha(m) - \beta(m)/\ddot{a}_x}.$$

Say, for $m = 12$, we would have

$$P^{(12)} = kP, \text{ where the coefficient } k \approx \frac{1}{1 - 11/(24\ddot{a}_x)}. \quad (1.2.15)$$

Since $\ddot{a}_x \geq 1$ (why?), the maximal value of the "correction" coefficient k is not larger than $\frac{1}{1-11/24} \approx 1.85$, but this is not an accurate bound. As a rule, $\ddot{a}_x$ is essentially larger than 1, which we saw in many examples. For instance, for $\ddot{a}_x = 5$ we would have $k \approx 1.1$, and $\ddot{a}_x = 10$ would lead to $k \approx 1.05$. In any case, if the insured pays monthly, she/he should pay more. See also Exercise 24. □

1.2.5 Combinations of insurances

In practice, one can see a large number of different insurance forms—in particular, various combinations of classical insurances we considered above. Nevertheless, the rule for net premium calculations for all these policies is the same. One should compute the APV of what the company pays, i.e., $E\{Z\}$, and the APV of the premium annuity, i.e., $E\{Y_P\}$. Then the equality $E\{Z\} = E\{Y_P\}$ will be an equation for the premium.

EXAMPLE 1. (a) An n-year deferred whole life annuity is issued to (x) with the provision for unit death benefit in the case of death before n. For the premium annuity, we have $E\{Y_P\}=P\ddot{a}_{x:\overline{n}|}$, while $E\{Z\}={}_{n|}\ddot{a}_x+A^{\,1}_{x:\overline{n}|}=v^n\,{}_np_x\ddot{a}_{x+n}+A^{\,1}_{x:\overline{n}|}$. Thus, $P=\left({}_nE_x\ddot{a}_{x+n}+A^{\,1}_{x:\overline{n}|}\right)/\,\ddot{a}_{x:\overline{n}|}$. Certainly, we could also consider it as if the insured bought the life insurance and the deferred annuity separately. In this case, we would compute the two premiums separately and would add them up, which would take a bit longer.

(b) Now, in the same deferred annuity plan, let the death benefit (in the case of death before n) consist in the return of the accumulated premium with interest. For the APV of the premium annuity, we write the same as above: $E\{Y_P\} = P\ddot{a}_{x:\overline{n}|}$.

The APV of the return may be computed directly, but one may reason as follows. If the insurer had returned the premium in any case, the APV would have been equal to $P\ddot{a}_{x:\overline{n}|}$, that is, exactly to the APV of what the insured paid. As a matter of fact, the insurer does not return the premium if the insured attains the age of $x+n$. On the other hand, the present value of the series of unit payments until time n is $\ddot{a}_{\overline{n}|} = 1+...+v^{n-1} = \frac{1}{d}(1-v^n)$. (See also Section **0**.8.3.) Hence, the expected present value of the part of the premium, which the insurer does not return, is $P\cdot{}_np_x\cdot\ddot{a}_{\overline{n}|}$, and the APV of the return is $P(\ddot{a}_{x:\overline{n}|}-{}_np_x\cdot\ddot{a}_{\overline{n}|})$.

Thus, $E\{Z\} = v^n\,{}_np_x\ddot{a}_{x+n} + P(\ddot{a}_{x:\overline{n}|}-{}_np_x\cdot\ddot{a}_{\overline{n}|})$. Setting $E\{Y_P\} = E\{Z\}$, we readily compute that

$$P = \frac{v^n\,{}_np_x\ddot{a}_{x+n}}{{}_np_x\cdot\ddot{a}_{\overline{n}|}} = \frac{v^n\ddot{a}_{x+n}}{\ddot{a}_{\overline{n}|}}. \qquad (1.2.16)$$

See also Exercise 25.

(c) (A similar example is contained in [42, 5.6].) Let us consider the case when the company returns the premium without interest. This means that if $K=k<n$, the company pays $P(k+1)$ units at the end of the year of death. This amounts to an n-year term regularly increasing insurance with the APV per unit of the premium equal to $(IA)^{\,1}_{x:\overline{n}|}$. We considered the precise formula for $(IA)^{\,1}_{x:\overline{n}|}$ in Exercise **8**.45.

Thus, $P\ddot{a}_{x:\overline{n}|} = v^n\,{}_np_x\ddot{a}_{x+n} + P(IA)^{\,1}_{x:\overline{n}|}$, and

$$P = \frac{{}_nE_x\ddot{a}_{x+n}}{\ddot{a}_{x:\overline{n}|}-(IA)^{\,1}_{x:\overline{n}|}}. \quad \square$$

In conclusion, it makes sense again to emphasize that the premiums we considered above are net premiums. To take into account the risk that the insurer incurs, a security loading should be incorporated into the premium. An explicit (and already known to us) way to do that will be discussed in the next section. One implicit way is to compute $E\{Z\}$ and $E\{Y\}$ in the representation $P = E\{Z\}/E\{Y\}$ with different interest rates. Taking a larger rate for the premium annuity (in comparison with the rate for the insurance part), we make the denominator smaller, and hence the premium larger. This is similar to what we can see in any bank: the loan rate is higher than the deposit rate, i.e., the rate of return on customers' investments.

1.3 Accumulated values

In this section, we generalize the notion of accumulated value from Section **0**.8.4. The characteristic at which we will arrive may be interpreted as the expected accumulated value of a temporary annuity, *given that the annuitant survives* the term of the annuity. To explain what it means rigorously, we use the notion of net premium, and this is why we consider accumulated value in the current chapter concerning premiums.

We begin with an illustrative but somewhat non-rigorous way of reasoning. Assume that an individual aged x invests in a fund (for example, a pension fund) a unit of money at the beginning of each year during n years or until death. If Y is the present value of this series of payments, then, as we know, $E\{Y\} = \ddot{a}_{x:\overline{n}|}$. The value of the same investment from the standpoint of the time $t = n$ is $S = v^{-n}Y$ (see also Section **0**.8.4). If the future lifetime T of the investor were larger n with probability one, then Y would have been equal to $\ddot{a}_{\overline{n}|}$ and S would have been equal to $v^{-n}\ddot{a}_{\overline{n}|} = \ddot{s}_{\overline{n}|}$, in accordance with (**0**.8.4.2).

However, T may be less than n, and the expected value of S is

$$E\{S\} = \frac{1}{v^n}E\{Y\} = \frac{1}{v^n}\ddot{a}_{x:\overline{n}|}.$$

Suppose that the fund consists of the contributions of k independent individuals of the same type—namely, all investors begin to invest at the same time, having the same age x and the same distribution of the future lifetimes. Let S_i be the accumulated value for the ith investor at time $t = n$, and let $W = S_1 + ... + S_k$.

At the end of the term, the total accumulated sum W will be distributed between the investors who attain the age of $x+n$. So, an investor who survives the term will not get W/k, but rather W/N_k, where N_k is the number of investors who will be alive at the end of the term.

Thus, the amount obtained per survivor is

$$\widetilde{S} = \frac{S_1 + ... + S_k}{N_k}. \tag{1.3.17}$$

Because $E\{N_k\} = k \cdot {}_np_x$, if we replace the denominator and the numerator in (1.3.17) by their expected values, we will come to the quantity

$$\frac{kv^{-n}\ddot{a}_{x:\overline{n}|}}{k \cdot {}_np_x} = \frac{\ddot{a}_{x:\overline{n}|}}{v^n \cdot {}_np_x}.$$

The last characteristic is denoted by $\ddot{s}_{x:\overline{n}|}$. It is called the *actuarial accumulated value* at the end of the term of an n-year temporary annuity-due and is interpreted as the expected accumulated value that is available if the investor survives.

However, the expected value of the ratio of r.v.'s is *not* equal to the ratio of the expected values of these r.v.'s. Therefore, the characteristic at which we arrived is *not* equal to $E\{\widetilde{S}\}$, and the interpretation above remains unclear.

To make it more understandable (and more rigorous), let us reason in a slightly different way. Let a r.v. η be what a particular investor will get at the end of the term, and let T be the investor's future lifetime. Since $\eta = 0$ for $T < n$, we write

$$E\{\eta\} = E\{\eta \,|\, T \geq n\}P(T \geq n) + 0 \cdot P(T < n) = E\{\eta \,|\, T \geq n\} \cdot {}_np_x. \tag{1.3.18}$$

The present value of η is $v^n\eta$. Then the present value of the investor's loss is the r.v. $L = Y - v^n\eta$, where Y is the present value of what the investor will pay by time n.

Let us now view the unit investment paid at the beginning of each year as an annual premium paid in order to get the benefit η at the end of the term. *Assume that the benefit* η *corresponds to the case when the unit premiums above are benefit premiums.* In other words, *we define* η *as the benefit such that* $E\{L\} = 0$ *in the case of unit premiums.* Then we may write

$$0 = E\{L\} = E\{Y\} - v^nE\{\eta\} = \ddot{a}_{x:\overline{n}|} - v^nE\{\eta \,|\, T \geq n\}{}_np_x.$$

Hence,

$$E\{\eta \,|\, T \geq n\} = \frac{\ddot{a}_{x:\overline{n}|}}{v^n \cdot {}_np_x} = \ddot{s}_{x:\overline{n}|}.$$

So, the quantity $\ddot{s}_{x:\overline{n}|}$ is $E\{\eta \,|\, T \geq n\}$, where η is defined as above. Thus, $\ddot{s}_{x:\overline{n}|}$ is indeed the expected value of what the investor will get *if* she/he survives but in the case of benefit premiums.

1.4 Percentile premiums 1,2,3

1.4.1 The case of a single risk

In this section, we show that in the case of a single contract, the percentile premium principle may lead to results contradicting common sense. Namely, this principle may determine the same premium for two insurances among which one deals with larger losses than the other.

We consider particular examples below but it is important to understand that the existence of such examples is by no means surprising. We noted already that the percentile principle coincides, in essence, with the VaR (or quantile) criterion considered in Section **1**.1.2.2. We have shown in Example **1**.1.2.2-2 that the function $q_\gamma(X)$ is not strictly monotone in the sense that there exist pairs of r.v.'s X and Y such that $P(X \geq Y) = 1$, $P(X > Y) > 0$, but $q_\gamma(X) = q_\gamma(Y)$.

To make it obvious, consider one more example.

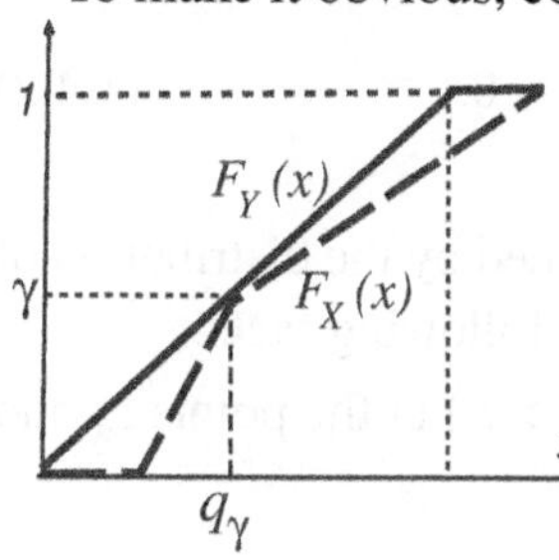

FIGURE 1.

EXAMPLE 1. Let r.v.'s X and Y be such that $P(X \geq Y) = 1$, and their d.f.'s are shown in Fig.1. The graph of $F_X(x)$ lies under the graph of $F_Y(x)$ since $P(X \leq x) \leq P(Y \leq x)$. Clearly, $P(X > Y) > 0$, since otherwise $P(X = Y) = 1$, and $F_X(x)$ would coincide with $F_Y(x)$. However, for the γ chosen in Fig.1, $q_\gamma(X) = q_\gamma(Y)$. $\square$

A more detailed discussion may be found in Section **1**.1.2.2.

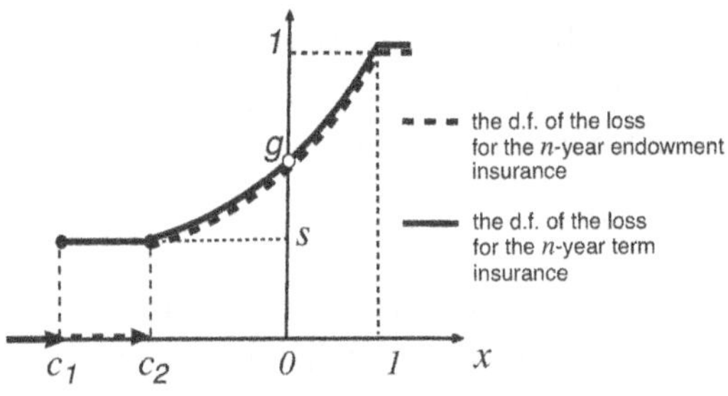

FIGURE 2.

The next example is similar to Example 1 and concerns some actual forms of insurance.

EXAMPLE 2[2]. For a (remaining) lifetime $T = T(x)$, consider an n-year term and an n-year endowment insurance on the fully continuous basis (for benefit and premium payments as well). For our example to be non-trivial, assume that $P(T > n) > 0$.

For the premiums below, we use the notation π instead of P, in order to avoid confusion with the probability symbol P.

Set $\delta = 0$. This does not change the essence of the matter but will make the example more illustrative. In this case, the present value of a unit payment at time t is one, and the present value of the premium accumulated by time t is πt.

Denote the losses for the insurances under consideration by $L_\pi^{(1)}$ and $L_\pi^{(2)}$, respectively. For the n-year term insurance, we have

$$L_\pi^{(1)} = L_\pi^{(1)}(T) = \begin{cases} 1 - \pi T & \text{if } T \leq n, \\ 0 - \pi n & \text{if } T > n. \end{cases} \tag{1.4.1}$$

The endowment insurance coincides with the term insurance if $T \leq n$, and it pays one unit when the insured attains the age of $x+n$. Thus,

$$L_\pi^{(2)} = L_\pi^{(2)}(T) = \begin{cases} 1 - \pi T & \text{if } T \leq n, \\ 1 - \pi n & \text{if } T > n. \end{cases}$$

The particular forms of these functions are not important for us (for $\delta > 0$ they will be more complicated), but what is important is that $L_\pi^{(1)}(T) = L_\pi^{(2)}(T)$ for $T \leq n$.

If $T > n$, then $L_\pi^{(1)}(T)$ takes only one value $c_1 = -\pi n$, and $L_\pi^{(2)}(T)$ takes only one value $c_2 = 1 - \pi n > c_1$. Thus,

$$L_\pi^{(2)} \geq L_\pi^{(1)}, \text{ and } P\left(L_\pi^{(2)} > L_\pi^{(1)}\right) = P(T > n) > 0. \tag{1.4.2}$$

The d.f.'s of the r.v.'s $L_\pi^{(1)}(T)$ and $L_\pi^{(2)}(T)$ are completely determined by the distribution of T; the typical graphs are shown in Fig.2. These graphs reflect the following facts:

(i) $L_\pi^{(1)}(0) = L_\pi^{(2)}(0) = 1$; (ii) the d.f.'s make jumps of $s = P(T > n)$ at the points c_1 and c_2, respectively; (iii) for $x > c_2$, the d.f.'s coincide.

[2]This example is based on the idea of Examples 6.2.3-4 in [19].

The requirement $P(L_\pi^{(1)}(T) > 0) \leq \gamma$ is equivalent to the inequality $P(L_\pi^{(1)}(T) \leq 0) \geq g$, where $g = 1-\gamma$. We see from the graphs that, if $g > s$, then the g-quantiles for both r.v.'s are the same. This means that for such g's (or γ's), the VaR approach identifies both insurances under consideration.

In light of (1.4.2), this contradicts common sense and says about the essential non-flexibility of the percentile approach.

▶ Let us state it in more detail. Consider **any** $\gamma \in (0, 1-s)$ and denote by t_γ the γ-quantile of T. Since $\gamma < 1-s = 1-P(T>n) = P(T \leq n)$, for the γ chosen we have $t_\gamma \leq n$.

From (1.4.1) it follows that for $L_\pi^{(1)}(T) > 0$ we should have $T \leq n$. In this case, $L_\pi^{(1)}(T) = 1-\pi T$, and hence the event $\{L_\pi^{(1)}(T) > 0\} = \{T < 1/\pi,\ T \leq n\}$. Thus, $P(L_\pi^{(1)}(T) > 0) \leq \gamma$ if and only if $P(T < 1/\pi,\ T \leq n) \leq \gamma$. Since $t_\gamma \leq n$, this is equivalent to $\frac{1}{\pi} \leq t_\gamma$, or

$$\pi \geq \pi_\gamma, \text{ where } \pi_\gamma = 1/t_\gamma.$$

Thus, π_γ is the minimal acceptable premium for the insurer.

Because $L_\pi^{(1)}(t) = L_\pi^{(2)}(t)$ for $t \leq n$, and we have chosen $t_\gamma < n$, the **same** π_γ is the minimal acceptable premium for the second insurance. ◀ □

1.4.2 The case of many risks. Normal approximation

The situation changes dramatically when we consider a portfolio of many risks. In this case, the total loss is the sum of the losses coming from separate contracts, and under mild conditions, for a large number of contracts, the distribution of the total loss is asymptotically normal. As we saw in Section **1**.1.2.5, when dealing only with normal r.v.'s, we cannot build examples similar to those from the previous section. So, for normal r.v.'s, "everything is fine", and we can use quantiles as a criterion.

The asymptotic normality of sums of r.v.'s takes place for a large class of dependencies between separate terms, but we will restrict ourselves to the independence case.

Consider a portfolio of n independent contracts of the *same* type, denoting by $L^{(i)}$, $i = 1, ..., n$, the loss corresponding to the ith contract. By assumption, the r.v. $L^{(i)}$ are i.i.d. Then, it is natural to assume that the premium rate is the same for all contracts. We will consider the continuous and discrete time cases simultaneously and denote the premium rate for both cases by π. Then

$$L^{(i)} = Z_i - \pi Y_i, \tag{1.4.3}$$

where Z_i and Y_i are the present values of the benefit payments and the premium annuity with unit rate, respectively, for the ith contract.

We will consider the case where Y_i is *consistent* with Z_i as it was defined in Section 1.2.2. In particular, this means that while the pairs (Z_i, Y_i) are independent (as was assumed), but "inside" each pair, for fixed i, the r.v.'s Z_i, Y_i are strongly dependent. More precisely, $Y_i = (1-Z_i)/\delta$ or $Y_i = (1-Z_i)/d$ in the continuous and discrete time case, respectively. From these relations, it follows that

$$L^{(i)} = Z_i\left(1+\frac{\pi}{\delta}\right) - \frac{\pi}{\delta} \text{ or } L^{(i)} = Z_i\left(1+\frac{\pi}{d}\right) - \frac{\pi}{d}, \tag{1.4.4}$$

depending on whether we consider the continuous or discrete time case. Below, we will use the symbol δ for both cases when it does not cause confusion.

Let us consider a γ-percentile premium; that is, let us set $L = L_n = L^{(1)} + ... + L^{(n)}$, and impose the condition

$$P(L_n > 0) \leq \gamma. \tag{1.4.5}$$

In order to use normal approximation, set $m = E\{L^{(i)}\}$, $\sigma^2 = Var\{L^{(i)}\}$, and

$$L_n^* = \frac{L_n - mn}{\sigma\sqrt{n}},$$

i.e., L_n^* is a normalized sum. Formally, we do not require the mean loss m to be negative. However, we should expect that this will be the case if we choose the premium large enough for the insurance to be profitable with a large probability.

Furthermore,

$$P(L_n > 0) = P\left(\frac{L_n - mn}{\sigma\sqrt{n}} > \frac{0 - mn}{\sigma\sqrt{n}}\right) = P\left(L_n^* > -\frac{m\sqrt{n}}{\sigma}\right).$$

Then, for (1.4.5) to hold, we should have $P\left(L_n^* \leq -\frac{m\sqrt{n}}{\sigma}\right) \geq 1 - \gamma$. Therefore, up to the accuracy of normal approximation, we should have $-\frac{m\sqrt{n}}{\sigma} \geq q_{1-\gamma,s}$, where $q_{1-\gamma,s}$ is the $(1-\gamma)$-quantile of the standard normal distribution. Eventually, we write it as

$$-m \geq \frac{q_{1-\gamma,s}\,\sigma}{\sqrt{n}}. \tag{1.4.6}$$

From (1.4.3) it follows that $m = A - \pi a$, where A and a are the respective expectations $E\{Z_i\}$ and $E\{Y_i\}$. Note that in our consistency case, $a = (1 - A)/\delta$.

From (1.4.4) we get that $\sigma = H\left(1 + \frac{\pi}{\delta}\right)$, where $H^2 = Var\{Z_i\}$. Then (1.4.6) may be rewritten as

$$\pi a - A \geq \frac{q_{1-\gamma,s}H}{\sqrt{n}}\left(1 + \frac{\pi}{\delta}\right).$$

Solving this inequality with respect to π, we obtain that

$$\pi \geq \pi_\gamma = \frac{A + \frac{q_{1-\gamma,s}H}{\sqrt{n}}}{a - \frac{q_{1-\gamma,s}H}{\delta\sqrt{n}}}, \tag{1.4.7}$$

provided that the denominator above is positive.

By construction, π_γ is the *minimal acceptable* γ*-percentile premium.*

Now, note that the benefit (net) premium is

$$\pi_{\text{net}} = \frac{A}{a}.$$

Let us look at (1.4.7). If $\gamma < \frac{1}{2}$, then the standard normal quantile $q_{1-\gamma,s} > 0$, and for $H > 0$,

$$\pi_\gamma > \frac{A}{a} = \pi_{\text{net}}.$$

We see also that

$$\pi_\gamma \to \pi_{\text{net}} \quad \text{as} \quad n \to \infty.$$

Both facts are quite natural and expected.

For the discrete time case, we should replace δ by d in (1.4.7).

EXAMPLE 1. We continue with Example 1.2.3-1 which was the last in the series of examples using the same data. In this example, we consider a 5-year endowment insurance for which the premium annuity is consistent with the insurance benefits. Time is discrete.

So far we know that, in the notation of this section, $A = A_{60:\overline{5}|} = 0.820$, $\pi_{\text{net}} = P_{60:\overline{5}|} = 0.182$, and $H^2 = Var\{Z\} = 0.0014$ (all with the accuracy chosen). So, $H \approx 0.037$.

The particular values above were obtained for $v = 0.96$. Hence, $d = 0.04$ and $a = \frac{1}{d}(1 - A) \approx 4.5$.

Consider $n = 100$ independent contracts of the same type. First, let $\gamma = 0.05$. Then $q_{1-\gamma,s} \approx 1.64$, and by (1.4.7),

$$\pi_\gamma \approx \frac{0.82 + 1.64 \cdot 0.037/\sqrt{100}}{4.5 - 1.64 \cdot 0.037/(0.04\sqrt{100})} \approx 0.190,$$

which is 0.008 larger than $\pi_{\text{net}} = 0.182$. For $\gamma = 0.01$ we should replace 1.64 above by $q_{0.99,s} \approx 2.33$, which leads to $\pi_{0.01} \approx 0.193$.

In our example, the percentile premium is close to the net premium, not only because n is large but also since the standard deviation H is small. We noticed earlier that for the data we use, the probability that the client dies within the term is small. Therefore, the insurance risk is not large. In the next example, the situation will be somewhat different.

EXAMPLE 2. Consider the whole life insurance on a fully continuous basis with the constant force of mortality $\mu = 0.04$ and $\delta = 0.06$. In this case, $A = \bar{A}_x = \mu/(\mu + \delta) = 0.4$, $a = \dfrac{1}{\mu + \delta} = 10$, and $\pi_{\text{net}} = \mu = 0.04$. The variance $H^2 = \dfrac{\mu}{\mu + 2\delta} - \left(\dfrac{\mu}{\mu + \delta}\right)^2 = 0.25 - 0.4^2 = 0.09$, and $H = 0.3$.

For $\gamma = 0.05$, we have

$$\pi_\gamma \approx \frac{0.4 + 1.64 \cdot 0.3/\sqrt{n}}{10 - 1.64 \cdot 0.3/(0.06\sqrt{n})} = \frac{0.4 + 0.492/\sqrt{n}}{10 - 8.2/\sqrt{n}}.$$

It is interesting how π_γ is changing in n, and how it is approaching $\pi_{\text{net}} = 0.04$. The table below illustrates the pattern:

n	100	150	200	1000	4000	10000
π_γ	0.0490	0.0472	0.0462	0.0427	0.0413	0.0408

Actually, the convergence is slow. □

The end of Route 1 ! Route 2 ⇒ page 523

1.5 Exponential premiums 3

The exponential premium is a premium based on the utility equivalence principle with an exponential utility function; see Section 1.1. Since this function is concave, the premium implicitly involves security loading. Consequently, such a premium should be larger than the net premium. However, unlike the percentile premiums considered in the previous section, exponential premiums turn out to be too large.

EXAMPLE 1. Consider a fully continuous whole life insurance on (x) with a constant force of mortality μ and interest rate $\delta = 0$. Certainly, such an example is artificial due to both conditions above, but it well illustrates the essence of the matter. The same phenomenon takes place in many much more realistic situations.

Let the death benefit be equal to C. Then, since $\delta = 0$, the loss $L_P = C - PT$, where P is a rate of continuous payment, and $T = T(x)$ is the future lifetime. Then the net premium

$$P_{\text{net}} = \frac{C}{E\{T\}}.$$

Consider the exponential premium. In our case, the left member of (1.1.5) is

$$E\{\exp\{\beta L_P\}\} = E\{\exp\{\beta(C - PT\}\} = e^{\beta C}E\{e^{-\beta PT}\} = e^{\beta C}M_T(-\beta P),$$

where $M_T(z)$ is the m.g.f. of the exponential r.v. T. Since $M_T(z) = 1/(1 - z/\mu)$,

$$E\{\exp\{\beta L_P\}\} = e^{\beta C}\frac{1}{1 + P\beta/\mu}.$$

Then the solution to the equation $E\{\exp\{\beta L_P\}\} = 1$ is

$$P = \frac{\mu}{\beta}\left(e^{\beta C} - 1\right). \tag{1.5.1}$$

Because $E\{T\} = 1/\mu$, we can rewrite (1.5.1) as

$$P = \frac{1}{\beta E\{T\}}\left(e^{\beta C} - 1\right) = P_{\text{net}} \cdot \frac{1}{\beta C}\left(e^{\beta C} - 1\right) = P_{\text{net}} \cdot k(\beta C),$$

where the function

$$k(z) = \frac{1}{z}\left(e^z - 1\right)$$

may be viewed as the security loading.

We see that the premium is not proportional to the sum insured, and it grows rapidly as C is increasing. If $\beta \to 0$, then $k(\beta C) \to 1$, and we come to the net premium, which has already been discussed at the end of Section 1.1. However, for a large β and/or C, the security loading may be very high. Say, $k(2) \approx 3.19$, so the premium should consist of more than 300% of the net premium.

The circumstance mentioned imposes a serious limitation on the application of the exponential approach.

One argument in favor of the exponential principle is that it may serve not for determining a premium for each particular case, but rather for figuring out a level after which the company should reinsure the risk. For example, H.U.Gerber in [43, p.51] writes that the issue we discuss "may be resolved by the following consideration: Assume that the insurer charges 250% of the net premium for all values of C: then the policies with a sum insured exceeding [the level corresponding to 250%] require reinsurance: policies with a lower sum insured are overcharged, which compensates for the relatively high fixed costs of these policies." □

In any case, in practice the most important and most frequently used approach is based on the equivalence principle. The net premium in this case serves as the starting point in premium calculations. A safety loading is either added to the net premium explicitly (for example, proceeding from a security level, which we did for large portfolios in Section 1.4.2), or it is incorporated implicitly by varying mortality and interest rates (see the remark at the end of Section 1.2.5 and Exercise 20).

---

2 RESERVES 2,3

2.1 Definitions and preliminary remarks

Consider an insurance on (x) with a level premium rate P. Denote by $\mathcal{E}_t$ the event that the policy is still in force at time t after the time of policy issue. For example, for the whole life insurance, $\mathcal{E}_t = \{T(x) > t\}$.

Our next step is to consider the company's future loss not only at $t = 0$, but at later time moments as well. Denote by ${}_tL = {}_tL_P$ the present value of the *future* loss of the company from the standpoint of time t, given that $\mathcal{E}_t$ occurred. The index P indicates that the loss so defined depends on the premium *chosen at the time of policy issue.* When it cannot cause misunderstanding, we will skip this index. The quantity

$$ {}_tV = {}_tV_P = E\{{}_tL_P \,|\, \mathcal{E}_t\}. $$

is said to be the *reserve* at time t. This is the expected value of what the company needs in order to fulfil its obligations after time t. For example, in the case of fully continuous whole life insurance, ${}_0V_P = \bar{A}_x - P\bar{a}_x$, while

$$ {}_tV_P = \bar{A}_{x+t} - P\bar{a}_{x+t}. \tag{2.1.1} $$

(Since $\mathcal{E}_t$ has occurred, the insured has attained the age of $x+t$. Consequently, when computing the *future* loss with respect to the time t, we do not take into account the premium amount received before this time. So to say, everything starts as from the very beginning, and we may view $x+t$ as the initial age.)

The counterpart of (2.1.1) for the discrete time case is

$$ {}_kV_P = A_{x+k} - P\ddot{a}_{x+k}. \tag{2.1.2} $$

Sometimes we will omit the index P in ${}_tV_P$ too.

Set $L_P = {}_0L_P$, $V_P = {}_0V_P$. Since $P(\mathcal{E}_0) = 1$, the conditional expectation $E\{L_P \,|\, \mathcal{E}_0\}$ equals the unconditional expectation $E\{L_P\}$.

The benefit premium P_{ben} is determined by the condition

$$V_{P_{\text{ben}}} = E\{L_{P_{\text{ben}}}\} = 0,$$

that is, we require the expected future loss with respect to the initial time to be zero. However, we *should not expect this property to hold at all time moments.* In other words, we should *not* expect that ${}_tV_{P_{\text{ben}}}$ equals 0 for all t.

Let us set ${}_tV_{\text{ben}} = {}_tV_{P_{\text{ben}}}$, the conditional expected reserve at time t corresponding to the benefit premium. We call such a reserve a *benefit reserve* or a *net premium reserve.*

As an example, let us consider the case (2.1.1). The benefit premium for this insurance is $P_{\text{ben}} = \overline{P}_x = \overline{A}_x / \bar{a}_x$, and hence

$${}_tV_{\text{ben}} = \overline{A}_{x+t} - \overline{A}_x \cdot \bar{a}_{x+t}/\bar{a}_x.$$

Replacing $\overline{A}_x$ by $1-\delta\bar{a}_x$, and $\overline{A}_{x+t}$ by $1-\delta\bar{a}_{x+t}$, after very simple algebra we get that

$${}_tV_{\text{ben}} = 1 - \frac{\bar{a}_{x+t}}{\bar{a}_x}. \tag{2.1.3}$$

Because it is usually the case (though not always (!), see Exercises **9**.4, **8**.1b) that $\bar{a}_{x+t} < \bar{a}_x$, in typical situations

$${}_tV_{\text{ben}} > 0 \text{ for } t > 0. \tag{2.1.4}$$

This is a desirable property, and it reflects the essence of the matter. In the case (2.1.4), the expected discounted loss of the insurer is positive, and hence the expected discounted loss of the insured is negative. Thus, on the average, the insured will get more than she/he will pay, which is a ground for the insured not to terminate the contract.

The traditional notation for the benefit reserves (or net premium reserves) inherits the same logic that was applied for the premium notation. For example,

${}_t\overline{V}_x$ is the benefit reserve at the moment t in the case of a fully continuous whole life insurance on (x);

${}_kV_x$ is the benefit reserve at an integer moment k in the case of a fully discrete whole life insurance;

${}_kV_{x:\overline{n}|}$ is the same for an n-year endowment;

${}_kV^{1}_{x:\overline{n}|}$ is the same for an n-year term insurance;

${}_kV(\overline{A}_{x:\overline{n}|})$ is the benefit reserve at a moment k in the case of an n-year endowment with the discrete type of premium payment and a benefit payable upon death.

Below, we will sometimes use this notation and sometimes prefer another symbolism, whichever is more convenient.

2.2 Examples of direct calculations

EXAMPLE 1 ([151, N6][3]). For a fully continuous whole life insurance on (40), you are given: the level annual premium is \$66 payable for the first 20 years; the death benefit is \$2000 for the first 20 years and \$1000 thereafter; $\delta = 0.06$; $1000\overline{A}_{50} = 333.33$; $1000\overline{A}^{\,1}_{50:\overline{10}|} = 197.81$; and $1000\,{}_{10}E_{50} = 406.57$. Calculate ${}_{10}\overline{V}$ for the premium given.

[3]Reprinted with permission of the Casualty Actuarial Society.

Since we compute the reserve needed when the insured attains the age of 50, the fact that now she/he is aged 40 does not matter. Assuming that the insured has attained the age of 50, we represent the insurance as a combination of two contracts with level payments: the whole life insurance with a benefit of \$1000, and the 10-year term insurance with the same benefit. The total APV of this combination is

$$1000\bar{A}_{50} + 1000\bar{A}^{\;1}_{50:\overline{10}|} = 333.33 + 197.81 = 531.14.$$

After $t = 10$, the insured pays only 10 years more. The APV of the premium annuity starting at $t = 10$ is the premium multiplied by $\bar{a}_{50:\overline{10}|} = \frac{1}{\delta}(1 - \bar{A}_{50:\overline{10}|})$. Now, $\bar{A}_{50:\overline{10}|} = \bar{A}^{\;1}_{50:\overline{10}|} + {}_{10}E_{50} = 604.38/1000$. So, $\bar{a}_{50:\overline{10}|} = \frac{1}{0.06}(1 - \bar{A}_{50:\overline{10}|}) \approx 6.593\bar{6}$. Thus, for $P = 66$, the APV of the premium annuity is $66 \cdot 6.593\bar{6} = 435.182$.

Eventually, ${}_{10}\bar{V} = 531.14 - 435.182 = 98.958$.

EXAMPLE 2. (The reader who skipped the multiple decrement scheme should omit this example.) A special fully discrete 3-year term insurance on (50) follows a double decrement model, with decrement 1 corresponding to accidental death and decrement 2 corresponding to all other causes. The probability distribution is specified by the following table:

x	50	51	52
$q_x^{(1)}$	0.004	0.004	0.004
$q_x^{(2)}$	0.02	0.03	0.04

The death benefit is 3000 for accidental deaths and 1000 for the second decrement; $v = 0.97$.

Find ${}_1V_{\text{ben}}$ and ${}_1V = {}_1V_P$ if starting from $t = 1$, the company charges the premium 10% larger than the benefit premium: $P = 1.1P_{\text{ben}}$.

First, we compute the APVs of the benefit and premium payments starting from $t = 1$. Denoting the present value of the corresponding benefit payment by Z_1, we have

$$E\{Z_1\} = v\left(3000q_{51}^{(1)} + 1000q_{51}^{(2)}\right) + v^2\left[1 - \left(q_{51}^{(1)} + q_{51}^{(2)}\right)\right]\left(3000q_{52}^{(1)} + 1000q_{52}^{(2)}\right) \approx 88.0033. \tag{2.2.1}$$

Let Y_1 be the present value of the premium payment at unit rate, starting from $t = 1$. Then

$$E\{Y_1\} = 1 + v\left[1 - \left(q_{51}^{(1)} + q_{51}^{(2)}\right)\right] = 1.93702. \tag{2.2.2}$$

The total APV of the benefit payment equals

$$E\{Z\} = v\left(3000q_{50}^{(1)} + 1000q_{50}^{(2)}\right) + v\left[1 - \left(q_{50}^{(1)} + q_{50}^{(2)}\right)\right]E\{Z_1\} \approx 114.3545,$$

while for the total premium with unit rate,

$$E\{Y\} = 1 + v\left[1 - \left(q_{50}^{(1)} + q_{50}^{(2)}\right)\right]E\{Y_1\} \approx 2.7361.$$

The benefit premium

$$P_{\text{ben}} = \frac{E\{Z\}}{E\{Y\}} \approx 41.7945.$$

The premium with the 10% loading is $P = 1.1P_{\text{ben}} \approx 45.9740$.

Thus, ${}_1V_{\text{ben}} = E\{Z_1\} - P_{\text{ben}} \cdot E\{Y_1\} \approx 88.003 - 41.7945 \cdot 1.93702 \approx 7.0464$, while ${}_1V_P = E\{Z_1\} - P \cdot E\{Y_1\} \approx 88.003 - 45.9740 \cdot 1.93702 \approx -1.04955748$. So, the reserve is negative, and for the insured it is reasonable not to renew the contract at time $t = 1$, if it is possible. □

2.3 Formulas for some standard types of insurance

In the "consistency case" described in Section 1.2.2, we can proceed as we did when deriving (2.1.3). Consider, for example, the fully discrete n-year endowment insurance. In this case, $P_{\text{ben}} = A_{x:\overline{n}|}/\ddot{a}_{x:\overline{n}|}$, and

$$ {}_kV_{\text{ben}} = A_{x+k:\overline{n-k}|} - P_{\text{ben}}\ddot{a}_{x+k:\overline{n-k}|} = A_{x+k:\overline{n-k}|} - A_{x:\overline{n}|}\frac{\ddot{a}_{x+k:n-k}}{\ddot{a}_{x:\overline{n}|}}. \tag{2.3.1} $$

On the other hand, $1 = A_{x:\overline{n}|} + d\ddot{a}_{x:\overline{n}|}$ for any x and n, which implies that $A_{x+k:\overline{n-k}|} = 1 - d\ddot{a}_{x+k:\overline{n-k}|}$ and $A_{x:\overline{n}|} = 1 - d\ddot{a}_{x:\overline{n}|}$. After substitution into (2.3.1), some like terms cancel, and following the traditional notation, we can write that

$$ {}_kV_{x:\overline{n}|} = 1 - \frac{\ddot{a}_{x+k:\overline{n-k}|}}{\ddot{a}_{x:\overline{n}|}}. \tag{2.3.2} $$

To consider the fully discrete whole life insurance, it suffices to let $n \to \infty$, which leads to

$$ {}_kV_x = 1 - \frac{\ddot{a}_{x+k}}{\ddot{a}_x}. \tag{2.3.3} $$

We have already obtained the similar formula for the fully continuous case in (2.1.3). In the traditional notation, we write it as

$$ {}_t\overline{V}_x = 1 - \frac{\overline{a}_{x+t}}{\overline{a}_x}. \tag{2.3.4} $$

The counterpart of (2.3.2) is derived absolutely in the same manner, and it looks as follows:

$$ {}_t\overline{V}_{x:\overline{n}|} = 1 - \frac{\overline{a}_{x+t:\overline{n-t}|}}{\overline{a}_{x:\overline{n}|}}. \tag{2.3.5} $$

EXAMPLE 1 ([159, N15][4]). You are given: An individual life aged 40 purchases a fully discrete whole life insurance policy with a death benefit of \$10,000; $i = 0.05$; $p_{40} = 0.98$; $\ddot{a}_{40} = 12$; premiums have been calculated according to the equivalence principle. Calculate the benefit reserve at time $t = 1$.

We have $\ddot{a}_{40} = 1 + v\,p_{40}\ddot{a}_{41}$, and $v = \frac{1}{1+i}$. Hence,

$$ \ddot{a}_{41} = \frac{12-1}{(1/1.05)\cdot 0.98} \approx 11.78. $$

[4]Reprinted with permission of the Casualty Actuarial Society.

Then, by (2.3.3), the benefit reserve

$$\approx 10{,}000 \times \left(1 - \frac{11.78}{12}\right) \approx 178.6.$$

EXAMPLE 2. Consider the fully continuous life insurance with the force of mortality $\mu(t) = \mu$.

(a) In the case of the whole life insurance $\bar{a}_x = 1/(\mu+\delta)$, and from (2.3.4) it follows that ${}_t\bar{V}_x = 0$ for all t. This is not surprising in view of the memoryless property and, as a consequence, of the fact that the benefit premium $\bar{P}_x = \mu$ and does not depend on x. [See Examples **9**.1.1 and 1.2.1-1.]

(b) We should not, however, expect the same for the n-year endowment insurance. For example, if t is close to n, the premium remaining to be paid is small (since the period $[t,n]$ is short), while the company still has an obligation to pay a unit of money, not depending on whether or not the insured survives n years after the time of policy issue.

Using, for example, (**9**.3.2.4), we have

$$\bar{a}_{x:\overline{n}|} = \int_0^n e^{-\delta t} e^{-\mu t}\, dt = \frac{1}{\mu+\delta}(1 - e^{-(\mu+\delta)n}),$$

and hence, by (2.3.5),

$${}_t\bar{V}_{x:\overline{n}|} = 1 - \frac{1 - e^{-(\mu+\delta)(n-t)}}{1 - e^{-(\mu+\delta)n}} = \frac{e^{(\mu+\delta)t} - 1}{e^{(\mu+\delta)n} - 1}.$$

As expected, when t runs from 0 to n, the last expression is strictly increasing from 0 (the reserve at the initial time) to 1 (the reserve before the final payment of one). □

2.4 Recursive relations

Let us consider a general fully discrete insurance on (x). We denote by P_k the premium paid at a time moment $k = 0, 1, \ldots$, and c_{k+1} denotes the benefit paid at the end of the period $[k, k+1]$ if the contract terminates in this period. All other notations are standard. Assume that the insured has attained the age of $x+k$. As in Section 2.1, set $\mathcal{E}_k = \{T(x) > k\}$.

We use the first step approach, considering two cases: when the insured lives at least one year more, and when she/he does not. The corresponding formula for total expectation looks as follows:

$$\begin{aligned} {}_kV = E\{{}_kL \,|\, \mathcal{E}_k\} &= E\{{}_kL \,|\, T(x) > k+1,\ \mathcal{E}_k\}P(T(x) > k+1 \,|\, \mathcal{E}_k\} \\ &\quad + E\{{}_kL \,|\, T(x) \le k+1,\ \mathcal{E}_k\}P(T(x) \le k+1 \,|\, \mathcal{E}_k\} \\ &= E\{{}_kL \,|\, T(x+k) > 1\}p_{x+k} + E\{{}_kL \,|\, 0 < T(x+k) \le 1\}q_{x+k}. \end{aligned}$$

If $T(x+k) > 1$, then there will be no benefit payment at the moment $k+1$, and before the receipt of the next premium P_{k+1}, the future loss at $t = k+1$ equals ${}_{k+1}L$. The discounted value of this loss is $v \cdot {}_{k+1}L$. Taking into account the premium P_k paid at the time $t = k$, we have

$$E\{{}_kL \,|\, T(x+k) > 1\} = E\{v \cdot {}_{k+1}L - P_k \,|\, T(x+k) > 1\} = v \cdot {}_{k+1}V - P_k.$$

If $T(x+k) \leq 1$ (the strict equality may be not taken into consideration), then there will be a payment of c_{k+1} at the moment $k+1$, and the contract will be terminated. Hence,

$$E\{_kL\,|\,T(x+k) \leq 1\} = vc_{k+1} - P_k.$$

Thus, $_kV = (v \cdot {}_{k+1}L - P_k)p_{x+k} + (vc_{k+1} - P_k)q_{x+k}$, which may be written as

$$_kV + P_k = v(c_{k+1} \cdot q_{x+k} + {}_{k+1}V \cdot p_{x+k}). \tag{2.4.1}$$

Proceeding from (2.4.1), one can use the forward and backward recursion. If the premiums are determined by the equivalence principle, we set $_0V = 0$ and may move forward starting from $t = 0$. If the contract can last at most n years, we can first calculate $_nV$ and move backward starting from $t = n$.

EXAMPLE 1. Let $p_x = 0.95$, the benefit payment at the first year (if any) be $10,000$, the premiums (which may vary) correspond to the equivalence principle, and $P_0 = 100$. Find the reserve at the beginning of the second year after the time of policy issue for $v = 0.97$.

The problem is simple. Since we proceed from the equivalence principle, $V_0 = 0$, and from (2.4.1) it follows that

$$0 + 100 = 0.97(1000 \cdot 0.05 + {}_1V \cdot 0.95).$$

Hence, $_1V \approx 55.89$.

EXAMPLE 2 ([152, N3][5]). For a fully discrete 10-year deferred whole life insurance of \$1000 on (40), you are given: $v = 0.95$, $p_{48} = 0.98077$, $p_{49} = 0.98039$, $A_{50} = 0.35076$. The annual benefit premium of \$23.4 is payable during the deferral period. Calculate $_8V$, the benefit reserve at time $t = 8$ right before the premium payment.

Since there is no premium payment after $t = 10$, the reserve $_{10}V = 1000A_{50}$. Since the company pays nothing if the insured dies within the interval $[49, 50)$, the payment $c_{10} = 0$, and in accordance with (2.4.1),

$$_9V = -P + v \cdot {}_{10}V \cdot p_{49} = -23.4 + 0.95 \cdot 0.98039 \cdot 1000 \cdot 0.35076 \approx 303.28752.$$

Similarly,

$$_8V = -P + v \cdot {}_9V \cdot p_{48} = -23.4 + 0.95 \cdot 0.98077 \cdot 303.2875166 \approx 259.18253. \ \square$$

Substituting $p_{x+k} = 1 - q_{x+k}$, we may rewrite (2.4.1) as

$$_kV + P_k = v({}_{k+1}V + (c_{k+1} - {}_{k+1}V)q_{x+k}). \tag{2.4.2}$$

The amount $c_{k+1} - {}_{k+1}V$ is called a *net amount at risk* in the period $[k, k+1]$.

EXAMPLE 3. (Similar examples are contained, for instance, in [19, Example 8.3.2], [154, N30], [154, N10].) A special fully discrete n-year term insurance on (x) pays a death benefit of one unit plus the benefit reserve at the end of the year of death, provided that the

[5]Reprinted with permission of the Casualty Actuarial Society.

insured dies before the time moment $x+n$. Given the probabilities q_x and a discount v, find the level benefit premium.

From the conditions of the problem it follows that ${}_nV = 0$, and $c_{k+1} = {}_{k+1}V + 1$ for $k = 0, ..., n-1$ if $K(x) < n$.

Then the formula (2.4.2) implies that ${}_kV + P = v({}_{k+1}V + q_{x+k})$ for $k = 0, ..., n-1$, or

$$ {}_kV = -P + v \cdot {}_{k+1}V + vq_{x+k}. \qquad (2.4.3) $$

Since we are looking for the benefit premium, ${}_0V$ should be equal to zero, so in this problem we may either move forward starting from ${}_0V = 0$, or move backward starting from ${}_nV = 0$. Let us choose the former way. Applying (2.4.3) consecutively in each step, we have

$$ \begin{aligned} 0 = {}_0V &= -P + v \cdot {}_1V + vq_x = -P + v(-P + v \cdot {}_2V + vq_{x+1}) + vq_x \\ &= -P - vP + v^2 \cdot {}_2V + v^2 q_{x+1} + vq_x \\ &= -P - vP + v^2(-P + v \cdot {}_3V + vq_{x+2}) + v^2 q_{x+1} + vq_x \\ &= -P - vP - v^2 P + v^3 \cdot {}_3V + v^3 q_{x+2} + v^2 q_{x+1} + vq_x = ... \\ &= -P - vP - ... - v^{n-1} + v^n \cdot {}_nV + v^n q_{x+n-1} + ... + v^2 q_{x+1} + vq_x = \\ &= -P(1 + ... + v^{n-1}) + v^n q_{x+n-1} + ... + v^2 q_{x+1} + vq_x, \end{aligned} $$

because ${}_nV = 0$. Solving it for P, we get that

$$ P = v \frac{q_x + vq_{x+1} + ... + v^{n-1} q_{x+n-1}}{1 + v + ... + v^{n-1}}. \qquad (2.4.4) $$

Certainly, we can write that the denominator equals $(1-v^n)/(1-v)$, but (2.4.4) reflects the logic of the answer: if in the numerator, we replace all q's by one, then we will come to the denominator. Consider two particular cases.

(a) Let the mortality force $\mu(x) = \mu$. Then all $q_x = q = 1 - e^{-\mu}$, and (2.4.4) leads to $P = vq$. In this case, from (2.4.3) it follows that ${}_kV = -vq + v \cdot {}_{k+1}V + vq = v \cdot {}_{k+1}V$, and because ${}_nV = 0$, we have ${}_kV = 0$ for all $k = 0, ..., n$. We could predict this proceeding from the memoryless property.

(b) Let $n = 2$. We know that ${}_0V = 0$ and ${}_2V = 0$. Let us find ${}_1V$. From (2.4.4) it follows that

$$ P = v \frac{q_x + vq_{x+1}}{1+v}. $$

Substituting it into (2.4.3) we have

$$ \begin{aligned} {}_1V = -P + vq_{x+1} &= -v \frac{q_x + vq_{x+1}}{1+v} + vq_{x+1} \\ &= v \cdot \frac{q_{x+1} - q_x}{1+v}. \end{aligned} $$

As was already discussed, if ${}_1V > 0$, the insured has an interest to renew the insurance. We see that this is the case if and only if $q_{x+1} - q_x > 0$. The former condition is equivalent to the growth of the mortality force. In reality, the last condition is true at least for large and moderate x's. □

3 EXERCISES

Section 1

1. In the situation of Section 1.2.1, show that if the benefit provided by an insurance contract increases by k percent, then the benefit premium increases in the same proportion, no matter whether the benefits are level or varying.

2. Write the traditional notation for all premiums considered in Table 1.2.1-1.

3. Consider the benefit premiums in the fully discrete case for (i) an n-year term insurance; (ii) an n-year pure endowment; (iii) the n-year endowment. Which premium is the largest? Write the relation between these three premiums. Clarify heuristically when the premium in the case (ii) is larger than in the case (i), and vice versa. Consider the same problem for the fully continuous case.

4. In each of the following, tell which quantity is larger: (i) $\overline{P}_x$ or P_x; (ii) $\overline{P}_{x:\overline{n}|}$ or $P_{x:\overline{n}|}$; (iii) P_x or $P_{x:\overline{n}|}$; (iv) $P^1_{x:\overline{n}|}$ or $P_{x:\overline{n}|}$.

5. Without any calculations, just proceeding from common sense, write the limits for all premiums in Table 1.2.1-1 as $\delta \to 0$. Clarify how to justify your answers rigorously.

6. As was told in Section **8**.2.4.1, the traditional notation for the APV of the pure endowment insurance is $A_{\substack{\;\;1\\x:\overline{n}|}}$ or ${}_nE_x$. The benefit premium is denoted by $P_{\substack{\;\;1\\x:\overline{n}|}}$.

   (a) Clarify why the formula $P_{x:\overline{n}|} = P^1_{x:\overline{n}|} + P_{\substack{\;\;1\\x:\overline{n}|}}$ is obvious from an economic point of view. Prove it by using Table 1.2.1-1.

   (b) Show that ${}_nP_x = P^1_{x:\overline{n}|} + P_{\substack{\;\;1\\x:\overline{n}|}}A_{x+n}$. (For the definition of ${}_nP_x$, see the remarks on notation in Section 1.2.1.)

7. A insurance company adds a 10% security loading to net premiums. You know that the company charges 20-year old clients the following annual premiums per unit benefit: (a) 0.044 for a 30-year payment whole life insurance; (b) 0.011 for a 30-year term insurance; (c) 0.055 for the 30-year endowment. Find A_{50}.

8. Consider two clients such that for the same δ's, the APV of the whole life annuity for the first client is larger than that for the second. Who should pay more for the whole life insurance?

9. A 50-year-old client buys a whole life insurance with unit benefit payable at the end of the year of death and with premiums paid at the beginning of each year. (a) Using the Illustrative Table, find the benefit premium. (b) Assume that, when calculating the actual premium, the actuary of the company adds 5% to the benefit premium. Estimate the probability that the company will make a profit dealing with 100 independent clients of the above type.

10. Using the Illustrative Table, find the benefit premiums in the case of 50-year old clients for the 30-year term life insurance.

11. Solve the problem of Example 1.2.1-3 for the case when $\mu(x) = 0.01$.

12. Solve the problem of Example 1.2.1-4 for the case when both independent lifetimes are uniformly distributed on $[0, 50]$. (*Advice*: Integrals to which you will come are standard, but computing them takes time. So, it is reasonable to use software.)

13. Find the formula for the variance of the loss in the case of a whole life insurance with fully continuous premiums and the lifetime uniformly distributed on $[0,\omega]$. (*Advice*: Look up Example **8**.1.1-1. Clearly, the answer should involve the age x and δ.)

14. Find concrete answers for the problem from Example 1.2.5-1a for $\delta = 0$ in terms of e_x and $_np_x$.

15. In the situation of Example 1.4.2-2, for the case of lifetimes uniformly distributed on $[0,50]$ and $\gamma = 0.05$, find the number n of contracts for which the security loading coefficient (with respect to the benefit premium) is not larger than 5%.

16. Similarly to what we did in Example 1.5-1, write an equation for the exponential premium in the case of uniformly distributed $T(x)$. Find an asymptotic approximation for the premium for large C.

17. One hundred 50-year-old clients buy a whole life insurance with a death benefit of \$10,000 payable at the end of the year of death and with premiums paid at the beginning of each year. Assume that, when calculating the actual premium, the actuary of the company adds k% to the benefit premium. Using the Illustrative Table for $\delta = 4\%$, find k for which the probability that the company will make a profit is greater than 0.95.

18. Consider two groups of clients of the same age x. In each group, the distribution of the future lifetime is the same. However, if $T^{(1)}(x)$ and $T^{(2)}(x)$ are the lifetimes for typical clients from the first and second group, respectively, then $P(T^{(1)}(x) \geq t) > P(T^{(2)}(x) \geq t)$ for ALL t. Which group is healthier? For each characteristic below, figure out for which group it will be larger: A_x, $\ddot{a}_x$, $A_{x:\overline{n}|}$, $\ddot{a}_{x:\overline{n}|}$, P_x, $P_{x:\overline{n}|}$.

19. Write an explicit formula for P_x in the exponential case.

20. For a rough estimate of the premium on the fully continuous basis for a whole life insurance contract, an actuary took as a force of mortality $\mu = 0.02$ and the interest rate $\delta = 0.06$ for the benefit payment. The actuary proceeded from the equivalence principle. However, when computing the APV of the premium annuity, the actuary used the quantity $k\delta$ as an interest rate, where k is a coefficient. (Look up the remark at the end of Section 1.2.5.) Should k be larger or smaller than one, in order to incorporate a safety loading into the premium? Find the premium and graph it as a function of k. Find k for which the premium is 10% larger than the net premium.

21. *Estimate* $P(\bar{A}_{30:\overline{35}|})$ in the situation of Exercise **8**.20. (*Hint*: One should be cautious when applying (**8**.2.1.7) to term insurances.)

22. On her thirtieth birthday, Mary decided to enter into a pension plan paying \$50,000 at the beginning of each year, starting from Mary's 65-year birthday. (a) Using the Illustrative Table for $\delta = 4\%$, find the premium Mary should pay if the plan adds 5% to the net premium. (b)* Recalculate the premium and benefits for the case when both are paid monthly. Clarify why the premium turned out to be smaller.

23. (a) Consider a fully continuous whole life insurance on a lifetime T. Denote by π the (level) premium rate, by δ the interest rate, by L the random loss of the company, and by $l(t)$ the loss of the company given $T = t$. Show that $l(t) = e^{-\delta t}(1 + \frac{\pi}{\delta}) - \frac{\pi}{\delta}$. Graph $l(t)$ and show that $P(L \geq 0) = P(T \leq t_0)$, where t_0 is a number such that $l(t_0) = 0$. Show that $\pi = \delta/(\exp\{\delta t_0\} - 1)$ and for $P(L \geq 0) = P(T \leq t_0) \leq \gamma$, we should choose

$$\pi \geq \frac{\delta \exp\{-\delta q_\gamma\}}{1 - \exp\{-\delta q_\gamma\}} = \frac{\delta}{\exp\{\delta q_\gamma\} - 1}, \tag{3.1}$$

where q_γ is the γ-quantile of T.

(b) Show that the counterpart of (3.1) in the fully discrete time case is

$$\pi > dv^{q_\gamma+1}/(1 - v^{q_\gamma+1}), \tag{3.2}$$

where q_γ is the γ-quantile of K, and v is the discount factor.

(c) Using the Illustrative Table, estimate the percentile premium for the whole life insurance on (50) for $\gamma = 0.05$. Compare it with the benefit premium computed in Exercise 9.

24.* Consider the situation of Example 1.2.4-1 and explain, from a heuristic point of view, why the insured should pay more if the premium payments are provided monthly. Why is the coefficient k in (1.2.15) large for $\ddot{a}_x = 1$?

25.* Explain from a common sense point of view that the premium for the plan in Example 1.2.5-1b must be larger that the premium for the usual n-year deferred annuity plan. Show that it follows from (1.2.16).

26.* By analogy with what we did in Section 1.3, derive the formula for the quantity $\bar{s}_{x:\overline{n}|}$, the actuarial accumulated value at the end of the term of an n-year temporary annuity in the case of continuous time.

*Section 2**

27. Write a general formula for the benefit reserves for an n-year term insurance in the fully continuous case. Write a precise expression for a constant mortality rate. Interpret the last result.

28. Solve the problem of Example 2.3-2b for the fully discrete case.

29. Without any calculations, simplify the formulas of Section 2.3 for the case $\delta = 0$.

30. Write and analyze the formula for the benefit reserves for deferred annuities.

31. Consider a life (x) with $q_x = 0.02$, $q_{x+1} = 0.03$, and $q_{x+2} = 0.04$. Let $v = 0.97$.

    (a) Find the level benefit premium and the benefit reserves for a special 3-year term insurance with payments $c_1 = 10$, $c_2 = 20$, and $c_3 = 15$.

    (b) Do the same for a 3-year endowment insurance with the death benefits as above and with a payment of 20 at the beginning of the 4th year if the insured attains age $x + 3$.

    (c) Do the same for a 3-year term insurance paying a death benefit of one unit plus half of the benefit reserve at the end of the year of death, provided that the insured dies within three years.

32. Using a spreadsheet technique, find the benefit premium and the benefit reserves for the insurance from Example **8**.4.1-3.

33. Let $r = 1 + i$ where i is an annual interest. Consider an n-year deferred whole life annuity on (x) with the provision for a death benefit equal to the benefit reserve, payable at the end of the year of death. Show that the benefit reserve

$$_kV = \frac{r^k - 1}{r^n - 1}\,\ddot{a}_{x+n}, \quad k \le n.$$

Chapter 11

Pension Plans

2,3

This chapter concerns pension models. We have already considered the simplest one in Section **10**.1.2.1 (see, in particular, the last position in Table 1 there), where the level deferred annuity may be viewed as future pension payments, and the temporal premium annuity as contributions to a retirement account. In this case, the benefit premium is a net contribution rate of the participant's payments for having a retirement annuity in the future, provided that the participant will survive the retirement age. As we saw in Table 1 mentioned, for example in the discrete time case, the net contribution rate (premium) per one unit of the future pension rate equals

$$P = \frac{{}_{n|}\ddot{a}_x}{\ddot{a}_{x:\overline{n|}}} = \frac{{}_{n}p_x \cdot v^n \cdot \ddot{a}_{x+n}}{\ddot{a}_{x:\overline{n|}}},$$

where x is the participant's age at the moment of entering the pension plan and n is the time to retirement.

In reality, the situation is much more complicated for many reasons. In particular, the future pension as well as the contributions to a retirement account may depend on the (changing in time) salary of the future retiree; the pension may depend on the age of the participant and may change in time due to inflation. This and other features will be reflected in models below.

When an actuary is analyzing a pension plan, her/his main task is to check the plan's solvency and provide conditions ensuring a certain balance between *future* benefit payments and *present* contributions to a retirement account. Due to the law of large numbers, such a balance may be achieved more easily if individual pension plans do not run separately but rather are parts of a common fund including many participants. The corresponding models in this case are more developed and are also considered below.

1 VALUATION OF INDIVIDUAL PENSION PLANS

We call a pension *plan* any arrangement of regular payments for life starting at a certain age. We use the term *individual plan* if we deal with one future retiree (usually, a participant of a plan), which we do in this section.

Actuaries distinguish two broad categories of pension plans: *defined benefit (DB)* and *defined contribution (DC) plans.*

In a DB plan, we start with a definition of the future (or projected) benefit; that is, what a worker (alone or with his/her spouse) can expect upon retirement.

A DC plan specifies a fixed contribution from a worker and/or his/her employer to a retirement account. In this case, the benefit is an annuity that can be purchased at the moment of retirement by the accumulated contributions made during the entire pre-retirement period.

1.1 DB plans

1.1.1 The APV of future benefits

Consider a participant age x with h years in service. Denote by r the minimal retirement age, and by y a *projected* time until the future retirement. In other words, the projected age at which the participant will retire is $x+y$, and we assume $x+y \geq r$.

Let $B = B(x,h,y)$ be the corresponding projected annual pension rate. It is worth emphasizing that in the current model the function $B(x,h,y)$ is predetermined. Note also that since r is the minimal retirement age, $B(x,h,y) = 0$ for $y < r-x$.

EXAMPLE 1. Suppose that the minimal retirement age is $r = 66$, a participant age $x = 33$ was hired 3 years ago, and her/his current annual salary is $w_0 = \$50,000$. Suppose the participant's salary grows at an annual rate of 2%. For simplicity, assume that this growth is running continuously; so, at age $x+y$, the participant's salary will equal $w_0\, e^{0.02y}$.

Assume also that the annual pension will amount to 2.5% of the annual salary at the moment of retirement multiplied by the number of years in service at the time of retirement. Then, for $y < 66-33 = 33$ the rate $B(x,h,y) = 0$, and for $y \geq 33$,

$$B(x,h,y) = 0.025\,(h+y)\,w_0\,e^{0.02y} = 0.025\,(3+y)\,50,000e^{0.02y} = 1250\,(3+y)\,e^{0.02y}. \ \square \tag{1.1.1}$$

Later, we consider other examples of the benefit rate function $B(x,h,y)$, but first let us find the APV of the future benefits given a function $B(\cdot)$.

There are four reasons for which a participant may leave the pension plan: withdrawal, disability, death, and retirement. We unify the first three decrements and denote by $\mu_x^{(d)}(y)$ the hazard rate corresponding to these decrements. Denote by $\mu_x^{(r)}(y)$ the hazard rate corresponding to the last factor: retirement itself, and set $\mu_x^{(\tau)}(y) = \mu_x^{(d)}(y) + \mu_x^{(r)}(y)$.

(The multiple decrement scheme is presented in detail in Section 7.2, but it is enough for the reader to look over the scheme at the end of Section 7.1.1. The only difference is that here we are dealing with the remaining life time.)

We assume the factors above to be acting independently. In particular, this means that the probability that the participant will not leave the plan before age $x+y$ is

$${}_yp_x^{(\tau)} = \exp\left\{-\int_0^y \mu_x^{(\tau)}(s)ds\right\},$$

and the probability that the participant will retire and this will happen in the time interval $[y, y+dy]$ is equal to

$${}_yp_x^{(\tau)}\,\mu_x^{(r)}(y)dy, \tag{1.1.2}$$

provided $y \geq r-x$.

Certainly, for $y < r-x$, the probability under discussion vanishes, and we set $\mu_x^{(r)}(y) = 0$ for $0 \leq y < r-x$.

The expression (1.1.2) needs to be clarified. Formally, it means that if a participant dies after the retirement age but before actually retiring, then the pension will not be paid. In other words, the participant does not have a beneficiary of her/his pension benefits.

To model a situation with a beneficiary, we should just interpret the parameters above in a different way. For example, we may define $\mu^{(r)}(y)$, for $y > r$, as a hazard rate corresponding to two possible decrements: retirement or death, and presuppose that once one of these decrements occurs, the pension payments will begin. Also, in this case, for $y > r$, the hazard rate corresponding to withdrawal may be set to be zero.

Let us denote by δ the risk-free interest rate, and by $a_z^{(r)}$ the expected present value of the pension annuity with a *unit rate* and starting at age z. (So, we regard $a_z^{(r)}$ as an APV from the standpoint of time z.)

In the no-beneficiary case, we may set $a_z^{(r)} = \bar{a}_z$, or $\ddot{a}_z$, or $\ddot{a}_z^{(12)}$ (monthly payments), depending on which pension annuity model we adopt.

If the participant has a beneficiary, in the case where the participant dies before retirement, we may view $a_z^{(r)}$ as the lump-sum paid to the beneficiary upon death of the participant.

In the model where the beneficiary is a spouse who will be receiving the pension if the participant dies, $a_z^{(r)}$ should involve survival probabilities for the spouse and be calculated differently depending on whether the participant or the spouse receives the pension. (We considered this case in Section 9.6.2.)

Proceeding from (1.1.2), we get that the APV of the pension benefits is equal to

$$\int_{r-x}^{\infty} e^{-\delta y} B(x,h,y)\, a_{x+y}^{(r)} \cdot {}_y p_x^{(\tau)}\, \mu_x^{(r)}(y) dy. \tag{1.1.3}$$

EXAMPLE 2. Let us revisit Example 1. Assume that the hazard rate $\mu_x^{(d)}(y)$ may be well approximated by a constant $\mu = 0.006$. Since $x = 33$, and the minimal retirement age $r = 66$, the probability that the participant will still be in the plan at the moment when retirement becomes possible, is $\exp\{-0.006 \cdot 33\} \approx 0.82$.

Next, we assume that the choice of the retirement age by the participant corresponds to the uniform distribution on the interval $[66,70]$. Then, for $y \in [33,37]$, the probability in (1.1.2) equals

$$\exp\left\{-\int_0^y (\mu_x^{(d)}(s) + \mu_x^{(r)}(s))ds\right\} \cdot \mu_x^{(r)}(y)dy = \exp\left\{-\int_0^y \mu_x^{(d)}(s)ds\right\}$$
$$\times \left[\mu_x^{(r)}(y) \cdot \exp\left\{-\int_0^y \mu_x^{(r)}(s)ds\right\}\right] dy = e^{-\mu y} \cdot \frac{1}{4} dy = e^{-0.006y} \cdot \frac{1}{4} dy.$$

(Since $\mu_x^{(r)}(y)$ corresponds to a uniform distribution, the factor in the brackets [ ] above is just $\frac{1}{4}$, the density of the distribution uniform on $[33,37]$.)

Suppose the free-interest rate $\delta = 0.04$. Then, in accordance with (1.1.3), the APV of the retirement annuity is

$$\int_{33}^{37} e^{-0.04y}\, 1250(3+y)\, e^{0.02y}\, a_{33+y}^{(r)} e^{-0.006y} \frac{1}{4} dy = 312.5 \int_{33}^{37} e^{-0.026y} (3+y)\, a_{33+y}^{(r)} dy.$$

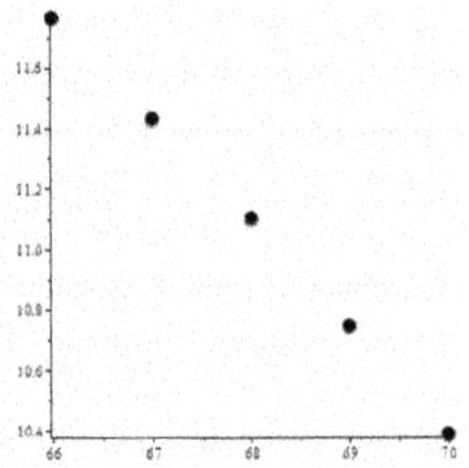

FIGURE 1.

It remains to choose an appropriate function $a^{(r)}_{33+y}$. It is noteworthy that it depends *only* on the survival probabilities for the population to which the participant (and the beneficiary if we consider the corresponding model) belong. Let us restrict ourselves to the no-beneficiary case and set $a^{(r)}_z = \overline{a}_z$, which may be considered an approximation for the APV with monthly payments. Let us proceed from the data from the Illustrative Table in Appendix, Section 3 and use the formulas $\overline{a}_x = \frac{1}{\delta}(1-\overline{A}_x)$ and $\overline{A}_x = \frac{e^{\delta}-1}{\delta}A_x$. The values of A_x are given in the table. Calculations lead to the values $11.76, 11.43, 11.10, 10.74, 10.39$ for $\overline{a}_x$ where $x = 66, ..., 70$ respectively; the corresponding plot is given in Fig.1. It looks practically linear, which allows us to accept—in this study example—the linear approximation

$$\overline{a}_x = 10.39 + 1.37\,\frac{70-x}{4} \quad \text{for } x \in [66, 70].$$

Thus, the desired APV equals

$$312.5\int_{33}^{37} e^{-0.026y}(3+y)\left(10.39+1.37\,\frac{37-y}{4}\right)dy \approx 211{,}652. \square$$

1.1.2 More examples of the benefit rate function $B(x,h,y)$

Most DB retirement plans involve the following three variables.

- The *salary of the participant.* It may be the salary at the moment of retirement or an average salary over the working period, computed with specified weights. Salaries corresponding to periods closer to the retirement are typically weighted more heavily. In the case of benefits not depending on salaries at all, one may proceed from a fixed "universal" salary equal to, say, a unit of money.

- The *accrual factor* that specifies which part of the salary contributes to the pension.

- The *number of years in service.*

Quite often, these three characteristics are multiplied. For instance, in Example 1.1.1-1, we multiplied the salary at the moment of retirement by the accrual factor $k = 0.025$, and the amount obtained was multiplied by the number of years in service at the moment of retirement.

The following is also worth emphasizing. Because we are talking about future benefits—and hence, about future salaries, we are dealing with *projected* (or estimated) salaries. Therefore, the function $B(x,h,y)$ represents *projected* (or estimated) retirement benefits.

Consider again a participant age x with h years in service. Denote by $w^{(a)}(x)$ the *actual* annual salary rate at age x, and by $w^{(e)}(x,y)$ the *estimated* annual salary rate for the same participant y years later. (Another notation uses the symbols $(AS)_x$ and $(ES)_{x+y}$, respectively; see, e.g., [19].)

One of the ways to estimate a future salary consists in introducing a *salary scale function* $S_{x,y}$ reflecting salary increases that are due to merit, seniority, inflation, and other factors.

One of possible examples is the function $S_{x,y} = e^{\beta(y-x)}\, s_{x,y}$, where the first factor reflects the salary growth due to inflation (at a rate of β), and $s_{x,y}$ is the factor representing the growth due to individual merit increases. The relation between the estimated and actual salary is given by the formula

$$w^{(e)}(x,y) = w^{(a)}(x)\,\frac{S_{x,y}}{S_{x,0}}. \tag{1.1.4}$$

To simplify examples and exercises below, we adopt the presentation

$$S_{x,y} = s_{x+y}, \tag{1.1.5}$$

where s_z is a scale function depending only on age.

Next, we consider several examples of the function $B(x,h,y)$. The first two are similar to Example 1.1.1-1.

1. The future pension rate is a fraction k_1 of the final salary rate, i.e.,

$$B(x,h,y) = k_1 w^{(e)}(x,y) = k_1 w^{(a)}(x)\,\frac{S_{x,y}}{S_{x,0}} = k_1 w^{(a)}(x)\,\frac{s_{x+y}}{s_x}. \tag{1.1.6}$$

2. Now, we include the number of years in service. The simplest way is to multiply the previous pension rate by the number of years in service, but certainly, in this case, the accrual factor may differ from k_1. So, using the symbol k_2, we write

$$B(x,h,y) = k_2\,(h+y)w^{(e)}(x,y) = k_2(h+y)w^{(a)}(x)\,\frac{s_{x+y}}{s_x}. \tag{1.1.7}$$

3. The next natural step is to replace the estimated salary at the moment of retirement by an average salary. It is quite common to replace $w^{(e)}(x,y)$ in (1.1.6) and/or (1.1.7) by the average of estimated salaries over the last m years prior to the projected retirement. For certainty, consider the case $m = 5$, which is common in practice. There are several ways to compute the average. First, we may replace $w^{(e)}(x,y)$ by

$$w^{(a)}(x)\,\frac{s_{x+y} + s_{x+y-1} + s_{x+y-2} + s_{x+y-3} + s_{x+y-4}}{5\,s_x}. \tag{1.1.8}$$

In a modified version, the last expression could be replaced by

$$w^{(a)}(x)\,\frac{0.5\,s_{x+[y]} + s_{x+[y]-1} + s_{x+[y]-2} + s_{x+[y]-3} + s_{x+[y]-4} + 0.5\,s_{x+[y]-5}}{5\,s_x}, \tag{1.1.9}$$

where $[y]$ is the integer part of y.

The motivation behind (1.1.9) is based on the assumption that retirement may occur at an intermediate moment of a year. So, instead of the five last years, we take the six years, but assign the weight 0.5 to the "end-point" years in the time interval considered.

Certainly, if $y < 5$, for the years that have already passed, estimated salaries should be replaced by actual salaries.

4. There is a version of the DB plan dealing with the average salary over the entire career. In this case the benefit is called a *career average benefit*.

5. Let us generalize the above definitions of average salary. To simplify the formula below, denote by $w(z)$ the salary of the participant at age z keeping in mind but suppressing in notation two cases. If z is less than or equal to the current age x, the symbol $w(z)$ denotes the actual salary, while if $z > x$, then $w(z)$ denotes the projected salary.

Recall that the number of years in service at age x was denoted by h, and for a given y, set $K = h + y$ (the projected total length of the in-service period). Then the average salary may be defined as

$$w_{\text{average}}(x,y) = \int_0^K q(s)w(x-h+s)\,ds, \tag{1.1.10}$$

where $q(s)$ is a weighting function. Since it is natural to place larger weights to years that are closer to the moment of retirement, we assume that $q(s)$ is a non-decreasing function. One of interesting particular examples is given in [94]:

$$q(s) = \gamma e^{\gamma(s-K)}, \tag{1.1.11}$$

where the parameter $\gamma > 0$.

In this case, we assign the weight $\gamma e^{-\gamma K}$ to the first salary earned (i.e., $w(x-h)$, usually, the lowest salary), and the weight γ to the last salary (i.e., $w(x+y)$, usually, the highest).

To understand (1.1.11), assume first that the salary is flat: $w(s)$ equals some number w_0. Then, by (1.1.10),

$$w_{\text{average}}(x,y) = w_0 \int_0^K \gamma e^{\gamma(s-K)}\,ds = w_0\,(1-e^{-\gamma K}), \tag{1.1.12}$$

which is close to the salary itself for large K.

Note also that the larger the parameter γ, the smaller the weights assigned to salaries early in the service term.

Suppose now that the salary is growing exponentially at a rate of β starting with a salary of w_0 at the age $x-h$. In other words, $w(x-h+s) = w_0 e^{\beta s}$. Substituting this and the formula for $q(s)$ from (1.1.11) into (1.1.10), we obtain that in this case,

$$w_{\text{average}}(x,y) = w_0 \frac{\gamma}{\gamma+\beta}\,(e^{\beta K} - e^{-\gamma K}). \tag{1.1.13}$$

Certainly, if $\beta = 0$ (the salary is flat), we come to (1.1.12).

Some particular examples may be found in Exercises 1-3.

1.2 DC plans

As was mentioned in the beginning of this section, for DC plans, we proceed from a given contribution rate.

1.2.1 Calculations at the time of retirement

For simplicity, we consider a continuous-time model. Let $[0,K]$ be a time interval. Consider a participant who entered a pension plan at time $t=0$ and retires at time $t=K$. All evaluations will be provided from the standpoint of the *retirement time*.

Let $w(t)$ be the salary rate at time $t \leq K$; that is, $w(t)dt$ is the salary received during the interval $[t, t+dt]$. Since the present time is the time of retirement, $w(t)$ is known. Let $c(t)$ be the contribution rate per unit of salary; that is, during a period $[t, t+dt]$ the participant contributed the amount $c(t)w(t)dt$ to the retirement account.

We adopt a model for which the contribution provided at time t is continuously increasing at a rate $\rho(t)$ due to investment. So, the contribution $c(t)w(t)dt$ increases to $e^{\rho(t)(K-t)}c(t)\times w(t)\,dt$ by time K. Consequently, the total retirement capital at the moment of retirement is equal to

$$C = \int_0^K c(t)w(t)\,e^{\rho(t)(K-t)}dt. \tag{1.2.1}$$

The participant is free to use this capital as she/he wishes. If the participant chooses to purchase a level life annuity with an APV of $a^{(r)}$, then the net pension annual rate equals

$$\pi = \frac{C}{a^{(r)}} = \frac{1}{a^{(r)}} \int_0^K c(t)w(t)\,e^{\rho(t)(K-t)}dt. \tag{1.2.2}$$

EXAMPLE 1. Consider an employee who is an active member of a pension plan, and assume that the contribution rate is constant: $c(t) = c$, as is the investment rate: $\rho(t) = \rho$. Assume also that the salary is growing at a constant rate β, that is, $w(t) = w_0 e^{\beta t}$. Suppose $\rho > \beta$. Then the accumulated capital equals

$$C = c\,w_0 \int_0^K e^{\beta t} e^{\rho(K-t)}dt = c\,w_0\,\frac{e^{\rho K} - e^{\beta K}}{\rho - \beta}. \tag{1.2.3}$$

Suppose the participant entered the plan at age 30 and retires at age 66; so, $K = 36$. Suppose further that the salary growth rate is $\beta = 0.02$; the investment rate is $\rho(t) = 0.08$; the employee was contributing to the retirement account at a rate of 3.5% and the employer was contributing for the participant at the same rate. That is, the total contribution rate is $c(t) = 0.07$. Then, by (1.2.3), it is easy to compute that the total accumulated capital per \$1 of the initial salary w_0 is equal to ≈ 18.38.

As for $a^{(r)}$, let us take $\bar{a}_{66}$ and adopt the value 11.76 obtained in Example 1.1.1-2. Then, the pension rate per \$1 of the initial salary is

$$\pi = \frac{18.38}{11.76} \approx 1.56.$$

For example, if the participant started at age 30 with a salary of \$40,000, then his (net) annual pension rate will amount to $\approx$ \$62,400 (while his annual salary at the moment of retirement is equal to $\$40,000\,e^{0.02\cdot 36} \approx \$82,177$).

The reader has probably noticed that while we set the investment rate $\rho = 0.08$, the value of $\bar{a}_{66}$ has been calculated proceeding from a risk-free rate of $\delta = 0.04$ (since we used the Illustrative Table). This is not meaningless: these characteristics are different, and the former is greater than the latter. Of course, the larger the investment rate is, the larger the pension. On the other hand, aggressive investment (hoping for a high rate) is risky.

In the model of this subsection, calculations are provided from the standpoint of the retirement time, so we do know what the investment rate *was*. It is easy to calculate that for $\rho = 0.06$, the pension rate would be \$39,385, and if $\rho = 0.04$ (the risk free rate), the pension decreases to \$25,788. □

1.2.2 Calculations at the time of entering a plan

We will use again a continuous time model and restrict ourselves to a particular scheme.

Suppose that at time $t = 0$, a participant age x enters a pension plan, and she/he is planning to retire at time K. Now, we compute the APV of the future contributions to the plan, and all evaluations will be provided from the standpoint of the *initial time*, i.e., $t = 0$.

Unlike what we did in Section 1.2.1, we will not include in our calculations the possibility of further growth of future contributions; for example, the contributions may go to a saving account.

Assume also that the plan includes a provision of returning a fraction d of the accumulated contributions with interest to the participant or her/his beneficiary in the case the participant leaves the plan before time K regardless of the reason.

Let $w(t)$ be the (projected) salary at time $t \in [0,K]$, and $c(t)$ be the (projected) contribution rate per unit of salary.

Denote by ${}_tp_x$ the probability that the participant will not leave the plan before time $t \in [0,K]$. In accordance with the general formula (**9**.1.1.7), without the returning-contribution-provision, the APV of the future contributions would be equal to the quantity

$$C_1 = \int_0^K e^{-\delta t} c(t) w(t) \cdot {}_tp_x dt,$$

where δ is a *risk-free* rate (rather than an investment rate ρ we used above).

To compute the APV of the return of contributions, we reason in the following way. If the plan returns the fraction d of the contributions in any case, such an APV would be exactly equal to $d \cdot C_1$. As a matter of fact, the plan does not return the contributions if the participant is still a member of the plan at time K, which occurs with probability ${}_Kp_x$.

Hence, the APV of the part of the contribution that the plan does not return is $d \cdot C_2$, where

$$C_2 = {}_Kp_x \int_0^K e^{-\delta t} c(t) w(t) dt.$$

Consequently, the APV of the contributions equals the quantity

$$\begin{aligned} C &= C_1 - d(C_1 - C_2) = (1-d)C_1 + dC_2 \\ &= (1-d) \int_0^K e^{-\delta t} c(t) w(t) \cdot {}_tp_x dt + d \cdot {}_Kp_x \int_0^K e^{-\delta t} c(t) w(t) dt. \end{aligned} \quad (1.2.4)$$

The next step may be either to find a contribution rate for which C is equal to the APV of a desired pension annuity or to figure out which annuity will be affordable for a given contribution rate.

EXAMPLE 1. For illustrative purposes, consider the "entirely exponential" case: $w(t) = w(0)\, e^{\beta t}$, and ${}_tp_x = e^{-\mu t}$, where μ is a leaving-hazard-rate. Let also $c(t) = c$, and $\delta > \beta$. Then

$$\begin{aligned} C &= c w_0 \left\{ (1-d) \int_0^K e^{-(\delta-\beta+\mu)t} dt + d \cdot e^{-\mu K} \int_0^K e^{-(\delta-\beta)t} dt \right\} \\ &= c w_0 \left\{ \frac{1-d}{\delta-\beta+\mu} \left(1 - e^{-(\delta-\beta+\mu)K}\right) + \frac{d}{\delta-\beta} e^{-\mu K} \left(1 - e^{-(\delta-\beta)K}\right) \right\}. \end{aligned} \quad (1.2.5)$$

As in Example 1.2.1-1, set $K=36$, $\beta=0.02$, $c=0.07$. Let $\delta=0.04$, $\mu=0.015$, and $d=0.9$. Substituting all of this into (1.2.5), we get

$$C \approx 1.085\, w_0. \tag{1.2.6}$$

Suppose that the retirement age is 66, and the participant enters the plan at age 30. As above, we use as an approximation the continuous-time APV $\bar{a}_{66}=11.76$ (see Example 1.1.1-2).

Next, we should recollect that we are dealing with a 36-year deferred annuity. If for the survival probability we accept the representation ${}_K p_x = e^{-\mu K}$ and for the discount—the quantity $e^{-\delta}$, then, the APV of the future pension annuity is

$${}_{36|}\bar{a}_{30} = {}_{36}p_{30}e^{-36\delta}\bar{a}_{66} = e^{-(0.015+0.04)36}\cdot 11.76 \approx 1.624.$$

Combining this with (1.2.6), we conclude that per one dollar of the initial salary w_0, the (net) pension annual rate is $\frac{1.085}{1.624} \approx 0.668$. Thus, for each one dollar of the initial salary, the participant will receive ≈ 66.8¢ of the annual pension paid in monthly portions (we consider $\bar{a}_{66}$ as an approximation of such a payment). For instance, for $w_0 = \$40,000$, the participant may hope for a (net) annual pension rate of $\approx \$26,720$. □

---

2 PENSION FUNDING. COST METHODS

This section differs from the previous in two respects. First, we consider a fund with many participants rather than one future retiree, and our main concern is the viability of the fund. Secondly, the models below are dynamic, and we pay attention to the evolution of funds in time. The model of Subsection 2.1 is similar to that in [19, Chapter 20].

2.1 A dynamic DB fund model

2.1.1 Modeling enrollment flow and pension payments

Consider a fund in which all participants enroll at age a and retire at age r. Suppose that in a time interval $[t, t+dt]$, the number of *new* entrants (age a) equals $n(t)dt$; in other words, $n(t)$ is the enrolling intensity at time t. Certainly, the actual process is random, so we view $n(t)$ as a mean characteristic.

To make the formulas below less cumbersome, we assume the existence of an initial time $t=0$ and set the function $n(t)=0$ for $t<0$.

Let $s_a(x)$, $x \geq a$, be the probability that a new entrant will be still a member of the fund at age x. We may view $s_a(x)$ as a survival function concerning remaining life, and we will call it as such, although $s_a(x)$ also concerns withdrawal from the plan for reasons other than death. By definition, $s_a(a)=1$. In the notation of Chapter 7, $s_a(x) = {}_{x-a}p_a$.

Now, let us consider the mean number of participants attaining age x (or more precisely, ages in $[x, x+dx]$) during the interval $[t, t+dt]$. These are those people who entered the fund $x-a$ years ago and survived to age x. So, the mean number in question is

$$n(t-(x-a))s_a(x)dt = n(t-x+a)s_a(x)dt. \tag{2.1.1}$$

(More precisely, the number of entrants with age in $[x, x+dx]$ is $(n(t-(x-a))s_a(x)dt)dx$. Since we set $n(t) = 0$ for $t < 0$, the last expression vanishes for $x > a+t$, which is natural: if the process starts at time $t = 0$, then at time t the maximal possible age for a participant is $a+t$.)

Suppose that at time t, all participants at age x have the same salary rate $w(x,t)$; that is, $w(x,t)dt$ is the salary earned during the interval $[t, t+dt]$. Thus, the mean salary depends only on age and current time. The latter type of dependence may reflect inflation and/or the common growth of productivity. A typical presentation is $w(x,t) = w(x)\,e^{\tau t}$, where τ is a rate due to the second type of dependency, and $w(x)$ reflects the dependence on age.

Then, the total annual salary rate at time t is

$$W_t = \int_a^r n(t-x+a)s_a(x)\,w(x,t)dx. \tag{2.1.2}$$

One may say that $W_t\,dt$ is the total payroll payment during the interval $[t, t+dt]$.

Next, we specify future pensions; so we consider a DB plan. Suppose that for a participant who retires at time t, the *initial* pension rate (that is, the rate at time t) is a fraction k of the salary at the moment of retirement, i.e., $kw(r,t)$.

Usually, the pension rate is not flat, and next we consider the pension rate $\pi(x,t)$ of a retiree age $x \geq r$ at time t. For such a retiree, the retirement occurred $x-r$ years ago; i.e., at time $t-(x-r) = t-x+r$. We assume that the pension rate $\pi(x,t)$ is equal to the initial pension multiplied by an adjustment factor $h(x)$ depending only on the age x and, naturally, such that $h(r) = 1$. Thus,

$$\pi(x,t) = \pi(r, t-x+r)h(x) = kw(r, t-x+r)h(x). \tag{2.1.3}$$

2.1.2 Normal cost

In this subsection, we restrict ourselves to the terminal funding case where a single contribution to the fund is made for participants who retire at time t. Our object of study is contribution rate, or in other terms *normal cost rate*, S_t (other notations used include $(NC)_t$ or ${}^TP_t$). More precisely, $S_t dt$ is the *required* contribution to the fund during the period $[t, t+dt]$ to ensure—on the average since we consider net characteristics—the payment of future pensions for *new* retirees, i.e., those who retire in the interval $[t, t+dt]$.

Let $\overline{a}_r^{(h)}$ be the APV of the pension annuity with the adjustment factor $h(\cdot)$ per *unit initial pension rate* payable continuously starting from the retirement age r. The APV $\overline{a}_r^{(h)}$ is calculated from the standpoint of the retirement time.

The probability that a retiree will attain age x is $s_a(x)/s_a(r)$ and if the initial pension rate is one, then the pension rate at age x is $1 \cdot h(x) = h(x)$. As usual, let δ be a risk-free interest rate. Then, in accordance with the general formula (**9**.1.1.7), making the variable change $y = x-r$ in the second step, we have

$$\overline{a}_r^{(h)} = \int_r^\infty e^{-\delta(x-r)}h(x)\frac{s_a(x)}{s_a(r)}dx = \int_0^\infty e^{-\delta y}h(r+y)\frac{s_a(r+y)}{s_a(r)}dy. \tag{2.1.4}$$

Let

$$\widetilde{s}_z(y) = \frac{s_a(z+y)}{s_a(z)}.$$

This is the probability that a participant will be still a member of the plan (perhaps as a retiree) at age $z+y$ given that this was true at age z. For us it is worth remembering that for $z \geq r$, the survival function $\widetilde{s}_z(y)$ is that of the remaining life and is based only on the mortality force after time z. Clearly, $\widetilde{s}_z(0) = 1$.

Set also $\widetilde{h}(y) = h(r+y)$, and note that $\widetilde{h}(0) = 1$.

So, with this new notation we can rewrite (2.1.4) as

$$\overline{a}_r^{(h)} = \int_0^\infty e^{-\delta y}\widetilde{h}(y)\widetilde{s}_r(y)dy. \tag{2.1.5}$$

By (2.1.1), the (mean) number of participants attaining the retirement age r in the time interval $[t, t+dt]$ is $n(t-r+a)s_a(r)dt$, and these people start to collect pensions at the initial rate $kw(r,t)$. Consequently, the normal cost $S_t = kw(r,t)n(t-r+a)s_a(r)\overline{a}_r^{(h)}$. Writing the factors that do not depend on t first, we have

$$S_t = ks_a(r)\overline{a}_r^{(h)}\,w(r,t)n(t-r+a). \tag{2.1.6}$$

(For $t < r-a$, this expression vanishes, which is understandable: there are no retirees yet.)

EXAMPLE 1 (*The "exponential" case.*) First, let $h(x) = e^{\alpha(x-r)}$, where α is a constant rate. Then $\widetilde{h}(y) = e^{\alpha y}$. It is natural to assume $\alpha \leq \delta$ (the pension growth rate is not larger than the risk free interest rate). Substituting into (2.1.5), we have

$$\overline{a}_r^{(h)} = \int_0^\infty e^{-(\delta-\alpha)y}\widetilde{s}_r(y)dy. \tag{2.1.7}$$

The r.-h.s. looks as a "standard" level annuity with a risk-free rate $\delta-\alpha$ and depends only on the survival function $\widetilde{s}_r(x)$. Denote the r.-h.s. of (2.1.7) by $\overline{a}^{(r)}$.

Next, suppose that $n(t)$ is growing exponentially, and set $n(t) = n_0e^{\omega t}$ for $t \geq 0$, where ω is the corresponding population growth rate. As above, $n(t) = 0$ for $t < 0$.

For the salary, we adopt the presentation $w(x,t) = w(x)\,e^{\tau t}$ already mentioned in Section 2.1.1.

Thus, for $t > r-a$,

$$S_t = kw(r)s_a(r)\overline{a}^{(r)}\,e^{\tau t}n_0e^{\omega(t-r+a)} = \left(kn_0w(r)s_a(r)\overline{a}^{(r)}e^{-\omega(r-a)}\right)\,e^{(\tau+\omega)t} = S_0\,e^{(\tau+\omega)t}. \tag{2.1.8}$$

Thus in the exponential case, the normal cost (i.e., the required contribution rate) is growing exponentially, and this growth is specified by the population and salary growth rate. As follows from (2.1.8), the coefficient $S_0 = kn_0w(r)s_a(r)\overline{a}^{(r)}e^{-\omega(r-a)}$.

Since in the case under consideration the growth of the number of new active participants as well as the salary growth are also exponential, one may hope that with appropriate contributions of active participants, the fund will be viable. □

2.1.3 The benefit payment rate and the APV of future benefits

At time t, the number of retirees ages from $[x, x+dx]$ equals $n(t-(x-a))s_a(x)dx$. Since we are talking about retirees, $x \geq r$ and these retirees retired $x-r$ years ago.

In accordance with (2.1.3), the initial pension rate for these people is $kw(r,t-x+r)$, and the current pension rate is $\pi(x,t) = kw(r,t-x+r)h(x)$. Hence, for $t \geq r-a$, the benefit payment rate is

$$B_t = \int_r^\infty n(t-x+a)s_a(x)\cdot\pi(x,t)\,dx = k\int_r^\infty n(t-x+a)s_a(x)w(r,t-x+r)h(x)\,dx. \quad (2.1.9)$$

(As a matter of fact, the integration runs only from r to $a+t$ as the integrand equals zero for $x > a+t$.) It is worthwhile to emphasize that B_t is the rate of *current* pension payment; that is, at time t.

Now, we turn to the APV of *future* benefit payments. Consider a retiree age x. Similarly to (2.1.5), for a *unit initial* pension rate, the APV of the *future (remaining)* pension annuity is

$$\bar{a}_x^{(h)} = \int_0^\infty e^{-\delta y}h(x+y)\tilde{s}_x(y)\,dy = \int_0^\infty e^{-\delta y}\tilde{h}(x-r+y)\tilde{s}_x(y)\,dy, \quad (2.1.10)$$

where as above, δ is a risk free rate. As has been already noticed, since $x \geq r$, the survival function $\tilde{s}_x(y)$ is based only on mortality.

Thus, the APV of the future pension payments is

$$\int_r^\infty n(t-x+a)s_a(x)\cdot kw(r,t-x+r)\bar{a}_x^{(h)}\,dx = k\int_r^\infty n(t-x+a)s_a(x)w(r,t-x+r)\bar{a}_x^{(h)}\,dx. \quad (2.1.11)$$

The actuarial notation for this quantity is $(rA)_t$.

EXAMPLE 1. (*The "exponential-uniform" case.*) Let us again consider the exponential case setting $h(x) = e^{\alpha(x-r)}$, $n(t) = n_0e^{\omega t}$, and $w(x,t) = w(x)e^{\tau t}$. However, for $s_a(x)$ we will accept a more realistic assumption than that of the exponential case, that is, where the mortality rate is constant. In Section **7**.1.6, we saw that for elderly people the distribution of the remaining life time was close to a uniform distribution. So, we assume that the survival function for the remaining life time of a retiree, i.e., $\tilde{s}_r(x)$, corresponds to the uniform distribution on an interval $[0,d]$. Say, if $r = 66$, as an approximation, we can set $d = 30$ certainly realizing that, as a matter of fact, people may live longer than 96 years. Thus for $x > r$, we will use the representation

$$s_a(x) = s_a(r)\left(1-\frac{x-r}{d}\right), \quad (2.1.12)$$

where $s_a(r)$ is a fixed quantity: the probability that a member chosen at random will attain the retirement age r. Then, assuming $t > (r-a)+d$, proceeding from (2.1.9), and making the variable change $y = x-r$, we have

$$B_t = k\int_r^{r+d} n_0e^{\omega(t-x+a)}s_a(r)\left(1-\frac{x-r}{d}\right)w(r)e^{\tau(t-x+r)}e^{\alpha(x-r)}\,dx \quad (2.1.13)$$

$$= kn_0s_a(r)w(r)\int_0^d e^{\omega(t-y-r+a)}\left(1-\frac{y}{d}\right)e^{\tau(t-y)}e^{\alpha y}\,dy$$

$$= kn_0s_a(r)w(r)e^{(\omega+\tau)t}e^{-\omega(r-a)}\int_0^d\left(1-\frac{y}{d}\right)e^{-(\omega+\tau-\alpha)y}\,dy \quad (2.1.14)$$

$$= \left(kn_0s_a(r)w(r)e^{-\omega(r-a)}b_1\right)e^{(\omega+\tau)t}, \quad (2.1.15)$$

where $b_1 = b_1(d,\omega,\tau,\alpha)$ is the value of the integral in (2.1.14). Note that b_1 does not depend on time t or the retirement age r. Thus, in the case under consideration, the benefit rate is growing exponentially, which might be expected. In Exercise 15, we discuss other aspects of (2.1.15).

Now, let us turn to the APV. We know that for a uniform r.v. X, the conditional distribution of X given $X \geq x$ is uniform. Hence, the survival function $\tilde{s}_x(y)$ corresponds to the distribution uniform on $[0, r+d-x]$. So, by virtue of (2.1.10), for $x \in [r, r+d]$, we have

$$\bar{a}_x^{(h)} = \int_0^{r+d-x} e^{-\delta y} e^{\alpha(y+x-r)} \left(1 - \frac{y}{r+d-x}\right) dy$$
$$= e^{\alpha(x-r)} \int_0^{d-(x-r)} e^{-(\delta-\alpha)y} \left(1 - \frac{y}{d-(x-r)}\right) dy = e^{\alpha(x-r)} b_2(x-r),$$

where for $s < d$

$$b_2(s) = b_2(s; d, \delta, \alpha) = \int_0^{d-s} e^{-(\delta-\alpha)y} \left(1 - \frac{y}{d-s}\right) dy.$$

For $s = d$, we may set $b_2(d) = \lim_{s \to d-} b_2(s) = 0$.

When substituting all of the above representations into (2.1.11), it is convenient to keep in mind that formulas (2.1.9) and (2.1.11) are similar. Using the variable change $y = x - r$ in the second step, we have

$$(rA)_t = k \int_r^{r+d} n_0 e^{\omega(t-x+a)} s_a(r) \left(1 - \frac{x-r}{d}\right) w(r) e^{\tau(t-x+r)} e^{\alpha(x-r)} b_2(x-r)\, dx \quad (2.1.16)$$
$$= k n_0 s_a(r) w(r) \int_0^d e^{\omega(t-y-r+a)} \left(1 - \frac{y}{d}\right) e^{\tau(t-y)} e^{\alpha y} b_2(y)\, dy$$
$$= k n_0 s_a(r) w(r) e^{(\omega+\tau)t} e^{-\omega(r-a)} \int_0^d \left(1 - \frac{y}{d}\right) e^{-(\omega+\tau-\alpha)y} b_2(y)\, dy \quad (2.1.17)$$
$$= \left(k n_0 s_a(r) w(r) e^{-\omega(r-a)} b_3\right) e^{(\omega+\tau)t}, \quad (2.1.18)$$

where $b_3 = b_3(d,\omega,\tau,\alpha)$ is the value of the integral in (2.1.17).

So, the APV is also growing exponentially at the same rate as in (2.1.15). We discuss the last formula in more detail in Exercise 15. □

2.2 More on cost methods

In a sense, the models of this section are simpler than the previous because the fund we consider does not accept new participants after it has been established at an initial time. On the other hand, we explore new details. Time below is discrete, and all plans are DB plans.

The models below are similar to some models in [4]; the reader may find more details there, though our exposition is somewhat different.

2.2.1 The unit-credit method

For an integer time moment t denote the set of all active (non-retired) participants by $\mathcal{A}_t$; the age of the j-th participant at time t by x_j (the second index t is suppressed); and the same for all participants retirement age by r. Suppose the fund was established at an initial time, the age of participant j at this time was a_j, and there were no enrollments after that. So, for active (non-retired) participants $a_j \le x_j < r$.

For simplicity, we assume all x's, a's, and r to be integers. To make notations less cumbersome (especially when it does not cause misunderstanding), we will omit the index j in x_j, a_j, and other individual characteristics.

For a participant, denote by $B(x)$ a *projected* pension rate that the participant has already earned by age x. (As a matter of fact, $B(x) = B_j(x_j)$; we suppress the index in notation.) The function $B(x)$ is non-decreasing, and it is natural to set $B(a) = 0$. We call $B(x)$ an *accrued benefit rate* at age x. So, $B(r)$ is the maximal benefit rate for which the participant can hope. The pension annuity is assumed to be level at the rate $B(r)$.

Thus, at age $x \le r$, the APV of the participant's *accrued benefit* is $B(x) \cdot v^{r-x} \cdot {}_{r-x}p_x \cdot \ddot{a}$, where as usual v is a discount, ${}_sp_x$ is the probability that the participant will be an active member s years later, and $\ddot{a}$ is the APV of the future pension annuity at a *unit* rate from the standpoint of the retirement time.

For example, $\ddot{a}$ may be $\ddot{a}_r^{(12)}$, the APV of the discrete time annuity paid *monthly* with a unit *annual* rate.

For simplicity, we call ${}_sp_x$ a survival function and assume it to be the same for all participants.

Thus, the APV of the total future pension benefits from the standpoint of time t is

$$\sum_{j \in \mathcal{A}_t} B_j(x_j) v^{r-x_j} {}_{r-x_j}p_{x_j} \ddot{a}. \tag{2.2.1}$$

For the fund to be viable (on the average), this quantity should equal the APV of the obligations of the fund; so we call the last expression an *accrued liability* and denote it by $(AL)_t$.

The accrued liability changes in time for two reasons: the participants are getting older, and the structure of the active group may change. As was told, we assume that there are no new entrants. However, we should take into account that (a) some contracts may be terminated due to death or withdrawal, (b) once participants reach the retirement age, they cease to be active members.

Regarding a particular participant, how much should be added to the fund to take into account these changes?

If participant j at age x_j is becoming one year older, then the accrued benefit rate will grow by $\Delta B_j = \Delta B_j(x_j) = B(x_j + 1) - B(x_j)$. However, this change will be needed only if the participant survives the retirement age, which will happen with probability ${}_{r-x_j}p_{x_j}$. Also, the payments will start when the participant attains age r. So, *on the average*, the amount that should be added to the fund is $\Delta B_j v^{r-x_j} {}_{r-x_j}p_{x_j} \ddot{a}$. We call this (net!) characteristic a *normal cost* of the plan at time t and denote it by $(NC)_{tj}$. So,

$$(NC)_{tj} = \Delta B_j v^{r-x_j} {}_{r-x_j}p_{x_j} \ddot{a}. \tag{2.2.2}$$

The total normal costs for the whole fund at time t are

$$(NC)_t = \sum_{j \in \mathcal{A}_t} (NC)_{tj}.$$

► Let us consider it in more detail, working with r.v.'s rather than with their expected values. Suppose that we are at time t; so the set $\mathcal{A}_t$ is fixed. Denote by $\mathcal{T} = \mathcal{T}_t$ the set of participants whose contracts will be terminated in the period $[t, t+1]$, and by $\mathcal{R} = \mathcal{R}_t$ the set of all participants who will retire at time $t+1$.

Let the r.v. $I_{j,t} = 1$ if the j-th contract will be terminated during the period mentioned, and $I_{j,t} = 0$, otherwise. Then the set $\mathcal{T} = \{j : I_{j,t} = 1,\ j \in A_t\}$ and $\mathcal{R} = \{j : x_j = r-1,\ I_{j,t} = 0,\ j \in A_t\}$. Clearly,

$$\mathcal{A}_t = \mathcal{A}_{t+1} + \mathcal{T}_t + \mathcal{R}_t,$$

where $\mathcal{A}_{t+1} = \{j : x_j < r-1, I_j = 0, j \in \mathcal{A}_t\}$. It is worthwhile to emphasize that from the standpoint of time t, the sets $\mathcal{A}_{t+1}$, $\mathcal{T}_t$, and $\mathcal{R}_t$ are random.

Let Y_j be the present value of the future pension payments for member j from the standpoint of the retirement time; so $E\{Y_j\} = \ddot{a}$. Let $\widetilde{I}_{j,t} = 1$ if participant j (of age x_j) is a member of the fund at the retirement age r, and $\widetilde{I}_{j,t} = 0$, otherwise.

Denote by Υ_t the (random) present value of the fund's liability at time t. (So, $E\{\Upsilon_t\} = (AL)_t$.) We have

$$\Upsilon_t = \sum_{j \in \mathcal{A}_t} B_j(x_j) v^{r-x_j} \widetilde{I}_{j,t} Y_j = \sum_{j \in \mathcal{A}_t} B_j(x_j+1) v^{r-x_j} \widetilde{I}_{j,t} Y_j - \sum_{j \in \mathcal{A}_t} \Delta B_j v^{r-x_j} \widetilde{I}_{j,t} Y_j. \tag{2.2.3}$$

Denote the second sum in the far r.-h.s. of (2.2.3) by G_{1t}. Furthermore,

$$\sum_{j \in \mathcal{A}_t} B_j(x_j+1) v^{r-x_j} \widetilde{I}_{j,t} Y_j = \sum_{j \in \mathcal{A}_{t+1}} + \sum_{j \in \mathcal{T}_t} + \sum_{j \in \mathcal{R}_t}. \tag{2.2.4}$$

The first sum above equals

$$v \sum_{j \in \mathcal{A}_{t+1}} B_j(x_j+1) v^{r-(x_j+1)} \widetilde{I}_{j,t} Y_j. \tag{2.2.5}$$

Let us observe that if $j \in \mathcal{A}_{t+1}$ (i.e, the participant is active at time $t+1$), then $\widetilde{I}_{j,t} = \widetilde{I}_{j,t+1}$, and the sum in (2.2.5) equals the r.v. Υ_{t+1}. Then, the whole expression is equal to $v\Upsilon_{t+1}$.

For $j \in \mathcal{T}$, the indicator $\widetilde{I}_{j,t} = 0$; so the second sum in (2.2.4) vanishes.

The third sum in (2.2.4) equals $v \sum_{j \in \mathcal{R}_t} B_j(x_j+1) v^{r-(x_j+1)} \widetilde{I}_{j,t+1} Y_j$. The last sum—denote it by G_{t2}—is the (random) present value from the standpoint of time $t+1$ of the total pension benefits to be paid for the participants who will retire at time $t+1$.

Collecting everything, we write

$$\Upsilon_t = v\Upsilon_{t+1} + vG_{t2} - G_{t1}. \tag{2.2.6}$$

Let i be the risk-free interest. Then $v = 1/(1+i)$, and we will rewrite (2.2.6) as

$$\Upsilon_{t+1} = (1+i)\,[\Upsilon_t + G_{t1}] - G_{t2}. \tag{2.2.7}$$

This is a balance equation. The accrued liability Υ_t may be viewed as the capital of the fund at time t needed for the fund to be viable. By time $t+1$, this capital will have grown to $(1+i)\Upsilon_t$, and the amount G_{t2} will be withdrawn for the new retirees to purchase the desired annuities. From (2.2.7), it follows that for the fund to be still viable at time $t+1$, the amount G_{t1} should be added at the beginning of the year t.

Certainly, rigorously speaking, all of this is impossible because from the standpoint of time t, the quantities G_{t1} and G_{t2} are random, and we do not know how much should be deposited. If we restrict ourselves to expected values, we will come to net characteristics. In a more sophisticated model, we would add to the expected values security loadings which depend on the lowest acceptable probability of viability.

The reader will readily double check that the expected value $E\{G_{t1}\}$ is eqaul to the *normal cost* $(NC)_t = \sum_{j\in\mathcal{A}_t}(NC)_{tj}$, where the normal cost $(NC)_{tj}$ is given in (2.2.2). ◄

Now, let us consider the relative growth of the normal cost for one participant during a year; that is,

$$\frac{(NC)_{(t+1)j}-(NC)_{tj}}{(NC)_{tj}}.$$

If the age of a participant at time t is x, then the age at time $t+1$ is $x+1$. To make the result more explicit, instead of the last expression, let us consider the infinitesimal characteristic $\frac{1}{(NC)_{tj}}\frac{d}{dx}(NC)_{tj}$. In view of (2.2.2),

$$\frac{1}{(NC)_{tj}}\frac{d}{dx}(NC)_{tj}=\frac{d\ln(NC)_{tj}}{dx}=\frac{d\ln\Delta B_j(x)}{dx}+\frac{d\ln{}_{r-x}p_x}{dx}+\frac{d\ln v^{r-x}}{dx}.$$

(The index j in x_j is suppressed.)

As usual, set $v=e^{-\delta}$ and ${}_{r-x}p_x=\exp\left\{-\int_x^r\mu(s)ds\right\}$. Then

$$\frac{d\ln(NC)_{tj}}{dx}=\frac{d\ln\Delta B_j(x)}{dx}+\mu(x)+\delta. \tag{2.2.8}$$

This is important: in the framework of the method under discussion, the normal costs are growing; moreover, they are growing faster than the accrued benefits. On the other hand, the latter is strongly connected with salary growth; so the contributions needed are growing faster than the salary.

In Section 2.2.2, we fix this problem to some extent by considering a model with level costs, but for now we will stick to the current actuarial method and explore the expected value of all accumulated normal costs prior to time t.

More specifically, we will show that as may be expected,

> In a balanced fund, the APV of accumulated prior normal costs equals the APV of accrued liability.

To this end, we recall the definition of an accumulated value from Section **10**.1.3; more precisely, the expected accumulated value of a temporary annuity, *given that the annuitant survives* the term of the annuity. The corresponding presentation obtained in Section **10**.1.3 was

$$\ddot{s}_{x:\overline{n}|}=\frac{\ddot{a}_{x:\overline{n}|}}{v^n\cdot{}_np_x}. \tag{2.2.9}$$

The term $\ddot{a}_{x:\overline{n}|}$ above concerns a level temporal annuity. To compute the APV of the accumulated *prior* individual normal costs (which are not level) for a participant at time t, we may use the same formula replacing the numerator by the APV of the annuity with payments equal to normal costs. Namely, for a particular participant, omitting the index j, we replace $\ddot{a}_{x:\overline{n}|}$ in (2.2.9) by $\sum_{k=0}^{t-1} v^k\, {}_kp_a\, (NC)_{(k)}$, where $(NC)_{(k)}$ is the normal cost for the participant under consideration at time $k < t$, and a is the age at enrollment. For the moment, let us denote the survival function by $s(x)$. Proceeding from (2.2.2), we have

$$\sum_{k=0}^{t-1} v^k\, {}_kp_a\, (B(a+k+1) - B(a+k))v^{r-(a+k)}\, {}_{r-(a+k)}p_{a+k}\, \ddot{a}$$

$$= \ddot{a} v^{r-a} \sum_{k=0}^{t-1} (B(a+k+1) - B(a+k))\, \frac{s(a+k)}{s(a)}\, \frac{s(r)}{s(a+k)}$$

$$= \ddot{a} v^{r-a} \frac{s(r)}{s(a)} \sum_{k=0}^{t-1} (B(a+k+1) - B(a+k)) = \ddot{a} v^{r-a} \frac{s(r)}{s(a)}\, B(a+t).$$

(We used "telescoping" and the fact that $B(a) = 0$. Note also that $B(a+t) = B(x)$.)

Thus, for the normal costs accumulated prior to time t, we use (2.2.9) replacing the numerator by the above value obtained, and the denominator by $v^{x-a}\, {}_{x-a}p_a$. Then, the APV in question is

$$\frac{1}{v^{x-a}\, {}_{x-a}p_a} \ddot{a} v^{r-a} \frac{s(r)}{s(a)}\, B(x) = B(x) v^{r-x} \frac{s(a)}{s(x)}\, \frac{s(r)}{s(a)}\, \ddot{a} = B(x)\, v^{r-x}\, {}_{r-x}p_x \ddot{a},$$

which is consistent with (2.2.1).

2.2.2 The entry-age-normal method

In the model of this subsection, we assume the contributions to be level in time. So, the notion of contribution is not derived from original premises of the model but is defined at the very beginning. Let c_j be the (level) one-period contribution of participant j. We again suppress j in the notation when considering a particular participant.

For each participant, we require the APV of the total payments (from the standpoint of the initial time) to be equal to the APV of future benefits. That is, we require

$$B(r)\, v^{r-a}\, {}_{r-a}p_a\, \ddot{a} = c\ddot{a}_{a:\overline{r-a}|}, \tag{2.2.10}$$

where $\ddot{a}_{a:\overline{r-a}|}$ is the APV of a unit-rate temporal annuity starting from age a and lasting no more than $r-a$ years, and $\ddot{a}$ is the same as in Section 2.2.1.

The quantity c above is a *normal cost*, and relation (2.2.10) is *a definition of normal cost* in the framework of the actuarial cost method under discussion.

We saw in Section 2.2.1 that under the unit-cost method, at any time t, the accrued liability equals the APV of *prior* costs. Now, we choose this property as a definition of accrued liability. That is, in accordance of general formula (2.2.9), the accrued liability at time t of a participant age x is

$$(AL)_t = c\ddot{s}_{a:\overline{x-a}|} = c\, \frac{\ddot{a}_{a:\overline{x-a}|}}{v^{x-a} \cdot {}_{x-a}p_a}. \tag{2.2.11}$$

To clarify this definition, we show that

> In a balanced fund, at any time, the accrued liability is equal to the APV of future benefits minus the APV of future costs.

In other words, from the standpoint of any time, *the APV of the prior costs plus the APV of future costs equal the APV of the future benefits.*

To prove this, let us recollect that for $x \le r$, we have $\ddot{a}_{a:\overline{x-a}|} = \ddot{a}_{a:\overline{r-a}|} - v^{x-a}\,{}_{x-a}p_a \ddot{a}_{x:\overline{r-x}|}$. Substituting c from (2.2.10) only in the first term below, we write

$$(AL)_t = c\frac{\ddot{a}_{a:\overline{x-a}|}}{v^{x-a}\cdot{}_{x-a}p_a} = c\frac{\ddot{a}_{a:\overline{r-a}|}}{v^{x-a}\cdot{}_{x-a}p_a} - c\frac{v^{x-a}\,{}_{x-a}p_a\ddot{a}_{x:\overline{r-x}|}}{v^{x-a}\cdot{}_{x-a}p_a}$$

$$= \frac{B(r)\,v^{r-a}\,{}_{r-a}p_a\,\ddot{a}}{\ddot{a}_{a:\overline{r-a}|}}\cdot\frac{\ddot{a}_{a:\overline{r-a}|}}{v^{x-a}\cdot{}_{x-a}p_a} - c\frac{v^{x-a}\,{}_{x-a}p_a\ddot{a}_{x:\overline{r-x}|}}{v^{x-a}\cdot{}_{x-a}p_a} = B(r)v^{r-x}\,{}_{r-x}p_x\ddot{a} - c\ddot{a}_{x:\overline{r-x}|},$$

which is exactly what we were trying to prove.

The above calculation concerned a single participant. Clearly, the same properties are true for total characteristics of the plan: for the total accrued liability we should add up the accrued liabilities for all participants, and the same is true for the total normal costs.

---

3 EXERCISES

Section 1

1. (a) A pension plan for a company's employees provides a (future) annual income at retirement equal to a fraction of the final salary times the number of years in service. Suppose the future salary is projected to grow linearly, i.e., as $w_0(1+\tau z)$, where w_0 is an initial salary, τ is a linear relative rate, and z is the number of in-service years. Considering the pension as a function of the time to retirement, with what type of functions are we dealing?

   (b) Consider two *new* employees of ages 25 and 30, with respective salaries \$50,000 and \$60,000. The salary growth rates mentioned above for these participants are different: 0.02 and 0.0125, respectively. The employees are friends and have decided to retire simultaneously. (i) At what time should they retire for their pensions to be the same? (ii) If the employees instead decided to retire at the same age, at what age should they retire to have the same pension?

2. The current salaries of two employees of age 40 and 45 are \$40,000 and \$60,000, respectively. Their salaries are growing exponentially with respective rates 4% and 3%. The projected pension is a fraction k of the final salary, and retirement is possible after age 60.

   (a) If the employees retire at the same age, what should this age be for their pensions to be the same?

   (b) If the employees retire at the same time, when should it happen for their pensions to be the same? If the answer is non-realistic, change somehow the employees's ages to make the answer realistic.

3. Suppose that (1.1.4)–(1.1.5) are true, and the function s_x is linear.

   (a) If $\frac{s_{2x}}{s_x} = 2$, what is s_0? If $\frac{s_{2x}}{s_x} \neq 2$, what is the structure (or type) of the function $\frac{s_{2x}}{s_x}$?

   (b) Let $w^{(e)}(40,5) = 1.1w^{(a)}(40)$. Find s_x. Explain why it suffices to have a solution up to a constant multiplier.

4. In the case of Example 1.1.1-2, find the probability that the participant will retire. Write a general formula for this probability.

5. (a) In the framework of Section 1.1.1, find a formula for the APV of future pension benefits for a plan for which all participants must retire at an age r. (b) What should we change in Example 1.1.1-2 for this example to satisfy the above condition? (c) Find the particular value of the APV in this case.

6. In the situation of Section 1.1.1, among two participants having identical salaries at each age, the first has a smaller death and withdrawal hazard rates for all ages. Do you expect the APV of pension benefits for the first participant to be larger? If not, which additional condition should be imposed for this to be true?

7. In the situation of Example 1.1.1-2, consider the case of constant hazard rates and the presence of a beneficiary (see remarks in Section 1.1.1). More specifically, assume that the decrement rate $\mu_x^{(d)}(y) = \mu$ for $y \leq r - x$ and $\mu_x^{(d)}(y) = 0$ for $y > r - x$; the retirement hazard rate (including the force of mortality) $\mu_x^{(r)}(y) = 0$ for $y < r - x$ and $\mu_x^{(r)}(y) = \mu_1$ for $y \geq r - x$. When computing the APV of the pension annuity from the standpoint of the retirement time, we proceed from a constant mortality rate μ_2.

   (a) How will (1.1.3) look in this case?

   (b) For the data in Example 1.1.1-2 with $\mu_1 = 0.1$ and $\mu_2 = 0.05$, find the particular value of the APV.

8. (a) Find the limit of the r.-h.s. of (1.1.13) as $\gamma \to \infty$. Comment on the result from an economic point of view. (b) For a finite γ, how should (1.1.13) be changed if instead of the initial salary w_0, we are given a salary w at age x. (c) Let $\gamma = 1$. What will change in (1.1.1) if we replace the final salary with the average salary in the sense of (1.1.13).

9. How would formula (1.2.3) change in the case $\rho = \beta$? Is the r.-h.s. of (1.2.3) still positive if $\rho < \beta$? Explain that the restriction $\rho \geq \beta$ is reasonable from an economic point of view. (*Hint*: For $\rho < \beta$, the investment growth is smaller than the salary growth.)

10. In the situation of Section 1.2, consider the case where the future retiree does not need more than an amount $\overline{w}$ for comfortable (in her/his opinion) living. More specifically, assume that the contribution rate is equal to a constant $c < 1$ up to the moment when the income remaining after the contribution in the retirement account reaches the level $\overline{w}$. After that the participant contributes the entire surplus into the retirement account. Assume also $w(t) = w_0 e^{\beta t}$ and $\rho(t) = \rho$. Write the integral in (1.2.1) in this case. When will it differ from (1.2.3)? You may also provide particular calculations of the integral which are simple but a bit cumbersome.

11. (a) Is the expression in (1.2.4) increasing in d? Justify the answer mathematically and give a common-sense explanation.

    (b) In Example 1.2.2-1, provide formulas and calculations for $d = 1$. Did you expect a larger or smaller pension rate in comparison with $d < 1$?

Section 2

12. Let $t \geq r-a$. (a) Find the payroll function W_t for a stationary population and a flat wage, and express your answer in terms of the person-years-characteristics T_x from Section **7**.1.5.1. (b) Find W_t in the exponential case $w(x,t) = w(x)e^{\tau t}$, $w(x) = w_0 e^{\beta(x-a)}$, $n = n_0 e^{\omega t}$, $s_a(x) = e^{-\mu(x-a)}$.

13. What is the r.-h.s. (2.1.7)—we denoted it by $\overline{a}^{(r)}$—equal to if $\alpha = \delta$?

14. This exercise concerns the model of Section 2.1. (a) Will the increase of the retirement age cause a decrease of the normal costs? (*Advice*: Analyze factors reflecting the influence of the retirement age in (2.1.8).)

    (b) Certainly, a growth of the enrollment intensity ω must lead to a growth of normal costs. Verify that this is reflected in (2.1.8).

15. (a) Analyze the dependence of the benefit rate in (2.1.15) on retirement age and enrollment intensity. (*Advice*: Regarding the latter parameter, it is more convenient to look at the original representation (2.1.13).)

    (b) Analyze the dependence of the APV in (2.1.18) on retirement age and enrollment intensity. (*Advice*: Regarding the latter parameter, it is more convenient to look at the original representation (2.1.16).)

16. Rework Example 2.1.3-1 for the simple case when the survival function $s_a(x) = \exp\{-\mu(x-a)\}$, $\tau = 0$, and $h(x) \equiv 1$.

17. Not providing complete calculations, say how the formulas and examples of Section 2.1 will change if the initial pension rate equals a fraction k of the salary at the moment of retirement times the number of years in service.

18. (a) Suppose that in (2.2.8), the projected pension rate $B(x)$ is a fraction k of the current salary $w(x)$, and $w(x)$ is growing exponentially (in discrete time); more precisely, $w(x+1) = w(x)(1+\beta)$. Provide the expression (2.2.8) in this case.

    (b) Do the same for the continuous-time approximation $w(x) = w(a)e^{\beta(x-a)}$.

19. In the framework of Section 2.2.1, prove that if ΔB is constant, then $(NC)_{(t+1)j} = (1+i)(NC)_{tj}/p_{x_j}$, where i is a one-period effective interest. Find the relation between the total normal costs $(NC)_{t+1}$ and $(NC)_t$, if we only include participants who will not retire in the next year.

20. In the framework of Section 2.2.2, find the one-period contribution c and accrued liability for a participant of age $x \leq r$ at time t if the pension rate equals a fraction k of the salary at the retirement time—denote this salary by $w(r)$, multiplied by the number of years in the plan. The mortality force is a constant μ; the risk-free rate is δ, instead of $\ddot{a}$ we use $\overline{a}_r$, and all other APV's of annuities are replaced by their continuous-time counterparts. Find also AL for the case $x = r$ and comment on the result.

Chapter 12

Risk Exchange: Reinsurance and Coinsurance

2,3

The process of redistribution of risk, starting with purchasing insurances by individual clients, continues at the next level: insurance companies redistribute the risk they incurred between themselves. Such a risk redistribution may be even more flexible than that at the first level: the companies may share individual risks in different ways or redistribute total accumulated risk. A common practice is to protect the portfolio against excessive claims although there are many other forms of reinsurance.

A company which reinsures a part of its risk plays the role of a *cedent*, while the company which assumes this part is a *reinsurer*. Section 1 concerns optimal forms of reinsurance and the amount to be retained from the standpoint of the cedent.

In Section 2, we consider the negotiation process between two companies sharing a risk. In this case, each company is a cedent and reinsurer simultaneously, and the result of the negotiation is based on principles equally acceptable to both companies.

At least theoretically, these principles are not necessarily connected with payments for reinsurance. The negotiation may be direct, and the companies may agree with a certain form of risk distribution without paying each other for reinsurance. However, when many companies are simultaneously involved in reinsurance, market price mechanisms are the most, if not the only, realistic mechanisms of exchanging risk.

Actual reinsurance practice contains various combinations of forms of reinsurance: mutual agreements on direct reinsurance, trading risks, some financial products as special options, futures, or bonds. (See, e.g., [28], [37, Section 8.7], and references therein.) However, in any case, this is a market where commodities to be exchanged are risks. We touch on this question in Section 3.

The exposition below may be called fragmentary. Our goal is not to build a comprehensive theory but rather to make the reader acquainted with some basic notions and to give some examples.

1 REINSURANCE FROM THE STANDPOINT OF A CEDENT

1.1 Some optimization considerations

We identify below risks and the r.v.'s of future payments and sometimes even use the term "risk" when talking about r.v.'s.

1.1.1 Expected utility maximization

Let X be the r.v. of the future payment of a company regarding either a particular policy or a risk portfolio, and c be the premium corresponding to risk X. The reinsurance procedure is specified by a *retention* function $R(x)$ in the following way. The company *retains* an amount $R(X)$ and purchases a reinsurance coverage for the risk $X_{\text{reins}} = X - R(X)$. The function $R(x)$ specifies a particular type of reinsurance.

Assume the following.

- The expected value
$$E\{X_{\text{reins}}\} = E\{X - R(X)\} = \lambda, \tag{1.1.1}$$
where λ is a fixed quantity. For example, the reinsurer agrees to reinsure risks only with a given mean value λ.

- The company pays the reinsurance premium
$$c_{\text{reins}} = (1+\theta_{\text{reins}})E\{X_{\text{reins}}\} = (1+\theta_{\text{reins}})\lambda,$$
where θ_{reins} is a fixed reinsurance loading coefficient.

- The optimal reinsurance corresponds to the maximization of the expected utility of the company's surplus (or wealth) for a given utility function $u(x)$ and an initial surplus w.

Then the expected utility of the company's future surplus is

$$E\{u(w+c-c_{\text{reins}}-R(X))\}. \tag{1.1.2}$$

Comparing (1.1.2) with the quantity

$$E\{u(w-G-\xi+r(\xi))\} \tag{1.1.3}$$

which we considered in Section **1**.5, we see that, although these quantities represent different economic situations, they are mathematically identical. It suffices to establish the correspondence $\xi = X$, $G = c_{\text{reins}} - c$, and

$$r(x) = x - R(x). \tag{1.1.4}$$

Moreover, the restriction (1.1.1) coincides with the restriction (**1**.5.1.1).

We proved in Section **1**.5 that if $u(\cdot)$ is an increasing and concave function, then the maximum of (1.1.3) is attained at the function

$$r(x) = r_d(x) = \begin{cases} 0 & \text{if } x \le d, \\ x-d & \text{if } x > d, \end{cases} \tag{1.1.5}$$

where d is specified by the condition (**1**.5.1.1) [or (1.1.1), which is the same]. Combining (1.1.4) and (1.1.5), we see that the maximum of (1.1.2) is attained at the function

$$R(x) = R_d(x) = \begin{cases} x & \text{if } x \le d, \\ d & \text{if } x > d. \end{cases} \tag{1.1.6}$$

Thus, the retained risk $R(X)$ is the result of truncation of the original risk at level d. This type of reinsurance is called *excess-of-loss reinsurance* if it concerns each contract separately, and *stop-loss reinsurance* when it is applied to the whole risk portfolio. In both cases, the company fixes a certain level of claims and reinsures the amount of the claim which exceeds this level.

As in Section **1**.5, it is worth noting that rule (1.1.6) is the same for all concave utility functions. So, to specify a particular reinsurance policy, we do not have to know the utility function of the company; actually it would be rather naive to think that such a function exists. We just assume that the preferences of the company are close to those based on expected utility maximization for some utility function, perhaps different in different situations.

EXAMPLE 1. Consider a homogeneous portfolio of n independent risks. The payment for each risk has an exponential distribution which we assume, without loss of generality, to be standard. The insurance security loading is 5%, the reinsurance loading is 10.5%.

The company decides to spend 20% of the premium on stop-loss reinsurance. This implies that, if X is the total risk (payment), and X_{reins} is the risk to be reinsured, then $(1+\theta_{\text{reins}})E\{X_{\text{reins}}\} = 0.2(1+\theta)E\{X\}$, where $\theta = 0.05$ and $\theta_{\text{reins}} = 0.105$ are the corresponding loading coefficients. Thus, $(1+0.105)E\{X_{\text{reins}}\} = 0.2(1+0.05)E\{X\}$, from which it follows that

$$E\{X_{\text{reins}}\} = kE\{X\}, \text{ where } k \approx 0.19. \tag{1.1.7}$$

In accordance with what we have proved above, the optimal risk to be reinsured is $X_{\text{reins}} = X - R_d(X)$.

To specify the level d, it is more convenient not to use (**1**.5.1.4) but rather write the expectation of R_d directly as follows:

$$E\{R_d(X)\} = \int_0^d x dF(x) + d \cdot P(X > d) = \int_0^d x dF(x) + d(1-F(d)), \tag{1.1.8}$$

where $F(x)$ is the d.f. of X.

In our case, $X = X_1 + \ldots + X_n$, where X_i is the payment corresponding to the ith risk and having the standard exponential distribution. Hence, F is the Γ-distribution with parameters $(1,n)$, and

$$\begin{aligned} E\{R_d(X)\} &= \frac{1}{\Gamma(n)} \int_0^d x \cdot x^{n-1} e^{-x} dx + d \cdot \frac{1}{\Gamma(n)} \int_d^\infty x^{n-1} e^{-x} dx \\ &= \frac{\Gamma(n+1)}{\Gamma(n)} \cdot \frac{1}{\Gamma(n+1)} \int_0^d x^{(n+1)-1} e^{-x} dx + d \cdot \frac{1}{\Gamma(n)} \int_d^\infty x^{n-1} e^{-x} dx = \\ &= n\Gamma(d; n+1) + d\,(1 - \Gamma(d; n)), \end{aligned} \tag{1.1.9}$$

where

$$\Gamma(t;n) = \frac{1}{\Gamma(n)} \int_0^t x^{n-1} e^{-x} dx,$$

the Γ-d.f. with parameters $(1,n)$. If n is an integer, $\Gamma(t;n)$ may be written explicitly; see (**4**.2.1.3) where the right member is the Γ-d.f. with parameters (λ, n).

Since $X_{\text{reins}} = X - R_d(X)$, from (1.1.7) it follows that $E\{R_d(X)\} = (1-k)E\{X\} = (1-k)n \approx 0.81n$. Combining it with (1.1.9), we get an equation for d:

$$n\Gamma(d;n+1) + d\,(1 - \Gamma(d;n)) = 0.81n. \tag{1.1.10}$$

With use of good software, it is not difficult to estimate the solution $d = d(n)$. For example, $d(10) \approx 8.789$, which the reader may verify on her/his own. (For instance, in Excel, $\Gamma(t;n)$ is in the list of functions; in Maple, the function $GAMMA(n,t) = \Gamma(n)[1 - \Gamma(t;n)]$.)

The expected profit after reinsurance is

$$\begin{aligned} 0.8(1+\theta)E\{X\} - E\{R_d(X)\} &= 0.8(1+\theta)E\{X\} - (1-k)E\{X\} \\ &\approx 0.8(1+0.05)E\{X\} - 0.81E\{X\} = 0.03E\{X\}. \end{aligned}$$

So, after the reinsurance operation, the average return (profit) is reduced from 5% (since $\theta = 0.05$) to 3%. This is the payment for stabilization.

It is interesting to compare it with the result for the excess-of-loss insurance where we truncate the payment r.v.'s for each risk separately. In this case, we should consider the equation (1.1.10) for $n = 1$, which corresponds to the standard exponential distribution. It is easy to verify (using just a calculator) that the solution is $d_1 \approx 1.660$. We see that $10d_1 > d(10)$, which is to be expected. It can be shown that the same is true for all n.

Note that in the excess-of-loss case the average return is, certainly, the same as in the stop-loss case:

$$\begin{aligned} 0.8(1+\theta)E\{X\} - \sum_{i=1}^{10} E\{R_{d_1}(X_i)\} &= 0.8(1+\theta)E\{X\} - \sum_{i=1}^{10}(1-k)E\{X_i\} \\ &= 0.8(1+\theta)E\{X\} - (1-k)E\{X\}, \end{aligned}$$

since $X = X_1 + ... + X_n$. So, the difference is in the level of stability.

In Exercise 1, we compute the variance of the retained payment in both cases, determine which is larger, and interpret the answer. □

1.1.2 Variance as a measure of risk

In this subsection we follow mainly the ideas from [11, Sec.5.1]; see also references therein.

It is interesting that we will come to the same reinsurance strategy (1.1.6) if we minimize the variance of the retained risk, $Var\{R(X)\}$, given the mean value $E\{R(X)\}$. Note that under condition (1.1.1), $E\{R(X)\} = m - \lambda$, where $m = E\{X\}$.

Proposition 1 *For any $R(x)$ with $0 \le R(x) \le x$, and*

$$E\{R(X)\} = \beta \tag{1.1.11}$$

for a fixed β,

$$Var\{R(X\} \ge Var\{R_d(X)\},$$

where $R_d(x)$ is defined in (1.1.6), and d is determined by the condition $E\{R_d(X)\} = \beta$.

A way to find d is given in (1.1.8); see also Section **1**.5. A proof of Proposition 1 is relegated to the end of this subsection.

In the next setup, the role of variance differs from what we considered above, which will lead to a different optimal strategy.

Assume that the cedent specifies somehow what the retained risk should be on the average, and what variance of the retained risk is acceptable. More precisely, the cedent set $E\{R(X)\} = \beta$, $Var\{R(x)\} = \sigma^2_{\text{ret}}$, where β and σ^2_{ret} are fixed numbers chosen by the cedent. (The index "ret" is the abbreviation of "retained".)

In turn, the reinsurer charges a reinsurance premium which is determined by the expected value of the risk reinsured and its variance. In other words, the reinsurance premium c_{reins} is a function of $E\{X_{\text{reins}}\}$ and $Var\{X_{\text{reins}}\}$.

In this case, $E\{X_{\text{reins}}\} = m - \beta$ and is also fixed. Consequently, in order to minimize the cost of reinsurance given β and σ^2_{ret}, the cedent should minimize $Var\{X_{\text{reins}}\}$.

Proposition 2 *Under the conditions stated above, $Var\{X_{reins}\}$ attains its minimum at the function*

$$R(X) = rX, \quad \text{where } r \text{ is some non-negative number.} \tag{1.1.12}$$

The reinsurance of the type (1.1.12) is called *proportional* or *quota share* reinsurance. Since $R(x) \leq x$, the *retention coefficient* $r \leq 1$.

Let us turn to proofs.

Proof of Proposition 1. Because $Var\{R(X)\} = E\{R^2(X)\} - (E\{R(X)\})^2$ and $E\{R(X)\}$ is fixed, instead of minimizing $Var\{R(X)\}$, we may minimize $E\{R^2(X)\}$. Now,

$$\begin{aligned} E\{R^2(X)\} &= E\{[R(X) - d + d]^2\} = E\{[R(X) - d]^2\} + 2dE\{R(X) - d\} + d^2 \\ &= E\{[R(X) - d]^2\} + 2dE\{R(X)\} - d^2. \end{aligned}$$

The last two terms are fixed, and d is specified by the condition $E\{R_d(X)\} = \beta$. Hence, it suffices to find the minimum of $E\{[R(X) - d]^2\}$. We have

$$E\{[R(X) - d]^2\} = E\{[R(X) - d]^2;\ X \leq d\} + E\{[R(X) - d]^2;\ X > d\} \tag{1.1.13}$$

$$\geq E\{[R(X) - d]^2;\ X \leq d\}. \tag{1.1.14}$$

(The symbol $E\{X; B\}$ means integration over a set B; formally, $E\{X; B\} = E\{X \,|\, B\}/P(B)$.)

From (1.1.13) and (1.1.6) it follows that

$$E\{[R_d(X) - d]^2\} = E\{[X - d]^2;\ X \leq d\} + 0 = E\{[X - d]^2;\ X \leq d\}.$$

From (1.1.14) we get that for *any* $R(X)$ such that $0 \leq R(X) \leq X$,

$$E\{[R(X) - d]^2\} \geq E\{[R(X) - d]^2;\ X \leq d\} \geq E\{[X - d]^2;\ X \leq d\} = E\{[R_d(X) - d]^2\}.$$

Comparing the left and the rightmost terms, we see that $R_d(X)$ minimizes $E\{[R(X) - d]^2\}$ over all $R(X)$ under consideration. ■

Proof of Proposition 2. We write

$$Var\{X_{\text{reins}}\} = Var\{X - R(X)\} = Var\{X\} + Var\{R(X)\}$$
$$-2\sqrt{Var\{X\} \cdot Var\{R(X)\}}Corr\{X, R(X)\} = \sigma_X^2 + \sigma_{\text{ret}}^2 - 2\sigma_X\sigma_{\text{ret}}Corr\{X, R(X)\},$$

where $\sigma_X^2 = Var\{X\}$, and $Corr\{X, R(X)\}$ is the correlation coefficient; see Section **0**.2.4.3. As was noted in Section **0**.2.4.3, the correlation attains its maximum (which is equal to one) when $R(X) = rX$ for some $r > 0$. ■

1.2 Proportional reinsurance: Adding a new contract to an existing portfolio

1.2.1 The case of a fixed security loading coefficient

Consider a portfolio whose risk is represented by a r.v. X with mean m and variance σ^2. Denote by c the total premium of the portfolio. We set

$$c = (1 + \theta)m \tag{1.2.1}$$

and assume that the above security loading coefficient θ is the same for all portfolios or separate risks under consideration.

Suppose that the portfolio has already gone through all preliminary adjustments such as reinsurance, an arrangement of the reserve fund, etc., so the company dealing with this portfolio is satisfied with its level of security (or at least the company optimized the portfolio riskiness).

As we have done repeatedly, let us write the probability that the payment will not exceed the total premium, which is

$$P(c - X \geq 0) = P(X - m \leq \theta m) = P\left(X^* \leq \theta\frac{m}{\sigma}\right) = P\left(X^* \leq \frac{\theta}{k}\right),$$

where the normalized r.v. $X^* = (X - m)/\sigma$, and $k = \sigma/m$, the coefficient of variation (see also Section **2**.3.1.1).

The loading coefficient θ is fixed. If X were normal, the probability we consider would have been completely determined by k. Hence, in this case we could view k as a risk measure: a smaller k would indicate a greater probability above.

If X is not normal but represents a large portfolio, we may assume X to be close to a normal r.v. by the CLT. Then k may be viewed as a characteristic of riskiness up to normal approximation.

Note that if X^* had had another standard distribution distinct from normal, we could reason similarly. However, such a generalization would be rather formal, so the main argument for choosing k as a risk measure is based on the assumption of approximate normality.

Suppose now that a new risk with a random future payment X_0 is added to the portfolio. Set $m_0 = E\{X_0\}$, $\sigma_0^2 = Var\{X_0\}$, and denote by c_0 the premium corresponding to the new risk. As we agreed, in our scheme, $c_0 = (1 + \theta)m_0$, where θ is the same as in (1.2.1).

The company is going to retain the risk rX_0, and reinsure $(1 - r)X_0$, where r is a *retention coefficient*. The goal is to find a "reasonable" $r \leq 1$. If this reasonable r turns out to be less than one: $r < 1$, then the company should reinsure a part of the new risk.

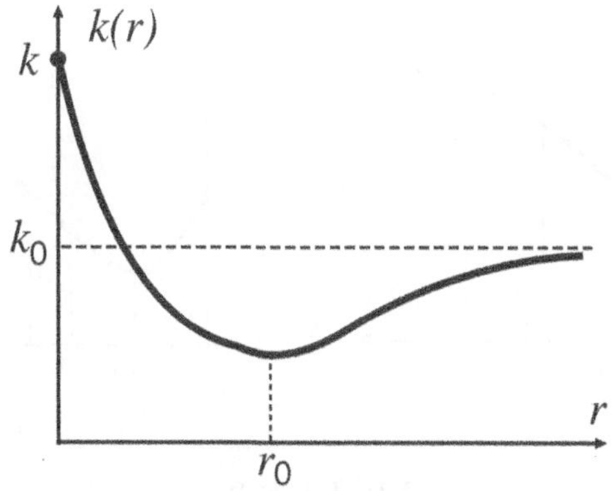

FIGURE 1.

Assume that X and X_0 are independent, and suppose that, in order to reinsure $(1-r)X_0$, the company should yield exactly the corresponding share of the premium, that is, $(1-r)c_0$. This is not a mild assumption, but we adopt it here.

In this case, the future payment for the new portfolio is the r.v. $X_r = X + rX_0$ with the expected value $m_r = m + rm_0$ and the variance $\sigma_r^2 = \sigma^2 + r^2\sigma_0^2$. The premium for the new portfolio is $c_r = c + rc_0 = (1+\theta)m + r(1+\theta)m_0 = (1+\theta)m_r$. Hence, the expected profit after reinsurance is equal to $c_r - m_r = \theta m_r = \theta(m + rm_0)$ and is increasing with r.

Thus, regarding the mean profit, the company wants r to be as large as possible. The principle we accept consists in choosing *the largest r for which the riskiness of the new portfolio is not larger than the riskiness of the original portfolio.* The riskiness in our scheme is completely characterized by the new coefficient of variation

$$k(r) = \frac{\sigma_r}{m_r} = \frac{\sqrt{\sigma^2 + r^2\sigma_0^2}}{m + rm_0}. \tag{1.2.2}$$

Thus, we choose the largest r for which $k(r) \le k$.

Elementary calculations show that

- $k(0) = k = \dfrac{\sigma}{m}$, $k(r) \to k_0 = \dfrac{\sigma_0}{m_0}$ as $r \to \infty$;
- $k'(r) \le 0$ for $r \le r_0 = \dfrac{\sigma^2 m_0}{\sigma_0^2 m}$, and $k'(r) > 0$ for $r > r_0$;
- the equation $k(r) = k$ has a positive solution iff $k < k_0$, and this solution is

$$r_1 = 2k^2 m \cdot \frac{1}{m_0(k_0^2 - k^2)}. \tag{1.2.3}$$

We consider three cases.

CASE 1: $k \ge k_0$; see Fig.1. We see that in this case, $k(r) \le k$ for all r. Consequently, the best $r = 1$, and reinsurance is not needed.

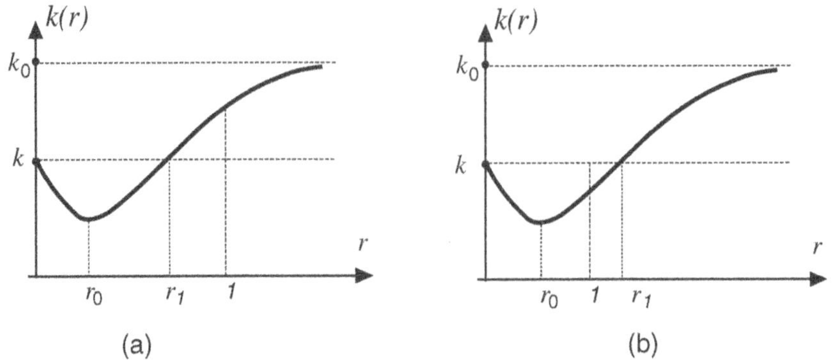

FIGURE 2.

CASE 2: $k < k_0$, and $r_1 < 1$; see Fig.2a. In this case, we choose $r = r_1$. We have $k(r_1) = k$, and r_1 is the biggest r for which $k(r) \leq k$. So, the riskiness of the new portfolio is the same as before, but because $r_1 > 0$, after adding the corresponding part of the new risk, the average profit of the portfolio has increased.

CASE 3: $k < k_0$, and $r_1 \geq 1$; see Fig.2b. In this case, we take $r = 1$, that is, reinsurance is again not needed.

EXAMPLE 1. Let the original portfolio consist of n i.i.d. risks X_i with $E\{X_i\} = \widetilde{m}$, $Var\{X_i\} = \widetilde{\sigma}^2$. Then $m = n\widetilde{m}$, $\sigma^2 = n\widetilde{\sigma}^2$, $k = \dfrac{\sigma}{m} = \dfrac{1}{\sqrt{n}}\dfrac{\widetilde{\sigma}}{\widetilde{m}} \to 0$ as $n \to \infty$. Hence, $k < k_0 = \dfrac{\sigma_0}{m_0}$ for sufficiently large n. The retention coefficient

$$r_1 = 2k^2 m \frac{1}{m_0(k_0^2 - k^2)} = \frac{2\widetilde{\sigma}^2}{\widetilde{m}} \frac{1}{(\sigma_0^2/m_0) - m_0\widetilde{\sigma}^2/(\widetilde{m}^2 n)} \sim \frac{2\widetilde{\sigma}^2 m_0}{\sigma_0^2 \widetilde{m}}. \tag{1.2.4}$$

Thus, for large n, reinsurance is needed if $2\widetilde{\sigma}^2 m_0 \leq \sigma_0^2 \widetilde{m}$.

Now let the new risk have the same parameters as the risks composing the original portfolio: $m_0 = \widetilde{m}$, $\sigma_0 = \widetilde{\sigma}$. Then, it is straightforward to calculate that $r_1 = \frac{2n}{n-1} > 1$, and no reinsurance is needed.

EXAMPLE 2. Let the new risk X_0 take on two values: v_0 (the sum insured) and 0 with probabilities q and $1-q$, respectively. The insurer wants to reinsure a part r of the payment v_0. In this case, the retained risk corresponds to a r.v. taking values $v_{\text{retained}} = rv_0$ and 0 with probabilities q and $1-q$, respectively. The part $(1-r)v_0$ is reinsured. We have $m_0 = v_0 q$, $\sigma_0 = v_0\sqrt{q(1-q)}$, the "reasonable" retention $r = r_1$, and in view of (1.2.4),

$$r_1 = 2k^2 m \frac{1}{\sigma_0^2/m_0 - k^2 m_0} = 2k^2 m \frac{1}{v_0[(1-q) - kq]} = 2k^2 m \frac{1}{v_0[1-(1+k^2)q]}.$$

Then the retained possible payment

$$v_{\text{retained}} = r_1 v_0 = 2k^2 m \frac{1}{1-(1+k^2)q} \sim 2k^2 m \tag{1.2.5}$$

for small q.

This is an old celebrated formula known from the beginning of the previous century, if not from an earlier time. The quantity m is interpreted as the net premium for the original portfolio and is denoted sometimes by P. So, the formula becomes

$$v_{\text{retained}} = 2k^2 P.$$

□

Route 2 ⇒ page 563

1.2.2 The case of the standard deviation premium principle

3

In this section, we follow mainly [123]. Consider the same problem with the following two changes. First, for a risk X with mean m and variance σ^2, we adopt, as the premium principle, the rule

$$c = m + \lambda\sigma, \tag{1.2.6}$$

where λ plays a role of loading. Second, we do not assume λ to be the same for all risks.

The reader who has not skipped Section **1**.4 will recognize the standard deviation premium principle in (1.2.6). However, to understand the material below, it suffices to view (1.2.6) as the definition of the coefficient

$$\lambda = (c-m)/\sigma.$$

As above, let X be the risk of the original portfolio. Then

$$P(c - X \geq 0) = P(X - m \leq \lambda\sigma) = P(X^* \leq \lambda).$$

Following the same logic, we adopt λ as a stability characteristic and λ^{-1} as a risk measure.

Consider a new risk X_0 with mean m_0 and variance σ_0^2. Denote by c_0 the premium coming from the new risk. In accordance with (1.2.6), set $\lambda_0 = (c_0 - m_0)/\sigma_0$. Then

$$c_0 = m_0 + \lambda_0\sigma_0.$$

We retain all assumptions on X and X_0 and reinsurance cost, made in the previous section. Then for the risk $X_r = X + rX_0$, we have again $m_r = m + rm_0$, $\sigma_r^2 = \sigma^2 + r^2\sigma_0^2$, and the premium

$$\begin{aligned} c_r &= c + rc_0 = m + \lambda\sigma + r(m_0 + \lambda_0\sigma_0) = m_r + \lambda\sigma + r\lambda_0\sigma_0 = m_r + \frac{\lambda\sigma + r\lambda_0\sigma_0}{\sigma_r}\sigma_r \\ &= m_r + \lambda(r)\sigma_r, \end{aligned}$$

where

$$\lambda(r) = \frac{\lambda\sigma + r\lambda_0\sigma_0}{\sigma_r} = \frac{\lambda\sigma + r\lambda_0\sigma_0}{\sqrt{\sigma^2 + r^2\sigma_0^2}}. \tag{1.2.7}$$

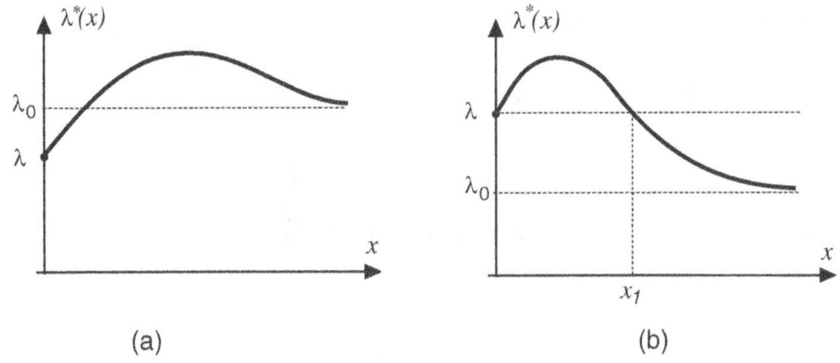

FIGURE 3.

Let $a = \sigma_0/\sigma$. To make (1.2.7) easier to explore, set $x = ra$. Then, dividing the numerator and the denominator in (1.2.7) by σ, we have

$$\lambda(r) = \frac{\lambda + \lambda_0 x}{\sqrt{1+x^2}}.$$

Denote the function on the right by $\lambda^*(x)$. Then

$$\lambda(r) = \lambda^*(ra).$$

It is straightforward to verify that

$$\lambda(0) = \lambda^*(0) = \lambda, \;\; \lambda^*(x) \to \lambda_0 \text{ as } x \to \infty, \text{ and}$$
$$\lambda^*(x) = \lambda \;\text{ for } x = x_1 = \frac{2\lambda_0\lambda}{\lambda^2 - \lambda_0^2} \;\text{ if } \lambda > \lambda_0.$$

The larger $\lambda(r)$, the less risky the new portfolio. We are looking for the largest $r \leq 1$ for which $\lambda(r) \geq \lambda$. Let

$$r_1 = \frac{x_1}{a} = \frac{2\lambda_0\lambda\sigma}{\sigma_0(\lambda^2 - \lambda_0^2)}.$$

CASE 1: $\lambda \leq \lambda_0$; see Fig.3a. Then $\lambda(r) \geq \lambda$ for all r. Consequently, the best r equals 1, and reinsurance is not needed.

CASE 2: $\lambda > \lambda_0$; see Fig.3b. In this case, if $r_1 < 1$, we choose $r = r_1$, and if $r_1 \geq 1$, we choose $r = 1$.

EXAMPLE 1. Let the original portfolio consists of n i.i.d. risks X_i with $E\{X_i\} = \widetilde{m}$, $Var\{X_i\} = \widetilde{\sigma}^2$. We assume that for each separate risk, the premium $\widetilde{c} = \widetilde{m} + \widetilde{\lambda}\widetilde{\sigma}$, where $\widetilde{\lambda}$ is a loading for separate risks. Then, since $m = n\widetilde{m}$, $\sigma^2 = n\widetilde{\sigma}^2$, we can write that $\lambda = \dfrac{c-m}{\sigma} = \dfrac{n(\widetilde{m} + \widetilde{\lambda}\widetilde{\sigma}) - n\widetilde{m}}{\widetilde{\sigma}\sqrt{n}} = \widetilde{\lambda}\sqrt{n}$. Then

$$r_1 = \frac{2\lambda_0\lambda\sigma}{\sigma_0(\lambda^2 - \lambda_0^2)} = \frac{2\lambda_0\widetilde{\sigma}}{\sigma_0} \cdot \frac{\widetilde{\lambda}n}{\widetilde{\lambda}^2 n - \lambda_0^2} \sim \frac{2\lambda_0\widetilde{\sigma}}{\widetilde{\lambda}\sigma_0} \tag{1.2.8}$$

for large n. Thus, for large n, reinsurance is needed if $2\lambda_0\tilde{\sigma} \leq \tilde{\lambda}\sigma_0$.

Now let the loading for the new risk be the same as for the separate risks in the original portfolio: $\lambda_0 = \tilde{\lambda}$. Then reinsurance is reasonable if $\sigma_0 \geq 2\tilde{\sigma}$. In particular, if the standard deviation of the new risk equals the standard deviation of separate risks in the original portfolio, there is no need for reinsurance. □

1.3 Long-term insurance: Ruin probability as a criterion 2,3

In the case of long-term insurance, security requirements should concern the behavior of the insurance process in the long run. For example, one can choose, as a characteristic of riskiness, the ruin probability for a large or infinite horizon.

Certainly, whatever risk measure we choose, we do not have to proceed basing solely on this measure. Usually a particular strategy of managing a risk portfolio is a trade-off between security and future profit, for example, the expected profit. If the ruin probability were the only criterion the company follows, it would not take any risk at all. Then the ruin probability would be equal to zero, but the profit would be zero too.

If we choose ruin probability as a risk measure, the goal of reinsurance is to reduce this probability to a certain chosen level. There are two difficulties, one essential and one technical, in this problem.

First, the cost of reinsurance may be high. In this case, since the payment for reinsurance goes from the original premium, the part of the premium that remains may turn out to be low. Then the purchase of reinsurance may lead not to a smaller but to a higher ruin probability. We will see it in examples below. Hence, we should first figure out the limits of reasonable reinsurance.

Secondly, the ruin probability depends on the interior characteristics of the insurance process and on the initial surplus. When buying a reinsurance, we change the former; however, we could instead of purchasing a reinsurance, increase the initial surplus. Consequently, the purchase of reinsurance is an additional compromise between the level of the surplus and the size of the risk retained. To make the problem simpler and solutions to be unified for all possible sizes of the initial surplus, we can proceed not from the ruin probability itself but from Lundberg's upper bound $e^{-\gamma u}$, where u stands for the initial surplus and γ for the adjustment coefficient; see (**6**.2.1.4). Then we can choose, as a security characteristic, the adjustment coefficient γ and work with it independently of u.

We will restrict ourselves to two particular examples.

1.3.1 An example with proportional reinsurance

Consider the discrete time model of Section **6**.2.2.2. Let X be a separate claim, $m = E\{X\}$, $\sigma^2 = Var\{X\}$, and c be the corresponding premium. For the adjustment coefficient γ, we use the representation

$$\gamma = \frac{2(c-m)}{\sigma^2}, \tag{1.3.1}$$

which is precise if X is normal and, as was shown in Section **6**.2.2.2, may serve as a good approximation in many other situations.

Consider proportional reinsurance, denoting by r the retention coefficient. The retained risk amounts to the r.v. $X_r = rX$. Note that, if X is normal, then X_r is also normal, and we can apply (1.3.1) to it.

Assume that the price for reinsurance is specified by a reinsurance loading θ_{reins}. Then given r, the company pays for reinsurance, per each claim X, the price $(1+\theta_{\text{reins}})E\{X - rX\} = (1+\theta_{\text{reins}})(1-r)m$. Consequently, if θ is the security loading of the original insurance, the premium retained after reinsurance is

$$c_r = (1+\theta)m - (1+\theta_{\text{reins}})(1-r)m. \tag{1.3.2}$$

It is straightforward to verify that (1.3.2) may be written as

$$c_r = [r + \theta_{\text{reins}} r - \Delta]m,$$

where $\Delta = \theta_{\text{reins}} - \theta$.

It is natural to assume that $\Delta > 0$. Otherwise, the insurer would yield the whole risk to a reinsurer, keeping the remaining premium as a pure profit.

We have $E\{X_r\} = m_r = rm$, $Var\{X_r\} = r^2\sigma^2$, and by virtue of (1.3.1), the adjustment coefficient for the retained insurance is equal to

$$\gamma_r = \frac{2(c_r - m_r)}{\sigma_r^2} = \frac{2[\theta_{\text{reins}} r - \Delta]m}{r^2\sigma^2}. \tag{1.3.3}$$

For $r = 1$ (no reinsurance), the adjustment coefficient

$$\gamma_1 = \gamma = \frac{2\theta m}{\sigma^2}.$$

Combining it with (1.3.3), we can write that

$$\gamma_r = \gamma g(r), \tag{1.3.4}$$

where the function

$$g(r) = \frac{1}{\theta} \cdot \frac{\theta_{\text{reins}} r - \Delta}{r^2}.$$

Clearly, $g(1) = 1$. The reader will readily verify that $g(r)$ attains its unique maximum at the point

$$r_0 = \frac{2\Delta}{\theta_{\text{reins}}} = \frac{2(\theta_{\text{reins}} - \theta)}{\theta_{\text{reins}}}, \tag{1.3.5}$$

and $g(r)$ is decreasing at a point r iff $r > r_0$.

The relation $r_0 \geq 1$ is equivalent to

$$\theta_{\text{reins}} \geq 2\theta. \tag{1.3.6}$$

We interpret this instance as a high reinsurance cost. In this case, $g(r) \leq 1$ for all $r \leq 1$, and reinsurance does not lead to a higher stability.

If $r_0 < 1$, taking an r between $[r_0, 1]$, we get a higher adjustment coefficient, which we interpret as the case of higher stability. The maximum is attained at $r = r_0$, but this is also the case of the lowest (for r from $[r_0, 1]$) expected income.

EXAMPLE 1. Let $\theta = 0.1$. In view of (1.3.6), reinsurance makes sense only if $\theta_{\text{reins}} < 0.2$. Say, for $\theta_{\text{reins}} = 0.15$, we will have $r_0 = \frac{2}{3}$, and one third of risk may be reinsured. The choice of a particular r between $\frac{2}{3}$ and 1 should be determined by a compromise between stability requirements and the mean profit. □

1.3.2 An example with excess-of-loss insurance

Consider now the continuous time model of Section **6**.2.2.4. In this case, the adjustment coefficient γ is a solution to the equation

$$M_X(z) = 1 + cz, \tag{1.3.7}$$

where $c = (1+\theta)m$ is the premium per one claim, and X is the r.v. of one claim; see (**6**.2.2.22). For the rest, the notation is the same as in the previous section.

Let X be uniformly distributed on $[0,1]$. This is, of course, a particular case, but the fact that we chose the unit interval does not restrict generality.

Consider the instance of excess-of-loss insurance with a level $d \le 1$, which means the following.

- For each claim X the company covers the part amounting to the r.v.

$$X_d = \begin{cases} X & \text{if } X \le d, \\ d & \text{if } X > d. \end{cases}$$

- The company pays for such a reinsurance $(1+\theta_{\text{reins}})E\{X - X_d\}$ per one claim.

Since X is uniform on $[0,1]$, the r.v. X_d assumes the value d with probability $1-d$. It is not difficult to verify that the conditional distribution of X_d, given $X \le d$, is uniform on $[0,d]$. Making use of it, one can calculate that

$$E\{X_d\} = m_d = d - \frac{1}{2}d^2, \tag{1.3.8}$$

$$E\{X - X_d\} = \frac{1}{2}(1-d)^2,$$

$$Var\{X_d\} = \sigma_d^2 = \frac{1}{12}d^3(4-3d). \tag{1.3.9}$$

Consequently, the premium per one claim after reinsurance is equal to

$$c_d = (1+\theta)E\{X\} - (1+\theta_{\text{reins}})E\{X - X_d\} = \frac{1}{2}[(1+\theta) - (1+\theta_{\text{reins}})(1-d)^2]. \tag{1.3.10}$$

The m.g.f.

$$M_{X_d}(z) = \int_0^d e^{zx}dx + e^{zd}(1-d) = \frac{1}{z}\left(e^{zd}[1+(1-d)z] - 1\right). \tag{1.3.11}$$

Given d, the adjustment coefficient γ_d is a solution to the equation

$$M_{X_d}(z) = 1 + c_d z.$$

Substituting (1.3.10) and (1.3.11), we come to the equation

$$e^{zd}[1+(1-d)z] = 1 + z + \frac{1}{2}[(1+\theta) - (1+\theta_{\text{reins}})(1-d)^2]z^2. \tag{1.3.12}$$

The l.-h.s. of the equation is an exponential function multiplied by a linear function, and the r.-h.s. is a quadratic function. Certainly, this equation cannot be solved analytically, but solving it numerically for particular values of the parameters does not present any difficulty. A reasonable procedure would be as follows.

For given values of θ and θ_{reins}, we consider a sequence of d's running from 1 to 0 in small steps. For each d, we find numerically a solution γ_d to (1.3.12). We stop the procedure when the sequence γ_d starts to decrease.

We will not carry out this procedure here, but consider a simpler approximation based on (1.3.1). As we noted in Section **6**.2.4, (1.3.1) may be not a bad approximation in the continuous time case also. [See in this section the comparison of the equations (**6**.2.4.1) and (**6**.2.4.2).]

Thus, we use the approximation

$$\gamma_d \approx \frac{2(c_d - m_d)}{\sigma_d^2} = 24q(d), \text{ where} \tag{1.3.13}$$

$$q(d) = \frac{1}{d^3(4-3d)}\left(\frac{1}{2}[(1+\theta) - (1+\theta_{\text{reins}})(1-d)^2] - d + \frac{d^2}{2}\right).$$

The last formula follows from (1.3.8), (1.3.9), and (1.3.10).

The function $q(d)$ is easy to calculate. The table below contains its values for θ=0.1, $\theta_{\text{reins}} = 0.3$.

d	1	0.9	0.8	0.7	0.69	0.68	0.67	0.66	0.65
$100q(d)$	5	5.11	5.37	5.60	5.6125	5.6207	5.6247	5.6238	5.617

We see that γ_d begins to decrease at $d \approx 0.67$.

However, this is a rough approximation. As we saw in Section **6**.2.4, the approximation (1.3.13) works well for small γ's. In our case, γ has an order of $24q(d) \approx 24 \cdot 0.056 = 1.344$, which is not small. So, for a more precise solution we should appeal to (1.3.12). □

---

2 RISK EXCHANGE AND RECIPROCITY OF COMPANIES

2.1 A general framework and some examples

As was noted in the beginning of this chapter, when talking about risk exchange, we should distinguish two different approaches (which, however, may be combined). In the first approach, risk (or random income) is considered a commodity which may be traded. The exchange mechanism in this case is a decentralized market mechanism which appears to be efficient in the case of equilibrium.

A classical example is a stock exchange market, but we can talk also about a reinsurance market where insurance companies exchange risks. We consider a corresponding model in Section 3.

The second approach concerns the situation when companies enter into direct negotiations about possibilities of risk exchange. The process of such negotiations is connected with the concept of a non-zero-sum or cooperative game.

A zero-sum game, say, in the situation of two players, is a game in which for any combination of the players' strategies, one player wins exactly the amount the player's opponent loses. In a non-zero-sum game, a gain by one player does not necessarily correspond with a loss of the other. There are strategies under which *both* players may benefit in comparison with what they had in their initial positions. The problem here consists in determining such strategies and choosing one of them. The corresponding theory is usually called the bargaining theory (see, e.g., [82], [97]).

The negotiations between two or more companies about sharing the risks they have may be considered a cooperative or non-zero-sum game. The companies analyze possible outcomes and establish principles which would lead to a particular exchange of risk, reasonable ("fair") from all companies' point of view.

EXAMPLE 1. We proceed from the scheme of Section 1.2.1. Two companies have initial portfolios whose risks amount to r.v.'s X_1, X_2 with means m_1, m_2 and variances σ_1^2, σ_2^2, respectively. The companies decide to share a new risk X_0 with a mean of m_0 and a variance of σ_0^2. The security loading is the same for all possible insurances.

Let r be the share of the new risk which goes to the first company and, accordingly, $(1-r)$ is the share of the second company. The companies negotiate on a reasonable choice of r.

Given r, the new risks for the companies are $X_{r1} = X_1 + rX_0$, $X_{r2} = X_2 + (1-r)X_0$.

Suppose, as in Section 1.2.1, that the companies choose variation coefficient (1.2.2) as a risk measure. Then the companies proceed from the respective coefficients

$$k_1(r) = \frac{\sqrt{\sigma_1^2 + r^2\sigma_0^2}}{m_1 + rm_0}, \quad k_2(1-r) = \frac{\sqrt{\sigma_2^2 + (1-r)^2\sigma_0^2}}{m_2 + (1-r)m_0}. \tag{2.1.1}$$

One of natural rules here could consist in the principle of equal riskiness for both companies, which leads to the equation

$$k_1(r) = k_2(1-r), \tag{2.1.2}$$

provided that a solution $r \in [0,1]$ exists; see also Fig.4.

Assume, for instance, that $m_1 = m_2$ and are much larger than m_0. Then the denominators in (2.1.1) are close to each other, and (2.1.2) may be approximated by the equation

$$\sigma_1^2 + r^2\sigma_0^2 = \sigma_2^2 + (1-r)^2\sigma_0^2.$$

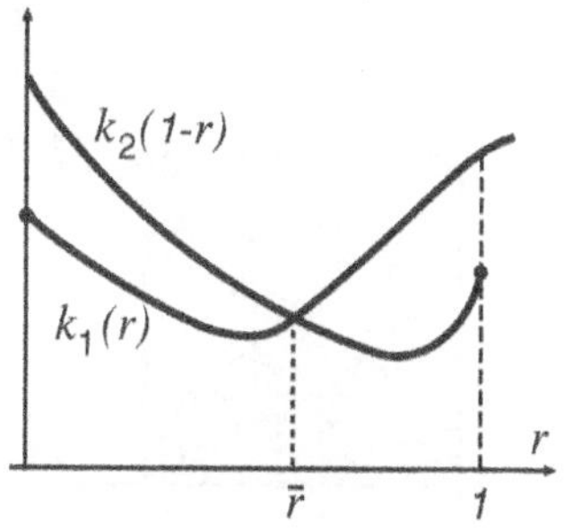

FIGURE 4.

A simple algebra leads to the solution

$$\bar{r} = \frac{\sigma_2^2 + \sigma_0^2 - \sigma_1^2}{2\sigma_0^2}. \tag{2.1.3}$$

If the two original portfolios have the same parameters (that is, $\sigma_1^2 = \sigma_2^2$, in addition to the equality of the mean values), then $\bar{r} = \frac{1}{2}$, which is natural. The example is continued in Exercise 8. □

Consider now a general scheme. To emphasize that it concerns not only insurance but a quite general situation, we will sometimes use the term 'participant' rather than 'company'.

The scheme involves n participants. The ith participant is characterized by a future random income Y_i.

When a participant is an insurance company, Y_i is the future surplus of the company, and we set $Y_i = c_i - X_i$, where X_i is the future payment provided by the company, and c_i is the premium collected, $i = 1, .., n$. In the general framework, the structure of Y_i is arbitrary.

Set $\mathbf{Y} = (Y_1, ..., Y_n)$. This is the vector of initial (random) incomes. Assume that the participants exchange somehow parts of their risks. In the case of insurance, it means that the companies exchange parts of their portfolios.

Denote by Z_i the income of the ith participant after exchange. The random vector (r.vec.) $\mathbf{Z} = (Z_1, ..., Z_n)$ specifies this exchange. We assume that not all types of exchange are allowed or possible, and the vector $\mathbf{Z}$ belongs to some set $\mathcal{Z}$ which represents all admissible types of exchange. For example, in many situations it is natural to assume that the total income does not change, so for all $\mathbf{Z} \in \mathcal{Z}$ we should have

$$\sum_{i=1}^{n} Z_i = \sum_{i=1}^{n} Y_i. \tag{2.1.4}$$

However, this condition may not exhaust all possible conditions for portfolios exchange.

EXAMPLE 2. Consider two companies with original risks X_1, X_2 and premiums c_1, c_2, respectively. The companies are negotiating about mutual reinsurance of their risks. Assume that the companies choose to restrict themselves to proportional reinsurance. Denote by r_1, r_2 the retention coefficients for the first and the second company, respectively. More precisely, the first company retains the share r_1 of its own risk, cedes the share $(1 - r_1)$ of its risk, and accepts the share $(1 - r_2)$ of the second company's risk. The respective numbers for the second company are r_2, $(1 - r_2)$, and $(1 - r_1)$.

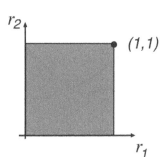

FIGURE 5.

In this case, the reinsurance procedure is specified by a pair $\mathbf{r} = (r_1, r_2)$ which is a point in the unit square $\mathcal{R} = \{0 \le r_1 \le 1, 0 \le r_2 \le 1\}$; see Fig.5.

Given $\mathbf{r}$, the first company covers the claim $r_1 X_1$ and the claim $(1 - r_2)X_2$ ceded by the second company. So, the total claim covered by the first company is $X_{\mathbf{r}1} = r_1 X_1 + (1 - r_2)X_2$. For the second company the corresponding claim is $X_{\mathbf{r}2} = (1 - r_1)X_1 + r_2 X_2$.

Certainly, the companies should somehow redistribute the premiums they have. The ways to do that may be different. We will consider some of them below but for now, not specifying the rule of premium distribution, we will just denote by $c_{\mathbf{r}i}$, the premium the ith company keeps after reinsurance. Certainly,

$$c_{\mathbf{r}1} + c_{\mathbf{r}2} = c_1 + c_2.$$

Then, the profit of the ith company after such an exchange is the r.v.

$$Z_{\mathbf{r}i} = c_{\mathbf{r}i} - X_{\mathbf{r}i}, \; i = 1, 2,$$

and the set $\mathcal{Z}$ consists of random vectors $\mathbf{Z_r} = (Z_{\mathbf{r}1}, Z_{\mathbf{r}2})$ where $\mathbf{r} \in \mathcal{R}$. □

Suppose now that each participant, when evaluating the quality of its position, proceeds from a function $U(Z)$ assuming real values and *perhaps different for different participants*, defined on a set of r.v.'s Z. The participant views $U(Z)$ as a characteristic of "quality" of the r.v. Z (from the participant's point of view). For example, if a participant is an expected utility maximizer, then

$$U(Z) = E\{u(Z)\}, \tag{2.1.5}$$

where $u(x)$ is the utility function of the participant. Another example concerns the mean-variance criterion for which

$$U(Z) = \tau E\{Z\} - Var\{Z\}, \tag{2.1.6}$$

and τ is a tolerance-to-risk coefficient, a weight the participant assigns to expectation; see Section **1**.1.2.5. Different participants may have different τ's. We remember that the criterion (2.1.6) is not applicable in the general case (see the section mentioned for detail) but, for example, when Z's are normal, the criterion is meaningful.

We call the value of $U(Z)$ a quality index of Z.

Denote by $U_i(Z)$ the quality-measure-function U of the ith participant. Then, before exchange, the quality index of the ith participant is equal to the number $V_{i0} = U_i(Y_i)$, and after exchange—to the number $V_i = U(Z_i)$. Set $\mathbf{V}^{(0)} = (V_{10}, ..., V_{n0})$, $\mathbf{V} = (V_1, ..., V_n)$.

The vector $\mathbf{V}$ is specified by the choice of $\mathbf{Z}$. If we denote by $\mathbf{U}(\mathbf{Z})$ the vector-function $(U_1(Z_1), ..., U_n(Z_n))$, then

$$\mathbf{V} = \mathbf{U}(\mathbf{Z}). \tag{2.1.7}$$

Thus, each r.vec. $\mathbf{Z}$ from the set $\mathcal{Z}$ of all admissible exchanges generates a point $\mathbf{V} = \mathbf{U}(\mathbf{Z}) = (U_1(Z_1), ..., U_n(Z_n))$ in $\mathbb{R}^n$; for $n = 2$, the point lies on a plane.

Denote by $\mathcal{V}$ the set of *all* possible points $\mathbf{V} = \mathbf{U}(\mathbf{Z})$ generated by the map (2.1.7). The set $\mathcal{V}$ is the image of $\mathcal{Z}$, and represents all possible positions the participants may attain. A typical picture is depicted in Fig.6.

The point $\mathbf{V}^{(0)}$, representing the initial position of the participants, is called a *status-quo point*. Usually it is an interior point of the set $\mathcal{V}$, as it is shown in Fig.6.

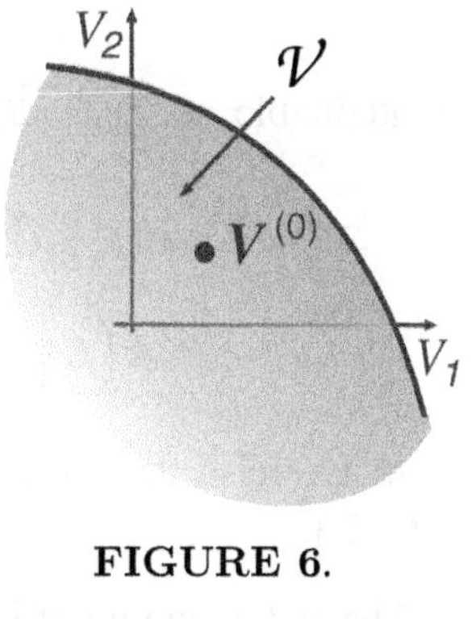

FIGURE 6.

EXAMPLE 3. This example is somewhat artificial but illustrative. Later, we will consider more realistic examples. Let $n = 2$, each participant is an expected utility maximizer with $u(x) = \sqrt{x}$; the initial incomes Y_1, Y_2 are positive r.v.'s such that $E\{u(Y_i)\} = E\{\sqrt{Y_i}\}$ is finite for $i = 1, 2$.

To make the example simpler, assume that the participants have a right to refuse a part of the income, say, choose $Z_1 = 0$, $Z_2 = 0$. Certainly, it will not correspond to optimal behavior and eventually such solutions will be rejected, but it is convenient to include these possibilities into original consideration. Such an assumption means that, instead of the condition $Z_1 + Z_2 = Y_1 + Y_2$, we will consider the condition

$$Z_1 + Z_2 \leq Y_1 + Y_2. \tag{2.1.8}$$

Assume that the class $\mathcal{Z}$ consists of all positive r.vec.'s $\mathbf{Z} = (Z_1, Z_2)$ for which (2.1.8) is true.

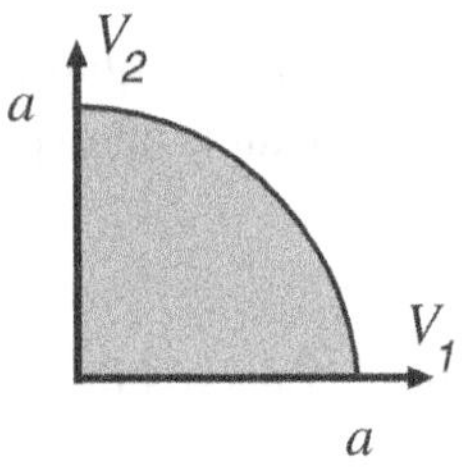

FIGURE 7.

In this case, $\mathcal{V}$ is the quarter of a disk, more precisely, $\mathcal{V}$ is the set

$$V_1^2 + V_2^2 \le a^2,\ 0 \le V_1 \le a,\ 0 \le V_2 \le a, \tag{2.1.9}$$

where $a = E\{\sqrt{Y_1+Y_2}\}$; see Fig.7.

Indeed, let us set $S = Y_1 + Y_2$, $V_i = E\{\sqrt{Z_i}\}$. By the Cauchy-Schwarz inequality (**0**.2.4.4),

$$V_i^2 = (E\sqrt{Z_i})^2 = \left(E\left\{\frac{\sqrt{Z_i}}{S^{1/4}}S^{1/4}\right\}\right)^2 \le E\left\{\left(\frac{\sqrt{Z_i}}{S^{1/4}}\right)^2\right\}E\left\{\left(S^{1/4}\right)^2\right\} = E\left\{\frac{Z_i}{\sqrt{S}}\right\}E\left\{\sqrt{S}\right\}.$$

Then, since $Z_1 + Z_2 \le Y_1 + Y_2 = S$,

$$\begin{aligned} V_1^2 + V_2^2 &\le E\left\{\frac{Z_1}{\sqrt{S}}\right\}E\left\{\sqrt{S}\right\} + E\left\{\frac{Z_2}{\sqrt{S}}\right\}E\left\{\sqrt{S}\right\} = E\left\{\sqrt{S}\right\}\left[E\left\{\frac{Z_1}{\sqrt{S}}\right\} + E\left\{\frac{Z_2}{\sqrt{S}}\right\}\right] \\ &= E\left\{\sqrt{S}\right\}E\left\{\frac{Z_1+Z_2}{\sqrt{S}}\right\} \le \left(E\left\{\sqrt{S}\right\}\right)^2 = a^2. \end{aligned}$$

To show that the boundary

$$\{V_1^2 + V_2^2 = a^2,\ 0 \le V_1 \le a,\ 0 \le V_2 \le a\} \tag{2.1.10}$$

is attainable, set $Z_1 = kS$, $Z_2 = (1-k)S$ for a constant $k \in [0,1]$. Then

$$\begin{aligned} V_1^2 + V_2^2 &= (E\{\sqrt{Z_1}\})^2 + (E\{\sqrt{Z_2}\})^2 = \left(E\left\{\sqrt{kS}\right\}\right)^2 + \left(E\left\{\sqrt{(1-k)S}\right\}\right)^2 \\ &= k\left(E\left\{\sqrt{S}\right\}\right)^2 + (1-k)\left(E\left\{\sqrt{S}\right\}\right)^2 = \left(E\left\{\sqrt{S}\right\}\right)^2 = a^2. \end{aligned}$$

In Exercise 9a, we show that any point satisfying (2.1.9) corresponds to a solution (Z_1, Z_2).

The status-quo point is $\mathbf{V}^{(0)} = (U_1(Y_1), U_2(Y_2)) = (E\{\sqrt{Y_1}\}, E\{\sqrt{Y_2}\}$. It may be shown that under mild conditions, it is located inside $\mathcal{V}$.

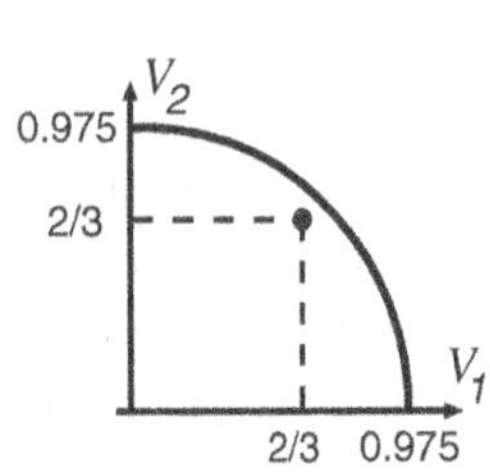

FIGURE 8.

Let us consider a particular example. Let Y_1, Y_2 be independent, and uniformly distributed on $[0,1]$. Then $E\{\sqrt{Y_i}\} = \int_0^1 \sqrt{x}dx = \frac{2}{3}$, and $\mathbf{V}^{(0)} = (\frac{2}{3}, \frac{2}{3})$.

The r.v. S has a triangular distribution with the density $f(x) = 1 - |x-1|$ for $x \in [0,2]$. (See Example **2**.2.1.1; the density (**2**.2.1.6) in this example may be represented as we wrote.) Then $a = E\{\sqrt{S}\} = \int_0^2 \sqrt{x}(1-|x-1|)dx = \frac{16}{15}\sqrt{2} - \frac{8}{15} \approx 0.975$. See Fig.8.

EXAMPLE 4. In the situation of Example 2, $\mathcal{V}$ consists of all points $(V_1(\mathbf{r}), V_2(\mathbf{r}))$, where $V_i(\mathbf{r}) = U_i(c_{\mathbf{r}i} - X_{\mathbf{r}i})$, and $\mathbf{r} \in \mathcal{R}$. The status-quo point corresponds to $\mathbf{r}^{(0)} = (1,1)$ (see also Fig.5), so $\mathbf{V}^{(0)} = (U_1(Y_1), U_2(Y_2)) = (U_1(c_1 - X_1), U_2(c_2 - X_2))$. We will consider this example later in greater detail. □

We continue to consider the general scheme. The participants enter into negotiations in order to agree on some rule of exchange $\mathbf{Z}$ appropriate for all participants.

The participants would act rationally if they rule out any agreement such that there exists another agreement which will give higher quality indices simultaneously to all participants.

Accordingly, we say that a point $\mathbf{V}$ from $\mathcal{V}$ is *Pareto optimal* if there is no point from $\mathcal{V}$ all of whose coordinates are not less than the corresponding coordinates of $\mathbf{V}$, and at least one coordinate is strictly larger. We denote the set of all Pareto optimal points by $\overline{\mathcal{V}}$, and the set of corresponding r.vec.'s $\mathbf{Z}$ by $\overline{\mathcal{Z}}$.

In the general picture in Fig.6, $\overline{\mathcal{V}}$ is the "North-East" boundary of $\mathcal{V}$. In the particular Example 3, this is the boundary (2.1.10): the quarter of the circle in Fig.7.

Certainly, we may restrict ourselves to points from $\overline{\mathcal{V}}$, and accordingly to exchanges (solutions) from $\overline{\mathcal{Z}}$.

Now, if the status-quo point $\mathbf{V}^{(0)}$ is an interior point of $\mathcal{V}$, there are points $\mathbf{V}$ all of whose coordinates are strictly larger than the corresponding coordinates of $\mathbf{V}$. This implies that all participants can simultaneously improve their quality indices in comparison with their initial positions.

Denote the set of all such points by $\mathcal{V}_0$ (the area "to the North-East" of $\mathbf{V}^{(0)}$ in Fig.9.), and the set of the corresponding r.vec.'s $\mathbf{Z}$ by $\mathcal{Z}_0$. Each point from $\mathcal{Z}_0$ generates a point from $\mathcal{V}_0$; in other words, $\mathcal{V}_0$ is the image of $\mathcal{Z}_0$.

It is clear that all participants may agree simultaneously only on a point from $\mathcal{V}_0$. On the other hand, it is reasonable to consider only points from $\overline{\mathcal{V}}$, so we should consider the intersection of these sets, that is, the set of all Pareto optimal points from $\mathcal{V}_0$.

Denote this set by $\overline{\mathcal{V}}_0$, and the set of corresponding r.vec.'s $\mathbf{Z}$ by $\overline{\mathcal{Z}}_0$. In Fig.9, the set $\overline{\mathcal{V}}_0$ is the part of the boundary to the "North-East" of $\mathbf{V}^{(0)}$.

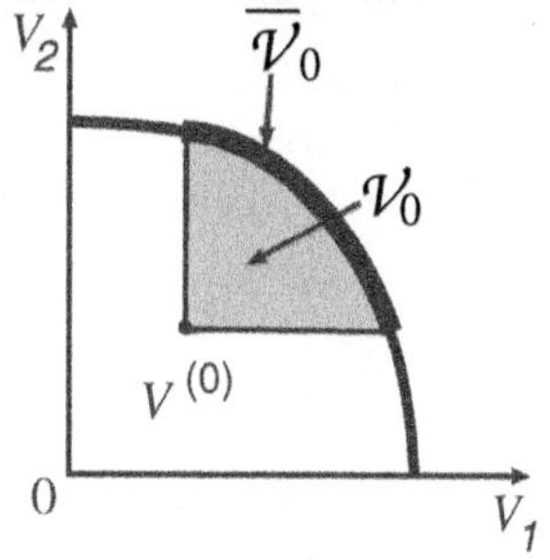

FIGURE 9.

We call points from $\overline{\mathcal{Z}}_0$ *Pareto optimal solutions* to the bargaining problem under consideration.

EXAMPLE 5. Consider the numerical sub-example of Example 3. We saw already that in this case, Pareto optimal solutions are vectors $\mathbf{Z} = (Z_1, Z_2) = (kS,\ (1-k)S)$, for $k \in [0,1]$. The Pareto optimal points $\mathbf{V} = (E\{\sqrt{kS}\}, E\{\sqrt{(1-k)S}\}) = E\{\sqrt{S}\}(\sqrt{k},\ \sqrt{(1-k)}) = a(\sqrt{k},\ \sqrt{(1-k)})$. On the other hand, for $\mathbf{V}$ to belong to $\overline{\mathcal{V}}_0$, both coordinates should be larger than $2/3$. So, we should have

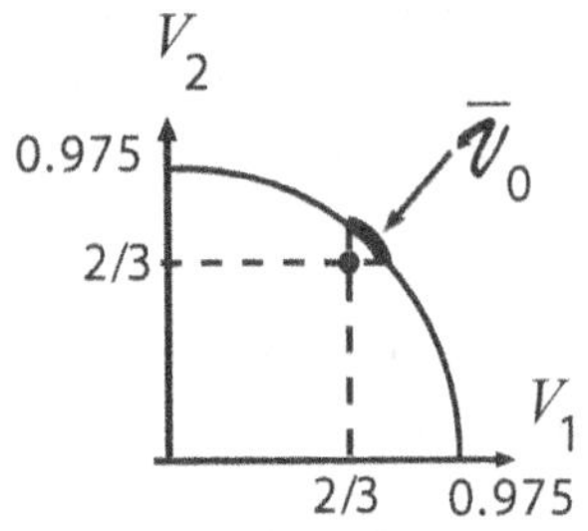

FIGURE 10.

$$a\sqrt{k} \geq \frac{2}{3},\ a\sqrt{1-k} \geq \frac{2}{3},$$

which results in

$$\frac{4}{9a^2} \leq k \leq 1 - \frac{4}{9a^2}.$$

Since $a = \frac{16}{15}\sqrt{2} - \frac{8}{15} \approx 0.975$, we come to the solutions

$$\mathbf{Z} = (kS,\ (1-k)S) \quad \text{for} \quad 0.467 \leq k \leq 0.532$$

up to the third digit. The set $\overline{\mathcal{V}}_0$ in this case is shown in Fig.10. □

Thus, for any particular problem, the task is to determine the set $\overline{\mathcal{Z}}_0$, and then, proceeding from some additional principles, to choose one particular Pareto solution.

These additional principles may be different and are strongly connected with the nature of the concrete problem under consideration. In the next sections, we consider some examples.

Next, we discuss one general approach to determining the set $\overline{\mathcal{Z}}_0$. We will do that at a heuristic level, skipping some formalities. The approach consists in maximizing the function

$$G(\mathbf{Z}) = a_1U_1(Z_1) + \ldots + a_nU_n(Z_n), \tag{2.1.11}$$

where a_i's are positive constants. Consider the set of all $\mathbf{Z}$'s for which $G(\mathbf{Z})$ is equal to some constant c. The image of each point $\mathbf{Z}$ from this set is a point $\mathbf{V} = (V_1, \ldots, V_n)$ for which

$$a_1V_1 + \ldots + a_nV_n = c. \tag{2.1.12}$$

Thus, the point $\mathbf{V}$ lies in the n-dimensional plane defined by (2.1.12).

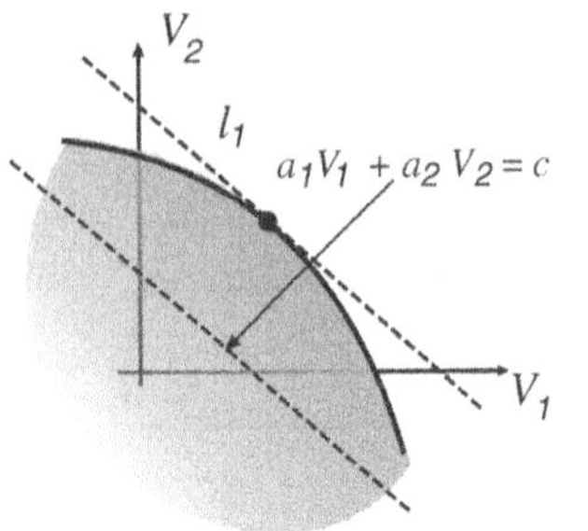

FIGURE 11.

In the two-dimensional case, this is a line, as is shown in Fig.11. To make the constant c (and, hence, the value of $G(\mathbf{Z})$) as large as possible, we should "move this line to the North-East" up to the moment when the line becomes the supporting line for the set $\mathcal{V}$, that is, when the whole set $\mathcal{V}$ is below the line. It is possible if the set $\mathcal{V}$ is convex. In Fig.11, this is the line l_1.

The intersection of this highest line with $\mathcal{V}$ will lie in $\overline{\mathcal{V}}$, the Pareto optimal boundary of $\mathcal{V}$. So, for a fixed vector of coefficients $\mathbf{a} = (a_1, a_2)$, we will get a Pareto optimal point from $\overline{\mathcal{V}}$. If $\mathcal{V}$ is convex, considering all possible vectors $\mathbf{a}$, we will come to all possible Pareto optimal points in $\overline{\mathcal{V}}$. The cases $a_1 = 0$ or $a_2 = 0$ will correspond to the end-points of the Pareto optimal set.

The n-dimensional case is similar. The only difference is that in this case we deal not with supporting lines but with supporting n-dimensional planes.

EXAMPLE 6. Consider the scheme of Examples 2 and 4. All vectors $\mathbf{Z}$ in this case may be identified with the vector $\mathbf{r} = (r_1, r_2)$ of retention coefficients. For each $\mathbf{r}$, we consider the point $\mathbf{V}(\mathbf{r}) = (V_1(\mathbf{r}), V_2(\mathbf{r})) = (V_1(r_1, r_2), V_2(r_1, r_2))$, as it was defined in Example 4.

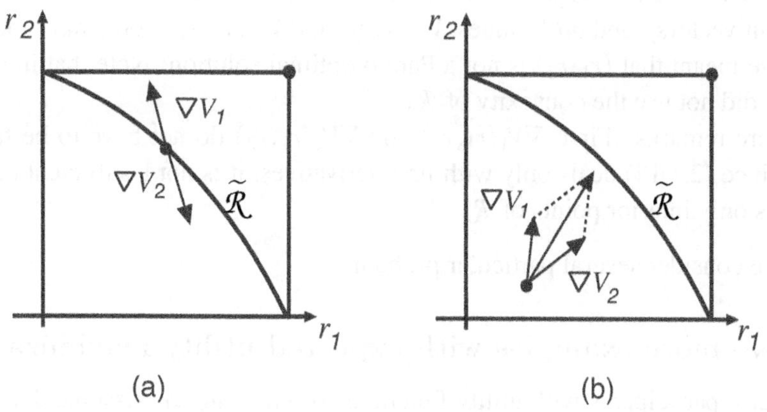

FIGURE 12.

Denote by $\overline{\mathcal{R}}$ the set corresponding to the set $\overline{\mathcal{Z}}$. The set $\overline{\mathcal{R}}$ is the set of all Pareto optimal solutions. Typically, this is a curve in $\mathcal{R}$, as it was shown in Fig.12. To find $\overline{\mathcal{R}}$, we should maximize the function

$$Q(r_1, r_2) = a_1 V_1(r_1, r_2) + a_2 V_2(r_1, r_2)$$

for all non-negative a_1, a_2.

Assume that $V_i(r_1, r_2), i = 1, 2$, are differentiable. Then we can make use of the gradients of the functions $V_1(r_1, r_2)$ and $V_2(r_1, r_2)$, that is, the vectors of partial derivatives

$$\nabla V_1 = \nabla V_1(r_1, r_2) = \left(\frac{\partial V_1(r_1, r_2)}{\partial r_1}, \frac{\partial V_1(r_1, r_2)}{\partial r_2} \right),$$
$$\nabla V_2 = \nabla V_2(r_1, r_2) = \left(\frac{\partial V_2(r_1, r_2)}{\partial r_1}, \frac{\partial V_2(r_1, r_2)}{\partial r_2} \right).$$

As we know from Calculus, the gradient ∇V_i points out in what direction we should move from the point (r_1, r_2) for the function $V_i(r_1, r_2)$ to change fastest. In particular, if we move in a direction such that the projection of ∇V_i on this direction is positive, then $V_i(r_1, r_2)$ increases.

If the maximum is attained at a point $\mathbf{r} = (r_1, r_2)$, and this point is an interior point of the square $\mathcal{R}$, then

$$\nabla Q = a_1 \nabla V_1 + a_2 \nabla V_2 = 0. \tag{2.1.13}$$

If the point $\mathbf{r}$ does not correspond to an end-point of the Pareto optimal set, both a_i's are positive, and $\nabla V_1 = -(a_2/a_1)\nabla V_2$. Thus, we have come to the condition

$$\nabla V_1 = -k \nabla V_2 \text{ for some } k > 0. \tag{2.1.14}$$

Condition (2.1.14) means that if an interior point $\mathbf{r}$ is Pareto optimal, the vectors $\nabla V_1(r_1, r_2)$ and $\nabla V_2(r_1, r_2)$ have opposite directions; see Fig.12a.

We could also come to the condition (2.1.14) reasoning in the following way. Assume that (r_1, r_2) is an interior point, $\nabla V_1, \nabla V_2$ are non-zero vectors, and the condition (2.1.14) is not true. Then—see also Fig.12b—we could move in the direction $\nabla V_1 + \nabla V_2$ (the sum of the gradient vectors), and *both* values, $V_1(r_1, r_2)$ and $V_2(r_1, r_2)$, would have increased. This would have meant that (r_1, r_2) is not a Pareto optimal solution. Note that in the reasoning above, we did not use the convexity of $\mathcal{R}$.

Two more remarks. First, $\nabla V_1(r_1, r_2)$ and $\nabla V_2(r_1, r_2)$ do not have to be tangent to $\widetilde{\mathcal{R}}$. Second, since (2.1.14) deals only with first derivatives, it is not a sufficient condition, and it concerns only interior points of $\mathcal{R}$. □

Next, we consider several particular problems.

2.2 Two more examples with expected utility maximization

Consider n participants with utility functions $u_1(x), ..., u_n(x)$. Assume that all functions $u_i(x)$ are sufficiently smooth, $u'(x) > 0$ and $u_i''(x) \leq 0$ for all x's. So, the participants are risk averters.

In accordance with the scheme of the previous section, to find the Pareto optimal solutions we should try to maximize the function

$$G(\mathbf{Z}) = \sum_{i=1}^{n} a_i E\{u_i(Z_i)\},$$

subject to the condition

$$\sum_{i=1}^{n} Z_i = \sum_{i=1}^{n} Y_i. \tag{2.2.1}$$

The following proposition belongs to K. Borch; see [16] and references therein. Some refinements and generalizations are obtained by H. Gerber in [42].

Proposition 3 *Let $\widetilde{\mathbf{Z}} = (\widetilde{Z}_1, ..., \widetilde{Z}_n)$ satisfy (2.2.1), and*

$$a_1 u_1'(\widetilde{Z}_1) = ... = a_n u_n'(\widetilde{Z}_n). \tag{2.2.2}$$

Then, for any $\mathbf{Z} = (Z_1, ... Z_n)$ satisfying (2.2.1),

$$G(\mathbf{Z}) \leq G(\widetilde{\mathbf{Z}}). \tag{2.2.3}$$

Proof. Denote by k the number $a_i u_i'(\widetilde{Z}_i)$ which, in accordance with (2.2.2), does not depend on i. Since the functions $u_i(x)$ are concave and smooth, for any $\widetilde{x}$ and x

$$u_i(x) \leq u_i(\widetilde{x}) + u_i'(\widetilde{x})(x - \widetilde{x}).$$

(See Section **0**.4.3.) Then

$$a_i u_i(Z_i) \leq a_i u_i(\widetilde{Z}_i) + a_i u_i'(\widetilde{Z}_i)(Z_i - \widetilde{Z}_i) = a_i u_i(\widetilde{Z}_i) + k(Z_i - \widetilde{Z}_i).$$

Consequently,

$$\sum_{i=1}^{n} a_i u_i(Z_i) \leq \sum_{i=1}^{n} a_i u_i(\widetilde{Z}_i) + k\Big(\sum_{i=1}^{n} Z_i - \sum_{i=1}^{n} \widetilde{Z}_i\Big) = \sum_{i=1}^{n} a_i u_i(\widetilde{Z}_i),$$

because $\mathbf{Z}$ and $\widetilde{\mathbf{Z}}$ both satisfy (2.2.1). Taking expectations we come to (2.2.3). ■

EXAMPLE 1 is practically the same as the corresponding example in [42]. Let $u_i(x) = -e^{-\beta_i x}$. In accordance with (2.2.2),

$$a_i u'(\widetilde{Z}_i) = k,$$

where k is a number not depending on i. From this it follows that $a_i\beta_i \exp\{-\beta_i\widetilde{Z}_i\} = k$, and

$$\widetilde{Z}_i = \frac{1}{\beta_i}C + b_i, \tag{2.2.4}$$

where $C = -\ln k$, $b_i = \frac{1}{\beta_i}\ln a_i\beta_i$. Since a_i is an arbitrary positive number, b_i is arbitrary number, positive or negative, for all i.

Recall that $\sum_{i=1}^n \widetilde{Z}_i = S$, where $S = \sum_{i=1}^n Y_i$. Summing up the left and the right sides of equation (2.2.4), we have

$$S = C\sum_{i=1}^n (1/\beta_i) + \sum_{i=1}^n b_i.$$

Solving it with respect to C and inserting into (2.2.4), we arrive at

$$\widetilde{Z}_i = \gamma_i S + c_i, \tag{2.2.5}$$

where

$$\gamma_i = \frac{1/\beta_i}{\sum_{i=1}^n (1/\beta_i)}, \tag{2.2.6}$$

$$c_i = b_i - \gamma_i \sum_{i=1}^n b_i. \tag{2.2.7}$$

Clearly $\gamma_i > 0$, and $\sum_{i=1}^n \gamma_i = 1$. Now, $\sum_{i=1}^n c_i = \sum_{i=1}^n b_i - (\sum_{i=1}^n \gamma_i)(\sum_{i=1}^n b_i) = \sum_{i=1}^n b_i - (\sum_{i=1}^n b_i) = 0.$

Thus,

$$\sum_{i=1}^n c_i = 0. \tag{2.2.8}$$

Moreover, since b_i are arbitrary numbers, c_i are arbitrary numbers for which (2.2.8) is true. Indeed, if we choose *arbitrary* b_i's for which $\sum_{i=1}^n b_i = 0$, by virtue of (2.2.7), $c_i = b_i$.

Thus, we can forget about b_i's, and claim that the set of all Pareto optimal solutions is described by (2.2.5) where

- γ_i is a certain non-random share of the total income S, which the ith participant will have. The share γ_i is completely specified by parameters β_i in accordance with (2.2.6).

- c_i's are arbitrary numbers for which (2.2.8) holds.

Let, for example, $n = 2$, $\beta_i = \beta$. Then, as is easy to check, all Pareto optimal solutions may be represented as follows:

$$\widetilde{Z}_1 = \frac{1}{2}S + c, \ \widetilde{Z}_2 = \frac{1}{2}S - c,$$

where c is an arbitrary constant.

To choose one Pareto optimal solution, the participants should choose a number c. Before the exchange, the participants may be not in equal positions because Y_1, Y_2 may have different distributions. Since in this case the participants share the total random income in equal proportions, the payment c is the only way to compensate for the inequality mentioned. For example, the participants may come to the agreement that, on the exchange, they will not make profit on the average. This amounts to the requirement $E\{\widetilde{Z}_1\} = E\{Y_1\}$, $E\{\widetilde{Z}_2\} = E\{Y_2\}$. Then the participants should choose

$$c = \frac{1}{2}\left(E\{Y_1\} - E\{Y_2\}\right). \ \square$$

Consider now the model of proportional mutual reinsurance as we built it in Examples 2.1-4 and 6, keeping the same notation. Thus, the quality-index-functions of two companies are

$$V_1(\mathbf{r}) = V_1(r_1, r_2) = E\{u_1(c_{\mathbf{r}1} - X_{\mathbf{r}1})\},\ V_2 = V_2(\mathbf{r}) = E\{u_2(c_{\mathbf{r}2} - X_{\mathbf{r}2})\},$$

where

$$X_{\mathbf{r}1} = r_1X_1 + (1 - r_2)X_2,\ X_{\mathbf{r}2} = (1 - r_1)X_1 + r_2X_2.$$

Let us assume that the part of a risk to be reinsured comes with the corresponding part of the premium. More precisely it means that, when ceding the part $(1 - r_1)X_1$, the first company yields also the premium $(1 - r_1)c_1$, and the same concerns the second company. So, the companies share their portfolios rather than their risks. Thus,

$$c_{\mathbf{r}1} = r_1c_1 + (1 - r_2)c_2,\ c_{\mathbf{r}2} = (1 - r_1)c_1 + r_2c_2. \tag{2.2.9}$$

The next example shows what we can expect in this case.

EXAMPLE 2. Let X_1, X_2 be independent and have the Γ-distributions with parameters (a_1, ν_1) and (a_2, ν_2), respectively. Set $m_i = E\{X_i\} = 1/a_i$. Assume that

$$u_1(x) = u_2(x) = -e^{-\beta x}.$$

We consider only the case $0 < \beta < \min(m_1^{-1}, m_2^{-1})$ because, as we will see, otherwise $V_1(\mathbf{r}), V_2(\mathbf{r})$ would not exist. We have

$$\begin{aligned} V_1(r_1, r_2) &= -E\{\exp\{-\beta[c_{\mathbf{r}1} - r_1X_1 - (1 - r_2)X_2]\}\} \\ &= -\exp\{-\beta c_{\mathbf{r}1}\}E\{\exp\{\beta[r_1X_1 + (1 - r_2)X_2]\}\} \\ &= -\exp\{-\beta c_{\mathbf{r}1}\}E\{\exp\{\beta r_1X_1\}\}E\{\exp\{\beta(1 - r_2)X_2\}\} \\ &= -\exp\{-\beta c_{\mathbf{r}1}\}\frac{1}{(1 - \beta m_1r_1)^{\nu_1}} \cdot \frac{1}{(1 - \beta m_2(1 - r_2))^{\nu_2}}. \end{aligned}$$

(For the m.g.f. of the Γ-distribution see Section **0**.4.3.5.)

Similarly,

$$V_2(r_1, r_2) = -\exp\{-\beta c_{\mathbf{r}2}\}\frac{1}{(1 - \beta m_1(1 - r_1))^{\nu_1}} \cdot \frac{1}{(1 - \beta m_2r_2)^{\nu_2}}.$$

Note now that to find Pareto optimal points, one does not need to know concrete values of the functions under consideration, but rather the areas where these functions are increasing or decreasing. Consequently, if instead of the functions $V_1(r_1,r_2)$, $V_2(r_1,r_2)$ we consider the quality indices $g(V_1(r_1,r_2))$, $g(V_2(r_1,r_2))$, where $g(x)$ is a strictly increasing function, then the set of Pareto optimal solutions will be the same. So, we may simplify the problem, taking $g(x) = \ln(-1/x)$, and instead of V_1, V_2, considering the functions

$$\begin{aligned} Q_1(r_1,r_2) &= g(V_1(r_1,r_2)) = \ln\left[\exp\{\beta c_{\mathbf{r}1}\}(1-\beta m_1 r_1)^{\nu_1}(1-\beta m_2(1-r_2))^{\nu_2}\right] \\ &= \beta c_{\mathbf{r}1} + \nu_1 \ln(1-\beta m_1 r_1) + \nu_2 \ln(1-\beta m_2(1-r_2)), \end{aligned} \tag{2.2.10}$$

$$\begin{aligned} Q_2(r_1,r_2) &= g(V_2(r_1,r_2)) = \ln\left[\exp\{\beta c_{\mathbf{r}2}\}(1-\beta m_1(1-r_1))^{\nu_1}(1-\beta m_2 r_2)^{\nu_2}\right] \\ &= \beta c_{\mathbf{r}2} + \nu_1 \ln(1-\beta m_1(1-r_1)) + \nu_2 \ln(1-\beta m_2 r_2). \end{aligned} \tag{2.2.11}$$

Taking into account (2.2.9), after simple algebra, we have

$$\nabla Q_1(r_1,r_2) = \left(\beta c_1 - \nu_1 \frac{\beta m_1}{1-\beta m_1 r_1},\ -\beta c_2 + \nu_2 \frac{\beta m_2}{1-\beta m_2(1-r_2)}\right),$$

$$\nabla Q_2(r_1,r_2) = \left(-\beta c_1 + \nu_1 \frac{\beta m_1}{1-\beta m_1(1-r_1)},\ \beta c_2 - \nu_2 \frac{\beta m_2}{1-\beta m_2 r_2}\right).$$

As we showed in Section 2.1, for (r_1,r_2) to be a Pareto optimal point, the vectors $\nabla Q_1, \nabla Q_2$ should have opposite directions. In this case the ratio of the coordinates for both vectors should be the same, which leads to the equation

$$\begin{aligned} &\left(c_1 - \nu_1 \frac{m_1}{1-\beta m_1 r_1}\right)\left(c_2 - \nu_2 \frac{m_2}{1-\beta m_2 r_2}\right) \\ &= \left(-c_1 + \nu_1 \frac{m_1}{1-\beta m_1(1-r_1)}\right)\left(-c_2 + \nu_2 \frac{m_2}{1-\beta m_2(1-r_2)}\right). \end{aligned} \tag{2.2.12}$$

(The l.-h.s. is the product of the first coordinate of the first vector and the second coordinate of the second vector, and the r.-h.s. is the product of the second coordinate of the first vector and the first coordinate of the second vector. We divided both sides by β^2.)

This is an equation for (r_1,r_2). To realize how the corresponding set can look, consider the case

$$m_1 = m_2,\ \beta m_i = 1, \text{ and } \quad c_i = (1+\theta)m_i.$$

Then m's will cancel, and (2.2.12) may be written as

$$\left((1+\theta) - \frac{\nu_1}{1-r_1}\right)\left((1+\theta) - \frac{\nu_2}{1-r_2}\right) = \left((1+\theta) - \frac{\nu_1}{r_1}\right)\left((1+\theta) - \frac{\nu_2}{r_2}\right). \tag{2.2.13}$$

First, let $\nu_1 = \nu_2 = \nu$. Setting in (2.2.13) $r_2 = 1 - r_1$, we come to an identity. Hence, the set

$$r_1 + r_2 = 1,\ 0 \le r_1 \le 1,\ 0 \le r_1 \le 1, \tag{2.2.14}$$

is one of the solutions of (2.2.13); see Fig.13. We skip calculations showing that other solutions, if any, do not fit the conditions of this problem.

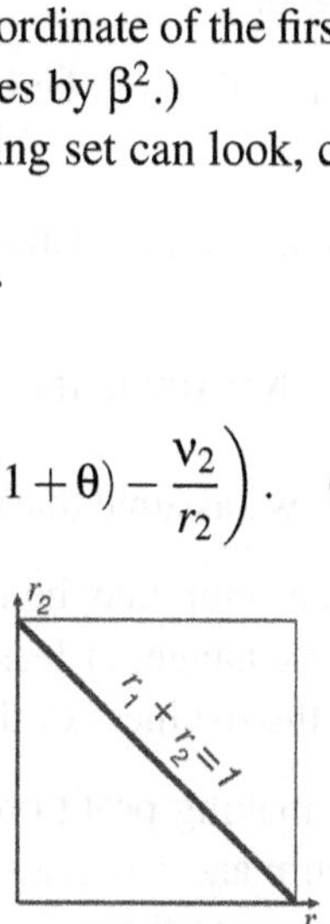

FIGURE 13.

If $\nu_1 \neq \nu_2$, the problem becomes more interesting. We depict the plots of the set (2.2.13), made by Maple, in Fig.14a for $\theta = 0.1$, $\nu_1 = 1$, $\nu_2 = 5$, and in Fig.14b for $\theta = 0.4$, $\nu_1 = 3$, $\nu_2 = 10$. In Exercise 11, we explain why in the last case the solution is close to (2.2.14).

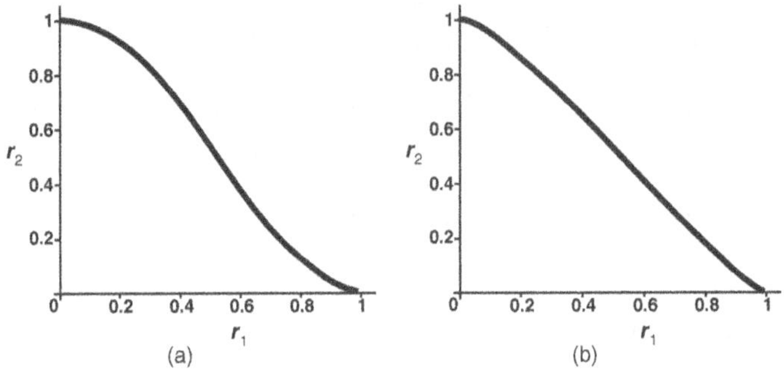

FIGURE 14.

In conclusion, note that since we used the "first-derivative-analysis", rigorously speaking, we still did not prove that the sets above are Pareto optimal. (This is similar to one-dimensional optimization problems: the fact that the first derivative equals zero at a point does not imply that this point is a maximizer or minimizer.) To complete the proof we should consider either the second derivatives of the functions (2.2.10)-(2.2.11), or the behavior of the gradients in a neighborhood of the sets above. Calculations are elementary, although they are lengthy, so we skip them also. □

2.3 The case of the mean-variance criterion

In this section we consider the case of proportional reinsurance described in Examples 2.1-2, 4, 6.

Assume X_1, X_2 to be independent, and set $m_i = E\{X_i\}$, $\sigma_i^2 = Var\{X_i\}$, $i = 1,2$. Suppose $\sigma_i > 0$.

We start with a relatively simple problem which corresponds to the case $\tau = 0$ in (2.1.6).

2.3.1 Minimization of variances

Here, we assume that the companies try to minimize the variances of their future profit.

Such a setup may be reasonable, for example, if the companies agree not to make profit on the exchange, at least, on the average, and under this condition, the companies try to reduce the riskiness of their portfolios.

(Not making profit on the average, certainly, amounts to the condition $E\{Z_{\mathbf{r}i}\} = E\{Y_i\}$ for each $\mathbf{r}$ and $i = 1,2$. In our case, this is true if when ceding the risk $(1 - r_i)X_i$, the ith company pays the *net* premium $(1 - r_i)m_i$, that is, the expected future payment for the risk ceded.

Indeed, for the first company this ends up with the premium $c_{\mathbf{r}1} = c_1 - (1-r_1)m_1 + (1-r_2)m_2$ (since the second company will pay the net premium $(1-r_2)m_2$ to the first). Then $E\{Y_{\mathbf{r}i}\} = c_{\mathbf{r}1} - E\{X_{\mathbf{r}1}\} = c_1 - (1-r_1)m_1 + (1-r_2)m_2 - r_1m_1 - (1-r_2)m_2 = c_1 - m_1 = E\{Y_1\}$. The same is true for the second company.)

Thus, the companies try to minimize the values of the functions

$$V_i(r_1, r_2) = Var\{Z_{\mathbf{r}i}\} = Var\{X_{\mathbf{r}i}\}.$$

Formally, if we set $\tau = 0$ in (2.1.6), we will come to $V(Y) = -Var\{Y\}$. In this case, it is convenient to omit the minus, and consider the corresponding minimization problem. Certainly, the problems of maximization $-Var\{Y\}$ and minimization of $Var\{Y\}$ are equivalent. Below, we will use the symbol $V_i(r_1, r_2)$ for the corresponding variance.

Since X_1, X_2 are independent and $X_{\mathbf{r}1} = r_1X_1 + (1-r_2)X_2$, $X_{\mathbf{r}2} = (1-r_1)X_1 + r_2X_2$, we have

$$V_1(r_1, r_2) = r_1^2\sigma_1^2 + (1-r_2)^2\sigma_2^2, \quad V_2(r_1, r_2) = (1-r_1)^2\sigma_1^2 + r_2^2\sigma_2^2. \tag{2.3.1}$$

To find Pareto optimal points, we again apply the method suggested in Section 2.1. In our case, it consists in minimization of the linear combination $a_1V_1(r_1, r_2) + a_2V_2(r_1, r_2)$ for positive a_1, a_2. So, we deal with a simple quadratic function which is concave upward and has a unique minimum; see also Exercise 3. For this reason, we restrict ourselves below to the first-derivative-analysis, looking for points where the gradients $\nabla V_1, \nabla V_2$ have opposite directions.

The same concerns the problem of portfolio redistribution we will consider in Section 2.3.2.

For the gradients we have

$$\nabla V_1 = 2\left(r_1\sigma_1^2,\ -(1-r_2)\sigma_2^2\right), \ \nabla V_2 = 2\left(-(1-r_1)\sigma_1^2,\ r_2\sigma_2^2\right).$$

If (2.1.14) is true, the ratio of the coordinates of ∇V_1 and ∇V_2 is the same, which amounts to

$$-\frac{r_1}{1-r_2} = -\frac{1-r_1}{r_2}.$$

This is equivalent to

$$r_1 + r_2 = 1; \tag{2.3.2}$$

see Fig.13.

Set $r_1 = r$. Then in the line (2.3.2), $r_2 = 1-r$, and

$$V_1(r, 1-r) = r^2[\sigma_1^2 + \sigma_2^2], \tag{2.3.3}$$

$$V_2(r, 1-r) = (1-r)^2[\sigma_1^2 + \sigma_2^2]. \tag{2.3.4}$$

For the first company, the smaller the r, the better. At the status-quo point, that is, at $\mathbf{r}^{(0)} = (1,1)$ the value $V_1(1,1) = \sigma_1^2$. Hence, the first company would accept reinsurance only if $V_1(r, 1-r) \le \sigma_1^2$. In view of (2.3.3), this corresponds to $r \le \sigma_1 \Big/ \sqrt{\sigma_1^2 + \sigma_2^2}$.

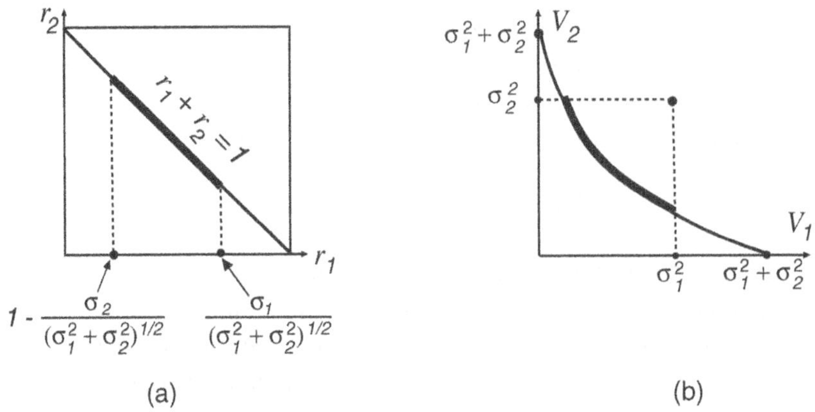

FIGURE 15.

Similarly, the second company would accept reinsurance if $1-r \leq \sigma_2 \Big/ \sqrt{\sigma_1^2+\sigma_2^2}$. Thus, the companies should choose the retention coefficients, from the segment

$$r_1+r_2=1, \quad 1-\frac{\sigma_2}{\sqrt{\sigma_1^2+\sigma_2^2}} \leq r_1 \leq \frac{\sigma_1}{\sqrt{\sigma_1^2+\sigma_2^2}}. \tag{2.3.5}$$

The reader is encouraged to show that the left member of (2.3.5) is not greater than the right member. Fig.15a depicts the set (2.3.5), Fig.15b - the corresponding set of Pareto optimal points in the plane (V_1, V_2). Since now we minimize the values of V_1 and V_2, the Pareto optimal points correspond to the "South-West" boundary in Fig.15b.

EXAMPLE 1. (a) Let $\sigma_1 = 3$, $\sigma_2 = 4$. Then (2.3.5) is equivalent to

$$r_1+r_2=1, \quad 0.2 \leq r_1 \leq 0.6; \tag{2.3.6}$$

(b) However, if $\sigma_1 = 10$, $\sigma_2 = 24$, then the set of Pareto optimal solutions is

$$r_1+r_2=1, \quad \frac{2}{13} \leq r_1 \leq \frac{5}{13}, \tag{2.3.7}$$

and does not include the middle point $(\frac{1}{2}, \frac{1}{2})$. □

Once the set of Pareto optimal solutions is determined, the companies should establish an additional principle leading to the choice of one solution from the set mentioned. Such a principle may require the volume of exchange to be, in some way or another, balanced.

Assume, for example, that the cost of reinsurance is proportional to the mean value of the reinsured risk. More precisely, for the risk $(1-r_1)X_1$ to be reinsured, the first company should pay to the second company the amount $(1+\theta_{\text{reins}})(1-r_1)m_1$, where θ_{reins} is a reinsurance loading coefficient. Respectively, the second company pays to the first the amount $(1+\theta_{\text{reins}})(1-r_2)m_2$. For the exchange to be balanced, we may require these two amounts to be equal (which is equivalent to the condition that the companies do not pay each other). Since in this case the factor $1+\theta_{\text{reins}}$ cancels, we have

$$(1-r_1)m_1 = (1-r_2)m_2. \tag{2.3.8}$$

Together with the condition $r_1 + r_2 = 1$, we obtain a simple solution

$$r_1 = \frac{m_1}{m_1 + m_2}, \quad r_2 = \frac{m_2}{m_1 + m_2}. \tag{2.3.9}$$

It is worth emphasizing two things.

First, we did not establish rule (2.3.8) in the very beginning. Only at the very end, when the set of Pareto optimal points has been determined, we impose condition (2.3.8) as an additional requirement.

Secondly, rule (2.3.8) may not lead to an appropriate solution.

EXAMPLE 2. Let $m_1 = 4$, $m_2 = 6$. Then $r_1 = 0.4$, $r_2 = 0.6$. In the situation of Example 1a, the point $(0.4, 0.6)$ belongs to the set of Pareto optimal solutions (2.3.6), so we can choose this point as a particular solution.

However, in the situation of Example 1b, the point $(0.4, 0.6)$ does not belong to the set (2.3.7), so in this case, (2.3.8) cannot serve as an additional condition. □

The last example points out a disadvantage of the expected value premium principle that does not take into account the risk to be carried by the reinsurer.

Suppose now that, when ceding the risk $(1 - r_i)X_i$, the ith company pays the net premium $(1 - r_i)m_i$ as we considered above, plus an additional security premium for reinsurance which is proportional to the standard deviation of the risk reinsured. More precisely, this additional payment is equal to $\lambda_{\text{reins}}(1 - r_i)\sigma_i$, where λ_{reins} is a loading coefficient. Such a rule corresponds to the standard deviation premium principle we considered in Section **10**.4; see also (**10**.4.2).

Assume now that the companies agree that these two additional premiums will compensate each other, so as a matter of fact, no company pays an additional reinsurance premium. This is equivalent to the condition

$$(1 - r_1)\sigma_1 = (1 - r_2)\sigma_2,$$

which together with $r_1 + r_2 = 1$, leads to

$$r_1 = \frac{\sigma_1}{\sigma_1 + \sigma_2}, \quad r_2 = \frac{\sigma_2}{\sigma_1 + \sigma_2}. \tag{2.3.10}$$

This solution looks quite natural. Note that, unlike the case of (2.3.9), the solution (2.3.10) always (!) belongs to the Pareto optimal set (2.3.5). We show this in Exercise 13. For particular examples, consider Exercise 14.

The end of Route 2

2.3.2 The exchange of portfolios

Let the ith portfolio be specified by the premium c_i and the future claim X_i. Now we assume that the part of a risk to be reinsured comes with the corresponding part of the premium. In other words, as in Section 2.2, we accept rule (2.2.9).

In this case, the profit of the first company is the r.v. $Z_{\mathbf{r}1} = c_{\mathbf{r}1} - X_{\mathbf{r}1} = r_1c_1 + (1-r_2)c_2 - r_1X_1 + (1-r_2)X_2 = r_1(c_1 - X_1) + (1-r_2)(c_2 - X_2)$. Thus,

$$Z_{\mathbf{r}1} = r_1Y_1 + (1-r_2)Y_2. \tag{2.3.11}$$

Similarly, the profit after reinsurance of the second company is

$$Z_{\mathbf{r}2} = (1-r_1)Y_1 + r_2Y_2. \tag{2.3.12}$$

Now we adopt the mean-variance criterion (2.1.6), assuming for simplicity that the tolerance-to-risk coefficient τ is the same for both companies. Suppose that X_1 and X_2, and hence Y_1 and Y_2, are independent. Then,

$$E\{Y_i\} = s_i = c_i - m_i, \text{ and } Var\{Y_i\} = \sigma_i^2.$$

By (2.3.11), (2.3.12),

$$\begin{aligned} V_1(r_1,r_2) &= \tau E\{Z_{\mathbf{r}1}\} - Var\{Z_{\mathbf{r}1}\} = \tau[r_1s_1 + (1-r_2)s_2] - [r_1^2\sigma_1^2 + (1-r_2)^2\sigma_2^2], \\ V_2(r_1,r_2) &= \tau E\{Z_{\mathbf{r}2}\} - Var\{Z_{\mathbf{r}2}\} = \tau[(1-r_1)s_1 + r_2s_2] - [(1-r_1)^2\sigma_1^2 + r_2^2\sigma_2^2]. \end{aligned} \tag{2.3.13}$$

Certainly, the problem includes that of the previous section. If $\tau = 0$, the functions V_i do not involve premiums, so any premium redistribution may be assumed, including the redistribution we considered in Section 2.3.1.

As in Section 2.3.1, we restrict ourselves to the first-derivative-analysis, that is, we will specify the set where ∇V_1 and ∇V_2 have opposite directions. See the remark directly following (2.3.1).

Differentiating V_i's, we get the gradients

$$\begin{aligned} \nabla V_1 &= \left(\tau s_1 - 2r_1\sigma_1^2,\ -\tau s_2 + 2(1-r_2)\sigma_2^2\right), \\ \nabla V_2 &= \left(-\tau s_1 + 2(1-r_1)\sigma_1^2,\ \tau s_2 - 2r_2\sigma_2^2\right). \end{aligned}$$

We will see that it is convenient to use the characteristics

$$\delta_1 = \frac{\tau s_1}{\sigma_1^2},\ \delta_2 = \frac{\tau s_2}{\sigma_2^2}, \tag{2.3.14}$$

and represent the gradients by the formulas

$$\nabla V_1 = \left(\sigma_1^2[\delta_1 - 2r_1], \sigma_2^2[2 - \delta_2 - 2r_2]\right), \nabla V_2 = \left(\sigma_1^2[2 - \delta_1 - 2r_1], \sigma_2^2[\delta_2 - 2r_2]\right). \tag{2.3.15}$$

Let first, $\delta_1 = \delta_2 = 1$. Then

$$\nabla V_1 = \left(\sigma_1^2[1 - 2r_1], \sigma_2^2[1 - 2r_2]\right),\ \nabla V_2 = \left(\sigma_1^2[1 - 2r_1], \sigma_2^2[1 - 2r_2]\right).$$

Thus, in this case, $\nabla V_1 = \nabla V_2$, and $\nabla V_1 = \nabla V_2 = 0$ at $\mathbf{r} = (r_1, r_2) = \left(\frac{1}{2}, \frac{1}{2}\right)$. Hence, both functions, V_1 and V_2, attain their maximum values at the *same* point $\mathbf{r} = \left(\frac{1}{2}, \frac{1}{2}\right)$.

Consequently, in this case the set of Pareto optimal solutions consists of one (!) point $\left(\frac{1}{2}, \frac{1}{2}\right)$ and for both companies the best solution is to divide the risks into equal shares. This is an extreme case.

Assume now that at least one δ is not equal to one. If (2.1.14) is true, the ratio of the coordinates of ∇V_1 and ∇V_2 should be the same, which implies that

$$\frac{\tau_1 s_1 - 2r_1\sigma_1^2}{-\tau_1 s_2 + 2(1-r_2)\sigma_2^2} = \frac{-\tau_2 s_1 + 2(1-r_1)\sigma_1^2}{\tau_2 s_2 - 2r_2\sigma_2^2}. \tag{2.3.16}$$

Straightforward algebraic manipulation shows that (2.3.16) is equivalent to the equation

$$r_1(1-\delta_2) + r_2(1-\delta_1) = 1 - \frac{1}{2}(\delta_1 + \delta_2). \tag{2.3.17}$$

Thus, the set of Pareto-optimal solutions is a subset of the *line* (2.3.17). Let us now consider some particular cases.

2.3.2.1 The symmetric case. Let $\delta_1 = \delta_2 = \delta \neq 1$. It is not difficult to verify that in this case δ in (2.3.17) cancels, and we again come to the segment

$$r_1 + r_2 = 1, \ 0 \leq r_1 < 1, \ 0 \leq r_2 < 1. \tag{2.3.18}$$

By (2.3.14), in our case,

$$\tau s_1 = \delta\sigma_1^2, \ \tau s_2 = \delta\sigma_2^2. \tag{2.3.19}$$

Let $r_1 = r$. Then, in accordance with (2.3.13), (2.3.15), and (2.3.19), in the segment (2.3.18),

$$V_1(r, 1-r) = \tau r[s_1 + s_2] - r^2[\sigma_1^2 + \sigma_2^2] = (\sigma_1^2 + \sigma_2^2)[\delta r - r^2], \tag{2.3.20}$$

$$\begin{aligned} V_2(r, 1-r) &= \tau(1-r)[s_1 + s_2] - (1-r)^2[\sigma_1^2 + \sigma_2^2] \\ &= (\sigma_1^2 + \sigma_2^2)[\delta(1-r) - (1-r)^2], \end{aligned} \tag{2.3.21}$$

$$\nabla V_1(r, 1-r) = \left(\sigma_1^2[\delta - 2r], \sigma_2^2[-\delta + 2r]\right) = (\delta - 2r)\cdot\left(\sigma_1^2, -\sigma_2^2\right), \tag{2.3.22}$$

$$\nabla V_2(r, 1-r) = \left(\sigma_1^2[2 - \delta - 2r], \sigma_2^2[\delta - 2 + 2r]\right) = (2r + \delta - 2)\cdot\left(-\sigma_1^2, \sigma_2^2\right). \tag{2.3.23}$$

Solutions we choose from (2.3.18) should satisfy the following two conditions.
(1) $\nabla V_1(r, 1-r)$ and $\nabla V_2(r, 1-r)$ must have opposite directions.
(2) The positions of both companies should be not worse than they were at the status-quo point, namely,

$$V_1(r, 1-r) \geq \tau s_1 - \sigma_1^2 = \sigma_1^2(\delta - 1), \ V_2(r, 1-r) \geq \tau s_2 - \sigma_2^2 = \sigma_2^2(\delta - 1). \tag{2.3.24}$$

In view of (2.3.22) and (2.3.23), the first condition holds if $(\delta - 2r)$ and $(2r + \delta - 2)$, as functions of r, have the same sign. It is not difficult to figure out (say, graphing these linear functions) that for $0 \leq r \leq 1$

(i) if $\delta \geq 2$, this is always true;
(ii) if $1 \leq \delta < 2$, this is true for $1 - \frac{\delta}{2} \leq r \leq \frac{\delta}{2}$;
(iii) if $0 \leq \delta < 1$, this is true for $\frac{\delta}{2} \leq r \leq 1 - \frac{\delta}{2}$.
Note that, if $\delta = 1$, we have only one point $r = \frac{1}{2}$, which we already observed above.

In view of (2.3.20) and (2.3.21), condition (2.3.24) is equivalent to the quadratic inequalities

$$(\sigma_1^2+\sigma_2^2)[\delta r-r^2]\geq\sigma_1^2(\delta-1),\ (\sigma_1^2+\sigma_2^2)[\delta(1-r)-(1-r)^2]\geq\sigma_2^2(\delta-1). \quad (2.3.25)$$

EXAMPLE 1. (a) Let $\sigma_1=1,\sigma_2=2,\delta=2$. These numbers determine all the necessary information about s_1,s_2,τ. Since $\delta=2$, the condition (i) holds automatically, and we should solve (2.3.25) which is equivalent to

$$5[2r-r^2]\geq 1,\ 5[2(1-r)-(1-r)^2]\geq 4.$$

The solution to these equations is

$$1-\frac{2}{\sqrt{5}}\leq r\leq\frac{1}{\sqrt{5}},$$

so approximately we deal with the segment

$$\{0.105\leq r_1\leq 0.447,\ r_2=1-r_1\}.$$

(b) Let $\sigma_1=1,\sigma_2=2,\delta=1.5$. Then we should solve the equations

$$5[1.5r-r^2]\geq 0.5,\ 5[1.5(1-r)-(1-r)^2]\geq 2.$$

The approximate solution here is $0.0699\leq r\leq 0.6531$. However, we should now take into account the requirement $1-\frac{\delta}{2}\leq r\leq\frac{\delta}{2}$, which in our case is $0.25\leq r\leq 0.75$. Hence, the eventual answer is

$$\{0.25\leq r_1\leq 0.6531,\ r_2=1-r_1\}.$$

The choice of one point requires an additional agreement similar to what we discussed in Section 2.3.1. □

2.3.2.2 A typical situation in the general case. Consider now the case when $\delta_1\neq\delta_2$. The main point here is that, by virtue of (2.3.17), Pareto optimal solutions lie in a line distinct from the diagonal of the square $\mathcal{R}$. Note also that, in view of (2.3.14) we can rewrite (2.3.13) as

$$V_1(r_1,r_2)=r_1\delta_1\sigma_1^2+(1-r_2)\delta_2\sigma_2^2-[r_1^2\sigma_1^2+(1-r_2)^2\sigma_2^2], \quad (2.3.26)$$
$$V_2(r_1,r_2)=(1-r_1)\delta_1\sigma_1^2+r_2\delta_2\sigma_2^2-[(1-r_1)^2\sigma_1^2+r_2^2\sigma_2^2]. \quad (2.3.27)$$

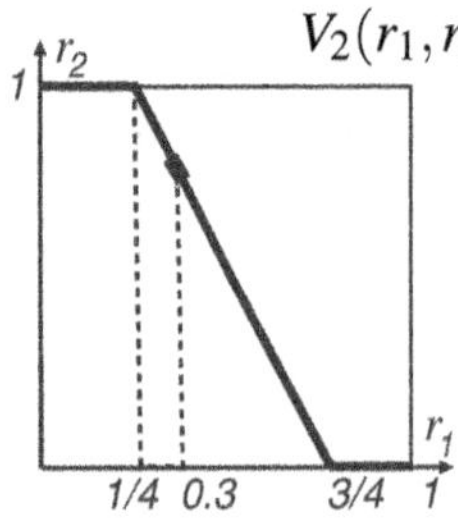

FIGURE 16.

General formulas here are a bit cumbersome, so we restrict ourselves to an example.

EXAMPLE 2. Let $\delta_1=5,\delta_2=9$. Substituting it into (2.3.17), we come to the line

$$4r_1+2r_2=3. \quad (2.3.28)$$

So, we should consider the segment connecting the points $(\frac{1}{4}, 1)$ and $(\frac{3}{4}, 0)$; see Fig.16.

Note that border points from the segments $\{r_2 = 1, 0 \le r_1 \le \frac{1}{4}\}$, and $\{r_2 = 0, \frac{3}{4} \le r_1 \le 1\}$, see again Fig.16, are also Pareto optimal. (The condition $\nabla V_1 = -k\nabla V_2$ applies only to interior points; the border or a part of it may be Pareto optimal in the absence of this condition.) To show this, we should consider the gradients along these segments. However, we will not do that since, as we will see, these segments should be excluded from consideration.

Consider the third, central, segment. Set again $r_1 = r$. Now $r \in [\frac{1}{4}, \frac{3}{4}]$. In accordance with (2.3.15), substituting r_2 from (2.3.28), we have

$$\begin{aligned} \nabla V_1 &= (\sigma_1^2[5-2r_1], \sigma_2^2[2-9-2r_2]) = (\sigma_1^2[5-2r], \sigma_2^2[-10+8r]) \\ &= (5-2r)\cdot(\sigma_1^2, -2\sigma_2^2), \\ \nabla V_2 &= (\sigma_1^2[2-5-2r_1], \sigma_2^2[9-2r_2]) = (2+3r)\cdot(-\sigma_1^2, 2\sigma_2^2). \end{aligned}$$

We see that $\nabla V_1, \nabla V_2$ have opposite directions for all $r \in [\frac{1}{4}, \frac{3}{4}]$.

Suppose now that $\sigma_1 = 1$, $\sigma_2 = 2$. Then, from (2.3.26), (2.3.27) it follows that in the line (2.3.28)

$$\begin{aligned} V_1(r_1, r_2) &= 5r_1 + 36(1-r_2) - [r_1^2 + 4(1-r_2)^2] \\ &= 5r + 18(4r-1) - [r^2 + (4r-1)^2] = -19 + 85r - 17r^2, \\ V_2(r_1, r_2) &= 5(1-r_1) + 36r_2 - [(1-r_1)^2 + 4r_2^2] \\ &= 5(1-r) + 18(3-4r) - [(1-r)^2 + (3-4r)^2] \\ &= 49 - 51r - 17r^2. \end{aligned}$$

At the status-quo point, $V_1(1,1) = \delta_1\sigma_1^2 - \sigma_1^2 = 4$, $V_2(1,1) = \delta_2\sigma_2^2 - \sigma_2^2 = 32$. Consequently, the restrictions $V_1(r_1, r_2) \ge V_1(1,1)$, $V_2(r_1, r_2) \ge V_2(1,1)$ are equivalent to the equations

$$-19 + 85r - 17r^2 \ge 4, \; 49 - 51r - 17r^2 \ge 32.$$

The solution to the system of these quadratic inequalities is

$$\frac{5}{2} - \frac{3}{34}\sqrt{629} \le r \le \frac{1}{2}\sqrt{13} - \frac{3}{2}.$$

Thus, the set of acceptable solutions for both companies is given approximately by

$$0.2871 \le r_1 \le 0.3027, \; 4r_1 + 2r_2 = 3$$

(again see Fig.16). The solution is close to the point $(0.3, 0.9)$, so in this case, additional considerations are not necessary. □

Other proofs, comments, and models of this type, and references may be found in [11, Chapter 5], [16], [31], [101].

3 REINSURANCE MARKET

3.1 A model of the exchange market of random assets

While in the previous sections we discussed direct risk exchange as a result of bargaining between companies, this section concerns the decentralized price mechanism of exchange. In this case, the participants of the market do not exchange their risks directly but trade them.

Constructions of this section serve merely to illustrate some basic ideas and concepts, and results below should not be, certainly, considered practical recommendations. Modern markets are complex and sophisticated mechanisms that use various financial tools, and a pure exchange model may serve only as a first approximation.

The classical model we discuss next is due to K. Arrow and H. Bühlmann (see, in particular, [5], [6], [21], [22]).

Consider the general framework of Section 2.1. Assume that the participants of the market are expected utility maximizers, and the ith participant has a utility function $\widetilde{u}_i(x)$ and an initial wealth s_i. Then for an additional income y (positive or negative) the utility of the ith participant's wealth is equal to $\widetilde{u}_i(s_i+y)$. Set $u_i(y)=\widetilde{u}_i(s_i+y)$. The function $u_i(y)$ is the utility function which incorporates the initial wealth.

As usual, we fix a sample space $\Omega=\{\omega\}$ and a probability measure $P(A)$ defined on events A from Ω. All r.v.'s under consideration are functions on Ω.

To simplify calculations, we assume below that the space Ω is discrete: $\Omega=\{\omega_1,\omega_2,...\}$, though, as a matter of fact, the results at which we will arrive are true in a much more general situation.

If we denote by $P(\omega_k)$ the probability of the outcome ω_k, then for a r.v. $Y=Y(\omega)$,

$$E\{Y\}=\sum_k Y(\omega_k)P(\omega_k). \tag{3.1.1}$$

As in Section 2.1, we fix the vector $\mathbf{Y}=(Y_1,...,Y_n)$ of the original r.v.'s of future incomes. Thus, the future wealth of the ith participant is s_i+Y_i, and its utility is $\widetilde{u}_i(s_i+Y_i)=u_i(Y_i)$. The r.v. Y_i may be called the initial endowment (without the fixed initial wealth s_i) of the participant i.

The participants of the market exchange the r.v.'s of their future incomes. Thus, the commodities of the market are random variables, and the main concept we should define and clarify is the price of a r.v. We assume that it is represented by a numerical-valued function $G(Y)$ assuming numerical values and defined on the class of r.v.'s under consideration.

We do not exclude r.v.'s taking negative values, interpreting it as the case of losses, and we do not exclude the case when the price assumes a negative value. If an investor accepts a risk which may bring only losses, then the investor should be paid for this. This means that the price of this risk is negative.

Unlike the price for a usual commodity, what should be called a price for a random income (say, a random asset) is a non-trivial question.

We suppose that the price function $G(Y)$ is linear, that is, for any two r.v.'s Y_1 and Y_2, and numbers α_1 and α_2

$$G(\alpha_1 Y_1 + \alpha_2 Y_2) = \alpha_1 G(Y_1) + \alpha_2 G(Y_2). \tag{3.1.2}$$

This condition looks reasonable although it involves some simplification. As a matter of fact, prices are not linear: usually the more you buy the less you pay for a unit of commodity.

We assume also that

$$G(Y) \geq 0, \text{ for any non-negative r.v. } Y, \text{ and} \tag{3.1.3}$$
$$G(1) = 1; \text{ more precisely, } G(Y) = 1, \text{ for any r.v. } Y \equiv 1. \tag{3.1.4}$$

Such a price function G may be represented as follows. Let $Q(A)$ be a probability measure defined on events A from Ω. This measure may be different from the original "actual" probability measure $P(A)$. Denote by $E_Q\{Y\}$ the expected value of Y with respect to measure Q. If Ω is discrete, and $Q(\omega_k)$ is the probability of ω_k with respect to measure Q, then

$$E_Q\{Y\} = \sum_k Y(\omega_k) Q(\omega_k) \tag{3.1.5}$$

(compare with (3.1.1)).

Consider the function

$$G(Y) = G_Q(Y) = E_Q\{Y\}. \tag{3.1.6}$$

This function satisfies properties (3.1.2)-(3.1.4) as any expectation with respect to some probability measure.

Moreover, it may be shown that under some mild conditions, any function $G(Y)$ with the properties (3.1.2)-(3.1.4) is equal to G_Q for *some* Q. In particular, it is true if the sample space Ω is finite.

So, *we adopt, as a definition of price, the representation (3.1.6).* In this case, the determination of a pricing procedure in the market amounts to the determination of a measure Q. If Ω is discrete, in view of (3.1.5), to determine Q, it suffices to specify the "probabilities" $Q(\omega_k)$.

It makes sense to emphasize again that the numbers $Q(\omega_k)$ differ from the actual probabilities $P(\omega_k)$. One may say that $Q(\omega_k)$ reflects how the participants of the market estimate the possibility of the outcome ω_k or, in other words, $Q(\omega_k)$ indicates the participants' beliefs regarding ω_k. These beliefs may differ from reality.

Given a pricing function $G(Y)$, the participant i sells the endowment Y_i for the price $G(Y_i)$ and purchases another r.v. W_i for the price $G(W_i)$. This is equivalent to the exchange Y_i for W_i for the price $G(W_i) - G(Y_i)$. Note that $G(W_i) - G(Y_i) = G(W_i - Y_i)$, since the function $G(\cdot)$ is linear.

We do not assume that $G(W_i) \leq G(Y_i)$, and the value $G(W_i - Y_i)$ may have any sign. If $G(W_i) > G(Y_i)$, we interpret it as if the participant borrowed some money to purchase W_i, or took this money from the initial wealth. On the other hand, $G(W_i - Y_i)$ may be negative, which means that after exchange, the participant received additional cash.

Eventually, the future income of the ith participant is

$$Z_i = W_i - G(W_i - Y_i), \tag{3.1.7}$$

and the expected utility is

$$V_i = E\{u(Z_i)\} = E\{u(W_i - G(W_i - Y_i))\}. \tag{3.1.8}$$

Certainly, the exchange should be balanced, that is,

$$\sum_{i=1}^{n} W_i = \sum_{i=1}^{n} Y_i. \tag{3.1.9}$$

Because the price function is linear, from (3.1.9) it follows that $\sum_{i=1}^{n} G(W_i) = \sum_{i=1}^{n} G(Y_i)$, or

$$\sum_{i=1}^{n} G(W_i - Y_i) = 0. \tag{3.1.10}$$

That is, the total payments are also balanced. Say, if $n = 2$, one participant pays, and the other receives.

From (3.1.10) it immediately follows that (3.1.9) is equivalent to the balance for the Z's:

$$\sum_{i=1}^{n} Z_i = \sum_{i=1}^{n} Y_i. \tag{3.1.11}$$

Indeed, by virtue of (3.1.7), $\sum_{i=1}^{n} Z_i = \sum_{i=1}^{n} W_i - \sum_{i=1}^{n} G(W_i - Y_i)$, while the last sum equals zero by (3.1.10).

Given a pricing function G, the ith participant determines the r.v. W_i which maximizes the expected utility (3.1.8). Denote the result of this maximization by W_{iG}. The r.v. W_{iG} is the demand of the participant i.

Note that W_{iG} is not what the participant will buy. Rather it is what she/he wants to buy; this is a demand. The point is that if a pricing procedure does not closely reflect the real situation in the market, demand may be not equal to supply. Thus, we should not expect that for an arbitrary G the demand quantities W_{iG} satisfy the balance requirement (3.1.9).

We call a function $G^*(Y)$ and a vector $\mathbf{W}^* = (W_1^*, ..., W_n^*)$ an *equilibrium price* function and an *equilibrium demand vector*, respectively, if

(a) $W_i^* = W_{iG^*}$ for all $i = 1, ..., n$, that is, for each participant, the demand is optimal with respect to the pricing function G^*,

(b) the supply and demand are balanced, i.e.,

$$\sum_{i=1}^{n} W_i^* = \sum_{i=1}^{n} Y_i.$$

As was noted, the last property is equivalent to the balance equation

$$\sum_{i=1}^{n} Z_i^* = \sum_{i=1}^{n} Y_i,$$

where $Z_i^* = W_i^* - G^*(W_i^* - Y_i)$. So we can talk about the equilibrium vector $\mathbf{Z}^* = (Z_1^*, ..., Z_n^*)$.

It may be proved that, under some mild conditions, the solution $\mathbf{Z}^*$ exists and is Pareto optimal; see, e.g., [81], [86]. This is a remarkable fact. First, this means that under equilibrium prices, the market pricing mechanism of exchange leads to a solution which cannot be improved simultaneously in favor of all participants. Second, the pricing mechanism is decentralized. To reach the equilibrium solution, the participants do not have to enter into direct negotiations. Each participant can make decisions separately from the others, maximizing just her/his own expected utility of the future income.

3.2 An example concerning reinsurance

The classical example we consider here is due to K. Borch; see, e.g., [16].

Suppose the participants of the market are insurance companies and $Y_i = -X_i$, where X_i is the original claim for the ith company. We interpret the quantity s_i from the general scheme above as the amount of the fund available to cover the claim X_i. In particular, we assume that s_i involves the premium paid. Then $s_i + Y_i = s_i - X_i$ is the future surplus of the ith company.

Denote by $\widetilde{X}_i$ the claim retained by the ith company. (In the notation of the previous section, $W_i = -\widetilde{X}_i$.) This means that the company purchases the reinsurance for $X_i - \widetilde{X}_i$ and pays for this the price

$$d_i = G(X_i - \widetilde{X}_i), \tag{3.2.12}$$

where $G(\cdot)$ is a pricing function. Note that the r.v. $X_i - \widetilde{X}_i$ may take on negative values (the company reinsures somebody's risk). Similarly, $G(X_i - \widetilde{X}_i)$ may be negative (overall, the company behaves as a reinsurer).

The expected utility of the ith company is

$$V_i = E\{\widetilde{u}(s_i - \widetilde{X}_i - d_i)\}. \tag{3.2.13}$$

The exchange is balanced if

$$\sum_{i=1}^{n} X_i = \sum_{i=1}^{n} \widetilde{X}_i. \tag{3.2.14}$$

Assume that the pricing function G is generated by a probability measure Q in the fashion of the previous section. Then, in accordance with (3.1.5),

$$d_i = E_Q\{X_i - \widetilde{X}\} = \sum_k [X_i(\omega_k) - \widetilde{X}(\omega_k)]Q(\omega_k) = \sum_k (x_{ik} - \widetilde{x}_{ik})q_k,$$

where $x_{ik} = X_i(\omega_k)$, $\widetilde{x}_{ik} = \widetilde{X}_i(\omega_k)$, and $q_k = Q(\omega_k)$.

Setting $p_k = P(\omega_k)$, and $t_{ik} = x_{ik} - \widetilde{x}_{ik}$, we rewrite (3.2.13) as

$$V_i = \sum_l p_l \widetilde{u}_i\left(s_i - \widetilde{x}_{il} - \sum_k (x_{ik} - \widetilde{x}_{ik})q_k\right). \tag{3.2.15}$$

(Certainly, the running indices in the exterior and interior sums should be denoted by different symbols.)

Our goal is to find an equilibrium price G and an equilibrium demand. Since G is completely characterized by the measure Q, our objects of study are the numbers q_k and $\widetilde{x}_{ik}$.

First, considering just one company, we fix i and maximize (3.2.15) with respect to the vector $(\widetilde{x}_{i1}, \widetilde{x}_{i2}, ...)$. For a fixed m, the partial derivative with respect to $\widetilde{x}_{im}$ is

$$\frac{\partial V_i}{\partial \widetilde{x}_{im}} = \frac{\partial}{\partial \widetilde{x}_{im}} \sum_{l \neq m} p_l \widetilde{u}_i \left(s_i - \widetilde{x}_{il} - \sum_k (x_{ik} - \widetilde{x}_{ik}) q_k \right) + \frac{\partial}{\partial \widetilde{x}_{im}} p_m \widetilde{u}_i \left(s_i - \widetilde{x}_{im} - \sum_k (x_{ik} - \widetilde{x}_{ik}) q_k \right)$$

$$= \sum_{l \neq m} p_l \widetilde{u}_i' \left(s_i - \widetilde{x}_{il} - \sum_k (x_{ik} - \widetilde{x}_{ik}) q_k \right) q_m + p_m \widetilde{u}_i' \left(s_i - \widetilde{x}_{im} - \sum_k (x_{ik} - \widetilde{x}_{ik}) q_k \right) (-1 + q_m)$$

$$= q_m \sum_l p_l \widetilde{u}_i' \left(s_i - \widetilde{x}_{il} - \sum_k (x_{ik} - \widetilde{x}_{ik}) q_k \right) - p_m \widetilde{u}_i' \left(s_i - \widetilde{x}_{im} - \sum_k (x_{ik} - \widetilde{x}_{ik}) q_k \right)$$

$$= q_m E \left\{ \widetilde{u}_i' \left(s_i - \widetilde{X}_i - d_i \right) \right\} - p_m \widetilde{u}_i' (s_i - \widetilde{x}_{im} - d_i).$$

Setting all partial derivatives equal to zero, we have a system of equations for the equilibrium prices and equilibrium demand vectors; namely

$$q_m E \left\{ \widetilde{u}_i' \left(s_i - \widetilde{X}_i - d_i \right) \right\} = p_m \widetilde{u}_i' (s_i - \widetilde{x}_{im} - d_i) \text{ for all } i = 1, ..., m, \text{ and } m = 1, ..., n. \quad (3.2.16)$$

EXAMPLE 1. Let

$$\widetilde{u}_i(x) = -\frac{1}{2}x^2 + x.$$

The function above is increasing for $x \leq 1$, so we should assume all r.v.'s $s_i - X_i$, and $s_i - \widetilde{X}_i - d_i$ to be bounded by one.

Suppose also, for the sake of simplicity, that all $s_i = 1$.

Since in our case $\widetilde{u}_i'(x) = 1 - x$, and $s_i = 1$, equations (3.2.16) may be written as

$$q_m \left(1 - E \left\{ \left(1 - \widetilde{X}_i - d_i \right) \right\} \right) = p_m (1 - (1 - \widetilde{x}_{im} - d_i)),$$

or

$$q_m (E\{\widetilde{X}_i\} + d_i) = p_m (\widetilde{x}_{im} + d_i). \quad (3.2.17)$$

Note that, in view of (3.2.14),

$$\sum_{i=1}^{n} d_i = 0 \quad (3.2.18)$$

(see also (3.1.10)). Set

$$S = X_1 + ... + X_n,$$

and observe that, by virtue of (3.2.14), $\sum_{i=1}^{n} \widetilde{x}_{im} = \sum_{i=1}^{n} x_{im} = s_m$, where s_m is the value $S = S(\omega)$ at $\omega = \omega_m$, that is, $s_m = S(\omega_m)$. Summing over i in both sides of (3.2.17) and taking into account (3.2.18), we get that

$$q_m E\{S\} = p_m s_m,$$

or

$$q_m = p_m \frac{s_m}{E\{S\}}. \quad (3.2.19)$$

Recalling the definition of q_m, we can write it in the following more explicit form:

$$Q(\omega_m) = P(\omega_m)\frac{S(\omega_m)}{E\{S\}}. \tag{3.2.20}$$

Thus, we have found the equilibrium pricing measure Q. For such a measure, the price of any r.v. Y is $G_Q(Y) = E_Q\{Y\} = \sum_m Y(\omega_m)Q(\omega_m) = \frac{1}{E\{S\}}\sum_m Y(\omega_m)S(\omega_m)P(\omega_m) = E\{YS\}/E\{S\}$. Thus,

$$G(Y) = G_Q(Y) = \frac{E\{YS\}}{E\{S\}}. \tag{3.2.21}$$

Note that we have assumed Ω to be discrete just to simplify proofs. Formula (3.2.21) looks meaningful in the general case, and it indeed may be obtained under rather weak conditions.

Furthermore, it turns out that with respect to the equilibrium price (3.2.21), the optimal r.v.

$$\widetilde{X}_i = r_i S, \tag{3.2.22}$$

where

$$r_i = \frac{E\{X_i S\}}{E\{S^2\}}. \tag{3.2.23}$$

To prove this, we should show that such a solution satisfies (3.2.17). First of all, observe that, by (3.2.21), for the r.v. (3.2.22),

$$d_i = G(X_i - \widetilde{X}_i) = \frac{1}{E\{S\}}(E\{X_i S\} - E\{\widetilde{X}_i S\}) = \frac{1}{E\{S\}}(E\{X_i S\} - r_i E\{S^2\})$$
$$= \frac{1}{E\{S\}}(E\{X_i S\} - E\{X_i S\}) = 0,$$

in view of (3.2.23).

So, $d_i = 0$, and we can write (3.2.17) as the equality $q_m E\{\widetilde{X}_i\} = p_m \widetilde{x}_{im}$. Making use of (3.2.19), we write it as $s_m E\{\widetilde{X}_i\} = \widetilde{x}_{im} E\{S\}$. Recalling the definitions of s_m and $\widetilde{x}_{im}$, we have

$$S(\omega_m)E\{\widetilde{X}_i\} = \widetilde{X}_i(\omega_m)E\{S\}.$$

Substituting $\widetilde{X}_i$ from (3.2.22), we come to an identity.

Thus, in our model, equilibrium reinsurance corresponds to the proportional reinsurance (3.2.22). Since $G(X_i - \widetilde{X}_i) = 0$, we have

$$G(X_i) = G(\widetilde{X}_i), \tag{3.2.24}$$

that is, each participant retains a claim, perhaps different from the original claim, but of the same price as the claim X_i.

The last property is due to the special choice of utility functions and parameters s_i. It is not very difficult to show that in the case of arbitrary quadratic utility functions, (3.2.22) continues to be true (with different r_i) but (3.2.24) may not hold (see [16]).

Let us look at the solutions obtained more closely. Assume X_i's are independent, and set $m_i = E\{X_i\}$, $\sigma_i^2 = Var\{X_i\}$, $m_S = m_1 + ... + m_n$, $\sigma_S^2 = \sigma_1^2 + ... + \sigma_n^2$. Then the equilibrium price of the original claim X_i will be

$$G(X_i) = \frac{E\{X_i S\}}{E\{S\}} = \frac{1}{m_S}\left(E\{X_i^2\} + \sum_{j \neq i} E\{X_i X_j\}\right) \tag{3.2.25}$$

$$= \frac{1}{m_S}\left(E\{X_i^2\} + m_i \sum_{j \neq i} m_j\right) = \frac{1}{m_S}\left(E\{X_i^2\} + m_i(m_S - m_i)\right)$$

$$= \frac{1}{m_S}\left(\sigma_i^2 + m_i m_S\right) = m_i + \frac{\sigma_i^2}{m_S}, \tag{3.2.26}$$

which looks simple and nice. The larger σ_i, the more one should pay to reinsure the risk. Using what we have already computed, for the retention coefficient r_i we have

$$r_i = \frac{E\{X_i S\}}{E\{S^2\}} = \frac{m_i m_S + \sigma_i^2}{m_S^2 + \sigma_S^2} = \frac{m_i + \sigma_i^2/m_S}{m_S + \sigma_S^2/m_S}, \tag{3.2.27}$$

which is also nice and meaningful. □

---

4 EXERCISES

Section 1

1. Show that in the situation of Example 1.1.1-1, for any identically distributed X_i's,

$$E\{R_{d(n)}(X)\} = nE\{R_{d(1)}(X_1) + ... + R_{d(1)}(X_n)\} = nE\{R_{d(1)}(X_1)\}$$

and give an interpretation. Does it means that $d(n) = nd(1)$? For independent X_i distributed as in Example 1.1.1-1, paralleling (1.1.9), show that

$$E\{R_d^2(X)\} = n(n+1)\Gamma(d; n+2) + d^2\left(1 - \Gamma(d; n)\right).$$

Set $n = 10$ and compute $Var\{R_{d(10)}(X)\}$ and $Var\{R_{d(1)}(X_1) + ... + R_{d(1)}(X_{10})\}$, the variance of the total payment in the case of excess-of-loss reinsurance. Interpret the result.

2. In the situation of Example 1.1.1-1, write equation (1.1.10) for the case when X_i's have the Γ-distribution with parameters $(1, \nu)$.

3. Problems below concern the situation of Section 1.2.1.

   (a) A company has a portfolio with a *net* premium (i.e., mean payment) of \$300,000, and a standard deviation of \$100,000. For a new risk, the net premium is \$100,000 and the standard deviation is σ_0. For what σ_0 is the reinsurance of the new risk reasonable? Find the retention coefficient for $\sigma_0 = \$90{,}000$.

   (b) Let the original portfolio consist of i.i.d. risks with a mean of $\widetilde{m}$, and a standard deviation of $\widetilde{\sigma}$.

i. Write the formula for the coefficient $2k^2m$. Does it depend on the number of risks?
ii. Let the number of the risks in the original portfolio be large. How will the formula (1.2.4) look in this case?
iii. In the situation of Example 1.2.1-2, let $q = 0.1$, $\widetilde{m} = 4$, $\widetilde{\sigma} = 1$, and let the number of risks in the original portfolio be $n = 100$. Find v_{retained}. For which q in this case does the approximation $2k^2m$ have an error less than 5%?

4.** In the situation of Example 1.2.2-1, give an economic explanation why (i) if λ_0 gets smaller, the need for reinsurance increases; (ii) the same is true if σ_0 is increasing.

5. In the scheme of Section 1.3.1, find the maximal value of γ_r.

6. In the scheme of Section 1.3.1, assume that X has the Γ-distribution with parameters (a, ν), $\theta = 0.05$, $\theta_{\text{reins}} = 0.08$. (a) Using (1.3.1) as an approximation, find the estimate of the retention coefficient maximizing the adjustment coefficient. How does your estimate depend on a and ν? Is it a property of the estimate or of the real maximizer? Find the adjustment coefficient for the optimal r. (b) Make a heuristic conjecture about the values of θ, a, and ν for which the approximation you use will work relatively well.

7. (a) In the scheme of Section 1.3.2, assume that X is exponentially distributed, $E\{X\} = 1$, $\theta = 0.05$, $\theta_{\text{reins}} = 0.06$. Given d, write an equation for the adjustment coefficient γ_d. Using software, estimate d for which γ_d attains its maximum.

   (b) For the same X as in (a), write the formula for the optimal r in the case of proportional reinsurance applied to each claim. Find the particular answer for the θ's above.

Sections 2 and 3

8. Write a condition under which solution (2.1.3) is meaningful.

9. Consider the situation of Example 2.1-3.

   (a) Show that any point from (2.1.9) corresponds to some vector (Z_1, Z_2) for which (2.1.8) is true. (*Advice*: Consider vectors $(k_1S,\ k_2S)$ for various choices of k_1 and k_2.)

   (b) Does one point from (2.1.9) correspond to one solution (Z_1, Z_2) or perhaps many?

   (c) Find the set $\mathcal{V}$ and the set of Pareto optimal points $\overline{\mathcal{V}}$ for $u(x) = x$. Show that in this case, the status-quo point is Pareto optimal, and consequently, the set $\overline{\mathcal{V}}_0$ consists of only one point.

   (d) Let Y_1, Y_2 be independent and uniformly distributed on $[0,1]$, and let $u(x) = x^{1/3}$. Show that in this case, $\overline{\mathcal{V}} = \{(V_1, V_2) : V_1^3 + V_2^3 = a^3,\ 0 \le V_1 \le a,\ 0 \le V_2 \le a\}$, where $a = E\{(Y_1 + Y_2)^{1/3}\}$. Find a and the set $\overline{\mathcal{V}}_0$.
   To solve this problem, we need the following generalization of the Cauchy-Schwarz inequality, which is called *Hölder's inequality*: $E\{\xi\eta\} \le (E\{\xi^p\})^{1/p}(E\{\eta^q\})^{1/q}$ for positive ξ, η and $p > 1, q > 1, \frac{1}{p} + \frac{1}{q} = 1$; (see, e.g., [38]). The Cauchy-Schwarz inequality corresponds to $p = q = 2$. For our problem we need Hölder's inequality for $p = 3$, $q = 3/2$, that is, $E\{\xi\eta\} \le (E\{\xi^3\})^{1/3}(E\{\eta^{3/2}\})^{2/3}$.

10. Find a solution to the problem from Example 2.2-1 for $n = 3$ and $\beta_i = \beta$. Assume that the r.v.'s Y_i are uniformly distributed on $[0, 2i]$ for $i = 1, 2, 3$, and the participants do not make an additional profit on the exchange on the average.

11. Show that for large ν_1, ν_2, the solution to (2.2.13) is approaching (2.2.14).

12. (a) Sketch the graph of the function $V(r_1, r_2) = a_1 V_1(r_1, r_2) + a_2 V_2(r_1, r_2)$ from Section 2.3.1. The graph is represented by a surface in the space (V, r_1, r_2). What is this surface called? Does this function have a minimum? Is it unique? In your answer, refer to known facts from Calculus.

    (b) Sketch the regions in the plane (r_1, r_2) for which $V_1(r_1, r_2) \leq V_1(1,1)$ and $V_2(r_1, r_2) \leq V_2(1,1)$.

13. Show that the solution (2.3.10) belongs to the set of admissible solutions (2.3.5).

14. Consider the solution (2.3.10) in the case of Examples 2.3.1-1ab. Demonstrate that this is a Pareto optimal solution.

15.** Let $V_1(r_1, r_2)$ and $V_2(r_1, r_2)$ be the functions defined Section 2.3.2. Analyze the function $V(r_1, r_2) = a_1 V_1(r_1, r_2) + a_2 V_2(r_1, r_2)$ in the spirit of Exercise 12a.

16.** Consider the scheme of Section 2.3.2.

    (a) Show that if $c_i = (1+\theta)m_i$, the set of Pareto optimal solutions depends only on $\tau\theta$ rather than on τ and θ separately.

    (b) Let X_1, X_2 have the Γ-distributions with parameters $(1, \nu_1)$ and $(2, \nu_2)$, respectively and $\tau\theta = 4$. Find the set of Pareto optimal solutions.

17.** The problem concerns Example 3.2-1.

    (a) Find the prices $G_Q(X_i)$ and r_i in the case of i.i.d. X_i's. Interpret your result. In particular, describe the behavior of $G_Q(X_i)$ for large n.

    (b) Find the formulas for $G_Q(X_i)$ in the case when $Corr\{X_i, X_j\} = \rho$ for $i \neq j$. Compare the results with the case $\rho = 0$. What would you get if $m_i = m$ and $\sigma_i = \sigma$? Interpret the result. In particular, consider the case of large n and the case of $n = 2$ and $\rho < 0$.

Appendix

1 SUMMARY TABLES FOR BASIC DISTRIBUTIONS

TABLE 1. Some basic distributions

Distributions		cdf	mean	variance	Some properties
DISCRETE:	probabilities				
Binomial	$\binom{n}{m} p^m q^{n-m}$, $m = 0, 1, ..., n$	A step function	np	npq	The distribution of "the number of successes"
Geometric: 1st version	pq^{m-1}, $m = 1, 2, ...$	A step function	$1/p$	q/p^2	1. The distribution of "the 1st success" 2. The lack of memory 3. $P(X > m) = q^m$
Geometric: 2nd version	pq^m, $m = 0, 1, 2, ...$	A step function	q/p	q/p^2	$P(X > m) = q^{m+1}$
Negative binomial: 1st version	$\binom{m-1}{\nu-1} p^\nu q^{m-\nu}$, $m = \nu, \nu+1, ...$	A step function	ν/p	$\nu q/p^2$	The distribution of "the νth success"
Negative binomial: 2nd version	$\binom{\nu+m-1}{m} p^\nu q^m$, $m = 0, 1, 2, ...$	A step function	$\nu q/p$	$\nu q/p^2$	
Poisson	$e^{-\lambda}\lambda^m/m!$, $m = 0, 1, 2, ...$	A step function	λ	λ	The sum of two independent Poisson r.v.'s with parameters λ_1 and λ_2 have the Poisson distribution with the parameter $\lambda_1 + \lambda_2$
CONTINUOUS:	pdf				
Uniform	$1/(b-a)$, $a \le x \le b$	$\frac{x-a}{b-a}$, $a \le x \le b$	$\frac{a+b}{2}$	$\frac{(b-a)^2}{12}$	All values from $[a, b]$ are equally likely
Exponential	ae^{-ax}, $x \ge 0$	$1-e^{-ax}$, $x \ge 0$	$1/a$	$1/a^2$	1. The lack of memory 2. $P(X > z) = e^{-az}$

TABLE 1. (Continued)

Distributions		cdf	mean	variance	Some properties
Γ-distribution	$a^{\nu}x^{\nu-1}e^{-ax}/\Gamma(\nu)$, $x \geq 0$	—	ν/a	ν/a^2	
Normal	$\frac{1}{\sqrt{2\pi}\sigma}e^{-\frac{(x-m)^2}{2\sigma^2}}$	$\Phi\left(\frac{x-m}{\sigma}\right)$	m	σ^2	The sum of independent (m_1, σ_1^2)- and (m_2, σ_2^2)-normal r.v.'s is $(m_1+m_2, \sigma_1^2+\sigma_2^2)$-normal

TABLE 2. Some basic m.g.f.'s

Distributions		Moment generating function $M(z)$
DISCRETE:	probabilities	
Binomial	$\binom{n}{m}p^m q^{n-m}$, $m = 0, 1, ..., n$	$(pe^z + q)^n$
Geometric: 1st version	pq^{m-1}, $m = 1, 2, ...$	$e^z p/(1-qe^z)$, exists for $z < \ln(1/q)$
Geometric: 2nd version	pq^m $m = 0, 1, 2, ...$	$p/(1-qe^z)$, exists for $z < \ln(1/q)$
Negative binomial: 1st version	$\binom{m-1}{\nu-1}p^{\nu}q^{m-\nu}$, $m = \nu, \nu+1, ...$	$[e^z p/(1-qe^z)]^{\nu}$, exists for $z < \ln(1/q)$
Negative binomial: 2nd version	$\binom{\nu+m-1}{m}p^{\nu}q^m$, $m = 0, 1, 2, ...$	$[p/(1-qe^z)]^{\nu}$, exists for $z < \ln(1/q)$
Poisson	$e^{-\lambda}\lambda^m/m!$, $m = 0, 1, 2, ...$	$\exp\{\lambda(e^z - 1)\}$
CONTINUOUS:	pdf	
Uniform	$1/(b-a)$, $a \leq x \leq b$	$\left(e^{zb} - e^{za}\right)/[z(b-a)]$
Exponential	ae^{-ax}, $x \geq 0$	$1/(1-z/a)$, exists for $z < a$
Γ-distribution	$a^{\nu}x^{\nu-1}e^{-ax}/\Gamma(\nu)$ $x \geq 0$	$[1/(1-z/a)]^{\nu}$, exists for $z < a$
Normal	$\frac{1}{\sqrt{2\pi}\sigma}\exp\left\{-\frac{(x-m)^2}{2\sigma^2}\right\}$	$\exp\left\{mz + \sigma^2 z^2/2\right\}$

2 TABLES FOR THE STANDARD NORMAL DISTRIBUTION

TABLE 1. The standard normal distribution function $\Phi(x)$.

x	0	0.01	0.02	0.03	0.04	0.05	0.06	0.07	0.08	0.09
0	0.5	0.504	0.508	0.512	0.516	0.5199	0.5239	0.5279	0.5319	0.5359
0.1	0.5398	0.5438	0.5478	0.5517	0.5557	0.5596	0.5636	0.5675	0.5714	0.5753
0.2	0.5793	0.5832	0.5871	0.591	0.5948	0.5987	0.6026	0.6064	0.6103	0.6141
0.3	0.6179	0.6217	0.6255	0.6293	0.6331	0.6368	0.6406	0.6443	0.648	0.6517
0.4	0.6554	0.6591	0.6628	0.6664	0.67	0.6736	0.6772	0.6808	0.6844	0.6879
0.5	0.6915	0.695	0.6985	0.7019	0.7054	0.7088	0.7123	0.7157	0.719	0.7224
0.6	0.7257	0.7291	0.7324	0.7357	0.7389	0.7422	0.7454	0.7486	0.7517	0.7549
0.7	0.758	0.7611	0.7642	0.7673	0.7704	0.7734	0.7764	0.7794	0.7823	0.7852
0.8	0.7881	0.791	0.7939	0.7967	0.7995	0.8023	0.8051	0.8078	0.8106	0.8133
0.9	0.8159	0.8186	0.8212	0.8238	0.8264	0.8289	0.8315	0.834	0.8365	0.8389
1	0.8413	0.8438	0.8461	0.8485	0.8508	0.8531	0.8554	0.8577	0.8599	0.8621
1.1	0.8643	0.8665	0.8686	0.8708	0.8729	0.8749	0.877	0.879	0.881	0.883
1.2	0.8849	0.8869	0.8888	0.8907	0.8925	0.8944	0.8962	0.898	0.8997	0.9015
1.3	0.9032	0.9049	0.9066	0.9082	0.9099	0.9115	0.9131	0.9147	0.9162	0.9177
1.4	0.9192	0.9207	0.9222	0.9236	0.9251	0.9265	0.9279	0.9292	0.9306	0.9319
1.5	0.9332	0.9345	0.9357	0.937	0.9382	0.9394	0.9406	0.9418	0.9429	0.9441
1.6	0.9452	0.9463	0.9474	0.9484	0.9495	0.9505	0.9515	0.9525	0.9535	0.9545
1.7	0.9554	0.9564	0.9573	0.9582	0.9591	0.9599	0.9608	0.9616	0.9625	0.9633
1.8	0.9641	0.9649	0.9656	0.9664	0.9671	0.9678	0.9686	0.9693	0.9699	0.9706
1.9	0.9713	0.9719	0.9726	0.9732	0.9738	0.9744	0.975	0.9756	0.9761	0.9767
2	0.9772	0.9778	0.9783	0.9788	0.9793	0.9798	0.9803	0.9808	0.9812	0.9817
2.1	0.9821	0.9826	0.983	0.9834	0.9838	0.9842	0.9846	0.985	0.9854	0.9857
2.2	0.9861	0.9864	0.9868	0.9871	0.9875	0.9878	0.9881	0.9884	0.9887	0.989
2.3	0.9893	0.9896	0.9898	0.9901	0.9904	0.9906	0.9909	0.9911	0.9913	0.9916
2.4	0.9918	0.992	0.9922	0.9925	0.9927	0.9929	0.9931	0.9932	0.9934	0.9936
2.5	0.9938	0.994	0.9941	0.9943	0.9945	0.9946	0.9948	0.9949	0.9951	0.9952
2.6	0.9953	0.9955	0.9956	0.9957	0.9959	0.996	0.9961	0.9962	0.9963	0.9964
2.7	0.9965	0.9966	0.9967	0.9968	0.9969	0.997	0.9971	0.9972	0.9973	0.9974
2.8	0.9974	0.9975	0.9976	0.9977	0.9977	0.9978	0.9979	0.9979	0.998	0.9981
2.9	0.9981	0.9982	0.9982	0.9983	0.9984	0.9984	0.9985	0.9985	0.9986	0.9986
3	0.9987	0.9987	0.9987	0.9988	0.9988	0.9989	0.9989	0.9989	0.999	0.999

TABLE 2. The quantiles of the standard normal distribution: $\Phi^{-1}(y)$.

y	0.8	0.85	0.9	0.91	0.92	0.93	0.94	0.95	0.96	0.97	0.98	0.99	1
$\Phi^{-1}(y)$	0.842	1.036	1.2816	1.341	1.4051	1.476	1.555	1.645	1.751	1.88079	2.05	2.33	infinity

3 ILLUSTRATIVE LIFE TABLE

TABLE 1. Illustrative Life Table; $\delta = 4\%$[1]

1	2	3	4	5	6	7	8	1
x	q_x	l_x	d_x	$\mu(x)$	$1000A_x$	$\ddot{a}_x$	$1000\,{}^2A_x$	x
0–1	0.00697	100 000	697	0.006994	62.91043	23.89891	13.946618	0–1
1–2	0.000473	99 303	47	0.000473	58.91851	24.00071	8.1953125	1–2
2–3	0.000322	99 256	32	0.000322	60.87854	23.95073	8.4085569	2–3
3–4	0.000242	99 224	24	0.000242	63.06097	23.89507	8.789316	3–4
4–5	0.000202	99 200	20	0.000202	65.40848	23.83520	9.2817204	4–5
5–6	0.000171	99 180	17	0.000171	67.88993	23.77191	9.8551417	5–6
6–7	0.000141	99 163	14	0.000151	70.50124	23.70532	10.506343	6–7
7–8	0.000141	99 149	14	0.000141	73.24761	23.63527	11.241791	7–8
8–9	0.000151	99 135	15	0.000151	76.10645	23.56237	12.038585	8–9
9–10	0.000141	99 120	14	0.000141	79.07307	23.48671	12.891885	9–10
10–11	0.000151	99 106	15	0.000151	82.17047	23.40771	13.826322	10–11
11–12	0.000151	99 091	15	0.000151	85.38548	23.32572	14.828768	11–12
12–13	0.000182	99 076	18	0.000192	88.73218	23.24037	15.914845	12–13
13–14	0.000222	99 058	22	0.000222	92.18848	23.15222	17.061767	13–14
14–15	0.000263	99 036	26	0.000263	95.74994	23.06139	18.264756	14–15
15–16	0.000343	99 010	34	0.000354	99.42113	22.96776	19.52857	15–16
16–17	0.000576	98 976	57	0.000566	103.1706	22.87214	20.818797	16–17
17–18	0.000677	98 919	67	0.000688	106.8667	22.77788	21.9895	17–18
18–19	0.000850	98 852	84	0.000850	110.6257	22.68201	23.159305	18–19
19–20	0.000942	98 768	93	0.000942	114.3878	22.58606	24.259035	19–20
20–21	0.000932	98 675	92	0.000933	118.2258	22.48818	25.361779	20–21
21–22	0.000984	98 583	97	0.000984	122.2323	22.38600	26.566503	21–22
22–23	0.000944	98 486	93	0.000935	126.3611	22.28070	27.822582	22–23
23–24	0.000945	98 393	93	0.000956	130.6971	22.17012	29.223142	23–24
24–25	0.000946	98 300	93	0.000947	135.2136	22.05494	30.740919	24–25
25–26	0.000927	98 207	91	0.000927	139.9180	21.93496	32.385796	25–26
26–27	0.000958	98 116	94	0.000959	144.8358	21.80954	34.188179	26–27
27–28	0.000918	98 022	90	0.000908	149.9322	21.67956	36.11216	27–28
28–29	0.000939	97 932	92	0.000940	155.2755	21.54329	38.236782	28–29
29–30	0.000991	97 840	97	0.001002	160.8241	21.40178	40.52005	29–30
30–31	0.001023	97 743	100	0.001024	166.5611	21.25547	42.946009	30–31
31–32	0.001065	97 643	104	0.001066	172.5120	21.10370	45.546363	31–32
32–33	0.001066	97 539	104	0.001057	178.6776	20.94646	48.326154	32–33
33–34	0.001180	97 435	115	0.001191	185.1007	20.78265	51.339598	33–34
34–35	0.001254	97 320	122	0.001244	191.7007	20.61433	54.499572	34–35
35–36	0.001368	97 198	133	0.001369	198.5195	20.44043	57.857616	35–36
36–37	0.001453	97 065	141	0.001454	205.5341	20.26153	61.392071	36–37
37–38	0.001568	96 924	152	0.001570	212.7786	20.07677	65.147237	37–38
38–39	0.001715	96 772	166	0.001727	220.2394	19.88650	69.113307	38–39
39–40	0.001915	96 606	185	0.001907	227.9031	19.69105	73.279881	39–40

[1] This table was prepared mainly by Sara Zarei as a part of her Master Thesis [148].

TABLE 1 (continued).

1	2	3	4	5	6	7	8	1
x	q_x	l_x	d_x	$\mu(x)$	$1000A_x$	$\ddot{a}_x$	$1000\,{}^2A_x$	x
40–41	0.002088	95 780	200	0.002076	235.7404	19.49117	77.616789	40–41
41–42	0.002249	95 580	215	0.002237	243.7926	19.28581	82.177482	41–42
42–43	0.00237	95 365	226	0.002367	252.0708	19.07469	86.981719	42–43
43–44	0.002631	95 780	252	0.002635	260.6175	18.85672	92.088931	43–44
44–45	0.002826	95 528	270	0.002830	269.3311	18.63449	97.383939	44–45
45–46	0.003065	95 258	292	0.003060	278.2828	18.40619	102.95937	45–46
46–47	0.003296	94 966	313	0.003312	287.4555	18.17226	108.80271	46–47
47–48	0.003508	94 653	332	0.003514	296.8694	17.93218	114.94751	47–48
48–49	0.003891	94 321	367	0.003899	306.5525	17.68522	121.43956	48–49
49–50	0.00413	93 954	388	0.004138	316.4033	17.43399	128.16161	49–50
50–51	0.004425	93 566	414	0.004435	326.5348	17.17561	135.26474	50–51
51–52	0.00482	93 152	449	0.004832	336.927	16.91057	142.73742	51–52
52–53	0.005005	92 703	464	0.005018	347.5323	16.6401	150.5311	52–53
53–54	0.005551	92 239	512	0.005566	358.5046	16.36027	158.85828	53–54
54–55	0.005843	91 727	536	0.005861	369.6364	16.07637	167.4679	54–55
55–56	0.006722	91 191	613	0.006734	381.1051	15.78388	176.60436	55–56
56–57	0.006613	90 578	599	0.006646	392.5751	15.49136	185.84031	56–57
57–58	0.007624	89 979	686	0.007642	404.6593	15.18317	196.0015	57–58
58–59	0.008343	89 293	745	0.008379	416.7269	14.87541	206.27453	58–59
59–60	0.00943	88 548	835	0.009475	428.9696	14.56318	216.92105	59–60
60–61	0.009748	87 713	855	0.009796	441.2068	14.25109	227.70509	60–61
61–62	0.01088	86 858	945	0.010940	453.8895	13.92764	239.25446	61–62
62–63	0.011907	85 913	1 023	0.011979	466.6098	13.60323	251.03263	62–63
63–64	0.012958	84 890	1 100	0.013031	479.4542	13.27565	263.16664	63–64
64–65	0.014095	83 790	1 181	0.014207	492.4442	12.94436	275.69957	64–65
65–66	0.015313	82 609	1 265	0.015420	505.5724	12.60955	288.63526	65–66
66–67	0.016473	81 344	1 340	0.016611	518.8371	12.27126	301.98609	66–67
67–68	0.018212	80 004	1 457	0.018392	532.3068	11.92773	315.86774	67–68
68–69	0.019619	78 547	1 541	0.019814	545.7582	11.58468	329.97319	68–69
69–70	0.021674	77 006	1 669	0.021913	559.3867	11.23711	344.59746	69–70
70–71	0.02364	75 337	1 781	0.023911	572.9602	10.89094	359.41414	70–71
71–72	0.02564	73 556	1 886	0.025975	586.5695	10.54386	374.56309	71–72
72–73	0.027668	71 670	1 983	0.028059	600.2583	10.19474	390.12187	72–73
73–74	0.030537	69 687	2 128	0.031013	614.0775	9.842311	406.18399	73–74
74–75	0.033275	67 559	2 248	0.033841	627.7719	9.493058	422.3752	74–75
75–76	0.036579	65 311	2 389	0.037265	641.4615	9.143927	438.88263	75–76
76–77	0.039779	62 922	2 503	0.040593	655.0211	8.798111	455.51941	76–77
77–78	0.043331	60 419	2 618	0.044298	668.5689	8.452597	472.4736	77–78
78–79	0.047214	57 801	2 729	0.048366	682.0779	8.108073	489.71345	78–79
79–80	0.052513	55 072	2 892	0.053962	695.5393	7.764762	507.23497	79–80
80–81	0.057608	52 180	3 006	0.059316	708.6238	7.431065	524.51174	80–81
81–82	0.062248	49 174	3 061	0.064293	721.4993	7.102696	541.80071	81–82
82–83	0.071455	46 113	3 295	0.074140	734.4118	6.773383	559.50566	82–83
83–84	0.073427	42 818	3 144	0.076266	746.2522	6.471414	575.79362	83–84
84–85	0.084892	39 674	3 368	0.088718	759.0123	6.145988	593.93351	84–85
85–86	0.093125	36 306	3 381	0.097756	770.5061	5.85286	610.31981	85–86
86–87	0.101898	32 925	3 355	0.107479	781.6138	5.569577	626.35592	86–87
87–88	0.111261	29 570	3 290	0.117961	792.3532	5.295686	642.04865	87–88
88–89	0.121195	26 280	3 185	0.129203	802.7426	5.030721	657.40544	88–89
89–90	0.131674	23 095	3 041	0.141201	812.8176	4.773775	672.46303	89–90

TABLE 1 (continued).

1	2	3	4	5	6	7	8	1
x	q_x	l_x	d_x	$\mu(x)$	$1000A_x$	$\ddot{a}_x$	$1000\,{}^2A_x$	x
90–91	0.142765	20 054	2 863	0.154059	822.6350	4.523398	687.29561	90–91
91–92	0.154383	17 191	2 654	0.167710	832.2601	4.277926	701.99372	91–92
92–93	0.166541	14 537	2 421	0.182198	841.8022	4.034571	716.72837	92–93
93–94	0.179267	12 116	2 172	0.197594	851.4107	3.789523	731.74768	93–94
94–95	0.192478	9 944	1 914	0.213833	861.2915	3.537528	747.41210	94–95
95–96	0.206227	8 030	1 656	0.230865	871.7577	3.270606	764.29360	95–96
96–97	0.220270	6 374	1 404	0.249053	883.2599	2.977261	783.24968	96–97
97–98	0.234809	4 970	1 167	0.267753	896.5109	2.639317	805.68180	97–98
98–99	0.249803	3 803	950	0.287594	912.5685	2.229794	833.74700	98–99
99–100	0.264984	2 853	756	0.308116	933.0991	1.706195	870.95040	99–100
100 +	1	2 097	2 097	1.000000	960.7894	1	923.11635	100 +

4 SOME FACTS FROM CALCULUS

4.1 The "little o and big O" notation

4.1.1 Little o

What we discuss below is not a new notion but a notation which turns out to be useful in many calculations. We start with a particular definition.

Denote by the symbol $o(x)$ (little o of x) *any* function $o(x) = \varepsilon(x)x$, where the function $\varepsilon(x) \to 0$ as $x \to 0$. In other words, $o(x) \to 0$ faster than x.

Another way to define $o(x)$ is to say that

$$\frac{o(x)}{x} \to 0 \quad \text{as} \quad x \to 0,$$

which is the same.

For example, $x^2 = o(x)$, and $x^{3/2} = o(x)$, while $\sqrt{x}$ is not $o(x)$.

Heuristically, the formula $x^2 = o(x)$ means that "x^2 is much smaller than x" for small x's.

Certainly, we can replace x by, say, x^2 and in this case, the expression $o(x^2)$ denotes any function such that $[o(x^2)/x^2] \to 0$. For example, $x^3 = o(x^2)$.

If a function

$$f(x) = 1 + 2x^2 + o(x^2) \quad \text{as} \quad x \to 0, \tag{4.1.1}$$

we can say that $f(x)$ converges to one, as $x \to 0$, at a rate of $2x^2$ up to a remainder $o(x^2)$ which is negligible in comparison with the term $2x^2$ for small x's. Formally, (4.1.1) means that $f(x) - 1 - 2x^2 = o(x^2)$, that is,

$$\frac{f(x) - 1 - 2x^2}{x^2} \to 0 \quad \text{as} \quad x \to 0.$$

EXAMPLE 1. At what rate does the function $f(x) = (1+4x^2)^3$ converge to one as $x \to 0$? By the formula $(1+a)^3 = 1 + 3a + 3a^2 + a^3$,

$$(1+4x^2)^3 = 1 + 3 \cdot 4x^2 + A(x),$$

where $A(x)$ is the sum of the terms containing higher powers of $4x^2$, that is, $(4x^2)^2$ and $(4x^2)^3$. The sum of all such terms converges to zero faster than x^2. Hence, the remainder $A(x)$ is $o(x^2)$. Thus,

$$(1+4x^2)^3 = 1 + 12x^2 + o(x^2).$$

If we want to approximate $f(x)$ for small x's with greater accuracy, we can write

$$(1+4x^2)^3 = 1 + 3 \cdot 4x^2 + 3 \cdot (4x^2)^2 + o(x^4) = 1 + 12x^2 + 48x^4 + o(x^4). \ \square$$

The reader may find more sophisticated examples connected with Taylor's expansion in Section 4.2.

It is worth emphasizing that $o(x)$ denotes not a particular function but a function (or *some* function) with the above property. Therefore, although we can write the expression $2o(x)$, it would not make much sense since multiplying $o(x)$ by 2 we get a function with the same property: it converges to zero faster than x. Thus, we can write $2o(x) = o(x)$. For the same reason, we may write that $o(x) + o(x) = o(x)$, etc.

EXAMPLE 2. Let us approximate the function $f(x) = (1+x)^4 + (1+3x^2)^3$ for small x's by a quadratic function. Similar to what we did in Example 2, and recalling that $(1+a)^4 = 1+4a+6a^2+4a^3+a^4 = 1+4a+6a^2+o(a^2)$, we have

$$f(x) = 1+4x+6x^2+o(x^2)+1+3\cdot 3x^2+o(x^2) = 2+4x+15x^2+o(x^2).$$

So, $f(x) = 2+4x+15x^2+$ the term which is negligible in comparison with the main term as $x \to 0$. □

The same relations may be established certainly not only for x's close to zero but to any point, or for $x \to \infty$. The general definition looks as follows.

Let us consider functions $f(x)$ and $g(x)$. We say that

$$f(x) = o(g(x)) \quad \text{as} \quad x \to x_0, \tag{4.1.2}$$

(f is a little "o" of g) if

$$\frac{f(x)}{g(x)} \to 0 \quad \text{as} \quad x \to x_0. \tag{4.1.3}$$

Here x_0 is any number, or ∞, or $-\infty$.

For example, $x^{-2} = o(x^{-1})$ as $x \to \infty$, or $(x-1)^3 = o((x-1)^2)$ as $x \to 1$.

For another good example, let us set $g(x) \equiv 1$ in (4.1.2). Then $f(x) = o(1)$. By definition (4.1.3), this relation means that $f(x) \to 0$ as $x \to x_0$. Therefore, the expression $o(1)$ denotes *any* vanishing function. For instance, instead of saying that $\ln(1+x) \to 0$ as $x \to 0$, we can say that $\ln(1+x) = o(1)$. Actually, it is not shorter but sometimes convenient.

4.1.2 Big O

For a $d > 0$, we call an interval $\Delta = (x_0 - d, x_0 + d)$ a neighborhood of a point x_0. We say that

$$f(x) = O(g(x)) \quad \text{as} \quad x \to x_0,$$

("f is a big O of g"), if there exists a constant C and a neighborhood Δ of x_0 such that for all $x \in \Delta$

$$|f(x)| \le C|g(x)|. \tag{4.1.4}$$

For instance, if we want to say that $f(x) = 1+x+x^2$ has the same order as $1+x$ for x close to zero, we can write that

$$1+x+x^2 = O(1+x) \tag{4.1.5}$$

as $x \to 0$. Certainly, (4.1.5) is not true for all x's because $\frac{1+x+x^2}{x} \to \infty$ as $x \to \infty$ but, for example, $1+x+x^2 \le 2(1+x)$ for $|x| \le 1/2$, which the reader can readily verify.

If $x_0 = \infty$ (or $x_0 = -\infty$), we say that $f(x) = O(g(x))$ as $x \to \infty$ (or $x \to -\infty$), if (4.1.4) is true for x's larger than some c_0 (or smaller than some negative c_0).

For example, $x + 3x^2 = O(x^2)$ as $x \to \infty$.

4.2 Taylor expansions

4.2.1 A general expansion

The *Taylor expansion* concerns approximations of sufficiently smooth functions $f(x)$ by polynomials in a neighborhood of a particular chosen point x_0. Without loss of generality, we can set $x_0 = 0$. If it is not so, it suffices to translate the origin to x_0, or more precisely, to consider instead of $f(x)$ the function $f_0(x) = f(x+x_0)$, and apply the Taylor expansion to the latter function.

Let $f(x)$ be $n+1$ times differentiable in a neighborhood of zero, that is, for $x \in \Delta = (-d,d)$ for some $d > 0$. Then the *Taylor formula* states the following: for all $x \in \Delta$,

$$f(x) = P_n(x) + R_n(x), \tag{4.2.1}$$

where the *Taylor polynomial* of the nth order

$$P_n(x) = \sum_{k=0}^{n} \frac{f^{(k)}(0)}{k!} x^k,$$

$f^{(k)}$ is the kth derivative of f, and the remainder

$$R_n(x) = \frac{f^{(n+1)}(c)}{(n+1)!} x^{n+1} \tag{4.2.2}$$

for some c between 0 and x.

The polynomial $P_n(x)$ is the best approximating polynomial in the sense that the first n derivatives of $P_n(x)$ and $f(x)$ at the origin coincide; i.e., $P^{(k)}(0) = f^{(k)}(0)$, $k = 0, 1, ..., n$.

The remainder $R_n(x)$ serves as an estimate of the accuracy of the polynomial approximation.

If we do not need to estimate the accuracy, we do not have to assume existence of the derivative $f^{(n+1)}$, and may restrict ourselves to the formula

$$f(x) = \sum_{k=0}^{n} \frac{f^{(k)}(0)}{k!} x^k + o(x^n). \tag{4.2.3}$$

This is true if the nth derivative $f^{(n)}(x)$ exists and is continuous in a neighborhood of zero. The notation $o(\cdot)$ is explained in Section 4.1.

If f has all derivatives, then under some additional conditions, we can set $n = \infty$, and write

$$f(x) = \sum_{k=0}^{\infty} \frac{f^{(k)}(0)}{k!} x^k. \tag{4.2.4}$$

For example, this is true if $|f^{(n)}(x)| \leq M^n$ for some M and all n in the neighborhood where we approximate the function.

4.2.2 Some particular expansions

All expansions below are verified by making use of the general formulas (4.2.3) and (4.2.4).

The exponential function: for all x's,

$$e^x = 1 + x + \frac{x^2}{2!} + \ldots = \sum_{k=0}^{\infty} \frac{x^k}{k!}. \tag{4.2.5}$$

As $x \to 0$, it often suffices to consider the first three terms, that is, the approximation

$$e^x = 1 + x + \frac{x^2}{2} + o(x^2). \tag{4.2.6}$$

The logarithmic function. For $\ln(1+x)$, the expansion

$$\ln(1+x) = \sum_{k=1}^{\infty} (-1)^{k-1} \frac{x^k}{k!} \tag{4.2.7}$$

is true for $-1 < x \leq 1$. In particular,

$$\ln(1+x) = x - \frac{x^2}{2} + o(x^2) \quad \text{as} \quad x \to 0. \tag{4.2.8}$$

From the last relation it follows that

$$\ln(1+x) \sim x \text{ as } x \to 0,$$

where, as usual, the symbol $a(x) \sim b(x)$ means that $\frac{a(x)}{b(x)} \to 1$.

The power function. Next, we consider the function $(1-x)^{-\alpha}$ for $\alpha > 0$. The Taylor expansion here is true for $|x| < 1$, and is given by the formula

$$(1-x)^{-\alpha} = \sum_{m=0}^{\infty} \binom{-\alpha}{m} (-x)^m = \sum_{m=0}^{\infty} \binom{\alpha+m-1}{m} x^m, \tag{4.2.9}$$

where for any real r

$$\binom{r}{k} = \frac{r(r-1)\cdots(r-k+1)}{k!}.$$

[We defined it also in (3.1.3).] The second equality in (4.2.9) is based on the formula

$$\binom{-\alpha}{m} = (-1)^m \binom{\alpha+m-1}{m},$$

which is true since

$$\binom{-\alpha}{m} = \frac{(-\alpha)(-\alpha-1)\cdots(-\alpha-m+1)}{m!} = \frac{(-1)^m(\alpha)(\alpha+1)\cdots(\alpha+m-1)}{m!}$$
$$= (-1)^m \binom{\alpha+m-1}{m}.$$

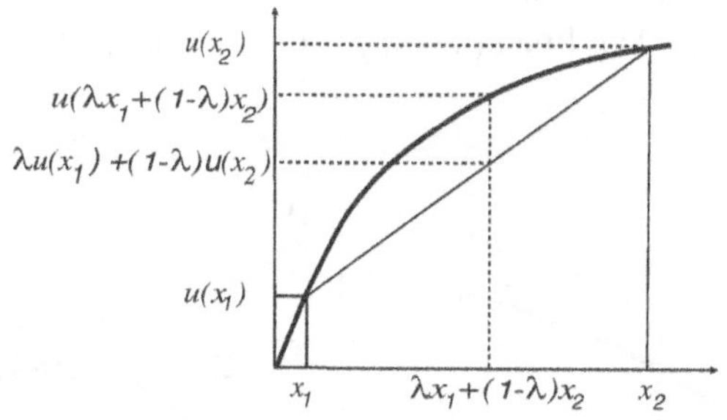

FIGURE 17. The first definition of a concave function.

For $\alpha = 1$, the coefficient $\binom{\alpha+m-1}{m} = \binom{m}{m} = 1$, and for $|x| < 1$ we have

$$\frac{1}{1-x} = 1+x+x^2+\ldots = \sum_{k=0}^{\infty} x^k. \tag{4.2.10}$$

We see that this is just the formula for the geometric series, which can certainly be derived without Taylor's expansion from the well known formula

$$1+x+x^2+\ldots+x^n = \frac{1-x^{n+1}}{1-x}. \tag{4.2.11}$$

The last formula is true for all $x \neq 1$.

4.3 Concavity

In introductory Calculus courses, a concave function u is often defined as a function for which $u''(x) \leq 0$. For us, this definition is somewhat restrictive. The definition below does not contradict the above definition if u is twice differentiable.

Definition. We say that a function $u(x)$ defined on an interval I is *concave* if for any $x_1, x_2 \in I$ and any $\lambda \in [0,1]$,

$$\lambda u(x_1) + (1-\lambda)u(x_2) \leq u(\lambda x_1 + (1-\lambda)x_2). \tag{4.3.1}$$

See, as an illustration, Fig.1.

It is known that $u(x)$ so defined is continuous at any interior point of I; see, e.g., [142]. The following proposition may be viewed as another definition of concavity.

Proposition 1 *A function $u(x)$ is concave on an interval I if and only if for any interior point $x_0 \in I$ there exists a number c, perhaps depending on x_0, such that for any $x \in I$*

$$u(x) - u(x_0) \leq c(x - x_0). \tag{4.3.2}$$

Note that Proposition 1 does not presuppose that the number c is unique. If $u(x)$ is differentiable at x_0, then (4.3.2) is true for $c = u'(x_0)$; see Fig.2i. However, if u is not smooth at x_0, then there may be many c's for which (4.3.2) is true; see Fig. 2ii. A proof may be also found in [142]. The line $c(x-x_0)$ is called a *support* of $u(x)$ at x_0.

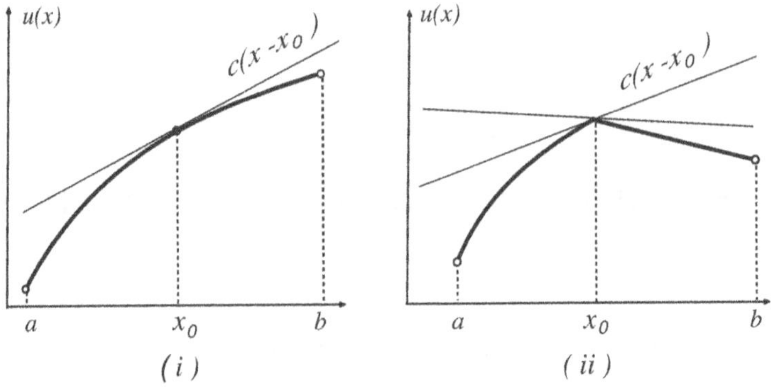

FIGURE 18. The second definition of a concave function.

We call a function $u(x)$ *convex* if (4.3.1) (or (4.3.2)) is true with the replacement of the inequality sign $\leq$ by the sign $\geq$. Note that there is no need to explore convex function separately because, if a function $u(x)$ is convex, then the function $-u(x)$ is concave.

References

The list below should not be considered a bibliography. It is a list of books and papers to which we refer in this book.

[1] Albanese, C., Credit Exposure, Diversification Risk and Coherent VaR, *Working Paper, Department of Mathematics, University of Toronto,* September, 1997.

[2] Allais, M., Le comportement de l'homme rationnel devant le risque: Crtique des postulats et axiomes de l'Ecole Americaine, *Econometrica*, 21, 1953.

[3] Arias, E., United States Life Tables, 2002, *Monthly vital statistics report,* vol. 53, no. 6, Hyattsville, Maryland: National Center for Health Statistics, November, 2004.

[4] Anderson, A.W., *Pension Mathematics for Actuaries*, 2nd edition, ACTEX Publications, Winsted, Connecticut, 1992. 13, 1, 35, 1993.

[5] Arrow, K.J., *Essays in the Theory of Risk Bearing*, Markham, Chicago, 1971.

[6] Arrow, K. J., and Hahn, F.H., *General Competitive Analysis*, San Francisco: Holden-Day, 1971.

[7] Arrow, K.J., Optimal insurance and generalized deductibles, *Scandinavian Actuarial Journal*, 1, 1974.

[8] Artzner, P., Delbaen, F., Eber, J.-M., and Heath, D., Coherent measures of Risk, *Math. Finance,* 9, 3, 1999.

[9] Artzner, P., Applications of coherent measures to capital requirements in insurance, *North American Actuarial Journal*, 3, 2, 1999.

[10] Asmussen, S., *Ruin Probability,* World Scientific, Singapore, River Edge, N.J., 2000.

[11] Beard, R.E., Pentikäinen, T., and Pesonen, E., *Risk Theory,* Chapman & Hall, London, NewYork, 1984.

[12] Bening, V.E. and Korolev, V.Yu., *Generalized Poisson Models and Their Applications in Insurance and Finance*, VSP, Utrecht, 2002.

[13] Bernoulli, D., Specimen Theoriae novae de mensura sortis, *Commentarii Academiae Scientiarum Imperialis Petropolitanae*, Tomus 5 (Papers of Imperial Academy of Sciences in Petersburg, v.5), p.175-192, 1738; translated into English in *Econometrica,* 22, N1, 22-36, 1954.

[14] Bhattacharya, R.N. and Rao, R.R., *Normal Approximation and Asymptotic Expansion*, SIAM, Fla, 2010.

[15] Borch, K., The rationale of the mean-standard deviation analysis: comment, *American Economic Review*, 64, 428, 1974.

[16] Borch, K., *The Mathematical Theory of Insurance*, D.C. Heath and Company, Lexington, Massachusetts, 1974.

[17] Borch, K., *Economics of Insurance*, North-Holland, Elsevier Science Publishers B.V., Amsterdam, New York, North-Holland, 1990.

[18] Borg, S., *On Optimal Dividend Payoffs: Karl Borch's Model*, Master Thesis, San Diego State University, 2003.

[19] Bowers, N.L., Herber, H.U., Hickman, J.C., Jones, D.A., and Nesbot, C.J., *Actuarial Mathematics,* Society of Actuaries, Schaumburg, Illinois 1997.

[20] Boyle, P.P., Karl Borch's Research Contributions to Insurance, *The Journal of Risk and Insurance*, 62, 2, 307, 1990.

[21] Bühlmann, H., An economic premium principle, *Astin Bullitin*, 11, 52, 1980.

[22] Bühlmann, H., The general economic premium principle. *Astin Bullitin*, 14, 13-21, 1984.

[23] Bühlmann, H., Gagliardi, B., Gerber, H., and Straub, E., Some inequalities for stop-loss premium, *Astin Bullitin*, 9, 75, 1979.

[24] Chew, S.H., Axiomatic utility theories with the betweeness property, *Annals of Operation Research*, 19, 273, 1989.

[25] Chew, S.H. and MacCrimmon, K.R., Alpha-nu choice theory: A generalization of expected utility theory, *Working paper* no. 669, *University of British Columbia*, Vancouver, 1979.

[26] Chew, S.H. and MacCrimmon, K.R., Alpha utility theory, lottery composition, and the Allais paradox, *Working paper* no. 686, *Faculty of Commerce, University of British Columbia*, Vancouver, 1979.

[27] Chow, Y.S. and Teicher, Henry, *Probability Theory, Independence, Interchangeability, Martingales*, Springer-Verlag, New York, 1978.

[28] Cowley, A. and Cummins, J.D., Securitization of Life Insurance and Liabilities. *The Journal of Risk and Insurance*, 72, 193, 2005.

[29] Cvitanic, J. and Zapatero, F., *Introduction to the Economics and Mathematics of Financial Markets*, MIT Press, Cambridge, Massachusets, 2004.

[30] Daniel, J.W., Multi-state transition models with actuarial applications, Study Notes, SOA and CAS; *http://casact.org/library/studynotes/daniel.pdf*, 2004.

[31] Daykin, C.D., Pentikäinen, T., and Pesonen, M. *Practical risk theory for actuaries,* Monographs on Statistics and Applied Probability, 53. Chapman and Hall, Ltd., London, 1994.

[32] Dekel, E., An axiomatic characterization of preferences under uncertainty: Weakening the independence axiom, *Journal of Economic Theory*, 40, 304, 1986.

[33] Denuit, M., Dhaene, J., Goovaerts M., and Kaas, R., *Actuarial Theory for Dependent Risks*, J.Wiley, New York, 2005.

[34] DeVylder, F., Martingales and ruin in a dynamic risk process, *Scandinavian Actuarial Journal*, 217, 1977.

[35] Doob, J.L., *Stochastic Processes*, Wiley, New York, 1990; the first edition – 1953.

[36] Durrett, R., *Essentials of Stochastic Processes*, Springer, New York, 1999.

[37] Embrechts, P., Klüppelberg, G., and Mikosch, T., *Modeling Extremal Events for Insurance and Finance*, Springer, Berlin, New York, 1997.

[38] Feller, W., *Introduction to Probability Theory*, 2, Wiley, New York, 1968.

[39] Fishburn, P.C., *Non-linear Preference and Utility Theory*, John Hopkins University Press, Baltimore, 1988.

[40] Folder, L., Expected utility and continuity, *Review of Economic Studies*, 39, 4, 1972.

[41] Gerber H.U., *Martingales in Risk Theory,* Mitt. Ver. Schweiz. Vers. Math. 73, 205, 1973.

[42] Gerber H.U., *An Introduction to Mathematical Risk Theory*, S.S. Heubner Foundation Monographs, University of Pennsylvania, 1979.

[43] Gerber, H.U. (with exercises contributed by Samuel H. Cox), *Life Insurance Mathematics*, 3rd ed., Springer, Berlin, New York, 1997.

[44] Gerber, H.U. and Shiu, E.S., On the time value of ruin, *North American Actuarial Journal*, 2, 1, 48, 1998.

[45] Gikhman, I.I. and Skorohod, A.V., *Introduction to the Theory of Random Processes,* W. B. Saunders Co., Philadelphia, 1969.

[46] Gihman, I.I. and Skorohod, A.V., *The Theory of Stochastic Processes,* Springer-Verlag, New York, 1974-1979.

[47] Gikhman, I.I. and Skorohod, A.V., *Controlled Stochastic Processes*, Springer, Berlin, 1979.

[48] Gollier, C. and Schlesinger, H., Arrow's theorem on optimality of deductibles: a stochastic dominance approach, *Economic Theory*, 7, 359-363, 1996.

[49] Goovaerts, M.G., De Vylder, F., and Haezendonck J., *Insurance Premiums,* North-Holland, Amsterdam, 1984.

[50] Grandel, J., *Aspects of Risk Theory*, Springer-Verlag, Berlin, New York, Heildeberg, 1991.

[51] Gradshteyn, I.S. and Ryzhik, I.M., *Table of Integrals, Series, and Products*, Academic Press, San Diego, 2000.

[52] Green, J. and Jullien, B. Ordinal independence in non-linear utility theory, *Quarterly Journal of Economics,* 102, 4, 785, 1988.

[53] Gupta, A.K. and Varga, T., *An Introduction to Actuarial Mathematics*, Kluwer Academic, Dordrecht, Boston, 2002.

[54] Haezendonck, J. and Goovaerts, M.J., A new premium calculations principle based on Orlicz norm, *Insurance: Mathematics and Economics*, 1, 41, 1982.

[55] Hardy, G.H, Littlewood, J.E., and Polya, G. *Inequalities*, 2d ed., Cambridge University Press, Cambridge, 1959.

[56] Hastie, R. and Dawes, R.M., *Rational Choice in an Uncertain World,* Sage, Thousand Oaks, Calif., 2001.

[57] van Heerwaarden, A.E., Kaaas, R., and Goovaerts, M.J., Properties of Essher premium calculation principle, *Insurance: Mathematics and Economics,* 8, 261, 1989.

[58] van Heerwaarden, A.E., Kaaas, R., and Goovaerts, M.J., Optimal resinsurance in relation of ordering of risks, *Insurance: Mathematics and Economics,* 8, 11, 1989.

[59] Hoel, P.G., Port, S.C., and Stone, C.J, *Introduction to Probability Theory*, Hougton Mifflin Company, Boston, 1971.

[60] Holton, G.A., *Value-at-Risk*, Academic Press, Amsterdam, Boston, 2003.

[61] Horn, R.A. and Johnson, C.A., *Matrix Analysis*, Cambridge University Press, Cambridge, 1985.

[62] Hull, J.C., *Options, Futures, and Other Derivatives*, Prentice Hall, New Jersey, 2002.

[63] Ibragimov, I.A., and Linnik, Yu.V., *Independent and Stationary Sequences of Random Variables*, Wolters-Noordhoff, Groningen, 1971.

[64] *Introduction to RiskMetrics*^{TM}. Morgan Guaranty Trust Company, 1995.

[65] Jensen, J.L.W.V., Sur les fonctions convexes et les ingalits entre les valeurs moyennnes, *Acta Math* 30, 175-193, 1906.

[66] Jorion, P., *Value at Risk: the New Benchmark for Controlling Market Risk,* McGraw-Hill, New York, 1997.

[67] Kaas, R., Goovaerts M., and Dhaene, J., *Modern Actuarial Theory*, Kluwer, Dordrecht, The Netherlands, London, 2001.

[68] Kahneman, D. and Tversky, A., Prospect Theory: Analysis of decision under risk, *Econometrica*, 47, 263, 1979.

[69] Kalashnikov, V.V., *Geometric Sums: Bounds for Rare Events with Applications in Risk Analysis, Reliability, Queueing,* Kluwer, Dordrecht, London, 1997.

[70] Kallenberg, O., *Foundations of Modern Probability*, Springer, New York, 2000.

[71] Karni, E., Optimal insurance: A non-expected utility analysis, in: G. Dione, ed., *Contributions to Insurance Economics*, Kluwer Academic Publishes, Dordrecht, London, 1992.

[72] Keeney, R.L. and Raiffa, H., *Decisions with Multiple Objectives: Preferences and Value Trade-offs*, Wiley, New York, 1976.

[73] Kiruta, A.Ya., Rubinov, A.M., and Yanovskaya, E.B., *Optimal Choice of Distributions in Complex Social-Economic Systems (Probability Approach)* (in Russian), Nauka, Moscow, 1980.

[74] Klugman, S.A., Panjer, H.H., and Willmot, G.E., *Loss Models: from Data to Decisions,* 2nd ed., Wiley-InterScience, New Jersey, 2004.

[75] Kolmogorov, A.N., *Foundations of the Theory of Probability*, Chelsea Pub. Co., New York, 1956.

[76] Kolmogorov, A.N. and Fomin, S.V., *Elements of the Theory of Functions and Functional Analysis,* Graylock Press, Rochester, New York, 1957-65.

[77] Korolev, V.Yu. and Shorgin, S.Ya., On the absolute constant in the remainder term estimate in the central limit theorem for Poisson random sums, in: *Probabilistic methods in Discrete Mathematics,* VSP, Utrecht, 1997, 305

[78] Kremer, E., On robust premium principles, *Insurance: Mathematics and Economics,* 5, 271, 1986.

[79] Lay, D., *Linear Algebra and its Applications,* Addison-Wesley, Reading, Massachusets, 2003.

[80] LeCam, L., An approximation theorem for the Poisson binomial distribution, *Pacific J. Math.*, 10, 1181, 1960.

[81] LeRoy, S. and Werner, J., *Principles of Financial Economics*, Cambridge University Press, 2001.

[82] Luce, R.D. and Raifa, H., *Games and Decisions: Introduction and Critical Survey,* John Wiley, New York, Wiley 1957; the last edition – Dover, 1989.

[83] Luce, R.D., *Utility of Gains and Losses*, Erlbaum, Mahwah, New Jersey, 2000.

[84] Machina, M.J., "Expected utility" analysis without the independence axiom, *Econometrica*, 50, 277, 1982.

[85] Machina, M.J., Choice under uncertainty: problems solved and unsolved, *Economic Perspectives*, 1, 121, 1987.

[86] Majumdar, M., *Equilibrium, Welfare, and Uncertainty: Beyond Arrow-Debreu,* Routledge, 2009

[87] Majumdar, M. and Rotar, V., Equilibrium prices in a random exchange economy with dependent summands, *Economic Theory,* 15, 531-550, 2000.

[88] Majumdar, M. and Rotar, V., Some general results on equilibrium prices in large random exchange economies, *Annals of Operation Research,* 114, 245-261, 2002.

[89] Marshall, A.W. and Olkin, I., A multivariate exponential distribution, *Journal of the American Statistical Association*, 62, 30, 1967.

[90] Marshall, A.W. and Olkin, I., Families of multivariate distributions, *Journal of the American Statistical Association*, 83, 834, 1988.

[91] Meyers, G., Coherent measures of risk: An exposition for the lay actuary, *http://casact.org/pubs/forum/00sforum/meyers/Coherent%20Measures%20of%20Risk.pdf*

[92] Michel, R., On Berry-Esseen results for the compound Poisson distribution, *Insurance: Mathematics and Economics*, 13, 1, 35, 1993.

[93] Mikosch, T., *Non-Life Insurance Mathematics. An Introduction with Stochastic Processes,* Springer, Berlin, Heildeberg, New York, 2004.

[94] Milevsky, M.A., *The Calculus of Retirement Income,* Cambridge University Press, New York, 2006.

[95] Minc, H., *Nonnegative Matrices,* Wiley, New York, 1998.

[96] von Neumann, J. and Morgenstern, O., *The Theory of Games and Economic Behavior,* Princeton University Press, New York, 1947.

[97] Owen, G., *Game Theory*, 3rd ed., Academic Press, San Diego,1995.

[98] Panjer, H. and Willmot, G., *Insurance Risk Models*, Society of Actuaries, Schaumburg, Illinois, 1992.

[99] Panjer, H. (ed.) et al., *Financial Economics*, The Actuarial Foundation, 1998.

[100] Petrov, V.V., *Limit Theorems of Probability Theory. Sequences of Independent Random Variables*, Clarendon Press, Oxford, 1995.

[101] Pesonen, Matti, Optimal reinsurance, *Scandinavian Actuarial Journal*, 2, 65, 1984.

[102] Pitman, J., *Probability*, Springer-Verlag, New York, 1993.

[103] Prelec, D., The Probability Weighting Function, *Econometrica*, 66, No. 3, 497-527,1998.

[104] Presman, E., On the approximation in variation of the distributions of a sum of independent Bernoulli random variables by the Poisson law, *Theory Probab.Applic.*, 18, 2, 39, 1983.

[105] Promislow, S.D., *Fundamentals of Actuarial Mathematics*, 2nd edition, Willey, 2010.

[106] Quiggin, J., *Generalized Expected Utility Theory*, Kluwer, Boston, 1993.

[107] Quiggin, J., A theory of anticipated utility, *Journal of Economic Behavior and Organization*, 3, 324, 1982.

[108] Quiggin, J., Stochastic dominance in regret theory, *Review of Economic Studies,* 57, 2, 503, 1989.

[109] Raviv, A., The design of an optimal insurance policy, *American Economic Review,* 69, 84-96, 1979.

[110] Reich, A., Properties of premium calculation principles, *Insurance: Mathematics and Economics,* 5, 97, 1986.

[111] Renyi, A., *Foundations of Probability*, Holden-Day, San Francisco, 1970.

[112] Revuz, D. and Yor, M., *Continuous Martingales and Brownian Motion*, Springer-Verlag, Berlin, New York, 1991.

[113] Rinott, Y. and Rotar, V., On Edgeworth expansions for dependency-neighborhoods-chain structures and Stein's Method, *Probability Theory and Related Fields*, 126, 528-570, 2003.

[114] Roell, A., Risk aversion in Quiggin and Yaari's rank-order model of choice under uncertainty, *Economic Journal*, 97, 143, 1987.

[115] Ross, S.M., *Stochastic Processes,* Wiley, New York, 1996.

[116] Ross, S.M., *A First Course in Probability*, 6th ed., Prentice Hall, Upper Saddle River, New Jersey, 2002.

[117] Ross, S.M., *Introduction to Probability Models*, 6th ed., Academic Press, San Diego, London, Boston, 1997.

[118] Rotar, G.V., A problem of control of reserve, *Theory Probab. Appl.*, 17, 3, 597, 1972.

[119] Rotar, G.V., On a problem of control of reserve, *Economics and Mathematical Methods*, 12, 733, 1976.

[120] Rotar, V.I., *Probability Theory*, World Scientific, Singapore, River Edge, New Jersey, 1997.

[121] Rotar, V.I., *Actuarial Models: The Mathematics of Insurance*, CRC Press, 2006.

[122] Rotar, V.I., *Probability and Stochastic Modeling*, CRC Press, 2013.

[123] Rotar, V.I. and Shorgin, S.Ya., On reinsurance of risks and a retention value, *Economics and Mathematical Methods*, 32, 4, 124, 1996.

[124] Royden, H.L., *Real Analysis*, 3rd ed., Prentice Hall, New Jersey, 1988.

[125] Schlesinger, H., Insurance demand without the expected-utility paradigm, *The Journal of Risk and Insurance*, 64, 1, 19-39, 1997.

[126] Schmidt, K.E., Positive homogeneity and multiplicativity of premium principles on positive risk, *Insurance: Mathematics and Economics,* 8, 1315, 1989.

[127] Senatov, V.V., *Normal Approximation: New Results, Methods, and Problems*, VSP, Utrecht, Boston, 1998.

[128] Shiganov, I.S., Refinement of the upper bound of the constant in the central limit theorem, *J. Soviet Mathematics*, 35, 3, 2545, 1986.

[129] Shiryaev, A.N., *Probability*, Springer, New York , New York, 1996.

[130] Shiryaev, A.N., *Essentials of Stochastic Finance: Facts, Models, Theory*, World Scientific, River Edge, New Jersey, 1999.

[131] Sholomitskii, A.G., *Choice Under Uncertainty and Modeling of Risk* (in Russian), GU-VSE (The High Economic School), Moscow, 2005.

[132] Shorgin, S., Approximation of a generalized binomial distribution, *Theory Probab. Appl.*, 22, 4, 867, 1977.

[133] Smolyak, S.A., On the dispersion of efficiency in the case of uncertainty (in Russian), in: *Methods and Models of Stochastic Optimization*, p.181-212, Moscow, CEMI, 1983.

[134] Smolyak, S.A., *Estimation of Efficiency of Investment Projects in the Case of Risk and Uncertainty* (in Russian), Nauka, Moscow, 2002.

[135] Stampfli, J. and Goodman, V., *The Mathematics of Finance, Modeling, and Hedging*, Brooks/Cole, Pacific Grove, CA 2001.

[136] Stewart, J., *Single Variable Calculus*, 5th ed., Thompson-Brooks/Cole, Belmont, California, 2003.

[137] Stewart, J., *Multivariable Calculus*, 5th ed., Thompson-Brooks/Cole, Belmont, California, 2003.

[138] Tapiero, C., Zuckerman, D., and Kahane, Y., Optimal Investment-Dividend Policy of an Insurance Firm under Regulation, *Scandinavian Actuarial Journal,* 65, 1983.

[139] Taylor, H. and Karlin, S., *An Introduction to Stochastic Modeling*, 3rd ed., Academic Press, San Diego, London, Boson, 1998.

[140] Wakker, P.P., *Additive Representations of Preferences*, Kluwer, Dordrecht, Boston, 1989.

[141] Waldmann, K.-H., On optimal dividend payments and related problems, *Insurance: Mathematics and Economics,* 7, 237, 1988.

[142] Webster, R., *Convexity*, Oxford University Press, Cambridge [Eng.], 1994.

[143] *Webster's New Universal Unabridged Dictionary*, Barnes and Noble Books, 2003.

[144] Winterfeldt, von D. and Edwards, W. , *Decision Analysis and Behavioral Research*, Cambridge University Press, Cambridge [Cambridgeshire], New York 1986.

[145] Williams, R.J.,*Introduction to Mathematics of Finance*, American Mathematical Society, 2006.

[146] Yaari, M.E., The dual theory of choice under risk, *Econometrica*, 55, 95-115, 1987.

[147] Yates, J.F., *Judgment and Decision Making*, Prentice-Hall, New Jersey, 1990.

[148] Zarei, S., *Survival Distributions and Analytic Laws of Mortality*, Master Thesis, San Diego State University, 2006.

[149] Zilcha, I. and Chew, S.H., Invariance of the efficient sets when the expected utility hypothesis is realxed, *Journal of Economic Behavior and Organization*, 13, 125-131, 1990.

[150] The Casualty Actuarial Society and the Canadian Institute of Actuaries, Exam 3, Fall 2003, *http://casact.org/admissions/studytools/exam3/.*

[151] The Casualty Actuarial Society and the Canadian Institute of Actuaries, Exam 3, Fall 2004, *http://casact.org/admissions/studytools/exam3/.*

[152] The Casualty Actuarial Society and the Canadian Institute of Actuaries, Exam 3, Spring 2004, *http://casact.org/admissions/studytools/exam3/.*

[153] The Casualty Actuarial Society and the Canadian Institute of Actuaries, Exam 3, Spring 2005, *http://casact.org/admissions/studytools/exam3/.*

[154] The Society of Actuaries, Exam 3, Fall 2004, *http://www.soa.org/STATIC/examinations.html.*

[155] The Society of Actuaries, Exam 3, Fall 2005, *http://www.soa.org/STATIC/examinations.html.*

[156] The Casualty Actuarial Society and the Canadian Institute of Actuaries, Exam 3L, Fall 2011, *http://www.casact.org/admissions/studytools/exam3/.*

[157] The Casualty Actuarial Society and the Canadian Institute of Actuaries, Exam 3L, Spring 2012, *http://www.casact.org/admissions/studytools/exam3/.*

[158] The Casualty Actuarial Society and the Canadian Institute of Actuaries, Exam 3L, Fall 2012, *http://www.casact.org/admissions/studytools/exam3/.*

[159] The Casualty Actuarial Society and the Canadian Institute of Actuaries, Exam 3L, Spring 2013, *http://www.casact.org/admissions/studytools/exam3/.*

Answers to Exercises

Possible additional remarks and errata will be posted in *http://actuarialtextrotar.sdsu.edu*

CHAPTER 0

3. $E\{Z_2^4 \mid Z_1\} = (1-Z_1^2)^2/5.$

4. (b) $f_{X_1 \mid S}(x \mid s) = \dfrac{\Gamma(\nu_1+\nu_2)}{\Gamma(\nu_1)\Gamma(\nu_2)} \dfrac{1}{s} \left(\dfrac{x}{s}\right)^{\nu_1-1} \left(1-\dfrac{x}{s}\right)^{\nu_2-1}$; $E\{X_1 \mid S\} = \frac{\nu}{\nu+1} S.$

5. $E\{Z_3^4 \mid Z_1, Z_2\} = (1-Z_1^2-Z_2^2)^2/5.$

6. $f(y \mid x) = 2(x+y)/(2x+1)$, $E\{Y \mid X\} = \frac{3X+2}{3(2X+1)}$. At $x=0$.

7. $f(y \mid x) = \frac{2y}{x^2}$ for $0 \le y \le x$, $E\{Y \mid X\} = 2X/3.$

9. $\dfrac{m_2}{m_1+m_2}.$

12. (a) $E\{Y\} = \frac{n}{2}$, $Var\{Y\} = \frac{n(n+2)}{12}$. (b) $E\{Y\} = p\lambda$, $Var\{Y\} = p\lambda$.
(c) $E\{Y\} = \lambda/2$, $Var\{Y\} = \frac{\lambda}{2}\left(1+\frac{\lambda}{6}\right)$; $P(Y=0) = \frac{1}{\lambda}\left(1-e^{-\lambda}\right)$ and $P(Y=1) = \frac{1}{\lambda}\left(1-e^{-\lambda}(1+\lambda)\right)$.

CHAPTER 1

2. 3.

3. (a) $[0, 0.7)$, $[0.8, 1]$. (b) $m \ge 1$; $\gamma \le \gamma_0 \approx 0.8$. (c) $m \ge 1$, $\gamma \le 1-\gamma_0 \approx 0.2$ (d) $\gamma \ge 0.5$.

4. (a) $[0, 0.1) \cup [0.4, 0.7) \cup [0.8, 1]$. (2) $\gamma < 1$.

5. The investment into one asset is more profitable for $\gamma > \gamma_0 \approx 0.68$.

13. No; 4.

17. 0.183.

18. (c) 25. (d) 8.05. (e) 50.7.

19. (a) ≈ 0.31.

23. No.

24. No.

31. $k \ge \$79$.

36. $-\beta$.

38. The first.

39. Worse.

45. Figure (a).

46. (b) For example, $(0.2, 0.5, 0.3)$.

47. (c) For example, $(0.1, 0.3, 0.3, 0.3)$.

49. $(0.1+0.1t, 0.5-0.2t, 0.4+0.1t)$ for $-1 \le t \le 2.5$.

50. (a) Yes. (b) $(\frac{1}{2},\frac{1}{4},\frac{1}{4})$ is better than $(\frac{3}{5},\frac{1}{10},\frac{3}{10})$. (c) No. However, this does not mean that Fred is a risk averter.

52. For example, $(\frac{7}{20},\frac{13}{20})$.

55. $c \approx 0.433$.

58. $\frac{1}{2}(1+2d-\sqrt{1+2d})$.

60. $(0.1)^{\beta/2}$.

61. (a) $\approx 0.69m$; (b) $\approx 0.59m$.

CHAPTER 2

2. (a) No if the number is positive. If the number is negative, then the c.v. changes the sign. (b) No. (c) Exponential. (d-ii) ≈ 0.246.

3. For all $c < a$.

5. (c) Heavy-tailed.

8. Only a.

9. (a) 1.5, 97.75. (b) 1.354, 83.905. (c) 1.35, 83.2275.

10. (a) ≈ 0.267. (b) ≈ 0.0134.

11. (a) $\approx 57.4\%$. (b) ≈ 0.918. (c) \$0.82, \$3.96.

12. ≈ 106.3.

14. (b) $E\{Y\} = \frac{q}{a}e^{-ad}(1+ad)$, $E\{Y^2\} = \frac{2q}{a^2}e^{-ad}\left(1+ad+(ad)^2\right)$.

15. (a) $\dfrac{\theta}{(\alpha-1)(1+d/\theta)^{\alpha-1}}$. (b) $\dfrac{3.12}{2(1+d/3.12)^2}$.

18. $P(X=1)=\frac{1}{6}$, $P(X=2)=\frac{7}{12}$, $P(X=3)=\frac{1}{4}$.

20. (a) $f_S(x) = 2(e^{-x}-e^{-2x})$.
(b) $f_S(x) = \frac{x}{2}$ if $0 \le x \le 1$, $f_S(x) = \frac{1}{2}$ if $1 \le x \le 2$, $f_S(x) = \frac{1}{2}(3-x)$ if $2 \le x \le 3$.

22. (a) 4. (c) ≈ 0.567, ≈ 0.433.

25. (a) 0.5. (b) ≈ 0.303.

26. ≈ 0.187.

27. (b) ≈ 0.772. (c) $C_1 = \frac{4^6}{5!}$, $C_2 = \frac{4^9}{8!}$.

33. Figure (b).

38. $P(X \le 3) = 1$ and $P(X=3) = 0.2$.

42. $f(x) = \dfrac{a^4}{9}\left(\dfrac{1}{6}x^3 + \dfrac{a^2}{30}x^5 + \dfrac{a}{6}x^4\right)e^{-ax}$.

46. Yes.

47. (a) $\theta \approx 0.105$. (b) 0.074. (c) 0.113.

48. (b) For the monetary unit equal to \$1000, the expected value for a separate policy is 0.05, the variance is 0.0975. For the total payment $E\{S\} = 100$, $Var\{S\} = 195$. (d) $\theta \approx 0.229$. (e) $\theta \ge 0.281$. (f) $\theta \approx 0.162$.

50. (a) $\theta \ge 0.063$. (b) $\theta_{n0} \approx 0.049$, $\theta_n \approx 0.058$. The premium for the first portfolio ≈ 1.058, for the second ≈ 1.116.

51. (a) $\theta_{\text{heuristic}} \approx 0.088$. (b) $\theta \ge 0.119$.

52. (a) 2255.

53. (a) $\frac{6}{5}\sqrt{\frac{3}{5}}$.

56. $\nu = \frac{1000}{9}$, $a = \frac{2}{3}$, $z = -\frac{200}{3}$.

57. $\approx 0.0665\frac{1}{\sqrt{n}}$.

60. (1) For the expected value principle, $\pi(X) = \pi(Y) = 1+\theta$.
(2) For the variance principle, $\pi(X) = 1+\lambda$, $\pi(Y) = 1+\lambda/3$.
(3) For the standard deviation principle: $\pi(X) = 1+\lambda$, $\pi(Y) = 1+\lambda/\sqrt{3}$.
(4) For the mean value principle, $\pi(X) = \sqrt{2}$, $\pi(Y) = \frac{2}{\sqrt{3}}$.
(5) For the exponential principle, $\pi(X) = \frac{1}{\beta}\ln\frac{1}{1-\beta}$ (we consider only $\beta < 1$), while $\pi(Y) = \frac{1}{\beta}\ln\left(\frac{e^{2\beta}-1}{2\beta}\right)$. In this case, $\pi(X) \geq \pi(Y)$.
(6) For the Escher principle, $\pi(X) = 2$, and $\pi(Y) = \frac{2}{e-1} \approx 1.16$. In this case, $\pi(X) > \pi(Y)$.

61. Indeed, in this case, the equation (4.10) is equivalent to $E\{X^\alpha\} = P^\alpha$, which leads to (4.3).

CHAPTER 3

4. For the same premium, the joint insurance is more profitable.

6. (a) $\lambda = 18.92, \bar{q}_n \approx 0.02695$. (b) $P(N=15) \approx 0.066$, $P(N \leq 15) \approx 0.2200$.
(c) Fifteen claims or less happen approximately once each five years.

10. $E\{\Lambda \,|\, N=6\} \approx 5.0598$.

11. The mean of Λ given $N=5$ is 4.2. The probability under consideration is 0.762.

12. (a) ≈ 245.62. (b) Not small. (c) ≈ 0.159.

13. (a) 2, 3. (b) $P(N \leq 4) = \sum_{m=0}^{4}\binom{3+m}{m}(\frac{2}{3})^4(\frac{1}{3})^m$. This probability is not small.

14. ≈ 0.994.

15. (a) Yes, if $p_1 \neq 0$.

20. For the zero-truncated version, $p_0 = 0$ and we deal with the geometric distribution.

21. ≈ 0.117.

22. (a) 425, 1275. (b) $\frac{245}{3}$; $\frac{61495}{9}$. (c) 588; 89964.

24. < 16.54.

27. (a) $\approx 0.0047, \approx 0.0444, \approx 0.0720$. (b) e^{-12}; ≈ 0.00038.

30. $P(S=0) = 0.01$; $f_S(x) = (0.18+0.81x)e^{-x}$ for $x > 0$.

31. 45.

32. (b) 400, 1600.

33. (a) ≈ 0.5375. (b) 0.58356. (d) 13,000; 3,700,000; $\exp\left\{10e^{100z}+40e^{300z}-50\right\}$.

34. (b) $\frac{35000}{3}$; $29\cdot 10^5$. (c) $\frac{10}{3}$, $\frac{80}{3}$, 20.

35. $E\{S\} = 11{,}500$, $Var\{S\} = 2{,}766{,}666.\bar{6}$.

36. $E\{S\} = 14{,}000$, $Var\{S\} = 4{,}166{,}666.\bar{6}$.

39. (b) $\theta \approx q_{\beta s}\sqrt{\frac{2}{\lambda}}$.
(c)

$$\theta \approx q_{\beta s}\frac{1}{\sqrt{\lambda}}\sqrt{1+\frac{1}{3}\left(\frac{b-a}{b+a}\right)^2},$$

and $c = (1+\theta)\frac{a+b}{2}\lambda$.

40. (b) $\theta \approx q_{\beta s}\frac{1}{\sqrt{\lambda}}\sqrt{\frac{1+p}{1-p}}$.

(c)

$$\theta \approx q_{\beta s}\frac{1}{\sqrt{\lambda}}\sqrt{\frac{p}{1-p}}\sqrt{e^{b^2}-1+\frac{1}{p}},$$

and $c = (1+\theta)e^{a+b^2/2}\frac{\lambda(1-p)}{p}$.

41. $\theta \geq c \approx 0.15$.

CHAPTER 4

5. $E\{X_t\} = \exp\{b^2t/2\}$, $Var\{X_t\} = \exp\{b^2t\}(\exp\{b^2t\}-1)$.

9. (a) $\approx 9.36 \cdot 10^{-14}$, that is, a very small number; ≈ 0.208. (b) ≈ 0.082, (c) ≈ 0.265. (d) 0.8, 0.4.

10. (a) $\approx 6.14 \cdot 10^{-6}$; 0.999. (b) ≈ 0.368. (c) ≈ 0.857. (d) 2, 1.

11. (a) 0.2. (b) ≈ 0.132. (c) ≈ 0.812. (The time unit is a day.)

12. (b) $E\{N_tN_{t+s}\} = \lambda t + \lambda^2 t(t+s)$, $Corr\{N_t, N_{t+s}\} = \sqrt{\frac{t}{t+s}}$. (c) $t+\frac{m}{\lambda}$, $\frac{m}{\lambda^2}$.

15. $\frac{80}{3}$, ≈ 0.6713.

16. (c) ≈ 232.484. (d) ≈ 0.317.

18. No. Yes. Yes.

19. ≈ 0.135.

20. $t < c \approx 0.206$.

21. $t < c \approx 1.280$.

25. (a) ≈ 309, ≈ 464. (b) ≈ 0.159. (c) ≈ 0.114.

26. 3650; ≈ 24333; ≈ 0.879.

27. 945.

28. (a) 165. (b) $u \geq 1.4$.

29. (a) Yes. (b) Yes. (c) Yes. (d) ≈ 0.147. (e) 0.639.

32. For Example 1, $\mathcal{P}^2 = \begin{Vmatrix} 0.505 & 0.4 & 0.095 \\ 0.465 & 0.44 & 0.095 \\ 0.4425 & 0.46 & 0.0975 \end{Vmatrix}$.

For Example 2, $p_{10}^{(2)} = 0$, $p_{11}^{(2)} = 1$, $p_{01}^{(2)} = p_{x+t}q_{x+t+1} + q_{x+t}$, $p_{00}^{(2)} = p_{x+t} \cdot p_{x+t+1}$.

33. For Example 1, 0.18; for Example 2, $p_{x+t} \cdot q_{x+t+1}$; for Example 4, $p^2(1-p)^2$.

35. ≈ 6.6795, ≈ 2.

36. ≈ 18.809.

38. ≈ 9.13.

41. ≈ 13, ≈ 58.33.

44. $\frac{40}{3}$, $12.9\bar{3}$.

49. ≈ 0.189.

50. ≈ 1.14, ≈ 0.159.

51. ≈ 0.038, ≈ 0.124, ≈ 0.443, ≈ 0.269, ≈ 0.126.

CHAPTER 5

2. ≈ 0.921.

3. $\sqrt{5}$.

7. ≈ 0.52.

8. ≈ 0.670.

11. Any $g(t)$ such that $g(t)t^{-1/2} \to \infty$.

14. ≈ 0.683.

26. The probability that at least one student will loose is ≈ 0.673. This probability is large, so the professor should warn student about this.

28. $\frac{1}{2}$, $\frac{1}{16}$, $\frac{1}{256}$.

30. ≈ 0.393.

31. No.

32. The price will never drop by 10% with probability ≈ 0.608. It will happen within one year with probability ≈ 0.277.

33. ≈ 0.667.

CHAPTER 6

9. ≈ 0.163.

10. ≈ 0.092.

11. $\gamma = a\left(1 - \dfrac{1+\sqrt{1+8(1+\theta)}}{4(1+\theta)}\right)$.

13. (a) ≈ 0.005. (b) $u \geq 57.62$. (c) $\theta \approx 0.234$.

14. (a) $e^{2z} + 2e^{3z} + e^{4z} = 4 + 13.2z$. (b) No. (c) $\gamma \in [0.59, 0.6]$, while approximation (2.2.25) gives $\gamma \approx 0.063$. (d) $u \geq 59.5$. (e) $\gamma \in [0.06, 0.061]$. The rest is practically the same since the solution does not differ much from the previous.

15. $\gamma \in [1.27, 1.28]$. Approximation (2.2.15) leads to 1.2.

16. $\theta \approx \frac{1.25}{u}$.

18. (b) If $P(X = a) = 1$, then F_1 is the distribution uniform on $[0, a]$. (c) $f_1(x) = \frac{5}{11}g_1(x) + \frac{4}{11}g_2(x) + \frac{2}{11}g_3(x)$, where $g_1(x)$, $g_2(x)$, and $g_3(x)$ are the uniform densities on $[0, 1]$, $[1, 2]$, and $[2, 3]$, respectively.

22. $\gamma = 1/2$, $\theta = 3/7$.

23. (b) $\Psi(u) \approx 0.93e^{-0.061u}$.

25. $u = 6, ..., 11$.

CHAPTER 7

4. No, because this function is not monotone.

5. A solution exists if the original probability multiplied by 1.1 does not exceed one. A solution is not unique. One of examples of a new mortality force is the force that is $\ln 1.1$ less than the original mortality provided that such a change does not lead to negative values).

6. $s(x) = \frac{1}{1+x}$.

8. $s(x)=\exp\left\{-\int_0^x \frac{du}{(\omega-u)^\alpha}\right\}=\exp\left\{-\frac{1}{(\alpha-1)\omega^{\alpha-1}}\left(\frac{1}{(1-x/\omega)^{\alpha-1}}-1\right)\right\}$ for $\alpha>1$. For $\alpha=1$, the distribution is uniform on $[0,\omega]$.

9. $s(x)=\left(1-\frac{x}{\omega}\right)^\alpha$ for $x\leq\omega$ and $\alpha>0$.

11. (a) 0.125. (b) ≈ 0.674.

14. ≈ 0.0956.

15. 75; ≈ 0.327; $\frac{50}{3}$.

16. (a) (i) If we consider just the ratio of the expected values of male and female survivors, then the estimate of the proportion is approximately $1:1.06$. (ii) ≈ 0.17.

17. $E\{\mathcal{L}(60)\}\approx 63.25$; $Var\{\mathcal{L}(60)\}\approx 23.25$.

19. ≈ 0.94.

20. The latter.

22. $P(T(20)>t)\approx 0.27e^{-t/50}+0.73e^{-t/80}$.

23. ≈ 32.28.

24. $\frac{50}{3}$.

29. 4900.

31. The answer to the second question is positive.

33. ${}_{40}p_{40}\approx 0.54$; $\mu(20)\approx 0.000932$; ${}_{10|10}q_{40}\approx 0.06$; ${}_{2|}q_0\approx 0.00032$, ${}_{2|}q_{20}\approx 0.000932$.

35. $a\approx 12.2$; about 6% of the people survived 90 years, and 1.2% of the total group of 100,000.

37. For $s(60.5)$, we have ≈ 0.827496 or 0.8275; for $s(61.3)$, we have ≈ 0.823497 or 0.8235. For $\mu(60.5)$, we have ≈ 0.00604231 or 0.0060422, so the difference is negligible. Next, $p_{60.25}\approx 0.9939667$.

39. $A\approx 0.00151$, $B\approx 0.0000215$, $\alpha=0.142$.

40. For $t>1/(\sqrt{2}-1)$.

41. ${}_{2|}q_{50}^{(1)}=0.60536$. $P(J=1\,|\,2<T\leq 3)=0.7$.

43. ${}_{2|}q_{70}\approx 0.115$; $Var\{\mathcal{D}_{71}^{(2)}\}=1.50144$; $P(\mathcal{D}_{71}^{(2)}=3)\approx 0.1649$, while the Poisson approximation gives ≈ 0.1662.

44. The estimates for $q_x^{(j)}$.

46. $f_{TJ}(t,j)=e^{-0.15t}\cdot\frac{j}{20}$; $f_T(t)=0.15e^{-0.15t}$; $P(J=j)=\frac{j}{3}$.

47. $f_{TJ}(t,j)=\frac{1}{20}jt^{j-1}\exp\{-\frac{1}{20}(t+t^2)\}$; $f_T(t)=\frac{1}{20}(1+2t)\exp\{-\frac{1}{20}(t+t^2)\}$; $P(J=1)\approx$ 0.175; $P(J=2)\approx 0.825$.

48. ${}_{10}p_x\approx 0.544$; $P(J=1)=\frac{2}{3}$.

49. ${}_{10}p_x\approx 0.273$; $P(J=1)\approx 0.303$.

55. ≈ 0.9053.

57. ${}_{20}p_{50:40}\approx 0.577$; ${}_{50}p_{50:40}=0$; ${}_{20}p_{\overline{50:40}}\approx 0.946$; ${}_{50}p_{\overline{50:40}}\approx 0.408$.

58. ${}_{20}p_{50:40}\approx 0.7236$, ${}_{50}p_{\overline{50:40}}\approx 0.9826$.

61. ≈ 0.426.

63. ${}_{15}q_{25:20}\approx 0.02225$. In the case of the absence of common shock, ${}_{15}q_{25:20}$ would be ≈ 0.01489, that is, about 30% less. Next, ${}_{15}q_{\overline{25:20}}\approx 0.00752$.

CHAPTER 8

2. (a) Yes. (c) When *delta* is close to zero or/and x is large.

3. ≈ 0.048.

6. $\frac{1}{12}$.

7. (b) $\overline{A}_{90} \approx 0.855$.

9. ≈ 0.25.

10. $\approx \$53,000$.

11. $P(Z \leq x) = x^{\mu/\delta}$. The distribution is uniform if $\mu = \delta$.

14. (a) ≈ 0.267. (b) ≈ 0.262.

18. (a) Non-positively correlated. (b) Non-positively correlated.

19. $\frac{u-sw}{1-s}$.

20. (a) $A_{30:\overline{35}|} \approx 0.386$; $A^{1}_{30:\overline{35}|} \approx 0.054$, and ${}_{35|}A_{30} \approx 0.146$. (b) ≈ 0.395 per \$1 of benefit.

21. $\overline{A}_{20} \approx 0.223$, $Var\{Z\} \approx 0.076$.

22. ≈ 0.032.

24. $\overline{A}_{50:\overline{20}|} \approx 0.545$; $\overline{A}^{\ \ 1}_{50:\overline{20}|} \approx 0.275$.

26. ≈ 0.2002. item[**28.**] 0.

31. None. item[**33.**] (a) All, under the condition $P(0 < \Psi < \infty) > 0$. item[**34.**] Under-estimate. item[**36.**] No.

38. $\mu_2 = \frac{0.02\delta}{\delta-0.01}$ for $\delta > 0.01$.

40. (a-i) $(\overline{IA})_{60} = (1-(40\delta+1)e^{-40\delta})/(40\delta^2)$.
(a-ii) $(\overline{IA})_{60} = \mu/(\mu+\delta)^2$.
(a-iii) ≈ 5.50, ≈ 7.45. (b-i) $(\overline{DA})_{60:\overline{20}|} = (20\delta - 1 + e^{-20\delta})/(40\delta^2)$.
(b-ii) $(\overline{DA})_{60:\overline{20}|} = \mu[20(\mu+\delta) - 1 + e^{-20(\mu+\delta)}]/(\mu+\delta)^2$.

42. ≈ 0.038.

43. ≈ 0.839.

44. ≈ 7.070.

46. APV≈ 0.93677, the probability under discussion is 0.06.

47. ≈ 19.44.

49. ≈ 0.580.

50. ≈ 5.744.

53. $\overline{A}_{x:y} \approx 0.965$; $Var\{Z\} \approx 0.000775$.

54. $(\overline{IA})_{x:y} \approx 0.342$; $Var\{Z\} \approx 0.048$.

55. ≈ 2.8681.

56. (i) $\frac{a}{\delta+2a}$; (ii) ≈ 0.347.

57. (a) ≈ 0.853. (b) ≈ 1.415. (c) $0.281(c_1+c_3)+0.145(c_2+c_4)$.

CHAPTER 9

1. ≈ 41.957; ≈ 26.292.

6. (a) Y is uniform on $[0, 1/\delta]$. (b) The d.f. $F_Y(x) = 1-(1-\delta x)^{\mu/\delta}$ for $x \leq 1/\delta$.

7. $P(Y \leq x) = -\frac{\ln(1-\delta x)}{\delta c}$ for $x \leq 1/\delta$.

9. Yes.

10. No.

20. (a) All. b) As $\delta \to 0$,

$$\begin{aligned}
\lim \bar{a}_x &= \overset{\circ}{e}_x, \\
\lim \ddot{a}_x &= e_x + 1, \\
\lim \bar{a}_{x:\overline{n}|} &= \overset{\circ}{e}_x - {}_n p_x \cdot \overset{\circ}{e}_{x+n}, \\
\lim \ddot{a}_{x:\overline{n}|} &= e_x + {}_n q_x - {}_n p_x \cdot e_{x+n}, \\
\lim {}_{m|}\bar{a}_x &= {}_m p_x \cdot \overset{\circ}{e}_{x+m}, \\
\lim {}_{m|}\ddot{a}_x &= {}_m p_x (e_{x+m} + 1).
\end{aligned}$$

25. $\bar{A}^{\,1}_{x:\overline{20}|} \approx 0.355$; ${}_{20}p_x \approx 0.836$.

26. $0.96\bar{6}$.

27. $c \approx \$31,866$.

28. (a) ≈ 58.82. (b) ≈ 1670.38.

29. For \$100,000 as a unit of money, (a) ≈ 27.07, ≈ 57.24. (b) ≈ 20.77, ≈ 34.35.

30. $(1 + 200\delta^2 - 20\delta - e^{-20\delta})/(200\delta^3)$.

31. $\approx 92,300$, $\approx 9,080,000$.

32. ≈ 21.983.

33. ≈ 18.75.

35. (a) ≈ 2.132. (b) ≈ 5.78.

40. (a) $\approx \$127,200$. (b) $\approx \$132,600$. (c) \$69,900.

41. $2.84c + 0.66\tilde{c}$.

43. (a) ≈ 10.81; ≈ 17.58. (b) ≈ 9.98; ≈ 16.74.

44. $\bar{a}_{\overline{x:y}} - \bar{a}_{x:y}$.

45. $\mu/[(\mu+\delta)(2\mu+\delta)]$.

46. $\mu_2/[(\mu_1+\delta)(\mu_1+\mu_2+\delta)]$.

CHAPTER 10

2. The results are given in the table below.

Whole life insurance with a benefit payable at the moment of death	$\overline{P}_x$ and $P(\overline{A}_x)$
Whole life insurance with a benefit payable at the end of the year of death	$\overline{P}(A_x)$ and P_x
n-Year term life insurance with a benefit payable at the moment of death	$\overline{P}^{1}_{x:\overline{n}\vert}$ and $P(\overline{A}^{1}_{x:\overline{n}\vert})$
n-Year endowment life insurance with a benefit payable at the moment of death	$\overline{P}_{x:\overline{n}\vert}$ and $P(\overline{A}_{x:\overline{n}\vert})$
m-Year payment whole life insurance with a benefit payable at the moment of death	${}_m\overline{P}_x$ and ${}_mP(\overline{A}_x)$
m-Year payment n-year term life insurance with a benefit payable at the moment of death	${}_m\overline{P}^{1}_{x:\overline{n}\vert}$ and ${}_mP(\overline{A}^{1}_{x:\overline{n}\vert})$
m-Year payment n-year term endowment life insurance with a benefit payable at the moment of death	${}_m\overline{P}_{x:\overline{n}\vert}$ and ${}_mP(\overline{A}_{x:\overline{n}\vert})$
n-Year pure endowment life insurance	$\overline{P}_{\substack{\ \ 1\\x:\overline{n}\vert}}$ and $P_{\substack{\ \ 1\\x:\overline{n}\vert}}$
n-Year deferred whole life annuity-due	$\overline{P}({}_{n\vert}\ddot{a}_x)$, and $P({}_n\ddot{a}_x)$

5. The results are given in the table below.

Whole life insurance with a benefit payable at the moment of death	$\overline{P}=1/\overset{\circ}{e}_x,\ P=1/(e_x+1)$
Whole life insurance with a benefit payable at the end of the year of death	$\overline{P}=1/\overset{\circ}{e}_x,\ P=1/(e_x+1)$
n-Year term life insurance with a benefit payable at the moment of death	$\overline{P}={}_nq_x/(\overset{\circ}{e}_x-{}_np_x\,\overset{\circ}{e}_{x+n})$, $P={}_nq_x/(e_x+{}_nq_x-{}_np_xe_{x+n})$
n-Year endowment life insurance with a benefit payable at the moment of death	$\overline{P}=1/(\overset{\circ}{e}_x-{}_np_x\,\overset{\circ}{e}_{x+n})$, $P=1/(e_x+{}_nq_x-{}_np_xe_{x+n})$
m-Year payment whole life insurance with a benefit payable at the moment of death	$\overline{P}=1/(\overset{\circ}{e}_x-{}_mp_x\,\overset{\circ}{e}_{x+m})$, $P=1/(e_x+{}_mq_x-{}_np_xe_{x+m})$

m-Year payment n-year term life insurance with a benefit payable at the moment of death	$\overline{P} = {}_nq_x/(\overset{\circ}{e}_x - {}_mp_x\,\overset{\circ}{e}_{x+m})$, $P = {}_nq_x/(e_x + {}_mq_x - {}_np_xe_{x+m})$
m-Year payment n-year term endowment life insurance with a benefit payable at the moment of death	$\overline{P} = 1/(\overset{\circ}{e}_x - {}_mp_x\,\overset{\circ}{e}_{x+m})$, $P = 1/(e_x + {}_mq_x - {}_np_xe_{x+m})$
n-Year pure endowment life insurance	$\overline{P} = {}_np_x/(\overset{\circ}{e}_x - {}_np_x\,\overset{\circ}{e}_{x+n})$, $P = {}_np_x/(e_x + {}_nq_x - {}_np_xe_{x+n})$
n-Year deferred whole life annuity-due	$\overline{P} = {}_np_x \cdot (e_{x+n}+1)/(\overset{\circ}{e}_x - {}_np_x\,\overset{\circ}{e}_{x+n})$, $P = {}_np_x \cdot (e_{x+n}+1)/(e_x + {}_nq_x - {}_np_xe_{x+n})$

6. ${}_nP_x = \dfrac{A^1_{x:\overline{n}|}}{\ddot{a}_{x:\overline{n}|}} + \dfrac{{}_nE_x}{\ddot{a}_{x:\overline{n}|}}A_{x+n} = \dfrac{A^1_{x:\overline{n}|}}{\ddot{a}_{x:\overline{n}|}} + P_{\underset{x:\overline{n}|}{\ \ 1}}A_{x+n}.$

7. $3/4$.

8. The second.

9. (a) $P_{50} \approx 0.0190$. (b) $P(L_n < 0) \approx 0.747$.

10. $P^1_{50:\overline{30}|} \approx 0.013$.

11. $P \approx 16.657$.

12. (a) $\overline{P}_{x:y} \approx 0.050$. (b) $\overline{P}_{\overline{x:y}} \approx 0.013$. (c) $P \approx 0.0196$.

13. $Var\{L_P\} = (1-e^{-s})\,(s+se^{-s}-2+2e^{-s})\,/[2(e^{-s}-1+s)^2]$.

15. $n \geq 1580$.

16. $P \sim \dfrac{1}{a\beta}e^{\beta C} \sim P_{\text{net}}\dfrac{1}{2\beta C}e^{\beta C}$ as $C \to \infty$.

17. $\geq 13.3\%$.

18. The premiums are smaller for the first group.

19. $v(1-q)$.

20. $k > 17/15$.

21. ≈ 0.0186.

22. (a) \$7407 per year. (b) \$7279 per year.

23. (c) ≈ 0.080.

31. (a) ${}_0V = 0$, ${}_1V \approx 0.2647$, ${}_2V \approx 0.1364$, ${}_3V = 0$.
(b) ${}_0V = 0$, ${}_1V \approx 6.4239$, ${}_2V \approx 12.9054$, ${}_3V = 20$.
(c) ${}_0V = 0$, ${}_1V \approx 0.0098$, ${}_2V \approx 0.01$.

CHAPTER 11

1. (a) A quadratic function. (b) (i) 40 years later. (ii) No solution.

2. (a) ≈ 65.65. (b) 40.55 years later, which is too much. However, if the participants had been twenty years younger, it would be realistic.

3. (a) 0. A hyperbola. (b) $s_x = b(10+x)$, where b is an arbitrary multiplier on which the ratio s_{x+y}/s_x does not depend.

4. ≈ 0.81; $\int_r^\infty y p_x^{(\tau)} \mu_x^{(r)}(y)dy$.

5. (a) $v^{r-x}\,{}_{r-x}p_x^{(\tau)} B(x,h,r-x)\,\overline{a}_r$. (b) \$224,385.

6. An additional condition should concern $\mu_x^{(r)}(y)$. For example, we may require $\mu_x^{(r)}(y)$ for the first participant to be not smaller then for the second for all y's.

7. (a) $\frac{\mu_1}{\mu_2+\delta} e^{-\mu(r-x)} \int_{r-x}^{\infty} e^{-(\delta+\mu_1)y} B(x,h,y)\,dy$. (b) ≈ 8014.

8. (a) $w_0 e^{\beta K}$. This is just the salary at the moment of retirement, and we take into account only this salary once the corresponding weight γ approaches infinity.

9. $cw(0)Ke^{\rho K}$.

10. The moment when the remaining income reaches the level $\overline{w}$ is

$$\bar{t} = \frac{1}{\beta} \ln\left(\frac{\overline{w}}{w(0)(1-c)}\right).$$

The problem makes sense if $\bar{t} < K$. The accumulated capital

$$C = \frac{w(0)e^{\rho K}}{\rho-\beta}\left(c + (1-c)e^{-(\rho-\beta)\bar{t}} - e^{-(\rho-\beta)K}\right) - \frac{\overline{w}e^{\rho K}}{\rho}\left(e^{-\rho\bar{t}} - e^{-\rho K}\right). \quad (3)$$

Any expression of the type of $e^{s\bar{t}}$ may be also written as

$$\left(e^{\beta\bar{t}}\right)^{s/\beta} = \left(\frac{\overline{w}}{w(0)(1-c)}\right)^{s/\beta},$$

though probably it makes sense to leave (3) as is.

11. (a) Expression (1.2.4), as a function of d, is non-increasing. (b) \$25,789.

12. (a) $w_0(T_a - T_r)$.

(b) $$\frac{w_0 n_0}{\omega+\mu-\beta} e^{(\omega+\tau)t}\left(1 - e^{-(\omega+\mu-\beta)(r-a)}\right).$$

13. $\overset{\circ}{e}_r$.

14. (a) Not necessarily.

18. (a) $\ln(1+\beta)+\mu(x)+\delta$. (b) $\beta+\mu(x)+\delta$.

19. $(NC)_{t+1} = (1+i)(NC)_t$.

20. $\frac{kw(r)(r-a)e^{-(\delta+\mu)(r-a)}}{1-e^{-\mu(r-a)}}$ and $\frac{kw(r)(r-a)}{\mu+\delta} \cdot e^{-(\delta+\mu)(r-x)} \frac{1-e^{-\mu(r-a)}}{1-e^{-\mu(x-a)}}$.

CHAPTER 12

1. $Var\{R^2_{d(10)}(X)\} \approx 1.38$, $Var\{R^2_{d(1)}(X_1)\} \approx 0.33$, and hence $10Var\{R^2_{d(1)}(X_1)\} \approx 3.3$. The stop-loss reinsurance gives the same result as the excess-of-loss reinsurance on the average, but the riskiness of the former is less than that of the latter reinsurance.

3. (a) $\sigma_0 > 88,192$. If $\sigma_0 = 90,000$, then $r_1 \approx 0.953$.
(b-ii) $v_{\text{retained}} \approx 2k^2 m/(1-q)$. (b-iii) $v_{\text{retained}} \approx 0.55$. For the second question, $q < 0.048$.

5. $\gamma_{r_0} = m\theta^2_{\text{reins}}/(2\sigma^2\Delta)$.

6. (a) $r_0 \approx \frac{3}{4}$, $\gamma_{r_0} \approx 0.10\bar{6}a$.

7. (b) $r_0 \approx 0.33$.

9. (d) $\mathbf{V} = (k^{1/3}, (1-k)^{1/3})E\{S^{1/3}\}$ for $0.453 \leq k \leq 0.547$.

10. $E\{\tilde{Z}_i\} = 2 + c_i$, where $c_1 = -1$, $c_2 = 0$, $c_3 = 1$.

16. (b) $7r_1 + 3r_2 = 5$.

17. (a) $G(X_i) = m + \frac{\sigma^2}{nm}$, $r_i = \frac{1}{n}$. (b) $G(X_i) = m_i + \frac{1}{m_S}\left(\sigma_i^2(1-\rho) + \rho\sigma_i\eta_S\right)$, where $\eta_S = \sum_{j=1}^{n} \sigma_j$.

Index